THE
PHYSIOLOGY
OF FISHES

Fourth Edition

CRC
MARINE BIOLOGY
SERIES

The late Peter L. Lutz, Founding Editor
David H. Evans, Series Editor

PUBLISHED TITLES

Biology of Marine Birds
E.A. Schreiber and Joanna Burger

Biology of the Spotted Seatrout
Stephen A. Bortone

The Biology of Sea Turtles, Volume II
Peter L. Lutz, John A. Musick, and Jeanette Wyneken

*Early Stages of Atlantic Fishes: An Identification Guide for the
Western Central North Atlantic*
William J. Richards

The Physiology of Fishes, Third Edition
David H. Evans and James B. Claiborne

Biology of the Southern Ocean, Second Edition
George A. Knox

Biology of the Three-Spined Stickleback
Sara Östlund-Nilsson, Ian Mayer, and Felicity Anne Huntingford

Biology and Management of the World Tarpon and Bonefish Fisheries
Jerald S. Ault

Methods in Reproductive Aquaculture: Marine and Freshwater Species
Elsa Cabrita, Vanesa Robles, and Paz Herráez

*Sharks and Their Relatives II: Biodiversity, Adaptive Physiology,
and Conservation*
Jeffrety C. Carrier, John A. Musick, and Michael R. Heithaus

Artificial Reefs in Fisheries Management
Stephen A. Bortone, Frederico Pereira Brandini, Gianna Fabi,
and Shinya Otake

Biology of Sharks and Their Relatives, Second Edition
Jeffrey C. Carrier, John A. Musick, and Michael R. Heithaus

The Biology of Sea Turtles, Volume III
Jeanette Wyneken, Kenneth J. Lohmann, and John A. Musick

The Physiology of Fishes, Fourth Edition
David H. Evans, James B. Claiborne, and Suzanne Currie

THE
PHYSIOLOGY
OF FISHES

Fourth Edition

Edited by **David H. Evans**
James B. Claiborne
Suzanne Currie

CRC Press is an imprint of the
Taylor & Francis Group, an   informa business

MIX
Paper from
responsible sources
FSC® C014174

CRC Press
Taylor & Francis Group
6000 Broken Sound Parkway NW, Suite 300
Boca Raton, FL 33487-2742

© 2014 by Taylor & Francis Group, LLC
CRC Press is an imprint of Taylor & Francis Group, an Informa business

No claim to original U.S. Government works

Printed on acid-free paper
Version Date: 20130524

International Standard Book Number-13: 978-1-4398-8030-2 (Hardback)

Library of Congress Cataloging-in-Publication Data

The physiology of fishes / edited by David H. Evans, James B. Claiborne, and Suzanne Currie. -- Fourth edition.
    pages cm. -- (CRC marine biology series)
    Includes bibliographical references and index.
    ISBN 978-1-4398-8030-2
    1. Fishes--Physiology. I. Evans, David H. (David Hudson), 1940- II. Claiborne, James B. III. Currie, Suzanne.

QL639.1.P49 2014
597.01--dc23   2013019480

**Visit the Taylor & Francis Web site at
http://www.taylorandfrancis.com**

**and the CRC Press Web site at
http://www.crcpress.com**

Contents

Preface

As was noted in the preface to the first edition, the idea for a single-volume on fish physiology dates from 1969, when DHE taught a graduate course in that subject for the first time at the University of Miami. The primary literature was often hard to find or too technical, and Hoar and Randall's Fish Physiology series (Academic Press) was just starting to be published as single volumes devoted to specific areas of research. Forty-four years later, the success of the first three editions of *The Physiology of Fishes* suggests that a single volume is still timely; hence, this fourth edition.

Because we felt that the utility and scientific interest of this fourth edition would be maximized by fresh approaches, and topics, we secured new authors for all but one of the chapters, and they come from three continents and five countries. In addition to completely new treatments on many of the original topics, we now have chapters on such timely topics as muscle plasticity, metabolism and membranes, oxygen sensing, endocrine disruption, pain, cardiac regeneration, and neuronal regeneration.

As in the past, we hope that this volume will be of interest to aquatic biologists, ichthyologists, fisheries scientists, and, of course, to comparative physiologists who still want to learn more about the physiological strategies that are unique to fishes, and those shared with other organisms.

Editors

David H. Evans, PhD, is professor emeritus of biology at the University of Florida (UF) and adjunct professor at the Mt. Desert Island Biological Laboratory (MDIBL). He received his AB in zoology from DePauw University, Indiana, in 1962 and his PhD in biological sciences from Stanford University, California, in 1967. He held postdoctoral positions in the Biological Sciences Department of the University of Lancaster, United Kingdom, and the Groupe de Biologie Marine du C.E.A., Villefranche-sur-mer, France, in 1967–1968. In 1969, he joined the Department of Biology at the University of Miami, Florida, as assistant professor and served as professor and chair from 1978 to 1981, when he became professor of zoology at UF. He also served as chair of zoology at UF from 1982 to 1985 and again from 2001 to 2006.

Dr. Evans served as director of the MDIBL, Salisbury Cove, Maine, from 1983 to 1992, as well as director of the MDIBL's Center for Membrane Toxicity Studies from 1985 to 1992. He has also served on a White House, Office of Science and Technology Policy, Acid Rain Peer Review Panel from 1982 to 1984 and the Physiology and Behavior Panel of the National Science Foundation from 1992 to 1996. He is a member of the Society for Experimental Biology, Sigma Xi, and the American Physiological Society. He has served on the editorial boards of *The Biological Bulletin*, *Journal of Experimental Biology*, *Journal of Experimental Zoology*, *Journal of Comparative Physiology*, and *American Journal of Physiology (Regulatory, Comparative, Integrative Physiology)*.

Dr. Evans received the University of Miami, Alpha Epsilon Delta, Premedical Teacher of the Year Award in 1974; the University of Florida and College of Liberal Arts and Sciences Outstanding Teacher Awards in 1992; the UF Teacher-Scholar of the Year Award in 1993; the Florida Blue Key Distinguished Faculty Award in 1994; the UF Professorial Excellence Program Award in 1996; the UF Chapter of Sigma Xi Senior Research Award in 1998; a UF Research Foundation Professorship in 2001; the August Krogh Distinguished Lectureship from the American Physiological Society in 2008; and the William S. Hoar Lectureship from the Department of Zoology at the University of British Columbia in 2011. In 1999, he was elected a fellow of the American Association for the Advancement of Science.

Dr. Evans has presented over 20 invitational lectures at international meetings and has published over 130 papers and book chapters. In addition to continuous funding from the National Science Foundation from 1970 to 2010, he has received grants from the National Institute of Environmental Health Sciences and the American Heart Association. Since retirement in 2007 and the closing of his laboratory in 2010, he has been writing the comprehensive history of the MDIBL.

James B. Claiborne, PhD, is a professor emeritus of biology at Georgia Southern University. He received his BS from Florida State University in 1977 and his PhD from the University of Miami (under the direction of Dr. David Evans) in 1981. He conducted postdoctoral research with Dr. Norbert Heisler at the Max Planck Institute for Experimental Medicine, Göttingen, Germany, and the Zoological Station of Naples, Naples, Italy. He joined the faculty at Georgia Southern as an assistant professor in 1983, served as acting department chair from 1987 to 1988, was promoted to full professor in 1996, and was named professor emeritus in 2012.

Dr. Claiborne has been a visiting scientist at the Mount Desert Island Biological Laboratory (MDIBL), Salisbury Cove, Maine, since 1986 and has served as editor of the *Bulletin of the Mt. Desert Island Biological Laboratory* and on the MDIBL Board of Trustees for many years. He has been a member of the American Physiological Society, the Society for Experimental Biology, and Sigma Xi. He has served on numerous external review panels for the National Science Foundation and was the recipient of the decennially awarded John Olin Eidson Presidential Award for Excellence in 2003.

He is a two-time winner of the GSU Award for Excellence in Research (1987, 1998) and was recognized by the biology department with a Parrish Scholar's Award in 2010.

"J.B." has taught a wide array of courses (14) during his career, and has mentored more than 40 high school, undergraduate, and graduate research students in his laboratory at GSU and at the MDIBL. He has published more than 100 refereed papers, short communications, book chapters, and edited volumes, several of which were coauthored with his students. His research and teaching activities have been funded through numerous research and Research Experiences for Undergraduates grants from the National Science Foundation from 1986 through 2014. He currently serves as managing member at Vote-now.com LLC.

Suzanne Currie, PhD, is a professor of biology and the Harold Crabtree Chair in Aquatic Animal Physiology at Mount Allison University in Sackville, New Brunswick, Canada. She received her BSc (honors) in biology from Acadia University in Wolfville, Nova Scotia, and then went on to earn an MSc and PhD in biology at Queen's University in Kingston, Ontario. She was the Charles and Katherine Darwin Research Fellow at the University of Cambridge during her postdoctoral studies in the Department of Zoology.

Dr. Currie's research is focused on the strategies aquatic animals use to cope with environmental stress. She is interested in understanding how physiological responses to changing environments are integrated with fish behavior and cell biology. With her students, postdoctoral fellows, and national and international collaborators, she studies a variety of freshwater and marine species in temperate and tropical environments. In the last few years, she has particularly enjoyed her collaborative ecophysiological field work in the tropics.

Dr. Currie is an active member of the Canadian Society of Zoologists and currently serves as second vice president. She is a member of the Society for Experimental Biology, the American Physiological Society, and the American Fisheries Society. She has also served on several national committees and panels for the Natural Sciences and Engineering Research Council of Canada. Dr. Currie serves on the editorial boards of *Physiological and Biochemical Zoology*, *Comparative Biochemistry and Physiology*, and *Bulletin of the Mount Desert Island Biological Laboratory* and is a review editor for *Frontiers in Aquatic Physiology*.

Contributors

James S. Ballantyne
Department of Integrative Biology
University of Guelph
Guelph, Ontario, Canada

Victoria A. Braithwaite
Department of Ecosystem Science and
    Management and Biology
Center for Brain, Behavior and Cognition
Penn State University
University Park, Pennsylvania

Suzanne Currie
Department of Biology
Mount Allison University
Sackville, New Brunswick, Canada

A. Kurt Gamperl
Ocean Sciences Centre
Memorial University of Newfoundland
St. John's, Newfoundland, Canada

Warren W. Green
Department of Biological Sciences
University of Windsor
Windsor, Ontario, Canada

Martin Grosell
Rosenstiel School of Marine and
    Atmospheric Science
University of Miami
Miami, Florida

Heather J. Hamlin
School of Marine Sciences
University of Maine
Orono, Maine

Pung-Pung Hwang
Institute of Cellular and Organismic Biology
Academia Sinica
Taipei, Taiwan, Republic of China

Ida Beitnes Johansen
Department of Biosciences
University of Oslo
Oslo, Norway

Michael G. Jonz
Department of Biology
University of Ottawa
Ottawa, Ontario, Canada

John E. Lewis
Department of Biology
and
Centre for Neural Dynamics
University of Ottawa
Ottawa, Ontario, Canada

Li-Yih Lin
Department of Life Science
National Taiwan Normal University
Taipei, Taiwan, Republic of China

Grant B. McClelland
Department of Biology
McMaster University
Hamilton, Ontario, Canada

Øyvind Øverli
Department of Animal and
    Aquacultural Sciences
Norwegian University of Life
    Sciences
Aas, Norway

Patricia M. Schulte
Department of Zoology
The University of British Columbia
Vancouver, British Columbia, Canada

Graham R. Scott
Department of Biology
McMaster University
Hamilton, Ontario, Canada

Holly A. Shiels
Faculty of Life Sciences
University of Manchester
Manchester, United Kingdom

Ruxandra F. Sîrbulescu
Laboratory of Neurobiology
Department of Biology
Northeastern University
Boston, Massachusetts

Christina Sørensen
Department of Biosciences
University of Oslo
Oslo, Norway

Viravuth P. Yin
Davis Center for Regenerative Biology and
    Medicine
Mount Desert Island Biological
    Laboratory
Salisbury Cove, Maine

Barbara S. Zielinski
Department of Biological Sciences
University of Windsor
Windsor, Ontario, Canada

Günther K.H. Zupanc
Laboratory of Neurobiology
Department of Biology
Northeastern University
Boston, Massachusetts

1 Muscle Plasticity

Grant B. McClelland and Graham R. Scott

CONTENTS

1.1 GENERAL INTRODUCTION

Muscle phenotype is the result of interactions between an animal's genotype and the environment. A single genotype can result in multiple phenotypes depending on environmental influences in what is known as phenotypic plasticity (Rezende et al., 2005) (Figure 1.1). In adult fishes, phenotypic plasticity occurs seasonally or with movement across environmental clines, and these changes are thought to be reversible. In contrast, environmental influences during ontogeny can lead to distinct and irreversible muscle phenotypes as an expression of different developmental trajectories (Spicer and Burggren, 2003) (Figure 1.1, Section 1.4). The importance of early life experience on the muscle phenotypes expressed later in life has long been recognized, but only recently has the influence of developmental exposures on muscle plasticity and swimming performance in adult fish been explicitly demonstrated (Scott and Johnston, 2012) (Figure 1.6). Skeletal muscles play numerous important roles in fishes, from powering locomotion to regulating whole-body metabolic homeostasis (Craig and Moon, 2011). Since muscles constitute a significant proportion of a fish's body mass, the ability to modulate the mechanical and metabolic properties of this tissue, in-line with environmental optima, is an important feature allowing species to adjust to changing environmental conditions. Phenotypic plasticity of the muscle is thus an important mechanism assuring that ecologically relevant tasks can still be performed in the face of environmental, mechanical, and/or energetic stress (Figure 1.1).

Fish are often faced with abiotic stressors in their aquatic habitat that they cannot avoid. Intra- and inter-specific variation in the magnitude of the response to a given stressor can reflect differences in the induction of "muscle genes" and/or the degree of nongenetic regulation (e.g., pH) of muscle phenotype. The magnitude of muscle plasticity can be enhanced, blunted, or even absent depending on the evolutionary forces that have shaped a particular species, which often results in

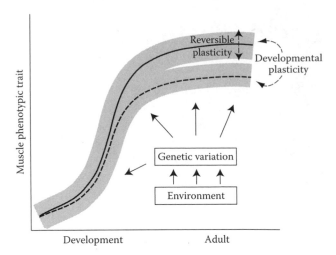

FIGURE 1.1 Muscle phenotype can change in early life through the expression of different developmental trajectories in the process of developmental plasticity. This irreversible developmental plasticity may impact upon the reversible phenotypic plasticity as adults. Environmental stress acts upon population genetic variation to affect different degrees of muscle plasticity at all stages. (Modified from Rezende, E.L. et al., *Revista Chilena de Historia Natural*, 78, 323, 2005; Garland, T. and Carter, P.A., *Ann. Rev. Physiol.*, 56, 579, 1994.)

a gradation of responses from non-, low-, or high responders. It has been suggested that the patchiness or "grain" of an organism's environment affects whether it is a specialist or generalist in terms of its acclimation response (Storz et al., 2010; Tattersall et al., 2012). An extreme example of a specialist strategy is exhibited by stenothermal Antarctic (*notothenioid*) fishes, which have lost the ability to mount a significant heat stress response when exposed to high temperatures (Hofmann et al., 2000), presumably because they are specialized to live in a "course-grained" environment in which temperature fluctuates little. The eurythermal killifish (*Fundulus heteroclitus*) is an example of a species with a generalist strategy that undergoes significant muscle plasticity with changing temperature (Fangue et al., 2009), reflecting the "fine-grained" nature of their thermal environments. Thus, the variability of any abiotic stressor in a species' natural environment may define the limits of their muscle phenotypic plasticity in response to that variable.

Muscle function affects locomotory performance, which can in many cases have profound impacts on whole-organismal performance and ultimately fitness (Garland and Carter, 1994). Changes in performance have therefore been regularly used as an outcome measure of muscle plasticity with prolonged environmental or energetic stress (McClelland et al., 2006; Scott and Johnston, 2012). Since many environmental variables can affect the metabolic and morphological characteristics of muscles that are responsible for power output and sustained locomotion, it is crucial that fish respond appropriately to stress. Although muscle plasticity can have a profound effect on fish survival, current understanding of the mechanistic underpinnings involved in remodeling of this tissue is somewhat limited.

1.2 MUSCLE ANATOMY OF ADULT FISHES

The structural and functional characteristics of fish muscle provide the means for suitable activity patterns. The different muscle fiber types in fishes have been characterized as fast, intermediate, and slow twitch. Unlike mosaic muscles in tetrapods, the three fiber types are anatomically separated (Figure 1.2), and the spatial orientation of the fast fibers is complex and distinct from other vertebrates. Fish white (fast) muscle is composed of type IIb fast fibers and supports burst locomotion. The fibers are arranged helically as hundreds of nested W-shaped blocks called myotomes, which are separated by sheets of connective tissue called myosepta. The myotomes run down the

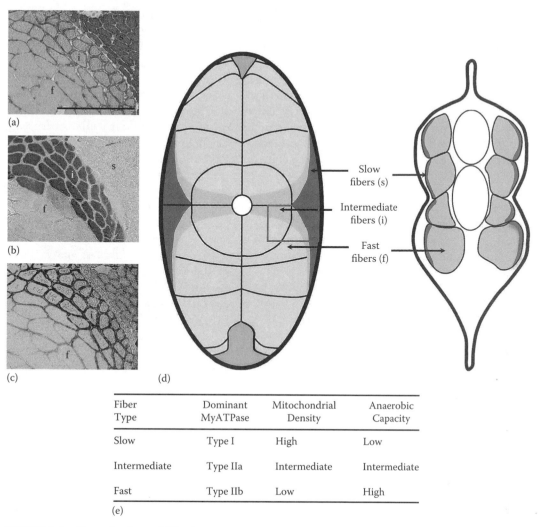

FIGURE 1.2 **(See color insert.)** The three main fiber types in the axial swimming muscle of fish are spatially segregated into distinct regions. Slow oxidative fibers (s) are located at the lateral periphery of the muscle, and in the adult fish, they are most abundant near the horizontal septum. Fast oxidative (intermediate) fibers (i) are located next to the slow fibers, and fast glycolytic fibers (f) make up the remainder of the axial musculature. (a–c) Representative histological sections from an adult zebrafish (Modified from Scott, G.R. and Johnston, I.A., *Proc. Natl. Acad. Sci. USA*, 109, 14247, 2012); (a) Slow fibers are identified using immunohistochemistry with the S58 antibody (scale bar is 50 μm). (b) Intermediate fibers are identified based on the activity of alkaline-resistant myosin-ATPase activity. (c) Slow and intermediate fibers both contain abundance mitochondria, reflected by the activity of succinate dehydrogenase activity. (d) Diagram of a transverse section through an adult fish (left) and an embryo during late segmentation (right) showing the arrangement of fiber types (the blue box indicates the area of the muscle shown in a–c). Serial sections of the same muscle fiber are represented with letters (s, i, f) in a–c. (e) summarizes the major phenotypic differences.

trunk of the body and are attached to posterior regions by tendons. In most fishes, the red (slow) muscle is situated superficially and parallel to the long axis of the body. It is composed of slow type I fibers (Figure 1.2) and supports axial undulatory locomotion (Alexander, 1969, 2003; Gemballa and Vogel, 2002). The distinct architecture of fast and slow fiber types ensures optimal changes in sarcomere length and relative shortening velocities to provide the tensions and power outputs needed for low and high swimming speeds, respectively (Rome, 1998).

White muscle is geared for high flux rates through glycolysis. The fibers of this muscle mass are larger than those in the red muscle and provide a higher power output (Rome, 2006). Red muscle fibers are smaller and owe their name to the high levels of myoglobin they possess to facilitate mitochondrial O_2 supply. They have greater DNA/g, greater mitochondrial densities, increased lipid stores, and a higher capacity for fatty acid oxidation, compared to the white muscle (Bone, 1966; Johnston, 1977; Leary et al., 2003; Morash et al., 2008). In most fish species, up to 85% of the trunk musculature and 60% of the body mass consist of white muscles (Sänger and Stoiber, 2001). The proportions of red muscle vary greatly between species, and some species have a third muscle type that is pink in appearance with intermediate contractile (type IIa myosin-ATPase) and metabolic properties (Figure 1.2). Different degrees of plasticity in these three muscle types occur because each detects and responds to stressors in distinct ways. Moreover, within a single muscle type, the response to one particular stressor may involve an increase in fiber size (i.e., hypertrophy), while the same muscle may respond to another stressor with a change in its metabolic phenotype but no hypertrophic growth (McClelland et al., 2006).

The spatial separation of muscle fiber types has facilitated the characterization of recruitment patterns with changes in swimming intensity (Johnston, 1980). As swimming speed increases, slow muscle fibers are initially recruited, followed by intermediate, and then finally fast muscle fibers at higher intensities. The absolute threshold swimming speed for the recruitment of different fiber types differs greatly between species and environmental conditions (Johnston, 1980). White muscle may be recruited during sustained swimming in some species but is restricted to unsustainable burst swimming in others, and these differences can be reflective of interspecific variation in the aerobic capacity and innervation patterns of this muscle type (Johnston, 1981). It stands to reason that variation in the response of white muscle to exercise training seen in different studies (Johnston and Moon, 1980; McClelland et al., 2006) is due to the choice of training speed, differences in muscle recruitment patterns, or the properties of a species' white muscle.

1.3 MECHANISMS OF MUSCLE PLASTICITY IN ADULTS

The focus of this chapter is on the events that lead to phenotypic changes in skeletal muscles; however, there are many interesting examples of plasticity in other muscle types, such as cardiac muscle (McClelland et al., 2005). Fish muscle is known to respond with hypertrophic and hyperplastic growth (increases in fiber size or number, respectively), through changes in metabolic phenotype, and via cell-structural changes such as membrane lipid composition (Hazel, 1995; Morash et al., 2008, 2009). Changes in muscle metabolic capacity can occur by dosage compensation where levels of existing proteins are increased or by the expression of different isoforms with distinct reaction kinetics. Mitochondria may change both quantitatively and qualitatively (e.g., changes in their capacity for O_2 consumption, lipid or carbohydrate oxidation, ATP synthesis, etc.). The contractile properties and myosin ATPase activity of muscle may also be modified by building more contractile machinery and/or by changing myosin heavy or light chain (MyHC, MyLC) isoform expression (Sänger, 1993) (Figure 1.3). These aspects of muscle plasticity have been studied for many years, but most of the molecular mechanisms responsible for these changes are not fully characterized. Our current knowledge is based in a limited number of species and only a few direct demonstrations of mechanism. Muscle research on fish often takes its cues from the mammalian literature with the *a priori* assumption that fish muscle responds similarly and through common mechanisms to environmental and metabolic perturbations. Some of these regulatory mechanisms appear to be conserved (McClelland et al., 2006), but there are many exceptions where fishes differ from mammals, for instance, (1) stressors such as cold can elicit a unique acclimation response in fish (McClelland et al., 2006; LeMoine et al., 2008; O'Brien, 2011), (2) some regulators central to muscle plasticity in mammals, such as peroxisome proliferator-activated receptor (PPAR)-gamma co-activator (PGC)-1α and myostatin (*mstn*), may not play a direct or the same role in fishes (McClelland et al., 2006; Johnston et al., 2008; LeMoine et al., 2008, 2010a,b), and (3) whole genome duplication (WGD) events in the course of fish evolution may have increased the number and types of regulators involved in muscle plasticity (Rescan et al., 2001; LeMoine et al., 2010a,b). WGD events can create distinct protein isoforms that are differentially expressed in different conditions (Morash et al., 2010; Morash and McClelland, 2011), thus providing a means to "fine tune" the acclimation response. However, not all fish lineages have retained the same paralogues in their genome, suggesting that there may be species-specific responses to the same stressor. Fishes may exhibit immediate genomic responses to stress (LeMoine et al., 2010a), but may also take several weeks of acclimation to show changes in gene expression for regulators of muscle plasticity (O'Brien, 2011). Nevertheless, it is a starting point to first presume that homologous genes and proteins have similar functions in regulating muscle plasticity in fishes as in mammals (the major regulators studied in fishes and mammals are outlined in Figure 1.3).

To respond to environmental and energetic stress, skeletal muscle must detect the perturbation, transmit the signal to the cell's genetic and metabolic machinery, and initiate a compensatory response, bringing its phenotype closer to a functional optimum (Figure 1.3). Plasticity may be initiated endogenously by changes in energy status through changes in the so-called cellular energy sensors (reviewed in Hock and Kralli, 2009) or exogenously by hormones, neurotransmitters, or autocrine/paracrine agents that bind to membrane or intracellular receptors. Genomic and nongenomic responses to stress can be triggered by second messenger pathways or by direct transduction of energy imbalance to the transcriptional machinery. The sarcolemma contains receptors for the major growth regulators insulin-like growth factors (IGF)-I and II, myostatin, leptin, and adiponectin (Hock and Kralli, 2009; Johnston et al., 2011). Membrane-bound adrenergic receptors (e.g., β-AR) respond to changes in the adrenal and/or sympathetic nervous system stimulation (e.g., epinephrine and norepinephrine). Muscle cells are also responsive to deformation, and stretch receptors trigger the release of autocrine factors that interact with a separate set of membrane receptors (Gundersen, 2011). The high-energy phosphate molecules, ADP, AMP, and ATP, and those responsible for cellular reduction to oxidation (redox) status (e.g., NAD^+ and NADH) change in response to mismatches between energy supply and energy demand. These changes activate AMP-kinase (AMPK) and the NAD^+-dependent protein deacetylase (SIRT)-1, respectively, to initiate cell signaling and gene expression changes. Muscle cells also respond to changes in cellular oxygen tension through the prolyl-hydroxylase domain (PHD) proteins that regulate the hypoxia inducible factor (HIF)-1α, a highly conserved transcription factor that regulates hypoxia responsive genes (Rytkönen and Storz, 2011). In addition, changes in lipid metabolism may affect cytosolic free fatty acid (FFA) concentrations and these FFAs can serve as ligands for the PPAR family of nuclear receptors, inducing the transcription of a number of genes encoding for mitochondrial proteins (Kondo et al., 2010; Gundersen, 2011). Many of these mechanisms have not been directly or extensively studied in fish, but cellular fluctuations in reactive oxygen species (ROS) and nitric oxide (NO) have been implicated in some instances of muscle remodeling (O'Brien, 2011) (Figure 1.3). For example, ROS and NO can induce cellular metabolic remodeling in mammals (Hock and Kralli, 2009) and are affected by cold acclimation in fishes (Malek et al., 2004) and exercise in mammals (Leary and Shoubridge, 2003) (as discussed in the following).

Downstream of the cell's ability to detect stressors is the cellular signal transduction machinery. The molecular features responsible for triggering changes in muscle growth, mitochondrial biogenesis, and contractile kinetics in response to single stressors may be unique to that stressor or may involve overlapping downstream pathways (Figure 1.3). For example, low temperature, exercise, leptin, adiponectin, and hypoxia may all tip the balance between ATP supply and demand and activate AMPK (Johnston et al., 2011). Both muscle growth and exercise show signaling through the PI3-K to serine/threonine-specific protein kinase (Akt) pathway, which regulates the expression of the mammalian target of rapamycin (mTor). This stimulates translation and upregulation of the myogenic regulatory factor (myoD) inducing gene expression, but by phosphorylating the forkhead box O (FOXO) transcription factor, this pathway also suppresses protein breakdown (Johnston et al., 2011). This pathway appears to be very important for regulating muscle growth by favoring protein synthesis over protein degradation. Myostatin is a factor that binds to the activin receptor on the muscle cell membrane. Acting through the FOXO transcription pathway, myostatin suppresses muscle differentiation and

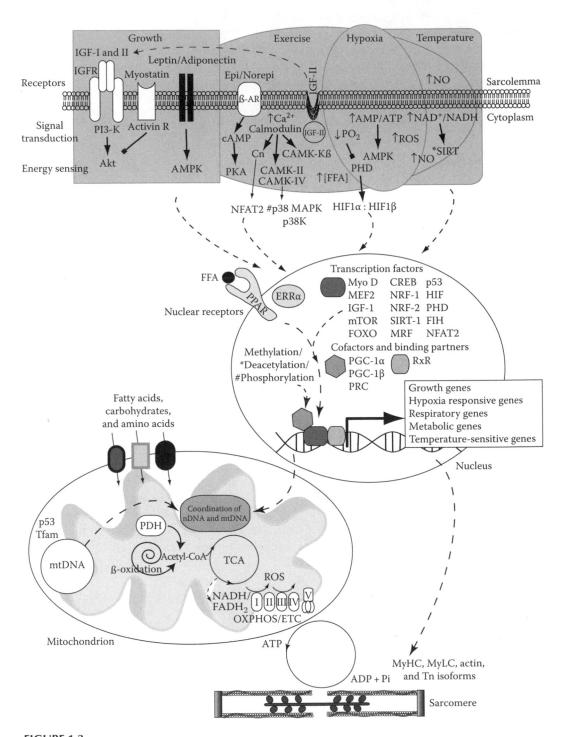

FIGURE 1.3

growth via regulation of various proteolytic functions by inhibiting Akt signaling (Carpio et al., 2009; Seiliez et al., 2011). Intracellular Ca^{2+} is a potent second messenger involved in triggering muscle plasticity by stimulating mitochondrial biogenesis. Ca^{2+} increases during excitation–contraction coupling but also with increases in oxidative stress (Schiaffino and Reggiani, 2011). Changes in Ca^{2+} are likely sensed by calmodulin to activate many downstream messenger proteins and kinases, principally the Ca^{2+}/calmodulin-dependent protein kinases (CAMK-II, IV, CAMKKβ) and the phosphatase calcineurin (Cn) (Hock and Kralli, 2009; Gundersen, 2011). Calcineurin appears to play an important role in regulating muscle fiber type in mammals by inducing the translocation of the calcineurin-dependent nuclear factor of activated T-cells (NFAT) to the nucleus, thus inducing the expression of slow MyHC genes (Pandorf et al., 2009). Calcineurin also activates the myocyte enhancer factor (MEF)-2 to affect type I MyHC expression (Pandorf et al., 2009; Schiaffino and Reggiani, 2011).

Muscles fibers respond to various stressors by enhancing their aerobic capacity through increased mitochondrial volume. Mitochondrial biogenesis has been extensively studied in mammals and research in this area on fishes has gained attention over the last decade. The building of mitochondrial volume involves the coordination of the nuclear and mitochondrial genomes. Only 13 of the proteins that make up mitochondria are encoded by mtDNA and all of these are for components of the electron transport chain (ETC). Moreover, the regulation of mtDNA is reliant on nuclear DNA (Scarpulla et al., 2012). For simplicity, regulation of the mitochondrial biogenesis pathway can be divided into the actions of three DNA-binding factors: (1) the PPARs, which regulate genes encoding for the fatty acid oxidation (β-oxidation) pathway; (2) estrogen-related receptors (ERR), which regulate genes encoding for the fatty acid oxidation pathway, the tricarboxylic acid cycle (TCA), and oxidative phosphorylation (OXPHOS); and (3) nuclear respiratory factors (NRF), which regulate respiratory genes for OXPHOS and the transcription factor A-mitochondrial (Tfam). NRF-1 and NRF-2 are implicated in regulating the expression of all 10 nuclear-encoded COX subunits and the expression of COX-17, a Cu chaperone

---

FIGURE 1.3 (continued) (See color insert.) The mechanisms of muscle plasticity may overlap in response to exercise, hypoxia, and temperature, and through signals for growth. Signals interact with membrane receptors (R) for insulin-like growth factor (IGFR), myostatin (activin R), leptin and adiponectin, and epinephrine (Epi) and norepinephrine (Norepi) (beta-adrenergic receptor, β-AR). Signal transduction pathways may act through cAMP or PI3-K to activate or inhibit protein kinase A (PKA), serine/threonine-specific protein kinase (Akt), or AMPK. Changes in intracellular calcium act through calmodulin to activate calcineurin (Cn), Ca^{2+}/calmodulin-dependent protein kinases (CAMK), PI3-K, nuclear factor of activated T-cells (NFAT), and MAPK. Reduction in intracellular PO_2 inhibits the prolyl-hydroxylase domain (PHD), leading to the stabilization of the hypoxia inducible factor (HIF)-1α, which dimerizes with the constitutively expressed HIF-1β subunit. Energy perturbations are "sensed" by changes in intracellular AMP/ATP, NAD^+/NADH, reactive oxygen species (ROS), and nitric oxide (NO). These can activate AMPK and NAD^+-dependent protein deacetylase (SIRT), but can also increase cytosolic free fatty acid concentrations (FFA) that can interact with nuclear receptors. The signal transduction and energy sensing responses affect gene transcription through the actions of a number of transcription factors: cAMP response element binding (CREB) protein, estrogen-related receptors (ERR), myogenic regulatory factor (MyoD), myocyte enhancer factor (MEF), mammalian target of rapamycin (mTOR), forkhead box O (FOXO), nuclear respiratory factor (NRF), myogenic regulatory factor (MRF), tumor suppressor protein (p53), factor inhibiting HIF (FIH), peroxisome proliferator-activated receptor (PPAR), transcription factor A-mitochondrial (Tfam), in coordination with important cofactors: PPAR-gamma coactivator (PGC)1α, PGC-1-related coactivator (PRC) and PGC1β, and binding partner, retinoic acid receptor (R×R). These transcription factors change the expression of genes that alter fiber size, metabolic machinery (e.g., glycolysis), mitochondrial biogenesis (e.g., tricarboxylic acid cycle [TCA], electron transport chain [ETC]), and sarcomere structure (e.g., myosin heavy chain and light chain [MyHC, MyLC] and troponin [Tn] isoforms). Arrows indicate an activation and blunt-ended lines indicate an inhibitory effect. (Based on data from fish and mammals, modified from Hock, M.B. and Kralli, A., *Annu. Rev. Physiol.*, 71, 177, 2009; Gundersen, K., *Biol. Rev.*, 86, 564, 2011; Johnston, I.A. et al., Molecular biotechnology of development and growth in fish muscle, in: Tsukamoto, K., Takeuchi, T., Beard, Jr. T.D., and Kaiser, M.J., Eds., *Fisheries for Global Welfare and Environment*, 5th World Fisheries Congress, Terrapub, pp. 241–262, 2008; Johnston, I.A. et al., *J. Exp. Biol.*, 214, 1617, 2011; Saleem, A. et al., *Exerc. Sport Sci. Rev.*, 39(4), 199, 2011; Scarpulla, R.C. et al., *Trend. Endocrinol. Metabol.*, 23(9), 459, 2012.)

and COX assembly factor (Hock and Kralli, 2009; Scarpulla et al., 2012). The actions of all of these DNA-binding factors are reliant on cofactors and binding factors to properly induce expression of target genes. The cofactor PGC-1α is now thought of as the master regulator of mitochondrial biogenesis in mammals (Scarpulla et al., 2012), as it drives most aspects of this process in concert with PGC-1β and PGC-related coactivator (PRC). However, the role of these cofactors in regulating mitochondrial biogenesis in the fishes has recently been questioned (LeMoine et al., 2010a,b; Bremer et al., 2012). PRC and PGC-1β are also growth-related factors that are essential for early embryonic development in mammals by promoting high rates of mitochondrial gene transcription. These factors respond dynamically to changes in cellular growth, differentiation, and energy metabolism. The actions of PGC-1α can be fine-tuned by posttranslational modification. Phosphorylation of PGC-1α occurs via p38-MAPK, Akt, and AMPK, methylation occurs by the protein arginine methyltransferase (PRMT)-1, and deacetylation by SIRT-1 can occur in response to energy depletion (Hock and Kralli, 2009). The induction of PGC-1α expression involves the upregulation and binding of transcription factors cAMP response element-binding (CREB) protein, MEF-2, PPARs, MyoD, and others in response to adrenergic stimulation, Ca^{2+} signaling via CAMK, and changes in cellular energy status (Hock and Kralli, 2009). Recently, p53, a tumor suppressor protein that is activated as part of cell stress response, has been implicated in the control of oxidative metabolism and mitochondrial biogenesis in mammals (Saleem et al., 2011). For example, p53 may increase PGC-1α expression and, in interacting with mtDNA, may help maintain genome integrity. Whether p53 functions in the same way in fishes is unknown. In the following sections, the mechanisms for muscle plasticity that have been investigated in fishes will be reviewed. Interestingly, of the factors examined so far, only NRF-1 has shown a clear correlation with the expression of genes encoding mitochondrial proteins in fish muscle (McClelland et al., 2006; LeMoine et al., 2008; Bremer and Moyes, 2011) (Figure 1.3).

1.3.1 ADULT MUSCLE GROWTH

The growth of muscles in adult fishes occurs as a part of their natural life cycle, in response to changes in nutrient availability, exercise, or abiotic features of their environment (Chauvigné et al., 2003; McClelland et al., 2006; Farrell, 2009; Palstra et al., 2010; Johnston et al., 2011). Muscle growth can occur by hyperplasia (increases in cell number) and hypertrophy (increases in cell size). Because muscle fibers are terminally differentiated and multinucleated cells, both types of growth require the recruitment of myogenic precursor cells (MPCs), which are distributed throughout the myotome as well as present in a dense collection in the external cell layer. In hyperplasia, MPCs fuse to form new multinucleated fibers. In hypertrophy, the MPCs are absorbed into existing cells and donate their DNA to the expanding fiber (Johnston et al., 2011). The addition of DNA probably serves to maintain the size of the myonuclear domain, thus preserving the number of nuclei for a given cytosolic volume, which would help avoid diffusion constraints (Johnston et al., 2011; Kinsey et al., 2011; Van der Meer et al., 2011). The activation and proliferation of MPCs in fish is possibly controlled by several factors such as IGF, GH, and myostatin (Johnston et al., 2011). Signaling by these factors affects the balance between anabolic and catabolic processes within muscle fibers, and growth occurs when protein synthesis outweighs degradation. The highly conserved myogenic regulatory factors (MRFs), myoD, myogenin, and other transcription factors play a central role in the differentiation and growth of muscle in both mammals and fishes by inducing muscle gene transcription (Johnston et al., 2008, 2011; Gundersen, 2011) (Figure 1.3).

The potential stimulation of growth by many factors has been explored in fish, including myostatin, follistatin, GH, IGF-I and II, and 11-keto-testosterone. In mammals, myostatin produced by the myocytes is a negative regulator of growth. It inhibits differentiation and thus limits hyperplastic muscle growth. This is supported by data on mutations in the myostatin gene, which lead to hypertrophic and hyperplastic muscle growth in dogs (Mosher et al., 2007) and other mammals. Trout express four myostatin genes, separated into the *mstn-1* and *mstn-2* clade, with four paralogues (*mstn-1a*, *mstn-1b* and *mstn-2a*, *mstn-2b*), which are differentially expressed in red and white muscles (Rescan et al., 2001). However, their ubiquitous tissue expression suggests a more diverse role in fishes compared to mammals (Carpio et al., 2009). In addition, changes in myostatin peptide did not correlate with muscle fiber hypertrophy in swim-trained trout (Martin and Johnston, 2005). The expression of the myostatin gene is regulated in mammals by the FOXO1 transcription factor. However, neither changes in the activity of FOXO1 nor another isoform, FOXO4, resulted in a change in *mstn-1a* or *mstn-1b* expression in a trout myocyte culture model (Seiliez et al., 2011). In contrast, reducing myostatin binding to its endogenous receptor by injecting a recombinant soluble version of the myostatin receptor (the activin IIB receptor) increased hyperplastic muscle growth in goldfish (Carpio et al., 2009). A similar role of myostatin in some fish and mammal species is supported by work with a myostatin knockdown zebrafish strain, which shows increased expression of MRFs myoD and myogenin and a 45% larger body weight (Lee et al., 2009). Based on these inconsistencies in the literature, the function and regulation of myostatin are still considered to be unresolved in fishes (Biga and Goetz, 2006; Seiliez et al., 2011).

Follistatin is a glycoprotein that can inhibit myostatin as well as other growth factors in mammals. Two paralogues of the follistatin gene exist in trout, and these genes are expressed in both fast and slow muscles in some species but have a more restricted tissue pattern in others (Medeiros et al., 2009; Johnston et al., 2011). The so-called six pack trout, because of their enlarged epaxial and hypaxial muscle mass, are thought to be the result of increased follistatin inactivation of myostatin (Medeiros et al., 2009).

Growth hormone (GH) can also suppress the actions of myostatin in mammals, but has varied effects on fishes, depending on the species. For instance, the external administration of GH caused only a modest stimulation of growth in zebrafish (Biga and Goetz, 2006). GH injection in trout upregulated *mstn1a* but downregulated *mstn1b* expression (Biga et al., 2004; Gahr et al., 2008), and GH decreased expressions of both genes in an *in vitro* trout myotube model (Seiliez et al., 2011). Moreover, transgenic salmon with an extra copy of the GH gene increase muscle mass (Devlin et al., 2001) by decreasing myostatin expression (Roberts et al., 2004).

Another mode of GH action is the stimulation of IGF-I production by the liver (Wood et al., 2005). The GH and IGF cell signaling systems appear to be conserved in vertebrates but some aspects of regulation are probably unique in fish, in part, due to the cyclical nature of growth (Johnston et al., 2008). IGF-I and II, the IGF-receptors, and several paralogues have been detected in fish (Chauvigné et al., 2003; Johnston et al., 2008) and are the main endocrine and autocrine regulators of skeletal muscle growth. In fact, the muscle growth response of fish to IGF is enhanced during periods of fasting when the density of IGF-R is increased (Chauvigné et al., 2003; Johnston et al., 2011). The signaling pathway from IGF-R to PI3-K and Akt may be conserved from fishes to mammals (Seiliez et al., 2011) (Figure 1.3). For example, treatment of cultured trout muscle cells with IGF-I leads to FOXO3 transcription factor phosphorylation, leading to the downregulation of expression of a downstream target gene, but surprisingly this did not include *mstn1a* or *mstn1b* (Seiliez et al., 2011). Exercise too can be a potent growth stimulus in fishes (see Section 1.3.3). The mechanism for this stimulation of growth probably involves the secretion of IGF from myocytes as they exert force. The IGF released binds to sarcolemmal IGF-R to stimulate protein synthesis and limit protein degradation.

Testosterone is another hormone that promotes protein synthesis in mammals, and the principle androgen in fishes, 11-keto-testosterone, also stimulates muscle growth in male trout (Thorarensen et al., 1996). However, the mechanisms for androgen action on fish muscle are unresolved. For example, testosterone administration did not affect protein turnover in isolated trout myocytes (Cleveland and Weber, 2011).

1.3.2 Temperature

Temperature is a potent stimulus for muscle plasticity in adult fishes. Some authors have even referred to temperature as the "abiotic master factor" because it is one of the most pervasive environmental stressors affecting the physiology of aquatic organisms (Brett, 1971). Not surprisingly, temperature is one of the most studied aspects of muscle plasticity in fishes. Acute exposure to cold decreases rates of diffusion and the catalytic rate of most enzymes. To maintain its function, muscle must either change the catalytic efficiency (i.e., express different isoforms) or change the amount of enzymes to compensate for the loss of kinetic energy (Tattersall et al., 2012). Muscle may also attempt to reduce cellular diffusion distances to account for the decelerating effect of low temperature on molecular motion (Guderley, 2004).

One of the ways fishes compensate for low environmental temperature is by increasing muscle mitochondrial volume (Johnston and Maitland, 1980; Dhillon and Schulte, 2011), mitochondrial oxidative capacity, and myoglobin content (Guderley, 2004) (Figure 1.7). Cold acclimation increases the components of the ETC, enzymes in the TCA, and enzymes involved in fatty acid oxidation (in some but not all species) (Johnston and Moon, 1980; McClelland et al., 2006). Chronic cold exposure also increases the mitochondrial ADP phosphorylation capacity by increasing the activity of the F1F0-ATPase. The greater mitochondrial membrane density that is afforded by the increase in mitochondrial volume also provides a conduit for intracellular oxygen diffusion, because oxygen solubility is higher in membrane lipids than aqueous cytoplasm (O'Brien, 2011). This combines with increases in muscle capillarity to improve oxygen diffusion capacity to mitochondria in the cold (Egginton and Sidell, 1989; Sidell, 1998). Fishes must also maintain contractile properties and membrane fluidity in the muscle with declining temperature (Johnston and Temple, 2002; McClelland, 2004; O'Brien, 2011). Cold exposure increases the proportion of red muscle mass in some fishes to compensate for the reduction in aerobic power output when the temperature is lowered (Johnston and Lucking, 1978). Cold also modifies the protein isoform complement of the contractile machinery (e.g., myHC composition) and increases the unsaturation index of sarcolemmal and mitochondrial membrane phospholipids (Hazel, 1995; Guderley, 2004).

Despite the wide interest in the physiological changes associated with temperature acclimation in fish, few of the underlying molecular mechanisms responsible for muscle remodeling are known. How muscle cells detect a drop in temperature to initiate the acclimation response is probably the least understood aspect of this process. Although no sarcolemmal thermal sensor has been identified, a drop in temperature may affect the cellular energy-sensing pathways that act via the sirtuins (SIRT), AMPK, ROS, thyroid hormone, and possibly NO (O'Brien, 2011). For example, if the thermal sensitivity of ATP production is greater than that of ATP consumption, then corresponding increases in cytosolic AMP would activate AMPK to compensate for this energy imbalance (Tattersall et al., 2012). AMPK may also increase NAD^+, thus activating SIRT-1 and subsequently the deacetylation of PGC-1α to promote mitochondrial biogenesis. However, the exact role of PGC-1α is currently unclear in the thermal response of fish muscle (discussed later). Alternatively, ROS are known to induce mitochondrial biogenesis in mammals (Scarpulla et al., 2012), and levels of ROS can increase with chronic cold in fish (Malek et al., 2004). These changes in oxidative stress may be a trigger for muscle remodeling. It has recently been proposed that ROS levels may increase due to cold-induced changes in membrane fluidity affecting ETC function or due to increased cellular O_2 concentration as O_2 solubility increases at low temperature (O'Brien, 2011). However, a direct link between changes in ROS and temperature and muscle remodeling in fish has not been demonstrated. In fact, cold exposures may even decrease ROS levels and induce the production of ROS-scavenging enzymes (Kammer et al., 2011). Another potential trigger for muscle remodeling is NO, known to stimulate mitochondrial biogenesis in mammals by its activation of PGC-1α (Leary and Shoubridge, 2003). Although many fish species express enzymes that synthesize NO, the NO synthases (NOS; O'Brien, 2011), some species (e.g., sticklebacks) do not express NOS in muscles nor show any changes in NO with cold acclimation (Mueller and O'Brien, 2011), but they readily remodel the muscle (Orczewska et al., 2010). This suggests that if NO does play a role in cold-induced mitochondrial biogenesis in fish muscle, it probably does so in a species-specific manner.

Although the ultimate trigger for cold-induced muscle remodeling has yet to be elucidated, some of the underlying molecular mechanisms that regulate the metabolic phenotype have been recently investigated. In contrast to mammals, past studies on fish show that the transcript changes for PGC-1α do not parallel changes in its putative target genes. In fact, with chronic cold exposure,

PGC-1α either does not change or declines in goldfish, zebrafish, and stickleback (McClelland et al., 2006; LeMoine et al., 2008; Orczewska et al., 2010; Bremer ad Moyes, 2011; Bremer et al., 2012). Structural differences in the PGC-1α gene between fishes and mammals, especially in upstream binding domains, may help explain this taxon-specific response (LeMoine et al., 2010b). However, a paralogue, PGC-1β, shows an increase in expression with cold exposure in goldfish muscle (LeMoine et al., 2008; Bremer and Moyes, 2011), and this response parallels changes in mRNA expression for the COX4-1 subunit of COX (Bremer and Moyes, 2011). This has led to the suggestion that it is PGC-1β, not PGC-1α, that serves as the master regulator for thermal remodeling in fish (Bremer and Moyes, 2011). However, lack of induction of PGC-1β in response to cold in another species casts some doubt on this hypothesis (Orczewska et al., 2010). That fishes respond to environmental stress by distinct molecular mechanisms is supported by data examining another DNA-binding factor that regulates muscle aerobic phenotype in mammals. The ERRα, which regulates many aspects of mitochondrial biogenesis in mammals, was not induced at the transcript or protein level after prolonged cold exposure in goldfish (Bremer et al., 2012). Moreover, cold acclimation in fish leads to a large decline in some of the nuclear receptors that regulate the fatty acid oxidation machinery in mitochondria of mammals. In zebrafish kept at 18°C for 4 weeks, the PPARα isoform declined but the PPARβ1 isoform showed no change in a mixed fiber-type muscle sample compared to 28°C controls (McClelland et al., 2006). PPARα was also found to decrease in the white muscle of cold-acclimated goldfish (Bremer et al., 2012), and the PPARβ isoform either increased or remained unchanged in this same species (LeMoine et al., 2008; Bremer et al., 2012). Similarly, medaka acclimated to 10°C showed no significant changes in the transcription factors PPARα1, PPARα2, PPARβ, and PPARγ in muscles compared to 30°C controls (Kondo et al., 2010). Cold acclimation data on R×R, an important DNA-binding partner that interacts with many of the nuclear receptors (Figure 1.3), has been equivocal (Bremer et al., 2012), making it hard to ascertain its role in fish muscle plasticity. Generally, the changes in the nuclear receptors do not suggest an important role in temperature-induced plasticity of muscle metabolic phenotype (Bremer et al., 2012). In contrast, NRF-1 expression is stimulated by cold exposure both at the transcript and protein levels (McClelland et al., 2006; LeMoine et al., 2008; Orczewska et al., 2010; Bremer et al., 2012), and this induction parallels increases in COX activity and COX subunit transcript expression (Orczewska et al., 2010; Bremer et al., 2012). These data support NRF-1 as a regulator of mitochondrial biogenesis whose function is conserved across taxa. Interestingly, the mitochondrial and nuclear genomes may respond distinctly to temperature. For example, genes from both the mitochondrial and nuclear genomes encode for the multisubunit F1F0-ATPase. In 10°C versus 30°C acclimated fishes, the transcription of the nuclear genes (α, β, γ F0-ATPase) increased twofold while transcripts for mitochondrial genes (e.g., ATPase 6–8) were six to seven times higher. The ratio of nuclear to mitochondrial DNA content was unaffected (Itoi et al., 2003). How this differential regulation of the two genomes occurs is unknown but certainly intriguing, given the tight coordination needed between them to ensure proper assembly of mitochondrial proteins.

The molecular mechanisms that regulate the changes in contractile phenotype and help counteract the declines in power output in the cold (Rome et al., 1984; Johnston et al., 1990; Johnston and Temple, 2002) have also been explored. Myosin ATPase activity increases with chronic cold (Johnston et al., 1990; Sidell, 1998), and changes in myosin heavy chain isoforms occur (Johnston et al., 2008) but with a temperature-dependent expression pattern (Watabe et al., 1992). The Ca^{2+}-binding protein parvalbumin has been implicated in variation in swimming performance in zebrafish (Seebacher and Walter, 2012), and calcium handling in muscles is temperature sensitive. Acclimation to cold in carp induces the transcription of parvalbumin and increases Ca^{2+}-ATPase and SR content (Johnston et al., 1990; Nelson et al., 2003), but some of these components show species-specific differences in temperature sensitivity (Erickson et al., 2005). Along with reduced diffusion distances from the sarcoplasmic reticulum to the sarcomere, these changes help decrease muscle twitch activation time in cold-acclimated fish (Johnston et al., 1990). The maximum rate of shortening (V_{max}) also increases from acute to chronic cold exposure, partly due to a change in myosin light chain (Crockford and Johnston, 1990) and myosin heavy chain isoform expressions (Watabe et al., 1992). However, it is important to note that temperature compensation in swimming performance is rarely complete.

1.3.3 EXERCISE

Skeletal muscle in adult fishes is highly responsive to repeated contractions, which induce significant changes in muscle growth and metabolic phenotype. In general, the response to chronic submaximal swimming is qualitatively similar to endurance exercise training in mammals, both resulting in the emergence of a more aerobic phenotype (Johnston and Moon, 1980; McClelland et al., 2006; LeMoine et al, 2010a; Martin-Perez et al., 2012). However, recent studies surveying global changes in gene expression and protein expression (Martin-Perez et al., 2012; Planas et al., 2013) demonstrate that exercise training affects many aspects of muscle physiology. Exercise-induced phenotype plasticity may be triggered by the cyclical changes in intracellular Ca^{2+}, metabolites (e.g., ADP, fatty acids), cytosolic PO_2, and mechanical stress with contractions (Gundersen, 2011). The signal transduction pathways involved in the exercise training response in mammals (Figure 1.3) include the CAMKs and p38MAPK and downstream activation of p38 kinase. The increased ATP consumption rates alter ratios of AMP to ATP, activating the cellular energy sensor AMPK, and changes in $NAD^+/NADH$ stimulate the SIRT pathway (Gundersen, 2011). Exercise may also increase the production of ROS and NO and increase circulating catecholamine concentrations. In mammals, these signals appear to all converge with the upregulation of PGC-1α to affect the binding of PPAR, NRF, and ERR to the target DNA (Hock and Kralli, 2009). Recently, the role of the many myokines (e.g., IL-6), synthesized and released by mammalian skeletal muscle in response to contraction, has been investigated as potential energy sensors that either act through AMPK or act as paracrine agents on peripheral organs (Pedersen, 2012). Very few of these factors have been examined in terms of muscle plasticity in fish. However, the phosphatase calcineurin does not seem to play a major role in muscle hypertrophy since its target protein NFAT2 decreases with exercise training in carp (Martin and Johnston, 2005).

Chronic swimming accelerates muscle growth in zebrafish and many salmonid species. Depending on the species, this increased growth can occur by muscle hypertrophy, muscle hyperplasia, or a combination of both (Palstra et al., 2010; Davison and Herbert, 2013). Enhanced growth may occur optimally at training speeds that coincide with a reduced cost of transport (COT) (Palstra et al., 2010), and thus variation between studies may reflect differences in relative training intensity (McClelland et al., 2006). Training, in turn, enhances swimming efficiency (lower COT) in some species, further improving growth (Brown et al., 2011). Not all fish species respond similarly, and the exercise-induced stimulation of growth appears to be more common in pelagic than benthic species (Davison and Herbert, 2013). Exercise in mammals stimulates IGF-I production, which is released by exocytosis to act on membrane IGF-R (Figure 1.3), and the same mechanism may occur in exercising fishes. Interestingly, exercise-trained zebrafish showed a reduction in transcript expression of an IGF receptor isoform (igf1rb) and a growth hormone receptor gene (ghrb) (Palstra et al., 2010). They also showed increases in expression of slow and fast myosin isoforms (smhc1 and myhz2) but no change in the growth factor myogenin (myog) mRNA expression (Palstra et al., 2010). Thus, it is yet unclear which molecular features are responsible for promoting muscle growth with exercise. Still, training does result in muscle hypertrophy in trout (Planas et al., 2013). This muscle growth was accompanied by changes in mRNA expression (as revealed by RNA-Seq) for many genes involved in muscle contraction, including α-actin, MyHC, MyLC, tropomyosin-α, Tn isoforms C, I, and T, as well as parvalbumin and creatine phosphokinase (CPK) (Planas et al., 2013). Moreover, gilthead sea bream (Sparus aurata) exercised for 4 weeks showed an increased expression of myomesin, a protein involved in sarcomere stabilization (Martin-Perez et al., 2012). It should be noted that transcriptional changes observed with exercise might not equate to differences in protein content (Martin-Perez et al., 2012; Planas et al., 2013). Variation in exercise performance may depend on differences in the kinetics of Ca^{2+} handling in muscle cells. For example, zebrafish with higher sustained or sprint swim performance also had higher mRNA expression of ryanodine receptor, Ca-ATPases, and in particular, parvalbumin (Seebacher and Walter, 2012). When parvalbumin was experimentally increased in muscles, zebrafish showed improvement of both sustained and sprint exercise performance (Seebacher and Walter, 2012). It is not surprising that exercise training in gilthead sea bream led to an increase in red muscle parvalbumin (Martin-Perez et al., 2012).

Early training studies on fish observed an increase in muscle aerobic capacity (Johnston and Moon, 1980; Farrell et al., 1991). However, not until recently was anything known regarding the molecular changes responsible for this change in metabolic phenotype. The transcriptional machinery of zebrafish muscle shows a robust response to even a single 3-h bout of swimming (LeMoine et al., 2010a). However, changes in mRNA expression of transcriptional regulators such as PGC-1α and PGC-1β either did not correlate with the expression of their target genes or did not respond at all to acute exercise (LeMoine et al., 2008). In contrast, zebrafish muscle responded to 4 weeks of swim training with an induction of NRF-1 mRNA and its downstream target gene citrate synthase (CS) (McClelland et al., 2006), in a similar manner as mammals. In contrast to mammals, zebrafish muscle showed a distinct response in PPARα, PPARβ1, and PGC-1α mRNAs, which either decreased or did not respond to training (McClelland et al., 2006). Even when the training period was extended to 8 weeks, PGC-1α expression was greatest early on in training in both muscle types, while only the red muscle showed a modest stimulation of PGC-1β expression in the first week (LeMoine et al., 2010a). Moreover, the expression of individual isoforms of PPARβ, β1 and β2, were not induced in either the red or white muscle by training (LeMoine et al., 2010a). These data suggest an indirect role for PGC-1α in the regulation of CS expression in fish and not as the putative "master regulator" of exercise-induced mitochondrial biogenesis as suggested for mammals. However, the supposed indispensable role of PGC-1α for mitochondrial biogenesis in mammals has recently been called into question (Rowe et al., 2012). In fish muscle, PGC-1α may be more important in coordinating the transcriptional regulation by the PPARs than by NRF-1, even though this transcription factor was more closely tied to changes in CS mRNA (McClelland et al., 2006). Currently, there is little information on how exercise may affect expression of other known regulators of muscle remodeling in fishes. PRC, SIRT, ERRα, and NRF-2 are involved in mitochondrial biogenesis in mammals, but their roles in exercise-induced muscle plasticity in fish are unclear.

Exercise can lead to reductions in myocyte PO_2 (Gundersen, 2011), and the role of HIF-1α in the endurance training response has recently been investigated in mammals. While the acute response to exercise results in an induction of HIF-1α expression and stabilization of the HIF protein (Ameln et al., 2005; Gundersen, 2011), endurance training reduces HIF-1 induction (Mason et al., 2004). Furthermore, muscle-specific HIF-1$^{-/-}$ knockout mice have more aerobic muscles than wild-type controls (Mason et al., 2004). However, since the level of HIF-1α is higher in glycolytic muscle fibers, these fibers may show a greater HIF response to exercise (Gundersen, 2011), suggesting that HIF plays a greater role in the remodeling of these fibers in mammals. Due to the evolutionarily conserved nature of the hypoxia response (Rytkönen and Storz, 2011), HIF-1α in fish may respond to stressors that induce cellular hypoxia, such as intense exercise, in a similar manner to mammals.

In adult fishes, exercise-induced muscle plasticity appears to be reversible. One study showed the effects of training to be reversed in half the time of training (Liu et al., 2009). Fish muscle can also respond to disuse with a drop in aerobic enzyme content (Urfi and Talesara, 1989). Interestingly, while myonuclei are added to growing muscle fibers during hypertrophy, they are not lost during atrophy in mammals, suggesting that the idea of constant myonuclear domain is more complex than once thought (Gundersen, 2011).

1.3.4 HYPOXIA

Many fishes live in O_2-variable environments where dissolved O_2 levels can change both spatially and temporally. Compensatory changes in skeletal muscle may serve to maintain ATP synthesis to match energy use with reduced O_2 availability. Some species respond to low cellular PO_2 with a corresponding drop in metabolic rate, adopting an oxyconformer strategy (e.g., goldfish, toadfish, brown bullhead) or by maintaining oxygen consumption over a range of environmental PO_2 as oxyregulators (Hughes, 1973). Nonetheless, there is a certain species-specific critical PO_2 (P_{crit}), where O_2 supply fails to meet O_2 demand, and oxygen consumption declines as a function of falling environmental PO_2. Oxyconformity of metabolic rate with PO_2 is common and is not unique to species that undergo whole-body hypometabolism. This drop in metabolic rate may be one of the triggers that change muscle phenotype in response to hypoxia (Gracey et al., 2001; Forgan and Forster, 2012).

Chronic hypoxia induces phenotypic changes in muscles that vary across species. While many respond with a coordinated increase in glycolytic capacity, a reduction in aerobic capacity, and even a lowering of mitochondrial content, other species increase mitochondrial content and capillary to fiber ratio, presumable to decrease diffusion distances. For example, hypoxia led to an increase in mitochondrial volume density and myoglobin content in carp, and lower postexercise muscle lactate in goldfish, suggesting an increase in muscle aerobic capacity in both species (Johnston and Bernard, 1984; Sänger, 1993; Fu et al., 2011). However, other species such as the tench (*Tinca tinca*) show a decline in mitochondrial content with hypoxia (Johnston and Bernard, 1982). Perhaps this represents differences in efficiency of mitochondrial respiration or, alternatively, a decline in oxygen delivery to muscles between species. Most likely some of the variation in response is due to differences in the experimentally induced severity of hypoxia and length of the acclimation period. Another source of variation may be the level of response between hypoxia-tolerant and hypoxia-intolerant species or between species that are oxyconformers and oxyregulators. Studies examining the global gene expression response to hypoxia reveal a robust response in fish muscle (Ton et al., 2003). However, early proteomic analysis of zebrafish muscle after hypoxia exposure showed only modest changes at the protein level (Bosworth et al., 2005). Recent technical advances using 2D-DIGE and MALDI-TOF/TOF MS have led to the detection of a greater number of proteins induced by hypoxia in zebrafish muscle (Chen et al., 2012). After only 48 h of hypoxia, protein levels of glycolytic enzymes triosephosphate isomerase B, fructose-bisphosphate aldolase, α-enolase, and pyruvate kinase were significantly increased. The protein contents of the β-oxidation enzymes β-hydroxyacyl-CoA dehydrogenase, pyruvate dehydrogenase, isocitrate dehydrogenase, and ATP synthase were all decreased. The structure of the sarcomeres may have also been modified with an increased expression of fast myosin heavy and light chains (e.g., MyHC4) (Chen et al., 2012). In contrast, proteomic analysis of muscle from 24-h hypoxic rainbow trout found that this species had a decrease in glycolytic enzymes (Wulff et al., 2012) similar to decreased activities in most glycolytic enzymes with 4 weeks of hypoxia seen in *Fundulus grandis* (Martínez et al., 2006). Hypoxia also reduces growth rates in many fish species (Martínez et al., 2006). For example, 28 days of hypoxia in killifish (*Fundulus heteroclitus*) led to lower muscle protein content and RNA to DNA ratio (Rees et al., 2012). The reductions in glycolytic enzyme specific activity and mitochondrial content seen in some species may reflect this hypoxia-mediated reduction in growth.

How vertebrate muscle detects and responds to hypoxia is still hotly debated. Hypoxia alters ATP/ADP in killifish (Richards et al., 2008) and likely stimulates AMPK to turn down anabolic processes. However, AMPK does not increase in hypoxic goldfish muscle (Jibb and Richards, 2008). The PHDs regulate HIF-1α in an oxygen-dependent manner, and HIF-1α responds very quickly to reduced cellular PO_2, making this pathway a likely candidate as a cellular hypoxia sensor. HIF regulates the expression of a large number of genes in hypoxia, including many responsible for angiogenesis and metabolism. In muscle, HIF stimulates the production of glycolytic enzymes, downregulates those for fat catabolism, and shifts the cells away from aerobic metabolism (Goda and Kanai, 2012). This may occur by HIF-1α suppression of mitochondrial biogenesis, possibly acting on c-Myc via the FOXO3A transcription factor (Goda and Kanai, 2012). Although direct connections between HIF and NRF-1 have not been established, HIF is known to downregulate PPARα in mammals (Mason and Johnson, 2007). Since HIF protein levels are higher in glycolytic fibers, and chronic hypoxia can lead to a slow-to-fast fiber transition in mammals (Pisani and Dechesne, 2005; Gundersen, 2011), fish white muscle may be more responsive then red muscle to low O_2. Interestingly, protein changes initiated by HIF signaling are thought to improve mitochondrial oxidation during hypoxia in mammals, through changes in the structure of COX by replacing the COX4-1 subunit isoform with the COX4-2 isoform (Fukuda et al., 2007). If the same occurs in fishes, it may explain why some species can reduce mitochondrial volume in oxygen poor waters but still maintain adequate muscle function. Currently, there is little evidence that HIF acts to stimulate mitochondrial biogenesis, making it an unlikely molecular player in those fish species that show increases in muscle mitochondrial volume with chronic hypoxia. However, only a few studies have examined the response of HIF to hypoxia in fish muscle (Kopp et al., 2011; Rimoldi et al., 2012).

1.4 MUSCLE PLASTICITY AT EARLY STAGES OF DEVELOPMENT

Muscle plasticity in juvenile and adult fishes is often reversible. The responses to changing environmental variables (e.g., temperature or oxygen, Sections 1.3.2 and 1.3.4) or physiological demands (e.g., exercise, Section 1.3.3) can generally be reversed in these animals if they return to their original state. In contrast, muscle plasticity is often irreversible if it occurs at embryonic and larval life stages (Johnston, 2006). Environmental stress during brief but critical windows in development can in fact have long-lasting effects on adult phenotype, because it can alter developmental trajectories (West-Eberhard, 2003; Beldade et al., 2011; Burggren and Reyna, 2011) (Figure 1.1). Embryos are also at the mercy of their environment, without the means to swim away from harsh conditions. A discussion of muscle plasticity would therefore be incomplete without appreciating its unique attributes in early life stages. This section therefore aims to distinguish developmental plasticity in the muscle of embryos from the mechanisms discussed earlier for adult fishes.

1.4.1 ONTOGENETIC MUSCLE GROWTH

Muscle changes rapidly and dramatically during ontogeny as the adult body plan becomes established (Currie and Ingham, 2001; Stellabotte and Devoto, 2007; Rescan, 2008; Johnston et al., 2011). Embryonic myogenesis (the creation of new muscle fibers) begins relatively early in development as pluripotent stem cells commit to a myogenic fate at the end of gastrulation, thus forming the MPC population from which the muscle arises. MyoD and myogenic factor 5 (Myf5), each part of the MRF gene family, are transcription factors that direct the development of the earliest MPCs (Hinits et al., 2009). MyoD and Myf5 expression is induced by hedgehog proteins secreted from the notochord (Coutelle et al., 2001). Their expression specifies the early MPCs to the myogenic lineage shortly before the development of the embryonic somites—the segments of mesoderm that arise in a rostral to caudal progression along the length of the neural tube, eventually forming the adult myotomes. These early MPCs are termed "adaxial cells" due to their location on both sides of the axial midline. Many adaxial cells differentiate and fuse together to form myotubes in an early step of fiber creation. They migrate to the lateral surface of the myotome, where they form a superficial layer of slow muscle fibers (Devoto et al., 1996). The migration of adaxial cells leaves a morphogenetic signal that patterns fast muscle fiber morphogenesis in the posterior region of the somite, driven by MyoD-induced activity of myogenin, another MRF (Henry and Amacher, 2004; Hinits et al., 2009). Concurrent with the morphogenesis of fast muscle is the movement of undifferentiated MPCs to form a cell layer that is external to the muscle fibers and just under the skin (Hollway et al., 2007; Stellabotte et al., 2007). These developmental events in early embryonic myogenesis result in the basic phenotype of the adult myotome (Figure 1.2): a lateral layer of slow muscle fibers, a deep and larger region of fast muscle fibers, and an external cell layer that contains undifferentiated MPCs (Hollway et al., 2007; Stellabotte et al., 2007).

A second wave of myogenesis begins later in embryonic development from MPCs in the external cell layer (Rowlerson and Veggetti, 2001; Stellabotte and Devoto, 2007; Johnston et al., 2011).

Stratified
hyperplasia

Mosaic
hyperplasia

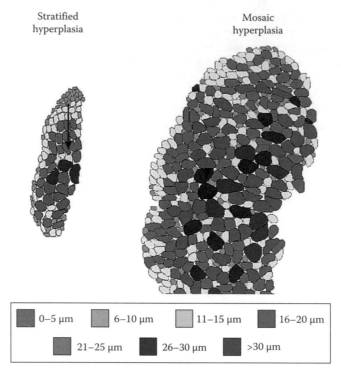

■ 0–5 μm	■ 6–10 μm	■ 11–15 μm	■ 16–20 μm
■ 21–25 μm	■ 26–30 μm	■ >30 μm	

FIGURE 1.4 **(See color insert.)** Myogenesis, the creation of new muscle fibers, occurs by multiple distinct processes. Stratified hyperplasia is the creation of new fibers from myogenic progenitor cells (MPC) located in distinct germinal zones. This leads to stratified variation in muscle fiber size between the younger and smaller fibers that are close to the germinal zones and the older and larger fibers located deep in the muscle. Myogenesis of this type begins in late embryogenesis and ends in the larval or early juvenile stage. Mosaic hyperplasia is the creation of new muscle fibers from MPCs that become interspersed throughout the fast muscle, which leads to a mosaic of muscle fiber sizes throughout the myotome (corresponding to differences in age). Camera lucida drawings of the fast muscle fibers from two zebrafish are shown: one at 7.5 mm total body length that is still undergoing myogenesis by stratified hyperplasia and one at 10 mm total body length that has begun mosaic hyperplasia. Fibers are color coded according to diameter size class. (Adapted from Johnston, I.A. et al., *J. Exp. Biol.*, 212, 1781, 2009. With permission.)

MPCs migrate from their source in the external cell layer to create distinct germinal zones of new fiber production in a layer between the fast and slow muscle and near the outer margins of the myotome. This new phase of myogenesis is termed "stratified hyperplasia" because of the distinct variation in muscle fiber size that is created—younger fibers that are closer to the germinal zones are much smaller (i.e., have not had as much time to grow and hypertrophy) than the older fibers located deeper in the muscle (Figure 1.4). In the fast muscle of many fish species, this mechanism of fiber production is only important until early juvenile stages of development. However, it continues throughout life to be the main mechanism for creating new slow and possibly intermediate muscle fibers in adults (Veggetti et al., 1990; Rowlerson et al., 1995) (note that the smallest slow fibers in Figure 1.2a are those closest to the boundary between the slow and fast muscle).

The final wave of myogenesis occurs in the fast muscle, beginning in larval stages and continuing until the fish reaches ~40% of its adult body length (Weatherley et al., 1988; Johnston, 2006). It is the MPCs distributed throughout the fast muscle, and not those in a germinal zone, that differentiate to form new fast fibers during this phase of growth. This process is therefore called "mosaic hyperplasia" because smaller younger fibers are interspersed among larger older fibers (Figure 1.4). The source of these MPCs is still unknown, but they may arise from MPCs in the external cell layer that migrate deep into the myotome (Johnston et al., 2011). This mechanism of fiber production is extremely important, resulting in the creation of the vast majority of fast fibers in many species (Weatherley et al., 1988), and continuing until much later in ontogeny in fish than in mammals and other tetrapods. Fish do eventually stop producing new fibers as a part of normal growth and before they reach their adult size. In zebrafish, the cessation of fiber recruitment is associated with an upregulation of mRNA transcripts for proteins involved in energy metabolism and cell communication, a repression of transcripts for contractile and structural proteins, and changes in a number of microRNAs (Johnston et al., 2009). Fiber hypertrophy, a process that occurs at all stages of ontogeny, then becomes the sole process that leads to further growth of muscle mass until adult size is reached. Preexisting muscle fibers absorb MPCs during hypertrophy, increasing the number of nuclei in each fiber as they grow (see Section 1.3.1). Nevertheless, mosaic hyperplasia can occur in adult fishes under some conditions, such as in response to injury (Rowlerson et al., 1997).

How plastic is muscle development? Any effect of environmental stress during early life on muscle phenotype will inevitably be overlaid on the predominant changes that occur as part of the basic ontogenic program. Molecular safeguards may in some situations make muscle development robust or insensitive to normal ranges of environmental variation (canalization), as observed for some other developmental systems (Braendle and Félix, 2008; Frankel et al., 2010). There are even some extreme examples of unique fish that can develop normally in some of the harshest habitats available to aquatic animals (Podrabsky and Culpepper, 2012). However, development is not impervious to environmental influences, and fishes in their native habitats experience a wide variety of stressors that could potentially induce developmental plasticity in the muscle. In general, developmental plasticity can result from environmentally induced changes in the relative temporal sequence of developmental events, which can alter the interactions between cells/tissues, or from persistent effects of the environment on each individual cell (West-Eberhard, 2003; Beldade et al., 2011). The remainder of this section will discuss the effects on early development of two more commonly studied stressors—temperature and hypoxia—to appreciate the mechanisms and consequences of muscle plasticity in early life stages.

1.4.2 TEMPERATURE

Temperature change has been shown to influence development and induce muscle plasticity by altering the relative timing of developmental events ("heterochrony"). Temperature has a strong effect on the overall pace of muscle development, as would be expected from its influence on biological rate processes in general. Most biological rates change two- to threefold over a 10°C range in temperature, and the same is true for the rate of somite formation during embryogenesis (e.g., Q_{10} is 2.8 from 20°C to 30°C in zebrafish) (Schröter et al., 2008). If the effect of temperature change on developmental rates is uniform across different components in the muscle, then muscle phenotype could be similar in fish that are compared at the same developmental stage (even though they may differ in age). For example, the effect of temperature on the rate of somitogenesis does not generally influence the final number of somites at hatching (Johnston et al., 1995; Vieira and Johnston, 1996; Galloway et al., 2006). However, temperature has been observed to cause heterochrony for a number of events in fish muscle development. Atlantic herring (*Clupea harengus*) experiences a relative delay (with respect to embryonic stage) in the expression of contractile proteins, the synthesis of contractile filaments, myofibril assembly, neuromuscular junction development, and the differentiation of myotubes into recognizable fiber types when embryos are raised at 5°C compared to 8°C or 12°C (Johnston et al., 1995, 1997, 2001). Heterochronies in the expression of genes involved in myogenesis, such as the MRF gene family members, could contribute to the disruption in relative timing of developmental events. In this regard, the onset of Myf5 and MRF4 mRNA induction is delayed relative to somite stage in Atlantic salmon (*Salmo salar*) embryos reared at 2°C compared to those reared at 8°C (Macqueen et al., 2007). In contrast, changing temperature does not affect the relative timing of MyoD or myogenin during embryonic development in a range of fish species (Temple et al., 2001; Hall et al., 2003; Cole et al., 2004; Galloway et al., 2006; Macqueen et al., 2007).

Embryonic temperature also has persistent effects on the muscle fibers themselves, at least some of which differ from the cellular effects of temperature change during adulthood. As described in Section 1.3.2, adjustment in the capacity for mitochondrial ATP production is an extremely important response to temperature change in the muscle of adult fishes. Cold acclimation induces mitochondrial biogenesis and angiogenesis in the slow and fast muscles of adults, but the embryos of some species starkly contrast this response. Colder temperatures during embryogenesis reduce mitochondrial abundance in the slow muscle fibers of plaice (*Pleuronectes platessa*) larvae (Brooks and Johnston, 1993) and in both the slow and fast fibers of larval herring (Vieira and Johnston, 1992). The mechanistic cause of this difference between embryos and adults is unclear, but it could relate to differences in how each stage responds to temperature. The pressure for adult fishes to sustain ATP turnover in the muscle so they can keep swimming in the cold should be a strong signal for mitochondrial biogenesis, and this signal is absent in embryos. In this regard, the intrinsic cellular effects of temperature may not differ between life stages, but the different behavioral responses to temperature may result in very different changes in the metabolic phenotype of muscle fibers.

The effects of temperature during early ontogeny lead to changes in muscle phenotype that can persist into adulthood. Numerous studies have found that temperature change solely during embryonic development is sufficient to alter the number and diameter of both slow and fast fibers in muscle myotomes at various life stages (Stickland et al., 1988; Vieira and Johnston, 1992; Brooks and Johnston, 1993; Hanel and Wieser, 1996; Rowlerson and Veggetti, 2001; Stoiber et al., 2002; Hall and Johnston, 2003). The critical window for this effect appears to occur in the first half of embryogenesis (until the eyed stage, ~47%–67% of the time to hatching in Atlantic salmon) (Macqueen et al., 2008). However, the consequence of temperature change during embryogenesis on muscle phenotype appears to change during ontogeny. For example, the final number of fast fibers was higher in zebrafish raised at an intermediate temperature (26°C) as embryos than in those raised at more extreme temperatures (22°C or 31°C), but the difference between treatment groups did not become apparent until after new fiber production ceased (Johnston et al., 2009) (Figure 1.5). Similarly, pearlfish (*Rutilus meidingeri*) raised at 13°C developed more slow and fast fibers than

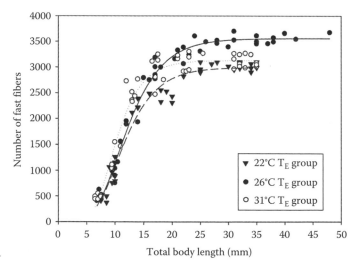

FIGURE 1.5 Embryonic temperature (T_E) affects the creation of new muscle fibers and the final number of fibers in the muscle myotome. Shown here is the number of fast muscle fibers in a transverse section through the myotome at 0.6 of total body length (near the rostral base of the anal fin), measured in zebrafish that were raised until hatching at one of three temperatures (22°C, 26°C, or 31°C) and then raised after hatching at a common temperature of 26°C. Muscle fibers are created during normal growth until the fish reach approximately 40% of adult body length, followed by a plateau in fiber number that differs between T_E groups. (Modified from Johnston, I.A. et al., *J. Exp. Biol.*, 212, 1781, 2009. With permission.)

those raised at 8.5°C or 16°C, but not until late juvenile and adult life stages (Steinbacher et al., 2011). In the latter study, having more muscle fibers was associated with having a larger pool of MPCs available to support growth, consistent with previous studies in Atlantic salmon (Johnston et al., 2003). The smaller pool of MPCs in fish raised at 16°C may have been caused by more MPCs exiting the cell cycle and differentiating during embryogenesis, the result of which would not become apparent until later life stages when MPC reserves became depleted (Steinbacher et al., 2011). The findings in pearlfish support the idea that embryonic temperature alters the number of fibers in fish muscle by altering the relative rates of MPC proliferation and differentiation, although other potential causes have not been excluded (Johnston, 2006).

Is muscle plasticity in response to temperature change during early development beneficial or is it a result of developmental disruption? Phenotypic plasticity is not always favorable (Ghalambor et al., 2007; Storz et al., 2010), and the responses to developmental temperature are by no means an exception to this general rule. The hypothesis that developmental plasticity in response to a particular environment always improves performance in that environment, broadly termed the beneficial acclimation hypothesis, is not universally supported in studies of thermal plasticity (Leroi et al., 1994; Huey et al., 1999). Muscle plasticity in response to embryonic temperature would support the beneficial acclimation hypothesis if it improved swimming performance—an ecologically important trait that influences fitness (Plaut, 2001)—at the same temperature as that experienced during embryogenesis. Some experiments have shown that embryonic temperature can influence routine swimming (Albokhadaim et al., 2007), the primarily aerobic critical swimming speed (U_{crit}) (Sfakianakis et al., 2011), or maximum burst swimming speed (Johnston et al., 2001; Burt et al., 2012) when measured at a single temperature in later life. However, there are few explicit tests of how embryonic temperature affects thermal sensitivity of swimming performance across a range of temperatures. One such example supports the hypothesis that developmental plasticity is beneficial. When zebrafish were raised until hatching at 22°C, 27°C, or 32°C and then reared at 27°C until adulthood, those fishes that were raised at a particular temperature also had the highest U_{crit} after acute transfer to that temperature (Figure 1.6a) (Scott and Johnston, 2012). The total area of each muscle fiber type was not different between embryonic temperature groups (see 27°C acclimation temperature in Figure 1.6c and d), but there were more slow fibers of smaller size in fishes raised at 27°C (Scott and Johnston, 2012). In contrast, there is another example suggesting that developmental plasticity is not beneficial in some species, and there may instead be an optimal developmental temperature that maximizes muscle function and swimming performance in later life. European sea bass (*Dicentrarchus labrax*) juveniles that were raised at 15°C until larval metamorphosis had more slow oxidative muscle and a higher U_{crit} after acute transfer to a range of temperatures (20°C, 25°C, and 28°C, but not at 15°C) than juveniles that had been raised at 20°C (Koumoundouros et al., 2009). The observation that juveniles raised at 20°C as embryos perform worse at 20°C than juveniles raised at 15°C contradicts the beneficial acclimation hypothesis.

The benefit of developmental muscle plasticity probably also depends on the thermal response in question. For the study discussed earlier, in which zebrafish were raised until hatching at 22°C, 27°C, or 32°C, the effect of embryonic temperature on the ability to acclimate to extreme temperatures (16°C or 34°C) did not always conform to the beneficial acclimation hypothesis (Scott and Johnston, 2012). The hypothesis was supported by the observation that zebrafish raised at 27°C and 32°C performed better after acclimation to warm temperatures than those raised at 22°C, but was not supported by the finding that zebrafish raised at 32°C acclimated better to the cold than those raised at 27°C (Figure 1.6b). The differences in U_{crit} after acclimation were associated with differences in muscle plasticity, reflected in both the total transverse area of swimming muscle (Figure 1.6c) and the relative abundance of aerobic (slow and intermediate) muscle fibers (Figure 1.6d). The differences in U_{crit} after cold acclimation were also associated with differences in gene expression in the fast muscle. The overall transcriptomic response to cold (detected using RNA-Seq) was accentuated in zebrafish raised at 32°C compared to those raised at 27°C, and the disparity between groups was caused largely by differences in expression of transcripts involved

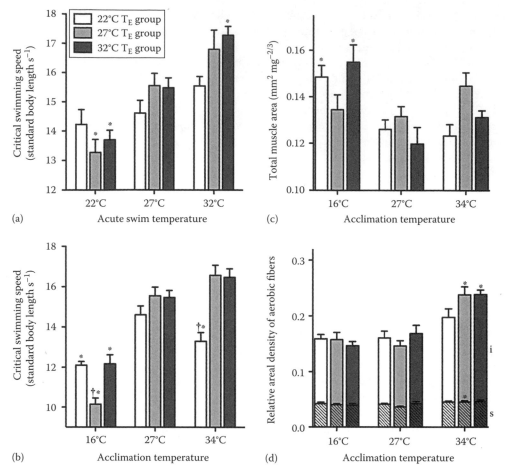

FIGURE 1.6 Embryonic temperature (T_E) treatment affects the temperature dependence of aerobic swimming performance and muscle phenotype in adult zebrafish. (a) T_E treatment reduced the sensitivity of critical swimming speed (U_{crit}) to acute temperature transfer at each group's respective T_E. T_E treatment also altered the thermal acclimation response of (b) U_{crit}, (c) the total transverse area of the swimming muscle (expressed relative to body mass$^{2/3}$ to account for isometric variation in body size), and (d) the relative proportion of muscle area composed of slow (s, hatched) and intermediate (i, unhatched) fiber types. *Represents a significant difference from U_{crit} at 27°C within each T_E group, and † represents a significant difference between T_E groups within each swim/acclimation temperature. (Adapted from Scott, G.R. and Johnston, I.A., *Proc. Natl. Acad. Sci. USA*, 109, 14247, 2012. With permission.)

in energy metabolism, angiogenesis, cell stress, muscle contraction and remodeling, and apoptosis (Scott and Johnston, 2012). The most parsimonious conclusion from the available literature appears to be that some thermal responses in the muscle of embryos are adaptive for future changes in temperature, but that some processes develop optimally within a specific narrow range of temperatures (e.g., acclimation capacity is optimized by warmer embryonic temperatures in zebrafish).

1.4.3 HYPOXIA

Oxygen deprivation has the potential to influence development in many fish species. Environmental hypoxia is perhaps the most obvious cause of cellular oxygen limitation, but it is not necessarily the most pervasive. The chorion and perivitelline fluid can be a significant barrier to oxygen diffusion that reduces cellular O_2 levels (Rombough, 1988; Ciuhandu et al., 2007), particularly when metabolic rate is high (e.g., at high temperatures). Regardless of its ultimate cause, cellular hypoxia often slows the overall pace of development, reducing the speed of somitogenesis and thus the formation of the axial muscle myotomes (Kinne and Kinne, 1962; Rombough, 1988; Schmidt and Starck, 2010). This is not always a consequence of developmental disruption *per se*, because some species such as the annual killifish (*Austrofundulus limnaeus*) and zebrafish can reversibly arrest development in hypoxia/anoxia at specific developmental stages (Padilla and Roth, 2001; Podrabsky and Culpepper, 2012). Nevertheless, hypoxia has been observed to cause developmental heterochronies in the muscle. Somite height was reduced across a range of developmental stages in zebrafish embryos raised at 0.83 mg O_2/L (~2.1 kPa) from fertilization until the end of segmentation (Schmidt and Starck, 2010). The development of the caudal vasculature was also hastened in zebrafish exposed to 10 kPa for the first 7 days after fertilization (Pelster, 2002). Several transcripts are differentially expressed in zebrafish embryos during hypoxia (5% O_2, ~5.1 kPa, from 48 to 72 h postfertilization), including an induction of glycolytic enzymes and stress response proteins and a repression of many tricarboxylic acid cycle enzymes, contractile proteins, globins, and proteins involved in transcription, translation, cell cycle regulation, and signaling (Ton et al., 2003). However, many of these transcriptional events are typical of the hypoxia response of many other species in juvenile and adult life stages, so it is unclear what role they play in regulating developmental heterochrony.

Hypoxia during embryonic development can also lead to changes in the number and size of muscle fibers. Both Atlantic salmon and rainbow trout had fewer muscle fibers at hatching when they were raised as embryos in water with low O_2 levels at 5°C (50% and 70% of air saturation, ~10.6 and ~14.8 kPa, respectively) (Matschak et al., 1997, 1998). The effect of hypoxia was eliminated when embryos were raised without a chorion, suggesting that cellular O_2 tensions were elevated after this barrier to diffusion was eliminated (Matschak et al., 1998). It has even been suggested that cellular O_2 depletion can explain the changes in muscle fiber number caused by warm temperatures, because warming decreases the oxygen content of water, increases metabolic O_2 demands, and thus reduces cellular O_2 levels (Matschak et al., 1995). The ultimate cause of changes in muscle fiber number during embryonic hypoxia is unclear but it could foreseeably involve a change in IGF signaling. HIF-1α interacts with IGF growth pathways in the muscle by regulating IGF expression and downstream signaling activity (Kajimura et al., 2006; Kamei et al., 2011), which could alter the relative rates of proliferation and differentiation of myoblasts (Ren et al., 2010).

Similar to the aforementioned discussion of developmental temperature, the hypothesis that developmental muscle plasticity in response to hypoxia is beneficial would be supported if it improves swimming performance in hypoxia. There have been few explicit tests of this hypothesis, but one relevant study found that aerobic swimming performance (U_{crit}) was impaired in both normoxia and hypoxia (1.5 mg O_2/L, ~3.8 kPa) in zebrafish that were raised in hypoxia from hatching to adulthood compared to those that were raised in normoxia (Widmer et al., 2006). This could suggest that developmental hypoxia is not beneficial for swimming performance, but rather that high oxygen levels are optimal for muscle performance and plasticity in later life. However, fishes were exposed to hypoxia for the entire period from fertilization to adulthood, so these observations could have been caused by plasticity in juvenile or adult life stages. Furthermore, it is unclear whether the effects of developmental hypoxia on swimming performance are mediated by muscle plasticity or by plasticity in other systems important for O_2 and nutrient transport during exercise (e.g., cardiorespiratory system) (Chapman et al., 2000; Pelster, 2002; Jonz and Nurse, 2005). If developmental muscle plasticity in response to hypoxia is not beneficial for swimming performance, it could still be important for coping with hypoxia in later life. For example, reductions in muscle aerobic capacity that are induced by embryonic hypoxia could be part of a suite of changes in the organism that improve hypoxia tolerance and minimize O_2 needs (Marks et al., 2005; Barrionuevo et al., 2010).

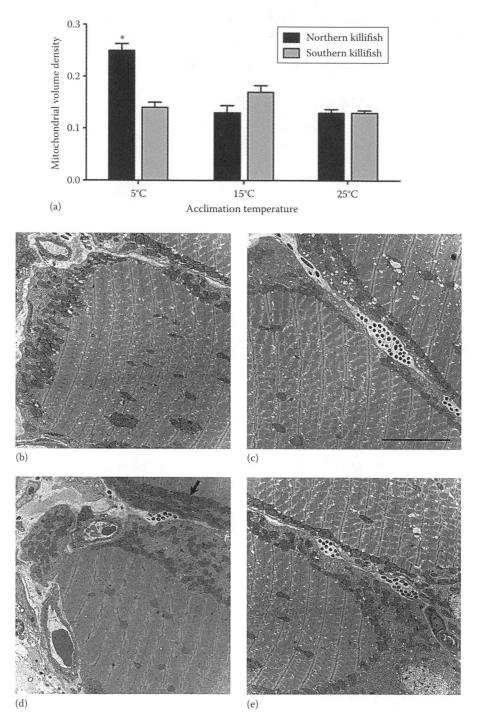

(a)

(b)

(c)

(d)

(e)

FIGURE 1.7 There are evolutionary differences in muscle plasticity in response to cold acclimation between killifish populations from a colder environment in the north and a warmer environment in the south. (a) Mitochondrial density (mitochondrial volume relative to total muscle volume) in the slow oxidative (red) swimming muscle increases in northern killifish in the cold (*) but not in southern killifish. (Data from Dhillon, R.S. and Schulte, P.M., *J. Exp. Biol.*, 214, 3639, 2011.) (b–e) Representative transmission electron micrographs from northern (b, d) and southern (c, e) killifish acclimated to 25°C (b, c) or 5°C (d, e). The arrow indicates the strong clustering of mitochondria near the cell membrane in cold-acclimated northern killifish (scale bar represents 10 nm).

1.5 EVOLUTION OF MUSCLE PLASTICITY

Can muscle plasticity evolve like many other forms of phenotypic plasticity (West-Eberhard, 2003; Beldade et al., 2011) or is it a fixed characteristic that is invariably essential for normal muscle function? Some insight into this question is provided by studies of intraspecific and interspecific variation in the cold acclimation response (Section 1.3.2). For example, northern populations of killifish (*Fundulus heteroclitus*) respond to cold acclimation by increasing the abundance and cristae surface density of mitochondria in the red and white muscles and by increasing the activities of citrate synthase and creatine kinase (but not lactate dehydrogenase) in the white muscle. In contrast, these responses are completely absent in southern killifish that do not as regularly experience cold temperatures in the wild (Dhillon and Schulte, 2011) (Figure 1.7). Furthermore, in a broad survey of many species that were wild-caught and sampled from the same lakes in summer and winter, some fishes increased cytochrome oxidase (COX) activity in the muscle in the cold (goldfish, northern redbelly dace, brook stickleback, pumpkinseed sunfish, black crappie, and central mudminnow), whereas some other fish did not (bluegill sunfish, largemouth bass, and northern pike) (Bremer and Moyes, 2011). There were no clear relationships between COX activity and the expression of genes that are believed to regulate mitochondrial biogenesis in mammals (NRF-1, NRF-2, and PGC-1α; see Section 1.3), so the cause of interspecific variation in this study was unclear. These findings suggest that muscle plasticity can indeed evolve, but the nature and conservation of the mechanisms involved have yet to be discovered.

1.6 CONCLUSIONS AND PERSPECTIVES

Skeletal muscle is a dynamic tissue that undergoes large changes in energy turnover in fishes, from rest to active swimming and burst escape responses. Due to its important role in whole animal metabolic homeostasis, it is not surprising that this tissue is highly responsive to energetic and environmental stresses. Muscle plasticity is an important means of adjusting to new physiological demands and to the unavoidable changes fishes often experience in their environment. In adults, muscle plasticity is often reversible, but in early life stages it can often have lifetime consequences on animal performance. We have begun to understand the molecular mechanisms of muscle plasticity, but our understanding is still in its natal stage and overly reliant on knowledge gleaned from research on mammals. Moreover, we are only just now gaining a greater appreciation of how embryonic exposures to stress affect adult physiology and muscle performance. Future research should aim to directly demonstrate the mechanisms of muscle plasticity in adults through the use of forward and reverse genetic techniques (e.g., transgenics, knockouts, morpholinos, and mutant screens) (Bower et al., 2012). We also need to further characterize how different developmental trajectories lead to changes in adult phenotypes and the capacity for phenotypic plasticity. Understanding how adult and developmental plasticity evolve will also be critical to appreciating the natural variation between fish species. These discoveries will be essential for predicting how fish will respond to changing environments in the future and provide useful information to help mitigate the potential negative effects on aquatic life.

ACKNOWLEDGMENTS

The research of G.B.M. and G.R.S. are supported by the Natural Science and Engineering Research Council of Canada Discovery Grant Program. The authors would like to thank D.J. Macqueen for excellent comments on an earlier version of this manuscript. I.A. Johnston and R.S. Dhillon kindly provided original images in Figures 2.4 and 2.7 (permission has been obtained where appropriate from the *Journal of Experimental Biology*, jeb.biologists.org).

REFERENCES

Albokhadaim I, Hammond CL, Ashton C, Simbi BH, Bayol S, Farrington S, and Stickland N. 2007. Larval programming of post-hatch muscle growth and activity in Atlantic salmon (*Salmo salar*). *J Exp Biol* 210: 1735–1741.

Alexander R. 1969. The orientation of muscle fibers in the myomeres of fish. *J Mar Biol Assoc UK* 49: 263–290.

Alexander R. 2003. *Principles of Animal Locomotion*. Princeton, NJ: Princeton University Press.

Ameln H, Gustafsson T, Sundberg CJ, Okamoto K, Jansson E, Poellinger L, and Makino Y. 2005. Physiological activation of hypoxia inducible factor-1 in human skeletal muscle. *FASEB J* 19: 1009–1011.

Barrionuevo WR, Fernandes MN, and Rocha O. 2010. Aerobic and anaerobic metabolism for the zebrafish, *Danio rerio*, reared under normoxic and hypoxic conditions and exposed to acute hypoxia during development. *Braz J Biol* 70: 425–434.

Beldade P, Mateus ARA, and Keller RA. 2011. Evolution and molecular mechanisms of adaptive developmental plasticity. *Mol Ecol* 20: 1347–1363.

Biga PR, Cain KD, Hardy RW, Schelling GT, Overturf K, Roberts SB, Goetz FW, and Ott TL. 2004. Growth hormone differentially regulates muscle myostatin-1 and -2 and increases circulating cortisol in rainbow trout (*Oncorhynchus mykiss*). *Gen Comp Endocrinol* 138: 32–41.

Biga PR and Goetz FW. 2006. Zebrafish and giant danio as models for muscle growth: Determinate vs. indeterminate growth as determined by morphometric analysis. *Am J Physiol Regul Integr Comp Physiol* 291: R1327–R1337.

Bone Q. 1966. On the function of the two types of myotomal muscle fibre in elasmobranch fish. *J Mar Biol Assoc UK* 46: 321–349.

Bosworth CA, Chou CW, Cole RB, and Rees BB. 2005. Protein expression patterns in zebrafish skeletal muscle: Initial characterization and the effects of hypoxic exposure. *Proteomics* 5: 1362–1371.

Bower NI, de la Serrana DG, Cole NJ, Hollway GE, Lee HT, Assinder S, and Johnston IA. 2012. Stac3 is required for myotube formation and myogenic differentiation in vertebrate skeletal muscle. *J Biol Chem* 287: 43936–43949.

Braendle C and Félix M-A. 2008. Plasticity and errors of a robust developmental system in different environments. *Dev Cell* 15: 714–724.

Bremer K, Monk CT, Gurd BJ, and Moyes CD. 2012. Transcriptional regulation of temperature-induced remodeling of muscle bioenergetics in goldfish. *Am J Physiol Regul Integr Comp Physiol* 303: R150–R158.

Bremer K and Moyes CD. 2011. Origins of variation in muscle cytochrome c oxidase activity within and between fish species. *J Exp Biol* 214: 1888–1895.

Brett JR. 1971. Energetic responses of salmon to temperature—Study of some thermal relations in physiology and freshwater ecology of Sockeye salmon (*Oncorhynchus nerka*). *Am Zool* 11: 99–113.

Brooks S and Johnston IA. 1993. Influence of development and rearing temperature on the distribution, ultrastructure and myosin sub-unit composition of myotomal muscle-fibre types in the plaice *Pleuronectes platessa*. *Mar Biol* 117: 501–513.

Brown EJ, Bruce M, Pether S, and Herbert NA. 2011. Do swimming fish always grow fast? Investigating the magnitude and physiological basis of exercise-induced growth in juvenile New Zealand yellowtail kingfish, *Seriola lalandi*. *Fish Physiol Biochem* 37: 327–336.

Burggren WW and Reyna KS. 2011. Developmental trajectories, critical windows and phenotypic alteration during cardio-respiratory development. *Respir Physiol Neurobiol* 178: 13–21.

Burt JM, Hinch SG, and Patterson DA. 2012. Developmental temperature stress and parental identity shape offspring burst swimming performance in sockeye salmon (*Oncorhynchus nerka*). *Ecol Freshw Fish* 21: 176–188.

Carpio Y, Acosta J, Morales R, Santisteban Y, Sanchéz A, and Estrada MP. 2009. Regulation of body mass growth through activin type IIB receptor in teleost fish. *Gen Comp Endocrinol* 160: 158–167.

Chapman LG, Galis F, and Shinn J. 2000. Phenotypic plasticity and the possible role of genetic assimilation: Hypoxia-induced trade-offs in the morphological traits of an African cichlid. *Ecol Lett* 3: 387–393.

Chauvigné F, Gabillard JC, Weil C, and Rescan PY. 2003. Effect of refeeding on IGFI, IGFII, IGF receptors, FGF2, FGF6, and myostatin mRNA expression in rainbow trout myotomal muscle. *Gen Comp Endocrinol* 132(2): 209–215.

Chen K, Cole RB, and Rees BB. 2012. Hypoxia-induced changes in the zebrafish (*Danio rerio*) skeletal muscle proteome. *J Prot* 78: 477–485.

Ciuhandu CS, Wright PA, Goldberg JI, and Stevens ED. 2007. Parameters influencing the dissolved oxygen in the boundary layer of rainbow trout (*Oncorhynchus mykiss*) embryos and larvae. *J Exp Biol* 210: 1435–1445.

Cleveland BM and Weber GM. 2011. Effects of sex steroids on indices of protein turnover in rainbow trout (*Oncorhynchus mykiss*) white muscle. *Gen Comp Endocrinol* 174(2): 132–142.

Cole NJ, Hall TE, Martin CI, Chapman MA, Kobiyama A, Nihei Y, Watabe S, and Johnston IA. 2004. Temperature and the expression of myogenic regulatory factors (MRFs) and myosin heavy chain isoforms during embryogenesis in the common carp *Cyprinus carpio* L. *J Exp Biol* 207: 4239–4248.

Coutelle O, Blagden CS, Hampson R, Halai C, Rigby PWJ, and Hughes SM. 2001. Hedgehog signalling is required for maintenance of myf5 and myoD expression and timely terminal differentiation in zebrafish adaxial myogenesis. *Dev Biol* 236: 136–150.

Craig PM and Moon TW. 2011. Fasted zebrafish mimic genetic and physiological responses in mammals: A model for obesity and diabetes? *Zebrafish* 8: 109–117.

Crockford T and Johnston IA. 1990. Temperature acclimation and the expression of contractile protein isoforms in the skeletal muscles of the common carp (*Cyprinus carpio* L.). *J Comp Physiol B* 160: 23–30.

Currie PD and Ingham PW. 2001. Induction and patterning of embryonic skeletal muscle cells in the zebrafish. In: Johnston IA (Ed.), *Muscle Development and Growth*. San Diego, CA: Academic Press, pp. 1–18.

Davison W and Herbert NA. 2013. Swimming-enhanced growth. In: Palstra AP and Planas JV (Eds.), *Swimming Physiology of Fish*. Heidelberg, Germany: Springer-Verlag, pp. 177–200.

Devlin RH, Biagi CA, Yesaki TY, Smailus DE, and Byatt JC. 2001. Growth of domesticated transgenic fish. *Nature* 409: 781–782.

Devoto SH, Melancon E, Eisen JS, and Westerfield M. 1996. Identification of separate slow and fast muscle precursor cells in vivo, prior to somite formation. *Development* 122: 3371–3380.

Dhillon RS and Schulte PM. 2011. Intraspecific variation in the thermal plasticity of mitochondria in killifish. *J Exp Biol* 214: 3639–3648.

Egginton S and Sidell BD. 1989. Thermal acclimation induces adaptive changes in subcellular structure of fish skeletal muscle. *Am J Physiol Regul Integr Comp Physiol* 256: R1–R9.

Erickson JR, Sidell BD, and Moerlanda TS. 2005. Temperature sensitivity of calcium binding for parvalbumins from Antarctic and temperate zone teleost fishes. *Comp Biochem Physiol A* 140: 179–185.

Fangue NA, Richards JG, and Schulte PM. 2009. Do mitochondrial properties explain intraspecific variation in thermal tolerance? *J Exp Biol* 212: 514–522.

Farrell AP. 2009. Environment, antecedents and climate change: Lessons from the study of temperature physiology and river migration of salmonids. *J Exp Biol* 212: 3771–3780.

Farrell AP, Johansen JA, and Suarez RK. 1991. Effects of exercise-training on cardiac performance and muscle enzymes in rainbow trout, *Oncorhynchus-mykiss*. *Fish Physiol Biochem* 9: 303–312.

Forgan LG and Forster ME. 2012. Oxygen dependence of metabolism and cellular adaptation in vertebrate muscles: A review. *J Comp Physiol B* 182: 177–188.

Frankel N, Davis GK, Vargas D, Wang S, Payre F, and Stern DL. 2010. Phenotypic robustness conferred by apparently redundant transcriptional enhancers. *Nature* 466: 490–493.

Fu SJ, Brauner CJ, Cao ZD, Richards JG, Peng JL, Dhillon R, and Wang YX. 2011. The effect of acclimation to hypoxia and sustained exercise on subsequent hypoxia tolerance and swimming performance in goldfish (*Carassius auratus*). *J Exp Biol* 15: 2080–2088.

Fukuda R, Zhang H, Kim JW, Shimoda L, Dang CV, and Semenza GL. 2007. HIF-1 regulates cytochrome oxidase subunits to optimize efficiency of respiration in hypoxic cells. *Cell* 129: 111–122.

Gahr SA, Vallejo RL, Weber GM, Shepherd BS, Silverstein JT, and Rexroad CE III. 2008. Effects of short-term growth hormone treatment on liver and muscle transcriptomes in rainbow trout (*Oncorhynchus mykiss*). *Physiol Genomics* 32: 380–392.

Galloway TF, Bardal T, Kvam SN, Dahle SW, Nesse G, Randøl M, Kjørsvik E, and Andersen Ø. 2006. Somite formation and expression of MyoD, myogenin and myosin in Atlantic halibut (*Hippoglossus hippoglossus* L.) embryos incubated at different temperatures: Transient asymmetric expression of MyoD. *J Exp Biol* 209: 2432–2441.

Garland T and Carter PA. 1994. Evolutionary physiology. *Annu Rev Physiol* 56: 579–621.

Gemballa S and Vogel F. 2002. Spatial arrangement of white muscle fibers and myoseptal tendons in fishes. *Comp Biochem Physiol* A133: 1013–1037.

Ghalambor CK, McKay JK, Carroll SP, and Reznick DN. 2007. Adaptive versus non-adaptive phenotypic plasticity and the potential for contemporary adaptation in new environments. *Funct Ecol* 21: 394–407.

Goda N and Kanai M. 2012. Hypoxia-inducible factors and their roles in energy metabolism. *Int J Hematol* 95: 457–463.

Gracey AY, Troll JV, and Somero GN. 2001. Hypoxia-induced gene expression profiling in the euryoxic fish *Gillichthys mirabilis*. *Proc Natl Acad Sci USA* 98: 1993–1998.

Guderley H. 2004. Metabolic responses to low temperature in fish muscle. *Biol Rev* 79: 409–427.

Gundersen K. 2011. Excitation-transcription coupling in skeletal muscle: The molecular pathways of exercise. *Biol Rev* 86: 564–600.

Hall TE, Cole NJ, and Johnston IA. 2003. Temperature and the expression of seven muscle-specific protein genes during embryogenesis in the Atlantic cod *Gadus morhua* L. *J Exp Biol* 206: 3187–3200.

Hall TE and Johnston IA. 2003. Temperature and developmental plasticity during embryogenesis in the Atlantic cod *Gadus morhua* L. *Mar Biol* 142: 833–840.

Hanel R and Wieser W. 1996. Growth of swimming muscles and its metabolic cost in larvae of whitefish at different temperatures. *J Fish Biol* 48: 937–951.

Hazel, JR. 1995. Thermal adaptation in biological membranes: Is homeoviscous adaptation the explanation? *Annu Rev physiol* 57(1): 19–42.

Henry CA and Amacher SL. 2004. Zebrafish slow muscle cell migration induces a wave of fast muscle morphogenesis. *Dev Cell* 7: 917–923.

Hinits Y, Osborn DPS, and Hughes SM. 2009. Differential requirements for myogenic regulatory factors distinguish medial and lateral somitic, cranial and fin muscle fibre populations. *Development* 136: 403–414.

Hock MB and Kralli A. 2009. Transcriptional control of mitochondrial biogenesis and function. *Annu Rev Physiol* 71: 177–203.

Hofmann GE, Buckley BA, Airaksinen S, Keen JE, and Somero GN. 2000. Heat-shock protein expression is absent in the Antarctic fish *Trematomus bernacchii* (family Nototheniidae). *J Exp Biol* 203: 2331–2339.

Hollway GE, Bryson-Richardson RJ, Berger S, Cole NJ, Hall TE, and Currie PD. 2007. Whole-somite rotation generates muscle progenitor cell compartments in the developing zebrafish embryo. *Dev Cell* 12: 207–219.

Huey RB, Berrigan B, Gilchrist GW, and Herron JC. 1999. Testing the adaptive significance of acclimation: A strong inference approach. *Am Zool* 39: 323–336.

Hughes GM. 1973. Respiratory responses to hypoxia in fish. *Am Zool* 13: 475–489.

Itoi S, Kinoshita S, Kikuchi K, and Watabe S. 2003. Changes of carp FoF1-ATPase in association with temperature acclimation. *Am J Physiol Regul Integr Comp Physiol* 284: R153–R163.

Jibb LA and Richards JG. 2008. AMP-activated protein kinase activity during metabolic rate depression in the hypoxic goldfish, *Carassius auratus*. *J Exp Biol* 211: 3111–3122.

Johnston IA. 1977. A comparative study of glycolysis in red and white muscles of the trout (*Salmo gairdneri*) and mirror carp (*Cyprinus carpio*). *J Fish Biol* 11: 575–588.

Johnston IA. 1980. Specialisations of fish muscle. In: Goldspink DF (Ed.), *Development and Specialisation of Muscle*, Vol. 7, Society for Experimental Biology. Cambridge, U.K.: Cambridge University Press, pp. 123–148.

Johnston IA. 1981. Structure and function of fish muscles. In: Day MH (Ed.), *Vertebrate Locomotion*. Symp *Zool Soc Lond* 48: 71–113.

Johnston IA. 2006. Environment and plasticity of myogenesis in teleost fish. *J Exp Biol* 209: 2249–2264.

Johnston IA and Bernard LM. 1982. Routine oxygen consumption and characteristics of the myotomal muscle in tench: Effects of long-term acclimation to hypoxia. *Cell Tissue Res* 227: 161–177.

Johnston IA and Bernard LM. 1984. Quantitative study of capillary supply to the skeletal muscles of Crucian carp *Carassius carassius* L.: Effects of hypoxic acclimation. *Physiol Zool* 57: 9–18.

Johnston IA, Bower NI, and Macqueen DJ. 2011. Growth and the regulation of myotomal muscle mass in teleost fish. *J Exp Biol* 214: 1617–1628.

Johnston IA, Cole NJ, Vieira VLA, and Davidson I. 1997. Temperature and developmental plasticity of muscle phenotype in herring larvae. *J Exp Biol* 200: 849–868.

Johnston IA, Fleming JD, and Crockford T. 1990. Thermal acclimation and muscle contractile properties in cyprinid fish. *Am J Physiol Regul Integr Comp Physiol* 259: R231–R236.

Johnston IA, Lee HT, Macqueen DJ, Paranthaman K, Kawashima C, Anwar A, Kinghorn JR, and Dalmay T. 2009. Embryonic temperature affects muscle fibre recruitment in adult zebrafish: Genome-wide changes in gene and microRNA expression associated with the transition from hyperplastic to hypertrophic growth phenotypes. *J Exp Biol* 212: 1781–1793.

Johnston IA and Lucking M. 1978. Temperature induced variation in the distribution of different types of muscle fibres in the goldfish (*Carassius carassius*). *J Comp Physiol* 124: 111–116.

Johnston IA, Macqueen DJ, and Watabe S. 2008. Molecular biotechnology of development and growth in fish muscle. In: Tsukamoto K, Takeuchi T, Beard TD Jr., and Kaiser MJ (Eds.), *Fisheries for Global Welfare and Environment*, 5th edn. World Fisheries Congress, Tokyo, Terrapub, pp. 241–262.

Johnston IA and Maitland B. 1980. Temperature acclimation in crucian carp, *Carassius carassius* L., morphometric analyses of muscle fibre ultrastructure. *J Fish Biol* 17: 3–125.

Johnston IA, Manthri S, Alderson R, Smart A, Campbell P, Nickell D, Robertson B, Paxton CG, and Burt ML. 2003. Freshwater environment affects growth rate and muscle fibre recruitment in seawater stages of Atlantic salmon (*Salmo salar* L.). *J Exp Biol* 206: 1337–1351.

Johnston IA and Moon TW. 1980. Endurance exercise training in the fast and slow muscles of a teleost fish (*Pollachius virens*). *J Comp Physiol* 135: 147–156.

Johnston IA and Temple GK. 2002. Thermal plasticity of skeletal muscle phenotype in ectothermic vertebrates and its significance for locomotory behaviour. *J Exp Biol* 205(pt 15): 2305–2322.

Johnston IA, Vieira VLA, and Abercromby M. 1995. Temperature and myogenesis in embryos of the Atlantic herring *Clupea harengus*. *J Exp Biol* 198: 1389–1403.

Johnston IA, Vieira VLA, and Temple GK. 2001. Functional consequences and population differences in the developmental plasticity of muscle to temperature in Atlantic herring *Clupea harengus*. *Mar Ecol Progr Ser* 213: 285–300.

Jonz MG and Nurse CA. 2005. Development of oxygen sensing in the gills of zebrafish. *J Exp Biol* 208: 1537–1549.

Kajimura S, Aida K, and Duan C. 2006. Understanding hypoxia-induced gene expression in early development: In vitro and in vivo analysis of hypoxia-inducible factor 1-regulated zebrafish insulin-like growth factor binding protein 1 gene expression. *Mol Cell Biol* 26: 1142–1155.

Kamei H, Ding Y, Kajimura S, Wells M, Chiang P, and Duan C. 2011. Role of IGF signaling in catch-up growth and accelerated temporal development in zebrafish embryos in response to oxygen availability. *Development* 138: 777–786.

Kammer AR, Orczewska JI, and O'Brien KM. 2011. Oxidative stress is transient and tissue specific during cold acclimation of threespine stickleback. *J Exp Biol* 214: 1248–1256.

Kinne O and Kinne EM. 1962. Rates of development in embryos of a cyprinodont fish exposed to different temperature-salinity-oxygen combinations. *Can J Zool* 40: 231–253.

Kinsey ST, Locke BR, and Dillaman RM. 2011. Molecules in motion: Influences of diffusion on metabolic structure and function in skeletal muscle. *J Exp Biol* 214: 263–274.

Kondo K, Misaki R, and Watabe S. 2010. Transcriptional activities of medaka *Oryzias latipes* peroxisome prolif-erator-activated receptors and their gene expression profiles at different temperatures. *Fish Sci* 76: 167–175.

Kopp R, Köblitz L, Egg M, and Pelster B. 2011. HIF signaling and overall gene expression changes during hypoxia and prolonged exercise differ considerably. *Physiol Gen* 43: 506–516.

Koumoundouros G, Ashton C, Sfakianakis DG, Divanach P, Kentouri M, Anthwal N, and Stickland NC. 2009. Thermally induced phenotypic plasticity of swimming performance in European sea bass *Dicentrarchus labrax* juveniles. *J Fish Biol* 74: 1309–1322.

Leary SC, Lyons CN, Rosenberger AG, Ballantyne JS, Stillman J, and Moyes CD. 2003. Fiber-type differences in muscle mitochondrial profiles. *Am J Physiol* 285: R817–R826.

Leary SC and Shoubridge EA. 2003. Mitochondrial biogenesis: Which part of "NO" do we understand? *BioEssays* 25: 538–541.

Lee CY, Hu SY, Gong HY, Chen MH, Lu JK, and Wu JL. 2009. Suppression of myostatin with vector-based RNA interference causes a double-muscle effect in transgenic zebrafish. *Biochem Biophys Res Commun* 387(4): 766–771.

LeMoine CMR, Craig PM, Dhekney K, Kim JJ, and McClelland GB. 2010a. Temporal and spatial patterns of gene expression in skeletal muscles in response to swim training in adult zebrafish (*Danio rerio*). *J Comp Physiol B* 180(1): 151–160.

LeMoine CM, Genge CE, and Moyes CD. 2008. Role of the PGC-1 family in the metabolic adaptation of gold-fish to diet and temperature. *J Exp Biol* 211: 1448–1455.

LeMoine CM, Lougheed SC, and Moyes CD. 2010b. Modular evolution of PGC-1alpha in vertebrates. *J Mol Evol* 70: 492–505.

Leroi AM, Bennett AF, and Lenski RE. 1994. Temperature acclimation and competitive fitness: An experimen-tal test of the beneficial acclimation assumption. *Proc Natl Acad Sci USA* 91: 1917–1921.

Liu Y, Cao ZD, Fu SJ, Peng JL, and Wang YX. 2009. The effect of exhaustive chasing training and detraining on swimming performance in juvenile darkbarbel catfish (*Peltebagrus vachelli*). *J Comp Physiol B* 179: 847–855.

Macqueen DJ, Robb D, and Johnston IA. 2007. Temperature influences the coordinated expression of myo-genic regulatory factors during embryonic myogenesis in Atlantic salmon (*Salmo salar* L.). *J Exp Biol* 210: 2781–2794.

Macqueen DJ, Robb DH, Olsen T, Melstveit L, Paxton CG, and Johnston IA. 2008. Temperature until the 'eyed stage' of embryogenesis programmes the growth trajectory and muscle phenotype of adult Atlantic salmon. *Biol Lett* 4: 294–298.

Malek RL, Sajadi H, Abraham J, Grundy MA, and Gerhard GS. 2004. The effects of temperature reduction on gene expression and oxidative stress in skeletal muscle from adult zebrafish. *Comp Biochem Physiol C* 138: 363–373.

Marks C, West TN, Bagatto B, and Moore FBG. 2005. Developmental environment alters conditional aggres-sion in zebrafish. *Copeia* 2005: 901–908.

Martin CI and Johnston IA. 2005. The molecular regulation of exercised-induced muscle fibre hypertrophy in the common carp: Expression of MyoD, PCNA and components of the calcineurin-signalling pathway. *Comp Biochem Physiol B* 142: 324–334.

Martínez ML, Landry C, Boehm R, Manning S, Cheek AO, and Rees BB. 2006. Effects of long-term hypoxia on enzymes of carbohydrate metabolism in the Gulf killifish, *Fundulus grandis*. *J Exp Biol* 209: 3851–3861.

Martin-Perez M, Fernandez-Borras J, Ibarz A, Millan-Cubillo A, Felip O, de Oliveira E, and Blasco J. 2012. New insights into fish swimming: A proteomic and isotopic approach in gilthead sea bream. *J Proteome Res* 11: 3533–3547.

Mason SD, Howlett RA, Kim MJ, Olfert IM, Hogan MC, McNulty W, Hickey RP, Wagner PD, Kahn CR, Giordano FJ, and Johnson RS. 2004. Loss of skeletal muscle HIF-1alpha results in altered exercise endurance. *PLoS Biol* 2: e288.

Mason S and Johnson RS. 2007. The role of HIF-1 in hypoxic response in the skeletal muscle. In: Roach R, Wagner PD, and Hackett P (Eds.), *Hypoxia and the Circulation*. New York: Springer.

Matschak TW, Hopcroft T, Mason PS, Crook AR, and Stickland NC. 1998. Temperature and oxygen tension influence the development of muscle cellularity in embryonic rainbow trout. *J Fish Biol* 53: 581–590.

Matschak TW, Stickland NC, Crook AR, and Hopcroft T. 1995. Is physiological hypoxia the driving force behind temperature effects on muscle development in embryonic Atlantic salmon (*Salmo salar* L.)? *Differentiation* 59: 71–77.

Matschak TW, Stickland NC, Mason PS, and Crook AR. 1997. Oxygen availability and temperature affect embryonic muscle development in Atlantic salmon (*Salmo salar* L.). *Differentiation* 61: 229–235.

McClelland GB. 2004. Fat to the fire: The regulation of lipid oxidation with exercise and environmental stress. *Comp Biochem Physiol B* 139: 443–460.

McClelland GB, Craig PM, Dhekney K, and Dipardo S. 2006. Temperature- and exercise-induced gene expression and metabolic enzyme changes in skeletal muscle of adult zebrafish (*Danio rerio*). *J Physiol* 577: 739–751.

McClelland GB, Dalziel AC, Fragoso NM, and Moyes CD. 2005. Muscle remodeling in relation to blood supply: Implications for seasonal changes in mitochondrial enzymes. *J Exp Biol* 208: 515–522.

Medeiros EF, Phelps MP, Fuentes FD, and Bradley TM. 2009. Overexpression of follistatin in trout stimulates increased muscling. *Am J Physiol Regul Integr Comp Physiol* 297: R235–R242.

Morash AJ, Bureau D, and McClelland GB. 2009. Dietary effects on the regulation of CPT I in rainbow trout (*Oncorhynchus mykiss*). *Comp Biochem Physiol B* 152: 85–93.

Morash AJ, Kajimura M, and McClelland GB. 2008. Intertissue regulation of carnitine palmitoyltransferase I (CPTI): Mitochondrial membrane properties and gene expression in rainbow trout (*Oncorhychus mykiss*). *Biochim Biophys Acta—Biomembranes* 1778: 1382–1389.

Morash AJ, Le Moine CMR, and McClelland GB. 2010. Genome duplication events have lead to a diversification in the CPT I gene family in fish. *Am J Physiol* 299: R579–R589.

Morash AJ and McClelland GB. 2011. Regulation of carnitine palmitoyltransferase (CPT) I during fasting in rainbow trout (*Oncorhynchus mykiss*) promotes increased mitochondrial fatty acid oxidation. *Physiol Biochem Zool* 84: 625–633.

Mosher DS, Quignon P, Bustamante CD, Sutter NB, Mellersh CS, Parker HG, and Ostrander EA. 2007. A mutation in the myostatin gene increases muscle mass and enhances racing performance in heterozygote dogs. *PLoS Genet* 3(5): 779–786.

Mueller IA and O'Brien KM. 2011. Nitric oxide synthase is not expressed, nor up-regulated in response to cold acclimation in liver or muscle of threespine stickleback (*Gasterosteus aculeatus*). *Nitric Oxide* 25: 416–442.

Nelson T, McEachron D, Freedman W, and Yang W-P. 2003. Cold acclimation increases gene transcription of two calcium transport molecules, calcium transporting ATPase and parvalbumin beta, in *Carassius auratus* lateral musculature. *J Ther Biol* 28: 227–234.

O'Brien KM. 2011. Mitochondrial biogenesis in cold-bodied fishes. *J Exp Biol* 214: 275–285.

Orczewska JI, Hartleben G, and O'Brien KM. 2010. The molecular basis of aerobic metabolic remodeling differs between oxidative muscle and liver of threespine sticklebacks in response to cold acclimation. *Am J Physiol Regul Integr Comp Physiol* 299: R352–R364.

Padilla PA and Roth MB. 2001. Oxygen deprivation causes suspended animation in the zebrafish embryo. *Proc Natl Acad Sci USA* 98: 7331–7335.

Palstra AP, Tudorache C, Rovira M, Brittijn SA, Burgerhout E, van den Thillart GE, Spaink HP, and Planas JV. 2010. Establishing zebrafish as a novel exercise model: Swimming economy, swimming-enhanced growth and muscle growth marker gene expression. *PLoS One* 5(12): e14483.

Pandorf CE, Jiang WH, Qin AX, Bodell PW, Baldwin KM, and Haddad F. 2009. Calcineurin plays a modulatory role in loading-induced regulation of type I myosin heavy chain gene expression in slow skeletal muscle. *Am J Physiol Regul Integr Comp Physiol* 297: R1037–R1048.

Pedersen BK. 2012. Muscular interleukin-6 and its role as an energy sensor. *Med Sci Sports Exerc* 44: 392–396.

Pelster B. 2002. Developmental plasticity in the cardiovascular system of fish, with special reference to the zebrafish. *Comp Biochem Physiol A* 133: 547–553.

Pisani DF and Dechesne CA. 2005. Skeletal muscle HIF1α expression is dependent on muscle fiber type. *J Gen Physiol* 126: 173–178.

Planas JV, Martín-Pérez M, Magnoni LJ, Blasco J, Ibarz A, Fernandez-Borras J, and Palstra AP. 2013. Transcriptomic and proteomic response of skeletal muscle to swimming-induced exercise in fish. In: Palstra AP and Planas JV (Eds.), *Swimming Physiology of Fish*. Heidelberg, Germany: Springer-Verlag.

Plaut I. 2001. Critical swimming speed: Its ecological relevance. *Comp Biochem Physiol A* 131: 41–50.

Podrabsky JE and Culpepper KM. 2012. Cell cycle regulation during development and dormancy in embryos of the annual killifish *Austrofundulus limnaeus*. *Cell Cycle* 11: 1697–1704.

Rees BB, Targett TE, Ciotti BJ, Tolman CA, Akkina SS, and Gallaty AM. 2012. Temporal dynamics in growth and white skeletal muscle composition of the mummichog *Fundulus heteroclitus* during chronic hypoxia and hyperoxia. *J Fish Biol* 81: 148–164.

Ren H, Accili D, and Duan C. 2010. Hypoxia converts the myogenic action of insulin-like growth factors into mitogenic action by differentially regulating multiple signaling pathways. *Proc Natl Acad Sci USA* 107: 5857–5862.

Rescan PY. 2008. New insights into skeletal muscle development and growth in teleost fishes. *J Exp Zool B Mol Dev Evol* 310: 541–548.

Rescan PY, Jutel I, and Ralliere C. 2001. Two myostatin genes are differentially expressed in myotomal muscles of the trout (*Oncorhynchus mykiss*). *J Exp Biol* 204: 3523–3529.

Rezende EL, Gomes FR, Ghalambor CK, Russell GA, and Chappell MA. 2005. An evolutionary frame of work to study physiological adaptation to high altitudes. *Rev. Chil. Hist. Nat.* 78: 323–336.

Richards JG, Sardella BA, and Schulte PM. 2008. Regulation of pyruvate dehydrogenase in the common killifish, *Fundulus heteroclitus*, during hypoxia exposure. *Am J Physiol Regul Integr Comp Physiol* 295: R979–R990.

Rimoldi S, Terova G, Ceccuzzi P, Marelli S, Antonini M, and Saroglia M. 2012. HIF-1α mRNA levels in Eurasian perch (*Perca fluviatilis*) exposed to acute and chronic hypoxia. *Mol Biol Rep* 39: 4009–4015.

Roberts SB, McCauley LA, Devlin RH, and Goetz FW. 2004. Transgenic salmon overexpressing growth hormone exhibit decreased myostatin transcript and protein expression. *J Exp Biol* 207: 3741–3748.

Rombough PJ. 1988. Respiratory gas exchange, aerobic metabolism and effects of hypoxia during early life. In: Hoar WS and Randall DJ (Eds.), *Fish Physiology*. Toronto, Ontario, Canada: Academic Press, pp. 59–161.

Rome LC. 1998. Matching muscle performance to changing demand. In: Weibel ER, Taylor CR, and Bolis L (Eds.), *Principles of Animal Design. The Optimization and Symmorphosis Debate*. Cambridge, U.K.: Cambridge University Press, pp. 103–113.

Rome LC. 2006. Design and function of superfast muscles: New insights into the physiology of skeletal muscle. *Annu Rev Physiol* 68: 193–221.

Rome LC, Loughna PT, and Goldspink G. 1984. Muscle fiber activity in carp as a function of swimming speed and muscle temperature. *Am J Physiol Regul Integr Comp Physiol* 247: R272–R279.

Rowe GC, El-Khoury R, Patten IS, Rustin P, and Arany Z. 2012. PGC-1α is dispensable for exercise-induced mitochondrial biogenesis in skeletal muscle. *PLoS ONE* 7: e41817.

Rowlerson A, Mascarello F, Radaelli G, and Veggetti A. 1995. Differentiation and growth of muscle in the fish *Sparus aurata* (L): II. Hyperplastic and hypertrophic growth of lateral muscle from hatching to adult. *J Muscle Res Cell Motil* 16: 223–236.

Rowlerson A, Radaelli G, Mascarello F, and Veggetti A. 1997. Regeneration of skeletal muscle in two teleost fish: *Sparus aurata* and *Brachydanio rerio*. *Cell Tissue Res* 289: 311–322.

Rowlerson A and Veggetti A. 2001. Cellular mechanisms of post-embryonic muscle growth in aquaculture species. In: Johnston IA (Ed.), *Muscle Development and Growth*. San Diego, CA: Academic Press, pp. 103–140.

Rytkönen KT and Storz JF. 2011. Evolutionary origins of oxygen sensing in animals. *EMBO Rep* 12: 3–4.

Saleem A, Carter HN, Iqbal S, and Hood DA. 2011. Role of p53 within the regulatory network controlling muscle mitochondrial biogenesis. *Exerc Sport Sci Rev* 39(4): 199–205.

Sänger AM. 1993. Limits to the acclimation of fish muscle. *Rev Fish Biol Fisheries* 3: 1–15.

Sänger AM and Stoiber W. 2001. Muscle fiber diversity and plasticity. In: Johnson I (Ed.), *Muscle Development and Growth*. London, U.K.: Academic Press, pp. 187–250.

Scarpulla RC, Vega RB, and Kelly DP. 2012. Transcriptional integration of mitochondrial biogenesis. *Trend Endocrinol Metabolism* 23(9): 459–466.

Schiaffino S and Reggiani C. 2011. Fiber types in mammalian skeletal muscles. *Physiol Rev* 91: 1447–1531.

Schmidt K and Starck JM. 2010. Developmental plasticity, modularity, and heterochrony during the phylotypic stage of the zebrafish, *Danio rerio*. *J Exp Zool B Mol Dev Evol* 314: 166–178.

Schröter C, Herrgen L, Cardona A, Brouhard GJ, Feldman B, and Oates AC. 2008. Dynamics of zebrafish somitogenesis. *Dev Dyn* 237: 545–553.

Scott GR and Johnston IA. 2012. Temperature during embryonic development has persistent effects on thermal acclimation capacity in zebrafish. *Proc Natl Acad Sci USA* 109: 14247–14252.

Seebacher F and Walter I. 2012. Differences in locomotor performance between individuals: Importance of parvalbumin, calcium handling and metabolism. *J Exp Biol* 215: 663–670.

Seiliez I, Sabin N, and Gabillard JC. 2011. FoxO1 is not a key transcription factor in the regulation of myostatin (mstn-1a and mstn-1b) gene expression in trout myotubes. *Am J Physiol Regul Integr Comp Physiol* 301: R97–R104.

Sfakianakis D, Leris I, and Kentouri M. 2011. Effect of developmental temperature on swimming performance of zebrafish (*Danio rerio*) juveniles. *Environ Biol Fish* 90: 421–427.

Sidell BD. 1998. Intracellular oxygen diffusion: The roles of myoglobin and lipid at cold body temperature. *J Exp Biol* 201: 1119–1128.

Spicer JI and Burggren WW. 2003. Development of physiological regulatory systems: Altering the timing of crucial events. *Zoology* 106: 91–99.

Steinbacher P, Marschallinger J, Obermayer A, Neuhofer A, Sänger AM, and Stoiber W. 2011. Temperature-dependent modification of muscle precursor cell behaviour is an underlying reason for lasting effects on muscle cellularity and body growth of teleost fish. *J Exp Biol* 214: 1791–1801.

Stellabotte F and Devoto SH. 2007. The teleost dermomyotome. *Dev Dyn* 236: 2432–2443.

Stellabotte F, Dobbs-McAuliffe B, Fernández DA, Feng X, and Devoto SH. 2007. Dynamic somite cell rearrangements lead to distinct waves of myotome growth. *Development* 134: 1253–1257.

Stickland NC, White RN, Mescall PE, Crook AR, and Thorpe JE. 1988. The effect of temperature on myogenesis in embryonic development of the Atlantic salmon (*Salmo salar* L.). *Anat Embryol (Berl)* 178: 253–257.

Stoiber W, Haslett JR, Wenk R, Steinbacher P, Gollmann H-P, and Sänger AM. 2002. Cellularity changes in developing red and white fish muscle at different temperatures: Simulating natural environmental conditions for a temperate freshwater cyprinid. *J Exp Biol* 205: 2349–2364.

Storz JF, Scott GR, and Cheviron ZA. 2010. Phenotypic plasticity and genetic adaptation to high-altitude hypoxia in vertebrates. *J Exp Biol* 213: 4125–4136.

Tattersall GJ, Sinclair BJ, Withers PC, Fields PA, Seebacher F, Cooper CE, and Maloney SK. 2012. Coping with thermal challenges: Physiological adaptations to environmental temperatures. *Compr Physiol* 2: 2151–2202.

Temple GK, Cole NJ, and Johnston IA. 2001. Embryonic temperature and the relative timing of muscle-specific genes during development in herring (*Clupea harengus* L.). *J Exp Biol* 204: 3629–3637.

Thorarensen H, Young G, and Davie PS. 1996. 11-Ketotestosterones stimulates growth of heart and red muscle in rainbow trout. *Can J Zool* 74: 912–917.

Ton C, Stamatiou D, and Liew CC. 2003. Gene expression profile of zebrafish exposed to hypoxia during development. *Physiol Genomics* 13: 97–106.

Urfi AJ and Talesara CL. 1989. Response of pectoral adductor muscle of *Channa punctata* to altered workload. *Indian J Exp Biol* 27: 668–669.

Van der Meer SFT, Jaspers RT, and Degens H. 2011. Is the myonuclear domain size fixed? *J Musculoskelet Neuronal Interact* 11: 286–297.

Veggetti A, Mascarello F, Scapolo PA, and Rowlerson A. 1990. Hyperplastic and hypertrophic growth of lateral muscle in *Dicentrarchus labrax* (L.). An ultrastructural and morphometric study. *Anat Embryol (Berl)* 182: 1–10.

Vieira VLA and Johnston IA. 1992. Influence of temperature on muscle-fibre development in larvae of the herring *Clupea harengus*. *Mar Biol* 112: 333–341.

Vieira VLA and Johnston IA. 1996. Muscle development in the tambaqui, an important Amazonian food fish. *J Fish Biol* 49: 842–853.

Watabe S, Hwang GC, Nakaya M, Guo XF, and Okamoto Y. 1992. Fast skeletal myosin isoforms in thermally acclimated carp. *J Biochem* 111: 113–122.

Weatherley AH, Gill HS, and Lobo AF. 1988. Recruitment and maximal diameter of axial muscle fibres in teleosts and their relationship to somatic growth and ultimate size. *J Fish Biol* 33: 851–859.

West-Eberhard MJ. 2003. *Developmental Plasticity and Evolution*. Oxford, U.K.: Oxford University Press.

Widmer S, Moore FBG, and Bagatto B. 2006. The effects of chronic developmental hypoxia on swimming performance in zebrafish. *J Fish Biol* 69: 1885–1891.

Wood AW, Duan C, and Bern HA. 2005. Insulin-like growth factor signaling in fish. *Int Rev Cytol* 243: 215–285.

Wulff T, Jokumsen A, Højrup P, and Jessen F. 2012. Time-dependent changes in protein expression in rainbow trout muscle following hypoxia. *J Prot* 75: 2342–2351.

2 Cardiovascular System

A. Kurt Gamperl and Holly A. Shiels

CONTENTS

2.1 INTRODUCTION

The cardiovascular system plays a critically important role in fishes, as it does in all vertebrates: it transports oxygen, carbon dioxide, metabolic substrates and wastes, hormones, and various immune factors throughout the body. As such, it is one of the most studied organ systems in fishes and thought to be key to understanding the evolution of this vertebrate group. The cardiovascular system also determines, in part, the ecological niche that can be occupied by many fish species and how they will/might be impacted by future abiotic and biotic changes. For example, over the past decade, it has been recognized that metabolic scope (the difference between resting and maximum metabolic rate) is a key determinant of fish thermal tolerance and that this parameter is largely determined by the capacity of the fish to deliver enough oxygen to the tissues to meet the rising metabolic demand as temperature increases (Wang and Overgaard, 2007; Farrell, 2009; Pörtner et al., 2010; Gamperl, 2011).

The cardiovascular system is composed of four essential components: a muscular pump, the heart; tubes through which the blood flows, blood vessels; a control system consisting of neural, endocrine, and paracrine factors; and a convective medium, the blood. In this chapter, we describe the anatomy and function of the fish heart and vasculature, the control systems that regulate the cardiovascular system, and how this system responds to meet biological demands and to maintain homeostasis in the face of challenging environmental conditions. However, we do not cover the blood's composition, cell types, or their functions, and instead point you to recent reviews by Farrell (2011a,b). Further, although there is a significant degree of variation in cardiovascular design and function among the various fish taxa, we largely focus on teleost species (i.e., advanced bony fishes) due to the limited space available and because this group's cardiovascular system is best understood.

2.2 GENERAL FEATURES OF THE CARDIOVASCULAR SYSTEM

The fish cardiovascular system is responsible for delivering blood, under pressure, to the tissues where the exchange of gases and various molecules occurs. The organ responsible for pumping blood around the body is the heart, and with the exception of air breathing fishes (see Burggren and Johansen (1986) and Burggren et al. (1997) for details on their circulatory design), most of the blood pumped by the heart flows through a single series circuit of vessels/structures: ventral aorta → afferent branchial arteries → the gill lamellae → efferent branchial arteries → dorsal aorta → arteries → arterioles → tissue capillary beds → venules → veins → major veins. The pressure that the heart is required to produce is largely determined by the amount of blood being pumped and the resistance to flow created by the vasculature. This relationship is described by the following equation:

$$P = Q \times R$$

where
  P is the arterial blood pressure (kPa)
  Q is the cardiac output (mL min^{-1})
  R is the vascular resistance (kPa mL^{-1} min^{-1}), and resistance in the total circulation or each segment of the circulation is in turn determined by frictional energy loss to the vessel walls and the blood's viscosity as described by the following equation:

$$R = \frac{(8\eta l)}{(\pi r^4)}$$

where
  l and r are the length and radius of the vessel
  η is the blood's viscosity

As seen from the second equation, (1) resistance to flow is lowest in the vessels with the largest radius (i.e., the arteries and veins that transport the blood to and from the tissues, respectively) and highest in the small vessels of the gills and tissues, and (2) the easiest way to control vascular resistance, and thus blood flow, is to change the radius of blood vessels; that is, resistance is inversely proportional to r to the fourth power, and if r doubles, R is reduced to 1/16th of its previous value. Thus, it is no surprise that the control of vascular resistance/blood pressure is largely at the level of the arterioles/metarterioles and that blood pressure is reduced dramatically in the capillaries (the site of gas/molecular diffusion) and in the venules/veins as compared to that measured on the arterial side of the circulation. Nonetheless, blood pressure in all vertebrates, including fishes, is not nearly as high as might be predicted, because the elements of the microcirculation (arterioles and capillaries) are arranged in parallel, and thus, the total cross-sectional area of this segment of the circulation is much greater than that of an individual vessel.

The general organization of the fish cardiovascular system as described earlier is shown in Figure 2.1. However, this schematic diagram also details a number of other features of fish circulatory

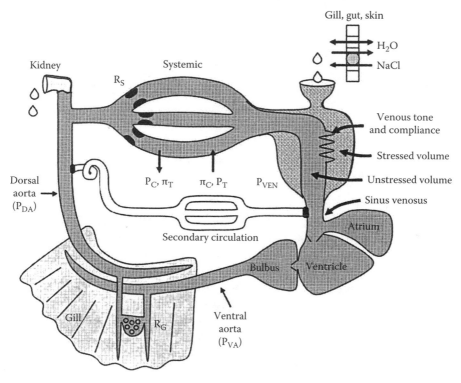

FIGURE 2.1 Schematic diagram of the teleost circulatory system. A single atrium and ventricle pump venous blood into a pulse-dampening, elastic, bulbus arteriosus and through the ventral aorta (P_{VA}, ventral aortic blood pressure) into the gill microcirculation. Blood leaving the gill enters the dorsal aorta (P_{DA}, dorsal aortic blood pressure) and carotid arteries, anteriorly, and is delivered to systemic tissues. The caliber of small vessels, mainly arteries and arterioles, determines the gill and systemic resistances (R_G and R_S) and regulates flow distribution to individual tissues and, along with cardiac output (Q), establishes ventral and dorsal aortic pressures. Venous return (P_{VEN}, central venous pressure) is affected by arteriolar resistance, the tone and compliance of the systemic veins, and by blood volume. Total blood volume (stressed + unstressed) is affected by water transfer across the gills, skin, and the gastrointestinal (GI) tract (gut), urine output by the kidney, and by Starling's forces, that is, capillary and interstitial tissue hydraulic (P_C and P_T, respectively) and oncotic (π_C and π_T, respectively) pressures that govern fluid exchange across the capillary endothelium. Venous return is directly coupled with Q. (Reproduced from Olson, K.R. and Farrell, A.P., The cardiovascular system, in Evans, D.H. and Claiborne, J.B., Eds., *The Physiology of Fishes*, Taylor & Francis, Boca Raton, FL, pp. 119–152, 2006. With permission.)

design/physiology. For example, there are other factors/mechanisms in addition to vascular resistance that control blood pressure, and there is an alternate route that plasma can take to get back to the heart. Blood pressure is also determined by total blood volume (30–80 mg kg^{-1} for teleosts and elasmobranchs; Olson, 1992) and the compliance (resistance of a hollow organ to a change in volume) of the vasculature. Blood volume is regulated between "stressed" and "unstressed" states by controlling (1) the amount of water/ions that move across the gill, gut, and skin epithelium and/or are eliminated at the kidneys and (2) fluid exchange across the capillary epithelium as determined by differences in hydrostatic and oncotic (i.e., due to protein concentration) pressures. Blood pressure at a particular blood volume is regulated by altering the compliance of the major veins (see Section 2.6). This control is very important in determining venous pressure and consequently ensuring that there is adequate filling of the heart. The secondary circulation (see Section 2.6.4) is an arterial–capillary–venous system that is in parallel with the primary circulation of fish. However, it has a very low hematocrit due to the very small anastomoses (openings, 2–20 μm) that connect it to the arteries and veins and is not thought to play a significant role in oxygen delivery/CO_2 transport. Instead, it has been suggested that it has nutritive, iono- or immunological functions (see Section 2.6.4).

2.3 CARDIAC MORPHOLOGY AND FUNCTION

2.3.1 HEART

In fishes, the heart is located ventrally and just anterior to the pectoral fins and peritoneal (gut) cavity and lies within the pericardial cavity (Figure 2.2a). This cavity is lined/formed by a membranous sac, the pericardium, that is comprised largely of epithelial-like (mesothelial) cells and connective tissue and folds inward to cover the surface of the heart as the epicardium. The pericardium secretes serous fluid that bathes the heart, and varies between species in its stiffness and in the degree that it is attached to adjacent structures (e.g., the esophagus, the pectoral girdle, and the pharyngeal muscles). In many species, it is sufficiently adhered to these structures to prevent large changes in its shape or in the volume of the pericardial cavity. This results in a reduction in pressure in the pericardial cavity during ventricular systole (contraction) that pulls on the thin-walled atrium, and thus, increases the pressure gradient from the sinus venosus/central veins to the atrium (Satchell, 1991). This change in pressure may contribute substantially to atrial filling in many fishes, an effect called *vis-a-fronte* ("force from in front") filling (see later text). A similar aspiration phenomenon occurs in elasmobranchs; however, this taxa, the sturgeons (order Acipenseriformes) and the hagfish (order Myxiniformes), also possess a pericardioperitoneal canal. This tube connects the pericardium with the peritoneum and allows the one-way flow of serous fluid from the pericardial to gut cavity during forceful contractions of the branchial muscles and heart (Shabetai et al., 1985; Abel et al., 1994; Head et al., 2001; Gregory et al., 2004). This movement of fluid augments the drop in pericardial pressure normally associated with ventricular contraction and may be required in these species to achieve large stroke volumes as pericardial volume can be 3.8 times that of the empty heart (Shabetai et al., 1985). This large difference in ventricular to pericardial volume would minimize the normal aspiration effect associated with ventricular contraction.

In fishes, the heart has classically been described as having four chambers (sinus venosus, atrium, ventricle, and bulbus arteriosus or conus arteriosus) (Farrell and Jones, 1992). However, it has recently been suggested that a more accurate description is that it has three chambers (sinus venosus, atrium, and ventricle) and an outflow tract (OFT); the OFT is located between the ventricle and the start of the ventral aorta (Icardo, 2012) (see later text).

2.3.1.1 Sinus Venosus

The sinus venosus is a thin-walled, compliant chamber that collects and stores blood prior to it entering the atrium. It is generally described as being composed of muscle and connective tissue.

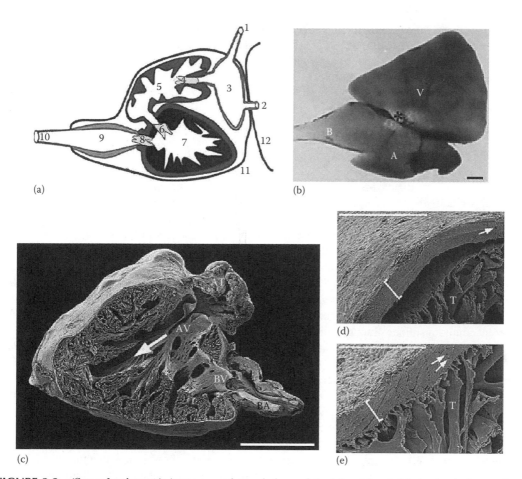

FIGURE 2.2 **(See color insert.)** Anatomy and morphology of the teleost heart. (a) Anatomical organization of the heart including the cardiac chambers and pericardial cavity. (1) Ductus Cuvier, (2) hepatic vein, (3) sinus venosus, (4) sinoatrial valve, (5) atrium, (6) atrioventricular valve, (7) ventricle, (8) bulboventricular valve, (9) bulbus arteriosus, (10) ventral aorta, (11) pericardium, and (12) peritoneum. (Modified from Farrell and Pieperhoff, 2011. With permission.) (b) Picture of the heart of the Atlantic horse mackerel, *Trachurus trachurus*. A, atrium; V, ventricle; and B, bulbus arteriosus. (Reproduced from Icardo, J.M., *Anat. Rec. A,* 288, 900, 2006. With permission.) (c) Scanning electron microscopy (SEM) image showing the left side of the adult sockeye salmon (*Oncorhynchus nerka*) heart. AT, atrium; AV, atrioventricular valve; BA, bulbus arteriosus; and BV, bulboventricular valve. Scale bar = 2 mm. (d, e) High magnification SEM images of the apex of the sockeye salmon heart showing two distinct muscle layers (compact and spongy myocardium). *Note*: separation of the layers in (d) is a processing artifact. The brackets show the thickness of the compact myocardium, T is the trabeculae of the spongy myocardium, and the arrows indicate small coronary vessels within the compact myocardium. Scale bar = 200 μm. c, d and e. (Reproduced from Pieperhoff, S. et al., *J. Anat.* 215, 536, 2009. With permission.)

However, the proportion of these two components in the chamber wall can vary widely among species, from mostly connective tissue in the zebrafish (*Danio rerio*) to mostly cardiac muscle tissue in the American eel (*Anguilla anguilla*) and to the replacement of cardiac muscle with smooth muscle in the common carp (*Cyprinus carpio*) (Yamauchi, 1980; Icardo, 2012). The sinus venosus is capable of rhythmic contractions, but its role in cardiac filling is thought to be minor because (1) there are few myocytes in the walls of the sinus venosus in most species (e.g., see Figure 4.2 in Genten et al., 2009) and (2) there are no valves to prevent retrograde flow into the large veins that drain into it (Farrell and Jones, 1992). Between the sinus venosus and atrium are the sinoatrial canal and the sinoatrial valve. The sinoatrial valve prevents backflow of blood from the atrium into the sinus venosus when the former chamber contracts. The sinoatrial region is also very important to cardiac function as it is the location of the heart's pacemaker cells. These specialized cells (see Section 3.3 for a further description), which are often organized into a ring around the sinoatrial confluence, set the contractile rhythm of the heart and are normally highly innervated by sympathetic and parasympathetic nerves that serve to increase and decrease heart rate, respectively (see Section 5.2).

2.3.1.2 Atrium

The atrium is a relatively thin-walled chamber that shows considerable variability in size and shape between species but is normally only 8%–25% of the mass of the ventricle (Santer, 1985; Farrell and Jones, 1992). The walls of the atrium are composed of a thick subepicardial layer of collagen, an external rim of myocardium, and a complex network of thin trabeculae (pectinate muscles) (Genten et al., 2009; Icardo, 2012). These thin (19–35 μm in diameter) trabeculae fan out from the atrioventricular orifice in a mesh-like arrangement (Santer, 1985), but do so in such a fashion as to leave a well-defined chamber lumen. This chamber expands greatly when it fills with blood due to the *vis-a-fronte* mechanism described earlier and/or the slightly positive mean circulatory filling pressure (0.1–0.2 kPa: Minerick et al., 2003; Sandblom and Axelsson, 2005, 2007a; Sandblom et al., 2009) that forces blood into the atria via *vis-a-tergo* ("force from behind") cardiac filling. At present, it is unclear to what extent these two mechanisms contribute to atrial filling in various species and under different conditions. For example, while opening of the pericardium, and thus elimination of *vis-a-fronte* filling, reduces maximum in situ cardiac output by 44% in the rainbow trout (*Oncorhynchus mykiss*; Farrell et al., 1988b), *in vivo* measurements suggest that *vis-a-tergo* is the primary filling mechanism in resting fish of this species (Minerick et al., 2003). Further, it is apparent that not all species (e.g., the eel; Franklin and Davie, 1991) are dependent on *vis-a-fronte* filling to achieve maximum levels of cardiac function.

When the atrium contracts, pressure in the chamber increases to 0.25–0.5 kPa (Satchell, 1991), and this is sufficient to open the atrioventricular valve and force blood into the relaxed ventricle. For many years, it was assumed that atrial contraction was the only way in which the ventricle was filled. However, there is now evidence that (1) ventricular filling is biphasic in both teleosts and elasmobranchs (Lai et al., 1996, 1998) and (2) the "early phase," which involves the passive flow of blood through the atrium based on the pressure gradient between the central veins and the relaxing ventricle, may be more important for ventricular filling than atrial contraction (the "late phase") in some species (e.g., the sea bass, *Paralabrax* sp.; Lai et al., 1998).

2.3.1.3 Ventricle

The ventricle is the largest of the heart chambers and shows the greatest variation in size, shape, and myocardial structure and vascularization. The shape of the ventricle is generally characterized as having three forms: tubular, pyramidal, or sac-like (Santer et al., 1983; Icardo, 2012). The tubular ventricle (often found in fishes with an elongated body form, e.g., eels) is the least common type, is almost circular in cross section, lacks an apex, and lies in the same longitudinal axis as the bulbus arteriosus.* Ventricles that are pyramidal in shape have the OFT and atrium connected to the chamber's base and a posteriorly pointed apex (see Figure 2.2a and b) and are generally found in fishes with an active lifestyle (e.g., salmonids, sharks, scombrids). Finally, the sac-like ventricle is the most common form and is rounded in shape with a blunt apex (e.g., cod, flounder). Fish hearts are also classified into four types based on whether the ventricle has a compact myocardial layer (e.g., see Figure 2.2d and e), the thickness (proportion) of this layer if present, and the extent of coronary artery vascularization (Davie and Farrell, 1991; Tota and Gattuso, 1996; Icardo, 2012). Type I hearts are found in approximately three quarters of all fishes, are composed entirely of "spongy" myocardium (i.e., loosely organized trabeculae), and either lack a coronary circulation or have it restricted to just under the epicardium. Thus, they depend exclusively on the venous blood percolating through the ventricle to supply the oxygen and nutritive requirements of the myocardium. In contrast, all the other heart types have some compact myocardium and some degree of coronary vascularization. This circulation brings highly oxygenated blood to the myocardium from the gills or the pectoral vasculature (Davie and Farrell, 1991). Type II hearts (e.g., Figure 2.2c through e) possess only a limited compact myocardial layer, and the coronary circulation is restricted to this myocardium. Type III hearts have less than 30% compact myocardium, but the coronary circulation now provides blood flow to both the compact and spongy muscle layers and possibly the atrium. Finally, Type IV hearts found in species like sharks and tunas have ventricles composed of >30% compact myocardium and a coronary circulation that not only serves the entire ventricle but also has an extensive network of capillaries in the atrium.

In fishes, relative ventricular mass (RVM, (ventricular mass/body mass $\times$ 100)) varies over a 10-fold range from ~0.03% (0.3 g kg body mass^{-1}) in flatfishes, to ~0.1% in the salmonids and elasmobranchs, and to 0.3%–0.4% in the tunas (Farrell and Jones, 1992). A large ventricular mass is most often, but not exclusively, related to the higher pressures that must be developed to perfuse the circulation of actively swimming species (e.g., up to ~8 kPa in salmonids (Kiceniuk and Jones, 1971) and ~12 kPa in tunas (Jones et al., 1993). For example, some Antarctic icefish lack hemoglobin in addition to having very low heart rates, and a very large RVM (0.3%–0.4%; Farrell and Jones, 1992) is needed to accommodate the large stroke volume required to meet the fish's oxygen demand. Other cold-stenothermal species such as the burbot (*Lota lota*) have an RVM (0.15%) nearly twofold greater than the RVM of other teleosts (Tiitu and Vornanen, 2002). This latter observation is likely due to the added work that the fish heart must do to pump the viscous blood at cold temperatures, because an increase in RVM (by 10%–30%) is seen in many, but not all, temperate teleosts when seasonally exposed or acclimated to cold temperatures (Driedzic et al., 1996). In addition to cold exposure, a number of other factors including stress (Johansen et al., 2011), exercise (Hochachka, 1961; Gallaugher et al., 2001), anemia (Siminot and Farrell, 2007; Powell et al., 2011, 2012), and sexual maturation (Clark and Rodnick, 1998) can result in cardiac enlargement (an increase in RVM). These increases in RVM are generally 10%–50%, but Clark and Rodnick (1998) reported that RVM increased by 2.4-fold in maturing male rainbow trout.

2.3.1.4 Outflow Tract

When blood is ejected from the ventricle during systole, it passes through a valve (called the bulboventricular valve in teleost fishes) and enters the OFT. Classically, this OFT was considered to be a conus arteriosus in elasmobranchs and primitive fishes and a bulbus arteriosus in more advanced bony fishes such as the teleosts. This led to the description of fish heart as having four chambers. However, recent work has shown that this schema needs to be revised. For example, a conal remnant exists at the junction between the ventricle and bulbus in numerous advanced teleosts (e.g., see Figure 4.10 in Genten et al., 2009), and a bulbar vessel is present between the conus and ventral aorta in sharks (Schib et al., 2002; Duran et al., 2008; Jones and Braun, 2011; Icardo, 2012). In fish with a bulbus arteriosus (e.g., Figure 2.2b), this structure dominates the OFT and is quite variable in shape, from tubular, to pear-shaped, to bulbus. The bulbus has walls composed mainly of vascular smooth muscle and connective tissue (elastin and collagen) (see Figure 4.9 in Genten et al., 2009) and a lumen that is often lined with ridges. This morphology gives the bulbus significant compliance (i.e., it is 30 times more distensible than the mammalian aorta) and allows it to expand greatly with blood during ventricular contraction. This increase in volume reduces pressure in the ventral aorta during ventricular ejection, and thus, may protect the delicate gill vasculature (Jones and Braun, 2011). In addition, elastic recovery of the bulbus during cardiac diastole allows for continued flow of blood into the ventral aorta and across the gills (improving gas exchange) during ventricular relaxation, and for blood pressure to be elevated throughout most of the cardiac cycle (Jones and Braun, 2011). In contrast to the bulbus, the conus arteriosus contains a large amount of compact myocardium (and thus is contractile), has one to several rows of leaf-like valves (Jones and Braun, 2011; Icardo, 2012), and its function is not well understood. For example, while some consider the conus to be a "muscular bulbus," it is unlikely to function in this way. The compact myocardium that predominates the conal wall is much stiffer than the smooth muscle in the bulbus; it contracts shortly (i.e., 250–500 ms) after the ventricle, and when the conus contracts, its valves stop the flow of blood from the heart (Jones and Braun, 2011).

2.3.2 CARDIOMYOCYTE

2.3.2.1 Gross Morphology

Across the ~28,000 extant fish species, gross cardiac morphology is quite diverse. In contrast, the morphology of the myocytes that make up these hearts is remarkably uniform and plays an important role in the organization of the various membrane systems and organelles that couple excitation with cellular contraction.

The majority of the atrium and ventricle are composed of working myocardial cells, the cardiomyocytes, whose contraction and relaxation facilitates cardiac pumping. These chambers are also composed of fibroblasts that are vital to the maintenance of the extracellular matrix, although very little is known about this abundant myocardial cell type in fishes. Finally, specialized cells of the cardiac conduction system are present in the fish heart as discussed in Section 3.3. In all fishes studied to date, the majority of working cardiomyocytes from the atria and the ventricle are spindle shaped with an extended length: width ratio, that is, ~90–180 μm long and 3–15 μm wide (Figure 2.3a through c). However, there is a small proportion of cells with a more sheet-like morphology (Figure 2.3d), and confocal z-stack imaging has shown that the thickness of these cells is usually <5 μm (Shiels and White, 2005; Shiels et al., 2011). These cell dimensions result in a ~3-fold

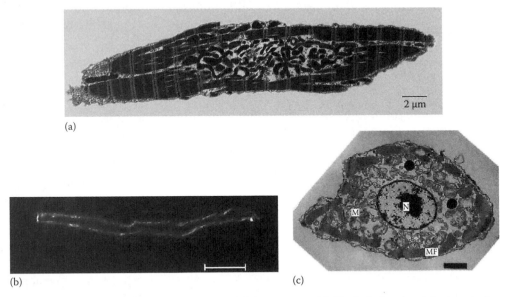

(a)

(b)   (c)

FIGURE 2.3 (See color insert.) Morphology of fish myocytes. (a) Single isolated atrial myocyte from a Pacific bluefin tuna (*Thunnus orientalis*); note the sarcomeric pattern of the myofibrils and the internal organization of the organelles along the longitudinal axis of the cells. (Di Maio, A., unpublished observation, with permission). (b) Confocal image of an adult zebrafish (*Danio rerio*) ventricular myocyte with the sarcolemma visualized with di-8-anepps (bar 20 μm). (Luxan, G. and Shiels, H., unpublished observation, with permission). (c) Electron micrographs of a ventricular myocyte from rainbow trout (*Oncorhynchus mykiss*) heart in cross section (M, mitochondria; MF, myofibrils; N, nucleus; dark droplets are lipid). Scale bar 2 μm. (Adapted from Vornanen, M., *J. Exp. Biol.*, 201, 533, 1998. With permission.)

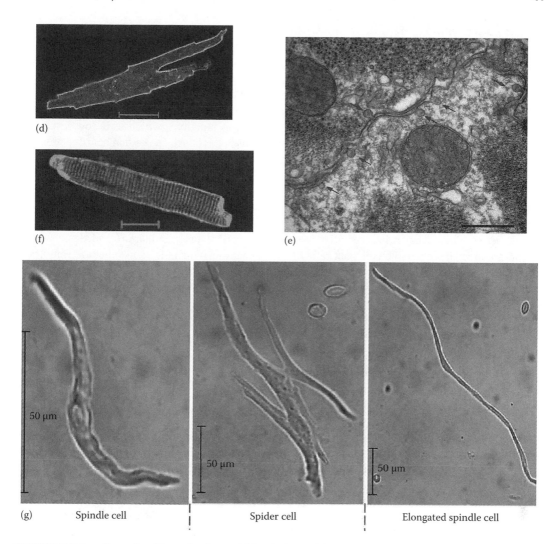

(d)

(f)

(e)

(g)   Spindle cell   Spider cell   Elongated spindle cell

FIGURE 2.3 (continued) **(See color insert.)** Morphology of fish myocytes. (d) Sheet-like ventricular myocyte from the rainbow trout labeled with both fluo-4 AM (cytosol) and di-8-ANEPPS (SL membrane). Notice the absence of T-tubular invaginations. (Adapted from Shiels, H.A. and White, E., *Am. J. Physiol.* 288, R1756, 2005. With permission.) (e) Bluefin tuna ventricle with arrows showing caveolae—small flask-like inward projections of the SL (bar 0.5 µm). (Di Maio, A., unpublished data. With permission.) (f) Pacemaker cells of the trout heart showing three distinct morphologies. (Adapted from Haverinen, J. and Vornanen, M., *Am. J. Physiol.*, 292, R1023, 2007. With permission.)

higher surface area to volume ratio in the fish myocyte compared with the mammalian myocyte (Figure 2.3a through d). Cardiomyocyte dimensions among different fish species are quite similar, despite large variations in body size and heart mass (e.g., compare zebrafish [Figure 2.3b] to bluefin tuna (*Thunnus orientalis*) [Figure 2.3a]). Thus, it is apparent that the fish heart can grow larger by increasing the number of cardiomyocytes (hyperplasia) in the myocardium. This regenerative potential has drawn much interest from cardiologists and the stem cell community. Nevertheless, growth/remodeling of the fish heart (e.g., see "RVM" in Section 3.2.1) is normally the result of both hyperplasia and hypertrophic myocyte growth (Farrell et al., 1988a; Clark and Rodnick, 1998; Klaiman et al., 2011).

2.3.2.2 Sarcolemmal Membrane and Cell-to-Cell Connections

The size and shape of the fish cardiomyocyte have significant implications for the organization of membrane structures and organelles (Vornanen, 1997). The sarcolemmal (SL) membrane of fish cardiomyocytes lack transverse-tubular (T-tubular) invaginations (compare Figure 2.3b and c with Figure 2.3f) (Santer, 1985; Di Maio and Block, 2008) but contain caveolae (Figure 2.3e; [Tiitu and Vornanen, 2002a]), which increase the cell surface area and may be important in cell signaling (Harvey and Calaghan, 2012). The SL also contains gap junctions, which form low resistance electrical pathways that conduct electrical impulses through the heart. In the fish heart, as in other vertebrate hearts, the gap junctions are formed by two hemichannels of connexon from opposing cells that come together to form a large conductance pore (connexin). This pore allows for the passage of ions, signaling molecules, and metabolites between cells (Severs et al., 2008), and recent work shows that fish connexins are remodeled during thermal acclimation, and thus, may compensate for thermally-induced changes in conduction velocity (Fenna, A. and Shiels, H., unpublished observations). There also exist specialized structures such as fascia adherens and desmosomes that do not electrically couple cells, but rather form strong mechanical linkages that are important in transmitting contractile force across the heart and may be involved in mechanical signaling (Pieperhoff et al., 2009). Finally, the SL membrane also contains ion channels and cell surface receptors that will be discussed in more detail under action potential (AP) (Section 2.4.1) and excitation–contraction (E–C) coupling (Section 2.4.2).

2.3.2.3 Organization of Intracellular Organelles

The myofilaments are generally located in a ring directly beneath the SL (Figure 2.3a and c) and are surrounded by the sarcoplasmic reticulum (SR). The peripheral location of myofilaments in fish cardiomyocytes is probably related to the lack of T-tubules, which requires that the contractile apparatus be brought in close proximity to the L-type calcium (Ca) channels embedded in the SL. The majority of fish hearts rely on transsarcolemmal Ca influx through these channels to a greater extent than the internal Ca stores of the SR for activating contraction. Indeed, SR membrane volume (relative to total nonmitochondrial cell volume) is less in fishes (ranging from 2% to 7%) than mammals (~12%). SR membrane volume has only been measured in a few fish species (perch, *Perca Flavescens* (4.5%–6%) (Bowler and Tirri, 1990); bluefin tuna (2%–3%) (Di Maio and Block, 2008)), and in tuna, atrial cardiomyocytes have a greater proportion of SR (3.2%) than ventricular cardiomyocytes (2.2%; Di Maio and Block, 2008). This is in line with observations from mammals (Bers, 2002). Environmental temperature can also alter SR density in fish. For example, cold acclimation increases SR density in bluefin tuna (Di Maio and Block, 2008; Shiels et al., 2011) and perch (Bowler and Tirri, 1990) hearts, and Shiels et al. (2011) showed that this adaptation improves Ca cycling in the cold. Similar changes have been seen in mammals during cold hibernation/torpor (Wang et al., 1997; Carey et al., 2003; Dibb et al., 2005).

Fish myocytes also contain mitochondria and glycogen/lipid droplets, which along with a single nucleus occupy the central space within the peripheral myofibril ring (see Figure 2.3c). The arrangement of intracellular mitochondria in rainbow trout myocytes is relatively chaotic, which is in contrast to the highly ordered parallel arrangement seen in mammalian ventricular myocytes (Birkedal et al., 2006).

2.3.3 Pacemaking and Cardiac Conduction

2.3.3.1 Pacemaker

The fish heart has an autonomous rhythm generated by a relatively small number (~28,000 in rainbow trout) of specialized cardiac myocytes located in the primary pacemaker region (Haverinen and Vornanen, 2007; Icardo, 2012). In most teleosts, the pacemaker cells are found in a specialized ring of tissue at the junction of the atrium and the sinus venous. This area of the myocardium is densely innervated by cholinergic and adrenergic neurons (Yamauchi, 1980) and regulated by hormonal factors that modulate the rate of pacemaker firing (see Section 2.3.3). The morphological and electrical properties of the fish primary pacemaker cells have only been studied thoroughly in the rainbow trout (Haverinen and Vornanen, 2007), although work on development of the conduction system has been well characterized in the embryonic zebrafish (Wong and Barnett, 2012). Similar to mammals, fish pacemaker cells show three distinct morphologies (Figure 2.3f) that give rise to fairly similar electrical profiles and the characteristic diastolic depolarization that causes APs. The diastolic depolarization in fish pacemaker cells could be the result of many interacting players including SL L- and T-type Ca channels, ion channels carrying the so-called funny (pacemaker) current (I_f), various repolarizing potassium (K) channels, and Ca transport mechanisms including the SR and the sodium (Na)-Ca exchanger (NCX). Haverinen and Vornanen (2007) show that the beating rate of all three cell types is modified by thermal acclimation, with cold-acclimated cells having a faster intrinsic rate compared with warm-acclimated cells (~44 bpm compared with ~38 bpm) at a common test temperature. Further, these authors report that these changes in intrinsic firing rate are unrelated to SR function, but are achieved, in part, via an increase in the delayed rectifier K current (I_{Kr}), which accelerates repolarization in the cold.

2.3.3.2 Conduction System

The electrical activity of the primary pacemaker needs to be carried throughout the heart to ensure the coordinated contraction of the cardiac chambers. In fishes, conduction system components, including fast conducting fibers, His bundles, and Purkinje cells have not been identified (Sedmera et al., 2003). It has been suggested that trabecular muscle bundles, which make up the spongy layer of the fish heart, provide a rapid conduction system and act as the functional correlate of the mammalian His–Purkinje system (Sedmera et al., 2003; Icardo and Colvee, 2011). Despite the absence of a morphologically defined conduction system, electrocardiograms (Patrick et al., 2010b) and optical mapping (Tsai et al., 2011) clearly show that electrical activity rapidly spreads across the fish heart and that the characteristic P, QRS, and T waves are readily identifiable (Figure 2.4a).

2.4 INTRINSIC CARDIAC CONTROL SYSTEMS

The remarkable variability in the function of fish hearts is achieved, in part, through a number of intrinsic control systems. These control systems allow the heart to adjust its function to changing demands and include variations in pacemaker firing rate and impulse conduction velocity, in cellular Ca cycling, in the way the myofilaments respond to Ca, and the release of hormones and paracrine factors (e.g., nitric oxide, NO). The next sections will investigate these processes in the fish heart.

2.4.1 ACTION POTENTIAL

The contraction of the fish heart is initiated and controlled by the AP, which occurs spontaneously in the pacemaker cells and is then carried through the working myocardium. The APs of fish atrial and ventricular myocytes (Figure 2.4b and c) are generally similar in shape to the APs in mammals and display a pronounced plateau phase (Nemtsas et al., 2010). Atrial and ventricular APs have been recorded from a number of fish taxa including zebrafish (Brette et al., 2008; Nemtsas et al., 2010), tuna (Galli et al., 2009a), flounder (*Pleuronectes flesus*) (Hoglund and Gesser, 1987), and a number of species from the orders Gadiformes, Cypriniformes, Perciformes, and Salmoniformes (Haverinen and Vornanen, 2009b). The atrial and ventricular myocytes of all fish species studied to date show a stable negative resting membrane potential (from −70 to −90 mV), that is maintained by diffusion of K out of the cell through inward rectifier (Kir2) channels. This outward current is called the inward rectifying K current, I_{K1}. In some species, mainly Cypriniformes, I_{K1} is modified

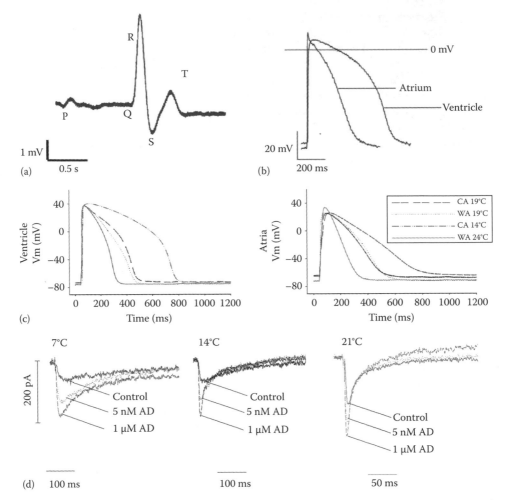

FIGURE 2.4 (See color insert.) Excitation of the fish heart. (a) Electrocardiogram from an isolated rainbow trout (*Oncorhynchus mykiss*) heart at 11°C showing atrial depolarization (P), ventricular depolarization (QRS), and ventricular repolarization (T). (Patrick, S. and Shiels, H., unpublished data.) (b) Action potentials of the rainbow trout heart recorded with microelectrodes from intact cardiac tissue at 4°C. Note the resting membrane potential (I_{K1}), the upstroke (I_{Na}), the plateau phase (I_{Ca}), and repolarization ($I_{Ks,r}$). (Adapted from Haverinen, J. and Vornanen, M., *J. Exp. Biol.,* 209, 549, 2006. With permission.) (c) Effect of thermal acclimation and acute temperature exposure on action potential (AP) characteristics in bluefin tuna (*Thunnus orientalis*) cardiomyocytes. Left is ventricle and right is atria, as indicated. Note the prolongation with cooling and acceleration with warming in both myocardial tissues. Also, note the lack of thermal compensation (compare CA–cold acclimated 19°C with WA–warm acclimated 19°C). (Adapted from Galli et al., *Am. J. Physiol.* 297, R502, 2009a. With permission.) (d) Effect of adrenergic stimulation (AD) and temperature (7°C, 14°C, and 21°C) on peak L-type Ca channel current (I_{Ca}) from trout atrial myocytes. Currents were elicited by 500 ms depolarizations from −70 to 0 mV using whole-cell patch clamp. Note the increase in current with increasing dose of AD. Also note the expanded timescale at 21°C. (Adapted from Shiels, H.A. et al., *Physiol. Biochem. Zool.,* 76, 816, 2003. With permission.)

via thermal acclimation to help maintain cellular excitability (Haverinen and Vornanen, 2009b). I_{K1} is usually smaller in atrial than ventricular myocytes. This means that less depolarizing current is required to initiate an AP in the fish atrium as compared to the ventricle, thus making it easier for the pacemaker cells to excite this tissue.

The rapid upstroke of the fish AP is produced by sodium (Na) influx (the Na current, I_{Na}) via Na channels (Na$_v$1 family) that open upon depolarization. omNav1.4a is the predominant Na$_v$ α subunit of the Na channel in the trout heart and is an ortholog of the mammalian skeletal muscle isoform (Haverinen et al., 2007). This partially explains the high tetrodotoxin (TTX) sensitivity of the trout (and other fish) Na channels, which are ~1000 times more sensitive to TTX than those of mammals (Vornanen et al., 2011). TTX is a neurotoxin that binds to voltage-gated Na channels preventing APs.

This Na-dependent depolarization of the SL opens the voltage-dependent calcium (Ca) channels (L-type Ca channels), which mediate Ca influx (the Ca current, I_{Ca}) and contribute to the plateau phase of the AP (Vornanen, 1997). Ca influx (I_{Ca}) during the AP plateau is particularly important in fish myocytes as it is the primary source of Ca for activation of the myofilaments (Hove-Madsen and Tort, 1998). L-type Ca channel expression is relatively insensitive to thermal acclimation (Vornanen, 1998), but Ca conductance through this channel is very sensitive to acute temperature changes (Shiels et al., 2000) and contraction frequency (Harwood et al., 2000). Both temperature-dependent changes in the duration of the AP plateau (Shiels et al., 2000) and adrenergic stimulation (Shiels et al., 2003) help to ensure this Ca influx is maintained when the thermal environment changes (see Sections 4.2 and 5.1).

The depolarization of the SL also causes slowly activating, voltage-dependent K channels to open. Enhanced K efflux together with decreasing Ca influx initiates SL repolarization and begins AP termination. There are two main slowly activating, voltage-dependent K currents in the fish heart, the rapid (I_{Kr}) and slow (I_{Ks}) components of the delayed rectifier (Vornanen et al., 2002a; Haverinen and Vornanen, 2009b; Hassinen et al., 2011). These currents are important for setting AP duration, and thus, are a focal site for modulation during thermal acclimation (Vornanen et al., 2002a). Repolarization also opens the Kir2 channels, allowing K to leave the cell on I_{K1}, which rapidly drives the membrane potential back to its stable resting voltage, ending the AP. I_{Kr} is usually greater in the atria than the ventricle, which accounts for the shorter AP duration in the atrium (Galli et al., 2009a; Haverinen and Vornanen, 2009b) and is strongly modified by thermal acclimation to compensate for the direct effect of temperature on AP duration.

Acute cooling increases AP duration in fish hearts, allowing more time for Ca influx, whereas acute warming shortens the AP, reducing the time for Ca influx. This can be seen in Figure 2.4c for both the atrial and ventricular APs of the bluefin tuna. These changes in AP duration compensate, in part, for the direct effect of temperature on I_{Ca} (Figure 2.4d); warm temperatures increase the amplitude but hasten inactivation of the fish I_{Ca} and acute cooling has the opposite effect (Aho and Vornanen, 1999; Shiels et al., 2000, 2001). The prolongation of the AP during acute cooling can also cause issues with restitution (Aho and Vornanen, 1999). Thus, temperate freshwater species (e.g., rainbow trout, burbot, perch, crucian carp (*Carassius carassius*), roach (*Rutilus rutilus*), and pike (*Esox lucius*)) acclimate to chronic cold by shortening the duration of their ventricular AP (Haverinen and Vornanen, 2009b). This mechanism occurs via modification of the I_{Kr} and I_{K1} (Vornanen et al., 2002a), and coupled with changes in Na currents (Haverinen and Vornanen, 2004, 2006) and remodeling of the SR (Shiels and Farrell, 1997; Shiels et al., 2010), allows the heart to beat faster and maintain excitability in the cold. However, this thermal compensation is not universal in fish, as it was not seen in tuna atria and ventricle (Figure 2.4c) (Galli et al., 2009a).

2.4.2 EXCITATION–CONTRACTION COUPLING

Excitation–contraction coupling (E-C coupling) is the process that drives contraction and relaxation of the working myocardium. It links SL excitation via the AP with mechanical contraction of the myofilaments, through the Ca transient. The time course of this process is shown in Figure 2.5a. The Ca transient ($\Delta[Ca^{2+}]_i$) is defined as the transient rise and fall in cytosolic Ca (from all external and internal Ca sources) during a contraction. Two consecutive Ca transients from a bluefin tuna ventricular myocyte are shown in Figure 2.5b. The rate at which the Ca transient rises and falls determines, in part, the rate of contraction and relaxation of the heart. The magnitude of the Ca transient is important in setting the strength of contraction and is used as an index of myocardial force (Yue, 1987).

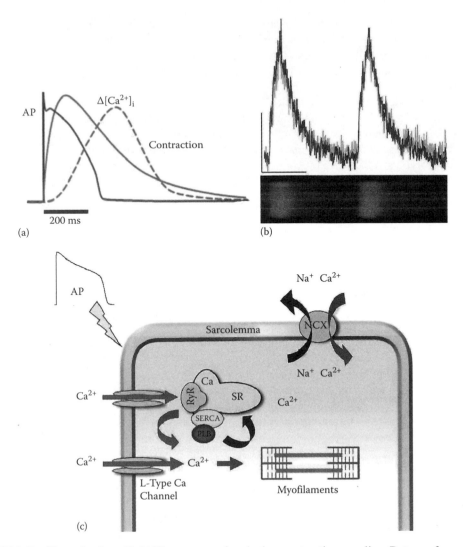

FIGURE 2.5 **(See color insert.)** (a) Time course of excitation–contraction coupling. Data are from a rabbit ventricular myocyte at 37°C. (Adapted from Bers, D.M., *Nature*, 415, 2002. With permission.) (b) The Ca transient measured across the width of a 14°C ventricular myocyte from Pacific bluefin tuna (*Thunnus orientalis*). Representative time courses (top) and corresponding raw line scan images (below) show temporal and spatial characteristics of Ca flux. Scale is 100 nM [Ca] by 1 s. Line scans are 2500 lines, 512 pixels. (Adapted from Shiels, H.A. et al., *Proc. Biol. Sci.*, 278, 18, 2011. With permission.) (c) Schema for excitation–contraction coupling in the fish cardiac myocyte. The figure shows the SL being excited by AP that opens L-type Ca channels, allowing Ca influx (arrows) down its concentration gradient. Ca can also enter the cell via reverse-mode Na–Ca exchange (NCX). Ca influx can trigger Ca release from the sarcoplasmic reticulum (SR) through ryanodine receptors (RyR). Together, these Ca influxes cause a transient rise in Ca that initiates contraction of the myofilaments. Relaxation occurs when Ca is removed from the cytosol (arrows) either back across the SL via forward-mode NCX or back into the SR via the SR Ca pump (SERCA) whose activity is regulated by phospholamban (PLB). (Figure composed by Dr G.L.J. Galli.)

The source of Ca can be of extracellular origin (coming in across the SL) or of intracellular origin (released from the SR). In fishes, in general, the extracellular Ca route is by far the most important (Tibbits et al., 1992) (see Figure 2.6). However, in certain species like trout and tuna, and under certain conditions, Ca release from the SR can play a significant role (see Figure 2.6). Figure 2.5c summarizes the ion channels, pumps, and organelles involved in E-C coupling in the fish heart.

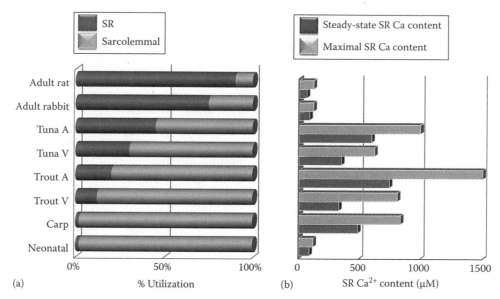

FIGURE 2.6　(See color insert.) (a) SR and SL Ca utilization during contraction. Note how transsarcolemmal Ca flux contributes more Ca to contraction than SR Ca release in fishes as compared with mammals. (b) Maximal and steady-state SR Ca content expressed as μM Ca/nonmitochondrial cell volume. Notice the incongruence between SR Ca utilization during contraction (a) and the Ca content of the SR (b) with a species. Both figures are compiled from a number of studies (see following text), using various methods to assess the source of Ca for contraction and the SR Ca contents. Thus, the figure is illustrative rather than quantitative. Data adapted from: Adult rat ventricle (a) (Negretti et al., 1993) (b) (Delbridge et al., 1997); adult rabbit ventricle (a) (Bassani et al., 1994) (b) (Satoh et al., 1996); tuna atrium, tuna ventricle (a and b) (Galli et al., 2011); trout atrium (a) (Shiels et al., 2002) (b) (Haverinen and Vornanen, 2009a,b); trout and carp ventricle (a and b) (Haverinen and Vornanen, 2009a,b); neonatal rabbit ventricle (Haddock et al., 1999). (Figure compiled with help of Dr G.L.J. Galli.)

The large surface area to volume ratio of fish myocytes, coupled with the large driving force for Ca influx, means that transsarcolemmal Ca results in a large change in cytosolic Ca. Between 20 and 80 μmol L^{-1} Ca (nonmitochondrial cell volume) can enter the fish myocyte (range depending on fish species and method of investigation) through L-type Ca channels and this is sufficient to initiate contraction of the myocyte (Vornanen et al., 2002b). The NCX can also bring Ca across the SL membrane, and thus, contribute to the Ca transient. The NCX transfers 3 Na and 1 Ca in opposite directions across the SL membrane. When the myocyte is at rest, the exchanger operates in forward mode and moves Ca out of the cell in exchange for Na. During the upsweep of the AP when the membrane is depolarized, the exchanger reverses direction and brings Ca into the cell and moves Na out (reverse-mode exchange). There are only a small range of membrane voltages (~0–10 mV) during a normal AP that favor Ca influx through the NCX (reverse mode), but due to the high surface area to volume ratio and high intracellular Na concentration of fish myocytes, (Birkedal and Shiels, 2007) reverse-mode NCX can play a large role in the systolic Ca transient in fish. Indeed, the Ca transient resulting from Ca influx through the NCX is alone sufficient to cause partial myofilament contraction in some fishes (trout (Hove-Madsen et al., 2000, 2003) and crucian carp (Vornanen, 1999)). In the cold stenothermic burbot, I_{Ca} is relatively small but I_{NCX} is relatively large and probably the dominant Ca influx pathway (Shiels et al., 2006b); whether this is a property of other cold stenotherms is not known. In most mammalian myocytes, Ca influx via I_{NCX} is not very important for bringing Ca into the cell.

Ca from the intracellular stores of the SR can also contribute to the Ca transient. The amount of SR Ca that contributes to the Ca transient (and thus contraction) is highly variable among species, tissue type, age, and environmental conditions and has been linked to cardiac performance and numerous cardiac pathologies (see Figure 2.6a). Myocardial force production in sluggish species like crucian carp shows no response to SR inhibition (Tiitu and Vornanen, 2001), whereas atrial and ventricular muscles of the rainbow trout show a 16% and 6% inhibition of force, respectively (Aho and Vornanen, 1999). Similar decreases have been seen in the Ca transient in rainbow trout ventricular myocytes after SR inhibition (Harwood et al., 2000). Highly active ectotherms rely more strongly on SR Ca stores than sedentary species. Tuna, which are considered to be teleost athletes, have a well-developed SR network (Di Maio and Block, 2008), and SR inhibition significantly decreases contractile force in bigeye tuna (*Thunnus obesus*) ventricle (by 30% (Galli et al., 2009b)), yellowfin tuna (*Thunnus albacares*) atria (by 50% (Shiels et al., 1999)), and skipjack tuna (*Katsuwonus pelamis*) atrium (by 30% (Keen et al., 1992)). Cold acclimation has also been shown to increase the response of fish heart muscle to SR inhibition (Keen et al., 1994; Shiels and Farrell, 1997). Myocardial force of cold stenothermic burbot shows a large decrease (~50% at 7°C) in force production after SR inhibition (Tiitu and Vornanen, 2002b). A link between phylogeny and SR dependence has also been suggested for neotropical freshwater fish species from the order Characiformes, superorder Ostariophysi, which utilize SR Ca during E-C coupling, independent of activity level or temperature (Costa et al., 2004; Rocha et al., 2007). The reasons underlying this phylogenetic SR dependence are unclear.

The majority of fish species studied to date do not rely strongly on SR Ca (Figure 2.6a). This is not due to lack of Ca in the fish SR, because studies with caffeine (which releases SR Ca stores) show that the fish SR holds very large amounts of Ca (Figure 2.6b). SR Ca content is determined by (1) the activity of the SR Ca-ATPase (SERCA) that pumps Ca from the cytosol into the SR lumen, (2) SR luminal Ca buffering (Korajoki and Vornanen, 2009), and (3) SR density (Di Maio and Block, 2008). Studies to date show SERCA activity is robust in active fishes like trout (Aho and Vornanen, 1998) and the tunas (Landeira-Fernandez et al., 2004). However, maximal or steady-state SR Ca content do not necessarily correlate with SERCA activity (Figure 2.6). Within the scombrids, mackerel (*Scomber japonicus*) SR Ca content is similar to Pacific bluefin tuna (Galli et al., 2011), and an inverse relationship between ryanodine sensitivity and SR Ca content has been reported for rainbow trout, burbot, and crucian carp (Haverinen and Vornanen, 2009a). This discrepancy may be underpinned by variable luminal Ca buffering within the SR of different ectothermic species (Korajoki and Vornanen, 2009).

Despite the high SR Ca content in fish (Figure 2.6b), it does not seem to be released from this cellular compartment during routine contractions. Ca is released from the SR through a process called Ca-induced Ca-release (CICR), whereby Ca influx across the SL membrane triggers the release of Ca from the SR through SR Ca release channels (ryanodine receptors). The role of CICR is hotly debated for fish cardiac myocytes. It is completely absent in some fish species, usually (but not always) low in sluggish fish like carp, and athletic fish with fast heart rates and high blood pressures (like salmon and tuna) often show more CICR. Indeed, the minor steady-state CICR in trout atria (Shiels, H.A. et al., 2002) has recently been shown to be increased via adrenergic stimulation (Brette, F., unpublished observations). In general, however, CICR is less important in fish E-C coupling than mammalian E-C coupling. Reasons may include a reduced Ca sensitivity of the fish ryanodine receptor to trigger Ca (Titu and Vornanen, 2003; Korajoki and Vornanen, 2009; Haverinen and Vornanen, 2009a), inadequate spatial organization of ryanodine receptors (Birkedal et al., 2009), or even the ryanodine receptor isoform itself (Shiels and White, 2005). Indeed, Ca sparks (spontaneous Ca release from clusters of ryanodine receptors) are infrequent in trout (Shiels and White, 2005) and zebrafish (Llach et al., 2011) despite high SR Ca loads. For a thorough discussion of the CICR in fish hearts and the role of the SR across vertebrates, please see Galli and Shiels (2012).

2.4.3 Ca and Myofilament Contraction

In fishes, as in all vertebrates, the contraction of the heart is permitted by Ca and powered by the generation of cross-bridges between the thick myosin filaments and the thin actin filaments.

Contraction is initiated by the Ca transient, which increases cytosolic Ca (from external and/or internal sources). This Ca binds to cardiac troponin C (cTnC), which in turn causes a conformational change that pulls cardiac troponin I (cTnI) away from the actin filament. This shifts tropomyosin (TM) from its groove in the actin filament, uncovering the myosin-binding sites on actin, and allowing for cross-bridge formation between actin and myosin. An increase in the amplitude of the Ca transient causes an increase in contractile force because more Ca means more Ca–cTnC interactions, and thus, more force producing cross-bridges.

The contractile apparatus in fishes has a number of interesting adaptations that permit function across a range of environmental conditions. Fish myofilaments are inherently more Ca sensitive than mammalian myofilaments (Churcott et al., 1994). This is in part due to modifications in the cTnC protein sequence, and Gillis and colleagues have shown that these modifications combat the reduction in myofilament Ca sensitivity that is known to occur in the cold (Gillis and Tibbits, 2002; Gillis et al., 2003).

2.4.4 CELLULAR LENGTH–TENSION RELATIONSHIP AND THE FRANK–STARLING RESPONSE

The force of contraction can be modulated by increasing Ca delivery to the myofilaments, but it can also be controlled by changing the way in which those myofilaments respond to Ca. Myofilament length has a profound impact on myofilament Ca sensitivity and force (tension) development as revealed by the cellular length–tension relationship (see Figure 2.7a). Myofilament length increases as the heart fills (stretches) during the cardiac cycle, and thus, stretch of the atrium and ventricle can play a significant role in regulating heart function through the Frank–Starling response in fishes

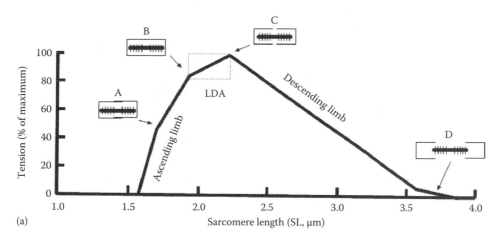

FIGURE 2.7 (a) Schematic diagram of the sarcomere length–tension relationship for mammalian cardiac muscle. The insets (A–D) show the role of myofilament overlap in the length-dependent increase in force. (A) This inset shows the position of actin and myosin at short sarcomere lengths (SLs) when myosin comes in contact with the Z-line and there is a rapid decline in tension as SLs decrease (to the left of the arrow). SLs to the left of inset (B) show how tension decreases when the thin filaments from opposite ends of the sarcomere overlap at the M-line. The region between (B) and (C) is the range of SLs where the potential availability of cross-bridges remains constant during sarcomere stretch because the central cross-bridge head-free zone of the myosin filament (M-line) is progressively uncovered. Inset (D) shows how tension declines toward zero when the sarcomere is stretched such that there is no overlap between thick and thin filaments. (Adapted from Shiels, H.A. and White, E., The effect of mechanical stimulation on vertebrate hearts: A question of class, in Kamkin, A. and Kiseleva, I., Eds., *Mechanosensitivity in Cells and Tissues Mechanosensitive Ion Channels*, Springer-Verlag, New York, pp. 323–342, 2008a; Shiels, H.A. and White, E., *J. Exp. Biol.*, 211, 2005, 2008b. With permission.)

(continued)

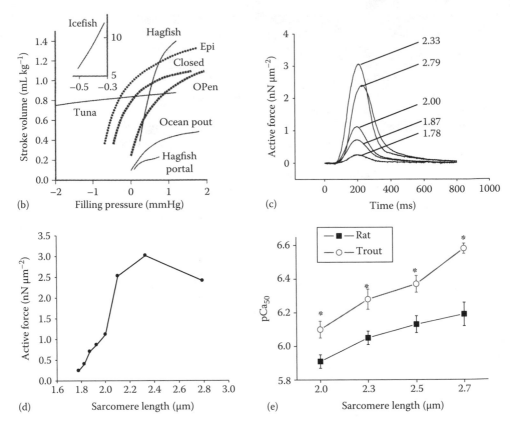

FIGURE 2.7 (continued) (b) *In vitro* Starling curves for various fish hearts. The dashed lines are data for the rainbow trout with "Epi," "Closed," and "Open," indicating hearts that were tested following epinephrine administration and those examined with an intact pericardium and one whose integrity had been breached, respectively. (Reproduced from Olson, K.R. and Farrell, A.P., The cardiovascular system, in Evans, D.H. and Claiborne, J.B., Eds., *The Physiology of Fishes*, Taylor & Francis, Boca Raton, FL, pp. 119–152, 2006. With permission.) (c) Contractile force produced by a single trout myocyte as it is stretched. The figure shows the contractions at each SL (indicated by numbers in μm) and how increasing SL increases force generation (up to a point). (d) SL–tension relationship from the same cell. Note the similarity between the SL–tension relationship in the single trout myocyte and the schematic diagram in part (a). (Part c and d: Adapted from Shiels, H.A. et al., *J. Gen. Physiol.*, 128, 37, 2006a. With permission.) (e) The SL dependency of myofilament Ca sensitivity in rat and trout permeabilized ventricular myocytes. The pCa for half maximal activation (pCa_{50}) is an index of myofilament Ca sensitivity of the contractile machinery. The * shows the data are significantly different. (Adapted from Patrick, S.M. et al., *J. Mol. Cell. Cardiol.* 48, 917, 2010a. With permission.)

(Farrell and Jones, 1992; Shiels, 2008) as it does in mammals (Allen and Kentish, 1985; Calaghan and White, 2005). However, the Frank–Starling response (i.e., ability of the heart to change its force of contraction and, therefore, stroke volume in response to changes in venous return/filling pressure) may be much more important to cardiac function in fishes as these taxa have the ability to modulate stroke volume to a greater degree than mammalian species. For example, it has been elegantly demonstrated through the use of in situ heart preparations, where the sinus venosus and ventral aorta are cannulated without disrupting the pericardium (Farrell et al., 1986; Graham and Farrell, 1989; Farrell, 1991; Mendonca et al., 2007), that small increments in filling pressure (0.2–0.3 kPa) result in very large (threefold) increases in stroke volume in fishes (e.g., see Figure 2.7b). This enhanced Frank–Starling response is partly because of the compliance of the fish's atrium and ventricle (Mendonca et al., 2007) as well as because of how sensitive the fish myocardium is to stretch. This stretched-induced increase in myocardial contractility ensures that even when the end-diastolic volumes of the ventricle are large, nearly all of the blood is normally ejected from the chamber (i.e., end-systolic volume is close to zero) (Franklin and Davie, 1992a).

The cellular mechanisms that allow the fish heart to accommodate large changes in stroke volume while retaining robust pumping ability (i.e., cardiac power) are only beginning to be understood. Figure 2.7c and d shows the increase in force with increased SL in a single myocyte from the rainbow trout heart (Shiels et al., 2006a). This work revealed a twofold extension of the functional ascending limb of the trout ventricular length–tension relationship as compared to that in the mammal, which is made possible by enhanced length-dependent myofilament Ca sensitivity (length-dependent activation; LDA (Patrick et al., 2010a)). This means that fish myofilaments are not only strongly activated by Ca, but that activation increases with length by a greater proportion in fish than in mammalian myofilaments. This feature is revealed by comparing the pCa_{50} values (the Ca concentration at which 50% of the myofilaments are activated) in permeabilized myocytes from fishes and mammals (Figure 2.7e). This figure shows that not only are the trout myofilaments more sensitive to Ca at any length than the rat, they also exhibit greater LDA (i.e., the slope of the fish line is greater than that of the mammal; Patrick et al., 2010a). These features make contractility of the fish heart very responsive to stretch and may underlie their robust Frank–Starling response. Fish myocytes also develop more titin-based passive tension than rat myocytes, which could account for the greater length-dependent activation (Patrick et al., 2010a). However, further studies are required to test this hypothesis. How length increases myofilament Ca sensitivity is still not completely understood, even for mammalian hearts (Moss and Fitzsimons, 2002). However, a more thorough discussion of these relationships and their impact on the passive properties of the fish heart is presented in Shiels and White (2008).

2.4.5 NATRIURETIC PEPTIDES AND NITRIC OXIDE

The mechanical performance and output of the fish heart is very strongly affected by filling pressure (preload) (see Section 2.4.4) and, under some circumstances, pressure in the ventral aorta (afterload). However, excessive preload and afterload (which may lead to elevated end-systolic and subsequently end-diastolic volumes) can stretch the myofibrils beyond their working length, and this can lead to cardiac distension and diminished pumping capacity (Olson and Farrell, 2006). On the other hand, chamber compliance (myocardial distensibility) is critical to ensuring adequate end-diastolic volumes/cardiac output. Thus, the heart has endocrine and paracrine mechanisms that help to modulate cardiac filling/function.

A number of natriuretic peptides have been definitively identified in fishes (e.g., atrial natriuretic peptide [ANP]; ventricular natriuretic peptide [VNP]; four cardiac natriuretic peptides [CNPs]; and brain natriuretic peptide [BNP]) (Johnson and Olson, 2008), and these are primarily synthesized by cardiomyocytes in response to cardiac distension (stretch) and subsequently released into the circulation. Once released, these hormones result in short-term decreases in blood pressure due to their potent direct vasodilatory effects and possibly their inhibitory effects on plasma levels of angiotensin II (Duff and Olson, 1986; Tsuchida and Takei, 1998; Farrell and Olson, 2000; Loretz and Pollina, 2000). These hormones also cause a more gradual decrease in blood volume due to their stimulation of diuresis and natriuresis (Loretz and Pollina, 2000). These changes culminate in reduced blood pressure (preload and afterload), and thus, return the chambers to a state of homeostasis with respect to end-diastolic volume (myocardial stretch). The protection afforded by these hormones against excessive preload and afterload is thought to be so important that it has recently been proposed as their main function in vertebrates—the "cardioprotective hypothesis" (Olson and Farrell, 2006; Johnson and Olson, 2008).

In addition to natriuretic peptides, the heart (endocardial epithelium, cardiac myocytes, and coronary vasculature) produces nitric oxide (NO), and this gasotransmitter has a number of paracrine/autocrine effects on heart function. For example, although it can have both negative (Atlantic salmon, *Salmo salar* (Gattuso, et al., 2002) and eel (Imbrogno, 2001; Pellegrino et al., 2003)) and positive (icefish, *Chionodraco hamatus* (Pellegrino et al., 2003; Cerra et al., 2009)) inotropic effects, it appears that the basal production of NO by the endocardial epithelium is critical to effective cardiac filling. In both the salmon and eel, blocking NO production resulted in a significant downward shift of the relationship between filling pressure and stroke volume/cardiac output (i.e., the Frank–Starling response). This result is consistent with experiments on mammalian hearts, which suggest that the endocardial epithelium could act as an important mechanosensor of luminar forces and that NO may enhance cardiac filling by modulating myocardial relaxation and diastolic tone (Brutsaert and Andries, 1992; Paulus et al., 1994; Shah et al., 1994; Angelone et al., 2012). Further, several NO-dependent effects have been identified when hearts are exposed to hypoxia. These include increases in coronary blood flow (Jensen and Agnisola, 2005) and cardiac O_2 utilization efficiency (Pedersen et al., 2010), and cardioprotection associated with the opening of ATP-sensitive (K_{ATP}) channels (Cameron et al., 2003).

2.5 EXTRINSIC CONTROL MECHANISMS

In addition to the control that central venous pressure (mean circulatory filling pressure) exerts on stroke volume through the Frank–Starling response (see previous sections), there are several other extrinsic control mechanisms that play a predominant role in regulating cardiac function. Cardiac output is the product of heart rate (f_H) and stroke volume (S_V), and thus, changes in either of these two parameters will affect blood flow to the tissues. Furthermore, any mechanisms that increase myocardial contractility (force development) will augment the heart's capacity to develop pressure, and thus, perfuse tissues where resistance may be considerable (e.g., the body musculature while swimming).

2.5.1 ROLE OF ADRENERGIC STIMULATION IN MODULATING CARDIAC FUNCTION

Adrenergic stimulation of the heart is a major extrinsic control system in fishes and is the result of catecholamines (epinephrine and norepinephrine) released from the chromaffin tissue or sympathetic nerve terminals (see next section) binding to cell surface adrenoreceptors on the cardiac myocytes and/or pacemaker cells. There is evidence in some species that stimulation of α-adrenoreceptors has a negative influence on heart rate and other aspects of cardiac function (Peyraud-Waitzenegger et al., 1980; Tirri and Ripatti, 1982). However, the main adrenergic influences on the fish heart appear to be mediated by stimulatory cardiac β-adrenoreceptors (βARs).

Currently, three types βARs have been identified in the fish heart (β_1, β_2, and β_3), although each of these has several subtypes and their coupling to/interaction with specific signaling pathways is not completely understood (Giltrow et al., 2011; Steel et al., 2011; Shiels, H.A. and Fenna, A., unpublished observations). In general (Steel et al., 2011), the β_1 and β_2 ARs are coupled to stimulatory G-proteins (G_s), and thus, mediate increases in heart rate and/or myocardial force development. After epinephrine or norepinephrine bind to these receptors, the α subunit of the G_s activates adenylyl cyclase (AC), resulting in increased levels of cyclic adenosine monophosphate (cAMP) and the subsequent activation of protein kinase A (PKA). Active PKA phosphorylates many cellular targets. In fishes, these targets include the L-type Ca channel, the phosphorylation of which increases Ca flux into the cell upon the arrival of an AP (Vornanen, 1998; Ballesta et al., 2012). For example, the effect of adrenergic stimulation on the trout atrial L-type Ca channel can be seen in Figure 2.4d at three temperatures (Shiels et al., 2003). This increased Ca influx can prolong the AP duration in fishes (Ballesta et al., 2012), enhance the amplitude of the systolic Ca transient (Llach et al., 2004), and in turn the force of myocyte contraction (Keen et al., 1993; Shiels and Farrell, 1997). This positive inotropic effect has been seen in numerous fish species (e.g., trout (Farrell et al., 1986; Keen et al., 1993; Farrell et al., 1996), eel (Franklin and Davie, 1992b), and tuna [Galli et al., 2009a]), and is thought to be a critical compensatory mechanism that enables the fish myocardium to maintain or increase myocardial power generation when faced with situations such as low temperatures (Keen et al., 1993; Shiels et al., 2003) and/or blood acidosis, hypoxia, and hyperkalemia (Hanson et al., 2006). In mammals, PKA also phosphorylates phospholamban, an accessory protein of SERCA. This reduces the inhibition of SERCA and increases Ca uptake into the SR, thereby increasing the rate of myocyte relaxation and the amount of Ca that can be released from the SR through CICR. Indirect studies have also shown that adrenergic stimulation decreases relaxation times in fish hearts and is SR dependent (Shiels and Farrell, 1997).

In contrast to the aforementioned findings, several species have now been identified (e.g., flounder, *Pleuronectes americanus* (Mendonca and Gamperl, 2009); Atlantic cod, *Gadus morhua* (Axelsson, 1988; Lurman et al., 2012); sea bass, *Dicentrarchus labrax* (Farrell et al., 2007) and tilapia, *Oreochromis* hybrid (Lague et al., 2012)) whose hearts show no, or very weak, responses to adrenergic stimulation. Although there are several possible explanations for this lack of responsiveness including ones related to differences in Ca handling (Mendonca et al., 2009; Lurman et al., 2012), one of the most plausible is that these hearts possess a large (significant) population of β_3 adrenoreceptors. These receptors have been found in the eel (Imbrogno et al., 2006) and trout (Nickerson et al., 2003) heart, have a much lower affinity for adrenaline than β_1/β_2 receptors (Gauthier et al., 2007), and mediate negative inotropy and lusitropy (i.e., their stimulation allows the heart to relax more slowly) through the induction of endothelial nitric oxide synthase-derived nitric oxide signaling (i.e., the NO-cGMP-PKG pathway) (Imbrogno et al., 2006). The significance/role of these receptors in modulating fish cardiac performance has only just begun to be understood (Petersen et al., 2013); however, it has been proposed that β_3 stimulation may serve to protect the myocardium by preventing excessive β_1/β_2 stimulation at high circulating catecholamine concentrations (Gauthier et al., 2007; Angelone et al., 2008; Mendonca et al., 2009).

2.5.2 Autonomic Nervous System and Control of Heart Rate

Myocardial function in fish is normally under control of the autonomic nervous system, which has efferent neural ganglions that synapse on the heart (Nilsson, 1983; Sandblom and Axelsson, 2011). However, the hagfish has an aneural heart (Jensen, 1961). Thus, this species relies exclusively on intrinsic or hormonal regulation of heart function and catecholamines stored within the heart tissue that can affect inotropy and chronotropy (Satchell, 1991). The teleost and elasmobranch hearts are innervated by cholinergic fibers that run in the vagus nerve and reach the sinus venosus via the ducts of Cuvier, and vagal nerve terminals are found in the sinus venosus, the pacemaker tissue, the walls of the atrium, and the AV junction, but not the ventricle (Santer, 1985). This cholinergic innervation is generally inhibitory and acts via muscarinic receptors in the pacemaker and atrial tissues to depress heart rate and reduce excitability. However, vagal innervation of the heart is stimulatory in the lamprey and elicits its effects through nicotinic receptors (Santer, 1985). The hearts of elasmobranchs and flatfish of the family Pleuronectidae appear to lack sympathetic innervation (Cobb and Santer, 1973; Donald and Campbell, 1982; Nilsson, 1994; Sandblom and Axelsson, 2011). However, the hearts of most fishes are also innervated by sympathetic nerve fibers that are normally excitatory and act via adrenergic receptors to increase myocardial contractility and heart rate (Nilsson, 1983; Sandblom and Axelsson, 2011). These sympathetic nerve fibers travel alongside the vagus to reach the heart, and nerve terminals have been found in all parts of the fish myocardium.

In fishes, the intrinsic rhythm of the pacemaker cells sets the basal rate at which the fish heart beats and varies significantly between species and directly with temperature and acclimation history (Farrell and Jones, 1992; Mendonca et al., 2007). For example, (1) at $10°C–12°C$, intrinsic heart rate for cyclostomes is ~15 bpm, while it ranges from 30 to 55 bpm for teleosts (Farrell and Jones, 1992); (2) intrinsic heart rate normally has a Q_{10} of ~2.0–2.5 (i.e., ~ doubles for every $10°C$ increase in temperature); and (3) rainbow trout hearts from acclimated at $4°C$ had an intrinsic heart rate of 61 bpm, whereas hearts from 17°C acclimated fishes contracted at 48 bpm (Aho and Vornanen, 2001). However, the rate at which the fish heart actually beats is rarely the intrinsic rate. This is because heart rate is also dependent upon the balance between cholinergic and sympathetic nervous stimulation, myocardial stretch, and circulating hormones (e.g., catecholamines) (Farrell and Jones, 1992). The cholinergic and adrenergic systems are antagonistic with regard to their effects on heart rate, with the former decreasing the rate at which the heart beats while the latter increases it. The contribution of these two systems to setting heart rate is estimated by calculating the cholinergic and adrenergic tones on the heart. This is accomplished by determining heart rate after applying pharmacological blockers that first eliminate the cholinergic influence on the heart (using atropine) and then the influence of the adrenergic nervous innervation (using bretylium); the heart rate (f_H) after applying both of these drugs is called the intrinsic heart rate (f_{Hint}) and the % cholinergic and adrenergic tones are calculated as [(f_H after atropine – f_H before atropine)/f_{Hint}] × 100 and [(f_H after atropine – f_{Hint})/f_{Hint} × 100)], respectively. It is difficult to compare cholinergic and adrenergic tones as they vary widely between species/studies (i.e., from ~10% to 80%) and are temperature dependent (e.g., see Table III in Farrell and Jones, 1992; Sandblom and Axelsson, 2011). For example, cold acclimation can increase (*Scyliorhinus canicula* (Taylor et al., 1977); rainbow trout (Wood et al., 1979)) or decrease (*Anguilla anguilla* (Siebert, 1979)) cholinergic tone, whereas adrenergic tone has been reported to decrease upon exposure to cold temperatures (Wood et al., 1979). Nonetheless, it appears that cholinergic tone dominates in fishes that live in cold temperatures (<2°C) (e.g., Antarctic species) and in fish species that live in warm (>20°C) temperatures, whereas adrenergic tone is often predominant in between these temperatures (Sandblom and Axelsson, 2011).

Variations in cholinergic tone, and thus heart rate, are also seen in response to changes in blood pressure and decreases in water or blood oxygen levels (hypoxia). These responses are primarily, but not exclusively, mediated by mechanoreceptors (baroreceptors) and chemoreceptors located in the gills (Sundin and Nilsson, 2002) and result in changes in heart rate via modulation of the heart's cholinergic nervous tone. For example, a baroreflex-mediated increase in heart rate, in concert with α-adrenergic constriction of the arterial and venous vasculature (Sandblom and Axelsson, 2005), reestablishes cardiovascular homeostasis following a fall in blood pressure (i.e., by increasing both heart rate and vascular resistance). These responses are opposite when blood pressure gets too high. The decrease in heart rate (bradycardia) that accompanies hypoxia is a well-studied phenomenon and is covered in more detail in Section 2.7.4.

Interestingly, it has been recognized for decades that there can be a 1:1 coupling between ventilation and heart rate in resting/quiescent elasmobranchs and teleosts (Jones et al., 1974; Taylor, 1992, 2011). This respiratory modulation of heart rate (cardiorespiratory synchrony [CRS]) is mediated predominantly by variations in inhibitory vagal tone imposed by the activity of cardiac preganglionic neurons within the medulla oblongata, although the neural basis of how ventilation modulates heart rate appears to vary between the major groups of fishes (Taylor, 2011). CRS has been hypothesized to optimize gas exchange under periods of respiratory distress or during energetic challenges by ensuring that ejection of blood from the heart is synchronized with the flow of fresh (oxygenated) water over the gills. However, there is little evidence to support this conclusion. For example, (1) cutting the vagus nerve (Atlantic cod; McKenzie et al., 2009) or injection with atropine (*Anguilla anguilla*; Iversen et al., 2010b) prevented cholinergically mediated bradycardia in response to hypoxia, but did not affect the fish's oxygen consumption down to PO_2 levels as low as 2.0 kPa; (2) Keen and Gamperl (2012) recently showed that there was little evidence of CRS in resting rainbow trout even at temperatures that approached the lethal temperature for this species; and (3) cholinergic tone is often reduced during stressful situations (Wood et al., 1979; Campbell et al., 2004), and thus, there is no mechanism to synchronize ventilation and cardiac output when CRS may be critical to supporting the fish's metabolic demand.

2.6 VASCULATURE

2.6.1 ARTERIES AND VEINS

In fishes, as in other vertebrates, the circulatory system consists of four basic categories of vessels (arteries, arterioles, capillaries, and veins), each with a different function. The arteries and veins are conductance vessels that distribute blood to and from the various tissues, and the general organization of these vessels in a teleost fish is shown in Figure 2.8. The heart pumps blood into the ventral aorta, which divides into four pairs of afferent branchial arteries that supply the gills with blood,

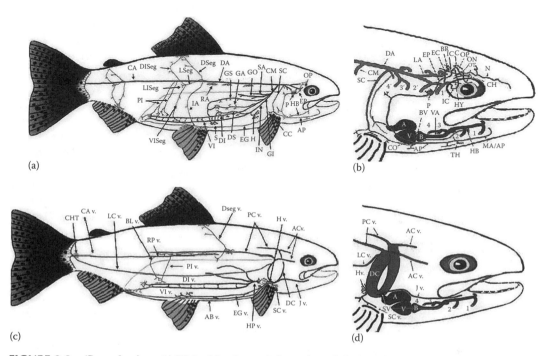

FIGURE 2.8 (See color insert.) Major blood vessels in a teleost fish. Arterial (panels a and b) and venous (panels c and d) circulations in the rainbow trout (*Oncorhynchus mykiss*). A, atrium; AB v., abdominal vein; AC v., anterior cardinal vein; AP, afferent pseudobranch artery; B, bulbus arteriosus; BL v., bladder vein; BR, brain; BV, bulboventricular valve; C, commissure vessel; CA, caudal artery; CA v., caudal vein; CAP, capillary network; CH, choroid artery; CHT, caudal heart; CM, celiacomesenteric artery; CO, coronary artery; DA, dorsal aorta; DC, ductus Cuvier; DI, dorsal intestinal artery; DI v., dorsal intestinal vein; DISeg, dorsal intersegmental artery; DS, duodenosplenic artery; DSeg, dorsal segmental artery; DSeg v., dorsal segmental vein; EC, external carotid artery; EG, epigastric artery; EG v., epigastric vein; EP, efferent pseudobranch artery; GA, gastric artery; GI, gastrointestinal artery; GO, gonadal artery; GS, gastrosplenic artery; H, hepatic artery; HB, hypobranchial artery; H v., hepatic vein; HP v., hepatic portal vein; HY, hyoidean artery; IA, intercostal artery; IC, internal carotid artery; ICos, intercostal artery; ICos v., intercostal vein; IN, intestinal artery; J v., jugular vein; LA, lateral aorta; LC v., lateral cutaneous vein; LISeg, lateral intersegmental artery; LSeg, lateral segmental artery; MA, mandibular artery; N, nasal artery; ON, orbito-nasal artery; OP, ophthalmic artery; OT, optic artery; P, pseudobranch; PC v., posterior cardinal vein; PI, posterior intestinal artery; PI v., post-intestinal vein; RA, renal artery; RP, renal portal; RP v., renal portal vein; S, spleen; SA, swim bladder artery; SC, subclavian artery; SC v., subclavian vein; SV, sinus venosus; TH, thyroidean artery; V, ventricle; VA, ventral aorta; VI, ventral intestinal artery; VI v., ventral intestinal vein; VISeg, ventral intersegmental artery; 1–4 afferent branchial arteries to gill arches 1–4; 1′–4′ efferent branchial arteries from gill arches 1–4. (Modified From Olson, K.R. and Farrell, A.P., Secondary circulation and lymphatic anatomy, in Farrell, A.P., Ed., *Encyclopedia of Fish Physiology, from Genome to Environment*, 1st edn., Academic Press, London, U.K.; Waltham, MA, pp. 1161–1168, 2011. With permission.)

and the majority of oxygenated blood is then directed into efferent branchial arteries (also four pairs), which give rise to the mandibular artery, hypobranchial artery, carotid arteries, hyoidean artery, and the dorsal aorta. The mandibular artery supplies the lower jaw and pseudobranch with blood. The hypobranchial artery eventually becomes the left and right coronary arteries (only in 20%–30% of fishes) that perfuse the fish heart. The carotid arteries (internal and external) and the hyoidean artery transport blood in an anterior direction and perfuse the anterior head, eyes, and brain. Finally, the remaining blood travels in a posterior direction just beneath the vertebral column, first in the dorsal aorta and then in the caudal artery. These two vessels give rise to all the major vessels that perfuse the fish's remaining organ systems. These vessels include the subclavian artery that perfuses the pectoral fin, pectoral girdle, and associated musculature and in some fishes connects into the hypobranchial circulation; the celiacomesenteric artery (or separate celiac and mesenteric arteries) that supplies the liver, stomach, spleen, intestine, gonads, and swim bladder through various vessels; the renal artery that perfuses the kidneys; a posterior intestinal artery that supplies blood to the posterior gut; and numerous segmental arteries that arise in association with each vertebrae and perfuse the red and white muscles and skin.

Blood is drained from the organs by veins that often bear the same name as the arteries that supply them. However, few of them drain directly into the central venous system. In the head, the jugular vein and anterior cardinal vein eventually carry venous blood to the sinus venosus and ductus Cuvier (see later text), respectively. Four cutaneous veins receive blood from the skin and eventually drain into the ductus Cuvier. These include the paired lateral cutaneous veins that lie just underneath the lateral line on each side of the fish, the dorsal vein that receives blood from the skin and the dorsal fins, and the abdominal vein that receives blood from the skin in the ventral part of the fish and often anastomoses with the epigastric and/or subclavian veins before draining into the ductus Cuvier. However, all other blood that is going back to the heart must pass through one of two portal systems; a portal system is defined as an arrangement by which blood collected from one set of capillaries passes through a large vessel or vessels and another set of capillaries before returning to the systemic circulation. The first portal system is the hepatic portal system, which is found in all vertebrates. In this system, blood draining from the liver, intestines, stomach, spleen, gonads, and swim bladder first passes through a short hepatic portal vein, through the sinusoids in the liver, and then travels a short distance in the hepatic vein before reaching the sinus venosus directly. The other portal system is the renal portal system, which is unique to fishes. Blood from the tail, trunk musculature, and skin returning through the caudal vein, most of the blood from the postabdominal region, and blood from the bladder and gonads, all eventually percolates through the renal parenchyma (tissue) and around the renal tubules. Thereafter, renal portal blood and postglomerular renal arterial blood are collected into the posterior cardinal vein(s) within the kidney, and this/these vessel(s) then carry the blood to the ductus Cuvier. As indicated earlier, the ductus Cuvier receives the majority of blood coming back to the heart. The ductus Cuvier is a large U-shaped vessel that lies between the peritoneal and pericardial cavities and is connected to the sinus venosus through an opening located at the bottom of the "U" (see Figure 2.8d).

In addition to providing the conduits through which blood flows to the tissues and returns to the heart, veins and arteries perform a number of other functions. For example, arterial walls are composed largely of elastin, collagen, and smooth muscle (Groman, 1982; Genten et al., 2009), and in addition to the OFT, they dampen the pressure pulse associated with ventricular contraction. Veins have ostial valves at junctions between regional (segmental) vascular beds and longitudinal veins that prevent backflow during contraction of the swimming muscles (Satchell, 1991). Finally, some arteries (Groman, 1982) have valves that prevent the backflow of blood during diastole.

Although the influence of nervous stimulation on conductance vessels of the arterial system appears to contribute little to regulating blood flow (Olson and Farrell, 2006), the smooth muscle of the ductus Cuvier and the central veins are responsive to several stimuli. These include catecholamines of nervous and/or humoral origin (acting via α-adrenergic mechanisms), angiotensin II, endothelin-1, and arginine vasotocin that are venoconstrictors, and atrial natriuretic peptide that has been identified as a potent venodilator (see Sandblom and Axelsson (2007b) for a review). These mechanisms allow venous tone, and thus, compliance to be modulated and ensure that *vis-a-tergo* filling (stroke volume) is appropriate to meet the heart's pumping requirements or to maintain blood pressure within an appropriate range.

2.6.2 ACCESSORY VENOUS PUMPS (HEARTS)

Fishes have relatively low arterial blood pressures (see Table 2.1) as compared to the other vertebrate groups, and thus, there is little potential energy in the venous circulation to get the blood back to the main (systemic) heart. Thus, several fishes have developed accessory pumps that help with venous return (Satchell, 1991, 1992; Farrell, 2011c). The hagfish has a portal heart that pumps blood from the gut (intestine) through the liver. This portal heart is composed of trabeculated cardiac muscle, generates its own rhythm (electrocardiogram), and has valves separating it from both its inflow and outflow vessels (Satchell, 1991; Davie et al., 1987). Both the hagfish and eel possess valved caudal hearts (although their morphology is quite different) whose change in volume/pumping is dependent upon the contraction of delicate sheets ("spindles") of skeletal muscle whose design and control are specific to this purpose (Kampmeier, 1969; Satchell, 1984, 1991). Finally, there are several other types of accessory hearts (valved chambers or vessels) that rely on the contraction of the ventilatory musculature (e.g., the cardinal hearts of hagfish and the branchial hearts found in many fish species) or tail musculature (e.g., the caudal hearts of sharks and eels) to enhance venous return. However, these latter three types differ from the former in that the musculature that compresses them performs other functions (e.g., in ventilation and locomotion), and their role in venous return is ancillary.

2.6.3 MICROCIRCULATION

The microcirculation consists of resistance vessels (small arteries, arterioles, and metarterioles), exchange vessels (sites), and venules and is where nutrients, ions, wastes, and gases are primarily exchanged between mediums (blood ⇔ water and blood ⇔ tissue). In fishes, there are two distinguishable microcirculations, one associated with the gill filaments and the other that is found in all other tissues. These are dealt with separately based on differences in morphology, function, and control.

2.6.3.1 Gill

The vascular anatomy of the gills (Figure 2.9a), which consists of both arterio-arterial and arteriovenous circulations, has been comprehensively reviewed by a number of authors, most recently by Olson (2002, 2011a). All of the fish's cardiac output is directed into the afferent branchial arteries (ABA), which delivers blood to, and distributes it along, the gill arch. Afferent filamental arteries (AFAs) then distribute the blood into afferent lamellar arterioles (ALAs) that deliver it to each respiratory lamella (L) where the exchange of molecules, ions, and gases occurs. Oxygenated blood leaves the lamellae through short efferent lamellar arterioles (ELAs) and from there passes into the efferent filamental arteries (EFAs). At this point, the blood can be directed in two ways. Approximately 90% of the blood leaving the gill lamellae enters the efferent branchial arteries (EBAs) (i.e., arterio-arterial circulation) that eventually connect up with the dorsal aorta/carotid arteries and allow this oxygenated blood to be distributed around the body (Ishimatsu et al., 1988). The remainder is directed into the arteriovenous circulation and then follows two potential routes: the blood can (1) leave the EFAs via short connecting vessels (arterio-arterial anastomoses, AVAs; Figure 2.9b) and enter the intralamellar (IL) vasculature or (2) enter the "nutrient circulation" of the gill via nutrient arteries (NA) that arise from the EFA or more commonly from the amalgamation of a large number of tortuous arterioles that arise from the EFA and EBA (Figure 2.9). In this chapter, these latter two circulations are considered to be part of the secondary circulation (see later text).

TABLE 2.1
Cardiovascular Variables in Various Fish Species at "Rest" and When Exposed to Variations in Temperature and Water Oxygen Levels (Hypoxia), Exercise, and Postfeeding

Species (Mass)	Condition (Swim Speed, Water PO_2, Temperature, or Feeding Level	Q	f_H	S_V	P_{VA}	P_{DA}	R_S	R_G
Cyclostomes								
Myxine glutinosa[a]	Rest (11°C)	8.7	22	0.41	7.8	5.8	0.67	0.23
(0.6–0.9 kg)								
Eptatretus cirrhatus[b]	Rest (17°C)	15.8	24.9	0.67	12	9.7	0.61	0.15
(1.4 kg)	70 mmHg	22.1	24.1	0.92	18.4	13.5	0.61	0.22
Elasmobranchs								
Hemiscyllium ocellatum[c]	Rest (28°C)	44.0	60	0.83		3.0	0.54	
(1.29)	45 mmHg	38	58	0.68		3.4	0.69	
	15 mmHg	27	40	0.68		2.7	0.81	
Aptychotrema rostrata[c] (1.54 kg)	Rest (28°C)	39	56	0.71		3.0	0.56	
	45 mmHg	32	42	0.77		2.8	0.69	
	15 mmHg	21	20	0.95		1.6	0.63	
Triakis semifasciata[d,e]	Rest (14°C–24°C)	33.1	51	0.77	47.7	32.3	0.98	0.47
(1.93 kg)	32–43 cm s^{-1}	56.2	55	1.02	58.1	36.2	0.64	0.39
	Postexercise	60.4	50	1.22	55	35.4	0.59	0.32
Chondrostei								
Acipenser transmontanus[f]	Rest (19°C)	25.8	43.2	1.4				
(2.53 kg)	6 h postfeeding (0.5% bwt)	32.3	51.3	1.4				
Teleosts								
Anguilla anguilla[g]	Rest (15°C)	11.5	37	0.29	38	23.3	2.03	1.28
(0.51 kg)	40 mmHg	7.2	23	0.31	29.6	15.7	2.18	1.93
A. australis[h]	Rest (17°C)	11.3	54	0.21	38.6	23.5	2.08	1.34
(0.62 kg)	15 cm s^{-1}	11.3	54	0.21	39	23.6	2.09	1.36
Cyprinus carpio[i]	Rest (6°C)	9	7	1.22				
	<10 mmHg	2	2	0.62				
	Rest (15°C)	15	17	0.77				
	<10 mmHg	3	3	0.65				
G. morhua[j]	Rest (11°C)	19.2	41	0.51	36.8	23.3	1.21	0.70
(0.4–1.3 kg)	35 mmHg	17	21	0.86	36	23.3	1.37	0.75
G. morhua[k]	Rest (10°C)	23.1	32.9	0.73				
(0.43–0.78 kg)	U_{crit}	44.5	46.3	0.99				
G. morhua[l]	Rest (10°C)	21.5	36.3	0.6				
(0.86–1.13 kg)	Rest (16°C)	32	56	0.6				
	Rest (20°C)	47	71	0.74				
Hemitripterus americanus[m]	Rest (11°C)	18.8	37	0.51	28.5	23.3	1.24	0.28
(0.7–1.4 kg)	30 cm s^{-1}	30.9	49	0.64	35.3	26.3	0.85	0.29
Oncorhynchus tshawytscha[n]	Rest (10°C)	35.8	57	0.63		3.2	0.71	
(0.35 kg)	U_{crit}	65.6	63	1.04		4.0	0.53	
Oncorhynchus mykiss[o]	Rest (10°C)	26.6	48.4	0.58		3.3	0.95	
(0.4–1.02 kg)	U_{crit}	48.7	66.9	0.73		4.1	0.68	
Ophiodon elongatus[p]	Rest (10°C)	11.2	29	0.38	38	28.3	2.53	0.87
(3.8–8.5 kg)	75 mmHg	9.9	28	0.37	39.2	30.2	3.05	0.91
	35 mmHg	7.7	12	0.7	33.8	23.9	3.10	1.29

TABLE 2.1 (continued)

Cardiovascular Variables in Various Fish Species at "Rest" and When Exposed to Variations in Temperature and Water Oxygen Levels (Hypoxia), Exercise, and Postfeeding

Species (Mass)	Condition (Swim Speed, Water PO$_2$, Temperature, or Feeding Level)	Q	f_H	S_V	P_{VA}	P_{DA}	R_S	R_G
Pagothenia borchgrevinki[q]	Rest (0°C)	29.6	11.3	2.16	28			0.95
(0.06 kg)	20 cm s^{-1}	51.8	21	2.16	28			0.54
Pleuronectes americanus[r]	Rest (4°C)	10	21	0.50				
	U$_{crit}$	25	31	0.80				
	Rest (10°C)	16	34	0.50				
	U$_{crit}$	39	52	0.74				
Pleuronectes americanus[s]	Rest (8°C)	12	36	0.35		2.5	2.18	
	Rest (16°C)	19	65	0.29		2.2	1.20	
	Rest (24°C)	30	65	0.44		1.9	0.83	
Dicentrarchus labrax[t]	20 cm s^{-1} (19°C)	43.4	64	0.85				
(0.33 kg)	3% bwt. meal (max.)	55.0	75	0.75				
	U$_{crit}$	90.6	106	0.89				
Thunnus albacares[u]	Spinalized (26°C)	115	97	1.3	89.7	32.6	0.28	0.50
(1.4 kg)	50 mmHg	74	71	1	90	33	0.44	0.77
Katsuwonus pelamis[u]	Spinalized (26°C)	132	126	1.08	87.3	40.2	0.30	0.36
(1.6 kg)	90 mmHg	105	86.2	1.22	87.3	36.2	0.34	0.49

Abbreviations: Q, cardiac output (mL min^{-1} kg^{-1}); f_H, heart rate (bpm); S_V, stroke volume (mL kg^{-1}); P_{VA} and P_{DA}, ventral and dorsal aortic pressures, respectively (mmHg); R_S and R_G, systemic and gill vascular resistances, respectively (mmHg mL^{-1} min^{-1} kg^{-1}).

Notes: [a]Axelsson et al. (1990), [b]Forster et al. (1992), [c]Speers-Roesch et al. (2012), [d]Lai et al. (1989), [e]Lai et al. (1990), [f]Grans et al. (2010), [g]Peyraud-Waitzenegger and Soulier (1989), [h]Davie and Forster (1980), [i]Stecyk and Farrell (2002), [j]Fritsche and Nilsson (1989), [k]Petersen and Gamperl (2010a), [l]Gollock et al. (2006), [m]Axelsson et al. (1989), [n]Gallaugher et al. (2001), [o]Thorarensen et al. (1996), [p]Farrell (1982), [q]Axelsson et al. (1992), [r]Joaquim et al. (2004), [s]Mendonca and Gamperl (2010), [t]Dupont-Prinet et al. (2009), [u]Bushnell and Brill (1992).

Blood flow through the gill, and thus pre- and post-branchial blood pressure, are modulated by changes in autonomic nervous system stimulation, nonadrenergic and noncholinergic (NANC) components, and various hormones (Nilsson and Sundin, 1998; Sundin and Nilsson, 2002). For example, in teleosts, catecholamines (of neuronal or humoral origin) increase the functional surface of the gill lamellae (without an increase in arterio-arterial resistance) by simultaneously constricting the efferent lamellar artery and dilating the afferent lamellar arterioles. These effects are mediated through α- and β-adrenoreceptors, respectively, with the former effect likely concentrated at the sphincter at the base of the EFAs. This lack of a change in overall gill resistance is in contrast to parasympathetic stimulation of the EFA sphincter and sympathetic (α-adrenergic-mediated) nervous constriction of the arteriovenous pathway. Both result in an increase in gill vascular resistance (ventral aortic blood pressure), with the latter also resulting in a redistribution of blood from the arteriovenous to the arterio-arterial pathway. With regard to NANC mediators of branchial blood flow, serotonin (5-HT) is probably the most studied/understood and appears to have both vasoconstrictory and vasodilatory effects. For example, serotonin constricts the arterio-arterial branchial vasculature in several fish species (Sundin et al., 1995a; Janvier et al., 1996; Sundin and Nilsson, 2002) and, in the Atlantic cod, also dilates the arteriovenous pathway and thus diverts blood away from the arterio-arterial circulation (Sundin, 1995a). However, there is recent evidence that other neurotransmitters including NO, vasointestinal peptide, and neuropeptide Y affect gill vascular resistance (Jonz and Zaccone, 2009).

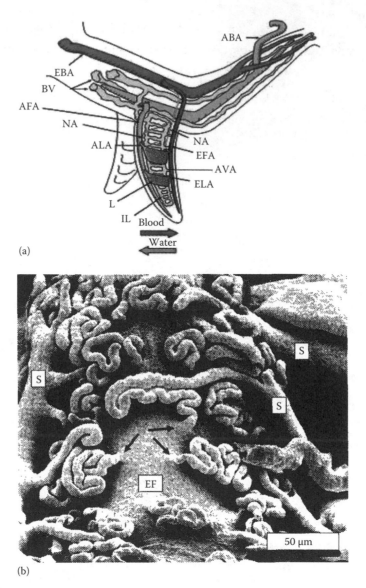

(a)

(b)

FIGURE 2.9   (See color insert.) Blood vessels in the fish gill. (a) The afferent brachial artery (ABA) delivers blood to, and distributes it along, the gill arch. Afferent filamental arteries (AFAs) distribute blood to the respiratory lamellae (L) via afferent lamellar arterioles (ALAs). Oxygenated blood leaves the lamellae through short efferent lamellar arterioles (ELAs) and from there passes into the efferent filamental arteries (EFAs) and then into the efferent branchial artery (EBA). Small nutrient arteries (NAs) originate from the EFA near the base of the filament and reenter the filament. The intralamellar vasculature (IL) forms an extensive sinus with the filament that ultimately drains into the branchial veins (BV) in the gill arch. (Modified from Olson, K.R. and Farrell, A.P., Secondary circulation and lymphatic anatomy, in Farrell, A.P., Ed., *Encyclopedia of Fish Physiology, from Genome to Environment*, 1st edn, Academic Press, London, U.K./Waltham, MA, pp. 1161–1168, 2011. With permission.) (b) Scanning electron micrograph of a vascular corrosion replica showing the origin of the circulation in the gill filament of the climbing perch, *Anabas testudineus*. Tortuous narrow-bore arterioles (arrows) arise from an EFA and anastomose to form progressively larger secondary arterioles (S) that ultimately perfuse the core of the gill filament and the arch support tissue. (Adapted from Olson, 1986. *J Exp Zool* 275: 172–185. With permission.)

Finally, the gasotransmitter hydrogen sulfide (H_2S) has been shown to constrict the gill vasculature (Dombrowski et al., 2004; Skovgaard and Olson, 2012), and there are a number of other substances that affect gill blood flow and vascular resistance in at least some fishes by constricting (e.g., adenosine, Cholecystokinin (CCK), isotocin, oxytocin, arginine vasotocin, neuropeptide Y, endothelin-1, prostaglandins) or dilating (e.g., atrial natriuretic peptides) various segments of this circulation (see Nilsson and Sundin (1998) and Sundin and Nilsson (2002) for reviews).

2.6.3.2 Other Tissues

In other fish tissues, the microcirculation consists of small resistance vessels (small arteries, arterioles, and metarterioles), capillaries, and venules (Satchell, 1991; Bushnell et al., 1992; Egginton and Syeda, 2011; Olson, 2011b). The resistance vessels range from 30 to 300 µm in diameter and have a considerable amount of smooth muscle in their walls to allow for control of perfusion/vascular resistance and a single layer of endothelial cells lining their lumen. However, the connections between the arterioles and capillaries (i.e., the metarterioles) have a reduced diameter (5–30 µm) and are encircled by a single smooth muscle cell (Davison, 1987), the precapillary sphincter, whose diameter is controlled by locally released substances (see later text). The capillaries, of which many arise from each arteriole, are thin tubes approx. 4–10 µm in diameter and 100–500 µm in length whose walls consist solely of a monolayer of endothelial cells. This morphology/architecture results in a vascular bed with large cross-sectional (which dramatically reduces blood flow velocity and increases blood transit time) and surface areas and a minimal barrier to diffusion. Thus, they are well suited to their primary function as the site of exchange of gases, metabolites, ions etc. Finally, venules are 8–100 µm in diameter, receive the blood from several capillaries, and carry it to larger veins. Although few details are available on their structure in fishes, the walls of mammalian venules are composed of an inner layer of squamous endothelial cells, a middle layer of muscle and elastic tissue, and an outer layer of fibrous connective tissue.

Resistance in the microcirculation of the tissues is modulated by the autonomic nervous system, hormones, and locally produced substances, although nervous innervation is restricted to the smooth muscle covering the arterioles (i.e., the precapillary sphincters are not innervated) (Olson, 2011b). In contrast to the gill vasculature, the microcirculation in other tissues does not receive cholinergic innervation, and sympathetic nerve stimulation via the activation of α-adrenoreceptors is a predominant mechanism controlling arteriolar (systemic) vascular resistance. However, it has also been shown that fish vessels are innervated by nerves that release NANC neurotransmitters (Johnsson et al., 2001), and thus, it has been suggested that substances such as NO, serotonin, vasoactive intestinal peptide, substance P, tachykinins, and others likely play a role in regulating tissue blood flow and resistance. Several hormones have also been implicated in controlling vascular resistance, including catecholamines, angiotensin II, arginine vasotocin, urotensin I and II, bradykinin, and adrenomedullin (see Olson and Farrell (2006) and Olson (2011b) for further details). Finally, it has been shown that the control of vascular resistance is quite sensitive to changes in the local chemical environment and that this allows tissue perfusion to match metabolic demand. For example, a decrease in tissue oxygen partial pressure (PO_2) or increases in H^+ and CO_2 produce vasodilation, and paracrine regulators such as adenosine, prostaglandins, endothelin-1, and the gasotransmitters carbon monoxide and H_2S appear to be important signaling molecules in fish vessels (Olson, 2011b). Of these paracrine regulators, H_2S has received the most attention, and although its effects on vascular resistance/blood pressure are variable, it has been suggested that H_2S serves as the oxygen "sensor" that initiates a cascade of events ultimately resulting in either hypoxic vasoconstriction or hypoxic vasodilation (Olson, 2008; Olson and Donald, 2009).

2.6.4 SECONDARY CIRCULATION

The secondary circulation is a vascular network of arteries, capillaries, and veins that arises from arterial–arterial anastomoses (e.g., see Figure 2.9b) but is in parallel with the primary circulation.

In bony fishes, there are two separate secondary circulations (Steffensen et al., 1986; Steffensen and Lomholt, 1992; Olson, 1996; Olson and Farrell, 2011). The first is associated with the systemic circulation, with anastomoses arising from the dorsal aorta and segmental arteries, and capillaries found in external tissues and those that line the body cavity (including the skin, fins, scales, buccal cavity, and peritoneum). The other "secondary circulation" is found in the gills and made up of NA and IL pathways/circulations (see Figure 2.9a and b). Blood entering the numerous IL vessels eventually enters the BV, and this oxygenated blood returns to the venous side of the heart. In contrast, blood entering the nutrient circulation provides blood to the tissue supporting the gill arch, the filamental adductor and abductor muscles, and the core of the gill filament. While both the nutrient and IL pathways begin as tortuous arterioles arising from arterial–arterial anastomoses associated with the efferent branchial and filamental arteries (Figure 2.9b), they do not exhibit all of the characteristics of a secondary circulation. This had led to significant and on-going debate, as to whether they are truly a secondary circulation, part of the primary circulation, or "lymphatic like" (Olson, 1996; Olson and Farrell, 2011).

At present, we have only a limited understanding of the function/physiology of the secondary circulations in fishes. However, there is information available on a number of its features. The volume of the secondary circulation is large, with estimates ranging from 50% to 150% of blood volume (Steffensen and Lomholt, 1992; Skov and Steffensen, 2003). The rate of flow through the secondary circulation appears to be small, with estimated flow rates in the rainbow trout (Steffensen and Lomholt, 1992) and the Atlantic cod (Skov and Steffensen, 2003) of 0.3% and 2.7% of cardiac output, respectively. Finally, although pressures have only been measured in the large veins of the secondary circulation, they are low, ranging from −0.4 to 0.4 kPa (Ishimatsu et al., 1992; Satchell, 1992). With regard to the function of the secondary circulation in fishes, a number of roles have been proposed, including gill hormone metabolism, immunological protection of the epithelium, the transepithelial exchange of water and/or ions, and that it is a prototype lymphatic system. However, it is clear that it does not have a role in oxygen transport/gas exchange. The entrance to the arterial–arterial anastomosis is surrounded by endothelial cells with long microvilli (Olson and Farrell, 2011), and they preclude the entry of most red blood cells (i.e., hematocrit in the secondary circulation has been estimated to be between 0.5% and 1.7% (Iwama et al., 1990; Lomholt and Steffensen, unpublished as cited in Steffensen and Lomholt, 1992)).

2.7 INTEGRATIVE CARDIAC FUNCTION

Control of the cardiovascular system is very complex, with cardiac output and blood pressure dependent upon a number of interrelated factors. A relatively simple schematic, and description, of the interrelationships between the main determinants of cardiac function and venous and arterial blood pressures are provided in Figure 2.10. This diagram does not capture the influence of numerous endocrine, paracrine, autocrine, and other nervous factors that control/mediate its function (e.g., see previous sections) or how cardiovascular variables are influenced by various biotic and abiotic factors. Thus, Table 2.1 provides data on cardiovascular variables in a number of fish species at rest, when swimming, after feeding, and when exposed to changes in temperature and water oxygen levels. The data presented is limited given space constraints and restricted to experiments where cardiac output and blood flows were directly measured using implanted flow probes to ensure accuracy, but provides a good qualitative and quantitative picture of the capacity of the fish cardiovascular system to respond to numerous challenges and the fish's metabolic demands. Nonetheless, it does not capture the effect of local changes in vascular resistance on tissue blood flow distribution (e.g., in resting and unfed fish, the percentage of cardiac output to the major organ systems is generally red muscle (1%–10%; Olson and Farrell, 1996), white muscle (25%–75%; Olson and Farrell, 1996), digestive organs (10%–40%; Seth and Axelsson, 2011), kidney (6%–13%, Olson and Farrell, 1996), and heart musculature (~1%, Axelsson and Farrell, 1993;

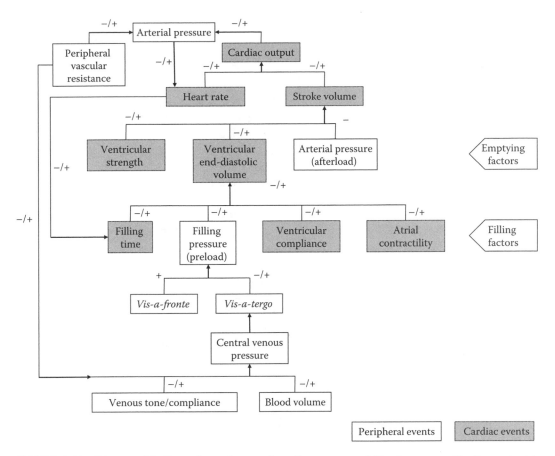

FIGURE 2.10 Direct and indirect determinants of cardiac output and blood pressure. Cardiac output is determined directly by heart rate and stroke volume. Heart rate is dependent on the intrinsic rate set by the pacemaker cells and sympathetic (positive chronotropic) and parasympathetic (negative chronotropic) influences; the latter includes negative feedback from arterial baroreceptors. Stroke volume is the difference between end-diastolic and end-systolic volumes. End-diastolic volume is highly dependent upon *vis-a-tergo* and *vis-a-fronte* filling mechanisms, ventricular compliance, the time available for filling, and possibly atrial contraction; with *vis-a-tergo* filling determined by venous pressure (preload), which is in turn dependent upon arterial vascular resistance, blood volume, and venous compliance. End-systolic volume is determined by ventricular inotropy, and in situations where ventral aortic pressure becomes excessive or the outflow tract is obstructed (e.g., see Gamperl and Farrell, 2004), afterload. Finally, arterial blood pressure is determined by cardiac output and resistance in the branchial and systemic vasculature. Shaded boxes indicate events/issues related to cardiac function, whereas white boxes indicate those that are peripheral to the heart. (+/−) indicates factors that can have a positive or negative effect. (+) indicates factors that have a positive influence on cardiac function. (−) indicates factors whose effects are predominantly negative (i.e., have a suppressive effect). (Modified from Olson, K.R. and Farrell, A.P., The cardiovascular system, in Evans, D.H. and Claiborne, J.B., Eds., *The Physiology of Fishes*, Taylor & Francis, Boca Raton, FL, pp. 119–152, 2006. With permission.)

Gamperl et al., 1994)) or provide comprehensive information of how these cardiovascular changes are mediated. Therefore, short descriptions follow on how fish cardiovascular function is altered by each of these conditions.

2.7.1 Exercise

Swimming in fishes can be loosely grouped into two categories: (1) aerobic swimming that can be sustained for prolonged periods and is often measured as the fish's critical swimming speed (U_{crit})

and (2) burst swimming that can only be maintained for a matter of seconds (minutes) and involves a large anaerobic component. Although the time course of changes in cardiovascular parameters is different between these two types of exercise (e.g., the peak in cardiac output (Q) occurs shortly after burst-type exercise finishes), the magnitude of changes in cardiovascular variables is generally similar (Farrell, 2011d), with the maximum increase in Q not normally more than 2–3 times the resting values (Table 2.1). However, fishes use heart rate f_H and S_V to varying degrees to elevate Q during exercise (Table 2.1). For example, while many fish species show percent increases in f_H that are only slightly higher than those measured for S_V, the sea bass (*Dicentrarchus labrax*) and Antarctic fish *Pagothenia borchgrevinki* increase Q almost exclusively by elevating f_H. Elasmobranchs increase Q by predominantly increasing S_V. In concert with an increase in the heart's pumping capacity are changes in vascular resistance and blood flow distribution. For example, both systemic and gill vascular resistance decrease (although the fall is generally greater in the former) and these changes partially offset the effect of the increase in Q on ventral and dorsal aortic pressures. The increase in f_H with exercise is mediated predominantly by a release of cholinergic inhibitory tone (Iversen et al., 2010a), although there can be some contribution of increased sympathetic stimulation (Axelsson et al., 1989). With regard to the increase in S_V with exercise, it is likely to be largely due to the increase in central venous and mean circulatory filling pressures that are associated with exercise (Sandblom et al., 2005, 2006) and result from (1) an increase in venous return due to skeletal muscle contractions physically compressing veins and massaging blood toward the heart and (2) decreased capacitance of the central veins due to α-adrenergic-mediated vasoconstriction (Sandblom et al., 2006; Farrell, 2011d). It is also possible that *vis-a-fronte* filling is enhanced when the heart is pumping maximally (Farrell, 2011d).

Blood flow distribution also changes considerably with exercise. Blood flow increases to both the red muscle (by 5- to 20-fold: Randall and Daxboeck, 1982; Kolok et al., 1993) and the heart/coronary artery (by 110% at 1 body length s^{-1}; Gamperl et al., 1995). In contrast, blood flow to both the white muscle (Randall and Daxboeck, 1982; Kolok et al., 1993) and GI tract (see later text) are diminished considerably.

2.7.2 FEEDING/DIGESTION

When food enters the GI tract, there is a substantial increase in the total oxygen consumption of fishes, which is referred to as specific dynamic action (SDA) or the heat increment of feeding (HIF). This increase in oxygen consumption is the result of digestion, absorption, and to a large extent protein metabolism, and blood flow to the stomach and intestine (gastrointestinal blood flow, GBF) must increase to meet this metabolic demand. In resting fishes fed meals of various sizes, GBF increases by 70%–150% and is a result of both an increased Q and an increase in the proportion of Q perfusing this vasculature. For example, Q increases by 15%–90% (mean 34%) and GBF as a percentage of Q increases by 20%–50% (mean 42%) (see Table 1 in Seth and Axelsson, 2011). The ability of fishes to increase/regulate GBF after a meal, however, depends on a number of interactive factors. Although hypoxia and exercise generally lead to large (50%–80%) decreases in GBF prior to feeding (Axelsson and Fritsche, 1991; Thorarensen et al., 1993; Axelsson et al., 2002; Altimiras et al., 2008), the data available for the sea bass shows that the proportion of Q perfusing the stomach and intestine postprandially is only affected by exercise, where it is decreased substantially (Axelsson et al., 2002; Altimiras et al., 2008; Dupont-Prinet et al., 2009). Further, the increase in blood flow (hyperemia) to the GI circulation is highly dependent upon diet composition, with diets high in lipid inducing a rapid, but short-lived, increase in blood flow, whereas high-protein diets result in a delayed and larger increase (Seth et al., 2009). With regard to the control of GBF postfeeding, it appears that both mechanical and chemical stimuli are involved. Distension of the stomach following ingestion of a meal results in an α–adrenoreceptor-mediated increase in systemic vascular resistance (including that of the GI circulation) and aortic blood pressure (P_{DA}), and this initial constriction of the GI vasculature is followed by a nutrient-induced GI dilation and increases in Q that result in postprandial hyperemia. Thus, the overall increase in R_{sys} (P_{DA}), combined with the reduction in resistance in the GI circulation, results in a redistribution of blood flow to the GI tract (Seth et al., 2008; Seth and Axelsson, 2009, 2011). What causes the nutrient-induced hyperemia is not well understood. However, (1) it is clear that the enteric nervous system (i.e., the intrinsic innervation of the gut) is much more important than extrinsic nervous regulation (sympathetic and parasympathetic nervous stimulation) in mediating postprandial GI hyperemia (Seth and Axelsson, 2010), and (2) it has been hypothesized that that the most probable trigger of postfeeding increases in GI blood flow is a change in the local chemical environment of the gut mucosa (e.g., a fall in oxygen tension/content that is accompanied by changes in H, ADP, and/or adenosine, and/or hyperosmolality due to an increase in [Na]) (Seth and Axelsson, 2010, 2011). Further, there is some evidence that hormones (e.g., cholecystokinin) also play a role in increases in GI blood flow following feeding (Seth et al., 2010).

2.7.3 TEMPERATURE

Temperature can have a profound effect on cardiovascular morphology and function, and throughout this chapter we have pointed out numerous ways in which cold acclimation/exposure affects aspects of the cardiovascular system. At the cellular level, the acute effects of temperature on Ca flux are to a large extent compensated for by corresponding changes in the shape of the AP (trout atria (Shiels et al., 2000); bluefin tuna ventricle (Shiels, H.A., Galli, G.L.J., and Block, B.A., unpublished data)). Moreover, thermal acclimation results in changes in the AP waveform, predominantly through changes in K channel function, that maintain excitability and Ca flux (Haverinen and Vornanen, 2009b). At the level of the heart, an increase in temperature elevates Q by ~1.8- to 2.3-fold for every 10°C (i.e., Q has a Q_{10} of ~2) until ~2°C–3°C before the fish's critical thermal maximum (CTM), at which point it crashes (see later text). We have learned a tremendous amount over the past few years about how this increase in Q is mediated and what ultimately limits Q when fishes are faced with rising temperatures (e.g., see Farrell (2009), Farrell et al. (2009), and Gamperl (2011) for recent reviews). When faced with an increase in temperature, fishes increase Q solely through an increase in f_H, while S_V remains virtually unchanged. This latter result is contrary to expectations based on the decrease in cardiac adrenergic sensitivity with temperature (Figure 2.4d) (Shiels et al., 2003; Farrell et al., 2007) and the negative effect that contraction frequency has on the isometric force generated by cardiac muscle preparations (Shiels et al., 2002). However, recent experiments by Syme et al. (submitted and unpublished) with ventricular muscle preparations from Atlantic cod and Atlantic salmon that are cycled around their resting length (i.e., similar to what the muscle experiences *in vivo*) fail to show a decrease in ventricular myocardial power with increased contraction frequency. This may suggest that myocardial stretch at high heart rates largely offsets the direct negative effects of contraction frequency on contractility.

Another novel finding is that fishes prevented from increasing f_H when exposed to acute increases in temperature by treatment with the pharmacological agent zetabradine (which inhibits the hyperpolarization-activated current of the heart's pacemaker cells) can reach similar maximal values of Q by increasing S_V (Gamperl et al., 2011; Keen and Gamperl, 2012). These studies demonstrate convincingly that fishes have the physiological capacity to increase S_V at high temperatures but that they preferentially increase f_H. Why fishes exclusively increase f_H when exposed to acute increases in temperature is not known, as there are several advantages to maintaining a low heart rate with respect to myocardial O_2 delivery and utilization (Farrell, 2007). However, it is unlikely to be due to the need for perfusion–ventilation matching at the gills or that contracting with a high rate and low strain, as opposed to a low rate and high strain, has performance advantages. Keen et al. (2012) failed to find any evidence of CRS in rainbow trout up to their CTM. Syme et al. (submitted) studied the ability of ventricular and atrial trabeculae from Atlantic cod acclimated to 10°C to sustain cyclical power production at 10°C and 20°C, and across a range of oxygen tensions (450% to 34% of air saturation), and showed that neither large strains nor high contraction rates convey a particular advantage with regard to the maintenance of myocardial contractility. It is possible that increases in f_H are inescapable because of the direct effects of temperature on the frequency of pacemaker discharge and that increasing cholinergic tone to slow the heart is not an option at high temperatures (Keen and Gamperl, 2012).

Maximum Q during an acute temperature challenge occurs ~2°C prior to the fish's CTM, at which point the heart becomes arrhythmic and Q falls dramatically (Gollock et al., 2006; Clark et al., 2008). What sets the maximum f_H in fishes when exposed to an acute temperature increase is not currently known. It is possible that the plateau in Q at high temperatures (despite the increase in metabolic demand), prior to loss of equilibrium, renders the brain hypoxic and subsequently results in neural dysfunction (Robertson, 2004) that negatively impacts heart rate. However, it is also plausible that temperature-induced increases in membrane fluidity and the inability of the pacemaker cells to conserve the functioning of important transmembrane ion channels precludes these cells from maintaining a normal cardiac rhythm (Lennard and Huddart, 1991). This issue needs to be addressed and could be done by measuring pacemaker potentials and ionic currents from isolated pacemaker cells of the fish heart as in Haverinen and Vornanen (2007). This technique would also be extremely useful for examining what factors mediate the large difference in maximum f_H between various fish species. For example, while the maximum f_H of the Atlantic cod (Gollock et al., 2006; Petersen and Gamperl, 2010a), flounder (Mendonca and Gamperl, 2009), and tilapia (even at >35°C; Maricondi-Massari et al., 1998) is only ~70 bpm, salmonids have maximum f_H values of ~120 bpm (Farrell, 2009), whereas f_H in the zebrafish (Gora and Burggren, 2012) and *Bathygobius soporator* (Rantin et al., 1998) can reach ~225 bpm.

2.7.4 HYPOXIA

In most fishes, with the exception of the hagfish, lungfishes, and some Antarctic fishes, brief (acute) exposure to low water oxygen levels (hypoxia) results in a cholinergically mediated bradycardia (decrease in heart rate). The extent of the decrease in f_H and the partial pressure of oxygen at which the bradycardia is initiated varies greatly between species and is temperature dependent (Gamperl and Driedzic, 2009). However, it is clear that the main purpose of the reduction in f_H is not improved oxygen uptake as the prevention of hypoxia-induced bradycardia via cutting the vagus nerve or pre-treatment with atropine does not affect fish oxygen consumption when exposed to graded hypoxia (McKenzie et al., 2009; Iversen et al., 2010b). Instead it appears that this bradycardia has several direct benefits for the heart related to oxygen delivery or the efficiency at which the myocardium utilizes oxygen (Farrell, 2007; Gamperl, 2011).

In those species where hypoxia-induced bradycardia occurs, all fishes exhibit an increase in stroke volume. However, there are three distinctly different patterns exhibited by fishes (Gamperl and Driedzic, 2009; Gamperl, 2011). In response pattern 1 (e.g., sea bass), S_V increases are either simultaneous with the onset of bradycardia or begin once bradycardia is initiated. However, these increases in S_V do not compensate for the decrease in f_H and so Q falls almost continuously (albeit slower than f_H) as hypoxia progresses. In response pattern 2 (e.g., dogfish shark, *Scyliorhinus canicula*; flounder), S_V increases when bradycardia is initiated and this increase maintains Q at, or near, normoxic levels until hypoxia becomes severe. In response pattern 3 (e.g., rainbow trout; Atlantic cod), S_V starts to increase well before the onset of hypoxic bradycardia (resulting in a significant increase in Q), and further increases in S_V once bradycardia is initiated maintain Q above normoxic levels until the fish reaches its hypoxic limit. At present, we do not have a good understanding of how the response patterns exhibited are related to differences in activity, the fish's lifestyle, and or hypoxia tolerance. However, there is strong evidence that the active regulation of venous tone and cardiac filling are important for regulating S_V in fishes during periods of oxygen shortage (Sandblom and Axelsson, 2004, 2005, 2006; Skals et al., 2006).

Accompanying hypoxia-related changes in Q, f_H and S_V are changes in gill and systemic vascular resistance, respectively. Most teleosts and elasmobranchs respond to severe hypoxia with an increase in branchial vascular resistance (R_{gill}), and this rise in R_{gill} can increase blood flow in the branchial vein considerably (by about 1.5-fold in cod; Sundin, 1995b). This redistribution of blood flow may potentially play a critical role in cardiac hypoxia tolerance by raising the oxygen content and partial pressure of the venous blood that supplies the heart (spongy myocardium). It is presently difficult to generalize about how systemic vascular resistance and aortic pressures are affected by hypoxia. However, it appears that elasmobranchs and sturgeons regulate systemic vascular resistance in a different way from teleosts. Systemic vascular resistance decreases in the dogfish (Butler and Taylor, 1975) and sturgeon (*Acipenser naccarii*) (Agnisola et al., 1999) at PO_2 levels less than 60 mmHg, while it increases in those teleosts examined to date (Holeton and Randall, 1967; Farrell, 1982; Chan, 1986).

Surprisingly, few studies have examined the effects of chronic hypoxia on *in vivo* cardiovascular function. Recently, Atlantic cod were shown to have higher values for f_H, but lower resting and maximum values for S_V and Q after hypoxic acclimation (40% air saturated seawater for >6 weeks) (Petersen and Gamperl, 2010a). The mechanisms that resulted in the resetting of f_H are unknown, but likely involve altered neural and/or hormonal control as the intrinsic f_H of in situ hearts from hypoxia-acclimated fish was not different (Petersen and Gamperl, 2010b). Potential explanations for the decline in pumping capacity of hearts from hypoxia-acclimated cod include (1) myocardial damage by constant exposure to low-oxygen conditions, (2) cardiac stunning, and (3) hypoxia-induced myocardial remodeling, without any damage or stunning (Petersen and Gamperl, 2010b). In support of this final explanation, acclimation of zebrafish and the cichlid *Haplochromis piceatus* to water of 50% O_2 content increased cardiac myocyte density (presumably through hyperplasia) and this resulted in a reduction in the size of the central ventricular cavity (Marques et al., 2008). The other species that has received considerable attention with regard to chronic hypoxia is the crucian carp. Crucian carp display a typical reflex bradycardia upon initial exposure to anoxia, with a large compensatory increase in S_V which, in turn, increases Q. However, by 48 h of anoxia, all cardiac parameters return to preanoxic levels and remain stable for at least 5 days (Stecyk et al., 2004). This long-term maintenance of cardiac function is thought to be important for transporting glucose from the liver to metabolically active tissues and ethanol from skeletal muscle (where it is produced) to the gills where it can be eliminated from the body.

Several strategies can enhance cardiac function during hypoxia, including reducing cardiac ATP demand, increasing cardiac anaerobic ATP production, and extending the lower limit at which oxygen can be extracted from the blood. These are reviewed extensively in Gamperl and Driedzic (2009) and Lague et al. (2012). At the cellular level, "channel arrest" has been hypothesized as a strategy for diminishing the metabolic cost of ion pumping. The channel arrest hypothesis (Hochachka, 1986) suggests that energy is conserved during anoxia by reducing the number or functionality of energy-consuming ion channels. However, studies in the anoxic crucian carp provide little evidence for channel arrest during anoxia (acute and chronic). For example, AP shape and I_{k1} are largely unaffected by acute or chronic anoxia (Vornanen and Tuomennoro, 1999; Paajanen and Vornanen, 2003), as is L-type Ca channel density (Vornanen and Paajanen, 2004). Clearly, other mechanisms must be employed to allow the carp heart to function long-term under anoxic conditions (see Stecyk et al. (2008) for a detailed review).

2.8 CONCLUDING REMARKS

In this chapter, we have summarized our current knowledge of fish cardiovascular function. However, this document is far from comprehensive and likely to be in need of revision/updating in the near future. This is because fish cardiovascular physiology is a very active field of scientific investigation, and it is likely that significant advances in our understanding of this critical organ system will continue at a rapid pace. The rapid growth of our knowledge in this area is partially due to the realization that cardiovascular function plays a critical role in determining fish thermal and hypoxia tolerance (i.e., how species may respond to accelerated climate change) (Wang and Overgaard, 2007; Farrell, 2009; Farrell et al., 2009; Gamperl and Driedzic, 2009), the surprising plasticity displayed by the fish cardiovascular system in dealing with biotic and abiotic challenges (Gamperl and Farrell, 2004; Franklin and Raverty, 2002), and the recognition that alterations in heart morphology/function can be impacted significantly by intensive aquaculture and have implications for fish welfare and profitability (Brocklebank et al., 2003; Mayer et al., 2011; Pombo et al., 2012). However, it is also due to the modern scientific approaches that are being applied to/adapted for use in fishes. These include functional genomics and proteomics (Haverinen and Vornanen, 2009b; Korajoki and Vornanen, 2009; Klaiman et al., 2011; Rytkonen et al., 2012), isolated microvessel techniques (Costa et al., 2012), confocal microscopy (Shiels et al., 2011), and cellular indices of myofilament Ca sensitivity (Patrick et al., 2010a) and ion flux (Haverinen and Vornanen, 2009a; Galli et al., 2011). A major challenge, however, is to integrate the information gleaned at the molecular and cellular levels with the growing body of data on heart and *in vivo* cardiovascular function.

ACKNOWLEDGMENTS

The authors acknowledge funding from the Natural Sciences and Engineering Council of Canada (Discovery and Accelerator grants to AKG) and the University of Manchester to HAS in support of the writing of this chapter. We also thank colleagues who provided access to unpublished data, including Dr. A. DiMaio for his EM images of the tuna heart, G. Luxan for his confocal images of zebrafish myocytes, and Dr. S. Patrick for ECG recordings from the trout heart. We also thank Mr. Gord Nash and Dr. G.L.J Galli for invaluable assistance with manuscript preparation.

REFERENCES

Abel DC, Lowell WR, and Lipke MA. 1994. Elasmobranch pericardial function. 3. The pericardioperitoneal canal in the horn shark *Heterodontus francisci*. *Fish Physiol Biochem* 13(3): 263–274.

Agnisola C, McKenzie DJ, Pellegrino D, Bronzi P, Tota B, and Taylor EW. 1999. Cardiovascular responses to hypoxia in the Adriatic sturgeon (*Acipenser naccarii*). *J Appl Ichthyol* 15: 67–72.

Aho E and Vornanen M. 1999. Contractile properties of atrial and ventricular myocardium of the heart of rainbow trout (*Oncorhynchus mykiss*): Effects of thermal acclimation. *J Exp Biol* 202: 2663–2677.

Aho E and Vornanen M. 2001. Cold acclimation increases basal heart rate but decreases its thermal tolerance in rainbow trout (*Oncorhynchus mykiss*). *J Comp Physiol B* 171: 173–179.

Allen DG and Kentish JC. 1985. The cellular basis of the length-tension relation in cardiac muscle. *J Mol Cell Cardiol* 9: 821–840.

Altimiras J, Claireaux G, Sandblom E, and Farrell AP. 2008. Gastrointestinal blood flow and postprandial metabolism in swimming sea bass *Dicentrarchus labrax*. *Physiol Biochem Zool* 81(5): 663–672.

Angelone T, Filice E, Quintieri AM, Imbrogno S, Recchia A, Puler E, Mannarino C, Pellegrino D, and Cerra MC. 2008. β3-Adrenoceptors modulate left ventricular relaxation in the rat heart via the NO-cGMP-PKG pathway. *Acta Physiol* 192: 1–11.

Angelone T, Gattuso A, Imbrogno S, Mazza A, and Tota B. 2012. Nitrite is a positive modulator of the Frank-Starling response in the vertebrate heart. *Amer J Physiol* 301: R1271–R1281.

Axelsson M. 1988. The importance of nervous and humoral mechanisms in the control of cardiac performance in the Atlantic cod *Gadus morhua* at rest and during non-exhaustive exercise. *J Exp Biol* 137: 287–301.

Axelsson M, Altimiras J, and Claireeaux G. 2002. Post-prandial blood flow to the gastrointestinal tract is not compromised during hypoxia in the sea bass *Dicentrarchus labrax*. *J Exp Biol* 205: 2891–2896.

Axelsson M, Davison W, Forster ME, and Farrell AP. 1992. Cardiovascular responses of the red-blooded Antarctic fishes, *Pagothenia bernacchii* and *P. borchgrevinki*. *J Exp Biol* 167: 179–201.

Axelsson M, Driedzic WR, Farrell AP, and Nilsson S. 1989. Regulation of cardiac output and gut blood flow in the sea raven, *Hemitripterus americanus*. *Fish Physiol Biochem* 6: 315–326.

Axelsson M and Farrell AP. 1993. Coronary blood flow in vivo in the coho salmon (*Oncorhynchus kisutch*). *Am J Physiol* 264: R963–R971.

Axelsson M, Farrell AP, and Nilsson S. 1990. Effects of hypoxia and drugs on the cardiovascular dynamics of the Atlantic hagfish *Myxine glutinosa*. *J Exp Biol* 151: 297–316.

Axelsson M and Fritsche R. 1991. Effects of exercise, hypoxia and feeding on the gastrointestinal blood flow in the Atlantic cod, Gadus morhua. *J. Exp. Biol.* 158: 181–198.

Ballesta S, Hanson LM, and Farrell AP. 2012. The effect of adrenaline on the temperature dependency of cardiac action potentials in pink salmon *Oncorhynchus gorbuscha*. *J Fish Biol* 80: 876–885.

Bassani JW, Bassani RA, and Bers DM. 1994. Relaxation in rabbit and rat cardiac cells: Species-dependent differences in cellular mechanisms. *J Physiol* 476: 279–293.

Bers DM. 2002. Cardiac excitation-contraction coupling. *Nature* 415: 198–205.

Birkedal R, Christopher J, Thistlethwaite A, and Shiels H. 2009. Temperature acclimation has no effect on ryanodine receptor expression or subcellular localization in rainbow trout heart. *J Comp Physiol B* 179: 961–969.

Birkedal R and Shiels HA. 2007. High [Na+]i in cardiomyocytes from rainbow trout. *Am J Physiol* 293: R861–R866.

Bowler K and Tirri R. 1990. Temperature dependence of the heart isolated from the cold or warm acclimated perch (*Perca fluviatilis*). *Comp Biochem Physiol A* 96: 177–180.

Brette F, Luxan G, Cros C, Dixey H, Wilson C, and Shiels HA. 2008. Characterization of isolated ventricular myocytes from adult zebrafish (*Danio rerio*). *Biochem Biophys Res Commun* 374: 143–146.

Brocklebank J and Raverty S. 2002. Sudden mortality caused by cardiac deformities following seining of preharvest farmed Atlantic salmon (*Salmo salar*) and by cardiomyopathy of postintraperitoneally vaccinated Atlantic salmon parr in British Columbia. *Can Vet J* 43: 129–130.

Brutsaert DL and Andries LJ. 1992. The endocardial endothelium. *Am J Physiol* 263: H985–H1002.

Burggren W, Farrell A, and Lillywhite H. 1997. Vertebrate cardiovascular systems. In: Dantzler WH (Ed.). *The Handbook of Physiology. Section 13. Comparative Physiology*, Vol. 1. New York: Oxford University Press, pp. 215–308.

Burggren WW and Johansen K. 1986. Circulation and respiration in lungfishes (Dipnoi). *J Morph Supp* 1: 217–236.

Bushnell PG and Brill RW. 1992. Oxygen transport and cardiovascular responses in skipjack tuna (*Katsuwonus pelamis*) and yellowfin tuna (*Thunnus albacares*) exposed to acute hypoxia. *J Comp Physiol B* 162: 131–143.

Bushnell PG, Jones DR, and Farrell AP. 1992. The arterial system. In: Hoar WS, Randall DJ, and Farrell AP (Eds.). *Fish Physiology, Volume XII, Part A, The Cardiovascular System*. San Diego, CA: Academic Press, Inc., pp. 89–139.

Butler PJ and Taylor EW. 1975. The effect of progressive hypoxia on respiration in the dogfish (*Scyliorhinus canicula*) at different seasonal temperatures. *J Exp Biol* 63: 117–130.

Calaghan SC and White E. 2005. Mechanical modulation of intracellular ion concentrations: Mechanisms and electrical consequences. In: Kamkin A and Kiseleva I (Eds.). *Mechanosensitivity in Cells and Tissues*. Moscow, Russia: Academia, pp. 55–89.

Cameron JS, Hoffman KE, Zia C, Hemmett HM, Kronsteiner A, and Lee CM. 2003. A role for nitric oxide in hypoxia-induced activation of cardiac K_{ATP} channels in goldfish (*Carassius auratus*). *J Exp Biol* 206: 4057–4065.

Campbell HA, Taylor EW, and Egginton S. 2004. The use of power spectral analysis to determine cardiorespiratory control in the short-horned sculpin *Myoxocephalus scorpius*. *J Exp Biol* 207(11): 1969–1976.

Carey HV, Andrews MT, and Martin SL. 2003. Mammalian hibernation: Cellular and molecular responses to depressed metabolism and low temperature. *Physiol Rev* 83: 1153–1181.

Cerra MC, Angelone T, Parisella ML, Pellegrino D, and Tota B. 2009. Nitrite modulates contractility of teleost (*Anguilla anguilla* and *Chionodraco hamatus*, i.e. the Antarctic hemoglobinless icefish) and frog (*Rana esculenta*) hearts. *Biochim Biophys Acta* 1787: 849–855.

Chan DKO. 1986. Cardiovascular, respiratory, and blood adjustments to hypoxia in the Japanese eel, *Anguilla japonica*. *Fish Physiol Biochem* 2: 179–193.

Churcott CS, Moyes CD, Bressler BH, Baldwin KM, and Tibbits GF. 1994. Temperature and pH effects on Ca^{2+} sensitivity of cardiac myofibrils: A comparison of trout with mammals. *Am J Physiol* 267: R62–R70.

Clark RJ and Rodnick KJ. 1998. Morphometric and biochemical characteristics of ventricular hypertrophy in male rainbow trout (*Oncorhynchus mykiss*). *J Exp Biol* 201: 1541–1552.

Clark TD, Sandblom E, Cox GK, Hinch SG, and Farrell AP. 2008. Circulatory limits to oxygen supply during an acute temperature increase in the Chinook salmon (*Onchorynchus tshawytscha*). *Am J Physiol* 295: R1631–R1639.

Cobb J and Santer R. 1973. Electrophysiology of cardiac function in teleosts: Cholinergically mediated inhibition and rebound excitation. *J Physiol* 230: 561–573.

Costa IASF, Hein TW, Secombes CJ, and Gamperl AK. 2012. Control of the steelhead trout (*Oncorhynchus mykiss*) coronary microcirculation: Temperature effects and role of the endothelium. Paper presented at Experimental Biology 2012. San Diego.

Costa MJ, Kalinin, AL, and Rantin FT. 2004. Role of the sarcoplasmic reticulum in calcium dynamics of the ventricular myocardium of *Lepidosiren paradoxa* (Dipnoi) at different temperatures. *J Therm Biol* 29: 81–89.

Davie PS and Farrell AP. 1991. The coronary and luminal circulations of the myocardium of fishes. *Can J Zool* 69: 1993–2001.

Davie PS and Forster ME. 1980. Cardiovascular responses to swimming in eels. *Comp Biochem Physiol* 67A: 367–373.

Davie PS, Forster ME, Davison W, and Satchell GH. 1987. Cardiac function in the New Zealand hagfish, *Eptatretus cirrhatus*. *Physiol Zool* 60: 233–240.

Davison W. 1987. Arterioles in the swimming muscles of the leatherjacket *Parika scaber* (Pisces: Balistidae). *Cell Tis Res* 248(3): 703–708.

Delbridge LM, Satoh H, Yuan W, Bassani JW, Qi M, Ginsburg KS, Samarel AM, and Bers DM. 1997. Cardiac myocyte volume, Ca^{2+} fluxes, and sarcoplasmic reticulum loading in pressure-overload hypertrophy. *Am J Physiol* 272: H2425–H2435.

Di Maio A and Block BA. 2008. Ultrastructure of the sarcoplasmic reticulum in cardiac myocytes from Pacific bluefin tuna. *Cell Tiss Res* 334: 121–134.

Dibb KM, Hagarty CL, Loudon ASI, and Trafford AW. 2005. Photoperiod-dependent modulation of cardiac excitation contraction coupling in the Siberian hamster. *Am J Physiol* 288: R607–R614.

Dombrowski RA, Russell MJ, and Olson KR. 2004. Hydrogen sulfide as an endogenous regulator of vascular smooth muscle tone in trout. *Am J Physiol* 286: R678–R685.

Donald J and Campbell G. 1982. A comparative study of the adrenergic innervation of the teleost heart. *J Comp Physiol* 147: 85–91.

Driedzic WR, Bailey JR, and Sephton DH. 1996. Cardiac adaptations to low temperature in non-polar teleost fish. *J Exp Zool* 275: 186–195.

Duff DW and Olson KR. 1986. Trout vascular and renal responses to atrial natriuretic factor and heart extracts. *Am J Physiol* 251: R639–R642.

Dupont-Prinet A, Claireaux G, and McKenzie DJ. 2009. Effects of feeding and hypoxia on cardiac performance and gastrointestinal blood flow during critical speed swimming in the sea bass *Dicentrarchus labrax*. *Comp Biochem Physiol A* 155: 233–240.

Duran AC, Fernandez B, Graimes AC, Rodriguez C, Arque JM, and Sans-Coma V. 2008. Chondrichthyans have a bulbus arteriosus at the arterial pole of the heart: Morphological and evolutionary implications. *J Anat* 213: 597–606.

Egginton S and Syeda F. 2011. Design and physiology of capillaries and secondary circulation. In: Farrell AP (Ed.). *Encyclopedia of Fish Physiology, from Genome to Environment*, 1st edn. London, U.K.: Academic Press, pp. 1142–1152.

Farrell AP. 1982. Cardiovascular changes in the unanaesthetized lingcod (*Ophiodon elongates*) during short-term, progressive hypoxia and spontaneous activity. *Can J Zool* 60: 933–941.

Farrell AP. 1991. From hagfish to tuna: A perspective on cardiac-function in fish. *Physiol Zool* 64: 1137–1164.

Farrell AP. 2007. A tribute to P. L. Lutz: A message from the heart—Why hypoxic bradycardia in fishes? *J Exp Biol* 210: 1715–1725.

Farrell AP. 2009. Environment, antecedents and climate change: Lessons from the study of temperature physiology and river migration of salmonids. *J Exp Biol* 212(23): 3771–3780.

Farrell AP. 2011a. Cellular composition of the blood. In: Farrell AP (Ed.). *Encyclopedia of Fish Physiology, from Genome to Environment*, 1st edn. London, U.K.: Academic Press, pp. 984–991.

Farrell AP. 2011b. *Encyclopedia of Fish Physiology, from Genome to Environment*, 1st edn. London, U.K.: Academic Press, 2272pp.

Farrell AP. 2011c. Accessory hearts in fishes. In: Farrell AP (Ed.). *Encyclopedia of Fish Physiology, from Genome to Environment*, 1st Edn. London, U.K.: Academic Press, pp. 1073–1076.

Farrell AP. 2011d. Integrated cardiovascular responses of fish to swimming. In: Farrell AP (Ed.). *Encyclopedia of Fish Physiology, from Genome to Environment*, 1st edn. London, U.K.: Academic Press, pp. 1215–1220.

Farrell AP, Axelsson M, Altimiras J, Sandblom E, and Claireaux G. 2007. Maximum cardiac performance and adrenergic sensitivity of the sea bass *Dicentrarchus labrax* at high temperatures. *J Exp Biol* 210: 1216–1224.

Farrell AP, Eliason EJ, Sandblom E, and Clarke TD. 2009. Fish cardiorespiratory physiology in an era of climate change. *Can J Zool* 87: 835–851.

Farrell AP, Gamperl AK, Hicks JMT, Shiels HA, and Jain KE. 1996. Maximum cardiac performance of rainbow trout (*Oncorhynchus mykiss*) at temperatures approaching their upper lethal limit. *J Exp Biol* 199: 663–672.

Farrell AP, Hammons AM, Graham MS, and Tibbits GF. 1988a. Cardiac growth in rainbow trout; *Salmo Gairdneri*. *Can J Zool* 66: 2368–2373.

Farrell AP, Johansen JA, and Graham MS. 1988b. The role of the pericardium in cardiac performance of the trout (*Salmo gairdneri*). *Physiol Zool* 61: 213–221.

Farrell AP and Jones DR. 1992. The heart. In: Hoar WS, Randall DJ, and Farrell AP (Eds.). *Fish Physiology, Volume XII, Part A, The Cardiovascular System*. San Diego, CA: Academic Press, Inc., pp. 1–88.

Farrell AP, MacLeod KR, and Chancey B. 1986. Intrinsic mechanical properties of the perfused rainbow trout heart and the effects of catecholamines and extracellular calcium under control and acidotic conditions. *J Exp Biol* 125: 319–345.

Farrell AP and Olson KR. 2000. Cardiac natriuretic peptides: A physiological lineage of cardioprotective hormones? *Physiol Biochem Zool* 73: 1–11.

Farrell AP and Pieperhoff S. 2011. Cardiac anatomy in fishes. In: Farrell AP (Ed.). *Encyclopedia of Fish Physiology, from Genome to Environment*, 1st edn. London, U.K.: Academic Press, pp. 998–1005.

Forster ME, Davison W, Axelsson M, and Farrell AP. 1992. Cardiovascular responses to hypoxia in the hagfish *Eptatretus cirrhatus*. *Respir Physiol* 88: 373–386.

Franklin CE and Davie PS. 1991. The pericardium facilitates pressure work in the eel heart. *J Fish Biol* 39: 559–564.

Franklin CE and Davie PS. 1992a. Dimensional analysis of the ventricle of an in situ perfused trout heart using echocardiography. *J Exp Biol* 166: 47–60.

Franklin CE and Davie PS. 1992b. Myocardial power output of an isolated eel (*Anguilla dieffenbachii*) heart preparation in response to adrenaline. *Comp Biochem Physiol* 101: 293–298.

Franklin CE, Davison W, and Seebacher F. 2007. Antarctic fish can compensate for rising temperatures: Thermal acclimation of cardiac performance in *Pagothenia borchgrevinki*. *J Exp Biol* 210: 3068–3074.

Fritsche R and Nilsson S. 1989. Cardiovascular responses to hypoxia in the Atlantic cod, *Gadus morhua*. *J Exp Biol* 48: 153–160.

Gallaugher PE, Thorarensen H, Kiessling AK, and Farrell AP. 2001. Effects of high intensity exercise training on cardiovascular function, oxygen uptake, internal oxygen transport and osmotic balance in Chinook salmon (*Oncorhynchus tshawytscha*) during critical speed swimming. *J Exp Biol* 204: 2861–2872.

Galli GLJ, Lipnick MS, and Block BA. 2009a. Effect of thermal acclimation on action potentials and sarcolemmal K+ channels from Pacific bluefin tuna cardiomyocytes. *Am J Physiol* 297: R502–R509.

Galli GL, Lipnick MS, Shiels HA, and Block BA. 2011. Temperature effects on Ca^{2+} cycling in scombrid cardiomyocytes: A phylogenetic comparison. *J Exp Biol* 214: 1068–1076.

Galli GLJ and Shiels HA. 2012. The sarcoplasmic reticulum in the vertebrate heart. In: Wang T and Sedmara D (Eds.). *Ontogeny and Phylogeny of the Vertebrate Heart Advances in Experimental Medicine and Biology*. Amsterdam, the Netherlands: Springer-Verlag, pp. 103–124.

Galli GLJ, Shiels HA, and Brill RW. 2009b. Temperature sensitivity of cardiac function in pelagic fishes with different vertical mobilities: Yellowfin tuna (*Thunnus albacares*), Bigeye tuna (*Thunnus obesus*), Mahimahi (*Coryphaena hippurus*), and Swordfish (*Xiphias gladius*). *Physiol Biochem Zool* 82: 280–290.

Gamperl AK. 2011. Integrated responses of the circulatory system: Temperature. In: Farrell AP (Ed.). *Encyclopedia of Fish Physiology, from Genome to Environment*, 1st edn. London, U.K.: Academic Press, pp. 1197–1205.

Gamperl AK, Axelsson M, and Farrell AP. 1995. Effects of swimming and environmental hypoxia on coronary blood flow in rainbow trout. *Am J Physiol* 269: R1258–R1266.

Gamperl AK and Driedzic WR. 2009. Cardiovascular responses to hypoxia. In: Richards JG, Farrell AP, and Brauner CJ (Eds.). *Hypoxia, Fish Physiology*, Vol. XXVII. London, U.K.: Academic Press, pp. 302–360.

Gamperl AK and Farrell AP. 2004. Cardiac plasticity in fishes: Environmental influences and intraspecific differences. *J Exp Biol* 207: 2539–2550.

Gamperl AK, Pinder AW, Grant RR and Boutilier RG. 1994. Influence of hypoxia and adrenaline administration on coronary blood flow and cardiac performance in seawater rainbow trout (*Oncorhynchus mykiss*). *J Exp Biol* 193: 209–232.

Gamperl AK, Swafford BL and Rodnick KJ. 2011. Elevated temperature, per se, does not limit the ability of rainbow trout to increase ventricular stroke volume. *J Thermal Biol* 36: 7–14.

Gattuso A, Mazza R, Imbrogno S, Sverdrup A, Tota B, and Nylund A. 2002. Cardiac performance in *Salmo salar* with infectious salmon anaemia (ISA): Putative role of nitric oxide. *Dis Aquat Org* 52: 11–20.

Gauthier C, Seze-Goismier C, and Rozec B. 2007. Beta 3-adrenoreceptors in the cardiovascular system. *Clin Hemorheol Microcirc* 37: 193–204.

Genten F, Terwinghe E, and Danguy A. 2009. *Atlas of Fish Histology*. Enfield, NH: Science Publishers, 215pp.

Gillis TE, Moyes CD, and Tibbits GF. 2003. Sequence mutations in teleost cardiac troponin C that are permissive of high Ca^{2+} affinity of site II. *Am J Physiol* 284: C1176–C1184.

Gillis TE and Tibbits GF. 2002. Beating the cold: The functional evolution of troponin C in teleost fish. *Comp Biochem Physiol A* 132: 763–772.

Giltrow E, Eccles PD, Hutchinson TH, Sumpter JP, and Rand-Weaver M. 2011. Characterisation and expression of β_1-, β_2- and β_3-adrenergic receptors in the fathead minnow (*Pimephales promelas*). *Gen Comp Endocr* 173: 483–490.

Gollock MJ, Currie S, Petersen LH, and Gamperl AK. 2006. Cardiovascular and haematological responses of Atlantic cod (*Gadus morhua*) to acute temperature increase. *J Exp Biol* 209: 2961–2970.

Gore M and Burggren WW. 2012. Cardiac and metabolic physiology of early larval zebrafish (*Danio rerio*) reflects parental swimming stamina. *Front Aq Physiol*. doi: 10.3389/fphys.2012.00035.

Graham MS and Farrell AP. 1989. The effect of temperature-acclimation and adrenaline on the performance of a perfused trout heart. *Physiol Zool* 62: 38–61.

Grans A, Olsson C, Pitsillides K, Nelson HE, Cech JJ, Jr., and Axelsson M. 2010. Effects of feeding on thermoregulatory behaviours and gut blood flow in white sturgeon (*Acipenser transmontanus*) using biotelemetry in combination with standard techniques. *J Exp Biol* 213: 3198–3206.

Gregory JA, Graham JB, Cech JJ, Jr., Dalton N, Michaels J, and Lai NC. 2004. Pericardial and pericardioperitoneal canal relationships to cardiac function in the white sturgeon (*Acipenser transmontanus*). *Comp Biochem Physiol A* 138: 203–213.

Groman DB. 1982. *Histology of the Striped Bass*. Bethesda, MD: American Fisheries Society, 116pp.

Haddock PS, Coetzee WA, Cho E, Porter L, Katoh H, Bers DM, Jafri MS, and Artman M. 1999. Subcellular $[Ca^{2+}]i$ gradients during excitation-contraction coupling in newborn rabbit ventricular myocytes. *Circ Res* 85: 415–427.

Hanson LM, Obradovich S, Mouniargi J, and Farrell AP. 2006. The role of adrenergic stimulation in maintaining maximum cardiac performance in rainbow trout (*Oncorhynchus mykiss*) during hypoxia, hyperkalemia and acidosis at 10°C. *J Exp Biol* 209: 2442–2451.

Harvey RD and Calaghan SC. 2012. Caveolae create local signalling domains through their distinct protein content, lipid profile and morphology. *J Mol Cell Cardiol* 52: 366–375.

Harwood CL, Howarth FC, Altringham JD, and White E. 2000. Rate-dependent changes in cell shortening; intracellular Ca^{2+} levels and membrane potential in single isolated rainbow trout (*Oncorhynchus mykiss*) ventricular myocytes. *J Exp Biol* 203: 493–504.

Haverinen J, Hassinen M, and Vornanen M. 2007. Fish cardiac sodium channels are tetrodotoxin sensitive. *Acta Physiol (Oxf)* 191: 197–204.

Hassinen M, Laulaja S, Paajanen V, Haverinen J, and Vornanen M. 2011. Thermal adaptation of the crucian carp (*Carassius carassius*) cardiac delayed rectifier current, IKs, by homomeric assembly of Kv7.1 subunits without MinK. *Am J Physiol* 301: R255–R265.

Haverinen J and Vornanen M. 2004. Temperature acclimation modifies Na+ current in fish cardiac myocytes. *J Exp Biol* 207: 2823–2833.

Haverinen J and Vornanen M. 2006. Significance of Na+ current in the excitability of atrial and ventricular myocardium of the fish heart. *J Exp Biol* 209: 549–557.

Haverinen J and Vornanen M. 2007. Temperature acclimation modifies sinoatrial pacemaker mechanism of the rainbow trout heart. *Am J Physiol* 292: R1023–R1032.

Haverinen J and Vornanen M. 2009a. Comparison of sarcoplasmic reticulum calcium content in atrial and ventricular myocytes of three fish species. *Am J Physiol* 297: R1180–R1187.

Haverinen J and Vornanen M. 2009b. Responses of action potential and K+ currents to temperature acclimation in fish hearts: Phylogeny or thermal preferences? *Physiol Biochem Zool* 82: 468–482.

Head BP, Graham JB, Shabetai R, and Lai NC. 2001. Regulation of cardiac function in the horn shark by changes in pericardial fluid volume mediated through the pericardioperitoneal canal. *Fish Physiol Biochem* 24: 141–148.

Hochachka PW. 1961. The effect of physical training on oxygen debt and glycogen reserves in trout. *Can J Zool* 39: 767–776.

Hochachka PW. 1986. Defense strategies against hypoxia and hypothermia. *Science* 231: 234–241.

Hoglund L and Gesser H. 1987. Electrical and mechanical activity in heart tissue of flounder and rainbow trout during acidosis. *Comp Biochem Physiol A* 87: 543–546.

Holeton GF and Randall DJ. 1967. Changes in blood pressure in the rainbow trout during hypoxia. *J Exp Biol* 46: 297–305.

Hove-Madsen L, Llach A, Tibbits GF, and Tort L. 2003. Triggering of sarcoplasmic reticulum Ca^{2+} release and contraction by reverse mode Na^+/Ca^{2+} exchange in trout atrial myocytes. *Am J Physiol* 284: R1330–R1339.

Hove-Madsen L, Llach A, and Tort L. 2000. Na(+)/Ca(2+)-exchange activity regulates contraction and SR Ca(2+) content in rainbow trout atrial myocytes. *Am J Physiol* 279: 1856–1864.

Hove-Madsen L and Tort L. 1998. L-type Ca^{2+} current and excitation-contraction coupling in single atrial myocytes from rainbow trout. *Am J Physiol* 275: 2061–2069.

Icardo JM. 2006. Conus arteriosus of the teleost heart: Dismissed, but not missed. *Anat Rec A* 288: 900–908.

Icardo JM. 2012. The teleost heart: A morphological approach. In: Sedmera D and Wang T (Eds.). *Ontogeny and Phylogeny of the Vertebrate Heart*. New York: Springer-Verlag, pp. 35–53.

Icardo JM and Colvee E. 2011. The atrioventricular region of the teleost heart. A distinct heart segment. *Anat Rec (Hoboken)* 294: 236–242.

Imbrogno S, Angelone T, Adamo C, Pulera E, Tota B, and Cerra M. 2006. Beta3-adrenoreceptor in the eel (*Anguilla anguilla*) heart: Negative inotropy and NO-cGMP-dependent mechanism. *J Exp Biol* 209: 4966–4973.

Imbrogno S, De Iuri L, Mazza R, and Tota B. 2001. Nitric oxide modulates cardiac performance in the heart of *Anguilla anguilla*. *J Exp Biol* 204: 1719–1727.

Ishimatsu A, Iwama GK, and Heisler N. 1988. In vivo analysis of partitioning of cardiac output between systemic and central venous sinus circuits in rainbow trout: A new approach using chronic cannulation of the branchial vein. *J Exp Biol* 137: 75–88.

Ishimatsu A, Iwama GK, Bentley TB, Heisler N. 1992. Contribution of the secondary circulation system to acid-base regulation during hypercapnia in rainbow trout (*Oncorhynchus mykiss*). *J Exp Biol* 170: 43–56.

Iversen NK, Dupont-Prinet A, Findorf I, Mckenzie DJ, and Wang T. 2010a. Autonomic regulation of the heart during digestion and aerobic swimming in the European sea bass (*Dicentrarchus labrax*). *Comp Biochem Physiol A* 156: 463–468.

Iversen NK, Mckenzie DJ, Malte H, and Wang T. 2010b. Reflex bradycardia does not influence oxygen consumption during hypoxia in the European eel (*Anguilla anguilla*). *J Comp Physiol B* 180(4): 495–502.

Iwama GK, Ishimatsu A, and Heisler N. 1990. The effect of environmental hypercapnia on the ionic and acid-base status of the fluid in the veno-lymphatic system of rainbow trout, *Oncorhynchus mykiss*. Abstract from an International Symposium "Mechanisms of Systemic Regulation in Lower Vertebrates: Respiration, Circulation, Ion Transfer and Metabolism." Gottingen, Germany.

Janvier JJ, Peyraud WM, Soulier P. 1996. Effects of serotonin on the cardio-circulatory system of the European Eel (Anguilla anguilla) in vivo. *J Comp Physiol* 166: 131–137.

Jensen D. 1961. Cardioregulation in an aneural heart. *Comp Biochem Physiol* 2: 181–201.

Jensen FB and Agnisola C. 2005. Perfusion of the isolated trout heart coronary circulation with red blood cells: Effects of oxygen supply and nitrite on coronary flow and myocardial oxygen consumption. *J Exp Biol* 208: 3665–3674.

Joaquim N, Wagner GN, and Gamperl AK. 2004. Cardiac function and critical swimming speed of the winter flounder (*Pleuronectes americanus*) at two temperatures. *Comp Biochem Physiol A* 138: 277–285.

Johansen IB, Lunde IG, Rosjo H, Christensen G, Nilsson GE, Bakken M, and Overli O. 2011. Cortisol response to stress is associated with myocardial remodeling in salmonid fishes. *J Exp Biol* 214: 1313–1321.

Johnson KR and Olson KR. 2008. Comparative physiology of the piscine natriuretic peptide system. *Gen Comp Endocr* 157: 21–26.

Johnsson M, Axelsson M, and Holmgren S. 2001. Large veins in the Atlantic cod (*Gadus morhua*) and the rainbow trout (*Oncorhynchus mykiss*) are innervated by neuropeptide-containing nerves. *Anat Embryol* 204: 109–115.

Jones DR and Braun MH. 2011. The outflow tract from the heart. In: Farrell AP (Ed.). *Encyclopedia of Fish Physiology, from Genome to Environment*, 1st edn. London, U.K.: Academic Press, pp. 1015–1029.

Jones DR, Brill RW, and Bushnell PG. 1993. Ventricular and arterial dynamics of anaesthetized and swimming tuna. *J Exp Biol* 182: 97–112.

Jones DR, Langille BL, Randall DJ, and Shelton G. 1974. Blood flow in dorsal and ventral aortas of the cod, *Gadus morhua*. *Am J Physiol* 226: 90–95.

Jonz MG and Zaccone G. 2009. Nervous control of the gills. *Acta Histochem* 111: 207–216.

Kampmeier OF. 1969. *Evolution and Comparative Morphology of the Lymphatic System*. Springfield, IL: Charles C Thomas, 620pp.

Keen JE, Farrell AP, Tibbits GF, and Brill RW. 1992. Cardiac physiology in tunas. 2. Effect of ryanodine, calcium, and adrenaline on force frequency relationships in atrial strips from skipjack tuna, Katsuwonus Pelamis. *Can J Zool* 70: 1211–1217.

Keen AN and Gamperl AK. 2012. Blood oxygenation and cardiorespiratory function in steelhead trout (*Oncorhynchus mykiss*) challenged with an acute temperature increase and zatebradine-induced bradycardia. *J Therm Biol* 37: 201–210.

Keen JE, Viazon DM, Farrell AP, and Tibbits GF. 1993. Thermal acclimation alters both adrenergic sensitivity and adrenoreceptor density in cardiac tissue of rainbow trout. *J Exp Biol* 181: 27–47.

Keen JE, Vianzon DM, Farrell AP, and Tibbits GF. 1994. Effect of temperature and temperature-acclimation on the ryanodine sensitivity of the trout myocardium. *J Comp Physiol B* 164: 438–443.

Kiceniuk JW and Jones DR. 1977. The oxygen transport system in trout (*Salmo gairdneri*) during sustained exercise. *J Exp Biol* 69: 247–260.

Klaiman JM, Fenna AJ, Shiels HA, Macri J, and Gillis TE. 2011. Cardiac remodeling in fish: Strategies to maintain heart function during temperature change. *PLoS ONE* 6: e24464.

Kolok AS, Spooner RM, and Farrell AP. 1993. The effect of exercise on the cardiac output and blood flow distribution of the largescale sucker *Catostomus macrocheilus*. *J Exp Biol* 183: 301–321.

Korajoki H and Vornanen M. 2009. Expression of calsequestrin in atrial and ventricular muscle of thermally acclimated rainbow trout. *J Exp Biol* 212: 3403–3414.

Lague SL, Speers-Roesch B, Richards JG, and Farrell AP. 2012. Exceptional cardiac anoxia tolerance in tilapia (*Oreochromis hybrid*). *J Exp Biol* 215: 1354–1365.

Lai NC, Graham JB, Bhargava V, and Shabetai R. 1996. Mechanisms of venous return and ventricular filling in elasmobranch fish. *Am J Physiol* 270(39): H1766–H1771.

Lai NC, Graham JB, Dalton N, Shabetai R, and Bhargava V. 1998. Echocardiographic and hemodynamic determinations of the ventricular filling pattern in some teleost fishes. *Physiol Zool* 71(2): 157–167.

Lai NC, Graham JB, Lowell WR, and Shabetai R. 1989. Elevated pericardial pressure and cardiac output in the leopard shark *Triakis semifasciata* during exercise: The role of the pericardioperitoneal canal. *J Exp Biol* 147: 263–277.

Lai NC, Shabetai R, Graham JB, Hoit BD, Sunnerhagen KS, and Bhargava V. 1990. Cardiac function of the leopard shark, *Triakis semifasciata*. *J Comp Physiol B* 160: 259–268.

Landeira-Fernandez A, Morrisette JM, Blank JM, and Block BA. 2004. Temperature dependence of Ca^{2+}-ATPase (SERCA2) in the ventricles of tuna and mackerel. *Am J Physiol* 286: R398–R404.

Lennard R and Huddart H. 1991. The effect of thermal stress on electrical and mechanical responses and associated calcium movements of flounder heart and gut. *Comp Biochem Physiol* 98A: 221–228.

Llach A, Huang J, Sederat F, Tort L, Tibbits G, and Hove-Madsen L. 2004. Effect of beta-adrenergic stimulation on the relationship between membrane potential, intracellular [Ca^{2+}] and sarcoplasmic reticulum Ca^{2+} uptake in rainbow trout atrial myocytes. *J Exp Biol* 207: 1369–1377.

Llach A, Molina CE, Alvarez-Lacalle E, Tort L, Benítez R, and Hove-Madsen L. 2011. Detection, properties, and frequency of local calcium release from the sarcoplasmic reticulum in teleost cardiomyocytes. *PLoS One* 6: e23708.

Loretz CA and Pollina C. 2000. Natriuretic peptides in fish physiology. *Comp Biochem Physiol* 125: 169–187.

Lurman GJ, Petersen LH, and Gamperl AK. 2012. Atlantic cod (*Gadus morhua* L.) in situ cardiac performance at cold temperatures: Long-term acclimation, acute thermal challenge and the role of adrenaline. *J Exp Biol* 215: 4006–4014.

Maricondi-Massari M, Kalinin AL, Glass ML, and Rantin FT. 1998. The effects of temperature on oxygen uptake, gill ventilation, and ECG waveforms in the Nile tilapia, *Oreochromis niloticus*. *J Therm Biol* 23: 282–290.

Marques IJ, Leito JTD, Spaink HP, Testerlink J, Jaspers RT, Witte F, van den Berg S, and Bagowski CP. 2008. Transcriptome analysis of the response to chronic constant hypoxia in zebrafish hearts. *J Comp Physiol* 178B: 77–92.

Mayer I, Meager J, Skjaeraasen JE, Rodewald P, Sverdup G, and Ferno F. 2011. Domestication causes rapid changes in heart and brain morphology in Atlantic cod (*Gadus morhua*). *Environ Biol Fish* 92: 181–186.

McKenzie DJ, Skov PV, Taylor EW, Wang T, and Steffensen JF. 2009. Abolition of reflex bradycardia by cardiac vagotomy has no effect on the regulation of oxygen uptake by Atlantic cod in progressive hypoxia. *Comp Biochem Physiol A* 153(3): 332–338.

Mendonca PC and Gamperl AK. 2009. Nervous and humoral control of cardiac performance in the winter flounder (*Pleuronectes americanus*). *J Exp Biol* 212: 934–944.

Mendonca PC and Gamperl AK. 2010. The effects of acute changes in temperature and oxygen availability on cardiac performance in winter flounder (*Pseudopleuronectes americanus*). *Comp Biochem Physiol A* 155: 245–252.

Mendonca PC, Genge AG, Deitch EJ, and Gamperl AK. 2007. Mechanisms responsible for the enhanced pumping capacity of the in situ winter flounder heart (*Pseudopleuronectes americanus*). *Am J Physiol* 293: R2112–R2119.

Minerick AR, Chang HC, Hoagland TM, and Olson KR. 2003. Dynamic synchronization analysis of venous pressure-driven cardiac output in rainbow trout. *Am J Physiol* 285: R889–R896.

Moss RL and Fitzsimons DP. 2002. Frank-Starling relationship: Long on importance, short on mechanism. *Circ Res* 90: 11–13.

Negretti N, O'Neill SC, and Eisner DA. 1993. The relative contributions of different intracellular and sarcolemmal systems to relaxation in rat ventricular myocytes. *Cardiovasc Res* 27: 1826–1830.

Nemtsas P, Wettwer E, Christ T, Weidinger G, and Ravens U. 2010. Adult zebrafish heart as a model for human heart? An electrophysiological study. *J Mol Cell Cardiol* 48: 161–171.

Nickerson J, Dugan S, Drouin G, Perry S, and Moon T. 2003. Activity of the unique beta-adrenergic Na^+/H^+ exchanger in trout erythrocytes is controlled by a novel $beta_3$-AR subtype. *Am J Physiol* 285: R526–R535.

Nilsson S. 1983. *Autonomic Nerve Function in the Vertebrates*. Berlin/Heidelberg, Germany; New York: Springer-Verlag, 276pp.

Nilsson S. 1994. Evidence for adrenergic nervous control of blood pressure in teleost fish. *Physiol Zool* 67: 1347–1359.

Nilsson S and Sundin L. 1998. Gill blood flow control. *Comp Biochem Physiol* 119(1): 137–147.

Olson KR. 1992. Blood and extracellular fluid volume regulation: Role of the renin-angiotensin system, kallikrein-kinin system, and atrial natriuretic peptides. In: Hoar WS, Randall DJ, and Farrell AP (Eds.). *Fish Physiology, Volume XII, Part B, The Cardiovascular System*. San Diego, CA: Academic Press, Inc., pp. 135–254.

Olson KR. 1996. Secondary circulation in fish: Anatomical organization and physiological significance. *J Exp Zool* 275: 172–185.

Olson KR. 2002. Vascular anatomy of the fish gill. *J Exp Zool* 293: 214–231.

Olson KR. 2008. Hydrogen sulfide and oxygen sensing: implications in cardiorespiratory control. *J Exp Biol* 211: 2727–2734.

Olson KR. 2011a. Branchial anatomy. In: Farrell AP (Ed.). *Encyclopedia of Fish Physiology, from Genome to Environment*, 1st edn. London, U.K.: Academic Press, pp. 1095–1103.

Olson KR. 2011b. Physiology of resistance vessels. In: Farrell AP (Ed.). *Encyclopedia of Fish Physiology, from Genome to Environment*, 1st edn. London, U.K.: Academic Press, pp. 1104–1110.

Olson KR and Donald JA. 2009. Nervous control of circulation—The role of gasotransmitters, NO, CO, and H_2S. *Acta Histochem* 111: 244–256.

Olson KR and Farrell AP. 2006. The cardiovascular system. In: Evans DH and Claiborne JB (Eds.). *The Physiology of Fishes*. Boca Raton, FL: Taylor & Francis, pp. 119–152.

Olson KR and Farrell AP. 2011. Secondary circulation and lymphatic anatomy. In: Farrell AP (Ed.). *Encyclopedia of Fish Physiology, from Genome to Environment*, 1st edn. London, U.K.: Academic Press, pp. 1161–1168.

Paajanen V and Vornanen M. 2003. Effects of chronic hypoxia on inward rectifier K(+) current (I(K1)) in ventricular myocytes of crucian carp (*Carassius carassius*) heart. *J Membr Biol* 194: 119–127.

Patrick SM, Hoskins AC, Kentish JC, White E, Shiels HA, and Cazorla O. 2010a. Enhanced length-dependent Ca^{2+} activation in fish cardiomyocytes permits a large operating range of sarcomere lengths. *J Mol Cell Cardiol* 48: 917–924.

Patrick SM, White E, and Shiels HA. 2010b. Mechano-electric feedback in the fish heart. *PLoS One* 5: e10548.

Paulus WJ, Vantrimpont PJ, and Shah AM. 1994. Acute effects of nitric oxide on left ventricular relaxation and diastolic distensibility in humans. *Circulation* 89: 2070–2078.

Pedersen CL, Faggiano S, Helbo S, Gesser H, and Fago A. 2010. Roles of nitric oxide, nitrite and myoglobin on myocardial efficiency in trout (*Oncorhynchus mykiss*) and goldfish (*Carassius auratus*): Implications for hypoxia tolerance. *J Exp Biol* 213: 2755–2762.

Pellegrino D, Acierno R, and Tota B. 2003. Control of cardiovascular function in the icefish *Chionodraco hamatus*: Involvement of serotonin and nitric oxide. *Comp Biochem Physiol A* 134: 471–480.

Petersen LH and Gamperl AK. 2010a. Effect of acute and chronic hypoxia on the swimming performance, metabolic capacity and cardiac function of Atlantic cod (*Gadus morhua*). *J Exp Biol* 213: 808–819.

Petersen LH and Gamperl AK. 2010b. In situ cardiac function in Atlantic cod (*Gadus morhua*): Effects of acute and chronic hypoxia. *J Exp Biol* 213: 820–830.

Petersen LH, Needham SL, Burleson ML, Overturf MD, and Huggett DB. 2013. Involvement of β_3-adrenergic receptors in in vivo cardiovascular regulation in rainbow trout (*Oncorhynchus mykiss*). *Comp Biochem Physiol* 164: 291–300.

Peyraud-Waitzenegger M, Barthelemy L, and Peyraud C. 1980. Cardiovascular and ventilator effects of catecholamines in unrestrained eels (*Anguilla anguilla* L.). *J Comp Physiol* 138: 367–375.

Peyraud-Waitzenegger M and Soulier P. 1989. Ventilatory and circulatory adjustments in the European eel (*Anguilla anguilla* L.) exposed to short term hypoxia. *Exp Biol* 48: 107–122.

Pieperhoff S, Bennett W, and Farrell AP. 2009. The intercellular organization of the two muscular systems in the adult salmonid heart, the compact and the spongy myocardium. *J Anat* 215: 536–547.

Pombo A, Blasco M, and Climent V. 2012. The status of farmed fish hearts: And alert to improve health and production in three Mediterranean species. *Rev Fish Biol Fish* 22: 779–789.

Portner HO. 2010. Oxygen- and capacity-limitation of thermal tolerance: A matrix for integrating climate-related stressor effects in marine ecosystems. *J Exp Biol* 213(6): 881–893.

Powell, MD, Burke MS and Dahle D. 2011. Cardiac remodelling, blood chemistry, haematology and oxygen consumption of Atlantic cod, Gadus morhua L., induced by experimental haemolytic anaemia with phenylhydrazine. *Fish Physiol Biochem* 37: 31–41.

Powell MD, Burke MS and Dahle D. 2012. Cardiac remodelling of Atlantic halibut Hippoglossus hippoglossus induced by experimental anaemia with phenylhydrazine. *J Fish Biol* 81: 335–344.

Randall DJ and Daxboeck C. 1982. Cardiovascular changes in the rainbow trout (*Salmo gairdneri* Richardson) during exercise. *Can J Zool* 60: 1135–1140.

Rantin FT, Gesser H, Kalinin AL, Guerra DR, De Freitas JC, and Driedzic WR. 1998. Heart performance, Ca^{2+} regulation and energy metabolism at high temperatures in *Bathygobius soporator*, a tropical marine teleost. *J Therm Biol* 23: 31–39.

Robertson RM. 2004. Modulation of neural circuit operation by prior environmental stress. *Integr Comp Biol* 44: 21–27.

Rocha ML, Rantin FT, and Kalinin AL. 2007. Importance of the sarcoplasmic reticulum and adrenergic stimulation on the cardiac contractility of the neotropical teleost *Synbranchus marmoratus* under different thermal conditions. *J Comp Physiol [B]* 177: 713–721.

Rytkonen KT, Renshaw GMC, Vaino PP, Ashton KJ, Williams-Pritchard G, Leder EH, and Nikinmaa M. 2012. Transcriptional responses to hypoxia are enhanced by recurrent hypoxia (hypoxic preconditioning) in the epaulette shark. *Physiol Genom* 44: 1090–1097.

Sandblom E and Axelsson M. 2004. Baroreflex mediated control of heart rate and vascular capacitance in trout. *J Exp Biol* 208: 821–829.

Sandblom E and Axelsson M. 2005. Effects of hypoxia on the venous circulation in rainbow trout (*Oncorhynchus mykiss*). *Comp Biochem Physiol* 140A: 233–239.

Sandblom E and Axelsson M. 2006. Adrenergic control of venous capacitance during moderate hypoxia in the rainbow trout (*Onchorhynchus mykiss*): Role of neural and circulating catecholamines. *Am J Physiol* 291: R711–R718.

Sandblom E and Axelsson M. 2007a. Venous hemodynamic responses to acute temperature increase in the rainbow trout (*Oncorhynchus mykiss*). *Am J Physiol* 292: R2292–R2298.

Sandblom E and Axelsson M. 2007b. The venous circulation: A piscine perspective. *Comp Biochem Physiol A* 148: 785–801.

Sandblom E and Axelsson M. 2011. Autonomic control of circulation in fish: A comparative view. *Auton Neurosci: Basic Clin* 165: 127–139.

Sandblom E, Axelsson M, and McKenzie DJ. 2006. Venous responses during exercise in rainbow trout, *Oncorhynchus mykiss*: α-adrenergic control and the antihypotensive function of the renin-angiotensin system. *Comp Biochem Physiol A* 144: 401–409.

Sandblom E, Cox GK, Perry SF, and Farrell AP. 2009. The role of venous capacitance, circulating catecholamines, and heart rate in the hemodynamic response to increased temperature and hypoxia in the dogfish. *Am J Physiol* 296: R1547–R1556.

Sandblom E, Farrell AP, Altimiras J, Axelsson M, and Claireaux G. 2005. Cardiac preload and venous return in swimming sea bass (*Dicentrarchus labrax* L.). *J Exp Biol* 208: 1927–1935.

Santer RM. 1985. Morphology and innervation of the fish heart. *Adv Anat Embryol Cell Biol* 89: 1–102.

Santer RM, Greer Walker M, Emerson L, and Witthames PR. 1983. On the morphology of the heart ventricle in marine teleost fish (Teleostei). *Comp Biochem Physiol* 76(3): 453–457.

Satchell GH. 1984. On the caudal heart of Myxine (Myxinoidea: Cyclostomata). *Acta Zool Stock* 65: 125–133.

Satchell GH. 1991. *Physiology and Form of Fish Circulation*. Cambridge, U.K.: Cambridge University Press, 256pp.

Satchell GH. 1992. The venous system. In: Hoar WS, Randall DJ, and Farrell AP (Eds.). *Fish Physiology, Volume XII, Part A, The Cardiovascular System*. San Diego, CA: Academic Press, Inc., pp. 141–183.

Satoh H, Delbridge LM, Blatter LA, and Bers DM. 1996. Surface:volume relationship in cardiac myocytes studied with confocal microscopy and membrane capacitance measurements: Species-dependence and developmental effects. *Biophys J* 70: 1494–1504.

Schib JL, Icardo JM, Duran AC, Guerrero A, Lopez D, Colvee E, de Andres AV, and Sans-Coma V. 2002. The conus arteriosus of the adult gilthead seabream (*Sparus auratus*). *J Anat* 201: 395–404.

Sedmera D, Reckova M, deAlmeida A, Sedmerova M, Biermann M, Volejnik J, Sarre A, Raddatz E, McCarthy RA, Gourdie RG, and Thompson RP. 2003. Functional and morphological evidence for a ventricular conduction system in zebrafish and Xenopus hearts. *Am J Physiol* 284: H1152–H1160.

Seth H and Axelsson M. 2009. Effects of gastric distension and feeding on cardiovascular variables in the shorthorn sculpin (*Myoxocephalus scorpius*). *Am J Physiol* 296: R171–R177.

Seth H and Axelsson M. 2010. Sympathetic, parasympathetic and enteric regulation of the gastrointestinal vasculature of rainbow trout (*Oncorhynchus mykiss*) under normal and postprandial conditions. *J Exp Biol* 213: 3118–3126.

Seth H and Axelsson M. 2011. Integrated responses of the circulatory system to digestion. In: Farrell AP (Ed.). *Encyclopedia of Fish Physiology, from Genome to Environment*, 1st edn. London, U.K.: Academic Press, pp. 1206–1214.

Seth H, Grans A, and Axelsson M. 2010. Cholecystokinin as a regulator of cardiac function and postprandial gastrointestinal blood flow in rainbow trout (*Oncorhynchus mykiss*). *Am J Physiol* 298: R1240–R1248.

Seth H, Sandblom E, and Axelsson M. 2009. Nutrient-induced gastrointestinal hyperemia and specific dynamic action in rainbow trout (*Oncorhynchus mykiss*)—Importance of proteins and lipids. *Am J Physiol* 296: R345–R352.

Seth H, Sandblom E, Holmgren S, and Axelsson M. 2008. Effects of gastric distension on the cardiovascular system in rainbow trout (*Oncorhynchus mykiss*). *Am J Physiol Regul Integr Comp Physiol* 294: R1648–R1656.

Severs NJ, Bruce AF, Dupont E, and Rothery S. 2008. Remodelling of gap junctions and connexin expression in diseased myocardium. *Cardiovasc Res* 80: 9–19.

Shabetai R, Abel DC, Graham JB, Bhargava V, Keyes RS, Witztum K. 1985. Function of the pericardium and pericardioperitoneal canal in elasmobranch fishes. *Am J Physiol* 248: H198–H207.

Shah AM, Spurgeon HA, Sollott SJ, Talo A, and Lakatta EG. 8-bromo-cGMP reduces the myofilament response to Ca^{2+} in intact cardiac myocytes. *Circ Res* 74(5): 970–978.

Shiels HA, Calaghan SC, and White E. 2006a. The cellular basis for enhanced volume-modulated cardiac output in fish hearts. *J Gen Physiol* 128: 37–44.

Shiels HA, Di Maio A, Thompson S, and Block BA. 2011. Warm fish with cold hearts: Thermal plasticity of excitation-contraction coupling in bluefin tuna. *Proc Biol Sci* 278: 18–27.

Shiels HA and Farrell AP. 1997. The effect of temperature and adrenaline on the relative importance of the sarcoplasmic reticulum in contributing Ca^{2+} to force development in isolated ventricular trabeculae from rainbow trout. *J Exp Biol* 200: 1607–1621.

Shiels HA, Freund EV, Farrell AP, and Block BA. 1999. The sarcoplasmic reticulum plays a major role in isometric contraction in atrial muscle of yellowfin tuna. *J Exp Biol* 202: 881–890.

Shiels HA, Paajanen V, and Vornanen M. 2006b. Sarcolemmal ion currents and sarcoplasmic reticulum Ca^{2+} content in ventricular myocytes from the cold stenothermic fish, the burbot (*Lota lota*). *J Exp Biol* 209: 3091–3100.

Shiels HA, Vornanen M, and Farrell AP. 2000. Temperature-dependence of L-type Ca^{2+} channel current in atrial myocytes from rainbow trout. *J Exp Biol* 203: 2771–2780.

Shiels HA, Vornanen M, and Farrell AP. 2002. Temperature dependence of cardiac sarcoplasmic reticulum function in rainbow trout myocytes. *J Exp Biol* 205: 3631–3639.

Shiels HA, Vornanen M, and Farrell AP. 2003. Acute temperature change modulates the response of I Ca to adrenergic stimulation in fish cardiomyocytes. *Physiol Biochem Zool* 76: 816–824.

Shiels HA and White E. 2005. Temporal and spatial properties of cellular Ca^{2+} flux in trout ventricular myocytes. *Am J Physiol* 288: R1756–R1766.

Shiels HA and White, E. 2008a. The effect of mechanical stimulation on vertebrate hearts: A question of class. In: Kamkin A and Kiseleva I (Eds.). *Mechanosensitivity in Cells and Tissues Mechanosensitive Ion Channels*. New York: Springer-Verlag, pp. 323–342.

Shiels HA and White E. 2008b. The Frank-Starling mechanism in vertebrate cardiac myocytes. *J Exp Biol* 211: 2005–2013.

Siebert H. 1979. Thermal adaptation of heart rate and its parasympathetic control in the European eel, *Anguilla anguilla* (L.). *Comp Biochem Physiol* 64C: 275–278.

Simonot, DL and Farrell AP. 2007. Cardiac remodelling in rainbow trout Oncorhynchus mykiss Walbaum in response to phenylhydrazine-induced anaemia. *J Exp Biol* 210: 2574–2584.

Skals M, Skovgaard N, Taylor EW, Leite CAC, Abe AS, and Wang T. 2006. Cardiovascular changes under normoxic and hypoxic conditions in the air-breathing teleost *Synbranchus marmoratus*: Importance of the venous system. *J Exp Biol* 209: 4167–4173.

Skov PV and Steffensen JF. 2003. The blood volumes of the primary and secondary circulatory system in the Atlantic cod *Gadus morhua* L., using plasma bound evans blue and compartmental analysis. *J Exp Biol* 206: 591–599.

Skovgaard N and Olson KR. 2012. Hydrogen sulfide mediates hypoxic vasoconstriction through a production of mitochondrial ROS in trout gills. *Am J Physiol* 303: R487–R494.

Speers-Roesch B, Brauner CJ, Farrell AP, Hickey AJ, Renshaw GM, Wang YS, and Richards JG. 2012. Hypoxia tolerance in elasmobranchs. II. Cardiovascular function and tissue metabolic responses during progressive and relative hypoxia exposures. *J Exp Biol* 215: 103–114.

Stecyk JAW and Farrell AP. 2002. Cardiorespiratory responses of the common carp (*Cyprinus carpio*) to severe hypoxia at three acclimation temperatures. *J Exp Biol* 205: 759–768.

Stecyk JAW, Galli GL, Shiels HA, and Farrell AP. 2008. Cardiac survival in anoxia-tolerant vertebrates: An electrophysiological perspective. *Comp Biochem Physiol C* 148: 339–354.

Stecyk JAW, Stensløkken K-O, Farrell AP, and Nilsson GE. 2004. Maintained cardiac pumping in anoxic crucian carp. *Science* 306: 77.

Steele SL, Yang X, Debiais-Thibaud M, Schwerte T, Pelster B, Ehkker M, Tiberi M, and Perry, SF. 2011. The in vivo and in vitro assessment of cardiac B-adrenergic receptors in larval zebrafish (Danio rerio). *J Exp Biol* 214: 1445–1457.

Steffensen JF and Lomholt JP. 1992. The secondary vascular system. In: Hoar WS, Randall DJ, and Farrell AP (Eds.). *Fish Physiology, Volume XII, Part A, The Cardiovascular System*. San Diego, CA: Academic Press, Inc., pp. 185–217.

Steffensen JF, Lomholt JP, and Vogel WOP. 1986. In vivo observations on a specialized microvasculature, the primary and secondary vessels in fishes. *Acta Zool (Stockh.)* 67: 193–200.

Sundin L. 1995a. Serotonergic vasomotor control in fish gills. *Braz J Med Biol Res* 28: 1217–1221.

Sundin L. 1995b. Responses of the branchial circulation to hypoxia in the Atlantic cod, *Gadus morhua*. *Am J Physiol* 268: R771–R778.

Sundin L and Nilsson S. 2002. Branchial innervation. *J Exp Zool* 293: 232–248.

Sundin L, Nilsson GE, Block M, and Lofman, CO. 1995. Control of gill filament blood flow by serotonin in the rainbow trout, *Oncorhynchus mykiss*. *Am J Physiol* 37: R1224–R1229.

Syme DA, Gamperl AK, Nash GW, and Rodnick KJ. In Press. Increased stiffness of ventricular tissue and decreased cardiac function in Atlantic cod (*Gadus morhua*) at high temperatures. *Am J Physiol*.

Taylor EW. 1992. Nervous control of the heart and cardiorespiratory interactions. In: Hoar WS, Randall DJ, and Farrell AP (Eds.). *Fish Physiol*. Vol. XIIb. San Diego, CA: Academic Press, pp. 343–387.

Taylor EW. 2011. Central control of cardiorespiratory interactions in fish. In: Farrell AP (Ed.). *Encyclopedia of Fish Physiology, from Genome to Environment*, 1st edn. London, U.K.: Academic Press, pp. 1178–1189.

Taylor EW, Short S, and Butler PJ. 1977. The role of the cardiac vagus in the response of the dogfish *Scyliorhinus canicula* to hypoxia. *J Exp Biol* 70: 57–75.

Thorarensen H, Gallaugher P, and Farrell AP. 1996. Cardiac output in swimming rainbow trout, *Oncorhynchus mykiss*, acclimated to seawater. *Physiol Zool* 69: 139–153.

Thorarensen H, Gallaugher PE, Kiessling AK, and Farrell AP. 1993. Intestinal blood flow in swimming Chinook salmon *Oncorhynchus tshawytscha* and the effects of haematocrit on blood flow distribution. *J Exp Biol* 179: 115–129.

Tibbits GF, Moyes CD, Hove-Madsen L, Hoar WS, Randall DJ, and Farrell AP. 1992. Excitation-contraction coupling in the teleost heart. In: Hoar WS, Randall DJ, and Farrell AP (Eds.). *The Cardiovascular System*. San Diego, CA: Academic Press, pp. 267–304.

Tiitu V and Vornanen M. 2001. Cold adaptation suppresses the contractility of both atrial and ventricular muscle of the crucian carp heart. *J Fish Biol* 59: 141–156.

Tiitu V and Vornanen M. 2002a. Morphology and fine structure of the heart of the burbot, a cold stenothermal fish. *J Fish Biol* 61: 106–121.

Tiitu V and Vornanen M. 2002b. Regulation of cardiac contractility in a stenothermal fish, the burbot (*Lota lota*). *J Exp Biol* 205: 1597–1606.

Tirri R and Ripatti P. 1982. Inhibitory adrenergic control of heart rate of perch (*Perca fluviatilis*) in vitro. *Comp Biochem Physiol* 73C: 399–401.

Tota B and Gattuso A. 1996. Heart ventricle pumps in teleosts and elasmobranchs: A morphodynamic approach. *J Exp Zool* 275: 162–171.

Tsai C-T, Wu C-K, Chiang F-T, Tseng C-D, Lee J-K, Yu C-C, Wang Y-C, Lai L-P, Lin J-L, and Hwang J-J. 2011. In-vitro recording of adult zebrafish heart electrocardiogram—A platform for pharmacological testing. *Clin Chim Acta* 412: 1963–1967.

Tsuchida T and Takei Y. 1998. Effects of homologous atrial natriuretic peptide on drinking and plasma ANG II level in eels. *Am J Physiol* 275: R1605–R1610.

Vornanen M. 1997. Sarcolemmal Ca influx through L-type Ca channels in ventricular myocytes of a teleost fish. *Am J Physiol* 41: R1432–R1440.

Vornanen M. 1998. L-type Ca^{2+} current in fish cardiac myocytes: Effects of thermal acclimation and beta-adrenergic stimulation. *J Exp Biol* 201: 533–547.

Vornanen M. 1999. Na+/Ca2(+) exchange current in ventricular myocytes of fish heart: Contribution to sarcolemmal Ca^{2+} influx. *J Exp Biol* 202: 1763–1775.

Vornanen M. 2006. Temperature- and Ca^{2+}-dependence of [3H]ryanodine binding in the burbot (*Lota lota* L.) heart. *Am J Physiol* 290: R345–R351.

Vornanen M, Hassinen M, and Haverinen J. 2011. Tetrodotoxin sensitivity of the vertebrate cardiac Na+ current. *Mar Drugs* 9: 2409–2422.

Vornanen M and Paajanen V. 2004. Seasonality of dihydropyridine receptor binding in the heart of an anoxia-tolerant vertebrate, the crucian carp (*Carassius carassius* L.). *Am J Physiol* 287: R1263–R1269.

Vornanen M, Ryokkynen A, and Nurmi A. 2002a. Temperature-dependent expression of sarcolemmal K+ currents in rainbow trout atrial and ventricular myocytes. *Am J Physiol* 282: R1191–R1199.

Vornanen M, Shiels HA, and Farrell AP. 2002b. Plasticity of excitation-contraction coupling in fish cardiac myocytes. *Comp Biochem Physiol A* 132: 827–846.

Vornanen M and Tuomennoro J. 1999. Effects of acute anoxia an heart function in crucian carp: Importance of cholinergic and purinergic control. *Am J Physiol* 46: R465–R475.

Wang SQ, Huang YH, Liu KS, and Zhou ZQ. 1997. Dependence of myocardial hypothermia tolerance on sources of activator calcium. *Cryobiology* 35: 193–200.

Wang T and Overgaard J. 2007. The heartbreak of adapting to global warming. *Science* 315(5808): 49–50.

Wood CM, Pieprzak P and Trott JN. 1979. The influence oftemperature and anaemia on the adrenergic and cholinergic mechanisms controlling heart rate in the rainbow trout. *Can J Zool* 57: 2440–2447.

Wong AFM and Barnett P. 2012. Basic cardiac development: The heart and its electrical components. In: Wang T and Sedmera D (Eds.). *Ontogeny and Phylogeny of the Vertebrate Heart*. New York: Springer-Verlag, pp. 177–206.

Yamauchi A. 1980. Fine structure of the fish heart. In: Bourne G (Ed.). *Heart and Heart-like Organs*, Vol. 1. New York: Academic Press, pp. 119–148.

Yue DT. 1987. Intracellular [Ca^{2+}] related to rate of force development in twitch contraction of heart. *Am J Physiol* 252: H760–H770.

3 Membranes and Metabolism

James S. Ballantyne

CONTENTS

3.1   INTRODUCTION

Fishes inhabit most aquatic environments on our planet, living and reproducing in extremes of temperature, salinity, pH, pressure, oxygen levels, and food availability. Adaptation to many of these situations involves enhancements or suppression of existing metabolic pathways coupled with membrane adaptation to maintain membrane properties via "homeoviscous adaptation." The study of metabolism involves not only an understanding of the biochemical pathways consuming or producing high-energy compounds but also comprehension of their compartmentation at the subcellular, cellular, tissue, and organ levels. The transport and communication between these compartments involve membranes with their associated protein and lipid components. Many metabolic processes span more than one subcellular compartment and relatively few studies have focused on how these are related. Studies of plasma turnover of metabolites do provide some insights into the overall metabolism of compounds but there are only a few of these available for fishes.

Membranes are the foci for connecting processes such as ion regulation and metabolism. Some proteins such as membrane-bound ion ATPases (e.g., Na$^+$/K$^+$ ATPase) provide the coupling. Since these processes are often studied in isolation the relationships between metabolism and ion transport are poorly understood in fishes.

Various fish groups represent key points in the metabolic evolution of the vertebrates. There are the jawless fishes (Agnatha) with relatively simple metabolic schemas, the urea-retaining chondrichthyans (sharks, skates, rays, and holocephalans) and coelacanths sharing an osmoconforming physiology, the air-breathing Dipnoi, and the vast side branch of the teleosts with their wide distribution. Aspects of the evolution of metabolism have been described in reviews of some phylogenetic groups (elasmobranchs (Ballantyne, 1997; Speers-Roesch et al., 2006a; Ballantyne and Robinson, 2010; Speers-Roesch and Treberg, 2010), lungfish (Ballantyne and Frick, 2011), sturgeon, and paddlefish (Singer and Ballantyne, 2004)). The current chapter outlines the basic organization of the important metabolic pathways and provides comparisons of the metabolic differences of the major fish groups where they occur.

The compartmentation of metabolism between tissues allows specialized functions to be localized and to evolve to become more efficient. This chapter brings together the available information on the metabolism of key groups of fishes and points out evolutionary trends and areas for further investigation. The goal is to summarize the basic metabolic schemas found in fishes and to rationalize the design of metabolic pathways in fishes in the context of compartmentation and evolution.

3.2   EVOLUTIONARY CONTEXT

Certain events early in the evolution of the vertebrates were critical in forming the raw material for improving metabolism. The evolution of the Metazoa with discrete tissues for specialized functions required specialized metabolism in those tissues. With specialization of tissues, specialized proteins were needed. Several whole-genome duplications allowed additional copies of genes to be

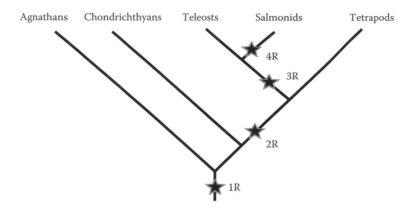

FIGURE 3.1 Diagram illustrating whole-genome duplications (1R, 2R, 3R, and 4R) in the evolution of fishes. Asterisk indicates the approximate evolutionary position of each duplication. (Based in part on Meyer, A. and Van de Peer, Y., *BioEssays*, 27, 937, 2005; Allendorf, F.W. and Thorgaard, G.H., Tetraploidy and the evolution of salmonid fishes, in: Turner, B.J. (ed.), *Evolutionary Genetics of Fishes*, Plenum Press, New York, pp. 1–53, 1984; Steinke, D. et al., *BMC Biology*, 4, 16, 2006.)

available for new compartmentalized functions (Figure 3.1). This allowed for better (faster) adaptation to new situations because entire pathways can be enhanced (van Hoek and Hogeweg, 2009). In support of this concept, model studies of yeast suggest that transporters and glycolytic genes are retained preferentially after whole-genome duplications (van Hoek and Hogeweg, 2009). The proliferation of tissue-specific isoforms of proteins happened extensively in the early evolution of the chordates (Iwabe et al., 1996) and likely was a major contributor to their success.

Tissue-specific isoform evolution has focused on three main tissues: liver, muscle, and neural tissue, while compartment-specific isoform evolution largely involves mitochondrial and cytosolic isoforms. There are at least three rationales for developing tissue- or compartment-specific isoforms. These are (1) protein net charge, (2) kinetic considerations, and (3) compartmental targeting. Neural tissue has a greater need for maintaining membrane potentials than other tissues. Isoforms specific to neural tissue have a high net negative charge (Fisher et al., 1980) to help maintain a high resting membrane potential (Merritt and Quattro, 2001). This would be a benefit in improving the speed of nerve conduction and muscle contraction. Evidence for the glycolytic enzyme triosephosphate isomerase (TPI) suggests that, after an initial gene duplication, selection was for a high overall net charge rather than specific amino acid substitutions in the neural isoform (Merritt and Quattro, 2001). High net negative charge has also been found in neural isoforms of aldolase in fishes (Merritt and Quattro, 2002).

The second rationale for tissue-specific protein isoforms involves enzyme kinetics and revolves around the concept that the K_m is similar to the in vivo concentration of its metabolite (Cleland, 1967). This allows optimal utilization of the enzyme catalytic capacity. Thus, as the concentrations of metabolites differ in different tissues, selection will act to adjust the kinetic properties of the tissue-specific isoforms to optimize tissue-specific enzyme function. This has been suggested to be the basis for the kinetic differences in liver and muscle isoforms of lactate dehydrogenase (LDH) and glyceraldehyde-3-phosphate dehydrogenase (G3PDH) (Senkbeil and White, 1978). Tissues such as white muscle with high anaerobic capacities experience increases in concentrations of glycolytic intermediates with contraction and thus require higher K_m values for those intermediates than would be seen in liver (Senkbeil and White, 1978).

The third rationale for isoform diversity involves the requirement that certain proteins must arrive in the correct compartment. While the need for subcellular compartment-specific isoforms could be driven by similar kinetic factors to those for tissue-specific isoforms, additionally most proteins in subcellular compartments are not made in that compartment and therefore must be imported. Getting the protein across the membrane requires special features. For example, mitochondrial proteins have to be imported into the mitochondria from cytosolic sites of synthesis. To facilitate import, there is a higher positive charge on mitochondrial protein isoforms to facilitate their transport into the negatively charged matrix side (Hartmann et al., 1991).

Thus, an overall trend toward improved tissue metabolism would involve tissue- and compartment-specific isoforms. The result would be faster nerve conduction, muscle contraction, and processing of metabolites. Overall, through the history of the vertebrates, there has been an increase in the number of tissue-specific isoforms (Fisher et al., 1980; Iwabe et al., 1996). For example, enolase, a glycolytic enzyme has three isoforms in higher vertebrates. The α isoform is found in nonneural tissue, the β form is found in muscle, and the γ form is found in nervous and neuroendocrine tissues (Tracy and Hedges, 2000). The α and β forms are the only forms found in chondrichthyans, dipnoans, and actinopterygians, while the γ form is found in mammals and birds (Tracy and Hedges, 2000). A subsequent duplication produced another α form in salmonids (Tracy and Hedges, 2000). Other duplications of individual genes have occurred in the evolution of the fishes and would have given rise to other protein isoforms with similar rationales for their kinetic design and targeting.

3.3   DIET AND DIGESTION

Metabolically, the diet of fishes must supply the energy requirements as well as the need for essential nutrients (amino acids, fatty acids, and vitamins). The composition of the diet and the way it is processed determine the raw materials available for metabolism involved in maintenance, growth, and reproduction. The majority of fish species are carnivorous, with only 5% of fish families having herbivorous representatives (Bone and Moore, 2008). In general, there is a decrease in the abundance of herbivorous fishes with increasing latitude (Meekan and Choat, 1997). In spite of this, the contribution of herbivorous and detritivorous species to the biomass of some ecosystems can be high (up to 50% [Bowen, 1987]). Although there are 30 families of freshwater herbivorous fishes (Okeyo, 1989) and numerous tropical marine species, little is known of their metabolism compared to that of carnivorous species.

Digestion has mechanical as well as biochemical components. The review by Clements and Raubenheimer (2006) provides a detailed analysis of the spectrum of dietary strategies and models for food processing. This chapter focuses primarily on the biochemical components of digestion. These consist of extracellular as well as intracellular contributions and can be of endogenous (from the fish) or exogenous (from gut microorganisms) origin. The importance of gut microorganisms in digestion varies with diet and is most important for food types for which the vertebrate genome lacks the genes for the digestive enzymes needed (e.g., cellulose).

Anatomically, the guts of fishes differ based on evolutionary, dietary, and environmental considerations. The most primitive fishes, the agnathans, lack a stomach (as do some teleosts, e.g., cyprinids), but most other groups have stomachs. In general, carnivorous fishes have shorter intestines than herbivores. In many carnivorous teleost species, diverticula of the intestine immediately posterior to the stomach, termed pyloric caeca, function to increase gut surface area (Clements and Raubenheimer, 2006). Pyloric caeca secrete digestive enzymes and do not function for microbial digestion (Clements and Raubenheimer, 2006). Another unusual anatomical feature of the digestive system, the spiral valve, is found in primitive jawed fishes including the elasmobranchs, coelacanths, bichirs, dipnoans, bowfins, gars, sturgeons, and the primitive teleosts, the osteoglossids. The spiral valve is a compact structure with a spiral internal design that increases the surface area for digestion. It is not known how its efficiency compares to that of the intestines of more advanced fish species.

The digestive system has the capacity to adapt anatomically and biochemically to environmental and dietary stresses (Buddington et al., 1987). For example, starvation and temperature have been shown to impact gut length (Clements and Raubenheimer, 2006) and the length of intestinal microvilli (Avella et al., 1992).

Similar to most animals, fishes display an elevated metabolic rate, termed specific dynamic action (SDA) associated with digestion (Smith et al., 1978; Tandler and Beamish, 1980; Brown and Cameron, 1991). Across fish species, SDA accounted for 7%–56% of the ingested energy with much of the variation due to diet (McCue, 2006). The lowest values (7%) are for carp eating a plant diet, while the highest values are for carnivorous diets (McCue, 2006). In spite of this, the SDA for protein is much lower in fishes than mammals due to the more efficient metabolism of protein (Smith et al., 1978). The protein component of SDA has been shown to be attributable to part of the cost of protein synthesis in fishes (Brown and Cameron, 1991).

The following sections outline the biochemical processes of digestion and assimilation of protein, lipid, and carbohydrate.

3.3.1 PROTEIN DIGESTION AND ASSIMILATION

Protein is an important component of the diet for the growth of all fishes with levels of 35%–45% required for optimal growth of salmonids (Anon, 1981) and 35%–55% for a range of other fish species including herbivorous and carnivorous species (Tacon and Cowey, 1985). This contrasts with the much lower requirements of mammals and birds (12%–25%) (Bowen, 1987), although the protein requirements for maintenance (no growth) of fishes may be similar to that of mammals (Bowen, 1987). The poor capacity of many fish species to utilize dietary carbohydrate may increase the need for dietary protein above that of mammals and birds (Bowen, 1987).

The efficiency of conversion of dietary protein into fish protein is high in fishes compared to that of mammals due to a slower turnover of existing protein (Huisman, 1976). Indeed, fishes are among the most efficient animals for converting feed into protein (Smith et al., 1978). Values of 2- to 20-fold greater than that for mammals and birds have been reported in fishes (Tacon and Cowey, 1985). Several factors contribute to this efficiency, including a lower mass-specific metabolic rate (due to the lack of need for heat production compared to mammals and birds), the lower rate of replacement of existing protein, lower costs of locomotion and reproduction, and lower cost of excreting nitrogen waste as ammonia rather than as compounds that require ATP for synthesis (e.g., urea and uric acid).

The amount of protein in the diet can influence the efficiency of its utilization. At least one marine herbivore (*Cebidichthys violaceus*) can grow on an entirely plant-based diet (10% protein by dry weight) due to higher (more efficient) protein utilization (Fris and Horn, 1993). Based on this and other studies, net protein utilization decreases with the protein content of the diet. Carnivorous fishes have a lower utilization of protein for growth than omnivorous species and both are lower than that of herbivorous species (Figure 3.2). Sustained swimming has been shown to improve utilization of dietary protein by increasing deposition in red and white muscles of trout (Felip et al., 2012).

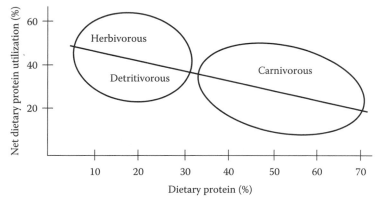

FIGURE 3.2 Relationship between protein content of diet and dietary protein utilization for carnivorous, detritivorous, and herbivorous fishes. (Based on Fris, M.B. and Horn, M.H., *J. Exp. Mar. Biol. Ecol.*, 166, 185, 1993.)

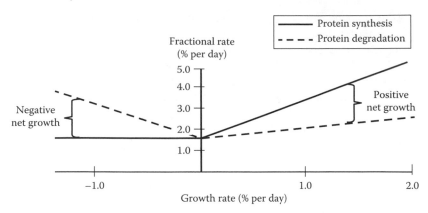

FIGURE 3.3 Relationship between rates of protein synthesis and degradation and growth rate in fed and starving fishes. (Based on Houlihan, D.F. et al., *Aquaculture*, 79, 103, 1989.)

Various studies have documented the rate of protein turnover, which influences the efficiency of deposition of dietary protein and the need for dietary protein. Protein turnover is higher in growing compared to mature fishes (Millward, 1989). The need for amino acids during starvation results in higher tissue protein breakdown such that net protein synthesis is reduced while protein degradation increases (Figure 3.3). In starving cod, *Gadus morhua*, a 60% rate of recycling (protein synthesis/protein breakdown × 100) of nitrogen has been reported (Houlihan et al., 1989). The breakdown of protein in different tissues differs substantially. Muscle protein is most stable, with 76% of muscle protein synthesized being retained compared to only 3%–5% of gill protein (Houlihan et al., 1986). Red muscle and heart also have higher protein turnover rates than white muscle (Houlihan et al., 1986; Millward, 1989).

Environmental parameters such as temperature and salinity influence protein dynamics in fishes. Temperature acclimation results in a compensatory increase in rates of protein synthesis in carp such that rates for cold-acclimated fishes are higher than those of warm-acclimated fishes when measured at the same temperature (Watt et al., 1988). While acclimation to higher temperatures increases the need for proteins, in some species this can be met by increased consumption (Anon, 1981; Tacon and Cowey, 1985). Acclimation to higher salinity increases the protein requirement of some salmonid species but not others (Tacon and Cowey, 1985). The basis for this has not been established.

The components of protein digestion for most fishes are similar to those of other vertebrates. Generally, endopeptidases including trypsin, chymotrypsin, pepsin, elastase, collagenase, and chymosin need to be activated after being secreted in an inactive form into the stomach or intestine. Many of these have acid pH optima. These enzymes cleave the protein into smaller peptides. Exopeptidases (carboxypeptidases A and B, leucine amino peptidase, dipeptidase, and tripeptidase) cleave off terminal amino acids or dipeptides that are then available for absorption. In the intestine, more alkaline conditions are consistent with the pH optima of carboxypeptidases A and B. Herbivorous fishes also have acid stomachs (pH 2.1–2.6) and pepsin activity (Ojeda and Caceres, 1995), as well as alkaline intestines and trypsin, chymotrypsin, and leucine aminopeptidase with alkaline pH optima (Sabapathy and Teo, 1995). The hydrolysis of dipeptides by brush-border enzymes has been reported (Bakke-McKellep et al., 2000).

More primitive fishes such as the hagfish, *Myxine glutinosa*, have trypsin, chymotrypsin, carboxypeptidase A, leucine amino peptidase, and dipeptidase but lack pepsin, carboxypeptidase B, or elastase (Nilsson and Fange, 1970), although this has not been verified through genome analysis. The pH of the guts of these fishes is not as acidic as that of more advanced fishes (pH 5.5–5.8) (Nilsson and Fange, 1970). Lampreys have trypsin and chymotrypsin (Cake et al., 1992), but whether they lack some of the other proteases has not been established. In the absence of a stomach, agnathans use caveolated cells in the anterior intestine for absorption of protein (Langille and Youson, 1985). Beginning with the most primitive jawed fishes, all of the proteases found in higher vertebrates are present. Trypsin and chymotrypsin have been found in chondrichthyans and teleosts (Zendzian and Barnard, 1967) as well as lungfish (Reeck et al., 1970). Elastase has been found in elasmobranchs (Nilsson and Fange, 1970).

The next component of digestion involves moving the resultant amino acids into the gut cells.

3.3.1.1 Amino Acid Transport in the Gut

The capacity for amino acid absorption from the gut compared to that of glucose is higher in carnivorous fishes than in omnivorous species and that of herbivorous fishes is lowest (Buddington et al., 1987). Amino acids and peptides are absorbed across the apical (brush-border) membrane into intestinal cells where some hydrolysis of the peptides may occur. The amino acids are then transported across the basolateral membrane into the blood. A variety of energy-dependent transporters using the sodium gradient are used with broad specificity for the various classes of amino acids. Some passive diffusion and paracellular transport may also occur (Clements and Raubenheimer, 2006).

Studies of the amino transport in the guts of fishes are limited to a few species. Even fewer studies have characterized specific carriers. In amino acid transport studies, one of the issues that must be considered is that transport of one amino acid is affected by the presence of other amino acids either in competition or via allosteric mechanisms. To understand the net transport occurring in tissues such as the intestine, mixtures of amino acids resembling the diet should be considered. The use of intact gut sacs for such studies has the disadvantage of the confounding effects of unstirred layers distorting kinetic parameters. In spite of this, they provide some insight into the absorption of mixtures of amino acids more closely resembling the in vivo function of the gut. In one such study using gut sacs from Pacific bluefin tuna, *Thunnus orientalis*, alanine and glycine display the highest rates of amino acid absorption (Martinez-Montano et al., 2010). In the same study, nonessential amino acids were absorbed at twice the rate as essential amino acids. This contrasts with another study of rainbow trout, totoaba, *Totoaba macdonaldi*, and Pacific tuna in which essential amino acids were absorbed faster than nonessentials (Rosas et al., 2008). Clearly more studies are needed to fully understand these processes.

Amino acid uptake in the gut brush-border membrane is by sodium-dependent secondary active transport with the same carrier capable of transporting several types of amino acids (Balocco et al., 1993). Sodium-dependent amino acid transporters in the brush-border membrane of intestinal regions have been reported with stoichiometries of one or two sodiums per amino acid (Ahearn and Storelli, 1994). At least four separate Na-dependent carriers have been identified in eel intestine (Storelli et al., 1989). Sodium-independent mechanisms have been reported for alanine, glycine, and lysine as well as diffusional transport of all amino acids in eel brush border (Storelli et al., 1989). The K_m for transport of some amino acids are lower than those reported for mammals (Storelli et al., 1989), but methionine displays a higher K_m compared to mammals (Berge et al., 2004). As would be expected, amino acid absorption rates for tuna are higher than those for trout or totoaba (Rosas et al., 2008), perhaps due to a greater density of carriers. Starvation reduced the capacity of some essential amino acid intestinal transporters but not nonessential amino acid carriers (Avella et al., 1992). Nonsaturable uptake of some amino acids (e.g., proline) has been reported in more distal regions of the intestine of some fishes (Bakke-McKellep et al., 2000). There are large differences in the apparent K_m for transport of amino acids depending on the region of the intestine examined such that higher affinities occur in more distal regions of the intestine, although exceptions occur (Bakke-McKellep et al., 2000). Higher affinities in regions where the gut concentrations are low would enhance the efficiency of uptake. Changes in amino acid transport have been observed for fishes adapting to low temperatures and are attributed to changing membrane phospholipid fatty acid content (Smith and Kemp, 1971).

Much work needs to be done to establish the range of carriers, their density in membranes, their kinetic characteristics, as well as how they fit into the simultaneous absorption of mixtures of amino acids.

3.3.2 Carbohydrate Digestion

Digestion and digestibility of carbohydrate are of considerable interest in the aquaculture industry where cheap sources of acceptable nutrients (especially carbohydrate) are always an issue. With the help of endogenous enzymes and gut microorganisms, herbivorous fishes can use a variety of plant carbohydrates (Stone, 2003). A major issue in understanding carbohydrate digestion is the relative importance of gut microorganisms versus the capacity of the fish to produce the appropriate digestive enzymes itself. Carbohydrate digestion requires different enzymes depending on the carbohydrate source.

Starch and glycogen digestion requires enzymes capable of hydrolyzing the α-1–4 glucan bonds (amylases). Fishes lack the salivary amylase found in mammals, but intestinal α-amylase is produced in the exocrine pancreas of fishes (Krogdahl et al., 2005). Lampreys have amylase in the diverticula of the gut (Cake et al., 1992). The number of isoforms of this enzyme has not been established. Teleosts may have more than one α-amylase (Krogdahl et al., 2005). Activities of the enzyme are higher in herbivorous and omnivorous species compared to carnivorous species (Krogdahl et al., 2005) and change with carbohydrate content of the diet (Cowey and Walton, 1989).

The products of the amylase reaction are shorter glucose chains including disaccharides such as maltose, lactose, sucrose, and trehalose, which must be hydrolyzed by disaccharidases to monomers such as glucose, fructose, and galactose for absorption. Disaccharidases (maltase, lactase, trehalase, and sucrase) have been found in brush-border membranes of several segments of the intestines of herbivorous, omnivorous, and carnivorous species, but activities are highest in herbivorous species (Krogdahl et al., 2005). The poor ability to use sucrose or lactose correlates with low intestinal sucrase and lactase activities in sturgeon (Hung et al., 1989). Channel catfish cannot use dietary mono- and disaccharides (Wilson and Poe, 1987). There is considerable variation in the type of disaccharidase across species and in different regions of the gut. Higher activities are usually found in more proximal regions compared to distal regions (Krogdahl et al., 2005).

Cellulose is an abundant carbohydrate in many plants and requires cellulase for digestion. The cellulase activity of both marine and freshwater fishes is likely due at least in part to gut microflora (Stickney and Shumway, 1971; Lindsay and Harris, 1980). In freshwater fishes, diet correlates with cellulase activity, with cyprinids feeding on detritus having high activity while salmonids displaying no activity (Prejs and Blaszczyk, 1977). Elasmobranchs lack cellulase activity in the stomach (Stickney and Shumway, 1971).

Wood contains cellulose that can be used by a very small number of species with the requisite enzyme capability. Indeed, this has been found in Amazonian catfish such as *Panaque* that ingest large quantities of wood and can grow on a wood diet relying on a suite of microorganisms for digestion (Nelson et al., 1999). The ability to rely completely on wood digestion has been questioned (German, 2009), but gut microbes capable of N-fixation have been found in *Panaque* that would improve the nitrogen availability (McDonald et al., 2012).

Although little is known of the dynamics of the use of dietary carbohydrate, there is evidence some can be used immediately in muscle, and sustained exercise can reduce the hyperglycemia in high-carbohydrate diets (Felip et al., 2012).

3.3.2.1 Intestinal Glucose Transport

Sugars, especially glucose, must be transported into the gut cells. Maximal rates of transport of glucose are low compared to those of amino acids (Bakke-McKellep et al., 2000).

The transport of glucose into gut cells occurs across the apical (brush-border) membrane of the intestine via a sodium-dependent mechanism (Storelli and Verri, 1993). The sodium-dependent glucose transporter in the intestine has a higher affinity for glucose in carnivorous fishes compared to that of herbivorous and omnivorous species, presumably as an adaptation to low luminal glucose levels (Buddington et al., 1987). Herbivorous fishes such as carp, *Cyprinus carpio*, and omnivorous species such as tilapia, *Sarotherodon mossambicus*, can regulate intestinal glucose uptake, but carnivorous fishes such as *Oncorhynchus mykiss* are unable to regulate glucose absorption in the intestine (Ferraris and Diamond, 1997). Fish intestine displays low- and high-affinity glucose transporters as do mammals with the low-affinity carriers occurring predominantly in the more proximal regions where higher concentrations of the substrate would occur (Ahearn and Storelli, 1994). Transport into the blood across the basolateral membrane is via sodium-independent glucose carriers (GLUT). These are discussed in more detail later.

Stoichiometries of one sodium ion or two sodiums per glucose have been reported for various intestinal brush-border preparations (Ahearn and Storelli, 1994). While the 2:1 stoichiometry should make uptake more favorable at low glucose levels, correlations of stoichiometry with glucose affinity and glucose availability in the gut are not obvious (Ahearn and Storelli, 1994). Differing values for kinetic parameters have been attributed to differences in methodology. K_m values for vesicle-based methods tend to be lower than those determined using intact tissues due to the confounding effects of unstirred layers in the latter (Ahearn and Storelli, 1994). Based on values for vesicles, the K_m values are similar to those of mammals (Ahearn and Storelli, 1994). The kinetics of glucose transport in the intestine of Atlantic salmon have been determined and a K_m of 0.3–1.7 mM reported, varying with the region of intestine (Bakke-McKellep et al., 2000).

Nutrient transport in the intestine can be influenced by salinity. The potential difference of freshwater fish intestine is blood-side positive while marine fishes are blood-side negative (Ferraris and Ahearn, 1984). Luminal sodium levels also vary with salinity and the gradients for glucose uptake may be more favorable in marine teleost fishes that drink seawater, and glucose uptake rates are higher in marine than freshwater fishes (Ferraris and Diamond, 1997). This may not be due only to changing sodium availability. The Na^+ concentrations in intestine of seawater and freshwater rainbow trout were not substantially different (Dabrowski et al., 1986). The membrane lipids of seawater-adapted fishes incorporate more unsaturated fatty acids that would increase membrane fluidity, allowing for nonsodium-dependent mechanisms of glucose uptake (Ferraris and Diamond, 1997). Indeed, increased glucose uptake via nonsaturable mechanisms has been reported in seawater-adapted fishes compared to freshwater fishes (Ferraris and Ahearn, 1983).

3.3.3 Chitin Digestion

In species feeding extensively on crustaceans, chitin comprising 20% of the available energy can be an important energy source (Clements and Raubenheimer, 2006). Chitin is a polymer of β(1–4) linked N-acetyl glucosamines. Its breakdown requires chitinase that breaks the chitin into smaller pieces by attacking the β(1–4) linkages. Another enzyme exo-N-acetyl-β-D-glucosaminidase (NAGase) further degrades the fragments into NAG monomers. Although bacterial involvement in chitin digestion may occur, the enzymes chitinase and NAGase are found in the fish genome (Krogdahl et al., 2005). Activities were low in the guts of hagfish, *M. glutinosa* (Fange et al., 1979). High chitinase and NAGase activities have been found in the stomachs of elasmobranchs and teleosts (Fange et al., 1979). Lysozyme has been implicated in chitin digestion but its importance has not been quantified (Krogdahl et al., 2005). The fate of the NAG resulting from chitin digestion has not been determined.

3.3.4 Lipid Digestion

Dietary lipid is an important energy source in fishes. There are three main forms of lipids (triacylglycerols, wax esters, and alkyldiacylglycerols) that can occur in the diet of fishes (Figure 3.4).

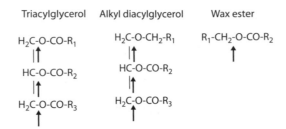

FIGURE 3.4 Structures of major storage lipids of fishes. Arrows indicate sites of lipase action. R_1, R_2, and R_3 indicate variable acyl chains.

While triacylglycerols are a common form of lipid, wax esters can occur at high levels in some prey such as marine copepods and other fishes, and alkyldiacylglycerols may also enter the diet of some fishes.

Lipid digestion in all fishes requires an emulsifier (bile salts or bile alcohols) to solubilize the lipids before the enzyme pancreatic lipase can begin the chemical breakdown of triacylglycerols. Bile salts are a vertebrate invention, and the types used tend to be a function of evolutionary position (Haslewood, 1967; Hagey et al., 2010). The types produced do not respond to changes in diet (Hagey et al., 2010).

Digestive lipases are secreted by the pancreas as in mammals and by the stomach or intestine in some species of fishes (Tocher, 2003). Lampreys have a lipase in the gut diverticula (Cake et al., 1992). The lipase from leopard shark, *Triakis semifasciata*, pancreas has an absolute requirement for bile salt (Patton et al., 1977). Cod also have a bile-activated lipase that can hydrolyze both phospholipids and triacylglycerols (Gjellesvik et al., 1989).

A colipase is needed to protect the lipase function from the bile salts. Lampreys have gall bladders and bile ducts, but these are lost in adults (Barrington, 1972; Cake et al., 1992). Adults apparently can still digest lipid (Youson et al., 1985). Pancreatic colipase has been identified in agnathans, elasmobranchs, and most other fish groups (Sternby et al., 1983). This coincides with the evolution of vertebrate bile salts (Sternby et al., 1983).

Fish lipases have less positional specificity than their mammalian counterparts. They cleave fatty acids from triacylglycerols at all three positions. The result is only fatty acids and glycerol being available for absorption, unlike the situation in mammals where 2-monoglyceride is also a major product (Figure 3.4) (Cowey and Sargent, 1977). The lipases act on wax esters as well (Bauermeister and Sargent, 1979a). More than one lipase is likely involved (Greene and Selivonchick, 1987). Fish digestive lipase can also hydrolyze triacylglycerols containing essential polyunsaturated fatty acids (PUFAs), unlike the mammalian enzyme (Cowey and Sargent, 1977). Wax esters and alkyldiacylglycerols may be hydrolyzed by the same lipase as triacylglycerols (Figure 3.4) (Tocher, 2003). The resulting fatty acids are absorbed and reesterified in intestinal cells (Bauermeister and Sargent, 1979b). The fatty alcohols from wax ester and alkyldiacylglycerol hydrolysis are oxidized to fatty acids before being reesterified to triacylglycerols in the intestine (Lie and Lambertsen, 1991). Wax esters are digested and the fatty alcohol oxidized to the corresponding fatty acid in the intestinal mucosa subsequently being esterified to triacylglycerol (Patton and Benson, 1975). The digestion of triacylglycerols is faster than that of wax esters even in fishes with significant wax ester content in the diet (Patton et al., 1975; Lie and Lambertsen, 1985). Alkyldiacylglycerols are digested more slowly than wax esters in cod, *G. morhua* (Lie and Lambertsen, 1991). Some digestion of wax esters occurs in the pyloric caeca (Bauermeister and Sargent, 1979a; Tocher and Sargent, 1984). The lipases from cod (Brockerhoff, 1966) and an elasmobranch (Brockerhoff and Hoyle, 1965) have a specificity for hydrolysis in the 1 and 3 positions, similar to the enzyme from mammals (Gjellesvik et al., 1989). There has been little recent work in fish lipases.

3.3.5 Dietary Vitamin Requirements

The need for many of the vitamins, identified as essential in higher vertebrates, is similar in fishes, with a few exceptions. Agnatha (Moreau and Dabrowski, 1998), chondrichthyans (Fracalossi et al., 2001; Nam et al., 2002; Maeland and Waagbo, 1998), dipnoans (Fracalossi et al., 2001), and chondrosteans (Dabrowski, 1994; Maeland and Waagbo, 1998) have the capacity for ascorbate synthesis, so do not need a dietary source. Many teleosts lack this capacity due to the loss of the enzyme gulonolactone oxidase (Dabrowski, 1990b). This includes representatives of the Osteoglossiformes (Fracalossi et al., 2001), Characiformes, Salmoniformes (Moreau and Dabrowski, 1996; Maeland and Waagbo, 1998), Clupeiformes (Fracalossi et al., 2001), Siluriformes (Moreau and Dabrowski, 1996), Cypriniformes (Fracalossi et al., 2001), Gymnotiformes (Fracalossi et al., 2001), Perciformes (Maeland and Waagbo, 1998; Fracalossi et al., 2001), Gadiformes (Maeland and Waagbo, 1998), Anguilliformes (Maeland and Waagbo, 1998), and Pleuronectiformes (Maeland and Waagbo, 1998). The loss of this capacity must be assumed to relate to the abundance of this vitamin in the diet. The absorption of ascorbate occurs mostly in the anterior intestine of carp (Dabrowski, 1990a) but is absorbed in the stomach and most of the intestine of rainbow trout (Dabrowski and Kock, 1989).

Other vitamins needed in the diet include thiamin, riboflavin, pyridoxine, B12, biotin, nicotinic acid, pantothenic acid, tocopherol (vitamin E), choline, and vitamin A (Anon, 1983). Folic acid can by synthesized in some species such as channel catfish, *Ictalurus punctatus* and *C. carpio* (Anon, 1983). Inositol can be synthesized in some species such as *I. punctatus* (Anon, 1983). Vitamin D3 can be synthesized in some species such as tilapia (Rao and Raghuramulu, 1996). Vitamin K can be synthesized by *C. carpio* (Anon, 1983).

The enzyme converting 25-hydroxycholecalciferol into 1,25-dihydroxycholecalciferol, 25-hydroxycholecalciferol-1-hydroxylase has been found in teleost fishes (Henry and Norman, 1975).

3.4 OXIDATIVE METABOLISM

Oxidative metabolism including the tricarboxylic acid (TCA) cycle and the electron transport chain is the mainstay of aerobic ATP production. Size affects aerobic metabolism in fishes in a similar way to that found in other vertebrates. In muscle, smaller fishes have higher aerobic enzyme activities than larger fishes (Somero and Childress, 1980; Moyes and Genge, 2010). The oxidation of energy sources such as glucose, amino acids, or lipids occurs in a tissue-specific manner in many cases with some evidence of an evolutionary trend from more generalized substrate preferences in some tissues to more specialized requirements in more advanced forms. Thus, the importance of various oxidative substrates in fish metabolism differs across taxonomic groups. The metabolism of some specific tissues have been reviewed (heart (Moyes, 1996), muscle (Dickson, 1996), liver (Moon, 2004), and brain (Soengas and Aldegunde, 2002)), and fish mitochondrial oxidative preferences have been described (Ballantyne, 1994). Carbohydrate is the preferred fuel for the hearts of hagfish (Sidell et al., 1984). In teleosts, red muscle and white muscle differ in their substrate preferences, with red muscle having a greater capacity for fatty acid and amino acid oxidation (Moyes et al., 1989). Elasmobranch red muscle has no capacity for lipid oxidation but can oxidize ketone bodies (Zammit and Newsholme, 1979a) and glutamine (Chamberlin and Ballantyne, 1992). In spite of the high levels of glutaminase, there are no in vivo studies demonstrating glutamine oxidation in muscle.

3.4.1 Mitochondrial Metabolism

Aside from differences in substrate preferences, mitochondria from red and white muscle and heart differ in their proton conductance with white muscle mitochondria being leakiest and heart the least leaky (Leary et al., 2003). Uncoupling protein 2 (UCP2) has been identified in carp and zebrafish, *Danio rerio* (Stuart et al., 1999), and uncoupling protein 1 (UCP1) has been identified in carp

(Jastroch et al., 2007). These UCPs do not have a thermogenic function in fishes, and although their function has not been established, they may help minimize superoxide production. A thermogenic role for proton leak in warm-bodied fishes has not been established (Ballantyne et al., 1992; Duong et al., 2006).

An important function of mitochondria is to facilitate balancing cytosolic redox. There are two main shuttles used to regenerate the NADH needed at the G3PDH reaction in glycolysis. Carp red and white muscle mitochondria balance cytosolic redox primarily with the malate/aspartate shuttle (Moyes et al., 1989), while mitochondria of more active species rely more on the α-glycerophosphate (α-GP) shuttle (Guppy and Hochachka, 1979). Hearts of air-breathing and water-breathing fresh-water fishes utilize the malate–aspartate shuttle, while the α-GP shuttle is only demonstrable in the heart of water-breathing species (Hochachka et al., 1979). The apparently faster rate of the α-glycerophosphate shuttle may be due to its simpler mechanism employing a membrane-bound mitochondrial α-glycerophosphate dehydrogenase as opposed to the complex malate–aspartate shuttle requiring two transporters, two isoforms of malate dehydrogenase (MDH), and two isoforms of aspartate aminotransferase.

The affinity of mitochondria for oxygen is high in some species that are tolerant of low oxygen levels. In goldfish red muscle, the K_m for oxygen is 0.09 μM (Bouwer and Van den Thillart, 1984). This is low compared to 0.5 μM in neuroblastoma and other isolated cells (Wilson et al., 1979; Scandurra and Gnaiger, 2010), 1 μM in isolated rat liver cells (Longmuir, 1957), and 0.5 μM in pigeon heart mitochondria (Oshino et al., 1974; Sugano et al., 1974).

3.5 CARBOHYDRATE METABOLISM

3.5.1 Aerobic Glycolysis

Glycolysis provides for more rapid ATP production than any other pathway. This makes it the main energy pathway for tissues capable of high activity such as muscle and nerves. This more than any other reason is why carbohydrate stores must be maintained even when dietary intake of carbohydrate ceases. By contrast with the negative scaling of aerobic enzymes in fish muscle, the glycolytic enzymes scale positively with body size (Somero and Childress, 1980). Thus, larger fishes have higher glycolytic enzyme activities per gram compared to smaller fishes. This relationship is influenced by a variety of factors including developmental, phylogenetic, diet, and environmental factors (Childress and Somero, 1990; Moyes and Genge, 2010).

Fishes are poor regulators of plasma glucose levels. It has been suggested that this in part may be due to the less than strict requirement of fish brains for glucose. There is a general trend toward higher glucose levels in more advanced fishes, with hagfish having the lowest levels and the ray finned fishes having the highest (Polakof et al., 2012) (Table 3.1). Plasma glucose levels are similar for carnivorous, herbivorous, and omnivorous fishes (Polakof et al., 2012) (Table 3.1). Herbivores and omnivores are more adept at regulating blood glucose than carnivores but plasma levels can rise many fold in all groups (Stone, 2003). Studies of the hepatectomized agnathan, the Pacific hagfish, *Eptatretus stouti*, indicate that the liver is not an important contributor to glucose homeostasis by contrast with higher vertebrates (Inui and Gorbman, 1978). The hormonal regulation of glucose levels has been reviewed (Polakof et al., 2012).

There is a wide range of responses of blood glucose to orally introduced glucose with some species displaying a sixfold increase (Polakof et al., 2012). It has been suggested that poor regulation of plasma glucose is due to the lack of an inducible glucokinase (Walton and Cowey, 1982). In mammals, when blood glucose levels rise after a high carbohydrate meal, glucokinase increases, reducing plasma levels. Subsequently, inducible glucokinase has been found in several fish species, including rainbow trout (see (Polakof et al., 2012) for review).

Glucose turnover in the plasma has been measured for several species, all of which are teleosts (Table 3.2). Starvation has little effect on glucose turnover in the short term (Bever et al., 1981) but

TABLE 3.1
Plasma Glucose and Nonesterified Fatty Acid Concentrations in Fish Species

Species	Glucose Concentration (mM)	Nonesterified Fatty Acids (μM)
Agnatha		
Lampetra fluviatilis	6–12 (Emelyanova et al., 2004)	
Petromyzon marinus (FW)	2.37 ± 0.18 (Foster et al., 1993)	1137 ± 167 (LeBlanc et al., 1995)
P. marinus (SW)	2.92 ± 0.47 (Foster et al., 1993)	
Myxine glutinosa	1.89 ± 0.12 (Emdin, 1982)	550 (Ballantyne, unpublished)
	1.5 (Larsson et al., 1976)	
Eptatretus stouti	0.92 (Plisetskaya et al., 1983)	
Chondrichthyans		
Chimaera monstrosa	1.3 (Larsson et al., 1976)	
Hydrolagus colliei		652 (Speers-Roesch et al., 2006b)
Squalus acanthias	3.5 (fed) (Oppelt et al., 1963)	415.5 ± 74.1 (Speers-Roesch et al., 2008)
	5.06 (fasted) (Oppelt et al., 1963)	399 + 20 (Ballantyne et al., 1993)
	2.8 (deRoos and deRoos, 1973)	
Etmopterus spinax	1.7 (Larsson et al., 1976)	
Chiloscyllium punctatum		171.6 ± 32.0 (Speers-Roesch et al., 2008)
Prionace glauca		193 ± 43 (Ballantyne et al., 1993)
Isurus oxyrinchus		203 ± 18 (Ballantyne et al., 1993)
Raja eglanteria	13.8 (fed) (Oppelt et al., 1963)	
Dasyatis violacea		287 (Ballantyne et al., 1993)
Leucoraja erinacea	2.8 (fed) (Grant et al., 1969)	572.6 ± 66.8 (Speers-Roesch et al., 2008)
	3.6 (fasted) (Grant et al., 1969)	
Raja radiata	2.2 (Grant et al., 1969)	
	1.1 (Larsson et al., 1976)	
Raja rhina		167.0 ± 20.4 (Speers-Roesch et al., 2008)
Raja ocellata	2.4 (Grant et al., 1969)	
Raja laevis	2.7 (Grant et al., 1969)	
Taeniura lymma		215.7 ± 27.6 (Speers-Roesch et al., 2008)
Himantura signifer (FW)		122.8 ± 40.9 (Speers-Roesch et al., 2008)
Potamotrygon motoro (FW)		105.7 ± 20.2 (Speers-Roesch et al., 2008)
Dipnoans		
Protopterus dolloi	1.2 (Frick et al., 2008a)	
Chondrosteans		
Acipenser transmontanus	5.2 (estimated from Figure in) (Hung, 1991)	
Acipenser fulvescens	4 (Gillis and Ballantyne, 1996)	2355 ± 396 (McKinley et al., 1993)
		2493 + 174 (Singer et al., 1990)
Acipenser brevirostrum (FW)	4.8 ± 0.6 (Jarvis and Ballantyne, 2003)	1061 ± 120 (Jarvis and Ballantyne, 2003)
Acipenser brevirostrum (20 ppt)	6.3 ± 0.6 (Jarvis and Ballantyne, 2003)	1225 ± 191 (Jarvis and Ballantyne, 2003)
Lepisosteus platyrhincus	9.0 ± 0.9 (Frick et al., 2007)	607 ± 28 (Frick et al., 2007)
Amia calva		758 (Singer and Ballantyne, 1991)
Teleosts		
Oncorhynchus mykiss	8.3 (Pottinger et al., 2003)	325.9 ± 36.4 (Pottinger et al., 2003)
O. mykiss	4.11 (Pottinger et al., 2003)	301.3 ± 13.9 (Pottinger et al., 2003)
Winter starved (120 days)		

(*continued*)

TABLE 3.1 (continued)
Plasma Glucose and Nonesterified Fatty Acid Concentrations in Fish Species

Species	Glucose Concentration (mM)	Nonesterified Fatty Acids (µM)
O. mykiss		1150 ± 180 (Dobson and Hochachka, 1987)
	0.9 (John et al., 1980)	253 ± 49 (John et al., 1980)
	2.93 ± 0.15 (van Raaij et al., 1996)	250 ± 20 (van Raaij et al., 1996)
		770 ± 14 (van Raaij, 1994)
		1523 ± 177 (Harrington et al., 1991)
Oncorhynchus nerka	4.9 (fed) (McKeown et al., 1975)	1298 ± 102 (fed) (McKeown et al., 1975)
	3.9 (starved 30 days) (McKeown et al., 1975)	1063 ± 31 (starved 30 days) (McKeown et al., 1975)
Oncorhynchus kisutch	3.29 ± 0.08 (Vijayan et al., 1993)	
Salvelinus fontinalis	4.2 ± 0.2 (Vijayan et al., 1990)	1000 + 50 (Vijayan et al., 1990)
Salvelinus alpinus	4.95 ± 0.27 (Bystriansky and Ballantyne, 2006)	2253 ± 280 (Bystriansky and Ballantyne, 2006)
		1214 ± 116 (Bystriansky et al., 2007)
		1601 ± 52 (Barton et al., 1995)
Lophius piscatorius	0.06 (Larsson et al., 1976)	
Sparus auratus	4.83 ± 0.12 (Sangiao-Alvarellos et al., 2003)	
Carassius auratus		950 ± 30 (van Raaij, 1994)
Cyprinus carpio	3.8 ± 0.16 (van Raaij et al., 1995)	770 ± 20 (van Raaij, 1994)
	4.62 ± 0.29 (van Raaij et al., 1996)	510 ± 30 (van Raaij et al., 1995)
		450 ± 30 (van Raaij et al., 1996)
Oreochromis mossambicus		300 ± 50 (van Raaij, 1994)
Gadus morhua	2.8 (Larsson et al., 1976)	
	8 (Claireaux and Dutil, 1992)	
Urophycis chuss		2374 ± 286 (Ballantyne et al., 1993)
Merluccius bilinearis		2030 (Ballantyne et al., 1993)
Scomber scombrus	2.4 (Larsson et al., 1976)	
Coryphaena hippurus (SW)	5.1 ± 0.3 (Morgan et al., 1996)	
Xiphias gladius		1284 ± 220 (Ballantyne et al., 1993)
Pleuronectes platessa	1.9 (Larsson et al., 1976)	
Dicentrarchus labrax	0.58 (fed) (Echevarria et al., 1997)	660 ± 50 (fed) (Echevarria et al., 1997)
	0.37 (starved 100 days) (Echevarria et al., 1997)	380 ± 20 (starved 100 days) (Echevarria et al., 1997)
Esox lucius		257 ± 21 (Ince and Thorpe, 1975)
Clupea harengus	3.7 (Larsson et al., 1976)	
Anguilla anguilla	2.36 (fed) (Larsson and Lewander, 1973)	384 ± 14 (fed) (Larsson and Lewander, 1973)
	1.8 (starved 145 days) (Larsson and Lewander, 1973)	545 ± 34 (starved 145 days) (Larsson and Lewander, 1973)

long-term studies of eels, *Anguilla rostrata*, show increases in turnover rate after 6 months of food deprivation and a large decrease in rate compared to controls in very long-term food deprivation (Cornish and Moon, 1985) (Table 3.2). For fed fishes, the values range from 0.34 to 42.2 µmol/min/kg (Table 3.2). The basis for this large range may be due to diet or other factors that remain to be determined. Investigations of glucose turnover in nonteleost groups would provide some evolutionary context for the importance of glucose in fishes.

TABLE 3.2
Turnover of Plasma Metabolites in Fishes

Substrate	Species	Turnover Rate (μmol/min/kg)	Reference
Glucose (fed)	*Paralabrus* spp.	2.2	Bever et al. (1981)
Glucose (starved)		1.9	Bever et al. (1981)
Glucose (fed)	*Anguilla rostrata*	42.2	Cornish and Moon (1985)
Glucose (starved 6 months)		76.7	Cornish and Moon (1985)
Glucose (starved 15 months)		5.0	Cornish and Moon (1985)
Glucose	*Oncorhynchus kisutch*	2.4	Lin et al. (1978)
Glucose	*Katsuwonus pelamis*	15.3	Weber et al. (1986)
Glucose normoxia	*Oncorhynchus mykiss*	10.7	Dunn and Hochachka (1987)
Glucose normoxia	*O. mykiss*	9	Haman and Weber (1996)
Glucose normoxia	*O. mykiss*	16.9	Haman et al. (1997)
Glucose hypoxia		10.6	Dunn et al. (1981)
Glucose unexercised	*Pleuronectes platessa*	5.7	Batty and Wardle (1979)
Glucose exercised		8.2	Batty and Wardle (1979)
Glucose	*Hoplias*	3.95	Machado et al. (1989)
Glucose	*Dicentrarchus labrax*	0.34	Garin et al. (1987)
Alanine	*O. mykiss*	3.32	Robinson et al. (2011)
Glutamine	*O. mykiss*	0.4	Robinson et al. (2011)
Lactate	*O. mykiss*	2.8	Dunn and Hochachka (1987)
Lactate	*A. rostrata*	0.15	Cornish and Moon (1985)
Lactate (rest)	*Ictalurus punctatus*	2.25	Cameron and Cech (1990)
Lactate (exercised)	*I. punctatus*	24.4	Cameron and Cech (1990)
Lactate	*K. pelamis*	112	Weber et al. (1986)
Glycerol	*O. mykiss*	8.1	Bernard et al. (1999)
Palmitate	*O. mykiss*	1.09	Bernard et al. (1999)
Palmitate	*O. mykiss*	0.48	Weber et al. (1986)
NEFA	*O. mykiss*	4.85	Bernard et al. (1999)
Triacylglycerol	*O. mykiss*	24	Magnoni et al. (2008)

Glycogen phosphorylase in fish exists in an unphosphorylated inactive a form and an active phosphorylated b form as in other vertebrates. Teleosts have three tissue-specific isozymes of glycogen phosphorylase with similar tissue distribution as found in amphibians and mammals (Yonezawa and Hori, 1979). Agnathans have only a single gene product (Yonezawa and Hori, 1977), while elasmobranchs have two isozymes, a muscle and liver form that differ in their affinities for glycogen and AMP (Yonezawa and Hori, 1979). Thus, two glycogen phosphorylase gene duplications have occurred in the evolution of fishes. Heart and muscle isoforms of carp have high and low affinities for AMP in the b form as in mammals (Yonezawa and Hori, 1979). Glycogen phosphorylase is activated by inosine monophosphate (IMP) in carp (Schmidt and Wegener, 1990). IMP is produced from ATP as it is removed from the adenylate pool during exhaustive exercise as part of the purine nucleotide cycle (see later text). It thus serves as a signal of low ATP and ADP pools.

Once in the cells, exogenous glucose is phosphorylated to glucose 6-phosphate so that it can enter glycolysis. This is catalyzed by hexokinase. Hexokinase exists in four forms: HKI, HKII, HKIII, and HKIV. HKIV is glucokinase, an isoform with a higher K_m for glucose that acts as a glucosensor. Glucokinase has been detected in fish liver, pancreas, and brain (Polakof et al., 2012). High-carbohydrate diet results in elevated activity and mRNA for glucokinase in liver in rainbow trout (Kirchner et al., 2008).

Glucokinase (GK) is regulated by glucokinase regulatory protein (GKRP) in mammals and a similar protein has been detected in fishes (Polakof et al., 2009). Binding of GKRP to GK increases when fishes are fed a high-carbohydrate diet and GK activity increases (Polakof et al., 2009). GK occurs in significant activity in liver, kidney, white muscle, gill, and brain, but not heart of rainbow trout (Soengas et al., 2006). Glucose can be sensed in parts of the brain of mammals with GK playing a role (Levin et al., 2004), and there is evidence this occurs in fishes.

Elasmobranch and trout PFK are inhibited by citrate unlike the situation in invertebrates where no inhibition occurs (Newsholme et al., 1977). Increases in concentrations of glycolytic intermediates such as G6P, F6P, and FBP occur in exercising carp muscle along with activation of PFK and PK (Driedzic and Hochachka, 1976).

Tuna display very high activities of glycolytic enzymes such as LDH, which occurs at the highest levels reported in any animal (5700 units per gram) (Guppy and Hochachka, 1978). Kinetically, the enzyme is not unusual (Guppy and Hochachka, 1978). The pH optimum for this enzyme is low (6.0–6.5) (Guppy and Hochachka, 1978). An analysis of seven sequential glycolytic enzymes that differ in their thermal sensitivity in tuna white muscle found no evidence for adjustment in enzyme activity along a thermal gradient (Fudge et al., 2001).

3.5.2 ANAEROBIC GLYCOLYSIS

Under hypoxic or anoxic environmental conditions and periods of high rates of exercise, the capacity of mitochondria to produce ATP is reduced or eliminated. Pyruvate dehydrogenase (PDH) controls carbohydrate oxidation and is decreased in activity during hypoxia similar to the situation in mammals (Richards et al., 2008). The decrease in activity is due to the accumulation of products (NADH and acetyl CoA) and an activation of one isoform of PDH kinase (PDK-2) that acts by phosphorylating PDH (Richards et al., 2008). During hypoxia, creatine phosphate, cytosolic $[NAD^+]/[NADH]$, $[ATP]/[ADP]$, glycogen levels, and intracellular pH fall and lactate accumulates (Richards et al., 2008). In most fishes, anaerobic glycolysis results in lactate accumulation. This can only proceed until osmotic or pH problems slow or stop metabolism.

Fishes retain more lactate intracellularly than higher vertebrates (Gleeson, 1996). Tuna accumulate lactate to very high (e.g., 84 mM) levels in white muscle during burst swimming (Guppy and Hochachka, 1979). This lactate can either be oxidized or used for gluconeogenesis. Thus, plasma lactate turnover rates are generally lower than those for glucose under the same conditions and range from 0.15 in eel to 112 µmol/min/kg for tuna, *Katsuwonus pelamis* (Table 3.2).

Lactate transport into and out of fish muscle fibers is carrier mediated (Wang et al., 1997; Laberee and Milligan, 1999), with at least two carriers involved. A monocarboxylate carrier is involved in transport and efflux involves an anion antiporter (Laberee and Milligan, 1999).

Fish muscle retains most of the lactate, and lactate transport activity in white muscle is 1/10 that of amphibians (Wang et al., 1997). Lactate influx is via the monocarboxylate carrier (33%), a lactate/Cl^-/HCO_3^- antiporter (45%), and simple diffusion (22%) (Wang et al., 1997). Efflux rates are 500-fold lower than influx in rainbow trout white muscle and largely due to noncarrier mediated diffusion (Sharpe and Milligan, 2003). Succinate and alanine are accumulated in addition to lactate in red and white muscles of crucian carp, *Carassius carassius*, at oxygen levels below 40 mmHg (Johnston, 1975).

3.5.2.1 Ethanol Production

In yeast, ethanol is produced from glucose with pyruvate being decarboxylated to acetaldehyde, which is then converted to ethanol via alcohol dehydrogenase. In bacteria, acetate is converted to acetaldehyde via acetaldehyde dehydrogenase and then acetaldehyde is converted to ethanol via alcohol dehydrogenase (van Waarde, 1991). Fishes use a different mechanism.

A few fish species deal with anoxia and the resultant acidosis due to lactate accumulation by producing a less acidic end product, ethanol. Two species of the genus *Carassius*, the common

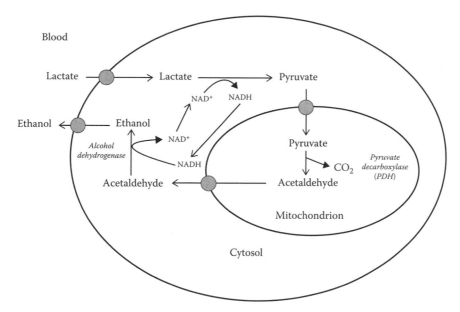

FIGURE 3.5 Pathway for ethanol production from lactate in red muscle.

goldfish, *Carassius auratus*, and the crucian carp, *C. carassius*, produce significant amounts of ethanol under anoxic conditions such as those encountered under ice in winter. In goldfish, ethanol is produced primarily from lactate. The likely pathway involves lactate produced by anoxic brain, liver, and kidney being transported in the blood to the red muscle where the lactate enters the cell. Cytosolic lactate is then converted to pyruvate, which enters the mitochondria on the monocarboxylate carrier. This pyruvate is decarboxylated to acetaldehyde by a pyruvate decarboxylase that may be part of the PDH complex that has undergone partial dissociation (Mourik et al., 1982). This decarboxylation apparently bypasses acetate production (van Waarde, 1991) (Figure 3.5). The acetaldehyde leaves the mitochondria by an as yet unknown mechanism and is converted to ethanol by a cytosolic alcohol dehydrogenase balancing the redox couple with LDH (Mourik, 1982; Mourik et al., 1982). Ethanol is freely diffusible and exits the mitochondria and the cell, equilibrating across tissues and fluid body compartments to be excreted at the gills to the environment.

Acetaldehyde dehydrogenase is not found in red muscle but is found in high activities in other tissues (brain, liver, kidney, heart, gill, and intestine) (Nilsson, 1988). This tissue-specific compartmentation is not found in trout or mammals where aldehyde dehydrogenase and alcohol dehydrogenase colocalize (Nilsson, 1988). There are two isoforms of ethanol dehydrogenase in cyprinids with a liver form involved in dealing with dietary ethanol and a muscle form designed for ethanol production. The liver form has a high affinity for ethanol to optimize flux in the ethanol to acetaldehyde directions, while the muscle form has a low affinity and is designed to use acetaldehyde for ethanol synthesis (Mourik, 1982; van Waarde, 1991).

In goldfish red muscle, ethanol levels can reach 4.6 mM (Shoubridge and Hochachka, 1980). The reduced acidosis (van Waarde et al., 1991) allows muscle activity under anoxic conditions with no accumulation of lactate. To accomplish this, in winter, crucian carp accumulate glycogen in their liver to higher levels than any other species (30% of liver wet weight) (Hyvarinen et al., 1985). Low-temperature acclimation results in increased levels of red muscle alcohol dehydrogenase in goldfish (van Waarde et al., 1991). Levels of ethanol accumulate to 1.48 mM in the bitterling, *Rhodeus amarus*, after 2 h of anoxia, but this ethanol is not produced in muscle (Wissing and Zebe, 1988).

Until recently, ethanol accumulation under anoxic conditions was thought to be confined to a very small number of freshwater species. Recent studies indicate that marine fishes, vertically migrating through the oxygen minimum layer, use ethanol production as strategy to prolong survival under anoxic conditions (Torres et al., 2012).

3.5.3 GLUCOSE TRANSPORT

A wide range of glucose transporters are used to move glucose into or out of cells. Transport into cells is not sodium dependent as in the intestine but is usually via facilitated diffusion on carriers designated GLUT. Thirteen GLUT transporters are expressed in zebrafish with some forms having paralogs (Tseng et al., 2009). Functionally, only a few of these have been characterized (Polakof et al., 2012). GLUT1 is found in the brain (Hall et al., 2005). GLUT2 has been reported in liver of rainbow trout (Panserat et al., 2001) and functions to keep liver cells at equilibrium with plasma glucose levels (Polakof et al., 2012). mRNA for GLUT2 has also been found in high levels in brain, heart, intestine, and kidney, but only low levels in muscle of sea bass, *Dicentrarchus labrax* (Terova et al., 2009).

Several of the GLUT carriers are regulated during periods of hypoxia. In response to hypoxia, GLUT2 mRNA is upregulated in liver of *D. labrax* (Terova et al., 2009) and GLUT1 mRNA is upregulated in gill of cod (Hall et al., 2005). By contrast, GLUT1 and GLUT3 mRNA decreases in spleen of cod under hypoxia (Hall et al., 2005). Insulin increases mRNA for GLUT1 and GLUT4 and translocation of existing GLUT4 to the plasma membrane in salmonids (Polakof et al., 2012).

3.5.4 GLUCONEOGENESIS

Gluconeogenesis is required to maintain, augment, or replenish carbohydrate (glucose and glycogen) reserves. Maintaining glycogen reserves is important following depletion of carbohydrate reserves during periods of starvation, hypoxia/anoxia, or exercise. In order to make glucose or glycogen, some reactions of the glycolytic pathway must be reversed or bypassed. The glycolytic enzymes hexokinase, PFK, and glycogen phosphorylase are bypassed by glucose 6-phosphatase, FBPase, and glycogen synthase, respectively. Pyruvate kinase is bypassed in one of several ways in liver or kidney but may be reversed in muscle (see later text).

Gluconeogenesis involves the conversion of carbon skeletons derived from amino acids, glycerol, volatile fatty acids (VFAs), or lactate into glucose or glycogen. Gluconeogenesis is an energy-intensive process, and since it involves several redox reactions (dehydrogenases), redox balance must be maintained. These requirements constrain the design of the pathways that can be used. Gluconeogenesis can only occur under aerobic conditions in which the mitochondria can supply the ATP needed. Mitochondria are also involved in certain gluconeogenic pathways as discussed later. In fishes, it occurs primarily in liver and kidney (Knox et al., 1980; Mommsen et al., 1985; Suarez and Mommsen, 1987), although it is important in muscle as well.

The bypass of pyruvate kinase involves pyruvate carboxylase that converts pyruvate to oxaloacetate and phosphoenolpyruvate carboxykinase (PEPCK) that converts oxaloacetate to pyruvate. Pyruvate carboxylase is always mitochondrial in mammals and fishes (Suarez and Mommsen, 1987). In fishes, PEPCK can be cytosolic or mitochondrial or both (Hers and Hue, 1983). For example, PEPCK is 30% cytosolic and 70% mitochondrial in *K. pelamis* (Buck et al., 1992). PEPCK is mitochondrial in liver of salmonids (Mommsen et al., 1985), plaice (Moon and Johnston, 1980), and elasmobranchs (Johnson et al., 1990) unlike the situation in mammals and birds (Scrutton and Utter, 1968). Fish brain seems incapable of gluconeogenesis based on the absence of PEPCK and FBPase (Moon and Foster, 1995). An exception is the lamprey with detectable levels of PEPCK and FBPase in the brain (Moon and Foster, 1995).

The various gluconeogenic pathways identified in fishes are discussed in the following sections.

3.5.4.1 Gluconeogenesis from Lactate

Lactate produced in muscle during anaerobic muscle activity or by other tissues in low-oxygen environments can contribute to gluconeogenesis in several ways (Figures 3.6 through 3.8). Two of these

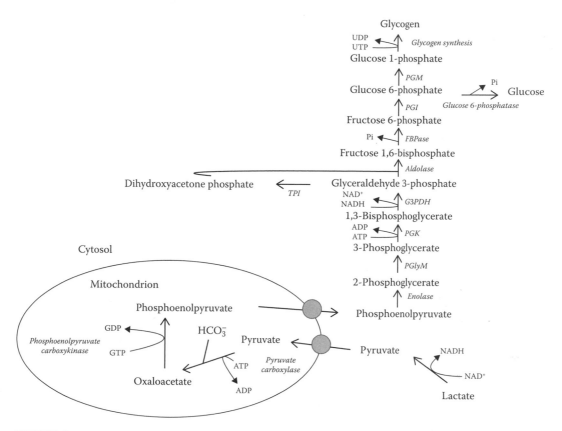

FIGURE 3.6 Diagram of pathways involved in gluconeogenesis from lactate in liver via mitochondrial PEPCK. Some intermediates have been omitted for clarity. *Abbreviations*: PGM, phosphoglucomutase; PGI, phosphoglucose isomerase; FBPase, fructose bisphosphatase; G3PDH, glyceraldehyde-3-phosphate dehydrogenase; PGK, phosphoglycerate kinase; PGlyM, phosphoglycerate mutase; TPI, triosephosphate isomerase.

pathways involve PEPCK (Figures 3.6 and 3.7). In this situation, lactate is converted to pyruvate, which then enters the mitochondria on the monocarboxylate carrier where it is converted to oxaloacetate via pyruvate carboxylase (Figure 3.6). Mitochondrial PEPCK converts the oxaloacetate to phosphoenolpyruvate, which exits the mitochondria also on the monocarboxylate carrier where it enters the gluconeogenic pathway. Cytosolic LDH balances redox at G3PDH. In general, high mitochondrial PEPCK correlates with high rates of gluconeogenesis from lactate, although exceptions exist (Suarez and Mommsen, 1987). If PEPCK is cytosolic, the fate of the oxaloacetate produced as described earlier is different (Figure 3.7). Transamination with glutamate via glutamate oxaloacetate transaminase (GOT) produces aspartate, which exits the mitochondria via the glutamate aspartate exchanger. The aspartate is converted back to oxaloacetate via cytosolic GOT. Cytosolic PEPCK then converts the oxaloacetate to phosphoenolpyruvate, which enters the gluconeogenic pathway. Redox balance is the same as for mitochondrial PEPCK via LDH.

In fish muscle, the lack of PEPCK (Knox et al., 1980) or sufficient liver capacity (Buck et al., 1992) requires a different mechanism for gluconeogenesis from lactate (Figure 3.8). In fishes, white muscle mitochondria are adequate to supply the ATP needed for gluconeogenesis from lactate (Moyes et al., 1992). There is little lactate oxidation in white muscle after exercise, and there is a quantitative conversion of muscle lactate back to glycogen (Moyes et al., 1992). Thus, fishes differ from mammals in the way lactate produced in muscle is used for gluconeogenesis. The Cori cycle, in which lactate produced in muscle is transported in the blood to the liver where gluconeogenesis occurs, is not important in fishes (Milligan, 1996). The mechanism by which muscle glycogen is

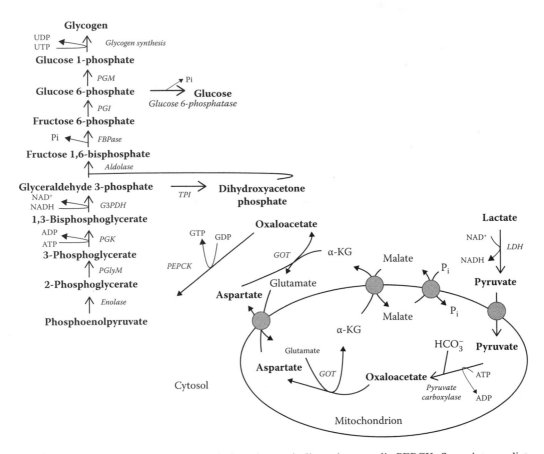

FIGURE 3.7 Diagram of gluconeogenesis from lactate in liver via cytosolic PEPCK. Some intermediates have been omitted for clarity. Metabolites in bold type indicate carbon flow. *Abbreviations*: GOT, glutamate oxaloacetate transaminase; PEPCK, phosphoenolpyruvate carboxykinase; others as for Figure 3.6.

replenished after exercise involves lactate but the pathway has not been unequivocally established even after decades of research. The lack of pyruvate carboxylase and PEPCK in most fish muscles requires lactate carbon be converted to glucose via reversal of pyruvate kinase (Moyes et al., 1992). Exceptions may be marlin, *Makaira nigricans* (Suarez et al., 1986), and the small-spotted catshark, *Scyliorhinus canicula* (Crabtree et al., 1972), where significant muscle PEPCK has been reported, and goldfish, where all gluconeogenic enzymes have been found (Van den Thillart and Smit, 1984). The restoration of muscle glycogen after exercise is faster in small fishes compared to large (Goolish, 1991). Goolish estimated 1–3 h for a small fish (4 cm length) compared to 10 days for a whale shark, *Rhincodon typus*, sized fish (800 cm). Interestingly, PK reversal has since been shown to be a viable pathway for gluconeogenesis in mammalian heart and skeletal muscle (Dobson et al., 2002; Jin et al., 2009). Hagfish accumulate lactate in the plasma after exhaustive exercise, but it is not known what mechanisms are used for gluconeogenesis from lactate (Davison et al., 1990).

3.5.4.2 Gluconeogenesis from Alanine

Amino acids such as alanine may be important gluconeogenic substrates when excess dietary protein is available or during nonfeeding spawning migrations when tissue protein is degraded and the resultant amino acids are needed to maintain glucose availability. In Pacific salmonids, a high rate of gluconeogenesis from amino acids, particularly alanine, occurs in liver (French et al., 1983). This would resemble the glucose–alanine cycle of starving mammals in which muscle alanine is transported to liver and kidney, where it is converted to glucose and returned to the plasma for transport

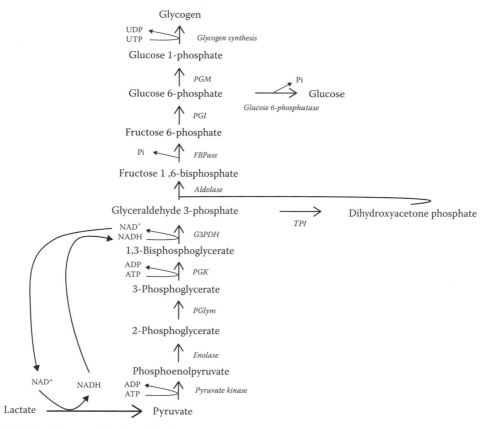

FIGURE 3.8 Diagram of pathway involved in gluconeogenesis from lactate in fish muscle. Some intermediates have been omitted for clarity. Abbreviations as for Figure 3.6.

back to muscle (Ruderman, 1975; Perriello et al., 1995). In this pathway (Figure 3.9), alanine is transaminated to pyruvate via cytosolic alanine aminotransferase (AAT) (Mommsen et al., 1985). The pyruvate enters the mitochondria on the monocarboxylate carrier and is converted to oxaloacetate via pyruvate carboxylase. Mitochondrial MDH converts the oxaloacetate to malate, which exits the mitochondria via the malate–phosphate exchanger. Cytosolic MDH converts the malate back to oxaloacetate, which is then converted to phosphoenolpyruvate via cytosolic PEPCK. The PEP enters the gluconeogenic pathway. Redox balance at G3PDH is provided by the cytosolic MDH.

The subcellular distribution of PEPCK can change in response to changing carbon sources for gluconeogenesis. For example, in plaice, PEPCK switches from primarily mitochondrial in fed fishes to primarily cytosolic in starved fishes (Moon and Johnston, 1980). This allows alanine mobilized from muscle protein to be used for gluconeogenesis.

3.5.4.3 Gluconeogenesis from Serine

Serine is the most important gluconeogenic amino acid in trout liver (French et al., 1981; Hansen and Abraham, 1989) and little skate, *Leucoraja erinacea*, kidney (Mommsen and Moon, 1987). Gluconeogenesis from serine occurs via serine pyruvate transaminase rather than serine dehydratase (Walton and Cowey, 1979, 1981; Mommsen and Moon, 1987) (Figure 3.10). In skate hepatocytes, rates of gluconeogenesis from serine are similar to those with lactate in liver, while in kidney, gluconeogenesis from serine was three times higher than that from glycerol in *A. rostrata* (Renaud and Moon, 1980a). The transporter for 2-phosphoglycerate out of the mitochondria has not been identified, and it is not known how redox at G3PDH is balanced.

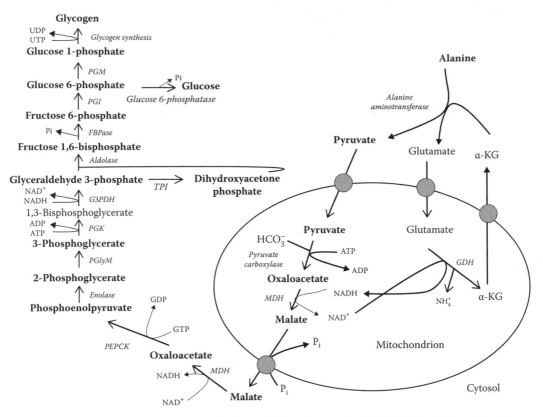

FIGURE 3.9 Diagram of gluconeogenesis from alanine via cytosolic PEPCK. Metabolites in bold type indicate carbon flow. Some intermediates have been omitted for clarity. *Abbreviations*: GDH, glutamate dehydrogenase; MDH, malate dehydrogenase; others as for Figure 3.6.

3.5.4.4 Gluconeogenesis from Glycerol

Glycerol derived from the action of lipases on triacylglycerols can be used as a gluconeogenic substrate (Figure 3.11). Glycerol kinase is required to phosphorylate the glycerol for entry into the gluconeogenic pathway at glyceraldehyde-3-phosphate. This bypasses all dehydrogenases eliminating the need for redox balance. A comparison of the rates of glucose production by isolated hepatocytes of *A. rostrata* indicate that the rates from glycerol are almost twice the rates from lactate or alanine (Renaud and Moon, 1980a) similar to the relative rates in elasmobranch hepatocytes (Mommsen and Moon, 1987). This pathway may be particularly important during spawning migrations when lipids are used as an energy source for red muscle contraction. Gluconeogenesis from glycerol is an important source of glycogen in muscle of spawning lamprey *Lampetra fluviatilis*, and muscle not liver is the site of synthesis (Savina and Wojtczak, 1977). This may not be the case in all fishes. Glycerol kinase has been demonstrated in liver of brook charr, *Salvelinus fontinalis* (Vijayan et al., 1990), but is very low in activity of red or white muscle of fishes (Newsholme and Taylor, 1969).

3.5.4.5 Gluconeogenesis from Volatile Fatty Acids

VFAs produced by gut microorganisms in herbivorous fishes are available for lipogenesis or gluconeogenesis. The enzyme required for activation of the VFA is acetyl CoA synthetase. Acetyl CoA synthetase is found in high activities in gut, kidney, and liver of temperate marine herbivorous fishes

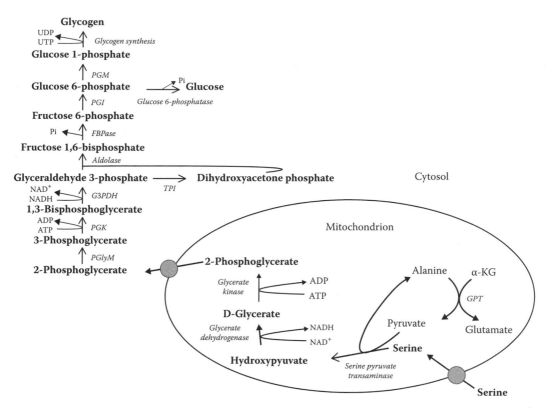

FIGURE 3.10   Diagram of pathways involved in gluconeogenesis from serine. Metabolites in bold type indicate carbon flow. Some intermediates have been omitted for clarity. *Abbreviations*: GPT, glutamate pyruvate transaminase; others as for Figure 3.6. (Based in part on Walton, M.J. and Cowey, C.B., *Comp. Biochem. Physiol.*, 62B, 497, 1979.)

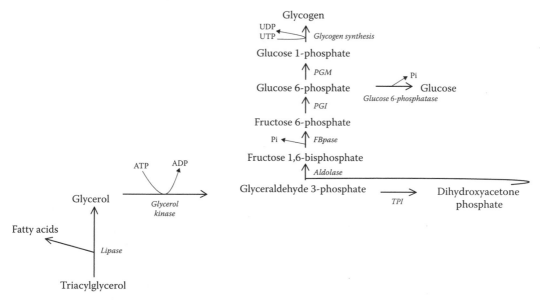

FIGURE 3.11   Diagram of pathway involved in gluconeogenesis from glycerol. Some intermediates have been omitted for clarity.

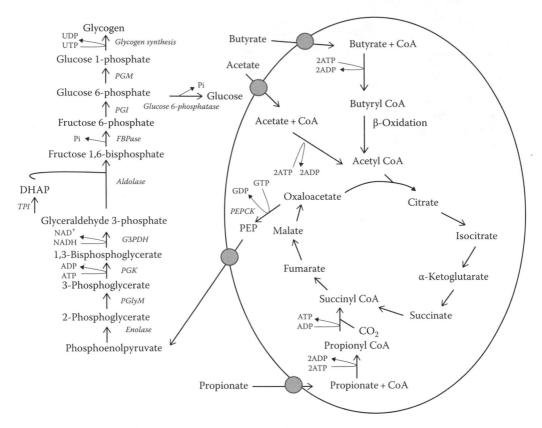

FIGURE 3.12 Possible gluconeogenic pathways for the volatile fatty acids, acetate, butyrate, and propionate, in herbivorous fish liver. Some intermediates have been omitted for clarity. Abbreviations as for Figure 3.6.

(Clements et al., 1994). The relative contribution of VFA to lipogenesis or gluconeogenesis is not known. Figure 3.12 illustrates the possible pathways involved in funneling carbon from VFAs produced by microorganisms in the guts of herbivorous fishes from glucose and glycogen.

3.5.5 GLYCOGEN

Some tissues can function aerobically using exogenous glucose, but for rapid or anoxic ATP production, glycogen is needed. Due to its smaller size, liver contains less than 10% of the glycogen found in muscle (Moyes et al., 1992). In burst exercise, glycogen is the main energy source for muscle and not exogenous glucose (Moyes et al., 1992). Resting glycogen levels vary from tissue to tissue and species to species and with environmental conditions. Table 3.3 indicates the range of levels found. Glycogen levels are usually highest in liver but agnathan liver levels are lower than heart (Table 3.3). Liver glycogen is generally higher in more anoxia-tolerant species such as carp where levels can be 10% of liver weight (Navarro and Gutierrez, 1995). Highest levels of glycogen in the brain for any vertebrate is for crucian carp in winter (3.3% of wet tissue weight) (Vornanen and Paajanen, 2006) (Table 3.3). Very high levels in liver are found in species that estivate. In liver of lungfish, glycogen is 9.6% of liver weight (Janssens, 1964).

3.5.6 GLYCEROL PRODUCTION

Glycerol may need to be produced in fishes consuming large amounts of wax esters since these are usually converted to triacylglycerols and this requires glycerol. The glycerol in these cases is largely

TABLE 3.3
Glycogen Content of Fish Tissues

Species and Tissue	Glycogen Content (μmol Glycosyl Units per g)	Reference
Petromyzon marinus muscle resting	14	Boutilier et al. (1993)
P. marinus liver	14	Boutilier et al. (1993)
P. marinus larva brain	45 ± 7	Rovainen et al. (1969)
P. marinus brain SW	2.61 ± 0.03	Foster et al. (1993)
P. marinus brain FW	0.34 ± 0.09	Foster et al. (1993)
Myxine glutinosa heart	20.45 ± 2.33	Sidell et al. (1984)
Protopterus dolloi liver	155	Frick et al. (2008a)
P. dolloi kidney	10	Frick et al. (2008a)
P. dolloi heart	50	Frick et al. (2008a)
P. dolloi muscle	5	Frick et al. (2008a)
Oncorhynchus mykiss liver	184 ± 36	Parkhouse et al. (1988)
O. mykiss hepatocytes	327	Moon (1998)
O. mykiss muscle	17	Richards et al. (2002)
O. mykiss muscle	7–10	Frolow and Milligan (2004)
O. mykiss white muscle	23 ± 1	Parkhouse et al. (1988)
O. mykiss white muscle	9.9 ± 0.9	Dobson and Hochachka (1987)
O. mykiss red muscle	18.1 ± 2.5	Parkhouse et al. (1988)
Oncorhynchus kisutch liver	166 ± 47	Vijayan et al. (1993)
Salmo salar muscle (cultured)	22.3 ± 2.2	McDonald et al. (1998)
S. salar muscle (wild)	12.0 ± 1.2	McDonald et al. (1998)
Carassius carassius brain (winter)	204	Vornanen and Paajanen (2006)
C. carassius brain (summer)	12.9	Vornanen and Paajanen (2006)
Coryphaena hippurus liver	84 ± 4	Morgan et al. (1996)
Sparus auratus seawater gill	0.14 ± 0.03	Sangiao-Alvarellos and others (2003)
Sparus auratus seawater liver	628 ± 36	Sangiao-Alvarellos and others (2003)
Sparus auratus seawater brain	0.11 ± 0.01	Sangiao-Alvarellos and others (2003)
Perca fluviatilis liver	602 ± 60	Mehner and Wieser (1994)
P. fluviatilis muscle	0.6 ± 0.5	Mehner and Wieser (1994)
Periophthalmodon schlosseri liver	300	Ip et al. (2006)
Periophthalmodon schlosseri muscle	30	Ip et al. (2006)
Boleophthalmus boddarti liver	85	Ip et al. (2006)
Boleophthalmus boddarti muscle	15	Ip et al. (2006)
Ictalurus hepatocytes	546	Moon (1998)
Anguilla rostrata hepatocytes	230	Moon (1998)
A. rostrata hepatocytes	200	Foster and Moon (1989)
Katsuwonus pelamis muscle	145 ± 34	Arthur et al. (1992)
K. pelamis red muscle	35.3 ± 4.1	Guppy and Hochachka (1979)
K. pelamis white muscle	49.2 ± 9.2	Guppy and Hochachka (1979)
Gadus morhua liver	31 ± 5	Claireaux and Dutil (1992)
G. morhua muscle	13 ± 2	Claireaux and Dutil (1992)
G. morhua heart	14 ± 5	Claireaux and Dutil (1992)

derived from carbohydrate sources via glycerol-3-phosphate dehydrogenase (Henderson and Tocher, 1987). Glycerol is also produced in large amounts in some northern fish species such as rainbow smelt, *Osmerus mordax*, and surf perch, *Hypomesus pretiosus*, as a colligative antifreeze molecule (Raymond, 1992). In rainbow smelt, glycerol can accumulate to more than 400 mM (Raymond, 1992)-but the sources of the carbon skeletons are the amino acids alanine and glutamate (Raymond and Driedzic, 1997) via G3PDH and glycerol-3-phosphatase rather than via polyol dehydrogenase (Driedzic et al., 1998). Alanine and glutamate carbon is used for glycerol synthesis in rainbow smelt at cold temperatures (Raymond and Driedzic, 1997).

3.6   NITROGEN METABOLISM

The nitrogen metabolism of fishes has several aspects that distinguish it from that of terrestrial vertebrates. These relate to the intake, metabolic need, and disposal of nitrogen. The chondrichthyans and the coelacanths maintain the osmolarity of all body fluid compartments similar to that of seawater. They do this by accumulating high levels of urea and methylamines. The need for nitrogen to synthesize these compounds has had profound effects on their metabolism (Ballantyne, 1997). Loss of urea from the gill of urea-retaining species represent a significant drain on nitrogen resources. The fact that all chondrichthyans are carnivorous may be due to the need for significant amounts of nitrogen for urea synthesis. Nitrogen metabolism of chondrichthyans has been reviewed (Ballantyne and Robinson, 2010).

More advanced fishes are osmoregulators and do not accumulate osmotically significant amounts of urea except for the estivating dipnoans. This would diminish their need for dietary nitrogen. The need for continuous mucus production may require a substantial part of ingested nitrogen (e.g., 11% (Fauconneau and Tesseraud, 1990)) (Figure 3.13). During reproductive migrations or seasonal food deprivation, amino acids may be mobilized from body protein as energy sources and for gluconeogenesis as described earlier. As aquatic organisms, fishes do not have the same problem as terrestrial vertebrates in disposing of excess nitrogen. The energetic efficiency in excreting

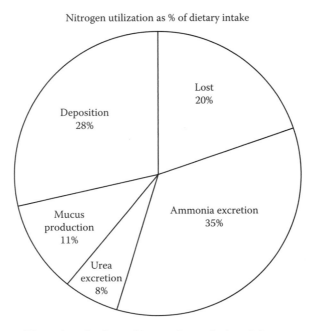

Nitrogen utilization as % of dietary intake

Lost 20%

Deposition 28%

Mucus production 11%

Urea excretion 8%

Ammonia excretion 35%

FIGURE 3.13   Diagram illustrating the fate of ingested protein in rainbow trout. (Based on data from Fauconneau, B. and Tesseraud, S.S., *Fish Physiol. Biochem.*, 8, 29, 1990.)

ammonia rather than the energetically costly urea or uric acid may play a role in the efficiency of fish metabolism. The potential for nitrogen limitation in herbivorous fishes needs further investigation, especially in light of the possibility of N-fixation by gut microorganisms in some species (McDonald et al., 2012).

3.6.1 Amino Acid Metabolism

Amino acids have various roles in fishes and these have been outlined previously (Ballantyne, 2001; Li et al., 2009). A major role is in protein synthesis. The diversion of amino acids toward protein synthesis rather than catabolism has been attributed to lower K_m values for amino acids in the amino acyl tRNA synthetases than for the catabolic amino acid enzymes (Walton and Cowey, 1982; Walton, 1985). The very low K_m for synthesis ensure first access of amino acids for protein synthesis. As amino acid levels rise above the amounts needed for growth and protein turnover, more amino acids are funneled for oxidation and gluconeogenesis via transaminases with relatively high K_m values. This view of amino acid metabolic regulation in fishes has not been challenged since it was first put forward (Walton and Cowey, 1982). Studies of the kinetics of fish aminoacyl tRNA synthetases and related enzymes are needed to validate it.

Generally, plasma levels of amino acids are lower than those of the intracellular compartment. Thus, uptake would require energy-dependent transport systems. Aside from a few studies of amino acid transport for cell volume regulation and transport in the gut, there is little known of amino acid transport into cells. Further work in this area is needed to establish the number, types, and substrate preferences of amino acid transporters in plasma membranes and subcellular membranes of various fish tissues. It is not known what the concentrations of amino acids are in other subcellular compartments such as the mitochondrial matrix. I am unaware of any studies of amino acid transport mechanisms in any fish mitochondria.

Protein breakdown during spawning migrations of Pacific salmonids involves elevated activities of acid and neutral proteases in muscle (Ando et al., 1986). Cathepsin L is a cysteine protease found in lysosomes. It is involved in the degradation of tissue protein and has been characterized in carp liver (Aranishi et al., 1997). Alanine is the main shuttle for nitrogen from muscle amino acids during migration (Mommsen et al., 1980). In situations of high amino acid catabolism such as starvation or a high-protein diet, activities of glutamate dehydrogenase (GDH) and glutamate pyruvate transaminase (GPT) increase in liver of rainbow trout (Sanchez-Muros et al., 1998).

The turnover of most amino acids in plasma is similar to that of mammals when corrected for the differences in metabolic rate (Robinson et al., 2011). Threonine turnover is unusually high perhaps due to its role in mucous synthesis (Figure 3.13) (Robinson et al., 2011). Based on a study of the turnover of 15 amino acids in the plasma of rainbow trout, the rate of disappearance of amino acids from the plasma (Rd) correlates with the molar proportion of amino acids in total body protein (Robinson et al., 2011). Such a correlation has been found in mammals (Lindsay, 1982; Obled and Arnal, 1991).

3.6.1.1 Essential Amino Acids

The proportions of essential amino acids needed in the diet are a function of the total body protein essential amino acid content and are similar across species (Tacon and Cowey, 1985). While the need for essential amino acids is similar for all fish groups (Dabrowski, 1993), essential fatty acid availability may vary widely depending on diet, temperature, and salinity.

The turnovers of essential amino acids in the plasma are similar to those of nonessential amino acids (Robinson et al., 2011). Fishes do not conserve the essential amino acid histidine with dietary restriction (Cowey et al., 1981). Histidine and its derivatives play a role in intracellular buffering in fishes, and levels are higher in tissues of high activity where acidosis may occur associated with lactate accumulation (van Waarde, 1988).

Arginine is considered an "indispensible" amino acid in fishes, although it can be synthesized (Cowey, 1994). Endogenous production of arginine by the kidney may play an important role in augmenting dietary sources of this amino acid (Buentello and Gatlin, 2001).

Taurine is important in cell volume regulation, as an antioxidant and in detoxification. Teleost fishes differ in their ability to synthesize taurine based on the presence or absence of cysteine sulfinate decarboxylase (CSD) in the liver (Yokoyama et al., 2001). Notably, tuna lack the capacity for taurine synthesis. Elasmobranchs also cannot synthesize taurine de novo (King et al., 1980). The need to synthesize taurine must relate to the taurine content of the diet.

3.6.1.2　Glutamine Metabolism

Glutamine is a very important amino acid due to its participation in numerous metabolic pathways (Ballantyne, 2001). The synthesis of glutamine is via glutamine synthetase. In chondrichthyans, a brain isoform is cytosolic and serves in ammonia detoxification. In liver, there is a mitochondrial isoform that is involved in urea synthesis. In teleosts such as the mudskipper, *Bostrichthys sinensis*, the liver has a cytosolic enzyme (Anderson et al., 2002).

Many of the roles of glutamine in fishes are similar to those of other vertebrates. The discovery that glutamine could be oxidized in red muscle and other extrahepatic tissues of fishes (Chamberlin et al., 1991; Chamberlin and Ballantyne, 1992) illustrated a major difference from the situation in mammals. In mammals, glutamine is synthesized in muscle and transported in the blood to tissues such as the intestine where it is oxidized as an energy source. In fishes, red muscle has high levels of phosphate-dependent glutaminase and mitochondria can oxidize glutamine in the absence of other substrates. Typically, mitochondrial oxidations require both two-carbon (e.g., acetyl CoA) and four-carbon (e.g., oxaloacetate) sources to maintain Krebs cycle. Glutamine oxidation in fish red muscle satisfies these requirements by using a mitochondrial form of malic enzyme (Figure 3.14). Glutamine enters the mitochondria via an as yet uncharacterized transporter and is deaminated to glutamate via phosphate-dependent glutaminase. Glutamate is deaminated to α-ketoglutarate via GDH. The α-ketoglutarate enters the TCA cycle and proceeds to form malate. One half of the

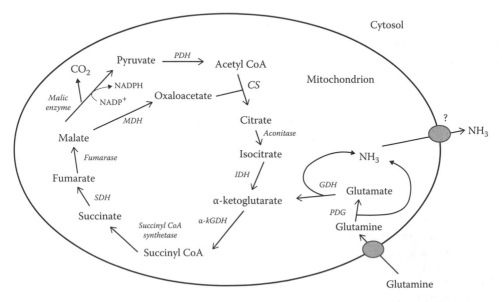

FIGURE 3.14 Pathways for glutamine oxidation in red muscle mitochondria. Some intermediates have been omitted for clarity. *Abbreviations*: PDG, phosphate-dependent glutaminase; CS, citrate synthase; IDH, isocitrate dehydrogenase; α-KGDH, α-ketoglutarate dehydrogenase; SDH, succinate dehydrogenase; MDH, malate dehydrogenase; PDH, pyruvate dehydrogenase; PDG, phosphate-dependent glutaminase; GDH, glutamate dehydrogenase.

malate is converted to oxaloacetate via MDH while the other half is converted to pyruvate via malic enzyme. The pyruvate is converted to acetyl CoA via PDH, which can condense with the oxaloacetate to form citrate. This system is similar to that used in some insect flight muscles that use proline as a sole fuel. The production of two ammonia molecules results from each glutamine oxidized. It is not known how ammonia leaves the mitochondria or the acid–base consequences of its production in the mitochondria. Malic enzyme is usually cytosolic and functions in lipogenesis (see later text). Changes in glutamine catabolism in red muscle during exercise in fishes have not been documented. Although there are no direct measurements of glutamine oxidation during exercise, GDH activity increases in red muscle and white muscle with exercise training in rainbow trout, and U_{crit} is correlated with heart GDH activity in rainbow trout (Farrell et al., 1991). Further studies are needed to establish the relative importance of glutamine in muscle compared to other oxidative substrates. The turnover of glutamine in plasma of trout is lower than that of mammals when adjusted for metabolic rate (Robinson et al., 2011).

Glutamine synthesis has been reported as a strategy for ammonia detoxification in tissues with glutamine synthetase such as brain (Chew et al., 2006). This is discussed in more details in Section 3.6.3.

3.6.1.3 Branched Chain Amino Acid Metabolism

Branched chain amino acid (BCA) metabolism in fishes is different in some respects from that of mammals. In mammals, most branched chain aminotransferase (BCAT) activity is found in muscle, so little BCA is removed from the portal blood by the liver and they can be taken up and used by extrahepatic tissues, especially muscle. Fishes have little white muscle BCAT activity (Cowey, 1993). BCAT has been detected in kidney, liver, and gill of lake charr, *Salvelinus namaycush*, with highest activity in posterior kidney (Hughes et al., 1983). Branched chain α-ketoacid dehydrogenase has been detected in salmonid liver using α-keto-isocaproate (Eisenstein et al., 1990). Activities are higher than those of rat and bird tissues (Hughes et al., 1983). The livers of fishes, therefore, play greater roles in BCA metabolism than in mammals. The basis for this is unknown.

3.6.1.4 Role of Amino Acids in Cell Volume Regulation

Amino acids play a role in cell volume regulation in all fishes (Ballantyne and Chamberlin, 1994). They are quantitatively more important in primitive fishes (agnathans, elasmobranchs, and coelacanths). Amino acids play a role in cell volume regulation of brain cells of the elasmobranch, *L. erinacea*, in which 50% of the free amino acids are taurine (Forster et al., 1978). There is a need to get rid of intracellular amino acids during hypoosmotic stress. Glutamate oxidation is sensitive to osmolarity in isolated mitochondria from elasmobranchs and teleosts with higher rates at low osmolarities in the elasmobranch (Ballantyne and Moon, 1986a).

Taurine plays an important role in cell volume regulation in elasmobranch (Goldstein et al., 1995) and teleost hearts (Vislie and Fugelli, 1975; Vislie, 1980), elasmobranch (Ballatori and Boyer, 1992a,b; Ballatori et al., 1994) and teleost liver cells (Michel et al., 1994), elasmobranch red blood cells (RBCs) (Goldstein and Brill, 1990; Haynes and Goldstein, 1993; Goldstein and Davis, 1994; Thoroed and Fugelli, 1994a; Perlman et al., 1996), and teleost RBCs (Fugelli and Thoroed, 1986; Thoroed and Fugelli, 1994b). Its transport is mediated by the band 3 protein that acts as a volume-activated channel that may transport other osmolytes (Perlman et al., 1996).

3.6.1.5 D-Amino Acids

D-amino acids are found in fish tissues in the 5–100 μM range (Kera et al., 1998). Levels are approximately 200 times lower than levels of the L forms, for example, serine in carp brain (Nagata et al., 1994). These are likely of dietary origin. D-amino acid oxidase occurs mainly in peroxisomes of liver, kidney, and intestine of marine and freshwater fishes (Sarpeitro et al., 2003). It catalyzes the oxidative deamination of a range of D-amino acids with the production of hydrogen peroxide and an imino acid. The enzyme is induced by dietary intake of D-amino acids (Sarpeitro et al., 2003). The enzyme accepts a wide range of D-amino acids.

3.6.2 PURINE NUCLEOTIDE CYCLE

The purine nucleotide cycle (Lowenstein, 1972) has been shown to operate in fish muscle where it functions to stabilize the adenylate energy charge (Mommsen and Hochachka, 1988). During exhaustive exercise, ATP decreases by 65% with a concomitant production of IMP and ammonia produced via AMP deaminase (Driedzic and Hochachka, 1976) (Figure 3.15). In the recovery period, AMP deaminase is not active and adenylosuccinate synthetase and lyase replenish ATP (Figure 3.15). The purine nucleotide cycle is more important in muscle where its capacity for ammonia production is similar to that of GDH (van Waarde, 1981). In trout white muscle, exhaustive exercise results in decreases in ATP concentration and increased levels of IMP and ammonia (Wang et al., 1994). The purine nucleotide cycle is not important in liver of fishes (van Waarde, 1981; Casey et al., 1983). The purine nucleotide cycle may operate in the deamination of aspartate under anoxic conditions in goldfish (van Waarde, 1983).

3.6.3 AMMONIA METABOLISM

Ammonia is produced in several reactions including histidine deaminase, urocanase, AMP deaminase, and most importantly GDH and glutaminase. There is a spike in plasma ammonia concentrations following a meal, indicating substantial amino acid catabolism at that time. Dietary amino acids not used for protein synthesis are deaminated and the ammonia excreted. Some of the enzymes of ammonia production are elevated at high dietary protein levels in some but not all species (Cowey et al., 1981).

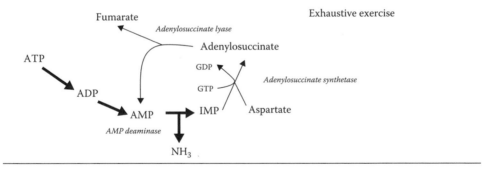

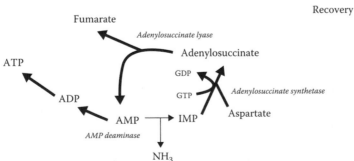

FIGURE 3.15 Diagram of the function of the purine nucleotide cycle in fish muscle. Some intermediates have been omitted for clarity. Bold arrows indicate main flux routes. (Based on Mommsen, T.P. and Hochachka, P.W., *Metabolism*, 37, 552, 1988.)

Two-thirds of ammonia excreted by carp originates in the liver and one-third in the kidney (Pequin and Serfaty, 1963). The carbon skeletons remaining are either oxidized or directed to gluconeogenic or lipogenic pathways. Ammonia is toxic to most fish species and is excreted mostly from the gills. In spite of this, plasma levels of ammonia are maintained within a relatively narrow range (0.1–0.6 mM) that differs little across species from widely divergent osmotic strategies and evolutionary groups (Table 3.4). The tight window of concentrations indicates a regulatory capacity for ammonia metabolism that has not been examined.

In some environmental situations, eliminating ammonia can be problematic. Environmental ammonia levels may be very high, making excretion against the concentration gradient difficult. Alkaline environments can also make excretion of ammonia difficult. A variety of strategies have evolved to deal with these situations.

The swamp eel, *Monopterus albus*, of Southeast Asia may be the fish most tolerant of environmental ammonia (120 mM for 144 h (Ip et al., 2004b)). It has the highest brain glutamine synthetase activities of any fish and detoxifies ammonia mostly in the liver by inducing higher activities of GS (Ip et al., 2004b).

TABLE 3.4
Plasma Levels of Ammonia in Fish Species

Species	Plasma Ammonia Concentration (mM)	Reference
Agnathans		
Petromyzon marinus	0.400	Wilkie et al. (1999)
P. marinus ammocoete	0.375	Wilkie et al. (2006)
P. marinus parasitic	0.700	Wilkie et al. (2006)
P. marinus migrating	0.250	Wilkie et al. (2006)
Chondrichthyans		
Squalus acanthias	0.125	Wood et al. (2010)
S. acanthias	0.081 + 0.013	Wood et al. (1995)
S. acanthias fed	0.113 + 0.028	Kajimura et al. (2008)
S. acanthias starved 56 days	0.007 + 0.003	Kajimura et al. (2008)
Scyliorhinus canicula	0.66 + 0.16	Robertson (1989)
Taeniura lymma	0.210 + 0.040	Ip et al. (2005a)
Potamotrygon sp. FW	0.306 + 0.061	Wood et al. (2002)
Potamotrygon motoro FW	0.250 + 0.04	Ip et al. (2003)
Himantura signifer FW	0.330 + 0.090	Ip et al. (2003)
Himantura signifer FW	0.170 + 0.030	Ip et al. (2005a)
Dipnoans		
Protopterus dolloi	0.163 + 0.015	Chew et al. (2003)
P. dolloi control	0.32 + 0.07	Ip et al. (2005b)
P. dolloi estivated 56 days	0.33 + 0.13	Ip et al. (2005b)
Teleosts		
Scophthalmus maximus	0.25	Rasmussen and Korsgaard (1998)
Dicentrarchus labrax	0.368 + 0.05	Echevarria et al. (1997)
Oncorhynchus mykiss	0.311 + 0.019	Wright et al. (1988)
Salmo salar	0.432	Knoph and Masoval (1996)
Cyprinus carpio	0.378 + 0.030	Ogata and Murai (1988)
Sander vitreus	0.145 + 0.039	Madison et al. (2009)
Takifugu rubripes	0.350	Nawata et al. (2010)

The excretion of ammonia under alkaline conditions is problematic because at pHs above the pK for ammonia the diffusion gradient for ammonia is not favorable. In response to alkaline environmental conditions, one species of tilapia, *Alcolapia grahami*, deals with the problem of nitrogen excretion by synthesizing and excreting urea (Randall et al., 1989). Other species in alkaline lakes excrete most of their nitrogen as ammonia with some urea excretion produced by uricolysis (Danulat and Kemper, 1992; McGeer et al., 1994). Under these conditions, ammonia may be excreted by energy-dependent transporters.

3.6.4 UREA SYNTHESIS

Urea plays different roles in different fish groups. It is an osmotically important compound in chondrichthyans and the coelacanths but plays an important role in ammonia detoxification during development and even in adult teleost fishes. Urea can be synthesized as part of several pathways. The pathways of uric acid breakdown produce urea at several enzyme-catalyzed reactions. As part of the pathway for uricolysis, urea is produced by allantoicase and ureidoglycollase. All the enzymes of uricolysis have been found in agnathans (Wilkie et al., 1999) and lungfishes (Brown et al., 1966). Earlier reports of the absence of uricase, allantoinase, and allantoicase in agnathans (e.g., the hagfish, *Bdellostoma cirrhatum*; Read, 1975) are apparently incorrect. This pathway is peroxisomal, and in most teleost fishes, urea is produced by this pathway (Wood, 1993). Urea is produced from dietary arginine via arginase (Gouillou-Coustans et al., 2002). Arginase produces urea either as part of the urea cycle or independently. Arginase in white muscle has been reported in some elasmobranchs such as the Pacific dogfish, *Squalus suckleyi* (Hunter, 1924), and *S. canicula* but not in rays (Connell, 1955). Thus, some elasmobranchs can synthesize urea in muscle but others cannot. The significance of this remains to be determined.

Marine chondrichthyans and coelacanths are osmoconformers and produce osmotically significant amounts of urea (~40% of osmolarity) via a functional urea cycle. The nitrogen donor is glutamine via carbamoyl phosphate synthetase III (CPSase III) and arginase is mitochondrial. The subcellular compartmentation of the urea cycle in elasmobranchs differs from that of mammals. Although arginase is mitochondrial in most fishes including elasmobranchs, bowfins, and teleosts (Felskie et al., 1998), it becomes cytosolic in the lungfishes, amphibia, reptiles, and mammals (Figure 3.16). Two genes coding for mitochondrial forms of arginase are found in teleosts such as the pufferfish, *Takifugu rubripes*, and four in rainbow trout (Wright et al., 2004; Wright, 2007). In elasmobranchs, urea synthesis via the urea cycle occurs in the liver. The rationale for a mitochondrial arginase is unknown but requires mitochondrial membrane transporters to facilitate urea efflux at rates matching synthetic rates. The carrier in *L. erinacea* liver mitochondria is energy dependent with a K_m of 0.34 mM (Rodela et al., 2008). Urea retention at the gills and kidneys also requires urea carriers. In the gills of elasmobranchs, a urea sodium antiporter in the basolateral membrane helps recover urea before it is lost to the environment (Fines et al., 2001). The gill basolateral membrane also has a high cholesterol content to reduce permeability to urea (Fines et al., 2001). The elasmobranch kidney has at least two urea carriers to recover urea in the filtrate before excretion (Morgan et al., 2003). One of these is sodium dependent with a very low K_m (0.70 mM) (Morgan et al., 2003). In elasmobranchs, synthesis of urea has been accompanied by complimentary membrane retention mechanisms.

Urea is produced in osmotically significant amounts by estivating lungfish. It should be appreciated that urea accumulation in estivation has at least three advantages. The first is an osmotic one in which accumulation of urea reduced the osmotic stress of drying out. The second is the acid–base advantage of consuming HCO_3^- while producing urea. The third is ammonia detoxification under conditions where its excretion is not possible due to lack of water.

The disposal of acid or base is easier in water than on land, so the urea cycle is thought not to be important in acid–base regulation in most fishes (Atkinson, 1992). Urea synthesis has been shown to not be important for acid–base regulation in the gulf toadfish, *Opsanus beta* (Barber

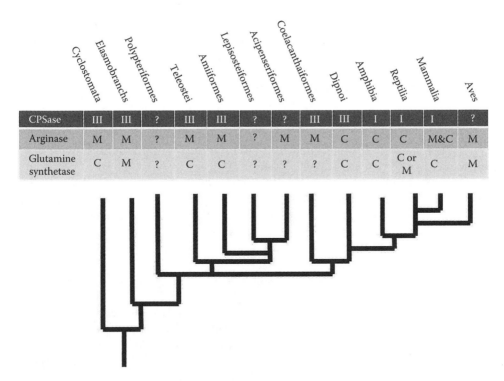

FIGURE 3.16 Evolution of carbamoyl phosphate synthetase (CPS) and the localization of CPSase III, arginase, and GS in various fishes and other vertebrate groups. (Based in part on Mommsen, T.P. and Walsh, P.J., *Science*, 243, 72, 1989; Loong, A.M. et al., *J. Comp. Physiol.*, 182B, 367, 2012; Campbell, J.W. et al., *Am. J. Physiol.*, 246, R805, 1984.)

and Walsh, 1993). Fishes in lakes with high bicarbonate levels that synthesize and excrete urea may obtain acid–base benefit under conditions where they are unable to excrete bicarbonate (Atkinson, 1992).

Hagfish lack a functional urea cycle (Read, 1975) and low CPSase III has been reported for lamprey and bowfin (Wright, 2007), implying little role for the urea cycle in these species. In adult teleost fishes, significant urea synthesis via the urea cycle with CPSase III only occurs in a few species. *A. grahami* from the alkaline lake Magadi uses urea as a mode of excretion of nitrogen due to unfavorable conditions for ammonia excretion (Wood et al., 1989). A marine teleost, the gulf toadfish, *O. beta*, produces urea for excretion of nitrogen (Walsh and Mommsen, 2001) perhaps due to the need to avoid producing ammonia that predators can detect. Some air-breathing teleost fishes such as the goby, *Mugilogobius abei*, have functional urea cycle in muscle, skin, and gill but not in liver as a mechanism for dealing with high levels of environmental ammonia (Iwata et al., 2000). This strategy is rare with most fishes using other mechanisms for detoxifying ammonia (Ip et al., 2004a).

Urea synthesis by the urea cycle plays an important role in teleost embryos (Depeche et al., 1979; Felskie et al., 1998). Urea levels up to 45 mM have been reported in embryos of live-bearing species with lower levels in embryos of egg-bearing species (Depeche et al., 1979). CPSase III is found in fish eggs (Depeche et al., 1979; Wright et al., 1995) but not in adult liver (Felskie et al., 1998). Urea production via a CPS III-containing urea cycle in early developmental stages may be associated with ammonia detoxification in both freshwater (Wright et al., 1995; Terjesen et al., 2001) and marine (Chadwick and Wright, 1999; Terjesen et al., 2000; Barimo et al., 2004) fish eggs. The fact that even pelagic eggs of marine fishes have a functional urea cycle calls into question the role in ammonia detoxification. It has been suggested that fingerling trout have a urea cycle for arginine synthesis (Chiu et al., 1986). This may apply to earlier developmental stages including embryos.

Some of the enzymes of the urea cycle may have other functions and display these in the absence of the other components of the cycle. CPSase III found in fish muscle may produce citrulline (Felskie et al., 1998) that has a role as a nitrogen shuttle from intestine to the kidney for arginine synthesis (Buentello and Gatlin, 2001). One isoform of arginase is cytosolic and may function in glutamate, proline, and polyamine synthesis (Ash, 2004).

3.6.5 METHYLAMINES

Methylamines such as trimethylamine N-oxide (TMAO) occur in high concentrations in marine elasmobranchs as well in deep sea teleosts (Kelly and Yancey, 1999; Yancey et al., 2001) and cold water teleosts (Raymond, 1998; Raymond and DeVries, 1998). Most freshwater fishes lack TMAO: exceptions being anadromous and catadromous species (Dyer, 1951).

Methylamines have several functions in fishes. They play a role in the metabolism of choline and thus impact membrane phospholipids and their metabolism. In chondrichthyans, they serve as counteracting solutes to attenuate the detrimental effects of urea on some proteins (Yancey and Somero, 1979). In deep sea fishes, TMAO counteracts the effects of high pressure (Gillett et al., 1997; Samerotte et al., 2007), while in cold water fishes, it acts as a colligative antifreeze (Raymond and DeVries, 1998).

In cold water Antarctic fishes, TMAO is synthesized via trimethylamine oxidase (TMAase) in liver (Raymond and DeVries, 1998). Some groups of elasmobranchs lack the capacity for TMAO synthesis from TMA or choline (Goldstein et al., 1967; Treberg et al., 2006). Synthesis of TMAO from TMA or choline has been demonstrated in nurse shark liver, *Ginglymostoma cirratum* (Goldstein and Funkhouser, 1972). Salmonids lack the ability to synthesize TMAO and must acquire it in the diet (Benoit and Norris, 1945). Kelp bass can synthesize TMAO from TMA in liver (Charest et al., 1988). Some exclusively freshwater species can synthesize TMAO from TMA (Baker et al., 1963), but this may be due to the action of a nonspecific oxidase.

β-alanine is synthesized in elasmobranch liver (King et al., 1980). Betaine is the main counteracting solute in elephant fish, *Callorhinchus milii* (Bedford et al., 1998). Sarcosine is synthesized in elasmobranch liver but at lower rates than β-alanine (King et al., 1980).

The interconversion of choline and betaine has been demonstrated in kelp bass, *Paralabrax clathratus*, and pink salmon, *Oncorhynchus gorbuscha*, and a low capacity to synthesize TMA and TMAO from either choline or betaine in both (Charest et al., 1988). Choline dehydrogenase and betaine aldehyde dehydrogenase activities have been detected in several tissues of marine elasmobranchs with highest activities in liver and kidney (Treberg and Driedzic, 2007). Both enzymes are mitochondrial (Treberg and Driedzic, 2007).

Under hypoosmotic conditions, methylamine levels decline and it has been suggested this is due to oxidation (Ballantyne, 1997). Sarcosine is oxidized by isolated liver cells and mitochondria of *L. erinacea* and the rate is influenced by osmolarity (Ballantyne et al., 1986). Sarcosine is oxidized to glycine via sarcosine oxidase in elasmobranch liver mitochondria and glycine is further catabolized via the glycine cleavage system (Moyes et al., 1986a). It has been suggested that mitochondria transport of sarcosine and glycine is sensitive to osmotic conditions that may be related to enhanced oxidation under hypoosmotic conditions (Moyes et al., 1986a).

3.7 LIPID AND KETONE BODY METABOLISM

In addition to their role as energy sources, lipids are essential components of membranes and are precursors for signaling molecules. It has been argued that, because many of the interactions between lipids and the proteins that are involved in their metabolism are reliant on hydrophobic interactions, there is a general, lower substrate specificity for fatty acid chains (Sargent et al., 1993). This is important in understanding not only the fatty acid composition of membrane phospholipids but also the array of signaling molecules of the eicosanoid pathways (e.g., the same desaturase can

FIGURE 3.17 Pathways of desaturation and elongation of fatty acids in marine and freshwater fishes. Thicker arrows indicate main pathways. Δ5, Δ6, and Δ9 refer to the respective desaturases.

act on n3, n6, n7, and n9 fatty acids and elongases have a similar wide range of possible substrates; Figure 3.17). Unlike the situation in mammals, highly unsaturated fatty acids (HUFAs) are needed as membrane components, especially at cold temperatures, in addition to their roles in cell signaling. The effects of temperature and solutes on hydrophobic interactions may also impact the transport and metabolism of lipids.

3.7.1 STORAGE LIPIDS

Lipids are stored as triacylglycerols, wax esters, or as alkyldiacylglycerols (Figure 3.4). The sites of storage differ somewhat among fish groups. Agnatha store lipid in liver and muscle (Kott, 1971; Fellows and McLean, 1982). In teleosts (Sheridan, 1994) and sturgeon (Lovern, 1932), lipids are stored in liver, mesenteric sites, and red muscle. Endogenous lipids are used in red muscle of some species since declines in muscle lipid have been found after exercise (e.g., in jack mackerel, *Trachurus symmetricus*; Pritchard et al., 1971). Elasmobranchs store large amounts of lipid in the liver, but since red muscle does not oxidize fatty acids in elasmobranchs little lipid is stored in this tissue (Ballantyne, 1997). While most fishes store lipid in the form of triacylglycerol, some groups like deep sea fishes also use wax esters or alkyldiacylglycerol for buoyancy (Morris and Culkin, 1989). All three of these lipids serve as energy sources. The three main storage forms are illustrated in Figure 3.4. Alkyldiacylglycerols are important in some elasmobranchs (Spener and Mangold, 1971). Wax esters may comprise up to 90% of total lipids in some lantern fishes (Myctophidae) (Nevenzel et al., 1969). In the coelacanths, 90% of fat and muscle lipids are wax esters but only 8% in liver and 15% in spleen (Nevenzel et al., 1966). In addition to their role in buoyancy, wax esters can be a major energy source for marine fishes (Cowey and Sargent, 1977). Mullets, *Mugil japonicus* and *M. cephalus* (Nevenzel, 1970), and gouramis (Sand et al., 1973) put wax esters in their eggs, synthesizing the required fatty alcohols by reducing fatty acids rather than using dietary sources (Sand et al., 1971). Fatty alcohols are incorporated into wax esters in the eggs of some tropical freshwater fishes such as the gourami, *Trichogaster cosby*, using dietary fatty alcohols that are first oxidized to fatty acids then reduced back to fatty alcohols (Sand et al., 1971). The body lipids of this species are largely triacylglycerols. The rationale for this strategy is not known.

3.7.2 LIPOGENESIS

When nonlipid dietary components are available in excess of needs for growth and maintenance, they can be converted to lipids and stored. The carbon source for lipogenesis would be excess dietary amino acids in carnivorous fishes or glucose in herbivorous species. It has been suggested that carnivorous animals including fishes preferentially divert gluconeogenic amino acids to carbohydrate synthesis rather than to lipid synthesis (Aster and Moon, 1981). Lipogenesis increases with high-carbohydrate diets in *I. punctatus* (Likimani and Wilson, 1982), rainbow trout (Brauge et al., 1995), coho salmon, *Oncorhynchus kisutch* (Lin et al., 1977a), and white sturgeon, *Acipenser transmontanus* (based on enzyme activities) (Fynn-Aikins et al., 1992). In white sturgeon, maltose and glucose were lipogenic (Hung et al., 1989). High-fat diets decreased lipogenic enzyme activity in coho salmon (Lin et al., 1977b) and rainbow trout (Jurss et al., 1985). There may be some environmental impacts on lipogenesis. Carp displayed higher rates of lipogenesis from alanine and glucose when cold acclimated (Shikata et al., 1995).

Liver is the main site of lipogenesis with capacities more than 10-fold higher than that of adipose tissue (Sargent et al., 1989; Segner and Bohm, 1994). Adipose tissue if present is more involved in uptake and esterification of fatty acids into triacylglycerols (Sargent et al., 1989). Although dietary lipid is a major source of lipid, some de novo synthesis (lipogenesis) occurs in fishes. Most lipogenesis occurs in liver but adipose tissue and ovaries also have lipogenic capacity (Henderson and Tocher, 1987).

Lipids are synthesized de novo via the same multienzyme complex fatty acid synthase (FAS) as in mammals (Sargent, 1989). This results in saturated fatty acids, which can then be desaturated as needed. Marine fishes consuming a diet high in lipid tend to have less de novo lipogenesis than freshwater fishes (Tocher, 2003).

De novo synthesis of fatty acids is a cytosolic process that requires NADPH that can be supplied by four cytosolic enzymes: G6PDH, 6-phosphoglycerate dehydrogenase (6PGDH), malic enzyme, and isocitrate dehydrogenase (IDH). Not all of these are required. One of the established pathways for lipogenesis is given in Figure 3.18. Carbon from glucose or amino acids enters the mitochondrion via pyruvate on the monocarboxylate carrier and is converted to both acetyl CoA and oxaloacetate via PDH and pyruvate carboxylase, respectively. The acetyl CoA and oxaloacetate combine to form citrate that exits the mitochondria on the tricarboxylate carrier (TCC). This cytosolic citrate is then converted to cytosolic acetyl CoA by the action of citrate lyase. Acetyl CoA carboxylase forms malonyl CoA by carboxylating acetyl CoA. The malonyl CoA is ultimately converted to fatty acids via fatty acid synthase. The need for NADPH by the cytosolic fatty acid synthase is met by cytosolic malic enzyme that takes oxaloacetate produced by citrate lyase and converts it to pyruvate, which reenters the cycle (Figure 3.18). The importance of citrate lyase has been questioned since activities are very low or absent in some species (e.g., rainbow trout) (Segner and Bohm, 1994). The pathway for lipogenesis in the absence of this enzyme has not been determined.

Elasmobranchs use a modification of this pathway (Figure 3.19), whereby cytosolic redox is balanced by an isoform of IDH that uses $NADP^+$ and NADPH (Saxrud et al., 1996). In elasmobranch liver, cytosolic $NADP^+$-requiring malic enzyme is absent (Zhou et al., 1995) and IDH is distributed equally between the cytosol and the mitochondria (Saxrud et al., 1996). Citrate is formed from α-ketoglutarate by a reversal of mitochondrial IDH (Figure 3.19). The resulting α-ketoglutarate reenters the mitochondria via a carrier that has not been established and contributes to citrate formation via a reversal of part of the TCA (IDH and aconitase) (Figure 3.19). This pathway may

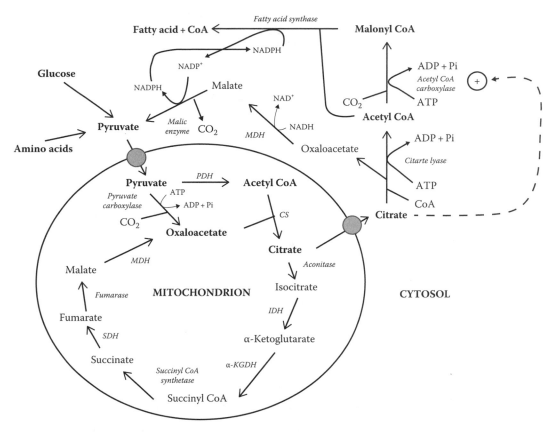

FIGURE 3.18 Diagram of the role of citrate in lipogenesis in teleost fish liver. Some intermediates have been omitted for clarity. Bold letters indicate carbon flow. Abbreviations as for Figure 3.14.

apply to nonelasmobranch species such as carp and goldfish (Saxrud et al., 1996). The absence of mitochondrial NAD-dependent IDH has been reported, but this is apparently due to methodological problems related to the lability of the enzyme. These problems have been resolved (Segner and Bohm, 1994).

In some species (e.g., eel), mitochondrial citrate is transported to the cytosol on the TCC where it activates acetyl CoA carboxylase and is converted to acetyl CoA via citrate lyase (Figure 3.18). The activity of the TCC increases in eels during a lipogenic phase in preparation for migration (Zara et al., 2000). The increase in TCC activity has been attributed to modifications of the lipid environment, in particular cardiolipin (Zara et al., 2000).

As would be expected, activities of malic enzyme, G6PDH, 6PGDH, citrate cleavage enzyme, and fatty acid synthase decreased after 3 weeks of starvation in coho salmon (Lin et al., 1977a). Malic enzyme, 6PGDH, and G6PDH are upregulated by elevated glucose or alanine concentrations similar to the situation in mammals (Sanden et al., 2003).

3.7.3 FATTY ACID DESATURATION AND ELONGATION

Fatty acids obtained in the diet may need to be modified to suit the needs of fishes for membrane modifications or eicosanoid production. This is accomplished by desaturases that introduce double bonds and elongases that add two carbon units to fatty acids. Fatty acids such as 16:0 and 18:0 can be desaturated to 16:1 and 18:1, respectively, by the action of a $\Delta 9$ fatty acid desaturase that requires NADPH and oxygen. This may be an important process in membrane adaptation during cold acclimation due to the lower melting points of 16:1 and 18:1. The absence of $\Delta 12$ and $\Delta 15$ fatty acid

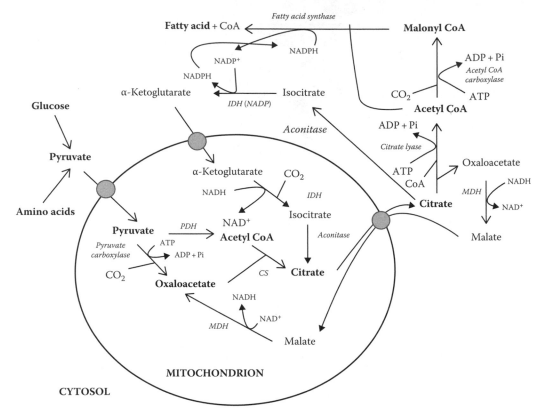

FIGURE 3.19 Diagram of lipogenesis without cytosolic malic enzyme such as may occur in elasmobranch liver. Metabolites in bold type indicate carbon flow. Abbreviations as for Figure 3.14. (Based in part on Saxrud, K.M. et al., *J. Exp. Zool.*, 274, 334, 1996.)

desaturases in animals means fishes cannot synthesize 18:2n6 and 18:3 n3, making these essential fatty acids in fishes. Deficiency of n3 fatty acids in rainbow trout results in higher hepatic respiration rates, reduced circulating hemoglobin, and increased muscle water (Castell et al., 1972). Increased mitochondrial osmotic fragility (faster swelling) was also observed in n3-deficient fishes (Castell et al., 1972). n6 deficiency does not elicit the same responses (Castell et al., 1972).

$\Delta 5$ and $\Delta 6$ desaturase activities occur in freshwater fishes and may be the same gene product based on characterization of a zebrafish gene product (Hastings et al., 2001). Marine fishes have a reduced or deficient $\Delta 5$ desaturase (Owen et al., 1975; Tocher et al., 1989) and thus have a limited capacity to elongate and desaturate 18:2n6 and 18:3n3 (Tocher, 2003), and this includes marine herbivorous species (Mourente and Tocher, 1993) (Figure 3.17).

3.7.4 LIPID TRANSPORT

As hydrophobic molecules, lipids need to be transported in ways other than in solution. This applies to plasma transport as well as transport through other fluid compartments including the cytosol. Fishes have the same plasma lipid transport systems, including nonesterified fatty acids (NEFAs) and various lipoproteins, as higher vertebrates with a few exceptions (Babin and Vernier, 1989). Lipoproteins are transported from the intestine via the lymphatic system as in mammals (Sheridan and Friedlander, 1985). Fishes have all of the same plasma lipoproteins as mammals, although the proportions of these vary (Sheridan, 1988; Babin and Vernier, 1989). The uptake of lipoproteins has not been extensively examined but elasmobranchs have an LDL receptor (Mehta et al., 1996).

Lipids may be transported as triacylglycerols on lipoproteins or as fatty acids after hydrolysis of the storage triacylglycerol. Triacylglycerol lipase is highest in liver and adipose tissue but significant activity occurs in other tissues where lipid is stored (Sidell and Hazel, 2002). The enzyme is activated by a cAMP-dependent phosphorylation (Michelsen et al., 1994). The activity of lipoprotein lipase increased in red muscle with endurance training in teleosts, although levels of specific plasma lipoproteins did not change (Magnoni and Weber, 2007). As suggested by studies of rainbow trout, the turnover of plasma TAG is very high in fishes compared to mammals and does not change with exercise (Magnoni et al., 2008).

The fastest delivery of lipid to tissues usually involves NEFAs. Plasma NEFAs are transported in the plasma bound to albumin in most species. Plasma NEFA concentrations do not change during exercise in sockeye salmon, *Oncorhynchus nerka* (McKeown et al., 1975), but plasma NEFA levels of migrating salmonids display sex-related differences (Ballantyne et al., 1996; Booth et al., 1999). Oleic acid is especially depleted in females, presumably for egg production (Ballantyne et al., 1996). Turnover of plasma NEFAs (palmitate), has only been determined for one species *O. mykiss* with a value of 1.09 µmol/min/kg (Bernard et al., 1999), about 10-fold lower than that of glucose (Table 3.2). Total NEFA turnover was about fivefold higher than that of palmitate alone (Table 3.2). Interestingly, triacylglycerol turnover in the same species is much higher (20-fold) (Magnoni et al., 2008). More work needs to be done to establish the roles of various plasma lipids in fishes since some groups such as elasmobranchs and a few teleosts lack albumin and thus have low circulating levels of NEFAs compared to other fish species (Zammit and Newsholme, 1979a; Speers-Roesch et al., 2008) (Table 3.1). Although plasma NEFAs are generally low in elasmobranchs (Speers-Roesch et al., 2008), a few species such as *L. erinacea* and *Squalus acanthias* have somewhat higher levels (Table 3.1). Further investigation of the lipid transport in these species seems warranted.

Albumin has been suggested to have been lost at least twice in the evolution of the teleost lineage (Metcalf et al., 1999b). Antarctic fishes from two orders lack albumin (Metcalf et al., 1999b). The weakening of hydrophobic interactions at very low temperatures may play a role in the loss of albumin in cold water species. Some teleosts such as *Esox lucius* have low levels (Ince and Thorpe, 1975) (Table 3.1), but the presence or absence of albumin has not been examined in these. In teleosts lacking albumin (e.g., *C. carpio, Dissostichus mawsoni, Anguilla australis*), high-density lipoprotein has been implicated in NEFA transport (De Smet et al., 1998; Metcalf et al., 1999a,b). In elasmobranchs, very-low-density lipoprotein (VLDL) and low-density lipoprotein (LDL) have been implicated (Metcalf and Gemmell, 2005), but the use of ketone bodies as an alternate transport form for lipid carbon also occurs (Ballantyne, 1997).

Fatty acids are transported into trout muscle cells via a saturable carrier similar to a mammalian fatty acid translocase (FAT/CD36) (Richards et al., 2004). The carrier has an affinity of 26 and 33 nM in trout red and white muscles, respectively (Richards et al., 2004). The maximal activity of the translocase is higher in red muscle than in white (Richards et al., 2004).

HOAD activity increases in red muscle and white muscle with exercise training in rainbow trout (Farrell et al., 1991). The transport of fatty acids within cells is handled by low-molecular-weight fatty acid binding proteins (FABPs) as in mammals (Tocher, 2003). In fishes, FABPs are tissue specific as in mammals and bind fatty acids with 1:1 stoichiometry (Londraville and Sidell, 1995). The FABP from shark liver resembles mammalian heart and adipose tissue FABP more than mammalian liver FABP (Bass and Kaur, 1989; Bass et al., 1991). Levels of muscle FABP increase with cold acclimation in striped bass, *Morone saxatilis* (Londraville and Sidell, 1996), reflecting increased utilization of lipids as energy sources.

3.7.5 FATTY ACID OXIDATION

Fatty acids are oxidized in both mitochondria and peroxisomes of fishes. Peroxisomal β-oxidation may serve to catabolize PUFAs that are poorly oxidized by mitochondria. The use of an acyl CoA oxidase in peroxisomes distinguishes the pathway from that of mitochondria. In fishes, liver peroxisomal oxidation can account for 30%–50% of lipid oxidation in cold water teleost species (Crockett and Sidell, 1993a,b). The preference for chain length and degree of unsaturation differs between these two systems with peroxisomes favoring longer chain HUFAs (Crockett and Sidell, 1993a). Some species (e.g., *M. glutinosa*) have negligible capacity for hepatic peroxisomal fatty acid oxidation (Moyes et al., 1991). Elasmobranchs on the other hand have significant hepatic peroxisomal fatty acid oxidative capacity (Moyes et al., 1991). Extrahepatic peroxisomal fatty acid oxidation is negligible in hagfish, elasmobranchs, and teleosts (Moyes et al., 1990). Fishes have at least three peroxisomal proliferator-activated receptor (PPAR) isoforms as in mammals (Ruyter et al., 1997; Leaver et al., 2005). The tissue distribution of the isoforms bears some similarity to that of mammals but there are discrete differences. The function of PPARs in regulating peroxisomal processes has not been established.

The main pathway for fatty acid oxidation is mitochondrial β-oxidation, and aside from differences in fatty acid preferences among tissues, there is no evidence it differs from that of higher vertebrates. This pathway is important in tissues such as red muscle and heart of teleosts. U_{crit} correlates with activity of a mitochondrial β-oxidative enzyme, hydroxyacyl CoA dehydrogenase (Farrell et al., 1991), implying it or the pathway may be limiting for cardiac performance. Elasmobranch liver mitochondria oxidize a range of acyl carnitines with 8, 10, 12, and 14 carbon chains being oxidized at faster rates than 16 or 18 carbon chains (Moyes et al., 1986b). This is similar to that observed for the freshwater lake charr (Ballantyne et al., 1989). In elasmobranchs, however, fatty acids can be incompletely oxidized to produce ketone bodies that are used as an energy source in extrahepatic tissues (see later text).

In isolated elasmobranch liver, mitochondria oxidation of fatty acids is enhanced by higher urea levels and inhibited by TMAO (Ballantyne and Moon, 1986b), but the physiological significance of this has not been established.

3.7.6 VOLATILE FATTY ACIDS

The guts of marine herbivorous fishes are anaerobic to provide the environment for anaerobic fermentation of a variety of marine algal complex carbohydrates to VFAs such as acetate, butyrate, propionate, valerate, and isovalerate with the help of an array of gut microorganisms (Clements et al., 1994). Blood levels of isovalerate can reach 4.4 mM in some marine herbivorous fishes: five times the pathological level in humans (Clements and Choat, 1995). VFAs have been detected in the guts of tropical marine herbivorous species such as *Kyphosus* spp. (Rimmer and Wiebe, 1987) as well as freshwater carnivorous and omnivorous species (Smith et al., 1996). Some of these may serve to provide the carbon skeletons for synthesis of the essential branched chain fatty acids.

Herbivorous teleosts that produce VFAs via gut fermentations transport these into the intestinal cells and then into the blood via specific anion exchangers using cytosolic bicarbonate ion (Titus and Ahearn, 1988, 1991). The kinetics of these carriers are such that the high affinity for bicarbonate on the blood side allows transport of bicarbonate against its concentration gradient into the cell from the blood in exchange for VFA moving down its concentration gradient into the blood. This differs from the mammalian model whereby unionized VFA diffuses out of the cells into the blood (Ahearn and Storelli, 1994).

3.7.7 EICOSANOID METABOLISM

The lack of precision in binding hydrophobic molecules such as the precursors and derivatives of the long chain n3 and n6 series fatty acids by enzymes creates a confusing spectrum of metabolites whose functions are largely unknown. Dietary lipids can influence the type of eicosanoids produced (Bell et al., 1992). Similarly, changes in membrane composition associated with the homeoviscous response to temperature alter the fatty acid pool available to eicosanoid-forming enzymes, resulting in different products that can have important consequences not the least of which involves the immune system (Bowden et al., 1996). The two enzymes cyclooxygenase (COX) and lipoxygenase occur in all fishes. Elasmobranchs have only a single cyclooxygenase (COX-1) (Yang et al., 2002). Whether this is due to the loss of an isoform is unknown since agnathans, which have a *COX-2* gene, have been suggested to possess an undiscovered *COX-1* gene (Havird et al., 2008). Teleosts have at least two cyclooxygenase genes due to genome duplications but some of these have been lost in certain groups (Nakamura and Nara, 2003; Havird et al., 2008). Resolvins are anti-inflammatory molecules derived from EPA and DHA by the action of COX-2. Resolvin E1 (Lu et al., 2007), D1 (Hong et al., 2005, 2007), and D5 (Hong et al., 2005) have been detected in trout.

Both 12- and 15-lipooxygenase have been characterized from fish gill. The 15-lipoxygenase hydroxylates 18-, 20-, and 22-carbon fatty acids (German and Creveling, 1990). 12-lipoxygenase in fish gill is cytoplasmic (German et al., 1986). Arachidonic and docosahexaenoic acid serve as substrates for the 12-lipoxygenase with similar affinities and V_{max} (German et al., 1986). Elasmobranchs lack the 12-lipoxygenase (Pettitt and Rowley, 1991). TXB_2 a nonfunctional product of COX-1 in mammals has been found in elasmobranchs (Cabrera et al., 2003), but its function is unknown.

Phosphatidylinositol of fishes usually has high levels of 20:4 n6 and may be the main source of this fatty acid for the prostaglandins and leukotrienes derived from this lipid (Sargent et al., 1993). The need for eicosanoids formed from 20:4n6 makes n6 series fatty acids essential for both marine and freshwater fishes (Sargent et al., 1993). The need for n6 series eicosanoids as part of the immune response of macrophages and neutrophils may explain the higher proportion of n6 series fatty acids in white cell membranes compared to red cell membranes (Sargent et al., 1993).

3.7.8 CHOLESTEROL METABOLISM

Cholesterol is an important component of some membranes and precursors for certain hormones. Fishes are capable of de novo cholesterol synthesis (Henderson and Tocher, 1987). HMG CoA reductase has been regarded as the rate-limiting enzyme in cholesterol synthesis in mammals. In fishes, increases in the activity of mevalonate-5-pyrophosphate decarboxylase (MVAPP decarboxylase) but not HMG CoA reductase with high dietary lipid may indicate this enzyme as rate limiting in cholesterol synthesis (Burgos et al., 1993). HMG CoA reductase activity increases in cold acclimation in carp even though the total cholesterol content of liver and cholesterol/phospholipid of microsomal membranes decreases in the cold (Teichert and Wodtke, 1992).

One of the intermediates in the de novo synthesis of cholesterol is squalene. It is found in high levels in some deep sea elasmobranchs (Heller et al., 1957; Malins and Wekell, 1969) and a few teleosts such as the eulachon, *Thaleichthys pacificus* (Ackman et al., 1968). It may be of dietary origin in the eulachon (Ackman et al., 1968). Elasmobranchs can synthesize squalene from acetate but not cholesterol (Diplock and Haslewood, 1965). The high levels of squalene in livers of elasmobranchs are thought to provide a buoyancy function (Phleger, 1991; Pelster, 1997).

3.7.9 KETONE BODY METABOLISM

In mammals and other higher vertebrates, ketone bodies are the only fuel that can replace glucose as an energy source in the brain during starvation. The ketone bodies acetoacetate and β-hydroxybutyrate (BHB) can be produced either by the incomplete β-oxidation of fatty acids, from amino acids such as tryptophan, phenylalanine, tyrosine, isoleucine, leucine, and lysine, or from VFAs. While all animals produce acetoacetate as part of β-oxidation of fatty acids, BHB can only be produced if the enzyme BHB dehydrogenase (BHBDH) is available.

The presence of BHBDH in fishes, however, varies considerably across fish groups. BHBDH has been detected in all of the primitive fishes that have been examined, including the sea lamprey, *Petromyzon marinus*, muscle (LeBlanc et al., 1995), elasmobranchs (Zammit and Newsholme, 1979a; Beis et al., 1980; Singer and Ballantyne, 1989; Ballantyne, 1997), lake sturgeon, *Acipenser fulvescens* (Singer et al., 1990), bowfins (*Amia calva*) (Singer and Ballantyne, 1991), gar, *Lepisosteus platyrhincus*

(Frick et al., 2007), and lungfish, *Protopterus dolloi* (Frick et al., 2008b). In teleosts, the situation is somewhat confusing with respect to BHBDH and BDH.

The earliest study of Jonas and Bilinski (1965) reporting BHB should be treated with suspicion since even the authors acknowledge interference of lactate with the assay of BHB. Subsequent studies of teleost fishes indicated the absence of β-hydroxybutyrate dehydrogenase in tissues and β-hydroxybutyrate in plasma in three species of teleost fishes (Zammit and Newsholme, 1979a,b). BHB could not be detected enzymatically or with NMR in carp (Segner, 1997). Red muscles of some teleosts have very low or undetectable levels of the enzyme (Zammit and Newsholme, 1979a; LeBlanc and Ballantyne, 1993), and isolated mitochondria from this tissue oxidize ketone bodies poorly (Moyes et al., 1989). It now appears that some teleost fishes do indeed have significant BHB concentrations in the plasma and tissue BHBDH activities. Detectable levels of BHBDH have been reported in tissues of seven teleost species (LeBlanc and Ballantyne, 1993; Segner, 1997). BHB has been detected in plasma of Atlantic salmon, *Salmo salar*, using enzymatic methods (Soengas et al., 1996). One study using an enzymatic assay reports BHB in trout and eel at levels similar to those of mammals (Phillips and Hird, 1977). Regardless of whether BHBDH is present, acetoacetate may still be an important metabolite in teleost fishes.

A further complication to understanding the role of ketone bodies in fishes is the finding of both D- and L-BHBDH in tissues of goldfish (LeBlanc and Ballantyne, 2000) and lungfish (Frick et al., 2008b). Both D- and L-BHBDH are cytosolic in liver and the D-form is mitochondrial in kidney (LeBlanc and Ballantyne, 2000). This somewhat resembles the situation in mammalian ruminants where both D- and L-BHBDH have been found with a cytosolic L-form and a mitochondrial D-form (Watson and Lindsay, 1972). The role of the cytosolic form in ruminants has been attributed to the use of butyrate from microorganisms as a carbon source for ketone body formation (Watson and Lindsay, 1972). Thus, in fishes capable of digesting plant carbohydrates, a system similar to that of ruminant mammals may prevail. Ketone body formation from butyrate has been demonstrated in liver slices of rainbow trout and eel, *A. australis* (Phillips and Hird, 1977). High levels of butyrate and other VFAs have been found in the guts of marine (Clements et al., 1994; Clements and Choat, 1995) and freshwater herbivorous/omnivorous fishes (Smith et al., 1996).

The group relying most on ketone bodies is the elasmobranchs. Elasmobranch fishes have high plasma levels of BHB under normal conditions and these increase during starvation (Zammit and Newsholme, 1979a; Wood et al., 2010). Tissue activities of BHBDH are high in liver, rectal gland, kidney, heart, and red muscle (Moon and Mommsen, 1987; Speers-Roesch et al., 2006a). BHB levels do not change during or after exhaustive exercise in elasmobranchs (Richards et al., 2003). BHB is oxidized by isolated liver mitochondria from *L. erinacea* (Moyes et al., 1986b). It has been suggested that the importance of ketone bodies in elasmobranchs is due to the reduced capacity for lipid oxidation in extrahepatic tissues (Ballantyne, 1997). The absence of albumin to carry non-esterified fatty acids in the plasma may be compensated for by using ketone bodies derived from liver lipids. Synthesis from amino acids may also occur in elasmobranchs. Ketone body formation from alanine has been demonstrated in elasmobranch liver mitochondria (Anderson, 1990). High activity levels of the enzyme hydroxymethylglutaryl CoA lyase (HMG CoA lyase) required for ketone body synthesis from amino acids have been reported in liver and kidney of an elasmobranch, indicating these tissues may be the main sites of ketogenesis from amino acids (Berges and Ballantyne, 1989).

An increased importance of ketone bodies during starvation is well established in mammals. There are some indications that this also occurs in fishes. Elasmobranchs display elevated plasma BHB and tissue BHBDH activities during starvation (Zammit and Newsholme, 1979a). Elevated BHBDH activities have been reported in brains of food-deprived rainbow trout (Howell, 1998) and Atlantic salmon (Soengas et al., 1996). By contrast, plasma BHB declined with food deprivation in *S. salar* (Soengas et al., 1996), and in cod, circulating levels of BHB do not change during prolonged starvation (Black and Love, 1986).

3.8   MEMBRANES AND MEMBRANE REMODELING

Aside from their role in compartmentation of metabolism, membranes provide barriers for the movement of water and solutes and serve as a matrix for membrane proteins. Their physical properties can impact the associated proteins and thus must be regulated. Some of the structural features of membranes relate to their location and role in mediating permeability. For example, the gill basolateral membrane of elasmobranchs has a very high cholesterol content that has been suggested to decrease the permeability of the gill to urea, preventing its loss to the environment (Fines et al., 2001). High levels of cholesterol have also been reported in swim bladder membranes (1:1 cholesterol:phospholipid) (Phleger et al., 1977), presumably for reducing permeability and loss of gases such as oxygen.

Cholesterol also plays a complex role in membrane structure and function. As a major component of detergent-resistant membrane regions termed "rafts," it contributes to linking related proteins for signal transduction and other functions. As a modulator of membrane fluidity, the amount of it in membranes may need to change to maintain membrane fluidity. Changes in temperature have been associated with increases, decreases, and no change in cholesterol content in various membranes (Crockett, 1998). The localization of membrane proteins such as Na^+ K^+-ATPase changes from a cholesterol-rich raft region to a nonraft region as part of a remodeling of the gill basolateral membrane during salinity acclimation in salmonids (Lingwood et al., 2005). During acclimation to warm temperatures, the cholesterol content of raft regions increases more than that of other membrane regions (Zehmer and Hazel, 2003).

A variety of factors impact membrane lipid properties and composition including diet, temperature, salinity, and pressure. An array of adaptive responses is available for restructuring membranes to maintain optimal properties and function. These include changes in phospholipid acyl chain saturation, position of double bonds, phospholipid head groups, and cholesterol content. Many of these have been demonstrated in various membranes of fishes (Table 3.5). Although cells have some autonomy in remodeling membranes, central control is needed for induction in the cold in carp (Macartney et al., 1996). Little is known of the regulation of membrane remodeling in fishes.

The need for membrane remodeling in response to temperature change has been extensively studied. One widely used mechanism for remodeling membrane phospholipid fatty acids is the Δ9 desaturase that introduces a double bond in the middle of 18-carbon saturated fatty acids, creating a monounsaturated fatty acid. The introduction of a single double bond reduces the melting point substantially, making it an effective mechanism for increasing membrane fluidity during cold acclimation. The activation of Δ9 desaturase involves an activation of the enzyme itself as well as increased enzyme synthesis due to increased mRNA levels (Trueman et al., 2000). Changes in acyl chains of phospholipids during cold acclimation is largely due to addition of monounsaturates to the sn-1 position in carp liver (Brooks et al., 2002), but more complex positional changes have been observed in carp intestinal microsomes (Miller et al., 1976). Two isoforms of Δ9 desaturase have been found in carp liver (Polley et al., 2003). Diet of saturated fatty acids causes an increase in transcripts of one isoform of the enzyme while low temperature induces the other (Polley et al., 2003).

Membrane remodeling in response to salinity change involves changes associated with the need to change gill transport proteins and permeability properties. In addition, changes in dietary lipids associated with each environment impact membrane composition. In the laboratory, salinity acclimation in salmonids has little effect on dietary lipid requirement; see Watanabe (1982) for review suggesting the changes observed in field studies are primarily dietary. All phospholipids from sea-run Arctic char red muscle mitochondria had much higher (3- to 20-fold) n3/n6 ratios than their landlocked freshwater conspecifics (Glemet et al., 1997).

In lab studies, dietary lipids influence the n3 and n6 content, the n3/n6 ratio, as well as the degree of unsaturation of erythrocyte membrane fatty acids (Leray et al., 1986). The cholesterol/phospholipid ratio seems to be defended in the face of changes in dietary lipid (Leray et al., 1986).

TABLE 3.5

Mechanisms of Homeoviscous Adaptation in Fish Membranes

Mechanism	Nature of Change	Membrane	Species	Reference
Cholesterol content	Increase in membrane content with increasing temperature	Liver mitochondria	*Cyprinus carpio*	Wodtke (1978)
	...	Microsomes	*C. carpio*	Wodtke (1983)
	...	Liver plasma membrane	*Oncorhynchus mykiss*	Robertson and Hazel (1995)
	...	Liver plasma membrane raft	*O. mykiss*	Zehmer and Hazel (2003)
	...	Kidney plasma membrane	*O. mykiss*	Robertson and Hazel (1995)
	...	Gill plasma membrane	*O. mykiss*	Robertson and Hazel (1995)
	...	Red blood cells	*C. carpio*	Wodtke (1983)
	...	Red blood cell plasma membrane	*Platichthys flesus*	Sorensen (1993)
	No change	Red blood cell plasma membrane	*O. mykiss*	Robertson and Hazel (1995)
	...	Intestinal brush-border membrane	*O. mykiss*	Crockett and Hazel (1995)
	...	Intestinal basolateral membrane	*O. mykiss*	Crockett and Hazel (1995)
	...	Sperm plasma membrane	*O. mykiss*	Labbe et al. (1995)
	...	Brain mitochondria outer membrane	*Carassius auratus*	Chang and Roots (1989)
PC/PE	Increase with increasing temperature	Liver mitochondria	*C. carpio*	Wodtke (1978)
	Decrease with increasing pressure	Liver mitochondria	*Anguilla*	Sebert et al. (1994)
n3/n6	n3/n6 lower at high temperature	Liver mitochondria	*C. carpio*	Wodtke (1978)
Unsaturation index	Lower with increasing temperature	Liver mitochondria	*C. carpio*	Wodtke (1978)
	...	Liver mitochondria	*Lepomis cyanellus*	Cossins et al. (1980)
	...	Brain synaptosomal membrane	*C. auratus*	Cossins et al. (1977)
	...	Red blood cell plasma membrane	*Platichthys flesus*	Sorensen (1993)
	...	Red muscle mitochondria	*C. carpio*	Wodtke (1983)
	...	Red blood cells	*C. carpio*	Wodtke (1983)
	...	microsomes	*C. carpio*	Wodtke (1983)
	...	Intestinal basolateral membrane	*C. carpio*	Lee and Cossins (1990)
	...	Intestinal mucosal membranes	*C. auratus*	Smith and Kemp (1971)

TABLE 3.5 (continued)
Mechanisms of Homeoviscous Adaptation in Fish Membranes

Mechanism	Nature of Change	Membrane	Species	Reference
	No change	Intestinal brush-border membrane	*C. carpio*	Lee and Cossins (1990)
	…	Intestinal basolateral membrane	*O. mykiss*	Crockett and Hazel (1995)
	…	Intestinal brush-border membrane	*O. mykiss*	Crockett and Hazel (1995)
	…	Gill mitochondria	*C. auratus*	Caldwell and Vernberg (1970)
	…	Gill mitochondria	*Ictalurus natalis*	Caldwell and Vernberg (1970)
Cardiolipin content of mitochondria	Increase in cold	Gill mitochondria	*C. auratus*	Caldwell and Vernberg (1970)
	Increase in cold	Gill mitochondria	*Ictalurus natalis*	Caldwell and Vernberg (1970)
Fatty acid elongation	Lower in warm	Liver microsomes	*Pimelodus maculatus*	de Torrengo and Brenner (1976)
Δ6 desaturase	Increase in cold with preference for n6 over n3	Liver mitochondria	*C. carpio*	Wodtke (1978)
	Decrease at high temperature	Liver microsomes	*Pimelodus maculates*	de Torrengo and Brenner (1976)
Δ9 desaturase	Induced in cold	Liver cells	*C. carpio*	Macartney et al. (1996); Tiku et al. (1996)

There may be a greater dietary need for n6 fatty acids compared to n3 fatty acids in marine and freshwater tropical fishes (Watanabe, 1982), presumably for membrane properties.

The mitochondrial membranes of elasmobranchs seem to differ from those of other fish groups in the high proportion of saturated fatty acids (Glemet and Ballantyne, 1996). This has been attributed to the effects of urea on membrane, but this has not been verified. There is some evidence that urea influences membrane fluidity in elasmobranchs and that this is not counteracted by TMAO (Barton et al., 1999). These effects may be mediated in part through effects on membrane proteins (Barton et al., 1999).

Plasmalogens are phospholipids with the fatty acid at the sn-1 position replaced by a fatty alcohol through an ether linkage. Most plasmalogens are ethanolamine based (40%–60%) with some choline-based forms (Chapelle, 1987). They have been implicated in mediating membrane dynamics via effects as antioxidants, mediators of membrane fluidity, and signaling (Nagan and Zoeller, 2001; Magnusson and Haraldsson, 2011), but relatively little is known of these lipids, especially in fishes. Plasmalogens can comprise 20%–30% of the phospholipids in the brain, nervous tissue, and gills of fishes (Chapelle, 1987).

3.9 STARVATION METABOLISM

Although some aspects of the effects of starvation on fish metabolism have been discussed earlier with respect to specific metabolic pathways, it is useful to take an overall look at the process in a few representative species. Some fishes are capable of living without food for very long periods and the way this is accomplished differs between species and from the pattern in mammals (see (Navarro and Gutierrez, 1995) for review).

Several terms have been used to describe the lack of food. "Fasting" is usually reserved for a voluntary cessation of feeding. In fishes, this may be associated with reproduction, overwintering, estivation, or migration. Studies of fasting in a laboratory situation may not duplicate natural conditions of temperature, photoperiod, or salinity and thus should be considered with caution.

"Food deprivation" is a term usually used in the context of laboratory studies. "Starvation" is the term that should be reserved for the state in which metabolic changes have occurred due to fasting or food deprivation.

The three phases of starvation that can be applied to birds and mammals (Wang et al., 2006) do not fit all fish groups. In mammals, in the initial phase, liver glycogen is depleted and mobilization of fatty acids from lipid depots occurs. This is followed by gluconeogenesis from glycerol (from lipolysis), some amino acids, as well as ketogenesis from lipids. Finally, adipose lipid stores are depleted and muscle protein is mobilized for gluconeogenesis until death. The metabolic changes associated with starvation in fishes differ among taxonomic groups (e.g., elasmobranchs vs. teleosts) and within taxonomic groups based on life history characteristics.

An important response to starvation is a decrease in the metabolic rate (Mehner and Wieser, 1994). This metabolic depression helps attenuate the effects of food deprivation. This strategy can be used only if starvation occurs in the absence of the need to remain active or migrate.

Thus, depending on the life history strategy of the species undergoing starvation, the duration and utilization of energy reserves of different species differ considerably.

One of the best studied species is the rainbow trout. Measurements of respiratory quotients and energy reserves of rainbow trout found that carbohydrate use increases from 25% to about 30% between days 5 and 15 (Lauff and Wood, 1996). FPBase activity increases in liver but not kidney (Soengas et al., 2006). Between days 5 and 15, protein use increased from about 15% to about 20% and lipid use declined from 68% to 37% (Lauff and Wood, 1996). Other indications of the decline in lipid use include decreased lipoprotein lipase activity in adipose tissue and liver and lower plasma VLDL and LDL after 8 weeks of starvation (Black and Skinner, 1986). The need for NADPH for growth and lipid synthesis diminishes with starvation, and the activities of the enzymes involved in its production decrease in activity in liver and adipose tissue but not kidney of rainbow trout (Barroso et al., 1993).

Carp have also been extensively examined with respect to starvation metabolism. In nature, carp do not migrate extensively and rely more extensively on their substantial glycogen reserves (Ince and Thorpe, 1976; Blasco et al., 1992a) than some other species. In carp, liver protein is conserved at least up to 67 days but not in muscle (Blasco et al., 1992b). Liver and muscle glycogen stays constant but lipid declines in both muscle and liver by 67 days (Blasco et al., 1992b). Gene expression in starving carp liver has been studied and indicates vitellogenin synthesis is the first pathway to be downregulated during starvation (Hung, 2005). This is followed by upregulation of proteolytic, glycolytic, and tricarboxylic acid cycle enzymes and downregulation of lipogenic and gluconeogenic enzymes (Hung, 2005).

In *G. morhua* at 9°C, liver reserves of lipid and glycogen and white muscle glycogen are depleted in first 3 months (Black and Love, 1986). At about 8 weeks, white muscle protein and red muscle glycogen and protein begin to be used with red muscle glycogen being depleted by about 7 months (Black and Love, 1986). Protein degrading enzymes in muscle lysosomes increase during starvation in both liver and muscle of cod (e.g., cathepsin D and cathepsin L) (Guderley et al., 2003) due to need for increased proteolysis.

Some species migrate long distances and do not feed. For example, European eels, *Anguilla anguilla*, migrate 5500 km without feeding (Van Ginneken et al., 2005). This remarkable feat is partly due to the superior efficiency of anguilliform locomotion (Van Ginneken et al., 2005). For these fishes, degrading protein would compromise the capacity to complete the migration. Lab studies mimicking such a migration show losses in body mass of about 20% over 173 days but no change in the proportions of carbohydrate, lipid, or protein (Van Ginneken et al., 2005). Thus, all energy reserves are used in proportion to their availability. The fishes are smaller but still capable of swimming. Studies of another species of eel, *A. rostrata*, show they rely largely on muscle protein for their energy needs during starvation and maintain liver glycogen for long periods (Dave et al., 1975; Moon and Johnston, 1980; Renaud and Moon, 1980b; French et al., 1983). In this study, fasting had no effect on activities of lipogenic enzymes (IDH, G6PDH, 6PGDH, ATP-citrate lyase, malic enzyme) (Aster and Moon, 1981).

3.10 CONCLUSIONS

Fishes are the earliest vertebrates and the nature of the metabolic changes that have occurred during their evolution has been part of the focus of this chapter. Many aspects of differences in metabolic organization of different fish groups may reflect their evolutionary history as much as the environmental constraints they face.

The increasing metabolic complexity of more advanced groups has occurred due to the recruitment of tissue-specific and subcellular isoforms made possible by several whole genome duplications at various times in the past. The compartmentation of pathways requires not only isoform diversity but also membranes with the appropriate structure for the transporters and receptors allowing communication between compartments. There may be a role for RNA editing in providing some isoform diversity in metabolism, but this has not been examined. Little is known of the membrane structural features needed for many of the membrane types found in fish cells. Similarly, there is a dearth of information on the nature of the transporters involved in amino acid transport into many cell types and organelles. Ultimately, the functional organization of the metabolic components of the genome of different fish groups should be mapped to establish how pathways are organized and regulated and how they have evolved. Fishes provide the greatest diversity of vertebrates for this endeavor.

REFERENCES

Ackman, R.G., R.F. Addison, and C.A. Eaton. 1968. Unusual occurrence of squalene in a fish, the eulachon *Thaleichthys pacificus*. *Nature* 220:1033–1034.

Ahearn, G.A. and C. Storelli. 1994. Use of membrane vesicle techniques to characterize nutrient transport processes of the teleost gastrointestinal tract. In *Biochemistry and Molecular Biology of Fishes* (Volume 3. Analytical techniques), P.W. Hochachka and T.P. Mommsen (eds.). Elsevier, Amsterdam, the Netherlands, pp. 513–524.

Allendorf, F.W. and G.H. Thorgaard. 1984. Tetraploidy and the evolution of salmonid fishes. In *Evolutionary Genetics of Fishes*, B.J. Turner (ed.). Plenum Press, New York, pp. 1–53.

Anderson, P.M. 1990. Ketone body and phosphoenolpyruvate formation by isolated hepatic mitochondria from *Squalus acanthias* (spiny dogfish). *J. Exp. Zool.* 254:144–154.

Anderson, P.M., M.A. Broderius, K.C. Fong, T.K.N. Tsui, S.F. Chew, and Y.K. Ip. 2002. Glutamine synthetase expression in liver, muscle, stomach and intestine of *Bostrichthys sinensis* in response to exposure to a high exogenous ammonia concentration. *J. Exp. Biol.* 205:2053–2065.

Ando, S., M. Hatano, and K. Zama. 1986. Protein degradation and protease activity of chum salmon (*Oncorhynchus keta*) muscle during spawning migration. *Fish Physiol. Biochem.* 1:17–26.

Anon, 1981. *Nutrient Requirements of Coldwater Fishes*. National Academy Press, Washington, DC, pp. 1–63.

Anon, 1983. *Nutrient Requirements of Warmwater Fishes and Shellfishes*. National Academy Press, Washington, DC, pp. 1–102.

Aranishi, F., H. Ogata, K. Hara, K. Osatomi, and T. Ishihara. 1997. Purification and characterization of cathepsin L from hepatopancreas of carp *Cyprinus carpio*. *Comp. Biochem. Physiol.* 118B:531–537.

Arthur, P.G., T.G. West, P.M. Schulte, and P.W. Hochachka. 1992. Recovery metabolism of skipjack tuna (*Katsuwonus pelamis*) white muscle: Rapid and parallel changes in lactate and phosphocreatine after exercise. *Can. J. Zool.* 70:1230–1239.

Ash, D.E. 2004. Structure and function of arginases. *J. Nutr.* 134:2760S–2764S.

Aster, P.L. and T.W. Moon. 1981. Influence of fasting and diet on lipogenic enzymes in the American eel, *Anguilla rostrata* LeSueur. *J. Nutr.* 111:346–354.

Atkinson, D.E. 1992. Functional roles of urea synthesis in vertebrates. *Physiol. Zool.* 65:243–267.

Avella, M., O. Blaise, and J. Berhaut. 1992. Effects of starvation on valine and alanine transport across the intestinal mucosal border in sea bass, *Dicentrarchus labrax. J. Comp. Physiol.* 162B:430–435.

Babin, P.J. and J. Vernier. 1989. Plasma lipoproteins in fish. *J. Lipid Res.* 30:467–489.

Baker, J.R., A. Struempler, and S. Chaykin. 1963. A comparative study of trimethylamine-N-oxide biosynthesis. *Biochim. Biophys. Acta* 71:58–64.

Bakke-McKellep, A.M., S. Nordrum, A. Krogdahl, and R.K. Buddington. 2000. Absorption of glucose, amino acids and dipeptides in the intestines of Atlantic salmon (*Salmo salar* L.). *Fish Physiol. Biochem.* 22:33–44.

Ballantyne, J.S. 1994. Fish mitochondria. In *Biochemistry and Molecular Biology of Fishes* (Volume 3. Experimental Techniques), P.W. Hochachka and T.P. Mommsen (eds.). Elsevier, Amsterdam, the Netherlands, pp. 487–502.

Ballantyne, J.S. 1997. Jaws, the inside story. The metabolism of elasmobranch fishes. *Comp. Biochem. Physiol.* 118B:703–742.

Ballantyne, J.S. 2001. Amino acid metabolism. In *Fish Physiology* (Volume 20. Nitrogen Excretion), P.A. Wright and P.M. Anderson (eds.). Academic Press, San Diego, CA, pp. 77–107.

Ballantyne, J.S. and M.E. Chamberlin. 1994. Regulation of cellular amino acid levels. In *Cellular and Molecular Physiology of Cell Volume Regulation*, K. Strange (ed.). CRC Press Inc., Boca Raton, FL, pp. 111–122.

Ballantyne, J.S., M.E. Chamberlin, and T.D. Singer. 1992. Oxidative metabolism in thermogenic tissues of the swordfish and mako shark. *J. Exp. Zool.* 261:110–114.

Ballantyne, J.S., D. Flannigan, and T.B. White. 1989. The effects of temperature on the oxidation of fatty acids, acyl carnitines and ketone bodies by mitochondria isolated from the liver of the Lake Charr, *Salvelinus namaycush. Can. J. Fish. Aquat. Sci.* 46:950–954.

Ballantyne, J.S. and N.T. Frick. 2011. Lungfish metabolism. In *Lungfish Biology*. CRC Press, Boca Raton, FL, pp. 301–335.

Ballantyne, J.S., H.C. Glemet, M.E. Chamberlin, and T.D. Singer. 1993. Plasma nonesterified fatty acids of marine teleost and elasmobranch fishes. *Mar. Biol.* 116:47–52.

Ballantyne, J.S., F. Mercure, M.F. Gerrits, G. Van Der Kraak, S.J. McKinley, D.W. Martens, S.G. Hinch, and R.E. Diewert. 1996. Plasma non-esterified fatty acid profiles in male and female sockeye salmon, *Oncorhynchus nerka*, during the spawning migration. *Can. J. Fish. Aquat. Sci.* 53:1418–1426.

Ballantyne, J.S. and T.W. Moon. 1986a. Solute effects on mitochondria from an elasmobranch (*Raja erinacea*) and teleost (*Pseudopleuronectes americanus*). *J. Exp. Zool.* 239:319–328.

Ballantyne, J.S. and T.W. Moon. 1986b. The effects of urea, trimethylamine oxide and ionic strength on the oxidation of acyl carnitines by mitochondria isolated from the liver of the little skate *Raja erinacea. J. Comp. Physiol.* 156:845–851.

Ballantyne, J.S., C.D. Moyes, and T.W. Moon. 1986. Osmolarity affects oxidation of sarcosine by isolated hepatocytes and mitochondria from a euryhaline elasmobranch. *J. Exp. Zool.* 238:267–271.

Ballantyne, J.S. and J.W. Robinson. 2010. Freshwater elasmobranchs: A review of their physiology and biochemistry. *J. Comp. Physiol.* 180B:475–493.

Ballatori, N. and J.L. Boyer. 1992a. Taurine transport in skate hepatocytes. I. Uptake and efflux. *Am. J. Physiol.* 262:G445–G450.

Ballatori, N. and J.L. Boyer. 1992b. Taurine transport in skate hepatocytes. II. Volume activation, energy and sulfhydryl dependence. *Am. J. Physiol.* 262:G451–G460.

Ballatori, N., T.W. Simmons, and J.L. Boyer. 1994. A volume-activated taurine channel in skate hepatocytes: Membrane polarity and role of intracellular ATP. *Am. J. Physiol.* 267:G285–G291.

Balocco, C., G. Boge, and H. Roche. 1993. Neutral amino acid transport by marine fish intestine: Role of the side chain. *J. Comp. Physiol.* 163B:340–347.

Barber, M.L. and P.J. Walsh. 1993. Interactions of acid-base status and nitrogen excretion and metabolism in the ureogenic teleost *Opsanus beta. J. Exp. Biol.* 185:87–105.

Barimo, J.F., S.L. Steele, P.A. Wright, and P.J. Walsh. 2004. Dogmas and controversies in the handling of nitrogenous wastes: Ureotely and ammonia tolerance in early life stages of the gulf toadfish, *Opsanus beta. J. Exp. Biol.* 207:2011–2020.

Barrington, E.J.W. 1972. The pancreas and intestine. In *The Biology of Lampreys*, Vol. 2, M.W. Hardisty and I.C. Potter (eds.). Academic Press, London, pp. 135–169.

Barroso, J.B., L. Garcia-Salguero, J. Peragon, M. De La Higuera, and J.A. Lupianez. 1993. Effects of long-term starvation on the NADPH production systems in several different tissues of rainbow trout (*Oncorhynchus mykiss*). In *Fish Nutrition in Practice*, S.J. Kaushik and P. Luquet (eds.). Institut national de la recherche agronomique, Paris, France, pp. 333–338.

Barton, K.N., M.M. Buhr, and J.S. Ballantyne. 1999. The effects of urea and trimethylamine N-oxide on fluidity of liposomes and erythrocyte membranes of an elasmobranch (*Raja erinacea*). *Am. J. Physiol.* 276:R397–R406.

Barton, K.N., M.F. Gerrits, and J.S. Ballantyne. 1995. Effects of exercise on plasma nonesterified fatty acids and free amino acid in Arctic char (*Salvelinus alpinus*). *J. Exp. Zool.* 271:183–189.

Bass, N.M. and S. Kaur. 1989. Elasmobranch liver contains a fatty acid binding protein (FABP) with primary structure related to mammalian heart FABP and myelin P2 protein. *Hepatology* 10:591-abs#91.

Bass, N.M., J.A. Manning, and C.A. Luer. 1991. Isolation and characterization of fatty acid binding protein in the liver of the nurse shark, *Ginglymostoma cirratum*. *Comp. Biochem. Physiol.* 98A:355–362.

Batty, R.S. and C.S. Wardle. 1979. Restoration of glycogen from lactic acid in the anaerobic swimming muscle of plaice, *Pleuronectes platessa* L. *J. Fish Biol.* 15:509–519.

Bauermeister, A.E.M. and J.R. Sargent. 1979a. Wax esters: Major metabolites in the marine environment. *TIBS* 4:209–211.

Bauermeister, A.E.M. and J.R. Sargent. 1979b. Biosynthesis of triacylglycerols in the intestine of rainbow trout (*Salmo gairdnerii*) fed marine zooplankton rich in wax esters. *Biochim. Biophys. Acta* 575:358–364.

Bedford, J.J., J.L. Harper, J.P. Leader, P.H. Yancey, and R.A.J. Smith. 1998. Betaine is the principle counteracting osmolyte in tissues of the elephant fish *Callorhinus millii* (Elasmobranchii, Holocephali). *Comp. Biochem. Physiol.* 119B:521–526.

Beis, A., V.A. Zammit, and E.A. Newsholme. 1980. Activities of 3-hydroxybutyrate dehydrogenase, 3-oxoacid CoA- transferase and acetoacetyl-CoA thiolase in relation to ketone-body utilisation in muscles of vertebrates and invertebrates. *Eur. J. Biochem.* 104:209–215.

Bell, J.G., J.R. Dick, J.R. Sargent, and A.H. McVicar. 1992. Dietary linoleic acid affects phospholipid fatty acid composition in heart and eicosanoid production by cardiomyocytes from Atlantic salmon (*Salmo salar*). *Comp. Biochem. Physiol.* 103A:337–342.

Benoit, G.J. and E.R. Norris. 1945. Studies on trimethylamine oxide. II. The origin of trimethylamine oxide in young salmon. *J. Biol. Chem.* 158:439–442.

Berge, G.E., M. Goodman, M. Espe, and E. Lied. 2004. Intestinal absorption of amino acids in fish: Kinetics and interaction of the in vitro uptake of L-methionine in Atlantic salmon (*Salmo salar* L.). *Aquaculture* 229:265–273.

Berges, J.A. and J.S. Ballantyne. 1989. 3-Hydroxy-3-methylglutaryl coenzyme A lyase from tissues of the oytser, *Crassostrea virginica*, the little skate, *Raja erinacea* and the lake charr, *Salvelinus namaycush*: A simplified spectrophotometric assay. *Comp. Biochem. Physiol.* 93B:583–588.

Bernard, S.F., S.P. Reidy, G. Zwingelstein, and J.M. Weber. 1999. Glycerol and fatty acid kinetics in rainbow trout: Effects of endurance swimming. *J. Exp. Biol.* 202:279–288.

Bever, K., M. Chenoweth, and A. Dunn. 1981. Amino acid gluconeogenesis and glucose turnover in kelp bass (*Paralabrax* sp.). *Am. J. Physiol.* 240:R246–R252.

Black, D. and R.M. Love. 1986. The sequential mobilisation and restoration of energy reserves in tissues of Atlantic cod during starvation and refeeding. *J. Comp. Physiol.* 156B:469–479.

Black, D. and E.R. Skinner. 1986. Features of the lipid transport system of fish as demonstrated by studies on starvation in the rainbow trout. *J. Comp. Physiol.* 156B:497–502.

Blasco, J., J. Fernandez, and J. Gutierrez. 1992a. Fasting and refeeding in carp, *Cyprinus carpio* L.: The mobilization of reserves and plasma metabolite and hormone variations. *J. Comp. Physiol.* 162B:539–546.

Blasco, J., J. Fernandez, and J. Gutierrez. 1992b. Variations in tissue reserves, plasma metabolites and pancreatic hormones during fasting in immature carp (*Cyprinus carpio*). *Comp. Biochem. Physiol.* 103A:357–363.

Bone, Q. and R.H. Moore. 2008. *The Biology of Fishes*, 3rd edn. Taylor & Francis Group, New York, pp. 1–478.

Booth, R.K., R.S. McKinley, and J.S. Ballantyne. 1999. Plasma non-esterified fatty acid profiles in wild Atlantic salmon during their freshwater migration and spawning. *J. Fish Biol.* 55:260–273.

Boutilier, R.G., R.A. Ferguson, R.P. Henry, and B.L. Tufts. 1993. Exhaustive exercise in the sea lamprey (*Petromyzon marinus*): Relationship between anaerobic metabolism and intracellular acid-base balance. *J. Exp. Biol.* 178:71–88.

Bouwer, S. and G. Van den Thillart. 1984. Oxygen affinity of mitochondrial state III respiration of goldfish red muscles: The influence of temperature and O_2 diffusion on K_m values. *Mol. Physiol.* 6:291–306.

Bowden, L.A., C.J. Restall, and A.F. Rowley. 1996. The influence of environmental temperature on membrane fluidity, fatty acid composition and lipoxygenase product generation in head kidney leucocytes of the rainbow trout, *Oncorhynchus mykiss. Comp. Biochem. Physiol.* 115B:375–382.

Bowen, S.H. 1987. Dietary protein requirements of fishes—A reassessment. *Can. J. Fish. Aquat. Sci.* 44:1995–2001.

Brauge, C., G. Corraze, and F. Medale. 1995. Effects of dietary levels of carbohydrate and lipid on glucose oxidation and lipogenesis from glucose in rainbow trout, *Oncorhynchus mykiss*, reared in freshwater or in seawater. *Comp. Biochem. Physiol.* 111A:117–124.

Brockerhoff, H. 1966. Digestion of fat by cod. *J. Fish. Res. Bd. Can.* 23:1835–1839.

Brockerhoff, H. and R.J. Hoyle. 1965. Hydrolysis of triglycerides by the pancreatic lipase of a skate. *Biochim. Biophys. Acta* 98:435–436.

Brooks, S., G.T. Clark, S.M. Wright, R.J. Trueman, A.D. Postle, A.R. Cossins, and N.M. Maclean. 2002. Electrospray ionisation mass spectrometric analysis of lipid restructuring in the carp (*Cyprinus carpio* L.) during cold acclimation. *J. Exp. Biol.* 205:3989–3997.

Brown, C.R. and J.N. Cameron. 1991. The relationship between specific dynamic action (SDA) and protein synthesis rates in the channel catfish. *Physiol. Zool.* 64:298–309.

Brown, G.W., J. James, R.J. Henderson, W.N. Thomas, R.O. Robinson, A.L. Thompson, E. Brown, and S.G. Brown. 1966. Uricolytic enzymes in liver of the dipnoan *Protopterus aethiopicus. Science* 153:1653–1654.

Buck, L.T., R.W. Brill, and P.W. Hochachka. 1992. Gluconeogenesis in hepatocytes isolated from the skipjack tuna (*Katsuwonus pelamis*). *Can. J. Zool.* 70:1254–1257.

Buddington, R.K., J.W. Chen, and J. Diamond. 1987. Genetic and phenotypic adaptation of intestinal nutrient transport to diet in fish. *J. Physiol.* 393:261–281.

Buentello, J.A. and D.M. Gatlin. 2001. Plasma citrulline and arginine kinetics in juvenile channel catfish, *Ictalurus punctatus*, given oral gabaculine. *Fish Physiol. Biochem.* 24:105–112.

Burgos, C., M.F. Zafra, M. Castillo, and E. Barcia-Peregrin. 1993. Effect of lipid content of diet on cholesterol content and cholesterogenic enzymes of European eel liver. *Lipids* 28:913–916.

Bystriansky, J.S. and J.S. Ballantyne. 2006. Anesthetization of Arctic char, *Salvelinus alpinus* (Linnaeus), with tricaine methanesulfonate or 2-phenoxyethanol for immediate blood sampling does not alter circulating plasma metabolite levels. *J. Fish Biol.* 69:613–621.

Bystriansky, J.S., N.T. Frick, and J.S. Ballantyne. 2007. Intermediary metabolism of Arctic char, *Salvelinus alpinus*, during short-term salinity acclimation. *J. Exp. Biol.* 210:1971–1985.

Cabrera, D.M., M.G. Janech, T.A. Morinelli, and D.H. Miller. 2003. A thromboxane A_2 system in the Atlantic stingray, *Dasyatis sabina. Gen. Comp. Endocrinol.* 130:157–164.

Cake, M.H., I.C. Potter, G.W. Power, and M. Tajbakhsh. 1992. Digestive enzyme activities and their distribution in the alimentary tract of larvae of the three extant lamprey families. *Fish Physiol. Biochem.* 10:1–10.

Caldwell, R.S. and F.J. Vernberg. 1970. The influence of acclimation temperature on the lipid composition of fish gill mitochondria. *Comp. Biochem. Physiol.* 34:179–191.

Cameron, J.N. and J.J. Cech. 1990. Lactate kinetics in exercised channel catfish *Ictalurus punctatus. Physiol. Zool.* 63:909–920.

Campbell, J.W., J.E. Vorhaben, and D.D. Smith. 1984. Hepatic ammonia metabolism in a uricotelic treefrog *Phyllomedusa sauvagei. Am. J. Physiol.* 246:R805–R810.

Casey, C.A., D.F. Perlman, J.E. Vorhaben, and J.W. Campbell. 1983. Hepatic ammoniagenesis in the channel catfish, *Ictalurus punctatus. Mol. Physiol.* 3:107–126.

Castell, J.D., R.O. Sinnhuber, D.J. Lee, and J.H. Wales. 1972. Essential fatty acids in the diet of rainbow trout (*Salmo gairdneri*): Physiological symptoms of EFA deficiency. *J. Nutr.* 102:87–92.

Chadwick, T.D. and P.A. Wright. 1999. Nitrogen excretion and expression of urea cycle enzymes in the Atlantic cod (*Gadus morhua* L.): A comparison of early life stages with adults. *J. Exp. Biol.* 202:2653–2662.

Chamberlin, M.E. and J.S. Ballantyne. 1992. Glutamine metabolism in elasmobranch and agnathan muscle. *J. Exp. Zool.* 264:269–272.

Chamberlin, M.E., H.C. Glemet, and J.S. Ballantyne. 1991. Glutamine metabolism in a Holostean fish (*Amia calva*) and a teleost (*Salvelinus namaycush*). *Am. J. Physiol.* 260:R159–R166.

Chang, M.C.J. and B.I. Roots. 1989. The lipid composition of mitochondrial outer and inner membranes from the brains of goldfish acclimated at 5°C and 30°C. *J. Thermal Biol.* 14:191–194.

Chapelle, S. 1987. Plasmalogens and *O*-alkylglycerophospholipids in aquatic animals. *Comp. Biochem. Physiol.* 88B:1–6.

Charest, R.P., M. Chenoweth, and A. Dunn. 1988. Metabolism of trimethylamines in kelp bass (*Paralabrax clathratus*) and marine and freshwater pink salmon (*Oncorhynchus gorbuscha*). *J. Comp. Physiol.* 158B:609–619.

Chew, S.F., T.F. Ong, L. Ho, W.L. Tam, A.M. Loong, K.C. Hiong, W.P. Wong, and Y.K. Ip. 2003. Urea synthesis in the African lungfish *Protopterus dolloi*—Hepatic carbamoyl phosphate synthetase III and glutamine synthetase exposure are upregulated by 6 days of aerial exposure. *J. Exp. Biol.* 206:3615–3624.

Chew, S.F., J.M. Wilson, Y.K. Ip, and D.J. Randall. 2006. Nitrogen excretion and defense against ammonia toxicity. In *Fish Physiology* (Volume 21. The Physiology of Tropical Fishes), A. Val, V. Almeida-Val, and D.J. Randall (eds.). Academic Press, New York, pp. 307–395.

Childress, J.J. and G.N. Somero. 1990. Metabolic scaling: A new perspective based on scaling of glycolytic enzyme activities. *Am. Zool.* 30:161–173.

Chiu, Y.N., R.E. Austic, and G.L. Rumsey. 1986. Urea cycle activity and arginine formation in rainbow trout (*Salmo gairdneri*). *J. Nutr.* 116:1640–1650.

Claireaux, G. and J. Dutil. 1992. Physiological response of the Atlantic cod (*Gadus morhua*) to hypoxia at various environmental salinities. *J. Exp. Biol.* 163:97–118.

Cleland, W.W. 1967. Enzyme kinetics. *Ann. Rev. Biochem.* 36:77–112.

Clements, K.D. and J.H. Choat. 1995. Fermentation in tropical marine herbivorous fishes. *Physiol. Zool.* 68:355–378.

Clements, K.D., V.P. Gleeson, and M. Slaytor. 1994. Short-chain fatty acid metabolism in temperate marine herbivorous fish. *J. Comp. Physiol.* 164B:372–377.

Clements, K.D. and D. Raubenheimer. 2006. Feeding and nutrition. In *The Physiology of Fishes*, D.H. Evans and J.B. Claiborne (eds.). CRC Press, Boca Raton, FL, pp. 47–82.

Connell, J.J. 1955. Arginase in elasmobranch muscle. *Nature* 175:562.

Cornish, I. and T.W. Moon. 1985. Glucose and lactate kinetics in American eel *Anguilla rostrata*. *Am. J. Physiol.* 249:R67–R72.

Cossins, A.R., M.J. Friedlander, and C.L. Prosser. 1977. Correlations between behavioral temperature adaptations of goldfish and the viscosity and fatty acid composition of their synaptic membranes. *J. Comp. Physiol.* 120A:109–121.

Cossins, A.R., J. Kent, and C.L. Prosser. 1980. A steady-state and differential polarised phase fluorometric study of the liver microsomal and mitochondrial membranes of thermally acclimated green sunfish (*Lepomis cyanellus*). *Biochim. Biophys. Acta* 599:341–358.

Cowey, C.B. 1993. Protein metabolism in fish. In *Aquaculture: Fundamental and Applied Research*, B. Lahlou and P. Vitiello (eds.). American Geophysical Union, Washington, D.C., pp. 125–137.

Cowey, C.B. 1994. Amino acid requirements of fish: A critical appraisal of present values. *Aquaculture* 124:1–11.

Cowey, C.B., D.J. Cooke, A.J. Matty, and J.W. Adron. 1981. Effects of quantity and quality of dietary protein on certain enzyme activities in rainbow trout. *J. Nutr.* 111:336–345.

Cowey, C.B. and J.R. Sargent. 1977. Lipid nutrition in fish. *Comp. Biochem. Physiol.* 57B:269–273.

Cowey, C.B. and M.J. Walton. 1989. Intermediary metabolism. In *Fish Nutrition*, J.E. Halver (ed.). Academic Press, Inc., New York, pp. 259–329.

Crabtree, B., S.J. Higgins, and E.A. Newsholme. 1972. The activities of pyruvate carboxylase, phosphoenolpyruvate carboxylase and fructose diphosphatase in muscles from vertebrates and invertebrates. *Biochem. J.* 130:391–396.

Crockett, E.L. 1998. Cholesterol function in plasma membranes from ectotherms: Membrane-specific roles in adaptation to temperature. *Am. Zool.* 38:291–304.

Crockett, E.L. and J.R. Hazel. 1995. Cholesterol levels explain inverse compensation of membrane order in brush border but not homeoviscous adaptation in basolateral membranes from the intestinal epithelia of rainbow trout. *J. Exp. Biol.* 198:1105–1113.

Crockett, E.L. and B.D. Sidell. 1993a. Peroxisomal beta-oxidation is a significant pathway for catabolism of fatty acids in a marine teleost. *Am. J. Physiol.* 264:R1004–R1009.

Crockett, E.L. and B.D. Sidell. 1993b. Substrate selectivities differ for hepatic mitochondrial and peroxisomal beta-oxidation in an Antarctic fish, *Notothenia gibberifrons*. *Biochem. J.* 289:427–433.

Dabrowski, K. 1990a. Absorption of ascorbic acid and ascorbic sulfate and ascorbate metabolism in common carp (*Cyprinus carpio* L.). *J. Comp. Physiol.* 160B:549–561.

Dabrowski, K. 1990b. Gulonolactone oxidase is missing in teleost fish. The direct spectrophotometric assay. *Biol. Chem. Hoppe-Seyler* 371:207–214.

Dabrowski, K. 1993. Ecophysiological adaptations exist in nutrient requirements of fish: True or false? *Comp. Biochem. Physiol.* 104A:579–584.

Dabrowski, K. 1994. Primitive Actinopterygian fishes can synthesize ascorbic acid. *Experientia* 50:745–748.

Dabrowski, K. and G. Kock. 1989. Absorption of ascorbic acid and ascorbic sulfate and their interaction with minerals in the digestive tract of rainbow trout (*Oncorhynchus mykiss*). *Can. J. Fish. Aquat. Sci.* 46:1952–1957.

Dabrowski, K., C. Leray, G. Nonnotte, and D.A. Colin. 1986. Protein digestion and ion concentrations in rainbow trout (*Salmo gairdnerii* Rich.) digestive tract in sea and freshwater. *Comp. Biochem. Physiol.* 83A:27–39.

Danulat, E. and S. Kemper. 1992. Nitrogenous waste excretion and accumulation of urea and ammonia in *Chalcalburnus tarichi* (Cyprinidae), endemic to the extremely alkaline Lake Van (Eastern Turkey). *Fish Physiol. Biochem.* 9:377–386.

Dave, G., M. Johansson-Sjobeck, A. Larsson, K. Lewander, and U. Lidman. 1975. Metabolic and hematological effects of starvation in the European eel, *Anguilla anguilla* L. I. Carbohydrate, lipid, protein and inorganic ion metabolism. *Comp. Biochem. Physiol.* 52A:423–430.

Davison, W., J. Baldwin, P.S. Davie, M.E. Forster, and G.H. Satchell. 1990. Exhausting exercise in the hagfish, *Eptatretus cirrhatus*: The anaerobic potential and the appearance of lactic acid in the blood. *Comp. Biochem. Physiol.* 95A:585–589.

De Smet, H., R. Blust, and L. Moens. 1998. Absence of albumin in the plasma of the common carp *Cyprinus carpio*: Binding of fatty acids to high density lipoproteins. *Fish Physiol. Biochem.* 19:71–81.

de Torrengo, M.P. and R.R. Brenner. 1976. Influence of environmental temperature on the fatty acid desaturation and elongation activity of fish (*Pimelodus maculatus*) liver microsomes. *Biochim. Biophys. Acta* 424:36–44.

Depeche, J., R. Gilles, S. Daufresne, and H. Chiapello. 1979. Urea content and urea production via the ornithine-urea cycle pathway during the ontogenic development of two teleost fishes. *Comp. Biochem. Physiol.* 63A:51–56.

deRoos, R. and C.C. deRoos. 1973. Elevation of plasma glucose levels by mammalian ACTH in the spiny dogfish (*Squalus acanthias*). *Gen. Comp. Endocrinol.* 21:403–409.

Dickson, K.A. 1996. Locomotor muscle of high performance fishes: What do comparisons of tunas with ectothermic sister taxa reveal. *Comp. Biochem. Physiol.* 113A:39–49.

Diplock, A.T. and G.A.D. Haslewood. 1965. Biosynthesis of cholesterol and ubiquinone in the lesser spotted dogfish. *Biochem. J.* 97:36P–37P.

Dobson, G.P., S. Hitchins, and W.E. Teague. 2002. Thermodynamics of the pyruvate kinase reaction and the reversal of glycolysis in heart and skeletal muscle. *J. Biol. Chem.* 277:27176–27182.

Dobson, G.P. and P.W. Hochachka. 1987. Role of glycolysis in adenylate depletion and repletion during work and recovery in teleost white muscle. *J. Exp. Biol.* 129:125–140.

Driedzic, W.R. and P.W. Hochachka. 1976. Control of energy metabolism in fish white muscle. *Am. J. Physiol.* 230:579–582.

Driedzic, W.R., J.L. West, D.H. Sephton, and J.A. Raymond. 1998. Enzyme activity levels associated with the production of glycerol as an antifreeze in liver of rainbow smelt (*Osmerus mordax*). *Fish Physiol. Biochem.* 18:125–134.

Dunn, J.F., W. Davison, G.M.O. Maloiy, P.W. Hochachka, and M. Guppy. 1981. An ultrastructural and histochemical study of the axial musculature in the African lungfish. *Cell Tissue Res.* 220:599–609.

Dunn, J.F. and P.W. Hochachka. 1987. Turnover rates of glucose and lactate in rainbow trout during acute hypoxia. *Can. J. Zool.* 65:1144–1148.

Duong, C.A., C.A. Sepulveda, J.B. Graham, and K.A. Dickson. 2006. Mitochondria proton leak rates in the slow, oxidative myotomal muscle and liver of the endothermic shortfin mako shark (*Isurus oxyrinchus*) and the ectothermic blue shark (*Prionace glauca*) and leopard shark (*Triakis semifasciata*). *J. Exp. Biol.* 209:2678–2685.

Dyer, W.J. 1951. Amines in fish muscle. VI. Trimethylamine oxide content of fish and marine invertebrates. *J. Fish. Res. Bd. Can.* 8:314–324.

Echevarria, G., M. Martinez-Bebia, and S. Zamora. 1997. Evolution of biometric indices and plasma metabolites during prolonged starvation in European sea bass (*Dicentrarchus labrax*, L.). *Comp. Biochem. Physiol.* 118A:111–123.

Eisenstein, R.S., R.H. Miller, G. Hoganson, and A.E. Harper. 1990. Phylogenetic comparisons of the branched-chain α-ketoacid dehydrogenase complex. *Comp. Biochem. Physiol.* 97B:719–726.

Emdin, S.O. 1982. Effects of hagfish insulin in the Atlantic hagfish, *Myxine glutinosa*. The in vivo metabolism of [^{14}C] glucose and [^{14}C] leucine and studies on starvation and glucose-loading. *Gen. Comp. Endocrinol.* 47:414–425.

Emelyanova, L.V., E.M. Koroleva, and M.V. Savina. 2004. Glucose and free amino acids in the blood of lampreys (*Lampetra fluviatilis* L.) and frogs (*Rana temporaria* L.) under prolonged starvation. *Comp. Biochem. Physiol.* 138A:527–532.

Fange, R., G. Lundblad, J. Lind, and K. Slettengren. 1979. Chitinolytic enzymes in the digestive system of marine fishes. *Mar. Biol.* 53:317–321.

Farrell, A.P., J.A. Johansen, and R.K. Suarez. 1991. Effects of exercise-training on cardiac performance and muscle enzymes in rainbow trout, *Oncorhynchus mykiss*. *Fish Physiol. Biochem.* 9:303–312.

Fauconneau, B. and S.S. Tesseraud. 1990. Measurement of plasma leucine flux in rainbow trout (*Salmo gairdneri* R.) using osmotic pump. Preliminary investigations of diet. *Fish Physiol. Biochem.* 8:29–44.

Felip, O., A. Ibarz, J. Fernandez-Borras, M. Beltran, M. Martin-Perez, J.V. Planas, and J. Blasco. 2012. Tracing metabolic routes of dietary carbohydrate and protein in rainbow trout (*Oncorhynchus mykiss*) using stable isotopes (^{13}C starch) and (^{15}N protein): Effects of gelatinization of starches and sustained swimming. *Br. J. Nutr.* 107:834–844.

Fellows, F.C.I. and R.M. McLean. 1982. A study of the plasma lipoproteins and the tissue lipids of the migrating lamprey, *Mordacia mordax*. *Lipids* 17:741–747.

Felskie, A.K., P.M. Anderson, and P.A. Wright. 1998. Expression and activity of carbamoyl phosphate synthetase III and ornithine urea cycle enzymes in various tissues of four fish species. *Comp. Biochem. Physiol.* 119B:355–364.

Ferraris, R.P. and G.A. Ahearn. 1983. Intestinal glucose transport in carnivorous and herbivorous marine fishes. *J. Comp. Physiol.* 152:79–90.

Ferraris, R.P. and G.A. Ahearn. 1984. Sugar and amino acid transport in fish intestine. *Comp. Biochem. Physiol.* 77A:397–413.

Ferraris, R.P. and J. Diamond. 1997. Regulation of intestinal sugar transport. *Physiol. Rev.* 77:257–302.

Fines, G.A., J.S. Ballantyne, and P.A. Wright. 2001. Active urea transport and an unusual basolateral membrane composition in the gills of a marine elasmobranch. *Am. J. Physiol.* 280:R16–R24.

Fisher, S.E., J.B. Shaklee, S.D. Ferris, and G.S. Whitt. 1980. Evolution of five multilocus isozyme systems in the chordates. *Genetica* 52–53:73–85.

Forster, R.P., J.A. Hannafin, and L. Goldstein. 1978. Osmoregulatory role of amino acids in brain of the elasmobranch, *Raja erinacea*. *Comp. Biochem. Physiol.* 60A:25–30.

Foster, G.D. and T.W. Moon. 1989. Insulin and the regulation of glycogen metabolism and gluconeogenesis in American eel hepatocytes. *Gen. Comp. Endocrinol.* 73:374–381.

Foster, G.D., J.H. Youson, and T.W. Moon. 1993. Carbohydrate metabolism in the brain of the adult lamprey. *J. Exp. Zool.* 267:27–32.

Fracalossi, D.M., M.E. Allen, L.K. Yuyama, and O.T. Oftedal. 2001. Ascorbic acid biosynthesis in Amazonian fishes. *Aquaculture* 192:321–332.

French, C.J., P.W. Hochachka, and T.P. Mommsen. 1983. Metabolic organization of liver during spawning migration of sockeye salmon. *Am. J. Physiol.* 245:R827–R830.

French, C.J., T.P. Mommsen, and P.W. Hochachka. 1981. Amino acid utilisation in isolated hepatocytes from rainbow trout. *Eur. J. Biochem.* 113:311–317.

Frick, N.T., J.S. Bystriansky, and J.S. Ballantyne. 2007. The metabolic organization of a primitive air-breathing fish, the Florida gar, (*Lepisosteus platyrhincus*). *J. Exp. Zool.* 307A:7–17.

Frick, N.T., J.S. Bystriansky, Y.K. Ip, S.F. Chew, and J.S. Ballantyne. 2008a. Carbohydrates and amino acid metabolism following 60 days of fasting and aestivation in the African lungfish, *Protopterus dolloi*. *Comp. Biochem. Physiol.* 151A:85–92.

Frick, N.T., J.S. Bystriansky, Y.K. Ip, S.F. Chew, and J.S. Ballantyne. 2008b. Lipid, ketone body and oxidative metabolism in the African lungfish, *Protopterus dolloi* during periods of fasting and aestivation. *Comp. Biochem. Physiol.* 151A:93–101.

Fris, M.B. and M.H. Horn. 1993. Effects of diets of different protein content on food consumption, gut retention, protein conversion, and growth of *Cebidichthys violaceus* (Girard), an herbivorous fish of temperature zone marine waters. *J. Exp. Mar. Biol. Ecol.* 166:185–202.

Frolow, J. and C.L. Milligan. 2004. Hormonal regulation of glycogen metabolism in white muscle slices from rainbow trout (*Oncorhynchus mykiss*). *Am. J. Physiol.* 287:R1344–R1353.

Fudge, D.S., J.S. Ballantyne, and E.D. Stevens. 2001. A test of biochemical symmorphosis in a heterothermic tissue: Bluefin tuna white muscle. *Am. J. Physiol.* 280:R108–R114.

Fugelli, K. and S.M. Thoroed. 1986. Taurine transport associated with cell volume regulation in flounder erythrocytes under anisosmotic conditions. *J. Physiol.* 374:245–261.

Fynn-Aikins, K., S.S.O. Hung, W. Liu, and H. Li. 1992. Growth, lipogenesis and liver composition of juvenile white sturgeon fed different levels of D-glucose. *Aquaculture* 105:61–72.

Garin, D., A. Rombaut, and A. Freminet. 1987. Determination of glucose turnover in sea bass *Dicentrarchus labrax*. Comparative aspects of glucose utilization. *Comp. Biochem. Physiol.* 87B:981–988.

German, D.P. 2009. Inside the guts of wood-eating catfishes: Can they digest wood? *J. Comp. Physiol.* 179B:1011–1023.

German, J.B., G.G. Bruckner, and J.E. Kinsella. 1986. Lipoxygenase in trout gill tissue acting on arachidonic, eicosapentaenoic and docosahexaenoic acids. *Biochim. Biophys. Acta* 875:12–20.

German, J.B. and R.K. Creveling. 1990. Identification and characterization of a 15-lipoxygenase from fish gills. *J. Agr. Food Chem.* 38:2144–2147.

Gillett, M.B., J.R. Suko, F.O. Santoso, and P.H. Yancey. 1997. Elevated levels of trimethylamine oxide in muscles of deep-sea gadiform teleosts: A high-pressure adaptation? *J. Exp. Zool.* 279:386–391.

Gillis, T.E. and J.S. Ballantyne. 1996. The effects of starvation on plasma free amino acid and glucose concentrations in lake sturgeon, *Acipenser fulvescens*. *J. Fish Biol.* 49:1306–1316.

Gjellesvik, D.R., A.J. Raae, and B.T. Walther. 1989. Partial purification and characterization of a triglyceride lipase from cod (*Gadus morhua*). *Aquaculture* 79:177–184.

Gleeson, T.T. 1996. Post-exercise lactate metabolism: A comparative review of sites, pathways, and regulation. *Ann. Rev. Physiol.* 58:565–581.

Glemet, H.C. and J.S. Ballantyne. 1996. A comparison of liver mitochondrial membranes from an agnathan (*Myxine glutinosa*), an elasmobranch (*Raja erinacea*) and a teleost fish (*Pleuronectes americanus*). *Mar. Biol.* 124:509–518.

Glemet, H.C., M.F. Gerrits, and J.S. Ballantyne. 1997. Membrane lipids of red muscle mitochondria from land-locked and sea-run Arctic char, *Salvelinus alpinus* L. *Mar. Biol.* 129:673–679.

Goldstein, L. and S. Brill. 1990. Isosmotic swelling of skate (*Raja erinacea*) red blood cells causes a volume regulatory release of intracellular taurine. *J. Exp. Zool.* 253:132–138.

Goldstein, L. and E.M. Davis. 1994. Taurine, betaine, and inositol share a volume-sensitive transporter in skate erythrocyte cell membrane. *Am. J. Physiol.* 267:R426–R431.

Goldstein, L., E.M. Davis-Amaral, P.C. Blum, and C.A. Luer. 1995. The role of anion channels in osmotically activated taurine release from embryonic skate (*Raja eglanteria*) heart. *J. Exp. Biol.* 198:2635–2637.

Goldstein, L. and D. Funkhouser. 1972. Biosynthesis of trimethylamine oxide in the nurse shark, *Ginglymostoma cirratum*. *Comp. Biochem. Physiol.* 42A:51–57.

Goldstein, L., S.C. Hartman, and R.P. Forster. 1967. On the origin of trimethylamine oxide in the spiny dogfish, *Squalus acanthias*. *Comp. Biochem. Physiol.* 21:719–722.

Goolish, E.M. 1991. Aerobic and anaerobic scaling in fish. *Biol. Rev.* 66:33–56.

Gouillou-Coustans, M.F., V. Fournier, R. Metailler, E. Desbruyeres, C. Huelvan, J. Moriceau, H. Le Delliou, and S.J. Kaushik. 2002. Dietary arginine degradation is a major pathway in ureagenesis in juvenile turbot (*Psetta maxima*). *Comp. Biochem. Physiol.* 132A:305–319.

Grant, W.C., F.J. Hendler, and P.M. Banks. 1969. Studies on blood-sugar regulation in the little skate *Raja erinacea*. *Physiol. Zool.* 42:231–247.

Greene, D.H.S. and D.P. Selivonchick. 1987. Lipid metabolism in fish. *Prog. Lipid Res.* 26:53–85.

Guderley, H., D. Lapointe, M. Bedard, and J.D. Dutil. 2003. Metabolic priorities during starvation: Enzyme sparing in liver and white muscle of Atlantic cod, *Gadus morhua* L. *Comp. Biochem. Physiol.* 135A:347–356.

Guppy, M. and P.W. Hochachka. 1978. Controlling the highest lactate dehydrogenase activity known in nature. *Am. J. Physiol.* 234:R136–R140.

Guppy, M. and P.W. Hochachka. 1979. Metabolic sources of heat and power in tuna muscles. II. Enzyme and metabolite profiles. *J. Exp. Biol.* 82:303–320.

Hagey, L.R., P.R. Moller, A.F. Hofmann, and M.D. Krasowski. 2010. Diversity of bile salts in fish and amphibians: Evolution of a complex biochemical pathway. *Physiol. Biochem. Zool.* 83:308–321.

Hall, J.R., R.C. Richards, T.J. MacCormack, K.V. Ewart, and W.R. Driedzic. 2005. Cloning of GLUT3 cDNA from Atlantic cod (*Gadus morhua*) and expression of GLUT1 and GLUT3 in response to hypoxia. *Biochim. Biophys. Acta* 1730:245–252.

Haman, F., M. Powell, and J. Weber. 1997. Reliability of continuous tracer infusion for measuring glucose turnover rate in rainbow trout. *J. Exp. Biol.* 200:2557–2563.

Haman, F. and J.M. Weber. 1996. Continuous tracer infusion to measure in vivo metabolite turnover rates in trout. *J. Exp. Biol.* 199:1157–1162.

Hansen, H.J.M. and S. Abraham. 1989. Compartmentation of gluconeogenesis in the fasted eel (*Anguilla anguilla*). *Comp. Biochem. Physiol.* 92B:697–703.

Harrington, A.J., K.A. Russell, T.D. Singer, and J.S. Ballantyne. 1991. The effects of tricaine methanesulfonate (MS-222) on plasma nonesterified fatty acids in rainbow trout, *Oncorhynchus mykiss*. *Lipids* 26:774–775.

Hartmann, C., P. Christen, and R. Jaussi. 1991. Mitochondrial protein charge. *Nature* 352:762–763.

Haslewood, G.A.D. 1967. Bile salt evolution. *J. Lipid Res.* 8:535–550.

Hastings, N., M. Agaba, D.R. Tocher, M.J. Leaver, J.R. Dick, J.R. Sargent, and A.J. Teale. 2001. A vertebrate fatty acid desaturase with $\Delta 5$ and $\Delta 6$ activities. *Proc. Natl. Acad. Sci. USA* 98:14304–14309.

Havird, J.C., M.M. Miyamoto, K.P. Choe, and D.H. Evans. 2008. Gene duplications and losses within the cyclooxygenase family of teleosts and other chordates. *Mol. Cell Biol.* 25:2349–2359.

Haynes, J.K. and L. Goldstein. 1993. Volume-regulatory amino acid transport in erythrocytes of the little skate, *Raja erinacea*. *Am. J. Physiol.* 265:R173–R179.

Heller, J.H., M.S. Heller, S. Springer, and E. Clark. 1957. Squalene content of various shark livers. *Nature* 179:919–920.

Henderson, R.J. and D.R. Tocher. 1987. The lipid composition and biochemistry of freshwater fish. *Prog. Lipid Res.* 26:281–347.

Henry, H. and A.W. Norman. 1975. The presence of renal 25-hydroxyvitamin-D-1-hydroxylase in species of all vertebrate classes. *Comp. Biochem. Physiol.* 50B:431–434.

Hers, H.G. and L. Hue. 1983. Gluconeogenesis and related aspects of glycolysis. *Ann. Rev. Biochem.* 52:617–653.

Hochachka, P.W., K.B. Storey, C.J. French, and D.E. Schneider. 1979. Hydrogen shuttles in air versus water breathing fishes. *Comp. Biochem. Physiol.* 63B:45–56.

Hong, S., Y. Lu, R. Yang, K.H. Gotlinger, N.A. Petasis, and C.N. Serhan. 2007. Resolvin D1, protectin D1, and related docosahexaenoic acid-derived products: Analysis via electrospray low energy tandem mass spectrometry based on spectra and fragmentation patterns. *J. Am. Soc. Mass Spec.* 18:128–144.

Hong, S., E. Tjonahen, E.L. Morgan, Y. Lu, C.N. Serhan, and A.F. Rowley. 2005. Rainbow trout, (*Oncorhynchus mykiss*) brain cells synthesize novel docosahexaenoic acid-derived resolvins and protectins—Mediator lipidomic analysis. *Prostaglandins Other Lipid Mediat.* 78:107–116.

Houlihan, D.F., S.J. Hall, and C. Gray. 1989. Effects of ration on protein turnover in cod. *Aquaculture* 79:103–110.

Houlihan, D.F., D.N. McMillan, and P. Laurent. 1986. Growth rates, protein synthesis, and protein degradation rates in rainbow trout: Effects of body size. *Physiol. Zool.* 59:482–493.

Howell, A.B. 1998. *Aquatic Mammals*. Dover Publications Inc., New York, pp. 1–338.

Hughes, S.G., G.L. Rumsey, and M.C. Nesheim. 1983. Branched-chain amino acid aminotransferase activity in the tissues of lake trout. *Comp. Biochem. Physiol.* 76B:429–431.

Huisman, E.A. 1976. Food conversion efficiencies at maintenance and production levels for carp, *Cyprinus carpio* L., and rainbow trout, *Salmo gairdneri* Richardson. *Aquaculture* 9:259–273.

Hung, C.Y. 2005. Survival strategies of common carp, *Cyprinus carpio*, during prolonged starvation and hypoxia, PhD thesis, City University of Hong Kong, Hong Kong, pp. 1–269.

Hung, S.S.O. 1991. Carbohydrate utilization by white sturgeon as assessed by oral administration tests. *J. Nutr.* 121:1600–1605.

Hung, S.S.O., F.K. Fynn-Aikins, P.B. Lutes, and R. Xu. 1989. Ability of juvenile white sturgeon (*Acipenser transmontanus*) to utilize different carbohydrate sources. *J. Nutr.* 119:727–733.

Hunter, A. 1924. Quantitative studies concerning the distribution of arginase in fishes and other animals. *Proc. R. Soc. Lond.* 97B:227–242.

Hyvarinen, H., I.J. Holopainen, and J. Piironen. 1985. Anaerobic wintering of crucian carp (*Carassius carassius* L.). I. Annual dynamics of glycogen reserves in nature. *Comp. Biochem. Physiol.* 82A:797–803.

Ince, B.W. and A. Thorpe. 1975. Hormone and metabolite effects on plasma free fatty acids in the northern pike, *Esox lucius* L. *Gen. Comp. Endocrinol.* 27:144–152.

Ince, B.W. and A. Thorpe. 1976. The effects of starvation and force-feeding on the metabolism of the Northern pike, *Esox lucius* L. *J. Fish Biol.* 8:79–88.

Inui, Y. and A. Gorbman. 1978. Role of the liver in regulation of carbohydrate metabolism in the hagfish, *Eptatretus stouti*. *Comp. Biochem. Physiol.* 60A:181–183.

Ip, Y.K., S.F. Chew, J.M. Wilson, and D.J. Randall. 2004a. Defences against ammonia toxicity in tropical air-breathing fishes exposed to high concentrations of environmental ammonia: A review. *J. Comp. Physiol.* 174B:565–575.

Ip, Y.K., C.B. Lim, and S.F. Chew. 2006. Intermediary metabolism in mudskippers, *Periophthalmodon schlosseri* and *Boleophthalmus boddarti*, during immersion or emersion. *Can. J. Zool.* 84:981–991.

Ip, Y.K., W.L. Tam, W.P. Wong, and S.F. Chew. 2005a. Marine (*Taeniura lymma*) and freshwater (*Himantura signifer*) elasmobranchs synthesize urea for osmotic water retention. *Physiol. Biochem. Zool.* 78:610–619.

Ip, Y.K., W.L. Tam, W.P. Wong, A.M. Loong, K.C. Hiong, J.S. Ballantyne, and S.F. Chew. 2003. A comparison of the effects of exposure to environmental ammonia on the Asian freshwater stingray *Himantura signifer* and the Amazonian freshwater stingray *Potamotrygon motoro*. *J. Exp. Biol.* 206:3625–3633.

Ip, Y.K., A.S.L. Tay, K.H. Lee, and S.F. Chew. 2004b. Strategies for surviving high concentrations of environmental ammonia in the swamp eel *Monopterus albus*. *Physiol. Biochem. Zool.* 77:390–405.

Ip, Y.K., P.J. Yeo, A.M. Loong, K.C. Hiong, W.P. Wong, and S.F. Chew. 2005b. The interplay of increased urea synthesis and reduced ammonia production in the African lungfish *Protopterus aethiopicus* during 46 days of aestivation in a mucus cocoon. *J. Exp. Zool.* 303A:1054–1065.

Iwabe, N., K.I. Kuma, and T. Miyata. 1996. Evolution of gene families and relationship with organismal evolution: Rapid divergence of tissue-specific genes in the early evolution of chordates. *Mol. Biol. Evol.* 13:483–493.

Iwata, K., M. Kajimura, and T. Sakamoto. 2000. Functional ureogenesis in the gobiid fish, *Mugilogobius abei*. *J. Exp. Biol.* 203:3703–3715.

Janssens, P.A. 1964. The metabolism of aestivating African lungfish. *Comp. Biochem. Physiol.* 11:105–117.

Jarvis, P.L. and J.S. Ballantyne. 2003. Metabolic responses to salinity acclimation in juvenile shortnose sturgeon *Acipenser brevirostrum*. *Aquaculture* 219:891–909.

Jastroch, M., J.A. Buckingham, M. Helwig, M. Klingenspor, and M.D. Brand. 2007. Functional characterization of UCP1 in the common carp: Uncoupling activity in liver mitochondria and cold-induced expression in the brain. *J. Comp. Physiol.* 177B:743–752.

Jin, E.S., A.D. Sherry, and C.R. Malloy. 2009. Evidence for reverse flux through pyruvate kinase in skeletal muscle. *Am. J. Physiol.* 296:E748–E757.

John, T.M., F.W.H. Beamish, and J.C. George. 1980. Diurnal variation in the effect of exogenous indoleamines on blood metabolite levels in the rainbow trout. *Comp. Biochem. Physiol.* 65C:105–109.

Johnson, W.V., J.R. Kemp, and P.M. Anderson. 1990. Purification and properties of mitochondrial phosphoenolpyruvate carboxykinase from liver of *Squalus acanthias*. *Arch. Biochem. Biophys.* 280:376–382.

Johnston, I.A. 1975. Anaerobic metabolism in the carp (*Carassius carassius* L.). *Comp. Biochem. Physiol.* 51B:235–241.

Jonas, R.E.E. and E. Bilinski. 1965. Ketone bodies in the blood of salmonid fishes. *J. Fish. Res. Bd. Can.* 22:891–898.

Jurss, K., T. Bittorf, and T. Vokler. 1985. Influence of salinity and ratio of lipid to protein in diets on certain enzyme activities in rainbow trout (*Salmo gairdneri* Richardson). *Comp. Biochem. Physiol.* 81B:73–79.

Kajimura, M., P.J. Walsh, and C.M. Wood. 2008. The spiny dogfish *Squalus acanthias* L. maintains osmolyte balance during long-term starvation. *J. Fish Biol.* 72:656–670.

Kelly, R.H. and P.H. Yancey. 1999. High contents of trimethylamine oxide correlating with depth in deep-sea teleost fishes, skates, and decapod crustaceans. *Biol. Bull.* 196:18–25.

Kera, Y., S. Hasegawa, T. Watanabe, H. Segawa, and R. Yamada. 1998. D-aspartate oxidase and free acidic D-amino acids in fish tissues. *Comp. Biochem. Physiol.* 119B:95–100.

King, P.A., C. Cha, and L. Goldstein. 1980. Amino acid metabolism and cell volume regulation in the little skate, *Raja erinacea*. II. Synthesis. *J. Exp. Zool.* 212:79–86.

Kirchner, S., S. Panserat, P.L. Lim, S. Kaushik, and R.P. Ferrari. 2008. The role of hepatic, renal and intestinal gluconeogenic enzymes in glucose homeostasis of juvenile rainbow trout. *J. Comp. Physiol.* 178B:429–438.

Knoph, M.B. and K. Masoval. 1996. Plasma ammonia and urea levels in the Atlantic salmon farmed in sea water. *J. Fish Biol.* 49:165–168.

Knox, D., M.J. Walton, and C.B. Cowey. 1980. Distribution of enzymes of glycolysis and gluconeogenesis in fish tissues. *Mar. Biol.* 56:7–10.

Kott, E. 1971. Liver and muscle composition of mature lampreys. *Can. J. Zool.* 49:801–805.

Krogdahl, A., G.I. Hemre, and T.P. Mommsen. 2005. Carbohydrates in fish nutrition: Digestion and absorption in post-larval stages. *Aqua. Nutr.* 11:103–122.

Labbe, C., G. Maisse, K. Muller, A. Zachaowski, S. Kaushik, and M. Loir. 1995. Thermal acclimation and dietary lipids alter the composition but not fluidity of trout sperm plasma membrane. *Lipids* 30:23–33.

Laberee, K. and C.L. Milligan. 1999. Lactate transport across sarcolemmal vesicles isolated from rainbow trout white muscle. *J. Exp. Biol.* 202:2167–2175.

Langille, R.M. and J.H. Youson. 1985. Protein and lipid absorption in the intestinal mucosa of adult lampreys (*Petromyzon marinus* L.) following induced feeding. *Can. J. Zool.* 63:691–702.

Larsson, A., M. Johansson-Sjobeck, and R. Fange. 1976. Comparative study of some haematological and biochemical blood parameters in fishes from the Skagerrak. *J. Fish Biol.* 9:425–440.

Larsson, A. and K. Lewander. 1973. Metabolic effects of starvation in the eel, *Anguilla anguilla* L. *Comp. Biochem. Physiol.* 44A:367–374.

Lauff, R.F. and C.M. Wood. 1996. Respiratory gas exchange, nitrogenous waste excretion, and fuel usage during starvation in juvenile trout, *Oncorhynchus mykiss. J. Comp. Physiol.* 165B:542–551.

Leary, S.C., C.N. Lyons, A.G. Rosenberger, J.S. Ballantyne, J. Stillman, and C.D. Moyes. 2003. Fibre type difference in muscle mitochondrial profiles. *Am. J. Physiol.* 285:R817–R826.

Leaver, M.J., E. Boukouvala, E. Antonopoulou, A. Diez, L. Favre-Krey, M.T. Ezaz, J.M. Bautista, D.R. Tocher, and G. Krey. 2005. Three peroxisome proliferator-activated receptor isotypes from each of two species of marine fish. *Endocrinology* 146:3150–3162.

LeBlanc, P.J. and J.S. Ballantyne. 1993. Beta-hydroxybutyrate dehydrogenase in teleost fish. *J. Exp. Zool.* 267:356–358.

LeBlanc, P.J. and J.S. Ballantyne. 2000. Novel aspects of the activities and subcellular distribution of enzymes of ketone body metabolism in liver and kidney of the goldfish, *Carassius auratus. J. Exp. Zool.* 286:434–439.

LeBlanc, P.J., T.E. Gillis, M.F. Gerrits, and J.S. Ballantyne. 1995. Metabolic organization of liver and somatic muscle of landlocked sea lamprey, *Petromyzon marinus*, during the spawning migration. *Can. J. Zool.* 73:916–923.

Lee, J.A.C. and A.R. Cossins. 1990. Temperature adaptation of biological membranes: Differential homeoviscous responses in brush-border and basolateral membranes of carp intestinal mucosa. *Biochim. Biophys. Acta* 1026:195–203.

Leray, C., G. Nonnotte, and L. Nonnotte. 1986. The effect of dietary lipids on the trout erythrocyte membrane. *Fish Physiol. Biochem.* 1:27–35.

Levin, B.C., V.H. Routh, L. Kang, N.M. Sanders, and A.A. Dunn-Meynell. 2004. Neuronal glucosensing. What do we know after 50 years. *Diabetes* 53:2521–2528.

Li, P., K. Mai, J. Trushenski, and G. Wu. 2009. New developments in fish amino acid nutrition: Towards functional and environmentally oriented aquafeeds. *Amino Acids* 37:43–53.

Lie, O. and G. Lambertsen. 1985. Digestive lipolytic enzymes in cod (*Gadus morhua*): Fatty acid specificity. *Comp. Biochem. Physiol.* 80B:447–450.

Lie, O. and G. Lambertsen. 1991. Lipid digestion and absorption in cod (*Gadus morhua*), comparing triacylglycerols, wax esters and diacylalkylglycerols. *Comp. Biochem. Physiol.* 98A:159–163.

Likimani, T.A. and R.P. Wilson. 1982. Effects of diet on lipogenic enzyme activities in channel catfish hepatic and adipose tissue. *J. Nutr.* 112:112–117.

Lin, H., D.R. Romsos, P.I. Tack, and G.A. Leveille. 1977a. Effects of fasting and feeding various diets on hepatic lipogenic enzyme activities in coho salmon (*Oncorhynchus kisutch* (Walbaum)). *J. Nutr.* 107:1477–1483.

Lin, H., D.R. Romsos, P.I. Tack, and G.A. Leveille. 1977b. Influence of dietary lipid on lipogenic enzyme activities in coho salmon, *Oncorhynchus kisutch* (Walbaum). *J. Nutr.* 107:846–854.

Lin, H., D.R. Romsos, P.I. Tack, and G.A. Leveille. 1978. Determination of glucose utilization in coho salmon [*Oncorhynchus kisutch* (Walbaum)] with (6-^{3}H)- and (U-^{14}C)-glucose. *Comp. Biochem. Physiol.* 59A:189–191.

Lindsay, D.B. 1982. Relationships between amino acid catabolism and protein anabolism in the ruminants. *Fed. Proc.* 41:2550–2554.

Lindsay, G.J.H. and J.E. Harris. 1980. Carboxymethylcellulase activity in the digestive tracts of fish. *J. Fish Biol.* 16:219–233.

Lingwood, D.D., G. Harauz, and J.S. Ballantyne. 2005. Regulation of fish gill Na$^+$-K$^+$-ATPase by selective sulfatide enriched raft partitioning during seawater adaptation. *J. Biol. Chem.* 280:36545–36550.

Londraville, R.L. and B.D. Sidell. 1995. Purification and characterization of fatty acid-binding protein from aerobic muscle of the Antarctic icefish *Chaenocephalus aceratus. J. Exp. Zool.* 273:190–203.

Londraville, R.L. and B.D. Sidell. 1996. Cold acclimation increases fatty acid-binding protein concentration in aerobic muscle of striped bass, *Morone saxatilis. J. Exp. Zool.* 275:36–44.

Longmuir, I.S. 1957. Respiration rate of rat-liver cells at low oxygen concentrations. *Biochem. J.* 64:378–382.

Loong, A.M., Y.R. Chng, S.F. Chew, W.P. Wong, and Y.K. Ip. 2012. Molecular characterization and mRNA expression of carbamoyl phosphate synthetase III in the liver of the African lungfish, *Protopterus annectens*, during aestivation or exposure to ammonia. *J. Comp. Physiol.* 182B:367–379.

Lovern, J.A. 1932. CCXXXV. Fat metabolism in fishes. II. The peritoneal, pancreatic and liver fats of the sturgeon (*Acipenser sturio*). *Biochem. J.* 26:1985–1988.

Lowenstein, J.M. 1972. Ammonia production in muscle and other tissues: The purine nucleotide cycle. *Physiol. Rev.* 52:382–414.

Lu, Y., S. Hong, R. Yang, J. Uddin, K.H. Gotlinger, N.A. Petasis, and C.N. Serhan. 2007. Identification of endogenous resolvin E1 and other lipid mediators derived from eicosapentaenoic acid via electrospray low-energy tandem mass spectrometry spectra and fragmentation mechanisms. *Rapid Commun. Mass Spectrom.* 21:7–22.

Macartney, A.I., P.E. Tiku, and A.R. Cossins. 1996. An isothermal induction of delta-9-desaturase in cultured carp hepatocytes. *Biochim. Biophys. Acta* 1302:207–216.

Machado, C.R., M.A.R. Garofalo, J.E.S. Roselino, I.C. Kettelhut, and R.H. Migliorini. 1989. Effect of fasting on glucose turnover in a carnivorous fish (*Hoplias* sp). *Am. J. Physiol.* 256:R612–R615.

Madison, B.N., R.S. Dhillon, B.L. Tufts, and Y.S. Wang. 2009. Exposure to low concentrations of dissolved ammonia promotes growth rate in walleye *Sander vitreus. J. Fish Biol.* 74:872–890.

Maeland, A. and R. Waagbo. 1998. Examination of the qualitative ability of some cold water marine teleosts to synthesize ascorbic acid. *Comp. Biochem. Physiol.* 121A:249–255.

Magnoni, L.J., E. Vaillancourt, and J.M. Weber. 2008. High resting triacylglycerol turnover of rainbow trout exceeds the energy requirements of endurance swimming. *Am. J. Physiol.* 295:R309–R315.

Magnoni, L.J. and J.M. Weber. 2007. Endurance swimming activates trout lipoprotein lipase: Plasma lipids as a fuel for muscle. *J. Exp. Biol.* 210:4016–4023.

Magnusson, C.D. and G.G. Haraldsson. 2011. Ether lipids. *Chem. Phys. Lipids* 164:315–340.

Malins, D.C. and J.C. Wekell. 1969. The lipid biochemistry of marine organisms. *Prog. Chem. Fats Lipids* 10:339–363.

Martinez-Montano, E., E. Pena, U. Focken, and M.T. Viana. 2010. Intestinal absorption of amino acids in the Pacific bluefin tuna (*Thunnus orientalis*): In vitro uptake of amino acids using hydrolyzed sardine muscle at three different concentrations. *Aquaculture* 299:134–139.

McCue, M.D. 2006. Specific dynamic action: A century of investigation. *Comp. Biochem. Physiol.* 144A:381–394.

McDonald, D.G., C.L. Milligan, W.J. McFarlane, S. Croke, S. Currie, B. Hooke, R.B. Angus, B.L. Tufts, and K. Davidson. 1998. Condition and performance of juvenile Atlantic salmon (*Salmo salar*): Effects of rearing practices on hatchery fish and comparison with wild fish. *Can. J. Fish. Aquat. Sci.* 55:1208–1219.

McDonald, R., H.J. Schreier, and J.E.M. Watts. 2012. Phylogenetic analysis of microbial communities in different regions of the gastrointestinal tract in *Panaque nigrolineatus*, a wood-eating fish. *PLoS One* 7:e48018.

McGeer, J.C., P.A. Wright, C.M. Wood, M.P. Wilkie, C.F. Mazur, and G.K. Iwama. 1994. Nitrogen excretion in four species of fish from an alkaline lake. *Trans. Am. Fish. Soc.* 123:824–829.

McKeown, B.A., J.F. Leatherland, and T.M. John. 1975. The effect of growth hormone and prolactin on the mobilization of free fatty acids and glucose in the Kokanee salmon, *Oncorhynchus nerka. Comp. Biochem. Physiol.* 50B:425–430.

McKinley, R.S., T.D. Singer, J.S. Ballantyne, and G. Power. 1993. Seasonal variation in plasma non-esterified fatty acids of lake sturgeon (*Acipenser fulvescens*) in the vicinity of hydro-electric facilities. *Can. J. Fish. Aquat. Sci.* 50:2440–2447.

Meekan, M.G. and J.H. Choat. 1997. Latitudinal variation in abundance of herbivorous fishes: A comparison of temperate and tropical reefs. *Mar. Biol.* 128:373–383.

Mehner, T. and W. Wieser. 1994. Energetics and metabolic correlates of starvation in juvenile perch (*Perca fluviatilis*). *J. Fish Biol.* 45:325–333.

Mehta, K.D., R. Chang, and J. Norman. 1996. *Chiloscyllium plagiosum* low-density lipoprotein receptor: Evolutionary conservation of five different functional domains. *J. Mol. Evol.* 42:264–272.

Merritt, T.J.S. and J.M. Quattro. 2001. Evidence for a period of directional selection following gene duplication in a neurally expressed locus in triosephosphate isomerase. *Genetics* 159:689–697.

Merritt, T.J.S. and J.M. Quattro. 2002. Negative charge correlates with neural expression in vertebrate aldolase isozymes. *J. Mol. Evol.* 55:674–683.

Metcalf, V.J., S.O. Brennan, G. Chambers, and P.M. George. 1999a. High density lipoprotein (HDL) and not albumin, is the major palmitate binding protein in New Zealand long-finned (*Anguilla dieffenbachii*) and short-finned eel (*Anguilla australis schmidtii*) plasma. *Biochim. Biophys. Acta* 1429:467–475.

Metcalf, V.J., S.O. Brennan, and P.M. George. 1999b. The Antarctic toothfish (*Dissostichus mawsoni*) lacks plasma albumin and utilizes high density lipoprotein as its major palmitate binding protein. *Comp. Biochem. Physiol.* 124B:147–155.

Metcalf, V.J. and N.J. Gemmell. 2005. Fatty acid transport in cartilaginous fish: Absence of albumin and possible utilization of lipoproteins. *Fish Physiol. Biochem.* 31:55–64.

Meyer, A. and Y. Van de Peer. 2005. From 2R to 3R: Evidence for a fish-specific genome duplication (FSGD). *BioEssays* 27:937–945.

Michel, F., B. Fossat, J. Porthe-Nibelle, B. Lahlou, and P. Saint-Marc. 1994. Effects of hyposmotic shock on taurine transport in isolated trout hepatocytes. *Exp. Physiol.* 79:983–995.

Michelsen, K.G., J.S. Harmon, and M.A. Sheridan. 1994. Adipose tissue lipolysis in rainbow trout, *Oncorhynchus mykiss*, is modulated by phosphorylation of triacylglycerol lipase. *Comp. Biochem. Physiol.* 107B:509–513.

Miller, N.G.A., M.W. Hill, and M.W. Smith. 1976. Positional and species analysis of membrane phospholipids extracted from goldfish adapted to different environmental temperatures. *Biochim. Biophys. Acta* 455:644–654.

Milligan, C.L. 1996. Metabolic recovery from exhaustive exercise in rainbow trout. *Comp. Biochem. Physiol.* 113A:51–60.

Millward, D.J. 1989. The nutritional regulation of muscle growth and protein turnover. *Aquaculture* 79:1–28.

Mommsen, T.P., C.J. French, and P.W. Hochachka. 1980. Sites and patterns of protein and amino acid utilization during the spawning migration of salmon. *Can. J. Zool.* 58:1785–1799.

Mommsen, T.P. and P.W. Hochachka. 1988. The purine nucleotide cycle as two temporally separated metabolic units: A study on trout muscle. *Metabolism* 37:552–556.

Mommsen, T.P. and T.W. Moon. 1987. The metabolic potential of hepatocytes and kidney tissue in the little skate, *Raja erinacea. J. Exp. Zool.* 244:1–8.

Mommsen, T.P. and P.J. Walsh. 1989. Evolution of urea synthesis in vertebrates: The piscine connection. *Science* 243:72–75.

Mommsen, T.P., P.J. Walsh, and T.W. Moon. 1985. Gluconeogenesis in hepatocytes and kidney of Atlantic salmon. *Mol. Physiol.* 8:89–100.

Moon, T.W. 1998. Glucagon: From hepatic binding to metabolism in teleost fish. *Comp. Biochem. Physiol.* 121B:27–34.

Moon, T.W. 2004. Hormones and fish hepatocyte metabolism: the good, the bad and the ugly. *Comp. Biochem. Physiol.* 139B:335–345.

Moon, T.W. and G.D. Foster. 1995. Tissue carbohydrate metabolism, gluconeogenesis and hormonal and environmental influences. In *Biochemistry and Molecular Biology of Fishes* (Volume 4. Metabolic Biochemistry), P.W. Hochachka and T.P. Mommsen (eds.). Elsevier Science, Amsterdam, the Netherlands, pp. 65–100.

Moon, T.W. and I.A. Johnston. 1980. Starvation and the activities of glycolytic and gluconeogenic enzymes in skeletal muscles and liver of the plaice, *Pleuronectes platessa. J. Comp. Physiol.* 136B:31–38.

Moon, T.W. and T.P. Mommsen. 1987. Enzymes of intermediary metabolism in tissues of the little skate, *Raja erinacea. J. Exp. Zool.* 244:9–15.

Moreau, R. and K. Dabrowski. 1996. The primary localization of ascorbate and its synthesis in the kidneys of Acipenserid (Chondrostei) and teleost (Teleostei) fishes. *J. Comp. Physiol.* 166B:178–183.

Moreau, R. and K. Dabrowski. 1998. Body pool and synthesis of ascorbic acid in adult sea lamprey (*Petromyzon marinus*): An agnathan fish with gulonolactone oxidase activity. *Proc. Natl. Acad. Sci. USA* 95:10279–10282.

Morgan, J.D., S.K. Balfry, M.M. Vijayan, and G.K. Iwama. 1996. Physiological responses to hyposaline exposure and handling and confinement stress in juvenile dolphin (Mahimahi: *Coryphaena hippurus*). *Can. J. Fish. Aquat. Sci.* 53:1736–1740.

Morgan, R.L., J.S. Ballantyne, and P.A. Wright. 2003. Urea transporter in kidney brush-border membrane vesicles from a marine elasmobranch, *Raja erinacea. J. Exp. Biol.* 206:3293–3202.

Morris, R.J. and F. Culkin. 1989. Fish. In *Marine Biogenic Lipids, Fats, and Oils*, R.G. Ackman (ed.). CRC Press, Boca Raton, FL, pp. 145–178.

Mourente, G. and D.R. Tocher. 1993. Incorporation and metabolism of ^{14}C-labelled polyunsaturated fatty acids in wild-caught juveniles of golden grey mullet, *Liza aurata*, in vivo. *Fish Physiol. Biochem.* 12:119–130.

Mourik, J. 1982. Anaerobic metabolism in red skeletal muscles of goldfish (*Carassius auratus* L.), PhD thesis, University of Leiden, Leiden, the Netherlands, pp. 1–163.

Mourik, J., P. Raeven, K. Steur, and A.D.F. Addink. 1982. Anaerobic metabolism of red skeletal muscle of goldfish, *Carassius auratus* (L.). *FEBS Lett.* 137:111–114.

Moyes, C.D., L.T. Buck, and P.W. Hochachka. 1990. Mitochondrial and peroxisomal fatty acid oxidation in elasmobranchs. *Am. J. Physiol.* 258:R756–R762.

Moyes, C.D., L.T. Buck, P.W. Hochachka, and R.K. Suarez. 1989. Oxidative properties of carp red and white muscle. *J. Exp. Biol.* 143:321–331.

Moyes, C.D. and C.E. Genge. 2010. Scaling of muscle metabolic enzymes: A historical perspective. *Comp. Biochem. Physiol.* 156A:344–350.

Moyes, C.D., T.W. Moon, and J.S. Ballantyne. 1986a. Osmotic effects on amino acid oxidation in skate liver mitochondria. *J. Exp. Biol.* 125:181–195.

Moyes, C.D., T.W. Moon, and J.S. Ballantyne. 1986b. Oxidation of amino acids, Krebs cycle intermediates, lipid and ketone bodies by mitochondria from the liver of *Raja erinacea*. *J. Exp. Zool.* 237:119–128.

Moyes, C.D., P.M. Schulte, and P.W. Hochachka. 1992. Recovery metabolism of trout white muscle: Role of mitochondria. *Am. J. Physiol.* 262:R295–R304.

Moyes, C.D., R.K. Suarez, G.S. Brown, and P.W. Hochachka. 1991. Peroxisomal beta-oxidation insights from comparative biochemistry. *J. Exp. Zool.* 260:267–273.

Moyes, C.M. 1996. Cardiac metabolism in high performance fish. *Comp. Biochem. Physiol.* 113A:69–75.

Nagan, N. and R.A. Zoeller. 2001. Plasmalogens: Biosynthesis and functions. *Prog. Lipid Res.* 40:199–229.

Nagata, Y., K. Horiike, and T. Maeda. 1994. Distribution of free D-serine in vertebrate brains. *Brain Res.* 634:291–295.

Nakamura, M.T. and T.Y. Nara. 2003. Structure, function, and dietary regulation of $\Delta 6$, $\Delta 5$, and $\Delta 9$ desaturases. *Ann. Rev. Nutr.* 24:345–376.

Nam, Y.K., Y.S. Cho, S.E. Douglas, J.W. Gallant, M.E. Reith, and D.S. Kim. 2002. Isolation and transient expression of a cDNA encoding L-gulono-γ-lactone oxidase, a key enzyme for L-ascorbic acid synthesis, from the tiger shark *Scyliorhinus torazame*. *Aquaculture* 209:271–284.

Navarro, I. and J. Gutierrez. 1995. Fasting and starvation. In *Biochemistry and Molecular Biology of Fishes* (Volume 4. Metabolic Biochemistry), P.W. Hochachka and T.P. Mommsen (eds.). Elsevier Science, Amsterdam, the Netherlands, pp. 393–434.

Nawata, C.M., S. Hirose, T. Nakada, C.M. Wood, and A. Kato. 2010. Rh glycoprotein expression is modulated in pufferfish (*Takifugu rubripes*) during high environmental ammonia exposure. *J. Exp. Biol.* 213:3150–3160.

Nelson, J.A., D.A. Wubah, M.E. Whitmer, E.A. Johnson, and D.J. Stewart. 1999. Wood-eating catfishes of the genus *Panaque*: Gut microflora and cellulolytic enzyme activities. *J. Fish Biol.* 54:1069–1082.

Nevenzel, J.C. 1970. Occurrence, function and biosynthesis of wax esters in marine organisms. *Lipids* 5:308–319.

Nevenzel, J.C., W. Rodegker, J.F. Mead, and M.S. Gordon. 1966. Lipids of the living coelacanth, *Latimeria chalumnae*. *Science* 152:1753–1755.

Nevenzel, J.C., W. Rodegker, J.S. Robinson, and M. Kayama. 1969. The lipids of some lantern fishes (Family: Myctophidae). *Comp. Biochem. Physiol.* 31:25–36.

Newsholme, E.A., P.H. Sugden, and T. Williams. 1977. Effect of citrate on the activities of 6-phosphofructokinase from nervous tissues from different animals and its relationship to the regulation of glycolysis. *Biochem. J.* 166:123–129.

Newsholme, E.A. and K. Taylor. 1969. Glycerol kinase activities in muscles from vertebrates and invertebrates. *Biochem. J.* 112:465–474.

Nilsson, A. and R. Fange. 1970. Digestive proteases in the cyclostome *Myxine glutinosa* (L.). *Comp. Biochem. Physiol.* 32:237–250.

Nilsson, G.E. 1988. A comparative study of aldehyde dehydrogenase and alcohol dehydrogenase activities in crucian carp and three other vertebrates: Apparent adaptations to ethanol production. *J. Comp. Physiol.* 158B:479–485.

Obled, C. and M. Arnal. 1991. Age-related changes in whole-body amino acid kinetics and protein turnover in rats. *J. Nutr.* 121:1990–1998.

Ogata, H. and T. Murai. 1988. Changes in ammonia and amino acid levels in the erythrocytes and plasma of carp, *Cyprinus carpio*, during passage through the gills. *J. Fish Biol.* 33:471–479.

Ojeda, F.P. and C.W. Caceres. 1995. Digestive mechanisms in *Aplodactylus punctatus* (Valenciennes): A temperate marine herbivorous fish. *Mar. Ecol. Prog. Ser.* 118:37–42.

Okeyo, D.O. 1989. Herbivory in freshwater fishes: A review. *Isr. J. Aq. Bamidegh* 41:79–97.

Oppelt, W.W., L. Bunim, and D.P. Rall. 1963. Distribution of glucose in the spiny dogfish (*S. acanthias*) and the brier skate (*R. eglanteria*). *Life Sci.* 7:497–503.

Oshino, N., T. Sugano, R. Oshino, and B. Chance. 1974. Mitochondrial function under hypoxic conditions: The steady states of cytochrome a + a3 and their relation to mitochondrial energy states. *Biochim. Biophys. Acta* 368:298–310.

Owen, J.M., J.W. Adron, C. Middleton, and C.B. Cowey. 1975. Elongation and desaturation of dietary fatty acids in turbot *Scophthalmus maximus* L., and rainbow trout, *Salmo gairdnerii* Rich. *Lipids* 10:528–531.

Panserat, S., E. Plagnes-Juan, and S. Kaushik. 2001. Nutritional regulation and tissue specificity of gene expression for proteins involved in hepatic glucose metabolism in rainbow trout (*Oncorhynchus mykiss*). *J. Exp. Biol.* 204:2351–2360.

Parkhouse, W.S., G.P. Dobson, and P.W. Hochachka. 1988. Organization of energy provision in rainbow trout during exercise. *Am. J. Physiol.* 254:R302–R309.

Patton, J.S. and A.A. Benson. 1975. A comparative study of wax ester digestion in fish. *Comp. Biochem. Physiol.* 52B:111–116.

Patton, J.S., J.C. Nevenzel, and A.A. Benson. 1975. Specificity of digestive lipases in hydrolysis of wax esters and triglycerides studied in anchovy and other selected fishes. *Lipids* 10:575–583.

Patton, J.S., T.G. Warner, and A.A. Benson. 1977. Partial characterization of the bile salt-dependent triacylglycerol lipase from the leopard shark pancreas. *Biochim. Biophys. Acta* 486:322–330.

Pelster, B. 1997. Buoyancy at depth. In *Fish Physiology* (Volume 16. Deep-Sea Fishes), D.J. Randall and A.P. Farrell (eds.). Academic Press Inc., New York, pp. 195–237.

Pequin, L. and A. Serfaty. 1963. L'excretion ammoniacale chez un teleosteen dulcicole: *Cyprinus carpio* L. *Comp. Biochem. Physiol.* 10:315–324.

Perlman, D.F., M.W. Musch, and L. Goldstein. 1996. Band 3 in cell volume regulation in fish erythrocytes. *Cell Mol. Biol.* 42:975–984.

Perriello, G., R. Jorde, N. Nurjhan, M. Stumvoll, G. Dailey, T. Jenssen, D.M. Bier, and J.E. Gerich. 1995. Estimation of glucose-alanine-lactate-glutamine cycles in postabsorptive humans: Role of skeletal muscle. *Am. J. Physiol.* 269:E443–E450.

Pettitt, T.R. and A.F. Rowley. 1991. Fatty acid composition and lipoxygenase metabolism in blood cells of the lesser spotted dogfish, *Scyliorhinus canicula. Comp. Biochem. Physiol.* 99B:647–652.

Phillips, J.W. and F.J.R. Hird. 1977. Ketogenesis in vertebrate livers. *Comp. Biochem. Physiol.* 57B:133–138.

Phleger, C.F. 1991. Biochemical aspects of buoyancy in fishes. In *Biochemistry and Molecular Biology of Fishes* (Volume 1. Phylogenetic and Biochemical Perspectives), P.W. Hochachka and T.P. Mommsen (eds.). Elsevier, New York, pp. 209–247.

Phleger, C.F., R.B. Holtz, and P.W. Grimes. 1977. Membrane biosynthesis in swimbladders of deep sea fishes *Coryphaenoides acrolepis* and *Antimora rostrata. Comp. Biochem. Physiol.* 56B:25–30.

Plisetskaya, E., W.W. Dickhoff, and A. Gorbman. 1983. Plasma thyroid hormones in cyclostomes: Do they have a role in regulation of glycemic levels? *Gen. Comp. Endocrinol.* 49:97–107.

Polakof, S., J.M. Miguez, and J.L. Soengas. 2009. A hepatic protein modulates glucokinase activity in fish and avian liver: A comparative study. *J. Comp. Physiol.* 179B:643–652.

Polakof, S., S. Panserat, J.L. Soengas, and T.W. Moon. 2012. Glucose metabolism in fish: A review. *J. Comp. Physiol.* 182B:1015–1045.

Polley, S.D., P.E. Tiku, R.T. Trueman, M.X. Caddick, I.Y. Morozov, and A.R. Cossins. 2003. Differential expression of cold- and diet-specific genes encoding two carp liver Δ9-acyl-CoA desaturase isoforms. *Am. J. Physiol.* 284:R41–R50.

Pottinger, T.G., M. Rand-Weaver, and J.P. Sumpter. 2003. Overwintering fasting and re-feeding in rainbow trout: Plasma growth hormone and cortisol levels in relation to energy mobilisation. *Comp. Biochem. Physiol.* 136B:403–417.

Prejs, A. and M. Blaszczzyk. 1977. Relationships between food and cellulase activity in freshwater fishes. *J. Fish Biol.* 11:447–452.

Pritchard, A.W., J.R. Hunter, and R. Lasker. 1971. The relation between exercise and biochemical changes in red and white muscle and liver in the jack mackerel, *Trachurus symmetricus. Fish. Bull.* 69:379–386.

Rao, D.S. and N. Raghuramulu, 1996. Lack of vitamin D3 synthesis in *Tilapia mossambica* from cholesterol and acetate. *Comp. Biochem. Physiol.* 114A: 21–25.

Randall, D.J., C.M. Wood, S.F. Perry, H. Bergman, G.M.O. Maloiy, T.P. Mommsen, and P.A. Wright. 1989. Urea excretion as a strategy for survival in a fish living in a very alkaline environment. *Nature* 337:165–166.

Rasmussen, R.S. and B. Korsgaard. 1998. Ammonia and urea in plasma of juvenile turbot (*Scophthalmus maximus* L.) in response to external ammonia. *Comp. Biochem. Physiol.* 120A:163–168.

Raymond, J.A. 1992. Glycerol is a colligative antifreeze in some northern fishes. *J. Exp. Zool.* 262:347–353.

Raymond, J.A. 1998. Trimethylamine oxide and urea synthesis in rainbow smelt and some other northern fishes. *Physiol. Zool.* 71:515–523.

Raymond, J.A. and A.L. DeVries. 1998. Elevated concentrations and synthetic pathways of trimethylamine oxide and urea in some teleost fishes of McMurdo Sound, Antarctica. *Fish Physiol. Biochem.* 18:387–398.

Raymond, J.A. and W.R. Driedzic. 1997. Amino acid are a source of glycerol in cold-acclimatized rainbow smelt. *Comp. Biochem. Physiol.* 118B:387–393.

Read, L.J. 1975. Absence of ureogenic pathways in liver of the hagfish *Bdellostoma cirrhatum*. *Comp. Biochem. Physiol.* 51B:139–141.

Reeck, G.R., W.P. Winter, and H. Neurath. 1970. Pancreatic enzymes of the African lungfish *Protopterus aethiopicus*. *Biochemistry* 9:1398–1403.

Renaud, J.M. and T.W. Moon. 1980a. Characterization of gluconeogenesis in hepatocytes isolated from the American eel, *Anguilla rostrata* LeSueur. *J. Comp. Physiol.* 135B:115–125.

Renaud, J.M. and T.W. Moon. 1980b. Starvation and the metabolism of hepatocytes isolated from the American eel, *Anguilla rostrata* LeSueur. *J. Comp. Physiol.* 135:127–137.

Richards, J.G., A. Bonen, G.J.F. Heigenhauser, and C.M. Wood. 2004. Palmitate movement across red and white muscle membranes of rainbow trout. *Am. J. Physiol.* 286:R46–R53.

Richards, J.G., G.J.F. Heigenhauser, and C.M. Wood. 2002. Glycogen phosphorylase and pyruvate dehydrogenase transformation in white muscle of trout during high-intensity exercise. *Am. J. Physiol.* 282:R828–R836.

Richards, J.G., G.J.F. Heigenhauser, and C.M. Wood. 2003. Exercise and recovery metabolism in the Pacific spiny dogfish *Squalus acanthias*. *J. Comp. Physiol.* 173B:463–474.

Richards, J.G., B.A. Sardella, and P.M. Schulte. 2008. Regulation of pyruvate dehydrogenase in the common killifish, *Fundulus heteroclitus*, during hypoxia exposure. *Am. J. Physiol.* 295:R979–R990.

Rimmer, D.W. and W.J. Wiebe. 1987. Fermentative microbial digestion of herbivorous fishes. *J. Fish Biol.* 31:229–236.

Robertson, J.C. and J.R. Hazel. 1995. Cholesterol content of trout plasma membranes varies with acclimation temperature. *Am. J. Physiol.* 269:R1113–R1119.

Robertson, J.D. 1989. Osmotic constituents of the blood plasma and parietal muscle of *Scyliorhinus canicula* (L.). *Comp. Biochem. Physiol.* 93A:799–805.

Robinson, J.W., D. Yanke, J. Mizra, and J.S. Ballantyne. 2011. Plasma free amino acid kinetics in rainbow trout (*Oncorhynchus mykiss*) using a bolus injection of ^{15}N-labelled amino acids. *Amino Acids* 40:689–696.

Rodela, T.M., J.S. Ballantyne, and P.A. Wright. 2008. Carrier-mediated urea transport across the mitochondrial membrane of an elasmobranch (*Raja erinacea*) and a teleost (*Oncorhynchus mykiss*) fish. *Am. J. Physiol.* 294:R1947–R1957.

Rosas, A., R. Vazquez-Duhalt, R. Tinoco, A. Shimada, L.R. Dabramo, and M.T. Viana. 2008. Comparative intestinal absorption of amino acids in rainbow trout (*Oncorhynchus mykiss*), totoaba (*Totoaba macdonaldi*) and Pacific bluefin tuna (*Thunnus orientalis*). *Aqua. Nutr.* 14:481–489.

Rovainen, C.M., O.H. Lowry, and J.V. Passonneau. 1969. Levels of metabolites and production of glucose in the lamprey brain. *J. Neurochem.* 16:1451–1458.

Ruderman, N.B. 1975. Muscle amino acid metabolism and gluconeogenesis. *Ann. Rev. Med.* 26:245–258.

Ruyter, B., O. Andersen, A. Dehli, A.K.O. Farrants, T. Gjoen, and M.S. Thomassen. 1997. Peroxisome proliferator activated receptors in Atlantic salmon (*Salmo salar*): Effects on PPAR transcription and acyl-CoA oxidase activity in hepatocytes by peroxisome proliferators and fatty acids. *Biochim. Biophys. Acta* 1348:331–338.

Sabapathy, U. and L.H. Teo. 1995. Some properties of the intestinal proteases of the rabbitfish, *Siganus canaliculatus* (Park). *Fish Physiol. Biochem.* 14:215–221.

Samerotte, A.L., J.C. Drazen, G.L. Brand, B.A. Seibel, and P.H. Yancey. 2007. Correlation of trimethylamine oxide and habitat depth within and among species of teleost fish: An analysis of causation. *Physiol. Biochem. Zool.* 80:197–208.

Sanchez-Muros, M.J., L. Garcia-Rejon, L. Garcia-Salguero, M. De La Higuera, and J.A. Lupianez. 1998. Long-term nutritional effects on the primary liver and kidney metabolism in rainbow trout. Adaptive response to starvation and a high-protein, carbohydrate-free diet on glutamate dehydrogenase and alanine aminotransferase kinetics. *Int. J. Biochem. Cell Biol.* 30:55–63.

Sand, D.M., J.L. Hehl, and H. Schlenk. 1971. Biosynthesis of wax esters in fish. Metabolism of dietary alcohols. *Biochemistry* 10:2536–2541.

Sand, D.M., C.H. Rahn, and H. Schlenk. 1973. Wax esters in fish: Absorption and metabolism of oleyl alcohol in the gourami (*Trichogaster cosby*). *J. Nutr.* 103:600–607.

Sanden, M., L. Froyland, and G.I. Hemre. 2003. Modulation of glucose-6-phosphate dehydrogenase, 6-phosphgluconate dehydrogenase and malic enzyme activity by glucose and alanine in Atlantic salmon, *Salmo salar* L. hepatocytes. *Aquaculture* 221:469–480.

Sangiao-Alvarellos, S., R. Laiz-Carrion, J.M. Guzman, M.P. Martin del Rio, J.M. Miguez, J.M. Mancera, and J.L. Soengas. 2003. Acclimation of *S. aurata* to various salinities alters energy metabolism of osmoregulatory and nonosmoregulatory organs. *Am. J. Physiol.* 285:R897–R907.

Sargent, J., R.J. Henderson, and D.R. Tocher. 1989. The lipids. In *Fish Nutrition*, J.E. Halver (ed.). Academic Press, Inc., New York, pp. 153–218.

Sargent, J.R. 1989. Ether-linked glycerides in marine animals. In *Biogenic Lipids, Fats and Oils*, Vol. I, R.G. Ackman (ed.). CRC Press, Inc., Boca Raton, FL, pp. 175–197.

Sargent, J.R., J.G. Bell, M.V. Bell, R.J. Henderson, and D.R. Tocher. 1993. The metabolism of phospholipids and polyunsaturated fatty acids in fish. In *Aquaculture: Fundamental and Applied Research*, B. Lahlou and P. Vitiello (eds.). American Geophysical Union, Washington, D.C., pp. 103–124.

Sarpeitro, M.G., T. Matsui, and H. Abe. 2003. Distribution and characteristics of D-amino acid and D-aspartate oxidases in fish tissues. *J. Exp. Zool.* 295A:151–159.

Savina, M.V. and A.B. Wojtczak. 1977. Enzymes of gluconeogenesis and the synthesis of glycogen from glycerol in various organs of the lamprey (*Lampetra fluviatilis*). *Comp. Biochem. Physiol.* 57B:185–190.

Saxrud, K.M., D.O. Lambeth, and P.M. Anderson. 1996. Isocitrate dehydrogenase from liver of *Squalus acanthias* (spiny dogfish) and citrate formation by isolated mitochondria. *J. Exp. Zool.* 274:334–345.

Scandurra, F.M. and E. Gnaiger. 2010. Cell respiration under hypoxia: Facts and artefacts in mitochondrial oxygen kinetics. *Adv. Exp. Med. Biol.* 662:7–25.

Schmidt, H. and G. Wegener. 1990. Glycogen phosphorylase in fish muscle: Demonstration of three intercovertible forms. *Am. J. Physiol.* 258:C344–C351.

Scrutton, M.C. and M.F. Utter. 1968. The regulation of glycolysis and gluconeogenesis in animal tissues. *Ann. Rev. Biochem.* 37:249–269.

Sebert, P., A. Meskar, B. Simon, and L. Barthelemy. 1994. Pressure acclimation of the eel and liver membrane composition. *Experientia* 50:121–123.

Segner, H. 1997. Ketone body metabolism in the carp *Cyprinus carpio*: Biochemical and ¹H NMR spectroscopic analysis. *Comp. Biochem. Physiol.* 116B:257–262.

Segner, H. and R. Bohm. 1994. Enzymes of lipogenesis. In *Biochemistry and Molecular Biology of Fishes* (Volume 3. Analytical Techniques), P.W. Hochachka and T.P. Mommsen (eds.). Elsevier, Amsterdam, the Netherlands, pp. 313–325.

Senkbeil, E. and H.B. White. 1978. Parallel evolution of pairs of dehydrogenase isoenzymes. *J. Mol. Evol.* 11:57–66.

Sharpe, R.L. and C.L. Milligan. 2003. Lactate efflux from sarcolemmal vesicles isolated from rainbow trout *Oncorhynchus mykiss* white muscle is via simple diffusion. *J. Exp. Biol.* 206:543–549.

Sheridan, M.A. 1988. Lipid dynamics in fish: Aspects of absorption, transportation, deposition and mobilization. *Comp. Biochem. Physiol.* 90B:679–690.

Sheridan, M.A. 1994. Regulation of lipid metabolism in poikilothermic vertebrates. *Comp. Biochem. Physiol.* 107B:495–508.

Sheridan, M.A. and J.K.L. Friedlander. 1985. Chylomicra in the serum of postprandial steelhead trout (*Salmo gairdneri*). *Comp. Biochem. Physiol.* 81B:281–284.

Shikata, T., S. Iwanaga, and S. Shimeno. 1995. Metabolic response to acclimation temperature in carp. *Fish. Sci.* 61:512–516.

Shoubridge, E.A. and P.W. Hochachka. 1980. Ethanol: Novel end product of vertebrate anaerobic metabolism. *Science* 209:308–309.

Sidell, B.D. and J.R. Hazel. 2002. Triacylglycerol lipase activities in tissues of Antarctic fishes. *Polar Biol.* 25:517–522.

Sidell, B.D., D.B. Stowe, and C.A. Hansen. 1984. Carbohydrate is the preferred metabolic fuel of the hagfish (*Myxine glutinosa*) heart. *Physiol. Zool.* 57:266–273.

Singer, T.D. and J.S. Ballantyne. 1989. Absence of extrahepatic lipid oxidation in a freshwater elasmobranch, the dwarf stingray *Potamotrygon magdalenae*: Evidence from enzyme activities. *J. Exp. Zool.* 251:355–360.

Singer, T.D. and J.S. Ballantyne. 1991. Metabolic organization of a primitive fish, the bowfin (*Amia calva*). *Can. J. Fish. Aquat. Sci.* 48:611–618.

Singer, T.D. and J.S. Ballantyne. 2004. Sturgeon and paddlefish metabolism. In *Sturgeons and Paddlefish of North America*, G.T.O. LeBreton, F.W.H. Beamish, and R.S. McKinley (eds.). Kluwer Academic Publishers, Norwell, MA, pp. 167–194.

Singer, T.D., V.G. Mahadevappa, and J.S. Ballantyne. 1990. Aspects of the energy metabolism in the lake stur-geon, *Acipenser fulvescens*: With special emphasis on lipid and ketone body metabolism. *Can. J. Fish. Aquat. Sci.* 47:873–881.

Smith, M.W. and P. Kemp. 1971. Parallel temperature-induced changes in membrane fatty acids and in the transport of amino acids by the intestine of goldfish (*Carassius auratus* L.). *Comp. Biochem. Physiol.* 39B:357–365.

Smith, R.R., G.L. Rumsey, and M.L. Scott. 1978. Heat increment associated with dietary protein, fat, carbohy-drate and complete diets in salmonids: Comparative energetic efficiency. *J. Nutr.* 108:1025–1032.

Smith, T., D.H. Wahl, and R.I. Mackie. 1996. Volatile fatty acids and anaerobic fermentation in temperate piscivorous and omnivorous freshwater fish. *J. Fish Biol.* 48:829–841.

Soengas, J.L. and M. Aldegunde. 2002. Energy metabolism of fish brain. *Comp. Biochem. Physiol.* 131B:271–296.

Soengas, J.L., S. Polakof, X. Chen, S. Sangiao-Alvarellos, and T.W. Moon. 2006. Glucokinase and hexokinase expression and activities in rainbow trout tissues: Changes with food deprivation and refeeding. *Am. J. Physiol.* 291:R810–R821.

Soengas, J.L., E.F. Strong, J. Fuentes, J.A.R. Veira, and M.D. Andres. 1996. Food deprivation and refeeding in Atlantic salmon, *Salmo salar*: Effects on brain and liver carbohydrate and ketone bodies metabolism. *Fish Physiol. Biochem.* 15:491–511.

Somero, G.N. and J.J. Childress. 1980. A violation of the metabolism-size scaling paradigm: Activities of gly-colytic enzymes in muscle increase in larger-size fish. *Physiol. Zool.* 53:322–337.

Sorensen, P.G. 1993. Changes of the composition of phospholipids, fatty acids and cholesterol from the eryth-rocyte plasma membrane from flounders (*Platichthys flesus* L.) which were acclimated to high and low temperatures in aquaria. *Comp. Biochem. Physiol.* 106B:907–923.

Speers-Roesch, B., Y.K. Ip, and J.S. Ballantyne. 2006a. Metabolic organization of freshwater, euryhaline, and marine elasmobranchs: Implications for the evolution of energy metabolism in sharks and rays. *J. Exp. Biol.* 209:2495–2508.

Speers-Roesch, B., Y.K. Ip, and J.S. Ballantyne. 2008. Plasma non-esterified fatty acids of elasmobranchs: Comparisons of temperate and tropical species and effects of environmental salinity. *Comp. Biochem. Physiol.* 149A:209–216.

Speers-Roesch, B., J.W. Robinson, and J.S. Ballantyne. 2006b. Metabolic organization of the spotted ratfish, *Hydrolagus colliei* (Holocephali: Chimaeriformes): Insight into the evolution energy metabolism in the chondrichthyan fishes. *J. Exp. Zool.* 306A:631–644.

Speers-Roesch, B. and J.R. Treberg. 2010. The unusual energy metabolism of elasmobranch fishes. *Comp. Biochem. Physiol. A Mol. Integr. Physiol.* 155:417–434.

Spener, F. and H.K. Mangold. 1971. The alkyl moieties in wax esters and alkyl diacyl glycerols of sharks. *J. Lipid Res.* 12:12–16.

Steinke, D., S. Hoegg, H. Brinkmann, and A. Meyer. 2006. Three rounds (1R/2R/3R) of genome duplications and the evolution of the glycolytic pathway in the vertebrates. *BMC Biol.* 4:16.

Sternby, B., A. Larsson, and B. Borgstrom. 1983. Evolutionary studies of pancreatic colipase. *Biochim. Biophys. Acta* 750:340–345.

Stickney, R.R. and S.E. Shumway. 1971. Occurrence of cellulase activity in the stomachs of fish. *J. Fish Biol.* 6:779–790.

Stone, D.A.J. 2003. Dietary carbohydrate utilization by fish. *Rev. Fish. Sci.* 11:337–369.

Storelli, C. and T. Verri. 1993. Nutrient transport in fish: Studies with membrane vesicles. In *Aquaculture: Fundamental and Applied Research*, B. Lahlou and P. Vitiello (eds.). American Geophysical Union, Washington, D.C., pp. 139–158.

Storelli, C., S. Vilella, M.P. Romano, M. Maffia, and G. Cassano. 1989. Brush-border amino acid transport mechanisms in carnivorous eel intestine. *Am. J. Physiol.* 257:R506–R510.

Stuart, J.A., J.A. Harper, K.M. Brindle, and M.D. Brand. 1999. Uncoupling protein 2 from carp and zebrafish, ectothermic vertebrates. *Biochim. Biophys. Acta* 1413:50–54.

Suarez, R.K., M.D. Mallet, C. Daxboeck, and P.W. Hochachka. 1986. Enzymes of energy metabolism and glu-coneogenesis in the Pacific blue marlin, *Makaira nigricans*. *Can. J. Zool.* 64:694–697.

Suarez, R.K. and T.P. Mommsen. 1987. Gluconeogenesis in teleost fishes. *Can. J. Zool.* 65:1869–1882.

Sugano, T., N. Oshino, and B. Chance. 1974. Mitochondrial functions under hypoxic conditions. The steady states of cytochrome c reduction and of energy metabolism. *Biochim. Biophys. Acta* 347:340–358.

Tacon, A.G.J. and C.B. Cowey. 1985. Protein and amino acid requirements. In *Fish Energetics: New Perspectives*, P. Tytler and P. Calow (eds.). The Johns Hopkins University Press, Baltimore, MD, pp. 155–183.

Tandler, A. and F.W.H. Beamish. 1980. Specific dynamic action and diet in largemouth bass, *Micropterus salmoides* (Lacepede). *J. Nutr.* 110:750–764.

Teichert, T. and E. Wodtke. 1992. Acyl-CoA:cholesterol acyltransferase and 3-hydroxy-3-methylglutaryl-CoA reductase in carp-liver microsomes: Effect of cold acclimation on enzyme activities and on hepatic and plasma lipid composition. *Biochim. Biophys. Acta* 1165:211–221.

Terjesen, B.F., T.D. Chadwick, J.A.J. Verreth, I. Ronnestad, and P.A. Wright. 2001. Pathways for urea production during early life of an air-breathing teleost, the African catfish *Clarias gariepinus* Burchell. *J. Exp. Biol.* 204:2155–2165.

Terjesen, B.F., I. Ronnestad, B. Norberg, and P.M. Anderson. 2000. Detection and basic properties of carbamoyl phosphate synthetase III during teleost ontogeny: A case study in the Atlantic halibut (*Hippoglossus hippoglossus* L.). *Comp. Biochem. Physiol.* 126B:521–535.

Terova, G., S. Rimoldi, F. Brambilla, R. Gornati, G. Bernardini, and M. Saroglia. 2009. In vivo regulation of GLUT2 mRNA in sea bass (*Dicentrarchus labrax*) in response to acute and chronic hypoxia. *Comp. Biochem. Physiol.* 152B:306–316.

Thoroed, S.M. and K. Fugelli. 1994a. Free amino compounds and cell volume regulation in erythrocytes from different marine fish species under hypoosmotic conditions: The role of a taurine channel. *J. Comp. Physiol.* 164B:1–10.

Thoroed, S.M. and K. Fugelli. 1994b. The Na^+-independent taurine influx in flounder erythrocytes and its association with the volume regulatory taurine efflux. *J. Exp. Biol.* 186:245–268.

Tiku, P.E., A.Y. Gracey, A.I. Macartney, R.J. Beynon, and A.R. Cossins. 1996. Cold-induced expression of Δ^9-desaturase in carp by transcriptional and posttranslational mechanisms. *Science* 271:815–818.

Titus, E. and G.A. Ahearn. 1988. Short-chain fatty acid transport in the intestine of a herbivorous teleost. *J. Exp. Biol.* 135:77–94.

Titus, E. and G.A. Ahearn. 1991. Transepithelial acetate transport in a herbivorous teleost: Anion exchange at the basolateral membrane. *J. Exp. Biol.* 156:41–61.

Tocher, D.R. 2003. Metabolism and functions of lipids and fatty acids in teleost fish. *Rev. Fish. Sci.* 11:107–184.

Tocher, D.R., J. Carr, and J.R. Sargent. 1989. Polyunsaturated fatty acid metabolism in fish cells: Differential metabolism of (n-3) and (n-6) series acid by cultured cells originating from a freshwater teleost fish and from a marine teleost fish. *Comp. Biochem. Physiol.* 94B:367–374.

Tocher, D.R. and J.R. Sargent. 1984. Studies on triacylglycerol wax ester and sterol ester hydrolases in intestinal caeca of rainbow trout (*Salmo gairdneri*) fed diets rich in triacylglycerols and wax esters. *Comp. Biochem. Physiol.* 77B:561–571.

Torres, J.J., M.D. Grigsby, and M.E. Clarke. 2012. Aerobic and anaerobic metabolism in oxygen minimum layer fishes: The role of alcohol dehydrogenase. *J. Exp. Biol.* 215:1905–1914.

Tracy, M.R. and S.B. Hedges. 2000. Evolutionary history of the enolase gene family. *Gene* 259:129–138.

Treberg, J.R. and W.R. Driedzic. 2007. The accumulation and synthesis of betaine in winter skate (*Leucoraja ocellata*). *Comp. Biochem. Physiol.* 147A:475–483.

Treberg, J.R., B. Speers-Roesch, P.M. Piermarini, Y.K. Ip, J.S. Ballantyne, and W.R. Driedzic. 2006. The accumulation of methylamine counteracting solutes in elasmobranchs with differing levels of urea: A comparison of marine and freshwater species. *J. Exp. Biol.* 209:860–870.

Trueman, R.J., P.E. Tiku, M.X. Caddick, and A.R. Cossins. 2000. Thermal thresholds of lipid restructuring and Δ9-desaturase expression in the liver of carp (*Cyprinus carpio*). *J. Exp. Biol.* 203:641–650.

Tseng, Y.C., R.D. Chen, J.R. Lee, S.T. Lin, and P.P. Hwang. 2009. Specific expression and regulation of glucose transporters in zebrafish ionocytes. *Am. J. Physiol.* 297:R275–R290.

Van den Thillart, G. and H. Smit. 1984. Carbohydrate metabolism of goldfish (*Carassius auratus* L.). Effects of long-term hypoxia-acclimation on enzyme patterns of red muscle, white muscle and liver. *J. Comp. Physiol.* 154B:477–486.

Van Ginneken, V.J.T., E. Antonissen, U.K. Muller, R. Booms, E. Eding, J. Verreth, and G. Van den Thillart. 2005. Eel migration to the Sargasso: Remarkably high swimming efficiency and low energy costs. *J. Exp. Biol.* 208:1329–1335.

van Hoek, M.J.A. and P. Hogeweg. 2009. Metabolic adaptation after whole genome duplication. *Mol. Biol. Evol.* 26:2441–2453.

van Raaij, M.T.M. 1994. The level and composition of free fatty acids in the plasma of freshwater fish in postabsorptive condition. *Comp. Biochem. Physiol.* 109A:1067–1074.

van Raaij, M.T.M., G.E.E.J.M. van den Thillart, M. Hallemeesch, P.H.M. Balm, and A.B. Steffens. 1995. Effect of arterially infused catecholamines and insulin on plasma glucose and free fatty acids in carp. *Am. J. Physiol.* 268:R1163–R1170.

van Raaij, M.T.M., G.E.E.J.M. van den Thillart, G.J. Vianen, D.S.S. Pit, P.H.M. Balm, and A.B. Steffens. 1996. Substrate mobilization and hormonal changes in rainbow trout (*Oncorhynchus mykiss*, L.) and common carp (*Cyprinus carpio*, L.) during deep hypoxia and subsequent recovery. *J. Comp. Physiol.* 166B:443–452.

van Waarde, A. 1981. Nitrogen metabolism in goldfish *Carassius auratus* (L.). Activities of transamination reactions, purine nucleotide cycle and glutamate dehydrogenase in goldfish tissues. *Comp. Biochem. Physiol.* 68B:407–413.

van Waarde, A. 1983. Aerobic and anaerobic ammonia production by fish. *Comp. Biochem. Physiol.* 74B:675–684.

van Waarde, A. 1988. Biochemistry of non-protein nitrogenous compounds in fish including the use of amino acids for anaerobic energy production. *Comp. Biochem. Physiol.* 91B:207–228.

van Waarde, A. 1991. Alcoholic fermentation in multicellular organisms. *Physiol. Zool.* 64:895–920.

van Waarde, A., I. De Graaf, G. Van den Thillart, and C. Erkelens. 1991. Acidosis (measured by nuclear magnetic resonance) and ethanol production in anoxic goldfish acclimated to 5 and 20°C. *J. Exp. Biol.* 159:387–405.

Vijayan, M.M., J.S. Ballantyne, and J.F. Leatherland. 1990. High stocking density alters the energy metabolism of brook charr, *Salvelinus fontinalis*. *Aquaculture* 88:371–381.

Vijayan, M.M., A.G. Maule, C.B. Schreck, and T.W. Moon. 1993. Hormonal control of hepatic glycogen metabolism in food-deprived, continuously swimming coho salmon (*Oncorhynchus kisutch*). *Can. J. Fish. Aquat. Sci.* 50:1676–1682.

Vislie, T. 1980. Cell volume regulation in isolated, perfused heart ventricle of the flounder (*Platichthys flesus*). *Comp. Biochem. Physiol.* 65A:19–27.

Vislie, T. and K. Fugelli. 1975. Cell volume regulation in flounder (*Platichthys flesus*) heart muscle accompanying an alteration in plasma osmolality. *Comp. Biochem. Physiol.* 52A:415–418.

Vornanen, M. and V. Paajanen. 2006. Seasonal changes in glycogen content and Na^+-K^+-ATPase activity in the brain of crucian carp. *Am. J. Physiol.* 291:R1482–R1489.

Walsh, P.J. and T.P. Mommsen. 2001. Evolutionary considerations of nitrogen metabolism and excretion. In *Fish Physiology* (Volume 20. Nitrogen Excretion), P.A. Wright and P.M. Anderson (eds.). Academic Press, San Diego, CA, pp. 1–30.

Walton, M.J. 1985. Aspects of amino acid metabolism in teleost fish. In *Nutrition and Feeding in Fish*, C.B. Cowey, A.M. Mackie, and J.G. Bell (eds.). Academic Press, Inc., New York, pp. 47–67.

Walton, M.J. and C.B. Cowey. 1979. Gluconeogenesis from serine in rainbow trout *Salmo gairdneri* liver. *Comp. Biochem. Physiol.* 62B:497–499.

Walton, M.J. and C.B. Cowey. 1981. Distribution and some kinetic properties of serine catabolizing enzymes in rainbow trout *Salmo gairdneri*. *Comp. Biochem. Physiol.* 68B:147–150.

Walton, M.J. and C.B. Cowey. 1982. Aspects of intermediary metabolism in salmonid fish. *Comp. Biochem. Physiol.* 73B:59–79.

Wang, T., C.C.Y. Hung, and D.J. Randall. 2006. The comparative physiology of food deprivation: From feast to famine. *Ann. Rev. Physiol.* 68:223–251.

Wang, Y., G.J.F. Heigenhauser, and C.M. Wood. 1994. Integrated responses to exhaustive exercise and recovery in rainbow trout white muscle: Acid-base, phosphagen, carbohydrate, lipid, ammonia, fluid volume and electrolyte metabolism. *J. Exp. Biol.* 195:227–258.

Wang, Y., P.M. Wright, G.J.F. Heigenhauser, and C.M. Wood. 1997. Lactate transport by rainbow trout white muscle kinetic characteristics and sensitivity to inhibitors. *Am. J. Physiol.* 272:R1577–R1587.

Watanabe, T. 1982. Lipid nutrition in fish. *Comp. Biochem. Physiol.* 73B:3–15.

Watson, H.R. and D.B. Lindsay. 1972. 3-Hydroxybutyrate dehydrogenase in tissues from normal and ketonaemic sheep. *Biochem. J.* 128:53–57.

Watt, P.W., P.A. Marshall, S.P. Heap, P.T. Loughna, and G. Goldspink. 1988. Protein synthesis in tissues of fed and starved carp, acclimated to different temperatures. *Fish Physiol. Biochem.* 4:165–173.

Weber, J.M., R.W. Brill, and P.W. Hochachka. 1986. Mammalian metabolic flux rates in a teleost: Lactate and glucose turnover in tuna. *Am. J. Physiol.* 250:R452–R458.

Wilkie, M.P., J.F. Claude, A. Cockshut, J.A. Holmes, Y.S. Wang, J.H. Youson, and P.J. Walsh. 2006. Shifting patterns of nitrogen excretion and amino acid catabolism capacity during the life cycle of the sea lamprey (*Petromyzon marinus*). *Physiol. Biochem. Zool.* 79:885–898.

Wilkie, M.P., Y. Wang, P.J. Walsh, and J.H. Youson. 1999. Nitrogenous waste excretion by the larvae of a phylogenetically ancient vertebrate: The sea lamprey (*Petromyzon marinus*). *Can. J. Zool.* 77:707–715.

Wilson, D.F., M. Erecinska, C. Drown, and I.A. Silver. 1979. The oxygen dependence of cellular energy metabolism. *Arch. Biochem. Biophys.* 195:485–493.

Wilson, R.P. and W.E. Poe. 1987. Apparent inability of channel catfish to utilize dietary mono- and disaccharides as energy sources. *J. Nutr.* 117:280–285.

Wissing, J. and E. Zebe. 1988. The anaerobic metabolism of the bitterling *Rhodeus amarus* (Cyprinidae, Teleostei). *Comp. Biochem. Physiol.* 89B:299–303.

Wodtke, E. 1978. Lipid adaptation in liver mitochondrial membranes of carp acclimated to different environmental temperatures. Phospholipid composition, fatty acid pattern, and cholesterol content. *Biochim. Biophys. Acta* 529:280–291.

Wodtke, E. 1983. On adaptation of biomembranes to temperature: Membrane dynamic and membrane functions. *J. Thermal Biol.* 8:416–420.

Wood, C.M. 1993. Ammonia and urea metabolism and excretion. In *The Physiology of Fishes*, D.H. Evans (ed.). CRC Press, Boca Raton, FL, pp. 379–425.

Wood, C.M., A.Y.O. Matsuo, R.J. Gonzalez, R.W. Wilson, M.L. Patrick, and A.L. Val. 2002. Mechanisms of ion transport in *Potamotrygon*, a stenohaline freshwater elasmobranch native to the ion-poor blackwaters of the Rio Negro. *J. Exp. Biol.* 205:3039–3054.

Wood, C.M., P. Part, and P.A. Wright. 1995. Ammonia and urea metabolism in relation to gill function and acid-base balance in a marine elasmobranch, the spiny dogfish (*Squalus acanthias*). *J. Exp. Biol.* 198:1545–1558.

Wood, C.M., S.F. Perry, P.A. Wright, H.L. Bergman, and D.J. Randall. 1989. Ammonia and urea dynamics in the Lake Magadi tilapia, a ureotelic teleost fish adapted to an extremely alkaline environment. *Resp. Physiol.* 77:1–20.

Wood, C.M., P.J. Walsh, M. Kajimura, G.B. McClelland, and S.F. Chew. 2010. The influence of feeding and fasting on plasma metabolites in the dogfish shark (*Squalus acanthias*). *Comp. Biochem. Physiol.* 155A:435–444.

Wright, P.A. 2007. Ionic, osmotic, and nitrogenous waste regulation. In *Fish Physiology* (Volume 26. Primitive Fishes), D.J. McKenzie, A.P. Farrell, and C.J. Brauner (eds.). Elsevier, Amsterdam, the Netherlands, pp. 283–318.

Wright, P.A., A. Campbell, R.L. Morgan, A.G. Rosenberger, and B.W. Murray. 2004. Dogmas and controversies in the handling of nitrogenous wastes: Expression of arginase Type I and II genes in rainbow trout: Influence of fasting on liver enzyme activity and mRNA levels in juveniles. *J. Exp. Biol.* 207:2033–2042.

Wright, P.A., A. Felskie, and P.M. Anderson. 1995. Induction of ornithine-urea cycle enzymes and nitrogen metabolism and excretion in rainbow trout (*Oncorhynchus mykiss*) during early life stages. *J. Exp. Biol.* 198:127–135.

Wright, P.A., C.M. Wood, and D.J. Randall. 1988. An in vitro and in vivo study of the distribution of ammonia between plasma and red cells of rainbow trout (*Salmo gairdneri*). *J. Exp. Biol.* 134:423–428.

Yancey, P.H., A.L. Fyfe-Johnson, R.H. Kelly, V.P. Walker, and M.T. Aunon. 2001. Trimethylamine oxide counteracts effects of hydrostatic pressure on proteins of deep-sea teleosts. *J. Exp. Zool.* 289:172–176.

Yancey, P.H. and G.N. Somero. 1979. Counteraction of urea destabilization of protein structure by methylamine osmoregulatory compounds of elasmobranch fishes. *Biochem. J.* 183:317–323.

Yang, T., S.J. Forrest, N. Stine, Y. Endo, A. Pasumarthy, H. Castrop, S. Aller, J.N. Forrest, J. Schnermann, and J. Briggs. 2002. Cyclooxygenase cloning in dogfish shark, *Squalus acanthias*, and its role in rectal gland Cl secretion. *Am. J. Physiol.* 283:R631–R637.

Yokoyama, M., T. Takeuchi, G.S. Park, and J. Nakazoe. 2001. Hepatic cysteinesulfinate decarboxylase activity in fish. *Aquaculture Res.* 32 (Suppl. 1):216–220.

Yonezawa, S. and S.H. Hori. 1977. Studies on phosphorylase isozymes in lower vertebrates. Purification and properties of lamprey phosphorylase. *Arch. Biochem. Biophys.* 181:447–453.

Yonezawa, S. and S.H. Hori. 1979. Studies on phosphorylase isozymes in lower vertebrates. Evidence for the presence of two isozymes in elasmobranchs. *Arch. Biochem. Biophys.* 197:149–157.

Youson, J.H., E.W. Sidon, W.D. Peek, and R.R. Shivers. 1985. Ultrastructure of the hepatocytes in a vertebrate liver without bile ducts. *J. Anat.* 140:143–158.

Zammit, V.A. and E.A. Newsholme. 1979a. Activities of enzymes of fat and ketone body metabolism and effects of starvation on blood concentrations of glucose and fat fuels in teleost and elasmobranch fish. *Biochem. J.* 184:313–322.

Zammit, V.A. and E.A. Newsholme. 1979b. The role of 3-oxo acid-CoA transferase in the regulation of ketogenesis in the liver. *FEBS Lett.* 103:212–215.

Zara, V., L. Palmieri, A. Giudetti, A. Ferramosca, L. Capobianco, and G.V. Gnoni. 2000. The mitochondrial tricarboxylate carrier: Unexpected increased activity in starved eels. *Biochem. Biophys. Res. Commun.* 276:893–898.

Zehmer, J.K. and J.R. Hazel. 2003. Plasma membrane rafts of rainbow trout are subject to thermal acclimation. *J. Exp. Biol.* 206:1657–1667.

Zendzian, E.N. and E.A. Barnard. 1967. Distributions of pancreatic ribonuclease, chymotrypsin, and trypsin in vertebrates. *Arch. Biochem. Biophys.* 122:699–713.

Zhou, X., J.J. Korte, and P.M. Anderson. 1995. Purification and properties of two malic enzyme activities in liver mitochondria of *Squalus acanthias* (spiny dogfish). *J. Exp. Zool.* 272:201–212.

4 Oxygen Sensing

Michael G. Jonz

CONTENTS

4.1 INTRODUCTION

Oxygen (O_2) is a gaseous molecule capable of accepting electrons and participating in oxidation–reduction reactions to promote the production of metabolic energy within cells. It is this dependence on O_2 for cellular energy that defines aerobic organisms and is a characteristic of most eukaryotes. Complex multicellular organisms, such as fishes, have evolved elaborate mechanisms through which uptake of O_2 across the gills, and subsequent delivery to the body tissues, can be regulated so as to defend the animal against hypoxic or ischemic damage. O_2 chemoreceptors, cells that display specific responses to chemical change, detect environmental or arterial hypoxia (low O_2) and initiate a suite of compensatory physiological adjustments, which may include increased frequency of gill ventilation (i.e., hyperventilation), increased ventilatory amplitude, decreased heart rate, vascular changes, and aquatic surface respiration (Perry et al., 2009; Milsom, 2012). These adaptations are fundamental for survival in environments with fluctuating O_2 availability. Physiological responses to hypoxia are exquisitely exemplified in fishes because they live in an aqueous environment where the solubility of O_2 is relatively low compared to that of air (Burleson et al., 1992). Thus, naturally occurring fluctuations in the partial pressure of O_2 (PO_2) will have profound physiological consequences, compared with air-breathing or terrestrial species.

A number of reviews have recently summarized the literature on chemoreceptor distribution and orientation, the hyperventilatory response to hypoxia, CO_2 sensitivity of gill chemoreceptors, gill neurotransmitters, and gill development, and the interested reader is directed to these comprehensive works for additional information (Perry et al., 2009; Brauner and Rombough, 2012; Milsom, 2012; Perry and Abdallah, 2012; Porteus et al., 2012). In this chapter, I present the most salient features from studies that have investigated O_2 chemoreceptors in fishes and contributed specifically to our understanding of O_2 "sensing" by these cells. This will include a morphological description of chemoreceptors at the cellular level, their biochemistry and innervation, a current working model for how chemoreceptors detect changes in O_2 and other chemical stimuli, and evidence from recent studies that are beginning to unravel the remarkable developmental transitions related to O_2 sensing in fishes.

4.2 RESPIRATORY CHEMORECEPTORS

4.2.1 Chemoreceptors Defined

The term "chemoreceptor" refers to a specialized cell type that is responsive to a chemical signal. Classical examples of chemoreceptors include those of the olfactory and taste organs, which mediate the sense of smell and taste, respectively. Such chemoreceptors may "sense" specific molecular or ionic species and elicit subsequent downstream effects that result in the perception of smell or taste (Farbman, 2000; Finger and Simon, 2000). The major difference between these types of chemoreceptors and respiratory chemoreceptors is that the latter perform an autonomic role, rather than a sense that can be perceived, and initiate respiratory, cardiac, and behavioral changes. Nevertheless, respiratory chemoreceptors respond to changes in the respiratory gases (O_2 and CO_2) as well as other signals, such as H^+ (i.e., a change in pH). As will be described in a later section, more recent evidence indicates the polymodal nature of respiratory chemoreceptors. It now appears that these chemoreceptors play other as yet undescribed roles.

In addition to chemical sensitivity, respiratory chemoreceptors generally have a number of typical morphological and physiological characteristics. These include (but are not limited to) exposure or access to a chemical environment, the ability to transduce a chemical signal into a change in membrane potential (V_m), the storage and release of neurotransmitter(s), and innervation by terminals of sensory neurons to relay the cell's response from the periphery to the central nervous system (CNS).

4.2.2 Distribution of Chemoreceptors in Fishes and Mammals

In vertebrates, respiratory chemoreceptors that are responsive to changes in O_2, CO_2, or H^+ may be found in the central or peripheral nervous system. In water-breathing fishes, however, there is little evidence for central O_2 and CO_2 sensing (Milsom, 2012). This chapter will therefore focus primarily on peripheral mechanisms of O_2 chemoreception that take place outside of the (CNS). These sites include the gills (or branchial arches) as well as extrabranchial regions.

In all embryonic vertebrates (i.e., amniotes and anamniotes), a series of bilateral aortic arches extend from the ventral aorta to paired dorsal aortae, forming the anterior arterial system (Figure 4.1a). In fishes, these aortic arches course through the pharyngeal arches and provide blood to the gills once they develop (Weichert, 1967). Although most vertebrates have six pairs of aortic arches during embryogenesis, the number of aortic arches is usually reduced by the time the adult stage is reached. In teleost fishes, the first two arches, the mandibular and the hyoid (I and II), degenerate and leave aortic arches III, IV, V, and VI to develop with the gills (i.e., gill arches 1, 2, 3, and 4; Figure 4.1b). By contrast, in mammals, aortic arch V is lost, in addition to I and II (Figure 4.1c). This leaves aortic arches III, IV, and VI that go on to form the internal carotid artery, the aortic arch proper, and

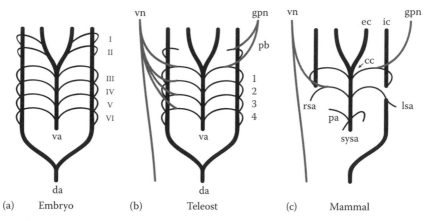

FIGURE 4.1 **(See color insert.)** The aortic arches and the distribution of chemoreceptors in teleosts and mammals. Associated innervations from the glossopharyngeal (gpn, IX) and vagus (vn, X) nerves are shown in red. Innervation is bilateral but is displayed individually for clarity. (a) The embryonic condition, with six aortic arches (I–VI) connecting the ventral aorta (va) to paired dorsal aortae (da). Numerals correspond to homologous structures in (b) and (c). (b) In teleosts, the first aortic arch is absent, the second is modified into the nonrespiratory pseudobranch (pb), and aortic arches III–VI become gill arches 1–4, respectively. The pb and first gill arch receive gpn innervation, whereas all four arches receive vn innervation. (c) In mammals, the ventral aorta is modified to the common carotid (cc), which bifurcates to the external and internal carotids (ec, ic, respectively). The right da is lost. The va is modified to become the systemic aorta (sysa). The site of the carotid body, the cc, receives gpn innervation. (From Zachar, P.C. and Jonz, M.G., *Respir. Physiol. Neurobiol.*, 184, 301, 2012a; With permission.)

pulmonary artery, respectively (Weichert, 1967). The organization of the aortic arches in vertebrates has therefore remained relatively conserved throughout phylogenesis. Likewise, the organization of the gills in fishes reflects a common phylogenetic or ancestral origin. As a beautiful example of this, the first gill arch (i.e., aortic arch III) in fishes is homologous to the internal carotid artery in adult mammals, a site just distal to the bifurcation of the common carotid artery. As will be discussed in the following sections, these regions are important because they represent sites of O_2 sensing. The mammalian carotid body is a paired organ that is associated with the internal carotid artery and is composed of chemoreceptive type I cells (also called glomus cells). In the field of O_2 sensing, much of the literature is dominated by studies carried out on the mammalian carotid body. The interested reader may consult a number of very good reviews on this subject (e.g., Gonzalez et al., 1994; Lahiri et al., 2006; Prabhakar, 2006; Kumar and Prabhakar, 2007; Fitzgerald et al., 2009; López-Barneo et al., 2009; Nurse, 2010; Peers et al., 2010). Because of its homology with the carotid body, the first gill arch in fishes has received considerable attention as the primary site of O_2 sensing. In fact, all of the gill arches have similar morphology and all appear to be a site of O_2 sensing. In the following sections, the gills, other extrabranchial sites, and their potential roles in O_2 sensing will be described.

4.3 GILLS

4.3.1 Gill Organization and Anatomy

Teleost fishes have four bilateral pairs of gill arches, and these contain numerous gill filaments that are further divided into two parallel hemibranchs. Each gill filament in turn gives rise to many lamellae, where gas exchange occurs during respiration. Blood flows into the gill filaments from the branchial arteries by way of the afferent filament arteries. It then courses through the vascular sinus of the lamellae and back to the systemic circulation by the efferent filament arteries. In this manner, blood within the gill vasculature generally moves opposite in direction to the flow of incident water over the gills, thereby maximizing gas exchange during ventilation.

The gill filaments and lamellae are covered by a thin epithelium composed of several cell types that provides a boundary between the external environment and the extracellular fluids. In addition to O_2 chemoreceptors, other important epithelial cell types include pavement cells, which mediate gas exchange across the lamellar epithelium, and ionoregulatory mitochondrion-rich cells, which control the ionic composition of the blood. The fine structure of the gill epithelium has been reviewed extensively (Laurent and Dunel, 1980; Hughes, 1984; Laurent, 1984; Olson, 2002; Wilson and Laurent, 2002; Evans et al., 2005a; Bailly, 2009; Jonz and Nurse, 2009; Dymowska et al., 2012).

4.3.2 Nerve Supply

In the embryonic vertebrate plan, four cranial nerves supply sensory and/or motor fibers to the pharyngeal region. The trigeminal nerve (V) innervates the mandibular arch; the facial nerve (VII) supplies the hyoid arch; the glossopharyngeal nerve (IX) innervates the third arch (i.e., first gill arch); and the vagus nerve (X) innervates the remaining posterior arches (Weichert, 1967; Gilbert, 2010). As development proceeds in fishes, the pharyngeal arches become specialized to support gills (see Figure 4.1), and branches of the glossopharyngeal and vagus nerves become the primary sources of innervation to the gill arches (Nilsson, 1984; Sundin and Nilsson, 2002).

In teleost fishes, the gill arches are innervated primarily by pretrematic (anterior to the gill slit) and posttrematic (posterior to the gill slit) branches of the glossopharyngeal and vagus nerves, which carry parasympathetic and motor (efferent) fibers to the gill vasculature and skeletal muscles and visceral sensory (afferent) fibers from chemoreceptors of the gill filaments (Figure 4.2a and b; Nilsson, 1984; de Graaf, 1990; Sundin and Nilsson, 2002). Most nerve fibers that enter the gill filaments and respiratory lamellae are carried by the branchial nerve, a large nerve trunk that runs the length of the arch and projects into the filaments at their base. This innervation can be described as extrabranchial or extrinsic, since the neuronal cell bodies of these nerve fibers reside outside the gill. In addition, intrabranchial or intrinsic innervation of the gills has also been observed. Several studies have demonstrated the presence of neuronal cell bodies (and their associated axons) in the gills in many fish species (Donald, 1984, 1987; Bailly and Dunel-Erb, 1986; Dunel-Erb and Bailly, 1986; Bailly et al., 1989, 1992; Dunel-Erb et al., 1989; de Graaf et al., 1990; Jonz and Nurse, 2003; Porteus et al., 2013).

4.3.3 Gill Neuroepithelial Cells

4.3.3.1 Morphology

The term "neuroepithelial cell" (NEC) in the present context refers specifically to a type of cell of epithelial origin possessing specialized sensory function. NECs include, among other cell types, taste cells, olfactory cells, mechanosensitive hair cells (such as in the cochlea), and respiratory chemoreceptors. NECs of the fish gill were first described in teleost fishes by Dunel-Erb et al. (1982) (Figure 4.3a), and three decades of morphological study since then using electron microscopy, immunohistochemistry, and confocal imaging have demonstrated that these cells display the structural characteristics typical of O_2 chemoreceptors (Bailly et al., 1992; Zaccone et al., 1997; Jonz and Nurse, 2003; Saltys et al., 2006; Vulesevic et al., 2006; Coolidge et al., 2008; Bailly, 2009; Regan et al., 2011; Tzaneva et al., 2011; Porteus et al., 2012, 2013). NECs reside in the gill filament epithelium (Figure 4.3b and c), where they may be equally exposed to the incident flow of water during ventilation and the arterial blood supply. All NECs are confined to the efferent aspects of the filament and lamellar epithelium. The "efferent" designation refers specifically to the flow of oxygenated arterial blood away from the gas exchange surfaces of the gills as it returns to the gill arches and systemic circulation. Within the primary epithelium, NECs are concentrated at the distal regions of the filaments (Bailly et al., 1992; Jonz and Nurse, 2003; Saltys et al., 2006; Coolidge et al., 2008; Porteus et al., 2012), while in the secondary epithelium, a population of NECs are usually present in the efferent epithelium of the lamellae. In zebrafish, lamellar NECs are more numerous in the

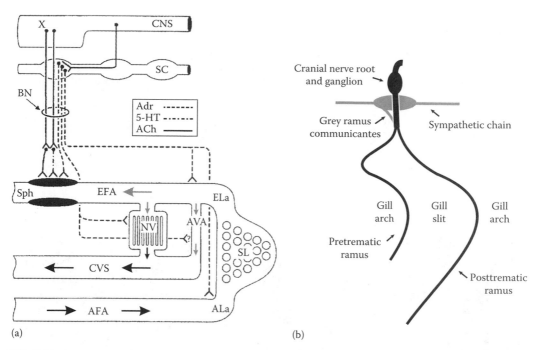

(a)   (b)

FIGURE 4.2 (a) Working model for innervation of the teleost gill by cranial and spinal autonomic nerves. The cranial ("parasympathetic") pathway includes nerve fibers traveling through the vagus (X) nerve and postganglionic neurons (indicated), while the spinal ("sympathetic") pathway travels through the sympathetic chain ganglia and the vagosympathetic trunk. Adrenergic, cholinergic, and serotonergic nerve fibers are indicated. A muscular "sphincter" is also shown at the proximal region of the efferent filament artery. Black arrows indicate the flow of deoxygenated blood and grey arrows indicate oxygenated blood. 5-HT, serotonergic nerve; ACh, cholinergic nerve; Adr, adrenergic nerve; AFA, afferent filament artery; ALa, afferent lamellar artery; AVA, arteriovenous anastomoses; BN, branchial nerve; CNS, central nervous system; CVS, central venous sinus; EFA, efferent filament artery; ELa, efferent lamellar artery; NV, nutritive vasculature; SC, sympathetic chain; SL, secondary (respiratory) lamellae; Sph, sphincter; X, vagus nerve. (Modified from Nilsson, S. and Sundin, L., *Comp. Biochem. Physiol. A Mol. Integr. Physiol.*, 119, 137, 1998. With permission.) (b) Simplified schematic of innervation to the gill region. The gills are innervated by nerve fibers (both sensory and motor) originating from the cranial nerves and by fibers that originate from the sympathetic chain and enter the cranial nerve trunk via grey rami communicantes. Each gill arch is innervated by a pretrematic ramus (anterior to the gill slit) and a posttrematic ramus (posterior to the gill slit) originating from different cranial nerve branches. Pretrematic rami carry sensory fibers, while posttrematic rami carry sensory and motor fibers. (From Jonz, M.G. and Zaccone, G., *Acta Histochem.*, 111, 207, 2009. With permission.)

proximal lamellae (i.e., lamellae nearest to the gill arch) compared to those lamellae located distally along the filaments (Jonz and Nurse, 2003). The location of gill NECs specifically to the distal filaments and proximal lamellae may provide further advantages for detection of changes in water PO_2. The distal regions of the filaments may receive a greater flow of incident water over the gills, thereby maximizing potential exposure to changes in water PO_2; while the lamellae that are found in a more proximal location along the filaments apparently receive a more continuous flow of inspired water during resting and hypoxic conditions (Hughes, 1972; Booth, 1978) and may provide a reliable measure of water PO_2. Although lamellar NECs have been shown to be exposed directly to the external environment (Jonz and Nurse, 2003), these cells have not yet been shown to respond directly to changes in PO_2. There is one notable exception to the typical distribution of NECs in the gill. Rainbow trout (*Oncorhynchus mykiss*) do not appear to possess NECs of the lamellae (Saltys et al., 2006; Porteus et al., 2012). While the reason for this is presently unclear, the absence of lamellar NECs in trout is presumably the result of evolutionary adaptation or natural history.

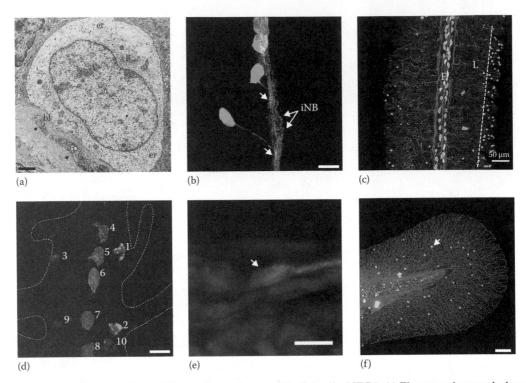

FIGURE 4.3 **(See color insert.)** Morphology of neuroepithelial cells (NECs). (a) Electron micrograph showing ultrastructural characteristics of a single NEC of the filament in trout (*Salmo gairdneri*). The NEC is apposed to the basal lamina (bl). Small arrows indicate tonofilaments of surrounding cells. Several nerve profiles reside beneath the basal lamina (arrowheads). er, rough endoplasmic reticulum; g, Golgi apparatus. Scale bar indicates 1 μm. (From Dunel-Erb, S. et al., *J. Appl. Physiol. Resp. Environ. Exercise Physiol.*, 53, 1342, 1982. With permission.) (b) Confocal micrograph of NECs and associated innervation in the gill filament of zebrafish (*Danio rerio*). NECs and nerve fibers were labeled by serotonin (5-HT) and zn-12 immunohistochemistry, respectively. This tissue was taken from a specimen that had been acclimated to chronic hypoxia. Note the long neuron-like processes that formed and extended terminals (arrows) to form contact with the intrabranchial nerve bundle (iNB) following acclimation. Scale bar indicates 10 μm. (Modified from Jonz, M.G. et al., *J. Physiol.*, 560, 737, 2004. With permission.) (c) Confocal image showing NECs of gill filament (F) and lamellae (L) of goldfish (*Carassius auratus*) labeled by 5-HT and zn-12 immunohistochemistry. This tissue was taken from a specimen that had been acclimated to 7°C. Note that following acclimation, NECs of the lamellae were redistributed toward the distal regions of lamellae, as indicated by the dashed line, and maintained their innervation. (Modified from Tzaneva, V. and Perry, S.F., *J. Exp. Biol.*, 213, 3666, 2010. With permission.) (d) In zebrafish, cells containing the vesicular acetylcholine transporter (VAChT) do not contain serotonin (5-HT) or the synaptic vesicle protein SV2. Confocal micrographs from an adult gill filament demonstrate triple immunohistochemical labeling with antibodies against VAChT, 5-HT, and SV2. 10 cells are indicated: cells 1–3 are VAChT-positive, cells 4–8 are 5-HT and SV2-positive, and cells 9 and 10 are only SV2-positive. A dashed outline indicates the position of the lamellae. Scale bar is 10 μm. (From Shakarchi, K. et al., *J. Exp. Biol.*, 216(5), 869, 2013. With permission.) (e) By contrast, in trout intrabranchial neurons were immunolabeled with both 5-HT and VAChT (arrow), but NECs were immunonegative for these markers. Scale bar is 10 μm. (From Porteus, C.S. et al., *Acta Histochem.*, 115, 158, 2013. With permission.) (f) Confocal image of NECs of the skin in developing zebrafish. NECs and associated innervation were labeled by 5-HT and zn-12 immunohistochemistry. Scale bar indicates 100 μm. (Modified from Coccimiglio, M.L. and Jonz, M.G., *J. Exp. Biol.*, 215, 3881, 2012. With permission.)

The size of NECs varies from species to species, where smaller fishes display gill NECs with a smaller cell diameter (Saltys et al., 2006), but the general morphology of gill NECs otherwise appears to be conserved across species. NECs are typically found as solitary cells and have similar morphology and distribution in the gill filaments on all gill arches and of all fish species studied to date (Table 4.1; and see references therein). The orientation of NECs within the epithelium can be categorized as one of two types. Previous studies have classified serotonin (5-hydroxytryptamine, 5-HT)-positive NECs of the filaments as "closed" to the external environment (i.e., oriented internally toward the arterial supply) and those NECs that are 5-HT-negative and immunoreactive for enkephalins as "open" and potentially exposed to the exterior (reviewed by Zaccone et al., 1994). However, this method of classification would not account for the 5-HT-positive NECs of the lamellae in zebrafish that are clearly of the open type (Jonz and Nurse, 2003). The orientation of NECs within the gill has important implications in determining the initiation of adaptive ventilatory and cardiac responses by hypoxia *versus* hypoxemia. The relationship between chemoreceptor orientation and cardioventilatory response is highly variable among species, but hypoxic hyperventilation is produced by stimulation of either internally or externally oriented chemoreceptors in the gills (Milsom, 2012).

NECs possess characteristics of cells involved in the secretion of neurochemicals, such as storage of the neurotransmitter 5-HT and, at the ultrastructural level, retention of cytoplasmic dense-cored vesicles (DCVs), and innervation by adjacent nerve fibers (Figure 4.3b and c; Dunel-Erb et al., 1982; Bailly et al., 1992; Jonz and Nurse, 2003; Bailly, 2009). NECs of the gill filaments and lamellae label strongly for antibodies against 5-HT and the synaptic vesicle protein SV2 (Jonz and Nurse, 2003; Saltys et al., 2006). Although 5-HT labeling appears to be diffuse throughout the cell, at high magnification a large nucleus is observed that is circumscribed by an intensely fluorescent 5-HT-positive cytoplasm, where the neurotransmitter is stored within vesicles. Consistent with previous ultrastructural reports of DCVs in gill NECs facing nerve profiles (Dunel-Erb et al., 1982; Bailly et al., 1992; Bailly, 2009), confocal images have confirmed that NECs of the gill filaments and lamellae in fishes are rich in SV2-positive synaptic vesicles within the basal cytoplasm that may allow for neurotransmitter release adjacent to nerve fibers during periods of hypoxia (Jonz and Nurse, 2003; see also Saltys et al., 2006).

Gill NECs undergo significant morphological modification following long-term changes in environmental conditions, such as available O_2 or temperature, and this plasticity may underlie important physiological changes that also occur under these conditions. In zebrafish maintained in chronic hypoxia, gill NECs containing 5-HT increased in size and extended neuron-like processes in the direction of adjacent nerve fibers (Figure 4.3b; Jonz et al., 2004). Furthermore, the density (i.e., number of cells per unit area) of nonserotonergic (but not serotonergic) NECs increased in zebrafish chronically exposed to hypoxia (Jonz et al., 2004); and the density of serotonergic NECs decreased in zebrafish following chronic exposure to hyperoxia (Vulesevic et al., 2006). Thus, environmental O_2 may affect changes in the size and number of gill NECs. The reason for the effects of hypoxia on proliferation of specifically nonserotonergic NECs is not explicitly clear, but one interpretation of these findings is that nonserotonergic NECs are proliferative cells that eventually give rise to NECs that will produce and retain 5-HT (Shakarchi et al., 2013). This is consistent with the original idea proposed by Pierre Laurent et al. that gill NECs displaying weak fluorescence for 5-HT may represent proliferative or differentiating NECs (Bailly et al., 1992). Similar SV2-positive NECs lacking 5-HT have also been found in the gills of goldfish (*Carassius auratus*), trout (*Oncorhynchus mykiss*), trairão (*Hoplias lacerdae*), traira (*Hoplias malabaricus*), and larvae of *Xenopus laevis* (Saltys et al., 2006; Coolidge et al., 2008).

In crucian carp and goldfish, the gills undergo extensive remodeling in response to ambient changes in O_2 and temperature. Prolonged exposure to temperatures below 15°C promotes production of an interlamellar cell mass (ILCM), which decreases the functional surface area of the gill, while hypoxia has the opposite effect and reduces the ILCM (Sollid et al., 2003, 2005; Tzaneva and Perry, 2010; Tzaneva et al., 2011; Nilsson et al., 2012). Such an adaptation may reduce the cost of osmoregulation across the

TABLE 4.1
Neurochemicals (or Other Cellular Markers) Identified in Gill Neuroepithelial Cells by Immunohistochemistry

	Species	Common Name	Reference
Neurotransmitter			
Serotonin	*Amia calva*	Bowfin	Goniakowska-Witalinska et al. (1995), Porteus et al. (2012)
	Anguilla anguilla	Eel	Zaccone et al. (1992)
	Blennius sanguinolentus	Blenny	Zaccone et al. (1992)
	Carassius auratus	Goldfish	Saltys et al. (2006), Coolidge et al. (2008), Tzaneva and Perry (2010), Porteus et al. (2012, 2013)
	Danio rerio	Zebrafish	Jonz and Nurse (2003, 2005), Jonz et al. (2004), Saltys et al. (2006), Vulesevic et al. (2006), Qin et al. (2010), Shakarchi et al. (2013)
	Dicentrarchus labrax	Sea perch	Bailly et al. (1992)
	Gadus morhua	Atlantic cod	Sundin et al. (1998)
	Heteropneustes fossilis	Indian catfish	Zaccone et al. (1992)
	Hoplias lacerdae	Trairão	Coolidge et al. (2008)
	Hoplias malabaricus	Traira	Coolidge et al. (2008)
	Ictalurus melas	Black bullhead	Bailly et al. (1992)
	Ictalurus nebulosus	Bullhead	Zaccone et al. (1992)
	Ictalurus punctatus	Channel catfish	Burleson et al. (2006)
	Kryptolebias marmoratus	Mangrove rivulus	Regan et al. (2011)
	Lepidosteus osseus	Gar	As cited by Zaccone et al. (1997)
	Micropterus dolomieui	Black bass	Bailly et al. (1992)
	Oncorhynchus mykiss	Rainbow trout	Bailly et al. (1992), Saltys et al. (2006), Coolidge et al. (2008), Zhang et al. (2011), Porteus et al. (2012)
	Oncorhynchus nerka	Sockeye salmon	Porteus et al. (2012)
	Oreochromis massambica	Tilapia	Bailly et al. (1992)
	Oreochromis niloticus	Nile tilapia	Monteiro et al. (2010)
	Oryzias latipes	Medaka	Porteus et al. (2012)
	Perca fluviatilis	Perch	Bailly et al. (1992)
	Protopterus annectens	African lungfish	Zaccone et al. (1992)
	Salmo trutta	Brown trout	Zaccone et al. (1992)
	Scyliorhinus stellaris	Dogfish	As cited by Zaccone et al. (1997)
Neuropeptides			
Endothelin	*Amia calva*	Bowfin	Goniakowska-Witalinska et al. (1995), Zaccone et al. (1996)
	Conger conger	Sea eel	Zaccone et al. (1996)
	Fundulus heteroclitus	Killifish	Hyndman et al. (2006)
	Heteropneustes fossilis	Indian catfish	Zaccone et al. (1996)
	Scyliorhinus canicula	Dogfish	Zaccone et al. (1996)
	Salmo trutta	Brown trout	Zaccone et al. (1996), Mauceri et al. (1999)
	Torpedo marmorata	Electric ray	Zaccone et al. (1996)
Enkephalins	*Amia calva*	Bowfin	Goniakowska-Witalinska et al. (1995)
	Anguilla anguilla	Eel	Zaccone et al. (1992)
	Blennius sanguinolentus	Blenny	Zaccone et al. (1992)
	Heteropneustes fossilis	Indian catfish	Zaccone et al. (1992)

TABLE 4.1 (continued)

Neurochemicals (or Other Cellular Markers) Identified in Gill Neuroepithelial Cells by Immunohistochemistry

	Species	Common Name	Reference
	Ictalurs nebulosus	Bullhead	Zaccone et al. (1992)
	Lampetra japonica	Lamprey	As cited by Zaccone et al. (1992)
	Lepidosteus osseus	Gar	As cited by Zaccone et al. (1997)
	Protopterus annectens	African lungfish	Zaccone et al. (1992)
	Salmo trutta	Brown trout	Zaccone et al. (1992)
Neuropeptide Y	*Oreochromis mossambica*	Tilapia	As cited by Zaccone et al. (1994)
PACAP 27 and 38[a]	*Heteropneustes fossilis*	Indian catfish	As cited by Zaccone et al. (2006)
	Pangasius hypothalamus	Vietnamese catfish	As cited by Zaccone et al. (2006)
Vasoactive intestinal polypeptide	*Heteropneustes fossilis*	Indian catfish	Zaccone et al. (2003)
	Salaria pavo	Blenny	As cited by Zaccone et al. (2006)
Biosynthetic enzymes			
Endothelial nitric oxide synthase[b]	*Heteropneustes fossilis*	Indian catfish	Zaccone et al. (2006)
	Lepidosteus osseus	Gar	As cited by Zaccone et al. (1997)
Neuronal nitric oxide synthase[b]	*Fundulus heteroclitus*	Killifish	Hyndman et al. (2006)
	Heteropneustes fossilis	Indian catfish	Mauceri et al. (1999), Zaccone et al. (2003)
Tyrosine hydroxylase[c]	*Ictalurus punctatus*	Channel catfish	Burleson et al. (2006)
Vesicular proteins			
Synaptic vesicle protein	*Danio rerio*	Zebrafish	Jonz and Nurse (2003), Jonz et al. (2004), Shakarchi et al. (2013)
	Carassius auratus	Goldfish	Saltys et al. (2006), Coolidge et al. (2008)
	Oncorhynchus mykiss	Rainbow trout	Coolidge et al. (2008)
	Hoplias lacerdae	Trairão	Coolidge et al. (2008)
	Hoplias malabaricus	Traira	Coolidge et al. (2008)
Vesicular acetylcholine transporter[d]	*Kryptolebias marmoratus*	Mangrove rivulus	Regan et al. (2011)
	Danio rerio	Zebrafish	Shakarchi et al. (2013)

[a] Pituitary adenylate cyclase-activating polypeptide.

[b] Biosynthetic enzyme involved in the production of nitric oxide.

[c] Biosynthetic enzyme involved in the production of catecholamines. Note that Porteus et al. (2012) found no evidence for tyrosine hydroxylase immunoreactivity in goldfish or trout gills.

[d] In both species, labeled cells were not shown to be serotonin immunoreactive. Also noteworthy is that Porteus et al. (2012) showed no immunohistochemical evidence of the vesicular acetylcholine transporter in goldfish or trout gills.

gill epithelium or perhaps prevent uptake of toxic substances or reduce infection (Nilsson et al., 2012). Importantly, covering of the gill surface may compromise the ability of NECs to sense changes in water PO_2. It is also evident that in goldfish acclimated to 7°C there is a redistribution of NECs, and their associated innervation, within the lamellae from a proximal (i.e., near the filament) to a more distal position (Figure 4.3c; Tzaneva and Perry, 2010), and this is correlated with production of the ILCM. Although in these studies there were no reported changes in filament NECs (which are considered to be the primary O_2 sensors) with acclimation, these results may suggest that lamellar NECs play a role in maintaining functional O_2 sensing responses during acclimatization to lower temperatures. Strictly speaking, however, lamellar NECs may not be essential for O_2 sensing. In embryonic and larval zebrafish, ventilatory responses to hypoxia develop several days before lamellae, or lamellar NECs, are present (Jonz and Nurse, 2005; Coccimiglio and Jonz, 2012; Shakarchi et al., 2013).

4.3.3.2 Biochemistry

There are numerous studies that have localized neurotransmitters, neuropeptides, and other proteins involved in neurosecretion in gill NECs. Studies that have used the combined techniques of immunohistochemistry and microscopy to characterize the biochemistry of NECs are summarized in Table 4.1. Most notably, the vast majority of studies indicates that NECs contain the neurotransmitter, 5-HT. A relatively small proportion of NECs, however, have been described in the gill filaments that label positive for antibodies against the synaptic vesicle protein SV2, but are negative for 5-HT (Jonz and Nurse, 2003; Saltys et al., 2006; Coolidge et al., 2008). As described earlier, 5-HT-negative NECs may represent a proliferative population of NECs that have not yet synthesized 5-HT. As an alternative explanation, these nonserotonergic NECs may retain other neurotransmitter(s) that have yet to be identified or that are not detectable by immunohistochemistry. This would suggest that other chemicals may play a role in O_2 sensing within the gill.

Cells containing acetylcholine (ACh) have also been localized in the gill. In the gills of zebrafish (Figure 4.3d) and the mangrove rivulus (*Kryptolebias marmoratus*), NEC-like cells were described that expressed the vesicular acetylcholine transporter (VAChT; Regan et al., 2011; Shakarchi et al., 2013), a transmembrane protein that mediates loading of ACh into secretory vesicles and that can be used as a convenient marker for cholinergic cells (Prado et al., 2002). In both studies, cells expressing the VAChT protein did not contain serotonin nor were they labeled by the SV2 antibody. However, Porteus et al. (2013) found no immunohistochemical evidence of the VAChT protein in the gills of goldfish or trout, suggesting that the presence of VAChT (and thus cholinergic NECs) may be species specific. Other candidates for neurotransmission in NECs may include nitric oxide or catecholamines, since there is evidence of nitric oxide synthase and tyrosine hydroxylase (the enzymes that synthesize these substances) in gill NECs (Mauceri et al., 1999; Zaccone et al., 2003; Burleson et al., 2006; Hyndman et al., 2006; Zaccone et al., 2006). However, no evidence of tyrosine hydroxylase was found in the gills of goldfish or trout (Porteus et al., 2013). In addition, a variety of neuropeptides, such as endothelin, enkephalins, neuropeptide Y, pituitary adenylate cyclase-activating polypeptide (PACAP), and vasoactive intestinal polypeptide (VIP), have been localized to NECs and may also be involved (Zaccone et al., 1992, 1996, 2003, 2006; Goniakowska-Witalinska et al., 1995; Mauceri et al., 1999; Hyndman and Evans, 2007).

4.3.3.3 Innervation

Most nerve fibers that enter the gill filaments and lamellae are carried by the branchial nerve. In addition, several studies have shown the presence of neuronal cell bodies within the gills (Figure 4.3e; Bailly and Dunel-Erb, 1986, 1989; Donald, 1984, 1987; Dunel-Erb and Bailly, 1986; Dunel-Erb et al., 1989; de Graaf et al., 1990; Jonz and Nurse, 2003; Bailly, 2009; Porteus et al., 2013). Evidence from earlier studies, in which electron microscopy was used to characterize ultrastructural features, and more recent confocal studies combined with immunohistochemistry have shown that 5-HT-containing NECs of the gill filaments receive innervation (Dunel-Erb et al., 1982; Bailly et al., 1992; Jonz and Nurse, 2003; Saltys et al., 2006; Zaccone et al., 2006; Bailly, 2009). Sources of this innervation include the extrabranchial (or extrinsic) population of nerve fibers that originate from neurons located outside the gill arches and from intrabranchial (or intrinsic) nerve fibers that originate from serotonergic neurons located within the gill filaments. NECs of the lamellae, on the other hand, receive only extrabranchial innervation, as has been shown in zebrafish and goldfish (Jonz and Nurse, 2003; Saltys et al., 2006; Tzaneva et al., 2011). While both populations of nerve fibers may convey sensory (afferent) signals away from NECs during stimulation, the extrabranchial innervation appears to initiate centrally mediated responses to hypoxia, such as hyperventilation and decreased heart rate, while the intrabranchial innervation may mediate local vascular changes within the gill that occur during hypoxia. The presence of nitrergic and serotonergic nerve fibers in the gills that contact NECs (Jonz and Nurse, 2003;

Zaccone et al., 2006) suggests that the response of O_2 chemoreceptors to hypoxia may potentially be influenced by efferent neurotransmission and, potentially, a mechanism of feedback.

4.3.3.4 Cellular Mechanisms of O_2 Sensing

The techniques of single-cell patch-clamp electrophysiology and Ca^{2+} imaging allow recording of ion channel activity and intracellular Ca^{2+} dynamics, respectively. Since the late 1980s (López-Barneo et al., 1988), these approaches have been important to our understanding of how type I cells of the mammalian carotid body can transform a chemical signal, such as hypoxia, into a meaningful cellular signal that can be passed on to the CNS to effect physiological change. However, only as recently as 2004 were these techniques applied to chemoreceptors of the gill. The field of O_2 sensing in fishes has therefore lagged behind by comparison to the much larger body of research on mammalian chemoreceptors. Nevertheless, the few studies exploring chemoreception by gill NECs that have emerged have revealed some very important clues about the role of NECs in fishes: (1) plasma membrane K^+ channels are involved in transduction of the hypoxic signal; (2) NEC stimulation results in a brief rise in intracellular Ca^{2+} ($[Ca^{2+}]_i$); (3) NECs are polymodal, capable of sensing or responding to multiple chemical signals; and (4) many of these characteristics, including species specificity, appear to be conserved between fishes and mammals. These concepts will now be explored in more detail in this and the following sections.

The patch-clamp technique has permitted characterization of O_2-sensitive ion channels, such as those that are permeable to K^+, Ca^{2+}, and Na^{2+}, in a variety of cell types in mammals, including carotid body type I cells, neurons, vascular smooth muscle cells, adrenal medullary chromaffin cells, and neuroepithelial bodies (NEBs) of the pulmonary epithelium (López-Barneo et al., 2001, 2009; Plant et al., 2002; Campanucci et al., 2003; Neubauer and Sunderram, 2004; Nurse, 2010; Peers et al., 2010; Domnik and Cutz, 2011). In fishes, patch-clamp recording of NECs has, so far, been applied only to isolated cells (Figure 4.4). These studies have demonstrated that gill NECs

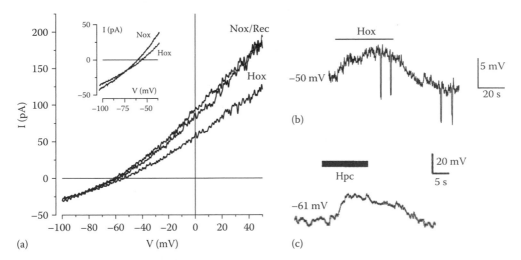

FIGURE 4.4 Physiological responses of NECs *in vitro* indicate that they are polymodal chemoreceptors. (a) Whole-cell recording from an O_2-sensitive NEC from zebrafish showing the current–voltage (I–V) relationship during exposure to normoxia (Nox, 150 mmHg), hypoxia (Hox, 25 mmHg), and after recovery in normoxia (Rec). Currents were evoked by changing the voltage from a holding potential of −60 mV to a ramp protocol between −100 and +50 mV. Inset, the average response of 10 such cells using the same experimental procedure. (b) Current-clamp recording of a reversible depolarization of membrane potential (V_m) during perfusion (bar) of a zebrafish NEC with hypoxic solution (25 mmHg). Resting V_m was ~ −50 mV. Downward deflections are due to experimental injection of 10 pA of current and indicate a greater voltage change (or membrane resistance) upon hypoxia due to ion channel inhibition. (c) Current-clamp recording of membrane depolarization during perfusion of a zebrafish NEC with hypercapnic solution (hpc, 7.5 mmHg). Resting V_m was ~ −61 mV.

(continued)

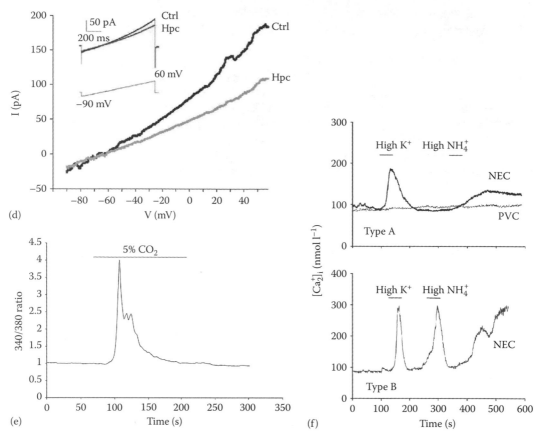

FIGURE 4.4 (continued)   (d) Average I–V relationship during exposure to hypercapnia from 12 NECs from zebrafish. Currents were evoked by changing the voltage from a holding potential of −60 mV to a ramp protocol between −90 and +60 mV (see lower inset). Upper inset shows a current trace from a single NEC. (e) Hypercapnic acidosis (5% CO_2 plus associated decrease in pH) increases the concentration of intracellular calcium $[Ca^{2+}]_i$ in isolated neuroepithelial cells (NECs) of zebrafish. A relative change in the 340/380 ratio from fura 2-AM experiments indicates an increase in $[Ca^{2+}]_i$. Bar represents the duration of the 5% CO_2 application. (f) In fura-2-based experiments, NECs from trout respond to high ammonia (NH_4^+) with an increase in $[Ca^{2+}]_i$. NECs were exposed briefly to 30 mM K^+ and 1 mM NH_4^+. Upper panel, a slow "type A" response to high K^+ and NH_4^+ in an NEC but not a pavement cell (PVC, for control). Lower panel, a fast "type B" response under the same conditions. (a and b: From Jonz, M.G. et al., *J. Physiol.*, 560, 737, 2004; c and d: Qin, Z. et al., *J. Physiol.*, 588, 861, 2010; e: Abdallah, S.J. et al., *Adv. Exp. Med. Biol.*, 758, 143, 2012; f: Zhang, L. et al., *J. Exp. Biol.*, 214, 2678, 2011. With permission.)

display physiological responses *in vitro* that are typical of O_2 chemoreceptors. As summarized in Table 4.2, membrane currents carried by K^+ have been described in gill NECs of three fish species: zebrafish *Danio rerio* (Jonz et al., 2004; Qin et al., 2010), channel catfish *Ictalurus punctatus* (Burleson et al., 2006), and goldfish *Carassius auratus* (Zachar and Jonz, 2012a). In all of these species, a decrease in PO_2 of the recording medium led to a reversible change in K^+ current across the plasma membrane. In zebrafish (e.g., Figure 4.4a), exposure of isolated NECs to 15–25 mmHg hypoxia reduced quinidine-sensitive background (or "leak") K^+ channels without significant effect on voltage-dependent K^+ channels, which are sensitive to the drugs tetraethylammonium (TEA) and 4-aminopyridine (4-AP) (Jonz et al., 2004; Qin et al., 2010). In channel catfish, a PO_2 of less than 10 mmHg induced either an increase or decrease in K^+ current, although the specific type of K^+ channel was not characterized in this species (Burleson et al., 2006). Furthermore, in goldfish,

TABLE 4.2
Summary of *In Vitro* Physiological Responses of Gill Neuroepithelial Cells (NECs) from Multiple Species

Species	Common Name	Stimulus or Drug	Effect	Reference
Carassius auratus	Goldfish	Hypoxia (25 mmHg)	None observed	Zachar and Jonz (2012)
		Anoxia (2–3 mmHg)	$\downarrow V_m$	Zachar and Jonz (2012)
		NaCN	$\downarrow V_m$	Zachar and Jonz (2012)
		TEA + 4-AP	$\downarrow K_{Ca}$	Zachar and Jonz (2012)
Danio rerio	Zebrafish	Hypoxia (25 mmHg)	$\downarrow V_m$	Jonz et al. (2004), Qin et al. (2010)
		Hypercapnia	$\downarrow V_m$	Qin et al. (2010)
			$\uparrow [Ca^{2+}]_i$	Abdallah et al. (2012), Abdallah, Perry and Jonz, unpublished observations
			$\uparrow [H^+]_i$	Abdallah, Perry and Jonz, unpublished observations
		TEA + 4-AP	$\downarrow K_v$	Jonz et al. (2004)
		Quinidine	$\downarrow K_B$, $\downarrow K_v$, $\downarrow V_m$	Jonz et al. (2004), Qin et al. (2010)
		4-AP	$\downarrow K_v$, $\downarrow V_m$	Qin et al. (2010)
		Acetazolamide	Reduced $\downarrow V_m$ response to hypercapnia	Qin et al. (2010)
Ictalurus punctatus	Channel catfish	Hypoxia (<10 mmHg)	$\downarrow$ and $\uparrow K^+$ current	Burleson et al. (2006)
		NaCN	$\downarrow K^+$ current	Burleson et al. (2006)
Oncorhynchus mykiss	Rainbow trout	High K^+	$\uparrow [Ca^{2+}]_i$	Zhang et al. (2011)
		NH_4^+	$\uparrow [Ca^{2+}]_i$	Zhang et al. (2011)

Note: Results were obtained from either patch-clamp electrophysiological recording or Fura-based calcium imaging. K_B, background K^+ channel; K_{Ca}, Ca^{2+}-dependent K^+ channel; K_v, voltage-dependent K^+ channel; NaCN, sodium cyanide; NH_4^+, ammonium ion; TEA, tetraethylammonium; V_m, membrane potential; $[Ca^{2+}]_i$, intracellular Ca^{2+} concentration; $[H^+]_i$, intracellular H^+ concentration; $\uparrow$ and $\downarrow$ indicate increase and decrease activity, respectively.

25 mmHg hypoxia had no observable effect on membrane currents or membrane potential (V_m), while anoxia (2–3 mmHg) and sodium cyanide (NaCN, which may mimic intracellular hypoxia or anoxia) induced pronounced plasma membrane depolarization (Zachar and Jonz, 2012a). One explanation for the lack of response of goldfish NECs to hypoxia is that during patch-clamp recording (in which the intracellular fluid is usually dialyzed with experimental media) an important cytosolic factor was lost, depressing the sensitivity of these cells to a much lower PO_2. Interestingly, gill NECs from zebrafish were similarly dialyzed during patch-clamp recording but had an intact response to hypoxia (Jonz et al., 2004).

Despite the difference in levels of PO_2 used in the aforementioned studies to induce a low O_2 challenge, the evidence suggests species-dependent mechanisms of O_2 sensing by gill NECs. Most notably, the response of zebrafish *versus* goldfish NECs to hypoxia may reflect evolutionary adaptation. That isolated goldfish NECs do not respond to low PO_2 until severe hypoxic or even anoxic levels, compared to zebrafish, may be related to differences in hypoxia tolerance in these species. Like the crucian carp (*Carassius carassius*; Vornanen et al., 2009), goldfish are anoxia tolerant (Bickler and

Buck, 2007) and have a blood oxygen affinity that is relatively high (P_{50} = 2.6 mmHg; Burggren, 1982) compared to other species, such as trout (P_{50} = 22.9 mmHg; Bushnell et al., 1984). Such an adaptation in goldfish may be beneficial during periods of prolonged environmental anoxia, when continuous stimulation of NECs would otherwise result in increased activation of sensory pathways, overall metabolic activity, and elevated ventilatory or behavioral responses without significant increase in O_2 uptake. Although goldfish do, indeed, hyperventilate in response to hypoxia, a statistically significant rise in ventilatory frequency above control first occurs at a relatively depressed water PO_2 of 75 mmHg (at 25°C; Tzaneva et al., 2011) compared to the hyperventilatory response in zebrafish, which is first significantly higher than control at a PO_2 of 110 mmHg (at room temperature; Vulesevic et al., 2006).

The regulation of leak K^+ channels by hypoxia appears to be a fundamental mechanism in fishes and mammals that has been relatively conserved and may have appeared early in vertebrate evolution. As indicated earlier, typically a decrease in activity of plasma membrane K^+ channels represents one of the first steps in the transduction of the hypoxic response in NECs. Indeed, similar K^+ channels are also O_2 sensitive in carotid body type I cells and are involved in hypoxic transduction (Buckler, 2007). These background K^+ channels are insensitive to changes in V_m and remain open under resting conditions. In this way, background channels impart a high permeability of the membrane to K^+ and contribute to setting the resting V_m and excitability of the cell (Lesage and Lazdunski, 2000; Goldstein et al., 2001; Lesage, 2003). The inhibition of background K^+ channels in gill NECs thus leads to membrane depolarization (Figure 4.4b and c) (Jonz et al., 2004; Qin et al., 2010). This "membrane hypothesis" in O_2 sensing was first proposed for type I cells and predicts that K^+ channel inhibition leads to a series of subsequent events, including Ca^{2+} influx through voltage-gated Ca^{2+} channels, release of neurotransmitters, and subsequent activation of afferent nerve fibers (López-Barneo et al., 1988, 2001, 2009; Montoro et al., 1996; Buckler, 2007; Nurse, 2010; Peers et al., 2010).

In gill NECs, similar downstream events appear to be involved in the transduction and signaling of hypoxia, as they are in mammalian chemoreceptors. Voltage-dependent Ca^{2+} channels of the plasma membrane are present in gill NECs of goldfish (Zachar and Jonz, unpublished observations), suggesting that these ion channels may increase the concentration of intracellular Ca^{2+} ($[Ca^{2+}]_i$) upon membrane depolarization by hypoxia (Figure 4.4b). Evidence from fura-based ratiometric Ca^{2+} imaging studies on NECs of trout (Zhang et al., 2011) and zebrafish (Abdallah et al., 2012) indicates that membrane depolarization (by high CO_2, high K^+, or NH_4^+) does, indeed, increase $[Ca^{2+}]_i$ (Figure 4.4e and f); however, direct evidence linking the activation of plasma membrane Ca^{2+} channels to increased $[Ca^{2+}]_i$ is not yet available. In addition, the regulation of $[Ca^{2+}]_i$ during NEC stimulation may be species and stimulus specific. Preliminary evidence indicates that, in goldfish NECs, hypoxic induction of synaptic vesicle recycling (a measure of neurosecretion) is inhibited by Cd^{2+} (Zachar and Jonz, unpublished observations), suggesting the involvement of plasma membrane Ca^{2+} channels in this species, while in zebrafish NECs, increased $[Ca^{2+}]_i$ induced by hypercapnia appears instead to originate from intracellular stores without significant involvement of the plasma membrane or extracellular Ca^{2+} (Abdallah et al., 2012; Abdallah, Perry and Jonz, unpublished observations). Discovery of the pathway(s) by which $[Ca^{2+}]_i$ is regulated in NECs may provide important information about how these cells respond to hypoxia (or anoxia) and how neurotransmitter release is controlled. Furthermore, in goldfish NECs, K^+ channels whose activity is dependent on intracellular Ca^{2+} (K_{Ca} channels) have been reported (Zachar and Jonz, 2012a). The role of K_{Ca} channels in goldfish NECs is not yet clear, but one possibility is that the hyperpolarizing current that they produce upon increased $[Ca^{2+}]_i$ (such as during membrane depolarization by severe hypoxia) may effectively decrease excitability (i.e., hyperpolarize the membrane by K^+ efflux), thereby suppressing the sensitivity of NECs to lower PO_2 levels. Although K_{Ca} channels have been reported, anecdotally, in gill cells isolated from zebrafish (Jonz, 2004), they do not appear to make a major contribution to whole-cell current in NECs of this species (Jonz et al., 2004; Qin et al., 2010). A current working model for a proposed O_2 sensing pathway by gill NECs is presented in Figure 4.5.

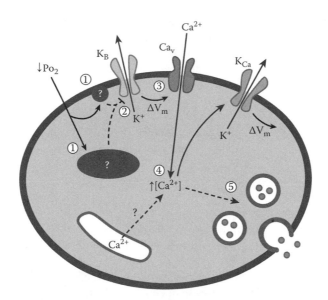

FIGURE 4.5 **(See color insert.)** A current working model of O_2 sensing in gill neuroepithelial cells (NECs). (1) A decrease in P_{O_2} is detected by an uncharacterized O_2 sensor that is membrane-delimited or cytosolic/mitochondrial. (2) Detection of hypoxia leads to inhibition of K^+ current through background K^+ (K_B) channels, leading to membrane depolarization (ΔV_m). (3) Voltage-dependent Ca^{2+} (Ca_V) channels are activated, permitting Ca^{2+} entry into the NEC. (4) Cytosolic Ca^{2+} increases, with intracellular stores possibly playing a role. (5) Ca^{2+} ions facilitate vesicle fusion and release of neurotransmitter. Ca^{2+} may also increase conductance through Ca^{2+}-activated K^+ (K_{Ca}) channels, as observed in goldfish NECs, modulating ΔV_m. Dotted lines represent presumptive mechanisms. (From Zachar, P.C. and Jonz, M.G., *Respir. Physiol. Neurobiol.*, 184, 301, 2012a; With permission.)

Despite significant advances in the characterization of the mechanisms of O_2 transduction and signaling in vertebrate chemoreceptors, the identification of the molecular "sensor" of O_2 in these cells (i.e., the mechanism by which membrane K^+ channels are regulated) remains controversial and several candidates have been considered, including the mitochondrion, AMP-activated protein kinase (AMPK), hydrogen sulfide (H_2S), reactive oxygen species (ROS), NADPH oxidase, and heme oxygenase-2 (HO-2) (Gonzalez et al., 2010; López-Barneo et al., 2010; Peers et al., 2010; Olson, 2011; Prabhakar, 2012). While many sensors have been proposed, the available evidence would seem to indicate that there does not appear to be a universal O_2 sensor that initiates the transduction of hypoxia in chemoreceptors of all vertebrates; although it has recently been suggested that AMPK may mediate hypoxic chemoreception regardless of species differences (in mammals) in K^+ channel expression (Peers et al., 2010). The "chemosome" hypothesis proposes that multiple sensors may mediate the O_2 sensing response within a single chemoreceptor (Prabhakar, 2006). The involvement of multiple sensors may then provide a means of setting the sensitivity of NECs to specific ranges of PO_2, thereby determining relative sensitivity or tolerance of an organism to hypoxia.

In fishes, only one candidate O_2 sensor has so far been studied. In trout, hydrogen sulfide (H_2S) may act as a sensor or transducer of O_2 and mediate responses to hypoxia in vascular smooth muscle of vertebrates, including fishes and mammals (Olson et al., 2006). The model proposes that decreased availability of O_2 during hypoxia leads to reduced oxidation of H_2S (an endogenous signaling molecule) and therefore its subsequent intracellular accumulation. H_2S has been shown to have effects on membrane potential similar to those of hypoxia (Olson et al., 2006). In isolated gill NECs of trout, both H_2S and hypoxia depolarized NECs by approximately 10 mV, suggesting that H_2S may contribute to molecular O_2 sensing in the gill (Olson et al., 2008). There is not yet evidence of how H_2S might regulate membrane conductance in NECs so as to result in depolarization, but evidence from studies on mammalian chemoreceptors indicates that H_2S can inhibit K_{Ca} channels

(Prabhakar, 2012). However, only leak K^+ channels have so far been reported to be O_2 sensitive in gill NECs (Jonz et al., 2004). An alternative candidate for a molecular O_2 sensor in these cells may therefore be AMPK since it has been shown to mediate membrane depolarization via inhibition of leak K^+ and K_{Ca} channels in the carotid body (Evans et al., 2005a; Wyatt et al., 2007). A potential model for O_2 sensing by AMPK in type I chemoreceptors of the mammalian carotid body is summarized by Peers et al. (2010). In this model, hypoxia inhibits mitochondrial oxidative phosphorylation, thus increasing the ratio of AMP:ATP. This would induce the activation of AMPK via stabilization of its phosphorylation by other protein kinases. Activated AMPK would then inhibit K_{Ca} and leak K^+ channels (possibly through phosphorylation) and induce membrane depolarization and Ca^{2+}-dependent neurosecretion.

4.3.3.5 Cellular Mechanisms of CO_2 Sensing

In addition to sensitivity to O_2, NECs are also responsive to changes in CO_2 and/or H^+ and may therefore mediate responses to hypercapnia. Elevated levels of water PCO_2, which will also liberate H^+, induce hyperventilation in fishes (Gilmour, 2001; Milsom, 2002, 2012; Perry and Abdallah, 2012). Moreover, in isolated NECs from the gills of zebrafish, patch-clamp studies indicate that high PCO_2 in the absence of pH change (i.e., isohydric hypercapnia) leads to K^+ channel inhibition and membrane depolarization (Figure 4.4c and d; Qin et al., 2010). Preliminary evidence from Ca^{2+} imaging studies, however, indicates that an increase in the extracellular H^+ concentration without an increase in PCO_2 (i.e., isocapnic acidosis) produces a rise in $[Ca^{2+}]_i$ (Abdallah et al., 2012; Abdallah, Perry and Jonz, unpublished observations). Figure 4.4e illustrates the rise in $[Ca^{2+}]_i$ produced by the decreased extracellular pH associated with high PCO_2. Elevated PCO_2 alone (i.e., isohydric hypercapnia) had no effect on $[Ca^{2+}]_i$ in zebrafish in these studies. These data are difficult to interpret and suggest that while changes in extracellular CO_2 mediate membrane depolarization in NECs, only extracellular H^+ can induce elevated $[Ca^{2+}]_i$ and therefore subsequent release of neurotransmitter. One potential problem of relating these two sets of data is that in the patch-clamp experiments (Qin et al., 2010) the cytoplasm of NECs was dialyzed with an intracellular potassium-based saline and may have resulted in the loss of an important cytosolic component, while in the Ca^{2+} imaging experiments the NEC cytoplasm was intact. An important next step in this area will be to verify the apparent uncoupling of membrane depolarization and Ca^{2+} signaling in NECs and whether CO_2 or H^+ (or both) act as the proximal signal in mediating the hyperventilatory response to hypercapnia.

4.3.3.6 Ammonia Sensing

In ammoniotelic teleosts, ammonia is a respiratory gas that is excreted across the gill epithelium during ventilation. In rainbow trout, elevated total ammonia (i.e., NH_3 gas plus ammonium ion, NH_4^+) in the blood plasma or water stimulated ventilation (Zhang and Wood, 2009; Zhang et al., 2011). Interestingly, it appears that in addition to O_2 and CO_2/H^+, gill NECs may also sense high ammonia. Bilateral ablation of the anterior gill arches in trout (i.e., arches I and II) reduced or eliminated the ventilatory response to external ammonia. In addition, NECs isolated from the gills responded to acute exposure of 1 mM NH_4Cl with an increase in $[Ca^{2+}]_i$ (Figure 4.4f), and this response was abolished in NECs isolated from trout chronically exposed to high external ammonia (Zhang et al., 2011).

4.3.3.7 Models of Neurotransmission

The mechanisms through which transmission of a chemoreceptor response to hypoxia results from interactions between gill NECs and sensory neurons, and perhaps between NECs, remain largely unknown. Earlier studies using electrophysiological recording of branchial nerves from isolated gill preparations demonstrated that, like hypoxia, the application of a number of neurochemicals induced an increase in discharge frequency from sensory fibers (Milsom and Brill, 1986; Burleson and Milsom, 1993). These studies were the first to show that ACh and nicotine each induced strong stimulation of nerve activity, while 5-HT and dopamine (DA) induced weak stimulation (Figure 4.6a and b; Burleson and Milsom, 1995a). Furthermore, ACh, nicotine, and 5-HT increased

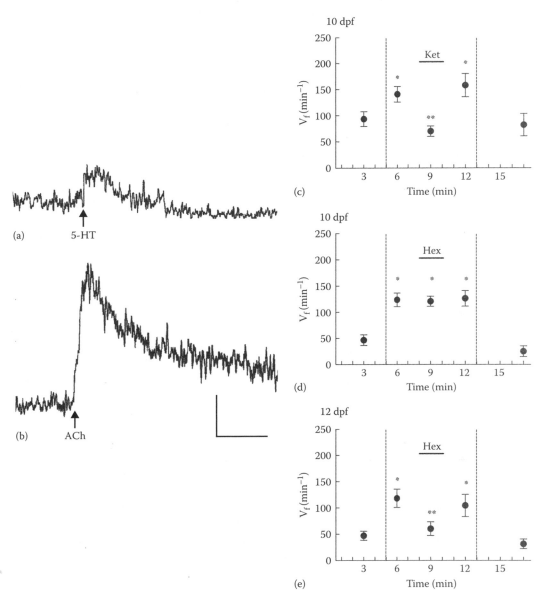

FIGURE 4.6 Serotonin (5-HT) and acetylcholine (ACh) are both important neurotransmitters involved in gill O_2 sensing. (a and b) In extracellular recordings from the gill nerves of trout, an increase in nerve activity was recorded upon stimulating an isolated gill arch with 5-HT (a) and ACh (b). The increase in nerve activity represents sensory nerve stimulation. The scale indicates 5 impulses/s (vertical bar) and 50 s (horizontal bar). (a and b: Modified from Burleson, M.L. and Milsom, W.K., *Respir. Physiol.*, 1003, 231, 1995a. With permission.) (c–e) In behavioral experiments, in which ventilation frequency (V_f, in min^{-1}) was monitored in developing zebrafish, application of hypoxia (from 5–13 min.) increased V_f. At 10 days postfertilization (dpf), 100 μM ketanserin (ket, a 5-HT receptor antagonist) reversed this response (c), while 100 μM hexamethonium (hex, a ACh receptor antagonist) was without effect (d). However, at 12 dpf hexamethonium inhibited the hypoxic ventilatory response (e). (From Shakarchi, K. et al., *J. Exp. Biol.*, 216(5), 869, 2013. With permission.)

gill ventilatory rate, while DA was without significant effect (Burleson and Milsom, 1995b). These studies provided evidence that ACh, 5-HT, and DA may have the potential to affect cardioventilatory responses to hypoxia. In the mammalian carotid body, ACh is a primary excitatory neurotransmitter, while DA and 5-HT are mediators of inhibition and modulation, respectively (Milsom and Burleson, 2007; Shirahata et al., 2007; Nurse, 2010).

In behavioral studies of developing zebrafish, in which anesthetized larvae were briefly exposed *in vivo* to a variety of neurochemicals, both serotonergic and cholinergic mechanisms contributed to the hyperventilatory response to hypoxia (Shakarchi et al., 2013). The exogenous application of 5-HT and ACh each induced hyperventilatory responses similar to those observed following the administration of hypoxic solution (25 mmHg). By contrast, DA reduced ventilatory frequency in zebrafish. Further pharmacological characterization using hypoxia combined with subsequent administration of either ketanserin or hexamethonium suggested that the ventilatory response to hypoxia included, at least in part, 5-HT_2 receptors and nicotinic ACh receptors (Figure 4.6c through e; Shakarchi et al., 2013). Although the cellular localization of these receptors was not investigated in this study, immunohistochemical evidence indicated that putative cholinergic NECs were present in the gill filaments in the vicinity of serotonergic NECs and that cholinergic NECs were not serotonergic. Although the details of cholinergic NECs (e.g., innervation, response to hypoxia) are not yet known, these data suggest that, in zebrafish, the cholinergic and serotonergic pathways of the hypoxic hyperventilatory response develop at different times and may be mutually exclusive.

In addition, in the amphibious fish, *K. marmoratus*, pre-exposure to 5-HT increased the sensitivity of emersion behavior (performed in order to promote a transition from gill to cutaneous respiration) when subjects were confronted with hypoxia, while ketanserin (a 5-HT_2 receptor antagonist) had the opposite effect; furthermore, ACh and hexamethonium (a nicotinic receptor antagonist) increased or decreased, respectively, the sensitivity of the emersion response to hypoxia (Regan et al., 2011).

In the mammalian carotid body, ACh is the favored candidate for mediating fast excitatory neurotransmission (along with adenosine triphosphate, ATP), and nicotinic ACh receptors have been localized to postsynaptic nerve terminals of sensory (petrosal) neurons and presynaptic type I cells (Shirahata et al., 2007; Nurse, 2010). Given the available evidence from studies in fishes (Burleson and Milsom, 1995a,b; Regan et al., 2011; Shakarchi et al., 2013), it appears that ACh may be equally important in mediating the hyperventilatory response to hypoxia. However, evidence from zebrafish indicates that in developing larvae a cholinergic contribution to the hyperventilatory response does not develop until after 12 days postfertilization (dpf), while a serotonergic pathway is present in the gill on or before 7 dpf (Shakarchi et al., 2013). 5-HT may thus have an additional role in O_2 sensing during early developmental stages. In the developing carotid body, 5-HT is present during embryonic stages in type I cells (Kameda, 2005), and there is evidence of ketanserin-sensitive 5-HT_{2A} receptors in the carotid body that participates in neuromodulation of the chemosensory response (Nurse, 2010). However, ionotropic 5-HT_3 receptors appear to mediate fast neurotransmission in the carotid body (Zhong et al., 1999; Nurse, 2010).

Direct evidence of the release of a neurotransmitter from gill NECs by hypoxia (or any other stimulus) has not yet been obtained. However, given that NECs are O_2 sensitive, are excitable and capable of dynamic changes in $[Ca^{2+}]_i$, and retain cytoplasmic synaptic vesicles, it is predicted that a hypoxic stimulus would induce such release with subsequent recycling of synaptic vesicles by endocytosis. Using an activity-dependent dye, Texas red hydrazide, a recent study found that NECs of trout did not take up the dye when exposed *in vivo* to hypoxia (45 mmHg) for 30 min, suggesting that recycling by endocytosis did not occur in these experiments (Porteus et al., 2013). By contrast, our preliminary evidence suggests that the recycling of synaptic vesicles in gill NECs stimulated by hypoxia can be observed *in vitro*. The activity-dependent styryl dye, FM1-43, has been used to demonstrate vesicular recycling in neurosecretory cells (Betz et al., 1996; Fukuda et al., 2003; Meyers et al., 2003; Zachar and Jonz, 2012b). In addition, FM1-43 was previously shown to be taken up by gill NECs in zebrafish following intraperitoneal injection (Jonz, 2004). In NECs isolated from the goldfish gill and then exposed to FM1-43, a brief stimulus of hypoxia (25 mmHg) induced uptake of the dye (Zachar and Jonz, unpublished observations), suggesting that hypoxia sufficiently stimulates vesicular release and recycling. These results may seem to be at odds with those described in a previous section, in which goldfish NECs did not respond to 25 mmHg hypoxia in patch-clamp recordings. However, it is important to note that unlike patch-clamp experiment, in which the intracellular medium is dialyzed, in FM1-43 experiments NECs are intact and retain their normal intracellular milieu. Sensitivity to hypoxia may be preserved in NECs when the normal composition of the cytosol is not disturbed.

4.4 EXTRABRANCHIAL CHEMORECEPTORS

4.4.1 Oropharyngeal Epithelium

There are numerous studies in the literature that have reported the preservation of physiological responses to hypoxia (and hypercapnia) following progressive or complete denervation of the cranial nerves, such as those that terminate within the gill arches. Under these circumstances, chemoreceptors of the denervated site(s), such as the gills, would be functionally ablated. Evidence from these studies was conveniently compiled in a comprehensive review by Milsom (2012). Predominantly, branchial chemoreceptors innervated by cranial nerves IX and X appear to initiate most cardioventilatory responses to hypoxia, such as decreased heart rate and increased ventilatory amplitude and frequency. However, in some species, extrabranchial chemoreceptors (i.e., outside of the gills) appear to contribute to a number of responses to hypoxia, including decreased heart rate in dogfish (*Scyliorhinus canicula*; Butler et al., 1977); increased gill amplitude in tambaqui (*Colossoma mesopotamicus*; Sundin et al., 2000), tench (*Tinca tinca*; Hughes and Shelton, 1962), sea raven (*Hemitripterus americanus*; Saunders and Sutterlin, 1971), traira (*Hoplias malabaricus*; Sundin et al., 1999), pacu (*Piaractus mesopotamicus*; Sundin et al., 2000), and bowfin (*Amia calva*; McKenzie et al., 1991); increased gill frequency in tench, sea raven, and bowfin (Hughes and Shelton, 1962; Saunders and Sutterlin, 1971; McKenzie et al., 1991); and aquatic surface respiration in tambaqui (*Colossoma mesopotamicus*; Florindo et al., 2006).

Although "extrabranchial" has not been clearly defined in most studies, there is ample evidence to suggest that O_2 chemosensory responses to hypoxia may originate from the oropharyngeal epithelium, which is additionally innervated by cranial nerves V and VII, in dogfish (Butler et al., 1977) and tambaqui (Sundin et al., 2000; Milsom et al., 2002; Florindo et al., 2006). Direct evidence for central O_2 chemoreceptors in fishes is lacking (Milsom, 2012). Unlike the gills, O_2-chemoreceptive cells of the extrabranchial oropharyngeal cavity have not been found. However, in zebrafish, serotonergic cells that are morphologically similar to NECs (i.e., innervation and apparent epithelial origin) occupy the oral aspects of the pharyngeal arches (Jonz and Nurse, 2005).

4.4.2 Skin

Additional evidence for the presence of extrabranchial O_2 chemoreceptors comes from the observation that developing zebrafish display behavioral and hyperventilatory responses to hypoxia before the gills are formed and in the complete absence of gill NECs (Jonz and Nurse, 2005; Coccimiglio and Jonz, 2012). In developing fishes, the skin is a major site for gas exchange (Rombough, 1988) and chemical sensing (Hansen et al., 2002; Northcutt, 2005). The skin may then be viewed as an optimal site for O_2 chemosensing in small organisms, such as developing zebrafish, that do not require a circulatory system to transfer respiratory gases between the external environment and tissues. In zebrafish embryos, serotonergic NECs of the skin first appear around 26 h postfertilization (hpf) and peak in number before 48 hpf (Figure 4.3f; Coccimiglio and Jonz, 2012). These cells retain synaptic vesicles and are innervated by catecholaminergic nerve terminals. The number of skin NECs, as identified by 5-HT immunohistochemistry, is markedly reduced by 7 dpf, and this developmental reduction can be postponed or eliminated if zebrafish larvae are acclimated to severe hypoxia (30 mmHg). Skin NECs have also been found in adults of the amphibious fishes, *K. marmoratus* (Regan et al., 2011), suggesting that in some species that retain cutaneous respiration skin O_2 sensing may be preserved. Although direct physiological evidence for O_2 sensitivity of skin NECs by patch-clamp or Ca^{2+} imaging is not yet available, the degradation of nerve terminals that contact skin NECs in zebrafish larvae effectively abolished the hyperventilatory response to hypoxia (Coccimiglio and Jonz, 2012).

4.5 DEVELOPMENT

4.5.1 DEVELOPMENT OF O_2 SENSING IN THE ZEBRAFISH GILL

Presently, evidence for the development of O_2 sensing, chemoreceptors, and associated neuronal pathways in fishes is available only for zebrafish. The zebrafish begins life as an organism that is tolerant to anoxia during early embryonic development but transitions to a hypoxia-sensitive larva between 2 and 3 dpf (Padilla and Roth, 2001; Jonz and Nurse, 2005; Turesson et al., 2006; Mendelsohn et al., 2008). The gills begin to develop at about 3 dpf and receive innervation around the same time (Kimmel et al., 1995; Higashijima et al., 2000). The gill filaments, where O_2-sensitive NECs are found, appear to arise from the pharyngeal ectoderm (Hogan et al., 2004). NECs first appear in gill primordia at 5 dpf and are predominantly innervated by 7 dpf (Jonz and Nurse, 2005). The hyperventilatory response to acute hypoxia in zebrafish begins at 3 dpf and peaks at 7 dpf, and this coincides with innervation of filament NECs (Jonz and Nurse, 2005; Turesson et al., 2006).

At approximately 7 dpf, zebrafish develop sensitivity to 5-HT and DA administered exogenously *in vivo*, which produce hyperventilation and hypoventilation, respectively (Shakarchi et al., 2013). This suggests that serotonergic and dopaminergic pathways in the gill develop first. At about 12 dpf, zebrafish develop sensitivity to the nicotinic ACh receptor antagonist, hexamethonium, while sensitivity to ACh develops a few days later, suggesting that an overlapping cholinergic pathway develops later and also mediates hypoxia-induced hyperventilation (Shakarchi et al., 2013).

4.5.2 EXTRABRANCHIAL TO BRANCHIAL O_2 SENSING

Gill NECs, however, may not significantly contribute to O_2-sensing responses before their innervation at 7 dpf since the hyperventilatory response to hypoxia develops earlier. These developmental changes suggest a temporal shift in O_2 sensing from an extrabranchial site elsewhere in the organism to the gills. As discussed in a previous section, a population of NEC-like cells of the skin was reported in zebrafish embryos (Jonz and Nurse, 2006; Coccimiglio and Jonz, 2012), and similar cells were reported in adults of the amphibious fish, *K. marmoratus* (Regan et al., 2011). Skin NECs, therefore, are good candidates for extrabranchial chemoreceptors that detect changes in water O_2 during the transition from anoxia tolerance to hypoxic sensitivity that occurs around the time of hatching. Indeed, that skin NECs are dramatically reduced in number by 7 dpf (when gill NECs are active) and are rarely observed in adults indicates that these cells play a role specifically in embryos and early larvae (Coccimiglio and Jonz, 2012). Since small organisms, such as zebrafish, do not require branchial gas exchange nor a circulatory system to survive (Rombough, 2007; Schwerte, 2009), it was suggested that the function of skin NECs may include coordination of hatching time or initiation of the development of functional ventilatory patterns responsive to hypoxia before gill O_2 sensors have matured (Coccimiglio and Jonz, 2012).

4.5.3 Renewal of O₂ Chemoreceptors

Recent evidence suggests that there may be a population of stem or progenitor cells in the gill filaments that may be related to O_2-chemoreceptive NECs, and this supports the initial ideas proposed by Laurent et al., in which NECs are continuously renewed by cells that do not contain 5-HT (Bailly et al., 1992). In developing zebrafish, nonserotonergic cells labeled with an antibody against a synaptic vesicle protein (SV2) are predominantly found in the gill arches, with very few of these cells in the gill filaments (Shakarchi et al., 2013), but in the adults, these cells occupy the gill filament epithelium in close proximity to serotonergic NECs (Jonz and Nurse, 2003). In addition, the acclimation of adult zebrafish to chronic hypoxia induced proliferation of these cells without effect on serotonergic NECs (Jonz et al., 2004). The developmental pattern of migration of SV2 cells from the gill arches to the filaments, and the control of cell number by hypoxia, would seem to suggest that these cells may play a role in the development or renewal of chemoreceptive NECs and maintenance of these populations during acclimation to hypoxia. Further evidence from developmental studies in zebrafish also indicates that the gene (*sall4*) encoding the stem cell transcription factor, called Sall4, is expressed in the gill arches and filaments of developing zebrafish (Jackson et al., 2013). These studies further support the presence of stem or progenitor cells in the gill.

SV2-positive cells of the gill arches in developing zebrafish may therefore be part of a population of progenitor cells that migrate from the gill arches into gill filament primordia, where they differentiate into NECs that will produce and store serotonin. In adult zebrafish, where the gill filaments are mature, SV2-positive cells reside adjacent to serotonergic NECs (Jonz and Nurse, 2003) and may contribute to maintaining this cell population. A similar arrangement can be found in the adult mammalian carotid body, where new type I cells can derive from adjacent progenitor type II cells (Pardal et al., 2007). While the carotid body is initially formed by the migration of sympathoadrenal progenitor cells from the superior cervical ganglion (Kameda, 2005), the embryonic origin of gill NECs is not yet understood.

4.6 CONCLUSION

O_2 chemoreceptors initiate autonomic reflexes and behavioral responses in fishes that lead to increased uptake and delivery of O_2 to systemic tissues, thus promoting survival in a hypoxic environment. Several decades of research have demonstrated the involvement of the gills in generating these responses, and it now seems clear that NECs of the gill filaments detect and transform the chemical signal of hypoxia. Considering the remarkable conservation of the distribution of NECs in the gills across all fish species, we understand relatively little about the significance of this distribution and how NECs convey information to the CNS regarding the level of O_2 in the environment or blood. At the cellular level, NECs from different species commonly utilize inhibition of membrane K^+ channels to generate changes in V_m during hypoxia, but we do not understand the expression of specific K^+ channel types across species and how this might be correlated with contribution to the transduction or modification of the hypoxic response. NECs are polymodal, suggesting that they may respond to multiple chemical stimuli and perhaps even fulfill multiple physiological roles in addition to cardiorespiratory regulation.

The zebrafish represents a particularly attractive model for pursuing studies of many areas of O_2 sensing, including intracellular mechanisms, neurotransmission, behavioral responses, and development. Further characterization of gill NECs with electrophysiological and fura-based Ca^{2+} imaging techniques may provide critical information about the specific roles of membrane ion channels and intracellular Ca^{2+} in O_2 sensing and hypoxia tolerance and may reveal an ubiquitous O_2 sensor in vertebrates. The use of mutations or gene knockdown studies that lead to perturbations in O_2 chemoreception by targeting specific proteins, such as membrane ion channels of NECs or postsynaptic receptors of sensory neurons, may lead to an improved understanding of how NECs communicate the hypoxic response within the gill. Continued studies of the role of the skin and gills in generating behavioral responses to hypoxia in zebrafish embryos may illuminate important features of the development of O_2 sensing in fishes. Generally, these studies will provide important information about O_2 chemoreception and respiratory regulation in fishes and how this system has evolved in vertebrates.

ACKNOWLEDGMENTS

I thank the Natural Sciences and Engineering Council of Canada for funding through a Discovery Grant (number 342303-2007) and the Canadian Foundation for Innovation and the Ontario Research Fund (grant number 16589).

REFERENCES

Abdallah SJ, Perry SF, Jonz MG. 2012. CO_2 signaling in chemosensory neuroepithelial cells of the zebrafish gill filaments: Role of intracellular Ca^{2+} and pH. *Adv Exp Med Biol* 758:143–148.

Bailly Y. 2009. Serotonergic neuroepithelial cells in fish gills: Cytology and innervation. In *Airway Chemoreceptors in the Vertebrates*, (eds. G. Zaccone, E. Cutz, D. Adriaensen, C.A. Nurse, and A. Mauceri). Enfield, NH: Science Publishers, pp. 61–97.

Bailly Y, Dunel-Erb S. 1986. The sphincter of the efferent filament artery in teleost gills: I. Structure and para-sympathetic innervation. *J Morphol* 187:219–237.

Bailly Y, Dunel-Erb S, Laurent P. 1992. The neuroepithelial cells of the fish gill filament: Indolamine-immunocytochemistry and innervation. *Anat Rec* 233:143–161.

Bailly Y, Dunel-Erb S, Geffard M, Laurent P. 1989. The vascular and epithelial serotonergic innervation of the actinopterygian gill filament with special reference to the trout, *Salmo gairderi*. *Cell Tissue Res* 258:349–363.

Betz WJ, Mao F, Smith CB. 1996. Imaging exocytosis and endocytosis. *Curr Opin Neurobiol* 63:365–371.

Bickler PE, Buck LT. 2007. Hypoxia tolerance in reptiles, amphibians, and fishes: Life with variable oxygen availability. *Annu Rev Physiol* 2007:145–170.

Booth, JH. 1978. The distribution of blood flow in the gills of fish: Application of a new technique to rainbow trout (Salmo gairdneri). *J Exp Biol* 73:119–129.

Brauner CJ, Rombough PJ. 2012. Ontogeny and paleophysiology of the gill: New insights from larval and air-breathing fish. *Respir Physiol Neurobiol* 1843:293–300.

Buckler KJ. 2007. TASK-like potassium channels and oxygen sensing in the carotid body. *Respir Physiol Neurobiol* 1571:55–64.

Burggren, W. 1982. "Air gulping" improves blood oxygen transport during aquatic hypoxia in the goldfish *Carassius auratus*. *Physiol Zool* 55:327–334.

Burleson ML, Milsom WK. 1993. Sensory receptors in the first gill arch of rainbow trout. *Respir Physiol* 931:97–110.

Burleson ML, Milsom WK. 1995a. Cardio-ventilatory control in rainbow trout: I. Pharmacology of branchial, oxygen-sensitive chemoreceptors. *Respir Physiol* 1003:231–238.

Burleson ML, Milsom WK. 1995b. Cardio-ventilatory control in rainbow trout: II. Reflex effects of exogenous neurochemicals. *Respir Physiol* 1013:289–299.

Burlseon ML, Mercer SE, Wilk-Blaszczak MA. 2006. Isolation and characterization of putative O_2 chemore-ceptor cells from the gills of channel catfish (*Ictalurus punctatus*). *Brain Res* 1092:100–107.

Burleson ML, Smatresk NJ, Milsom WK. 1992. Afferent inputs associated with cardioventilatory control in fish. In: *Fish Physiology*, Vol. XIIB, (eds. W.S. Hoar, D.J. Randall, and A.P. Farrell). San Diego, CA: Academic Press, pp. 389–426.

Bushnell PG, Steffensen JF, Johansen K. 1984. Oxygen consumption and swimming performance in hypoxia-acclimated rainbow trout *Salmo gairdneri*. *J Exp Biol* 113:225–235.

Butler PJ, Taylor EW, Short S. 1977. The effect of sectioning cranial nerves V,VII, IX and X on the cardiac response of the dogfish *Scyliorhinus canicula* to environmental hypoxia. *J Exp Biol* 69:233–245.

Campanucci VA, Fearon IM, Nurse CA. 2003. A novel O_2-sensing mechanism in rat glossopharyngeal neu-rones mediated by a halothane-inhibitable background K^+ conductance. *J Physiol* 548:731–743.

Coccimiglio ML, Jonz MG. 2012. Serotonergic neuroepithelial cells of the skin in developing zebrafish: Morphology, innervation and oxygen-sensitive properties. *J Exp Biol* 215:3881–3894.

Coolidge EH, Ciuhandu CS, Milsom WK. 2008. A comparative analysis of putative oxygen-sensing cells in the fish gill. *J Exp Biol* 211:1231–1242.

de Graaf PJF. 1990. Innervation pattern of the gill arches and gills of the carp (*Cyprinus carpio*). *J Morphol* 206:71–78.

Domnik NJ, Cutz E. 2011. Pulmonary neuroepithelial bodies as airway sensors: Putative role in the generation of dyspnea. *Curr Opin Pharmacol* 11:211–217.

Donald JA. 1984. Adrenergic innervation of the gills of brown and rainbow trout, *Salmo trutta* and *S. gaidneri*. *J Morphol* 182:307–316.

Donald JA. 1987. Comparative study of the adrenergic innervation of the teleost gill. *J Morphol* 193:63–73.

Dunel-Erb S, Bailly Y. 1986. The sphincter of the efferent filament artery in teleost gills: II. Sympathetic innervation. *J Morphol* 187:239–246.

Dunel-Erb S, Bailly Y, Laurent P. 1982. Neuroepithelial cells in fish gill primary lamellae. *J Appl Physiol Resp Environ Exercise Physiol* 53:1342–1353.

Dunel-Erb S, Bailly Y, Laurent P. 1989. Neurons controlling the gill vasculature in five species of teleosts. *Cell Tissue Res* 255:567–573.

Dymowska AK, Hwang PP, Goss GG. 2012. Structure and function of ionocytes in the freshwater fish gill. *Respir Physiol Neurobiol* 184:282–292.

Evans AM, Mustard KJ, Wyatt CN, Peers C, Dipp M, Kumar P, Kinnear NP, Hardie DG. 2005a. Does AMP-activated protein kinase couple inhibition of mitochondrial oxidative phosphorylation by hypoxia to calcium signaling in O_2-sensing cells? *J Biol Chem* 280:41504–41511.

Evans DH, Piermarini PM, Choe KP. 2005b. The multifunctional fish gill: Dominant site of gas exchange, osmoregulation, acid-base regulation, and excretion of nitrogenous waste. *Physiol Rev* 85:97–177.

Farbman AI. 2000. Cell biology of olfactory epithelium. In: *The Neurobiology of Taste and Smell*, (eds. T. E. Finger, W. L. Silver, and D. Restrepo). New York: Wiley & Sons, pp. 131–158.

Finger TE, Simon SA. 2000. Cell biology of taste epithelium. In: *The Neurobiology of Taste and Smell*, (eds. T. E. Finger, W. L. Silver, and D. Restrepo). New York: Wiley & Sons, pp. 287–314.

Fitzgerald RS, Eyzaguirre C, Zapata P. 2009. Fifty years of progress in carotid body physiology-invited article. *Adv Exp Med Biol* 648:19–28.

Florindo LH, Leite CAC, Kalinin AL, Reid SG, Milsom WK, Rantin FT. 2006. The role of branchial and oro-branchial O_2 chemoreceptors in the control of aquatic surface respiration in the neotropical fish tambaqui (*Colossoma macropomum*): progressive responses to prolonged hypoxia. *J Exp Biol* 209:1709–1715.

Fukuda J, Ishimine H, Masaki Y. 2003. Long-term staining of live Merkel cells with FM dyes. *Cell Tissue Res* 311:325–332.

Gilbert SF. 2010. *Developmental Biology*, 9th edn. Sunderland, MA: Sinauer Associates Inc.

Gilmour KM. 2001. The CO_2/pH ventilatory drive in fish. *Comp Biochem Physiol A Mol Integr Physiol* 130:219–240.

Goldstein SAN, Backenhauer D, O'Kelly I, Zilberberg N. 2001. Potassium leak channels and the KCNK family of two-P-domain subunits. *Nat Rev Neurosci* 2:1–11.

Goniakowska-Witalinska L, Zaccone G, Fasula S, Mauceri A, Licata A, Youson J. 1995. Neuroendocrine cells in the gills of the bowfin *Amia calva*. An ultrastructural and immunocytochemical study. *Fol Histochem Cytobiol* 33:171–177.

Gonzalez C, Almaraz L, Obeso A, Rigual R. 1994. Carotid body chemoreceptors: From natural stimuli to sensory discharges. *Physiol Rev* 74:829–898.

Gonzalez C, Agapito MT, Rocher A, Gomez-Niño A, Rigual R, Castañeda J, Conde SV, Obeso A. 2010. A revisit to O_2 sensing and transduction in the carotid body chemoreceptors in the context of reactive oxygen species biology. *Respir Physiol Neurobiol* 174:317–330.

Hansen A, Reutter K, Zeiske E. 2002. Taste bud development in the zebrafish, *Danio rerio*. *Dev Dyn* 223:483–496.

Higashijima S, Hotta Y, Okamoto H. 2000. Visualization of cranial motor neurons in live transgenic zebrafish expressing green fluorescent protein under the control of the islet-1 promoter/enhancer. *J Neurosci* 20:206–218.

Hogan BM, Hunter MP, Oates AC, Crowhurst MO, Hall NE, Heath JK, Prince VE, Lieschke GJ. 2004. Zebrafish *gcm2* is required for gill filament budding from pharyngeal ectoderm. *Dev Biol* 276:508–522.

Hughes, GM. 1972. Morphometrics of fish gills. *Resp Physiol* 14:1–25.

Hughes GM. 1984. General anatomy of the gills. In: *Fish Physiology*, Vol. XA, (eds. W.S. Hoar and D.J. Randall). San Diego, CA: Academic Press, pp. 1–72.

Hughes B, Shelton G. 1962. Respiratory mechanisms and their nervous control in fish. *Adv Comp Physiol Biochem* 1:275–364.

Hyndman KA, Evans DH. 2007. Endothelin and endothelin converting enzyme-1 in the fish gill: Evolutionary and physiological perspectives. *J Exp Biol* 210:4286–4297.

Hyndman KA, Choe KP, Havird JC, Rose RE, Piermarini PM, Evans DH. 2006. Neuronal nitric oxide synthase in the gill of the killifish, *Fundulus heteroclitus*. *Comp Biochem Physiol B Biochem Mol Biol* 144:510–509.

Jackson R, Braubach OR, Bilkey J, Zhang J, Akimenko MA, Fine A, Croll RP, Jonz MG. 2013. Expression of *sall4* in taste buds of zebrafish. *Dev Neurobiol* 73:543–558.

Jonz MG. 2004. Structure, function and development of O_2-sensitive neuroepithelial cells of the gills of zebrafish (*Danio rerio*). PhD thesis, McMaster University, Hamilton, Ontario, Canada.

Jonz MG, Nurse CA. 2003. Neuroepithelial cells and associated innervation of the zebrafish gill: A confocal immunofluorescence study. *J Comp Neurol* 461:1–17.

Jonz MG, Nurse CA. 2005. Development of oxygen sensing in the gills of zebrafish. *J Exp Biol* 208:1537–1549.

Jonz MG, Nurse CA. 2006. Ontogenesis of oxygen chemoreception in aquatic vertebrates. *Respir Physiol Neurobiol.* 154:139–152.

Jonz MG, Nurse CA. 2009. Oxygen-sensitive neuroepithelial cells in the gills of aquatic vertebrates. In: *Airway Chemoreceptors in the Vertebrates: Structure, Evolution and Function*, (eds. G. Zaccone, E. Cutz, D. Adriaensen, C.A. Nurse, and A. Mauceri). Enfield, NH: Science Publishers, pp. 1–30.

Jonz MG, Zaccone G. 2009. Nervous control of the gills. *Acta Histochem* 111:207–216.

Jonz MG, Fearon IM, Nurse CA. 2004. Neuroepithelial oxygen chemoreceptors of the zebrafish gill. *J Physiol* 560:737–752.

Kameda, Y. 2005. Mash1 is required for glomus cell formation in the mouse carotid body. *Dev Biol* 283:128–139.

Kimmel CB, Ballard WW, Kimmel SR, Ullmann B, Schilling TF. 1995. Stages of embryonic development of the zebrafish. *Dev Dyn* 203:253–310.

Kumar P, Prabhakar N. 2007. Sensing hypoxia: Carotid body mechanisms and reflexes in health and disease. *Respir Physiol Neurobiol* 157:1–3.

Laurent P. 1984. Gill internal morphology. In: *Fish Physiology*, Vol. XA., (eds. W.S. Hoar and D.J. Randall). San Diego, CA: Academic Press, pp. 73–183.

Laurent P, Dunel S. 1980. Morphology of gill epithelial in fish. *Am J Physiol Reg Integr Comp Physiol* 238:R147–R159.

Lahiri S, Roy A, Baby SM, Hoshi T, Semenza GL, Prabhakar NR. 2006. Oxygen sensing in the body. *Prog Biophys Mol Biol* 91:249–286.

Lesage F. 2003. Pharmacology of neuronal background potassium channels. *Neuropharmacology* 44:1–7.

Lesage F, Lazdunski M. 2000. Molecular and functional properties of two-pore-domain potassium channels. *Am J Physiol Renal Physiol* 279:F793–F801.

López-Barneo J, Pardal R, Ortega-Sáenz P. 2001. Cellular mechanism of oxygen sensing. *Annu Rev Physiol.* 63:259–287.

López-Barneo J, López-López JR, Ureña J, González C. 1988. Chemotransduction in the carotid body: K^+ current modulated by PO_2 in type I chemoreceptor cells. *Science* 241:580–582.

López-Barneo J, Nurse CA, Nilsson GE, Buck LT, Gassmann M, Bogdanova AY. 2010. First aid kit for hypoxic survival: Sensors and strategies. *Physiol Biochem Zool* 83:753–763.

López-Barneo J, Ortega-Sáenz P, Pardal R, Pascual A, Piruat JI, Durán R, Gómez-Díaz R. 2009. Oxygen sensing in the carotid body. *Ann NY Acad Sci* 1177:119–131.

Mauceri A, Fasulo S, Ainis L, Licata A, Lauriano ER, Martínez A, Mayer B, Zaccone G. 1999. Neuronal nitric oxide synthase (nNOS) expression in the epithelial neuroendocrine cell system and nerve fibers in the gill of the catfish, *Heteropneustes fossilis*. *Acta Histochem.* 101:437–448.

McKenzie DJ, Burleson ML, Randall DJ. 1991. The effects of branchial denervation and pseudobranch ablation on cardioventilatory control in an air-breathing fish. *J Exp Biol* 161:347–365.

Mendelsohn BA, Kassebaum BL, Gitlin JD. 2008. The zebrafish embryo as a dynamic model of anoxia tolerance. *Dev Dyn* 237:1780–1798.

Meyers JR, MacDonald RB, Duggan A, Lenzi D, Standaert DG, Corwin JT, Corey DP. 2003. Lighting up the senses: FM1-43 loading of sensory cells through nonselective ion channels. *J Neurosci* 23:4054–4065.

Milsom WK. 2002. Phylogeny of CO_2/H^+ chemoreception in vertebrates. *Respir Physiol Neurobiol* 131:29–41.

Milsom WK. 2012. New insights into gill chemoreception: Receptor distribution and roles in water and air breathing fish. *Respir Physiol Neurobiol* 184:326–339.

Milsom WK, Brill RW. 1986. Oxygen sensitive afferent information arising from the first gill arch of yellowfin tuna. *Respir Physiol* 66:193–203.

Milsom WK, Burleson ML. 2007. Peripheral arterial chemoreceptors and the evolution of the carotid body. *Respir Physiol Neurobiol* 157:4–11.

Milsom WK, Reid SG, Rantin FT, Sundin L. 2002. Extrabranchial chemoreceptors involved in respiratory reflexes in the neotropical fish *Colossoma macropomum* (the tambaqui). *J Exp Biol* 205:1765–1774.

Monteiro SM, Fontainhas-Fernandes A, Sousa M. 2010. An immunohistochemical study of gill epithelium cells in the Nile tilapia, *Oreochromis niloticus*. *Folia Histochem Cytobio* 48:112–121.

Montoro RJ, Ureña J, Fernández-Chacón R, Alvarez de Toledo G, López-Barneo J. 1996. Oxygen sensing by ion channels and chemotransduction in single glomus cells. *J Gen Physiol* 107:133–143.

Neubauer JA, Sunderram J. 2004. Oxygen-sensing neurons in the central nervous system. *J Appl Physiol* 96:367–374.

Nilsson GE, Dymowska A, Stecyk JA. 2012. New insights into the plasticity of gill structure. *Respir Physiol Neurobiol* 184:214–222.

Nilsson S. 1984. Innervation and pharmacology of the gills. In *Fish Physiology*, Vol. XA, (eds. W.S. Hoar and D.J. Randall). San Diego, CA: Academic Press, pp. 185–227.

Nilsson S, Sundin L. 1998. Gill blood flow control. *Comp Biochem Physiol A Mol Integr Physiol* 119:137–147.

Northcutt RG. 2005. Taste bud development in the channel catfish. *J Comp Neurol* 482:1–16.

Nurse CA. 2010. Neurotransmitter and neuromodulatory mechanisms at peripheral arterial chemoreceptors. *Exp Physiol* 95:657–667.

Olson KR. 2002. Vascular anatomy of the fish gill. *J Exp Zool* 293:214–231.

Olson KR. 2011. Hydrogen sulfide is an oxygen sensor in the carotid body. *Respir Physiol Neurobiol* 179:103–110.

Olson KR, Dombkowski RA, Russell MJ, Doellman MM, Head SK, Whitfield NL, Madden JA. 2006. Hydrogen sulfide as an oxygen sensor/transducer in vertebrate hypoxic vasoconstriction and hypoxic vasodilation. *J Exp Biol* 209:4011–4023.

Olson KR, Healy MJ, Qin Z, Skovgaard N, Vulesevic B, Duff DW, Whitfield NL, Yang G, Wang R, Perry SF. 2008. Hydrogen sulfide as an oxygen sensor in trout gill chemoreceptors. *Am J Physiol Regul Integr Comp Physiol* 295:R669–R680.

Padilla PA, Roth MB. 2001. Oxygen deprivation causes suspended animation in the zebrafish embryo. *Proc Natl Acad Sci USA* 98:7331–7335.

Pardal R, Ortega-Sáenz P, Durán R, López-Barneo J. 2007. Glia-like stem cells sustain physiologic neurogenesis in the adult mammalian carotid body. *Cell* 131:364–377.

Peers C, Wyatt CN, Evans AM. 2010. Mechanisms for acute oxygen sensing in the carotid body. *Respir Physiol Neurobiol* 174:292–298.

Perry SF, Abdallah S. 2012. Mechanisms and consequences of carbon dioxide sensing in fish. *Respir Physiol Neurobiol* 184:309–315.

Perry S, Jonz MG, Gilmour KM. 2009. Oxygen sensing and the hypoxic ventilatory response. In: *Fish Physiology*, vol. 27, (eds. J.G. Richards, A.P. Farrell, and C.J. Brauner). New York: Academic Press, pp. 193–253.

Plant LD, Kemp PJ, Peers C, Henderson Z, Pearson HA. 2002. Hypoxic depolarization of cerebellar granule neurons by specific inhibition of TASK-1. *Stroke* 33:2324–2328.

Porteus CS, Brink DL, Milsom WK. 2012. Neurotransmitter profiles in fish gills: Putative gill oxygen chemoreceptors. *Respir Physiol Neurobiol* 184:316–325.

Porteus CS, Brink DL, Coolidge EH, Fong AY, Milsom WK. 2013. Distribution of acetylcholine and catecholamines in fish gills and their potential roles in the hypoxic ventilatory response. *Acta Histochem* 115:158–169.

Prabhakar NR. 2006. O_2 sensing at the mammalian carotid body: Why multiple O_2 sensors and multiple transmitters? *Exp Physiol* 91:17–23.

Prabhakar NR. 2012. Hydrogen sulfide (H_2S): A physiologic mediator of carotid body response to hypoxia. *Adv Exp Med Biol* 758:109–113.

Prado MA, Reis RA, Prado VF, de Mello MC, Gomez MV, de Mello FG. 2002. Regulation of acetylcholine synthesis and storage. *Neurochem Int* 41:291–299.

Qin Z, Lewis JE, Perry SF. 2010. Zebrafish (*Danio rerio*) gill neuroepithelial cells are sensitive chemoreceptors for environmental CO_2. *J Physiol* 588:861–872.

Regan KS, Jonz MG, Wright PA. 2011. Neuroepithelial cells and the hypoxia emersion response in the amphibious fish *Kryptolebias marmoratus*. *J Exp Biol* 214:2560–2568.

Rombough PJ. 1988. Respiratory gas exchange, aerobic metabolism, and effects of hypoxia during early life. In *Fish Physiology*, vol. 11A, (eds. W.S. Hoar and D.J. Randall). New York: Academic Press, pp. 59–161.

Rombough, P. 2007. The functional ontogeny of the teleost gill: Which comes first, gas or ion exchange? *Comp Biochem Physiol A Mol Integr Physiol* 148:732–742.

Saltys HA, Jonz MG, Nurse CA. 2006. Comparative study of gill neuroepithelial cells and their innervation in teleosts and xenopus tadpoles. *Cell Tissue Res* 323:1–10.

Saunders RL, Sutterlin AM. 1971. Cardiac and respiratory response to hypoxia in the searaven, Hemepterus americanus, an investigation of possible control mechanism. *J Fisheries Res Board Can* 28:491–503.

Schwerte T. 2009. Cardio-respiratory control during early development in the model animal zebrafish. *Acta Histochem* 111:230–243.

Shakarchi K, Zachar PC, Jonz MG. 2013. Serotonergic and cholinergic elements of the hypoxic ventilatory response in developing zebrafish. *J Exp Biol* 216(5):869–880.

Shirahata M, Balbir A, Otsubo T, Fitzgerald RS. 2007. Role of acetylcholine in neurotransmission of the carotid body. *Respir Physiol Neurobiol* 157:93–105.

Sollid J, Weber RE, Nilsson GE. 2005. Temperature alters the respiratory surface area of crucian carp *Carassius carassius* and goldfish *Carassius auratus*. *J Exp Biol* 208:1109–1116.

Sollid J, De Angelis P, Gundersen K, Nilsson GE. 2003. Hypoxia induces adaptive and reversible gross morphological changes in crucian carp gills. *J Exp Biol* 206:3667–3673.

Sundin L, Nilsson S. 2002. Branchial innervation. *J Exp Zool* 293:232–248.

Sundin L, Holmgren S, Nilsson S. 1998. The oxygen receptor of the teleost gill? *Acta Zool* 79:207–214.

Sundin L, Reid SG, Rantin FT, Milsom WK. 2000. Branchial receptors and cardiorespiratory reflexes in the neotropical fish, Tambaqui (*Colossoma macropomum*). *J Exp Biol* 203:1225–1239.

Sundin LI, Reid SG, Kalinin AL, Rantin FT, Milsom WK. 1999. Cardiovascular and respiratory reflexes: The tropical fish, traira (*Hoplias malabaricus*) O_2 chemoresponses. *Respir Physiol* 116:181–199.

Turesson J, Schwerte T, Sundin L. 2006. Late onset of NMDA receptor-mediated ventilatory control during early development in zebrafish (*Danio rerio*). *Comp Biochem Physiol A Mol Integr Physiol* 143:332–339.

Tzaneva V, Perry SF. 2010. The control of breathing in goldfish (*Carassius auratus*) experiencing thermally induced gill remodelling. *J Exp Biol* 213:3666–3675.

Tzaneva V, Bailey S, Perry SF. 2011. The interactive effects of hypoxemia, hyperoxia, and temperature on the gill morphology of goldfish (*Carassius auratus*). *Am J Physiol Regul Integr Comp Physiol* 300:R1344–R1351.

Vornanen M, Stecyk JAW, Nilsson GE. 2009. The anoxia-tolerant crucian carp (*Carassius carassius* L.). In: *Fish Physiology*, vol. 27, (eds. J.G. Richards, A.P. Farrell, and C.J. Brauner). New York: Academic Press, pp. 398–443.

Vulesevic B, McNeill B, Perry SF. 2006. Chemoreceptor plasticity and respiratory acclimation in the zebrafish *Danio rerio*. *J Exp Biol* 209:1261–1273.

Weichert CK. 1967. *Elements of Chordate Anatomy*. New York: McGraw-Hill.

Wilson JM, Laurent P. 2002. Fish gill morphology: Inside out. *J Exp Zool* 293:192–213.

Wyatt CN, Mustard KJ, Pearson SA, Dallas ML, Atkinson L, Kumar P, Peers C, Hardie DG, Evans AM. 2007. AMP-activated protein kinase mediates carotid body excitation by hypoxia. *J Biol Chem* 282:8092–8098.

Zaccone G, Fasulo S, Ainis L. 1994. Distribution patterns of the paraneuronal endocrine cells in the skin, gills and the airways of fishes as determined by immunohistochemical and histological methods. *Histochem J* 26:609–629.

Zaccone G, Mauceri A, Fasulo S. 2006. Neuropeptides and nitric oxide synthase in the gill and the air-breathing organs of fishes. *J Exp Zool* 305A:428–439.

Zaccone G, Fasulo S, Ainis L, Licata A. 1997. Paraneurons in the gills and airways of fishes. *Microsc Res Tech* 37:4–12.

Zaccone G, Ainis L, Mauceri A, Lo Cascio P, Lo Giudice F, Fasulo S. 2003. NANC nerves in the respiratory air sac and branchial vasculature of the Indian catfish, *Heteropneustes fossilis*. *Acta Histochem* 105:151–163.

Zaccone G, Lauweryns S, Fasulo S, Tagliafierro G, Ainis L, and Licata A. 1992. Immunocytochemical localization of serotonin and neuropeptides in the neuroendocrine paraneurons of teleost and lungfish gills. *Acta Zool* 73:177–183.

Zaccone G, Mauceri A, Fasulo S, Ainis L, Lo Cascio P, Ricca MB. 1996. Localization of immunoreactive endothelin in the neuroendocrine cells of fish gill. *Neuropeptides*. 30:53–57.

Zachar PC, Jonz MG. 2012a. Neuroepithelial cells of the gill and their role in oxygen sensing. *Respir Physiol Neurobiol* 184:301–308.

Zachar PC, Jonz MG. 2012b. Confocal imaging of Merkel-like basal cells in the taste buds of zebrafish. *Acta Histochem* 114:101–115.

Zhang L, Wood CM. 2009. Ammonia as a stimulant to ventilation in rainbow trout *Oncorhynchus mykiss*. *Respir Physiol Neurobiol* 168:261–271.

Zhang L, Nurse CA, Jonz MG, Wood CM. 2011. Ammonia sensing by neuroepithelial cells and ventilatory responses to ammonia in rainbow trout. *J Exp Biol* 214:2678–2689.

Zhong H, Zhang M, Nurse CA. 1999. Electrophysiological characterization of 5-HT receptors on rat petrosal neurons in dissociated cell culture. *Brain Res* 816:544–553.

5 Intestinal Transport

Martin Grosell

CONTENTS

5.1 INTRODUCTION

All freshwater fishes regulate osmotic pressure and ionic concentrations above their ambient levels. In contrast, all three possible strategies for maintaining salt and water balance are found among fishes inhabiting marine environments. (1) Near osmoconformity/ionoconformity is found in the marine agnathan hagfishes, which are restricted to marine environments and do not regulate main electrolytes and osmotic pressure to any great extent (Morris, 1958). Hagfish do exert limited control over plasma ions by reducing Ca^{2+} and Mg^{2+} concentrations to ~50% of ambient with a resulting slight hyper-regulation of plasma Na$^+$ (Sardella et al., 2009). However, this control of plasma ionic composition is very minor compared to that displayed by elasmobranchs and teleosts. (2) Osmoconformity with regulation of main ions is seen in marine and some euryhaline elasmobranchs (Evans and Claiborne, 2008; Hazon et al., 2003) and in the lobe-finned coelacanth

(Griffith et al., 1974), which, in marine environments, maintain plasma osmolality slightly above that of the surrounding medium but NaCl concentrations at 30%–35% of ambient levels. (3) The most widespread strategy is osmoregulation, found in all teleosts (Evans and Claiborne, 2008; Hwang et al., 2011; Marshall and Grosell, 2006) and lamprey (Evans and Claiborne, 2008; Marshall and Grosell, 2006; Morris, 1958), which regulate Na^+ and Cl^- concentrations and osmotic pressure at ~150 mM and ~300 mOsm, respectively, regardless of ion concentrations in their surrounding environment.

Gas exchange by fishes is facilitated by a large gill surface area and close counter current contact between the blood and surrounding medium. While these properties are ideally suited for oxygen uptake from a medium with low oxygen solubility, they combine with large osmotic and ionic gradients across the branchial epithelium to result in diffusive salt loss and water movement in fishes that osmo- and ionoregulate.

While it has long been recognized that the gastrointestinal (GI) tracts of marine teleosts play vital roles in water balance (Smith, 1930), the GI tracts of marine elasmobranchs (Anderson et al., 2010) as well as the intestines of freshwater teleosts have more recently been implicated in osmolyte absorption associated with digestion (Wood and Bucking, 2011).

5.2 FRESHWATER FISHES

The following likely applies to freshwater lamprey, elasmobranchs, and teleosts, although studies on teleosts are by far the most abundant.

5.2.1 INTEGRATIVE PERSPECTIVE

To compensate for the excessive water gain, freshwater fishes produce copious amounts of urine. Although renal tubules in freshwater fishes are specialized for salt absorption, renal ion loss is unavoidable, which adds to the ion loss across the gill and must be compensated for by ion uptake. Ions are obtained from the diet (discussed later) and by active uptake across the branchial epithelium (discussed in the preceding chapter). For perspective, absolute ion uptake rates exhibit a strong dependence on size (Grosell et al., 2002) and correlate inversely with mass due to decreasing surface volume ratio as mass increases. An analysis, including 50 studies of Na^+ loss and uptake rates by freshwater organisms, reveals that 65% of the overall variation among organisms can be attributed to size (LOG uptake = 2.87–LOG[Mass]·0.274), where uptake rate and mass are in nmol/g/h and grams, respectively (Grosell et al., 2002). According to this relationship, a standard 1 g fish takes up Na^+ at a rate of ~750 nmol/g/h, while a 100 g fish takes up Na^+ at a rate of ~210 nmol/g/h.

5.2.2 DIETARY ION ABSORPTION

The majority of studies of osmoregulation by freshwater fishes have been performed on fasted animals to avoid fouling of the water in closed systems often employed for ion uptake experiments. As a consequence, little is known about the role of the GI tracts of freshwater fishes in salt and water balance. However, the potential role of dietary contributions to salt balance in freshwater fishes has received some recent attention (Bucking and Wood, 2006, 2007, 2008; Wood et al., 2010a), and the field was summarized in 2010 (Wood and Bucking, 2010). Natural diets for fish consisting of aquatic invertebrates contain 34–41 mmol Na^+/kg (Klinck et al., 2007). At a food consumption rate of 3% body mass per day and complete assimilation of ingested Na^+, dietary intake is equivalent to 20%–24% of the branchial uptake rates for a standard 100 g fish. Observations of no net gain of Na^+ by rainbow trout from ingested artificial diet (Bucking and Wood, 2006) suggest that little ingested Na^+ is assimilated. However, it should be noted that species-specific differences are likely to exist and that dietary Na^+ may play a more important role in waters with low Na^+ concentrations where branchial uptake may be limited. The freshwater stingray (*Potamotrygon sp.*) may serve as an example; branchial Na$^+$ uptake at ambient Na$^+$ concentrations is insufficient to offset the diffusive loss, leading to the suggestion that dietary Na$^+$ is of quantitative importance for salt balance in this freshwater elasmobranch (Wood et al., 2002).

Distinct from Na$^+$, 89% and 81% of ingested K$^+$ and Cl$^-$, respectively, are assimilated by rainbow trout, suggesting that dietary Cl$^-$ and especially K$^+$ may provide an important source of these electrolytes to freshwater fishes (Bucking and Wood, 2006). Uptake of Cl$^-$ from the diet appears to occur in exchange for HCO$_3^-$ secretion (Bucking and Wood, 2008; Cooper and Wilson, 2008; Taylor et al., 2007), which acts to neutralize acidic chyme from the stomach and alleviate, or at least reduce, the alkaline tide in fishes compared to that typical of most postprandial terrestrial vertebrates (Niv and Fraser, 2002). Interestingly, a number of fish species of distinct evolutionary origin (centracids, cyprinodons, and eels) have lost (or not developed) the ability to take up Cl$^-$ at the gill epithelium (Tomasso and Grosell, 2004). Obviously, for these species, dietary Cl$^-$ provides the sole source of compensation for renal and branchial Cl$^-$ loss.

The pathway for uptake of dietary K$^+$ in fishes is unknown.

5.3 MARINE TELEOST FISHES

Marine lampreys osmoregulate in ways very similar to marine teleosts. The following applies to teleosts and likely lamprey as well.

5.3.1 INTEGRATIVE PERSPECTIVE

The concentrated marine environment dictates ingestion of seawater to replace fluid lost by osmotic diffusion and low volumes of urine production. Marine teleosts generally drink 1–5 mL/kg/h (Carrick and Balment, 1983; Fuentes and Eddy, 1996, 1997a,b; Fuentes et al., 1996a,b; Grosell and Jensen, 1999; Grosell et al., 2004; Lin et al., 2002; Perrot et al., 1992; Shehadeh and Gordon, 1969; Takei and Tsuchida, 2000). The rate of seawater ingestion represents the most immediate level at which fluid absorption by the GI tract is controlled and is discussed in a separate section later. Of the seawater ingested, 70%–85% is absorbed by the GI tract, with the remaining fraction being released as rectal fluids of greatly altered chemistry (Genz et al., 2008; Hickman, 1968; Shehadeh and Gordon, 1969; Wilson et al., 1996). In contrast to the NaCl-rich ingested seawater, rectal fluids are low in NaCl with very high levels of Mg^{2+}, HCO$_3^-$, and SO$_4^{2-}$ (Table 5.1). As discussed in the following, the high levels of Mg^{2+} and SO$_4^{2-}$ stem from selective absorption of Na$^+$, Cl$^-$, K$^+$, and water, rendering these divalent ions greatly concentrated in the rectal fluids. In contrast, the high HCO$_3^-$ levels are the consequence of secondary active secretion by the intestine.

Water absorption from ingested seawater by the GI tract is a physiological feat accomplished by marine teleosts and only a few other vertebrates (lamprey, marine turtles, and likely saltwater crocodiles). The fluid absorption is driven by high rates of NaCl absorption; indeed, >95% of the NaCl ingested with seawater is absorbed by the GI tract. While this absorption of NaCl facilitates water uptake to offset the diffusive loss, it adds to the challenge of maintaining lower than ambient internal NaCl concentrations. In addition, the high concentrations of HCO$_3^-$ in the rectal fluids (Table 5.1) represent a substantial base loss from marine teleosts (Genz et al., 2008; Grosell et al., 1999; Wilson et al., 1996). The compensation for the excess NaCl gain and base loss (equivalent to acid gain) occurs in the form of NaCl and net acid secretion across the gill as discussed in the preceding chapter of this volume. Although intestinal absorption of Mg^{2+} and SO$_4^{2-}$ is extremely modest considering the gradients across the intestinal epithelium, some uptake is unavoidable and homeostasis of these divalent ions is achieved by renal excretion. Intestinal Ca^{2+} absorption is limited due to the formation of CaCO$_3$ precipitates in the intestinal lumen. The renal elimination of excess Ca^{2+} has long been assumed but urinary Ca^{2+} concentrations in

TABLE 5.1

Ionic Concentrations (mM), Osmotic Pressure (mOsm), and pH in Seawater and Fluids Obtained from Different Regions of the Gastrointestinal Tract

Parameter	Seawater	Medium				
		Gastric	Anterior	Mid	Posterior	Rectal
Na^+	420	191 (2)	81 (15)	70 (13)	63 (13)	50 (12)
K^+	8.9	3 (2)	13 (13)	10 (13)	9 (11)	7 (9)
Ca^{2+}	9.2	10 (2)	8 (14)	7 (14)	7 (12)	6 (11)
Mg^{2+}	48	26 (3)	72 (15)	87 (14)	82 (12)	119 (12)
Cl^-	489	268 (11)	115 (24)	109 (23)	96 (24)	97 (21)
SO_4^{2-}	28	20 (2)	76 (4)	116 (3)	93 (4)	101 (3)
HCO_3^- eqv.	2.3	5 (2)	45 14)	56 (16)	55 (15)	53 (11)
Osmotic pressure	1000	493 (2)	404 (6)	375 (4)	359 (6)	368 (5)
pH	7.9	8.2 (1)	8.0 (14)	8.2 (14)	8.3 (14)	8.2 (12)

Source: Walton Smith, F.G., *CRC Handbook of Marine Science*, Vol 1, CRC Press, Cleveland, OH, 1931; Bury, N.R. et al., *J. Exp. Biol.*, 204, 3779, 2001; Genz, J. and Grosell, M., *Comp. Biochem. Physiol. A*, 160(2), 156, 2011; Grosell, M. et al., *Am. J. Physiol.*, 293, R2099, 2007; Grosell, M. et al., *Fish Physiol. Biochem.*, 24, 81, 2001; Kirsch, R. and Meister, M.F., *J. Exp. Biol.*, 98, 67, 1982; McDonald, M.D. and Grosell, M., *Comp. Biochem. Physiol. A*, 143, 447, 2006; Parmelee, J.T. and Renfro, J.L., *Am. J. Physiol.*, 245, R888, 1983; Shehadeh, Z.H. and Gordon, M.S., *Comp. Biochem. Physiol.*, 30, 397, 1969; Sleet, R.B. and Weber, L.J., *Comp. Biochem. Physiol.*, 72A, 469, 1982; Smith, H.W., *Am. J. Physiol.*, 93, 480, 1930; Wilson, R.W. et al., *J. Exp. Biol.*, 199, 2331, 1996.

Note: >25 species of seawater-acclimated euryhaline and marine teleosts are represented (number of observations are indicated in parentheses).

marine teleosts are generally low (Grosell et al., 2001; Smith, 1930). Indeed, the mass balance calculations illustrated in Figure 5.1 suggest that elimination of excess Ca^{2+} may occur mainly across the gill epithelium.

It is challenging to achieve a complete overview of mass balance for salt, acid–base, and water exchange between marine fishes and their environment because three different organs (the GI tract, the gill, and the kidney) are intimately involved. An early account of intestinal and renal contributions for the winter flounder is still impressive but did not include acid–base equivalents (Hickman, 1968), while more recent examinations have included these parameters for branchial and intestinal fluxes (Genz et al., 2008; Wilson and Grosell, 2003) but not renal contributions. Simultaneous measurements of all relevant branchial, renal, and GI fluxes, including drinking rates, have not been performed on any species. However, most of these parameters have been measured on the gulf toadfish (*Opsanus beta*). Measurements of relevant fluxes on the gulf toadfish include branchial Na^+ and Cl^- extrusion rates (Kormanik and Evans, 1982), rectal and branchial acid–base flux measurements, drinking rates, rectal fluid, rectal ionic and $CaCO_3$ excretion (Genz et al., 2008) as well as urinary excretions, and composition of bladder urine (Genz et al., 2011; McDonald and Grosell, 2006). A mass balance for toadfish as an example of marine teleosts is presented in Figure 5.1.

As discussed in greater detail elsewhere (Grosell, 2011b), there is generally good agreement among the mass balance obtained from the aforementioned measurements on toadfish and a recent estimate based on average values from a range of marine teleosts (Grosell, 2007) and

(a)

	Ingestion	Branchial			Rectal	Renal
		Efflux	Influx	Net flux		
Volume	**2.6 mL/kg/h**			−1.8 mL/kg/h	−.07mL/kg/h	0.1 mL/kg/h
Na⁺	**1092**	**−7364**	6296	−1068	−19	−5
Cl⁻	**1274**	**−21597**	20394	−1203	−47	−24
Mg²⁺	**130**			5	−120	−15
SO₄²⁻	**78**			−9	−63	−6
Ca²⁺	**26**			−21.5	−4	−0.5
K⁺	**26**			−24.5	−1	−0.5
HCO₃⁻ eqv	**6**			<u>230</u>	−68	−0.3

(b)

FIGURE 5.1 (a) Gulf toadfish (*Opsanus beta*) with $CaCO_3$ precipitates voided with rectal fluids over a 48 h period. (Photo: Martin Grosell.) (b) Whole animal balance of water, acid–base equivalents, and main electrolytes (μmol/kg/h) in the gulf toadfish acclimated to natural seawater. Positive values indicate gain by the fish while negative values indicate loss and excretion. Values in **bold** represent measured values, underlined values represent estimated values (based on assumed Ca^{2+} and K^+ concentrations in urine of 5 mM and urine HCO_3^- concentrations of 3 mM). With the exception of branchial net flux of HCO_3^- equivalents (underlined), all branchial flux values are calculated from ingestion rates, and rectal and renal efflux rates assume steady state with respect to salt and water balance. Branchial net flux of HCO_3^- equivalents (uptake) is the sum of measured titratable nonintestinal acid excretion and ammonia excretion, with ammonia efflux contributing approximately 25% to the overall net acid excretion. (From Genz, J. et al., *J. Exp. Biol.*, 211, 2327, 2008.) Branchial unidirectional efflux rates for Na^+ and Cl^- were measured recently in toadfish in seawater (Grosell et al., unpublished) and are in agreement with earlier reports from this species and other marine teleosts. (From Grosell, M. and Wood, C.M., *J. Comp. Physiol. B*, 171, 585, 2001; Kormanik, G.A. and Evans, D.H., *J. Exp. Zool.*, 224, 187, 1982.) Branchial influx rates were calculated from the corresponding measured efflux rate and the calculated net flux rates. Drinking rates and loss via rectal fluids as well as renal losses were all measured (From Genz, J. et al., *J. Exp. Biol.*, 211, 2327, 2008; Genz, J. et al., *Am. J. Physiol.*, 300, R895, 2011), and ingestion of various ions was calculated from the measured drinking rate and an average assumed seawater composition (Table 5.1). See text for further details.

the early estimates based on winter flounder (Hickman, 1968). The gulf toadfish is atypical by having an aglomerular kidney and thus forming urine strictly by secretion. However, since urine flow and composition generally are indistinguishable among glomerular and aglomerular teleosts (Beyenbach, 2004; Beyenbach and Liu, 1996), the mass balance based on toadfish likely represented most marine teleosts.

5.3.2 DRINKING

Since drinking by marine fishes and subsequent intestinal water absorption were demonstrated in 1930 (Smith, 1930), it has been clearly shown that euryhaline fishes display 10- to 15-fold higher drinking rates in seawater than they do in freshwater (Carrick and Balment, 1983; Evans, 1967; Lin et al., 2002; Malvin et al., 1980; Perrot et al., 1992; Potts and Evans, 1967). Marine and euryhaline teleosts exhibit very similar drinking rates in the order of 1–5 mL/kg/h. In addition to ambient salinity, a range of factors affect drinking rate (reviewed by [Marshall and Grosell, 2006; Takei and Loretz, 2011]). The renin–angiotensin system with the terminally active angiotensin II (ANGII) acting on the medulla oblongata (the so-called swallowing center) is clearly involved in the regulation of drinking (Carrick and Balment, 1983; Fuentes and Eddy, 1996, 1998; Perrot et al., 1992; Takei and Tsuchida, 2000; Takei et al., 1979, 1988). Reduced blood volume and vasodilation (reduced blood pressure) result in elevated ANGII and potently stimulate drinking (Fuentes and Eddy, 1996; Hirano, 1974). Perhaps contrary to expectations are reports of elevated plasma osmolality reducing rather than stimulating drinking. However, it should be noted that the inhibitory response could be related to hypervolemia induced by NaCl infusions effectively masking the effect of increased osmotic pressure (Hirano, 1974; Takei et al., 1988). Clearly, additional work on this aspect of drinking control is required.

The active component of the kallikrein–kinin (KK) system, bradykinin (BK), and the atrial natriuretic peptide (ANP) both inhibit drinking by reducing plasma ANGII levels (Conlon, 1999; Gardiner et al., 1993; Takei et al., 2001; Tsuchida and Takei, 1998). However, at present, it is unknown if BK and ANP act solely via renin angiolensin system (RAS) or if they also act directly in some other way on the drinking reflex.

Although RAS responds to ambient salinity, hypovolemia, and hypotension to control drinking, it is clear that other factors are involved in the control of drinking. Convincing studies on the Japanese eel demonstrated that transfer to seawater stimulates drinking much faster than can be accounted for by hypovolemia and hypotension, and thus ANGII release, resulting from dehydration and thereby demonstrated that factors in addition to RAS control immediate changes in drinking rates. In contrast, distension of the stomach and increased salt concentrations in the intestinal lumen, both factors that would be associated with recent ingestion of seawater, act in a seemingly adaptive fashion to reduce drinking (Ando and Nagashima, 1996; Hirano, 1974). The drinking response to changes in these parameters occurs faster than blood pressure and blood volume changes pointing to sensory systems for ambient and intestinal lumen salt concentrations, ultimately affecting drinking. However, it remains to be established if the ANGII system is involved in this fast response. Elegant ion replacement studies have demonstrated that Cl⁻, rather than Na⁺, is the trigger for externally stimulated drinking and the negative feedback system associated with intestinal lumen salt concentrations (Ando and Nagashima, 1996; Hirano, 1974), but the nature of the Cl⁻ sensor remains unknown.

5.3.3 PROCESSING OF INGESTED FLUIDS

Due to the processing of ingested seawater that facilitates intestinal water absorption, the intestinal fluid chemistry is distinct from seawater and unique in composition (Table 5.1). When considering NaCl concentrations in fluids collected from various GI segments (Table 5.1), it is clear that the majority of the salt absorption occurs in the proximal segments and that these segments are largely impermeable to water. Although recent studies implicate the gastric mucosa of freshwater fishes in salt absorption from dietary sources (Wood and Bucking, 2010), the gastric mucosa of marine teleosts is generally assumed not to contribute to salt absorption by the GI tract. However, fluids sampled from the gastric lumen are much less concentrated than seawater (Kirsch and Meister, 1982; Parmelee and Renfro, 1983; Smith, 1930; Wilson et al., 1996), demonstrating the important role of the esophagus in desalinization of ingested seawater.

5.3.3.1 Na$^+$, Cl$^-$, and K$^+$ Absorption

More than 95% of the ingested NaCl is absorbed before rectal fluids are voided with distinct transport pathways utilized along the GI tract. The majority of the NaCl absorption occurs in the esophagus, which, combined with low water permeability of this tissue, is responsible for the marked reduction in osmotic pressure of the ingested seawater (~1000 mOsm) as it flows to the stomach (~500 mOsm) (Hirano and Mayer-Gostan, 1976; Kirsch and Meister, 1982; Parmelee and Renfro, 1983; Smith, 1930; Wilson et al., 1996). Both passive (along electrochemical gradients) and active absorption of Na$^+$ and Cl$^-$ occur in the esophagus in a 1:1 ratio (Grosell, 2006), with the passive component being quantitatively more significant (Esbaugh and Grosell, 2012). However, active absorption does occur and the absorption of Na$^+$ and Cl$^-$ is ouabain sensitive and Na$^+$ absorption is amiloride and 5-(N-Ethyl-N-isopropyl)amiloride (EIPA) sensitive, while furosemide is without effect on transport rates of both ions, eliminating the possibility of involvement of cotransporters, in the few species tested to date (Esbaugh and Grosell, 2012; Hirano and Mayer-Gostan, 1976; Parmelee and Renfro, 1983). Although early studies using amiloride were unable to distinguish between Na$^+$/H$^+$ exchangers (NHEs) and Na$^+$ channels, more recent work using the NHE-specific blocker EIPA clearly implicates NHEs in esophageal Na$^+$ absorption (Esbaugh and Grosell, 2012). Furthermore, cloning and expression measurements of NHEs in the Gulf toadfish revealed NHE2 as the likely candidate for the EIPA-sensitive Na$^+$ transport (Esbaugh and Grosell, 2012). However, it should be noted that complete inhibition of Na$^+$ transport was not achieved by EIPA or amiloride, suggesting additional uptake pathways. Considering that NHEs are involved in Na$^+$ uptake, it is tempting to suggest that anion exchangers might provide for Cl$^-$ uptake across the esophageal epithelium. However, the nature of Cl$^-$ uptake pathways in the esophagus remains unknown. There are no reports of K$^+$ concentrations in marine fishes' esophageal fluids, but two reports of low K$^+$ (2–3 mM) in gastric fluids, compared to seawater (~10 mM), of unfed fishes suggest modest esophageal and/or gastric absorption (Parmelee and Renfro, 1983; Smith, 1930) (Table 5.1).

Concentrations of Na$^+$ and Cl$^-$ in fluids from the anterior intestine are substantially lower than in the stomach with the absolute drop in concentrations being highest for Cl$^-$. Concentrations of both ions generally continue to drop with the lowest levels (50 and 82 mM for Na$^+$ and Cl$^-$, respectively) found in the posterior intestinal and rectal fluids. In contrast, K$^+$ levels in the anterior intestine are significantly higher than in gastric fluids but decrease along the more distal segments of the intestine (Table 5.1).

Coupled Na$^+$ and Cl$^-$ absorption via the apical Na$^+$:Cl$^-$ cotransporter NCC as well as coupled Na$^+$, Cl$^-$, and K$^+$ absorption via the absorption isoform of the Na$^+$:K$^+$:2Cl$^-$ cotransporter NKCC2 (Cutler and Cramb, 2002; Field et al., 1980; Frizzell et al., 1979; Halm et al., 1985a,b; Musch et al., 1982) relies on the inward directed electrochemical gradient for Na$^+$ and accounts for the majority of Na$^+$ uptake and 30%–70% of the intestinal Cl$^-$ absorption (Grosell, 2006) (Figure 5.2). The inward directed electrochemical gradient for Na$^+$ is established by the Na$^+$/K$^+$-ATPase (NKA), which is highly expressed in the lateral membranes of the enterocytes (Grosell, 2007) (Figure 5.3). The relative contribution of NKCC2 and NCC to Na$^+$ and Cl$^-$ absorption remains to be quantified, although thermodynamic considerations favor NCC in the more distal regions of the intestine (Loretz, 1995). Considering the stoichiometry of NKCC2 and the composition of seawater and intestinal fluids (Table 5.1), it is apparent that K$^+$ may be limiting for the activity of this apical transporter. However, considering the very low K$^+$ concentrations in the gastric fluids and considerably higher K$^+$ concentrations in the intestinal fluids, it seems likely that K$^+$ is secreted across the apical membrane to allow for continued operation of NKCC2. Indeed, Ba^{3+}-sensitive K$^+$ channels have been identified in the apical membrane of teleost intestinal epithelia (Musch et al., 1982).

Considering the inward-directed electrochemical gradient for Na$^+$, the possibility of Na$^+$ entering across the apical membrane via Na$^+$ channels exists but has yet to be thoroughly examined. Similarly, considering the alkaline luminal fluids and thus the outward-directed H$^+$ gradient across the apical membrane, apical NHE isoforms could contribute to Na$^+$ entry. At least for the euryhaline rainbow trout, mRNA expression of NHE3 is greatly increased upon transfer from freshwater to 65% seawater, suggesting a role for this transporter in seawater osmoregulation (Grosell et al., 2007). However, attempts to demonstrate a role for NHE3 in apical H$^+$ extrusion by isolated proximal intestinal segments showed no effect of amiloride (10^{-4} M) (Grosell et al., 2009a). In contrast to these data and in support of a role of NHEs are observations of increased intestinal base secretion by intact fish with amiloride (10^{-4} M) added to the solutions used to perfuse the intestinal lumen (Wilson et al., 1996). This increase in net base secretion is consistent with the inhibition of apical NHE exchange as it would reduce acid secretion into the intestinal lumen. The discrepancy between these *in vitro* and *in vivo* observations could perhaps indicate the NHE activity is most pronounced in the distal regions of the intestine, which were not considered in the study of isolated intestinal segments. The putative roles of apical, intestinal NHEs in seawater-acclimated rainbow trout and perhaps other euryhaline fishes await detailed

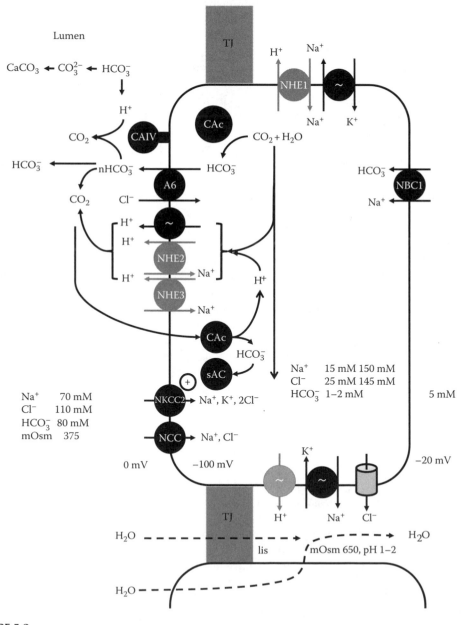

FIGURE 5.2

characterization. Recent work on the gulf toadfish, however, shows very low expression of the apical NHE2 and NHE3 in all intestinal segments, suggesting a limited, if any, role of these transporters in Na$^+$ uptake (Esbaugh and Grosell, 2012).

The net absorption of Cl$^-$, like Na$^+$, occurs against the transepithelial electrochemical gradient. Under *in vivo*-like conditions, the transepithelial potential (TEP) of the intestinal epithelium of marine fishes is −15 to −20 mV (luminal reference), and both luminal Na$^+$ and Cl$^-$ concentrations are typically around or less than 100 mM (Table 5.1). However, unlike Na$^+$, Cl$^-$ crosses the apical membrane against an electrochemical gradient. Cytosolic Cl$^-$ concentrations are above electrochemical equilibrium, and Cl$^-$ exits the epithelial cells across the basolateral membrane via K$^+$:Cl$^-$ cotransport (Smith et al., 1980) and Cl$^-$ channels, including cystic fibrosis transmembrane regulator (CFTR) paralogues (Loretz and Fourtner, 1988; Marshall et al., 2002b).

The outward-directed electrochemical gradient for Cl$^-$ across the apical membrane is overcome in part by NCC and NKCC2 as discussed earlier. Both these transporters contribute significantly to Cl$^-$ uptake, but Cl$^-$ uptake occurs at rates far greater than corresponding rates for Na$^+$. Excess Cl$^-$ uptake was first demonstrated for the southern flounder (Hickman, 1968) and have since been confirmed for all examined species (Grosell et al., 2001, 2005; Grosell and Jensen, 1999; Marshall et al., 2002a; Pickering and Morris, 1973; Shehadeh and Gordon, 1969; Skadhauge, 1974) (see [Grosell, 2006] for review). The excess Cl$^-$ uptake by the intestine is also illustrated by the data in Table 5.1; the difference between Na$^+$ and Cl$^-$ concentrations in the gastric fluids is 77 mM, while it is only 47 mM in the rectal fluids. This change in average Na$^+$:Cl$^-$ ratios along the intestine illustrates that net Cl$^-$ absorption exceeds Na$^+$ absorption by ~40%. Consistent with these observations is the demonstration of Na$^+$-independent Cl$^-$ uptake pathways (Grosell et al., 1999, 2001).

FIGURE 5.2 (continued)   Accepted (solid black) and putative (gray) transport processes in the intestinal epithelium of marine teleost fishes. (From Grosell, M., *Acta Physiol.*, 202, 421, 2011a.) Water transport, transcellular and/or paracellular, (dotted lines) is driven by active NaCl absorption, providing a hyperosmotic coupling compartment in the lateral interspace (*lis*). Entry of Na$^+$ across the apical membrane via cotransporters (NKCC2 and NCC) and extrusion across the basolateral membrane via Na$^+$/K$^+$-ATPase account for transepithelial Na$^+$ movement. In addition, recent evidence that NHE2 and NHE3 are expressed in the intestinal epithelium suggested that Na$^+$ uptake across the apical membrane may occur also via these transporters. Entry of Cl$^-$ across the apical membrane occurs via both cotransporters and Cl$^-$/HCO$_3^-$ exchange conducted by the SLC26a6 anion exchanger, while Cl$^-$ exits the cell via basolateral anion channels. Cellular HCO$_3^-$ for apical anion exchange is provided in part by HCO$_3^-$ entry across the basolateral membrane via NBC1 and in part by hydration of endogenous metabolic CO$_2$. Cytosolic carbonic anhydrase (CAc) found mainly in the apical region of the enterocytes facilitates CO$_2$ hydration. Protons arising from CO$_2$ hydration are extruded mainly across the basolateral membrane by a Na$^+$-dependent pathway and possibly by vacuolar H$^+$ pumps. Some H$^+$ extrusion occurs across the apical membrane via H$^+$ pumps and possibly via NHE2 and/or NHE3 and masks some of the apical HCO$_3^-$ secretion by HCO$_3^-$ dehydration in the intestinal lumen yielding molecular CO$_2$. This molecular CO$_2$ may diffuse back into the enterocytes for rehydration and continued apical anion exchange. Furthermore, molecular CO$_2$ from this reaction is rehydrated in the enterocytes and resulting HCO$_3^-$ is sensed by soluble adenylyl cyclase (sAC), which appears to stimulate ion absorption via NKCC2 (+). Conversion of HCO$_3^-$ to CO$_2$ in the intestinal lumen is facilitated by membrane-bound carbonic anhydrase, CA-IV, and possibly other isoforms, a process that consumes H$^+$ and thereby contributes to luminal alkalinization and CO$_3^{2-}$ formation. Titration of luminal HCO$_3^-$ and formation of CO$_3^{2-}$ facilitate formation of CaCO$_3$ precipitates to reduce luminal osmotic pressure and thus aid water absorption. SLC26a6, the electrogenic anion exchanger, exports nHCO$_3^-$ in exchange for 1Cl$^-$, and its activity is therefore stimulated by the hyperpolarizing effect of the H$^+$ pump. The apical electrogenic nHCO$_3^-$ exchanger (SLC26a6) and electrogenic H$^+$ pump constitute a transport metabolon perhaps accounting for the apparently active secretion of HCO$_3^-$ and the uphill movement of Cl$^-$ across the apical membrane. The indicated values for osmotic pressure and pH in the absorbed fluids are based on measured net movements of H$_2$O and electrolytes including H$^+$, but the degree of hypertonicity and acidity in *lis* are likely much less than indicated due to rapid equilibration with subepithelial fluid compartments. See text for further details.

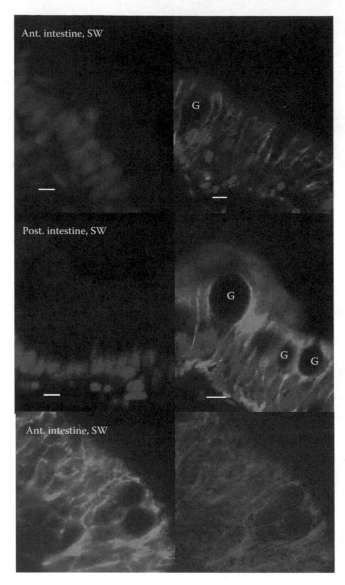

FIGURE 5.3 **(See color insert.)** V-type H⁺-ATPase in the anterior (top) and posterior intestine (middle) with negative controls (left panels) as well as Na⁺/K⁺-ATPase in the anterior intestine (bottom, left) and bright-field overlay (right) from toadfish held in seawater. Scale bar: 10 μ. (Images are from Guffey, S. et al., *Am. J. Physiol.*, 301, R1682, 2011.) (H⁺ pump) and (From Tresguerres, M. et al., *Am. J. Physiol.*, 299, R62, 2010) (Na⁺/K⁺-ATPase).

It is now well established that this Na⁺-independent Cl⁻ uptake is mediated by an apical anion exchanger, which at least in the euryhaline tiger puffer (*Takifugu rubripes*) and the gulf toadfish has been identified as a member of the SLC26 family of anion transporters, SLC26a6 (Grosell et al., 2009b; Kurita et al., 2008). The activity of this anion exchanger is, in part, responsible for the high luminal HCO_3^- concentrations and alkaline conditions (Table 5.1) as well as 30%–70% of the intestinal Cl⁻ uptake (Grosell, 2006). Interestingly, SLC20a6 from the tiger puffer and the toadfish is electrogenic, operating in an $nHCO_3^-/Cl^-$ mode (Grosell et al., 2009b; Kurita et al., 2008), which means that it is fueled in part by the apical, cytosolic negative membrane potential (~100 mV) (Loretz, 1995). This unusual stoichiometry accounts for Cl⁻ uptake and HCO_3^- secretion against seemingly unfavorable gradients for both ions (Grosell et al., 2009b). In addition to the electrogenicity of the teleost

SLC26a6, at least two other factors facilitate anion exchange, in the direction of Cl⁻ uptake, across the apical membrane despite apparent unfavorable gradients. Recent studies have demonstrated the presence and functional significance of a V-type H⁺-ATPase in the apical membrane in seawater-acclimated euryhaline and marine teleosts (Grosell et al., 2007, 2009a; Guffey et al., 2011) (Figure 5.3), and studies have demonstrated the presence of a membrane-bound carbonic anhydrase (CA) isoform (likely CA-IV) facing the intestinal lumen (Grosell et al., 2007, 2009a). Both the apical H⁺ pump and CA-IV facilitate continued anion exchange. H⁺ extrusion via an apical H⁺ pump acts to further polarize the apical membrane, which in itself provides driving force for the electrogenic SLC26a6. In addition, the secretion of acid inevitably will titrate off HCO_3^- in the boundary layer at the apical membrane, reducing the actual HCO_3^- gradient, which will create more favorable conditions for continued anion exchange. Last, but not least, the secretion of H⁺ into a high HCO_3^- environment will elevate the partial pressure of CO_2 (PCO_2), making molecular CO_2 available for diffusion back into the epithelial cells for hydration to (re)form HCO_3^- for continued anion exchange (Grosell et al., 2009b). Despite the activity of an apical H⁺ pump, the intestinal tissue performs high rates of net base secretion. The simultaneous H⁺ secretion accounts for 10%–30% of the HCO_3^- secretion depending on species, intestinal segment, and ambient salinity (Grosell et al., 2009a,b; Guffey et al., 2011). The apical, membrane-bound CA activity likely facilitates the titration of HCO_3^- in the boundary layer as H⁺ is being secreted by the H⁺ pump. In addition, the H⁺ arising from the formation of $CaCO_3^-$ precipitates (see "the following") will contribute to the HCO_3^- dehydration, both processes being facilitated by the CA-IV (Figure 5.2; [Grosell et al., 2009a,b]).

Two principal sources of HCO_3^- allow for the very high anion exchanges rates and thus HCO_3^- secretion by the intestinal epithelium. The hydration, facilitated by cytosolic CA (CAc) (Dixon and Loretz, 1986; Grosell et al., 2005, 2009a; Grosell and Genz, 2006; Wilson et al., 1996), of endogenous CO_2 from respiration of this metabolically active tissue contributes typically up to 50% of the apical HCO_3^- secretion (Ando and Subramanyam, 1990; Dixon and Loretz, 1986; Grosell et al., 2005; Grosell and Genz, 2006) and is, in at least one case, the sole source of cellular HCO_3^- for anion exchange (Grosell et al., 2009a). The elevated mRNA expression of CAc during seawater acclimation of rainbow trout and during acclimation to 60 ppt by the toadfish as well as elevated cellular CA enzymatic activity in both cases illustrates the significance of intestinal CAc for osmoregulation by seawater-acclimated euryhaline and marine teleosts (Grosell et al., 2007; Sattin et al., 2010). While a fraction of the H⁺ arising from the CO_2 hydration reaction is secreted across the apical membrane, the majority of the acid is secreted across the basolateral membrane via a Na⁺-dependent pathway, likely an NHE isoform (Grosell and Genz, 2006). High expression of NHE1 in all intestinal segments of the toadfish points to this basolateral isoform as the transporter responsible for the basolateral H⁺ secretion (Esbaugh and Grosell, 2012).

A member of the SLC4 family, NBC1 comprises the second source of HCO_3^- for apical secretion by facilitating import of extracellular HCO_3^- across the basolateral membrane of the enterocytes (Kurita et al., 2008; Taylor et al., 2010). As for the two CA isoforms discussed earlier, NBC1 displays elevated mRNA expression upon transfer to elevated salinity, illustrating the involvement of this transporter in hypo-osmoregulation by teleosts. Oocyte expression studies have revealed that NBC1 from marine teleosts is electrogenic (Kurita et al., 2008; Taylor et al., 2010), with saturation kinetics nearly identical to the intact intestinal epithelium, pointing to a rate-limiting role for NBC1 in transepithelial HCO_3^- transport by this tissue (Taylor et al., 2010). The affinity constant for the toadfish NBC1, as well as the basolateral step of the intact epithelium, is in the order of 10 mM HCO_3^-, meaning that transport rates by NBC1 and secretion rates by the epithelium are strongly influenced by variation in physiological plasma levels of HCO_3^- (3–7 mM). The aforementioned observations strongly suggesting a central role for NBC1 in intestinal HCO_3^- secretion have recently gained additional support by studies demonstrating that the marine fish NBC1 operates in a stoichiometry that allows for HCO_3^- import across the basolateral membrane of the enterocytes under physiological conditions (Chang et al., 2012).

The absorption of Na$^+$ (NCC, NKCC2, and possibly NHEs) and Cl$^-$ (NCC, NKCC2, and SLC26a6) across the intestinal epithelium occurs via multiple parallel pathways. The reason for this redundancy remains poorly understood but may reflect a need for salt and water absorption under a wide range of conditions imposed by the multiple functions of the intestinal epithelium. At least for Cl$^-$, information shedding light on this interesting phenomenon is emerging. First, Cl$^-$ absorption via anion exchange is relatively modest when luminal NaCl concentrations are high and thus favor Cl$^-$ absorption via NCC and NKCC2. In contrast, as luminal NaCl concentrations decrease, making cotransport less favorable, anion exchange appears to be stimulated (Grosell and Taylor, 2007). Second, high luminal HCO$_3^-$ concentrations, which present a challenge for continued anion exchange, stimulate NaCl absorption by cotransport pathways (Tresguerres et al., 2010). The activation of cotransport pathways by HCO$_3^-$ is mediated by soluble adenyl cyclase (sAC) and involves CA as well as the apical H$^+$ pump. The proposed mechanism involves creation of a high PCO$_2$ environment at the external surface of the apical membrane by the combined action of anion exchange and the apical H$^+$ pump followed by diffusion of CO$_2$ into the epithelial cells where CA facilitates hydration of CO$_2$ to form HCO$_3^-$, which is sensed by CAc, leading to the activation of NKCC2 in a cAMP-dependent manner as in other systems (Chen et al., 2000; Tresguerres et al., 2010) (Figure 5.2).

5.3.3.2 Divalent Ions

The major divalent ions in seawater, Mg^{2+} and SO$_4^{2-}$, have not received much attention from a marine fishes' osmoregulatory perspective despite being the most concentrated solutes in intestinal fluids (Table 5.1). However, recent studies have revealed that Mg^{2+} and SO$_4^{2-}$ concentrations, rather than NaCl concentrations, limit the survival of marine fishes at elevated salinities (60–90 ppt) (Genz et al., 2011). Both these divalent ions are left concentrated in the intestinal lumen as a function of selective absorption of Na$^+$, Cl$^-$, and water and reach concentrations of up to 150 and 120 mM, for Mg^{2+} and SO$_4^{2-}$, respectively (Table 5.1), ultimately preventing water absorption due to high luminal osmotic pressure. Despite the steep inward-directed gradient for both Mg^{2+} and SO$_4^{2-}$ (plasma concentrations <1 mM), the majority (~90%) of ingested Mg^{2+} and SO$_4^{2-}$ are eliminated with rectal fluids (Genz et al., 2008; Hickman, 1968; Wilson and Grosell, 2003). While the limited absorption, despite the steep gradients, is remarkable, some uptake of these divalent ions is unavoidable. Excess Mg^{2+} and SO$_4^{2-}$ entering the blood are eliminated by the kidney as evident from very high concentrations of both ions in marine teleost fishes' urine (McDonald, 2006; McDonald and Grosell, 2006). It seems that, at least for SO$_4^{2-}$, the limited intestinal uptake is not simply a product of passive barrier functions of the intestinal epithelium. Elegant studies have demonstrated that the intestinal epithelium of marine teleosts is capable of SO$_4^{2-}$ secretion via anion exchange for Cl$^-$ uptake to reduce overall net SO$_4^{2-}$ gain (Pelis and Renfro, 2003). The SLC26a6 isoform has recently been demonstrated to be involved in renal SO$_4^{2-}$ secretion in the euryhaline pufferfish under *in vivo*-like conditions (Kato et al., 2009). Considering the similarity between intestinal fluids and urine composition in marine teleosts and the presence of SLC26a6 in the intestinal epithelium, it seems likely that this transporter is involved in intestinal SO$_4^{2-}$ as well as HCO$_3^-$ secretion. Whether Mg^{2+} secretion, to reduce net absorption by the intestine of marine teleosts, occurs remains unknown.

Undoubtedly strong barrier properties of the intestinal epithelium against Mg^{2+} and SO$_4^{2-}$ uptake are involved in maintaining the high transepithelial gradients. The nature of these barrier properties awaits exploration, but recent studies demonstrating elevated mRNA expression of the tight junction proteins, claudin 3, 15, and 25b, in intestinal tissue of euryhaline fishes following transfer to elevated salinity may suggest involvement of these proteins in controlling the paracellular pathway and possibly uptake of divalent ions (Clelland et al., 2010; Tipsmark et al., 2010).

Concentrations of Ca^{2+} in intestinal fluids are typically <5mM as compared to 10 mM in seawater (Table 5.1). These low Ca^{2+} concentrations occur despite high rates of water absorption and are the product of intestinal Ca^{2+} absorption and the formation of CaCO$_3$ in the intestinal lumen (Genz et al., 2008; Hickman, 1968; Wilson and Grosell, 2003). The alkaline conditions in the intestinal lumen promote $CaCO_3$ formation such that macroscopic precipitates are present in intestinal fluids of fed as well as unfed marine fishes (Wilson et al., 2002; Wilson and Grosell, 2003) (Figure 5.1). It appears that the excess Ca^{2+} gained from intestinal uptake is eliminated across the gill epithelium, although a modest renal contribution cannot be dismissed (Grosell, 2011b) (discussed earlier).

5.3.3.3   Water Absorption

The absorption of Na^+ and Cl^- described earlier, along with adjustments of luminal osmotic pressure discussed later, provides driving forces for osmotic water movement despite the lack of net osmotic gradients or even slightly hyperosmotic conditions in the intestinal lumen (Table 5.1). Water absorption against osmotic pressure gradients has been demonstrated for the marine teleost intestinal epithelium (Genz et al., 2011; Skadhauge, 1974) and can be explained by high tonicity (>500 mOsm; [Grosell, 2006]) of fluids absorbed by the intestinal epithelium and the lateral interspace (*lis*) between enterocytes acting as a coupling compartment (Figure 5.2). The columnar enterocytes are elongated with large lateral membrane surface areas exposed to the lateral interspace. Densities of NKA and presumably other ion transport proteins are high in the lateral membrane (Grosell, 2007; Grosell et al., 2007; Guffey et al., 2011; Tresguerres et al., 2010) (Figure 5.3), and salt transport across the lateral membrane can be assumed to render fluids in the lateral interspace hyperosmotic, which will drive fluid movement into *lis*. Water movement in the direction from the intestinal lumen to the blood side of the epithelium is likely ensured by the barrier properties of the tight junctions located at the apical side of *lis* resulting in dispersal of salt, and thereby water, to the blood side of the epithelium.

At least for the killifish, *Fundulus heteroclitus*, water movement across the intestinal epithelium is uncoupled from the movement of polyethylene glycol (PEG), a paracellular marker, strongly suggesting that water movement is transcellular in this teleost (Wood and Grosell, 2012). It is unknown at present if this single study on killifish reflects teleosts in general, and clearly further studies are needed. Of the 13 documented aquaporins (AQPs), AQPe, AQP1, and AQP3 are all expressed in the intestinal epithelium and may play a role in water absorption (Aoki et al., 2003; Cutler et al., 2007; Lignot et al., 2002; Martinez et al., 2005a; Raldua et al., 2008). AQP1 shows elevated intestinal mRNA expression in euryhaline fishes following transfer to seawater, elevated AQP1 protein levels in the intestinal tissues from seawater-acclimated euryhaline fishes as well as localization in the apical and basolateral membranes. In addition, AQP1 confers water permeability in oocyte expression systems and is a likely candidate for the transcellular pathway for water movement across the marine teleost intestinal epithelium (Aoki et al., 2003; Martinez et al., 2005a,b; Raldua et al., 2008). However, it should be noted that mRNA expression of AQP1 in the esophagus has been reported to be increased following seawater transfer. This observation of elevated expression in a tissue known for its lack of water permeability may indicate that the presence of AQP1 is not associated with transcellular water movement but possibly plays roles in osmosensing and/or cell volume sensing as prolactin cells of the Mozambique tilapia (Watanabe et al., 2009). The area of intestinal AQP involvement in marine teleost osmoregulation, challenging as it may be, offers a fruitful area for further study.

5.3.3.4   Reduction of Intestinal Fluid Osmotic Pressure

In addition to the solute-coupled water absorption discussed earlier, marine teleost fishes employ at least two mechanisms, $CaCO_3$ precipitation and apical H^+ secretion, to reduce and control luminal osmotic pressure and thus facilitate continued water absorption in the intestine. The presence of $CaCO_3$ precipitates in the lumen of marine fishes is a consistent phenomenon contributing to a reduction in osmotic pressure by as much as 70–100 mOsm as HCO_3^- and Ca^{2+} are removed from solution (Grosell et al., 2009b; Wilson et al., 2002). This magnitude of osmotic pressure reduction is critical for water absorption as the maximal osmotic gradient against which the marine teleost intestine can absorb fluid is ~40 mOsm (Genz et al., 2011). Thus, without the reduction in osmotic pressure arising from $CaCO_3$ precipitation, water absorption and therefore survival in the marine environment would not be possible. It is unclear if the precipitation process is biologically controlled. However, precipitates from marine teleosts are unique in shape and composition, suggesting that they may not be formed strictly as a result of chemical conditions in the intestinal lumen (Genz et al., 2008; Perry et al., 2011; Salter et al., 2012). Interestingly, the activity of an extracellular, membrane-bound CA isoform (possibly CA-IV) likely facilitates the $CaCO_3$ precipitation reaction by catalyzing the titration of H^+ arising from the precipitation reaction (in bold) (Grosell et al., 2009a):

$$Ca^{2+} + HCO_3^- \rightarrow CaCO_3 + \mathbf{H^+ + HCO_3^-} \rightarrow \mathbf{CO_2}$$

Apical H^+-pump activity also contributes to reduce luminal osmotic pressure by converting HCO_3^- to CO_2 and thereby reducing the osmotic pressure. Estimates of apical H^+ secretion rates in marine teleosts are as high as 20%–30% of the corresponding HCO_3^- secretion rates, depending on species, intestinal segment, and ambient salinity (Grosell et al., 2009a,b; Guffey et al., 2011), but they likely underestimate the true contribution as pharmacological inhibition of H^+ efflux can be expected to simultaneously reduce HCO_3^- secretion (Grosell et al., 2009a). Postprandial base secretion in seawater-acclimated killifish is increased by 80% in the presence of proton pump inhibitors, demonstrating a very high rate of apical H^+ secretion following feeding (Wood et al., 2010a). Focusing on unfed marine teleosts, at luminal HCO_3^- concentrations of 50 mM (Table 5.1), a 20%–30% reduction in net HCO_3^- secretion rates due to apical H^+ extrusion would amount to a reduction in luminal osmotic pressure of 10–15 mM. While this reduction in osmotic pressure is modest compared to postprandial conditions and that arising from $CaCO_3$ precipitation, it is nevertheless significant considering the maximal osmotic pressure gradient (40 mOsm) against which the intestine can absorb water.

The $CaCO_3$ precipitation process and the apical H^+ secretion, both acting to reduce luminal osmotic pressure, may seem like conflicting processes since one relies on luminal HCO_3^- while the other acts to reduce HCO_3^- concentrations. However, it should be noted that Ca^{2+} rather than CO_3^{2-} is limiting for $CaCO_3$ formation in the intestinal lumen (Grosell et al., 2009a,b) and that titration of luminal HCO_3^- from apical H^+ secretion therefore can occur without impacting $CaCO_3$ precipitation rates. Indeed, both processes occur side-by-side *in vivo*. Interestingly, $CaCO_3$ precipitation and apical proton secretion both act to reduce luminal HCO_3^- concentrations, which will favor continued anion exchange and thus Cl^- (solute)-coupled water absorption. In addition, both $CaCO_3$ formation and titration of HCO_3^- due to apical H^+ secretion act to elevate luminal PCO_2 and thus the potential for CO_2 diffusion back into the enterocytes for the generation of HCO_3^- substrate for anion exchange. In addition to providing substrate for continued anion exchange, the diffusion of CO_2 back into the enterocytes and the subsequent hydration reaction to form HCO_3^- act to stimulate NaCl absorption via cotransport pathways (Tresguerres et al., 2010). Thus, the mechanisms in place to reduce luminal osmotic pressure and thereby promote osmotic water absorption operate in concert with the anion exchange and cotransport pathways contributing to solute-coupled water absorption.

5.3.4 Impact of Feeding on Intestinal Salt and Water Transport

The dual function of the marine teleost intestine, serving digestion and nutrient absorption as well as osmoregulatory functions, leads to potential conflicts between these functions. The ingestion of a natural meal may provide a source of "free water" in the case of a meal consisting of fish or other osmoregulating organisms, which would be a benefit for a marine teleost, but, on the other hand, may provide an additional salt load if the ingested meal consists of osmoconforming invertebrates. Experiments addressed this potential difference by feeding the gulf toadfish a diet consisting of fish and a diet consisting of squid, an osmoconformer (Taylor and Grosell, 2006b). The different osmoregulatory strategies of the prey organisms were evident from gastric fluids of the toadfish during the first 3–6 h postfeeding. After ingesting squid, gastric fluid osmotic pressure was much higher than after ingesting fish. However, this difference was not seen in the intestinal fluids as luminal osmotic pressure remained relatively constant over time between groups fed the two different diets nor was it reflected in the plasma osmotic pressure. Yet, a common observation in the two groups was an osmotic pressure exceeding the sum of electrolytes measured in the fluids for the first 36 h postfeeding. This observation of excess osmolytes was interpreted to reflect products of digestion, fatty acids, carbohydrates, and amino acids adding up as electrolytes for a total osmotic pressure of ~400 mOsm regardless of prey type. Despite this high luminal osmotic pressure, plasma osmolality remained constant ~320 mOsm (Taylor and Grosell, 2006b), suggesting that fluid postprandial absorption can occur against osmotic gradients as high as 80 mOsm, twice as high as the gradients at which fluid absorption can occur in the absence of organic osmolytes (Genz et al., 2011) such as amino acids, carbohydrates, and fatty acids. This suggestion ought to be tested.

Based on reductions in luminal Mg^{2+} and SO_4^{2-} in the intestine 24–48 h postfeeding, it appears that the intestine performs fluid secretion, presumably associated with the secretion of digestive enzymes. Concentrations of K^+ and Na^+ in intestinal fluids are modestly elevated following feeding, while Ca^{2+} is highly elevated, especially in fish fed a fish diet (Taylor and Grosell, 2006b). These changes in luminal fluid composition reflect the salt load associated with the diet. In contrast, luminal Cl^- drops precipitously from the gastric to the intestinal fluids and in the intestinal fluids for the first 48 h following a meal. In fact, Cl^- concentrations reach levels as low as 30 mM in the anterior intestine 24 h postfeeding. This drop in luminal Cl^- concentrations is paralleled by relatively low, but recovering, pH and total CO_2 levels and most likely reflects stimulated Cl^-/HCO_3^- exchange to buffer the acid chyme as it moves from the gastric to the intestinal lumen (Taylor and Grosell, 2006b). Increased intestinal HCO_3^- secretion, Cl^-, and water absorption following ingestion of a meal have since been confirmed for toadfish (Taylor and Grosell, 2009) and the euryhaline killifish (Wood et al., 2010a). Interestingly, the marine gulf toadfish and European flounder do not exhibit the postprandial alkaline tide typical of other vertebrates, which likely is associated with the high rates of intestinal HCO_3^- secretion postfeeding (Taylor et al., 2007; Taylor and Grosell, 2006b). Similar studies on freshwater as well as seawater-acclimated rainbow trout, however, have demonstrated postprandial alkaline tides (Bucking et al., 2009), suggesting perhaps that not all euryhaline teleosts are capable of perfect postprandial acid–base balance regulation via intestinal anion exchange.

The availability of commercially available dried, pelleted feeds is convenient for studies of dietary aspects of fish physiology if the species of interest accepts pelleted food. However, the low water content and associated high salt content of these artificial diets means that observations derived from studies using such diets may not be relevant to the biology and physiology of fishes in their natural environment. However, insight from such studies is of course directly relevant to species cultured on such diets. Pellet diets have been used to study intestinal processing of salts and water in seawater-acclimated rainbow trout (Bucking et al., 2009). These recent studies on rainbow trout confirm the observations on toadfish (discussed earlier) of fluid secretion by the intestine followed by salt as well as water absorption. Also in agreement with the toadfish study is the demonstration that Cl^- absorption exceeds that of all other ions, likely a consequence of anion exchange to neutralize the acid chyme. Furthermore, the study on rainbow trout indicated that, much like the case for freshwater-acclimated trout, the gastric epithelium appears to be an important site of ion uptake (Bucking et al., 2009).

5.3.5 Neuroendocrine Control and Regulation of Intestinal Transport Processes

The endocrine and neuroendocrine controls, including intracellular signaling pathways, of osmoregulatory intestinal processes are discussed in an excellent recent review (Takei and Loretz, 2011). Endocrine and neuroendocrine factors can be classified in two groups: (1) oligopeptides and amines (neurotransmitters), which are secreted nearly immediately after environmental change and decay quickly due to their short half-life, and (2) steroid hormones, which act at the transcriptional level to control long-term responses (Takei and Loretz, 2011).

A group of hormones or neurotransmitters acts to stimulate ion and water absorption across the intestinal epithelium of marine teleosts (Ando et al., 2003). Among these are somatostatin-related peptides (Uesaka et al., 1994), neuropeptide Y (Uesaka et al., 1996), urotensin II (Mainoya and Bern, 1984), and catecholamines (Ando and Hara, 1994). In contrast, natriuretic peptides (Ando et al., 1992; Loretz and Takei, 1997; O'Grady et al., 1985), guanylins (Takei and Yuge, 2007; Yuge et al., 2003; Yuge and Takei, 2007), vasointestinal peptide (Mainoya and Bern, 1984), acetylcholine, serotonin, and histamine (Mori and Ando, 1991) act to inhibit ion and water absorption by the intestinal epithelium of marine teleosts.

The inhibitory action of natriuretic peptides is likely through inhibition of the apical NKCC2 since inhibited Rb^+ (a K^+ maker) uptake in addition to Na^+ and Cl^- uptake was observed after addition (O'Grady et al., 1985). In contrast, the inhibitory action of vasointestinal peptide most likely is on the basolateral membrane to inhibit NaCl absorption (Mainoya and Bern, 1984; O'Grady, 1989). Guanylins act on the guanylyl cyclase-C receptor (GC-C) in the apical membrane to stimulate Cl^- secretion via the teleost CFTR homolog (Takei and Yuge, 2007; Yuge et al., 2003; Yuge and Takei, 2007). Guanylins and GC-C are upregulated upon transfer of euryhaline fishes to seawater, which seems counterintuitive, given the demand for absorption rather than secretion. It has been proposed that the activation of secretory Cl^- flux through CFTR is to provide luminal Cl^- for NKCC2 and for anion exchange and thus HCO_3^- secretion (Takei and Loretz, 2011), but an alternative explanation is that CFTR, activated by guanylins, acts to drive a secretory water flux, possibly to purge accumulated $CaCO_3$ precipitates with rectal fluids. Clearly, additional studies are needed to gain an integrative understanding of the role of these peptides.

The peptide hormone, prolactin (often associated with freshwater osmoregulation), and three steroid hormones, growth hormone, cortisol (often ascribed a role in seawater adaptation), and 17ß-estradiol, act on gene expression to cause long-term effects on epithelial function. A single study has addressed the impact of 17ß-estradiol on intestinal transport functions of seawater-acclimated euryhaline fishes to demonstrate reduced intestinal HCO_3^- secretion and NaCl absorption as well as increased Ca^{2+} uptake, the latter likely being a consequence of diminished $CaCO_3$ formation in the intestinal lumen (see above) (Whittamore et al., 2010).

Prolactin receptors are found in the intestinal epithelium and their expression is under the influence of ambient salinity (Huang et al., 2007; Pierce et al., 2007), but little is known about the role of this hormone in intestinal osmoregulation. A single study has reported elevated intestinal NKA activity in prolactin-treated seawater and freshwater-acclimated fishes perhaps acting to serve a role for intestinal ion uptake from dietary sources (Kelly et al., 1999), and it appears that some intestinal tight junction proteins (claudins) are influenced by prolactin (Tipsmark et al., 2010). While growth hormone is accepted as a seawater-adapting hormone in salmonids, little is known about osmoregulatory adjustments of the intestinal epithelium in response to this hormone. However, a single study has demonstrated elevated NKA activity in anterior regions of intestinal tissue from freshwater-acclimated fishes treated with growth hormone (Kelly et al., 1999), and, as for prolactin, it appears that growth hormone affects the expression of certain claudins (Tipsmark et al., 2010). Perhaps not surprising, growth hormone stimulates amino acid absorption (Takei and Loretz, 2011).

Cortisol is by far the most-studied endocrine effector of osmoregulation with several direct effects on the intestinal epithelium. Cortisol stimulates Na^+ and water transport of isolated intestinal tissue (Hirano and Utida, 1968) and stimulates NKA activity in cultured intestinal tissue as well as NKA activity and water absorption *in vivo* (Cornell et al., 1994; Veillette et al., 1995; Veillette and Young, 2004, 2005). In concert with elevated ion transport rate, cortisol appears to mediate enhanced capacity for transcellular water movement as AQP1, the most likely candidate for transcellular water movement (see above) across the intestinal epithelium, is stimulated by cortisol treatment (Deane et al., 2011; Martinez et al., 2005b). In addition, claudin (15 and 25b) expression is sensitive to cortisol, suggesting that this hormone may be involved in controlling paracellular permeability and epithelial integrity (Tipsmark et al., 2010).

5.3.6 Marine Teleost Acid–Base Balance Is Intimately Linked to Osmoregulation

As discussed earlier and displayed in Figure 5.1, intestinal transport processes serving osmo-regulation involve transepithelial secretion of HCO_3^- and H^+ extrusion across the basolateral membrane to the extracellular fluids with a resulting net acid gain. The secretion of acid across the basolateral membrane of the intestinal epithelium has been measured directly in the order of 0.30 $\mu mol/cm^2/h$ (Figure 5.4; [Grosell and Genz, 2006]), representing a substantial acid gain associated with water absorption. When compared to water absorption rates by the intestinal epithelium, these acid absorption rates equate to a theoretical pH in the absorbed fluids of <2 (Grosell and Genz, 2006). Subsequent simultaneous measurements of water and acid flux in isolated intestinal sac preparations confirmed absorption of highly acidic fluids (Grosell and Taylor, 2007). The aforementioned observations stem from toadfish, but absorption of acidic fluids appears to be a consistent trait among marine teleosts; a review (Grosell, 2006) of ion and water transport rates across isolated intestinal epithelia revealed differences across species between Na^+ and Cl^- absorption rates, with Cl^- absorption rates consistently exceeding Na^+ absorption rates. The apparent electrolyte charge imbalance of fluids absorbed by the intestinal epithelium is likely made up by absorption of H^+ arising from the CO_2 hydration reaction within the intestinal epithelium (Grosell, 2006; Grosell and Taylor, 2007). Thus, in addition to absorbing hypertonic fluids, which imposes a challenge for the gill to secrete excess NaCl, the intestine absorbs significant amounts of acid. The outcome, of this acid absorption, at least in toadfish, is an acidification of the blood at higher salinities (Genz et al., 2008). Intestinal acid absorption as part of marine osmoregulation is likely the explanation for a transient acidosis observed in euryhaline fish transferred from freshwater to elevated salinities (Maxime et al., 1991; Nonnotte and Truchot, 1990; Wilkes and Mcmahon, 1986), a situation that is character-ized by the onset of drinking, initiation of intestinal salt absorption, anion exchange, and thus acid absorption. The acidosis observed in fish transferred to higher salinities is modest in the context of the acid absorption rates discussed earlier, suggesting that compensation must occur. Indeed, studies on toadfish revealed a salinity-dependent secretion of acid, presumably by the gill epithelium (Genz et al., 2008), observations that have since been confirmed for the European flounder (Cooper et al., 2010), which show a tight correlation between the intestinal absorption of acid and the parallel branchial net acid excretion. The mechanisms by which the gill performs compensatory acid secretion remains to be fully characterized.

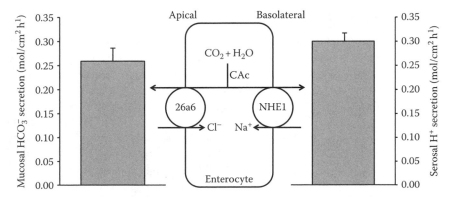

FIGURE 5.4 Apical base ($n = 42$) and basolateral acid ($n = 6$) secretion by isolated toadfish intestinal epithelium measured by pH-stat titration. Measurements were performed under HCO_3^--free conditions with endogenous CO_2 being the sole source of HCO_3^- secretion. As depicted in Figure 5.2, apical anion exchange is assumed to occur via SLC26a6, while basolateral and Na^+-dependent H^+ secretion likely involves NHE1. (From Grosell, M. and Genz, J., *Am. J. Physiol.*, 291, R1145, 2006.)

5.4 MARINE ELASMOBRANCHS AND OTHER OSMOCONFORMERS

Almost nothing is known about osmoregulation by coelacanths but their blood plasma is approximately isosmotic with seawater and of similar composition as elasmobranch plasma (Griffith et al., 1974). In addition, coelacanths possess what appears to be a salt secretion gland, the postanal gland, which is similar to the elasmobranch rectal gland at the structural and ultrastructural levels (Lemire and Lagios, 1979) and shares the high levels of NKA observed in the rectal glands of elasmobranchs (Griffith and Burdich, 1976). Given these similarities, the following discussion of osmolyte and water balance in elasmobranchs likely applies to coelacanths as well.

5.4.1 Integrative Perspective

Marine elasmobranchs display plasma osmolality slightly (~20 mOsm) above ambient (Holmes and Donaldson, 1969; Smith, 1930), and thus gain osmotically "free" water. Consequently, marine elasmobranchs show very low, although measurable, drinking rates (Anderson et al., 2007, 2010; De Boeck et al., 2001). With plasma Na^+ and Cl^- concentrations much below that of the surrounding seawater, urea (300–400 mM) is the main extracellular osmolyte. The denaturing effects of high urea concentrations are offset by high concentrations of methylamines, mainly trimethylamine oxide (TMAO). Glomerular filtration rates and flow rates of the hypoosmotic urine are relatively high compared to marine teleosts (Evans et al., 2004), and the kidney plays a central role in osmoregulation by reabsorbing mainly urea and TMAO, while the rectal gland secretes excess Na^+ and Cl^- resulting from the inward-directed gradients for these ions. The branchial epithelium plays an important role in osmoregulation by contributing to urea retention despite the substantial blood-to-water gradient (Fines et al., 2001; Part et al., 1998). Although the gill is normally not involved in salt secretion to any appreciable extent, it is playing a key role in maintaining acid–base balance (Tresguerres et al., 2005, 2006, 2007a,b).

5.4.2 Drinking

Contrary to common perception, elasmobranchs ingest seawater (De Boeck et al., 2001; Taylor and Grosell, 2006a) with drinking rates being controlled, at least in part, by extracellular fluid volume and the renin–angiotensin system (Anderson et al., 2001a,b, 2002a,b, 2006), as discussed for teleosts earlier.

5.4.3 Processing of Ingested Fluids

Almost nothing is known about the intestinal processing of the seawater ingested by elasmobranchs, but they clearly are capable of solute-coupled water absorption across various segments of the intestine (Anderson et al., 2007, 2010; Taylor and Grosell, 2006a). The absorption of Na^+ and Cl^- appears to be responsible for the solute-coupled water absorption, and substantial intestinal HCO_3^- secretion suggests that part of the intestinal Cl^- absorption occurs via anion exchange as is the case for teleosts (discussed earlier). However, nothing is known about the nature of the anion exchanger(s) involved and the possible roles for NCC and NKCC as discussed for teleosts earlier. It is interesting to note that intestinal anion exchange does not occur in crustaceans, even when induced to drink, but does occur in elasmobranchs, lamprey (but not hagfish), teleosts, and green turtle (Taylor, 2009). This phylogenetic distribution, however coarse, suggests a common vertebrate ancestral trait of intestinal anion exchange being involved in marine osmoregulation (Taylor, 2009; Taylor and Grosell, 2006a).

The main osmolyte in elasmobranchs, urea, appears to be lost into the proximal parts of the intestinal lumen and can reach concentrations as high as 375 mM in intestinal fluids (Anderson et al., 2007). Remarkably >90% of this urea is reabsorbed by the distal segments of the intestine such that urea concentrations fall as low as <22 mM in the rectal fluids. The urea transport mechanisms responsible for this impressive reabsorption remain unknown. However, high expression of a urea transporter and a member of the Rhesus-like ammonia transporters (Rhbg) suggest that these transporters might be involved in retention of nitrogen that would otherwise be lost with rectal fluids. Considering that many elasmobranchs are intermittent feeders and apparently, at least in some cases, are nitrogen limited (Wood et al., 2005, 2010b), it is perhaps not surprising that the intestine possesses a pronounced ability to reclaim urea.

5.4.4 IMPACTS OF FEEDING ON INTESTINAL OSMOLYTE AND WATER ABSORPTION

Sequential sampling of chyme from the intestinal lumen of the spiny dogfish revealed that the majority of fluid contained in a natural meal is absorbed by the intestine following a secretory component of the gastric mucosa, presumably associated with acid secretion (Wood et al., 2007). The osmolality of the chyme is elevated to be in equilibrium with the extracellular fluids by secretion of Na^+ and Cl^- in the gastric regions and substantial urea secretion in the proximal intestine. Interestingly, the intestine of the spiny dogfish (presumably representative of elasmobranchs in general) has high levels of glutamine synthetase and a set of ornithine–urea cycle enzymes (Kajimura et al., 2006) that would enable the intestine itself to produce, from ammonia resulting from digestion, urea (Wood et al., 2009) for absorption in the most distal segments of the intestine.

5.5 INTERACTIONS BETWEEN MARINE FISHES AND THE ENVIRONMENT

In addition to the obvious impact of ambient salinity on drinking (as mentioned earlier), the primary and most acute control of intestinal salt and water transport and the impact on specific intestinal transporters (discussed later), rapidly increasing atmospheric CO_2 levels, and ensuing ocean acidification have recently been reported to affect intestinal transport processes (Esbaugh et al., 2012; Heuer et al., 2012). While environmental impacts on fish physiology is intuitively easy to accept, a more surprising environment–organism interaction is represented by the recent discovery that fish impact the oceanic inorganic carbon cycle significantly (Perry et al., 2011; Salter et al., 2012; Wilson et al., 2009; Woosley et al., 2012). These fish–environment interactions are discussed in the following two sections.

5.5.1 IMPACT OF ENVIRONMENTAL FACTORS ON INTESTINAL OSMOREGULATORY TRANSPORT PROCESSES

Ambient salinity directly affects rates of seawater ingestion (discussed earlier) and has been demonstrated to affect the expression levels of a number of ion transport proteins and enzymes. For example, the intestine, which shows the highest expression of NKA of any of the osmoregulatory tissues (gill, kidney, and intestine) (Grosell et al., 1999; Hogstrand et al., 1999), also shows higher NKA expression in seawater-acclimated euryhaline fishes compared to freshwater-acclimated conspecifics. Not surprisingly, a similar trend has been observed for NKCC2 (Colin et al., 1985; Fuentes and Eddy, 1997a; Grosell et al., 2007; Jampol and Epstein, 1970; Kelly et al., 1999; Madsen et al., 1994; Seidelin et al., 2000). More recently, NHE3 expression has also been reported to be increased following seawater transfer (Grosell et al., 2007), although a functional role for intestinal NHE3 in osmoregulation of marine teleosts remains to be firmly established.

Increase in ambient salinity whether it is from freshwater to seawater or from seawater to higher salinities also affects components of the intestinal anion exchange pathway for Cl^- uptake. Basolateral NBC1, luminal CAc, apical CA-IV, SLC26a6, and the H^+ pump have all been demonstrated to respond to ambient salinity at the mRNA level and for CAc, CA-IV and the H^+ pump, also at the enzymatic activity level, in accordance with a role in osmoregulation (Grosell et al., 2007; Guffey et al., 2011; Kurita et al., 2008; Sattin et al., 2010; Taylor et al., 2010). The latter

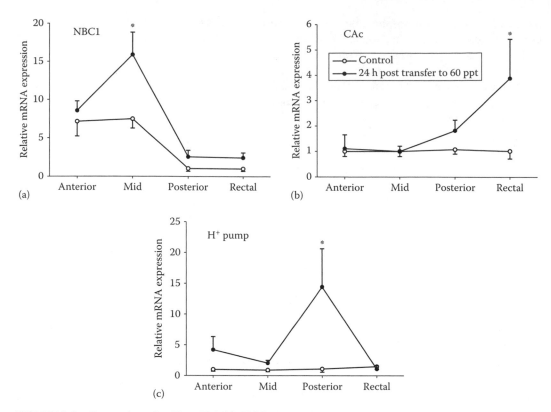

FIGURE 5.5 Expression of gulf toadfish Na:HCO_3^- cotransporter NBC1e (a), cytosolic carbonic anhydrase (CAc) (b), and H^+ pump (c) mRNA relative to elongation factor 1 alpha in anterior, mid, and posterior intestine and rectum following transfer from seawater (35 ppt) to hypersaline seawater (60 ppt) for 24 h. Values are normalized to a value of 1 for the tissue showing the lowest expression. Data is presented as mean ± SEM from $n = 8$ fish. "*" denotes statistically significant difference from seawater control. (From Guffey, S. et al., *Am. J. Physiol.*, 301, R1682, 2011; Sattin, G. et al., *Comp. Biochem. Physiol. A*, 156, 169, 2010; Taylor, J.R. et al., *J. Exp. Biol.*, 213, 459, 2010.)

studies demonstrated that not only does the intestine respond to ambient salinity but that different intestinal regions respond differently depending on the gene of interest (Figure 5.5). Unlike the euryhaline tiger puffer (Kurita et al., 2008), the gulf toadfish does not exhibit elevated expression of the anion exchanger SLC26a6 upon challenge to higher salinity (Grosell et al., 2009b), which might indicate that the anion exchanger is not rate limiting for intestinal HCO_3^- secretion in this species. Indeed, studies imply that, at least for the anterior intestine, NBC1 appears to be the rate-limiting step (Taylor et al., 2010). The basolateral NBC1, which is predominantly expressed in the more proximal regions of the intestine, shows strongly increased mRNA expression in the middle intestine upon transfer to 60 ppt, suggesting that transepithelial HCO_3^- secretion is prevalent in this region of the intestine in seawater as well as under extreme hypersaline conditions. Expressions of CAc and the H^+ pump are nearly constant along the GI tract of toadfish held in seawater. However, while CAc shows a gradually increasing expression (and activity) along the GI tract 24 h following transfer to 60 ppt, with the highest expression in the rectal tissue, the H^+ pump shows a marked increase in expression only in the posterior intestine. The elevated expressions of CAc and the H^+ pump in the distal regions of the intestine following exposure to elevated salinity imply that the hydration of endogenous CO_2 is the dominant source of HCO_3^- secretion in this region of the intestine. Clearly, different transport mechanisms are employed in different regions of the intestine, and at least for toadfish, it appears that the distal regions respond strongest to elevated ambient salinity. This observation may be related to a higher demand on the anterior intestine for digestive purposes, limiting its adaptability to high salinity levels.

Regardless of the reason(s) for the regional differences, they offer an exciting opportunity for comparative studies of epithelial transport physiology.

An additional and only recently recognized factor that affects intestinal transport physiology is elevated ambient CO_2. Levels as low as 1900 ppm CO_2 have recently been reported to influence fish acid–base physiology as well as intestinal transport processes (Esbaugh et al., 2012; Heuer et al., 2012). To place these levels in perspective, they are in line with predictions for year 2300 (Caldeira and Wickett, 2003) and, in fact, are exceeded by current day CO_2 levels in upwelling zones (Feely et al., 2008; Thomsen et al., 2010). In brief, these low levels of ambient CO_2 result in a modest respiratory acidosis rapidly compensated by retention of plasma HCO_3^-, returning blood pH to normal despite elevated PCO_2. Thus, fish acclimated to these modestly elevated ambient CO_2 levels display chronically elevated plasma HCO_3^- and PCO_2 (Esbaugh et al., 2012). Plasma HCO_3^- and PCO_2 both form substrates for intestinal HCO_3^- secretion (Grosell and Genz, 2006), and it is perhaps therefore not surprising that intestinal tissue exposed to serosal HCO_3^- and PCO_2 levels typical of CO_2-exposed fish demonstrates significantly elevated intestinal HCO_3^- secretion (>30%) when compared to tissue exposed to normal HCO_3^- and PCO_2 levels (Heuer et al., 2012). A qualitatively similar response was observed in midshipmen exposed to a much higher CO_2 level (5000 ppm) (Perry et 1., 2010).

Reduced luminal Cl^- concentrations in fish exposed to elevated ambient CO_2 further illustrate the activation of an anion exchange pathway by elevated plasma HCO_3^- and PCO_2 levels and point to a possible excess energy demand associated with ion transport by the intestine and the gill as elevated Cl^- uptake must be compensated by increased excretion across the gill (Heuer et al., 2012). Somewhat surprisingly, the increased intestinal base secretion observed in fish exposed to mild hypercapnia does not result in increased excretion of $CaCO_3$ precipitates. The likely explanation for the lack of elevated $CaCO_3$ production is that Ca^{2+} in the intestinal fluids becomes limiting for the $CaCO_3$ precipitation process (Heuer et al., 2012). It should be noted, however, that the aforementioned observations apply to unfed toadfish at 25°C and that other species, fed or unfed, under different conditions may respond differently. In light of the recently recognized role of teleosts in the oceanic inorganic carbon cycle (see below), the possibility that fishes may contribute by a greater amount under ocean acidification scenarios clearly warrants further attention.

5.5.2 Impact of Intestinal Processes on the Environment

Recent considerations of fish $CaCO_3$ production rates and the influence of temperature, fish size, and other factors combined with estimates of global fish biomass and size distribution revealed that teleost fishes contribute significantly to the oceanic inorganic carbon cycle (Wilson et al., 2009). Furthermore, the relatively high Mg^{2+} content of fish-produced $CaCO_3$, which stems from the high Mg^{2+} concentrations in intestinal fluids, has been suggested to result in relatively high solubility and thereby explains increases in titratable alkalinity much above the aragonite saturation depth (Wilson et al., 2009). Indeed, $CaCO_3$ precipitates produced by the gulf toadfish were recently found to be highly soluble, approximately two times higher than aragonite (Woosley et al., 2012). However, the unique crystal structure of fish-produced $CaCO_3$ precipitates have been documented from sediment samples, demonstrating that some of the precipitates produced by fishes can reach the ocean floor, at least in shallow water habitats (<30 feet) (Perry et al., 2011). While depth will obviously be a determining factor for the fate of the fish-produced $CaCO_3$, Mg^{2+} content, which likely influences the solubility (Woosley et al., 2012), will also be an important factor. In this context, a recent report of large species-specific differences in Mg^{2+} content of $CaCO_3$ precipitates (Salter et al., 2012) offers an exciting challenge as does the possible impact of feeding, temperature, and pH on Mg^{2+} content and thus solubility of these precipitates. Along with the impacts of ambient CO_2 on $CaCO_3$ production by marine fishes (discussed previously), answers to the question of sedimentation versus dissolution of fish-produced $CaCO_3$ will be of utmost importance for predicting pH and titratable alkalinity changes in our oceans during the coming decades of rapid climate change and ocean acidification.

5.6 FUTURE DIRECTIONS

The aforementioned discussions and syntheses are based almost exclusively on sedentary benthic species or standard euryhaline species like salmonids, killifish, and tilapia. An obvious question seems to be whether what we have learned from these species will apply to pelagic and truly marine species with higher metabolic demands. Studies on such species are logistically challenging but nevertheless are something worth pursuing.

While great strides have been made in the field of freshwater fish ontogeny of osmoregulation utilizing the zebrafish embryo model discussed extensively in the preceding chapter, nearly nothing is known about the onset of osmoregulation by larval marine fishes. It is unknown, perhaps with exception of a few species, when drinking commences and if the GI tract is open and functional for osmoregulatory purposes before it becomes active in nutrient assimilation.

Similarly, while the osmorespiratory compromise has been studied extensively in freshwater species, the interplay between gas exchange and osmoregulation in marine species is largely unexplored. Considering that pelagic species from tropical and subtropical regions produce small larvae and that these larvae will have extremely high metabolic rates while being exposed to water with low oxygen solubility, there is the potential for an extremely tight osmorespiratory compromise and likely extreme adaptations for gas exchange as well as osmoregulation.

ACKNOWLEDGMENTS

Grosell was supported by NSF (IOS 1146695) while writing this chapter.

REFERENCES

Anderson, W. G., Cerra, M. C., Wells, A., Tierney, M. L., Tota, B., Takei, Y., and Hazon, N. 2001a. Angiotensin and angiotensin receptors in cartilaginous fishes. *Comp Biochem Physiol A*. 128: 31–40.

Anderson, W. G., Dasiewicz, P. J., Liban, S., Ryan, C., Taylor, J. R., Grosell, M., and Weihrauch, D. 2010. Gastro-intestinal handling of water and solutes in three species of elasmobranch fish, the white spotted bamboo shark, *Chiloscyllium plagiosum*, little skate, *Leucoraja erinacea* and the clearnose skate, *Raja eglanteria*. *Comp Biochem and Physiol*. 155: 493–502.

Anderson, W. G., Pillans, R. D., Hyodo, S., Tsukada, T., Good, J. P., Takei, Y., Franklin, C. E., and Hazon, N. 2006. The effects of freshwater to seawater transfer on circulating levels of angiotensin II, C-type natriuretic peptide and arginine vasotocin in the euryhaline elasmobranch, *Carcharhinus leucas*. *Gen Comp Endocrinol*. 147: 39–46.

Anderson, W. G., Takei, Y., and Hazon, N. 2001b. The dipsogenic effect of the renin-angiotensin system in elasmobranch fish. *Gen Comp Endocrinol*. 124: 300–307.

Anderson, W. G., Takei, Y., and Hazon, N. 2002a. Osmotic and volaemic effects on drinking rate in elasmobranch fish. *J Exp Biol*. 205: 1115–1122.

Anderson, W. G., Taylor, J. R., Good, J. P., Hazon, N., and Grosell, M. 2007. Body fluid volume regulation in elasmobranch fish. *Comp Biochem Physiol*. 148: 3–13.

Anderson, W. G., Wells, A., Takei, Y., and Hazon, N. 2002b. The control of drinking in elasmobranch fish with special reference to the renin-angiotensin system. *Symp Soc Exp Biol*. 54: 19–30.

Ando, M. and Hara, I. 1994. Alteration of sensitivity to various regulators in the intestine of the eel following seawater acclimation. *Comp Biochem and Physiol*. 109A: 447–453.

Ando, M., Kondo, K., and Takei, Y. 1992. Effects of eel atrial natriuretic peptide on NaCl and water transport across the intestine of the seawater eel. *J Comp Physiol B*. 162: 436–439.

Ando, M., Mukuda, T., and Kozaka, T. 2003. Water metabolism in the eel acclimated to seawater: From mouth to intestine. *Comp Biochem Physiol B*. 136: 621–633.

Ando, M. and Nagashima, K. 1996. Intestinal Na^+ and Cl^- levels control drinking behavior in the seawater-adapted eel *Anguilla japonica*. *J Exp Biol*. 199: 711–716.

Ando, M. and Subramanyam, M. V. V. 1990. Bicarbonate transport systems in the intestine of the seawater eel. *J Exp Biol*. 150: 381–394.

Aoki, M., Kaneko, T., Katoh, F., Hasegawa, S, Tsutsui, N., and Aida, K. 2003. Intestinal water absorption through aquaporin 1 expressed in the apical membrane of mucosal epithelial cells in seawater-adapted Japanese eel. *J Exp Biol*. 206: 3495–3505.

Beyenbach, K. W. 2004. Kidneys sans glomeruli. *Am J Physiol*. 286: F811–F827.

Beyenbach, K. W. and Liu, P. L. F. 1996. Mechanism of fluid secretion common to aglomerular and glomerular kidneys. *Kidney Int*. 49: 1543–1548.

Bucking, C., Fitzpatrick, J. L., Nadella, S. R., and Wood, C. M. 2009. Post-prandial metabolic alkalosis in the seawater-acclimated trout: The alkaline tide comes in. *J Exp Biol*. 212: 2159–2166.

Bucking, C. and Wood, C. M. 2006. Gastrointestinal processing of monovalent ions (Na^+, Cl^-, K^+) during digestion: Implications for homeostatic balance in freshwater rainbow trout. *Am J Physiol*. 291: R1764–R1772.

Bucking, C. and Wood, C. M. 2007. Gastrointestinal transport of Ca^{2+} and Mg^{2+} during the digestion of a single meal in the freshwater rainbow trout. *J Comp Physiol. B* 177: 349–360.

Bucking, C. and Wood, C. M. 2008. The alkaline tide and ammonia excretion after voluntary feeding in freshwater rainbow trout. *J Exp Biol*. 211: 2533–2541.

Bury, N. R., Grosell, M., Wood, C. M., Hogstrand, C., Wilson, R. W., Rankin, J. C., Busk, M., Lecklin, T., and Jensen, F. B. 2001. Intestinal iron uptake in the European flounder (*Platichthys flesus*). *J Exp Biol*. 204: 3779–3787.

Caldeira, K. and Wickett, M. E. 2003. Anthropogenic carbon and ocean pH. *Nature*. 425: 365–1365.

Carrick, S. and Balment, R. J. 1983. The renin-angiotensin system and drinking in the euryhaline flounder, *Platichthys flesus*. *Gen Comp Endocrinol*. 51: 423–433.

Chang, M. H., Plata, C., Kurita, Y., Kato, A., Hirose, S., and Romero, M. F. 2012. Euryhaline pufferfish NBCe1 differs from nonmarine species NBCe1 physiology. *Am J Physiol*. 302: C1083–C1095.

Chen, Y., Cann, M. J., Litvin, T. N., Iourgenko, V., Sinclair, M. L., Levin, L. R., and Buck, J. 2000. Soluble adenylyl cyclase as an evolutionarily conserved bicarbonate sensor. *Science*. 289: 625–628.

Clelland, E. S., Bui, P., Bagherie-Lachidan, M., and Kelly, S. P. 2010. Spatial and salinity-induced alterations in claudin-3 isoform mRNA along the gastrointestinal tract of the pufferfish *Tetraodon nigroviridis*. *Comp Biochem Physiol A*. 155: 154–163.

Colin, D. A., Nonnotte, G., Leray, C., and Nonnotte, L. 1985. Na transport and enzyme activities in the intestine of the freshwater and sea-water adapted trout (*Salmo gairdnerii* R.). *Comp Biochem Physiol A*. 81: 695–698.

Conlon, J. M. 1999. Bradykinin and its receptors in non-mammalian vertebrates. *Regul Pept*. 79: 71–81.

Cooper, C. A., Whittamore, J. M., and Wilson, R. W. 2010. Ca^{2+}-driven intestinal HCO_{3-} secretion and $CaCO_3$ precipitation in the European flounder in vivo: Influences on acid-base balance regulation and blood gas transport. *Am J Physiol*. 298: R870–R876.

Cooper, C. A. and Wilson, R. W. 2008. Post-prandial alkaline tide in freshwater rainbow trout: Effects of meal anticipation on recovery from acid-base and ion regulatory disturbances. *J Exp Biol*. 211: 2542–2550.

Cornell, S. C., Portesi, D. M., Veillette, P. A., Sundell, K., and Specker, J. L. 1994. Cortisol stimulates intestinal fluid uptake in Atlantic salmon (*Salmo salar*) in the post-smolt stage. *Fish Physiol Biochem*. 13: 183–190.

Cutler, C. P. and Cramb, G. 2002. Two isoforms of the $Na^+/K^+/2Cl^{(-)}$ cotransporter are expressed in the European eel (*Anguilla anguilla*). *Biochim Biophys Acta*. 1566: 92–103.

Cutler, C. P., Martinez, A. S., and Cramb, G. 2007. The role of aquaporin 3 in teleost fish. *Comp Biochem Physiol A*. 148: 82–91.

De Boeck, G., Grosell, M., and Wood, C. 2001. Sensitivity of the spiny dogfish (*Squalus acanthias*) to waterborne silver exposure. *Aquat Tox*. 54: 261–275.

Deane, E. E., Luk, J. C. Y., and Woo, N. Y. S. 2011. Aquaporin 1a expression in gill, intestine, and kidney of the euryhaline silver sea bream. *Front Physiol*. 2: 1–9.

Dixon, J. M. and Loretz, C. A. 1986. Luminal alkalinization in the intestine of the goby. *J Comp Physiol*. 156: 803–811.

Esbaugh, A. and Grosell, M. 2012. Na^+ transport by the toadfish esophageal epithelium. *In preparation*.

Esbaugh, A. J., Heuer, R., and Grosell, M. 2012. Impacts of ocean acidification on respiratory gas exchange and acid-base balance in a marine teleost, *Opsanus beta*. *J Comp Physiol B*. 182: 921–934.

Evans, D. H. 1967. Sodium, chloride and water balance in the intertidal teleost, *Xiphister atropurpureus* III. The roles of simple diffusion, exchange diffusion, osmosis and active transport. *J Exp Biol*. 47: 525–534.

Evans, D. H. and Claiborne, J. B. 2008. Osmotic and ionic regulation in fishes. *In Osmotic and Ionic Regulation: Cells and Animals*. Evans, D. H. (Ed.), CRC Press, Boca Raton, FL, pp. 295–366.

Evans, D. H., Piermarini, P. M., and Choe, K. P. 2004. Homeostasis: Osmoregulation, pH regulation, and nitrogen excretion. In: *Biology of Sharks and Their Relatives,* J. C. Carrier, J. A. Musick, and M. R. Heithaus (Eds.), CRC Press, Boca Raton, FL, pp. 247–268.

Feely, R. A., Sabine, C. L., Hernandez-Ayon, J. M., Ianson, D., and Hales, B. 2008. Evidence for upwelling of corrosive "acidified" water onto the continental shelf. *Science*. 320: 1490–1492.

Field, M., Smith, P. L., and Bolton, J. E. 1980. Ion transport across the isolated intestinal mucosa of the winter flounder *Pseudopleuronectes americanus*: II. Effects of cyclic AMP. *J Membr Biol*. 53: 157–163.

Fines, G. A., Ballantyne, J. S., and Wright, P. A. 2001. Active urea transport and an unusual basolateral membrane composition in the gills of a marine elasmobranch. *Am J Physiol*. 280: R16–R24.

Frizzell, R. A., Smith, P. L., Field, M., and Vosburgh, E. 1979. Coupled sodium-chloride influx across brush border of flounder intestine. *J Membr Biol*. 46: 27–39.

Fuentes, J., Bury, N. R., Carroll, S., and Eddy, F. B. 1996a. Drinking in Atlantic salmon presmolts (*Salmo salar* L.) and juvenile rainbow trout (*Oncorhynchus mykiss* Walbaum) in response to cortisol and sea water challenge. *Aquaculture*. 141: 129–137.

Fuentes, J. and Eddy, F. B. 1996. Drinking in freshwater-adapted rainbow trout fry, *Oncorhynchus mykiss* (Walbaum), in response to angiotensin I, angiotensin II, angiotensin-converting enzyme inhibition, and receptor blockade. *Physiol Zool*. 69: 1555–1569.

Fuentes, J. and Eddy, F. B. 1997a. Drinking in Atlantic salmon presmolts and smolts in response to growth hormone and salinity. *Comp Biochem Physiol A*. 117: 487–491.

Fuentes, J. and Eddy, F. B. 1997b. Effect of manipulation of the renin-angiotensin system in control of drinking in juvenile Atlantic salmon (*Salmo salar* L) in fresh water and after transfer to sea water. *J Comp Physiol B*. 167: 438–443.

Fuentes, J. and Eddy, F. B. 1998. Cardiovascular responses in vivo to angiotensin II and the peptide antagonist saralasin in rainbow trout *Oncorhynchus mykiss*. *J Exp Biol*. 201: 267–272.

Fuentes, J., McGeer, J. C., and Eddy, F. B. 1996b. Drinking rate in juvenile Atlantic salmon, *Salmo salar* L fry in response to a nitric oxide donor, sodium nitroprusside and an inhibitor of angiotensin converting enzyme, enalapril. *Fish Physiol Biochem*. 15: 65–69.

Gardiner, S. M., Kemp, P. A., and Bennett, T. 1993. Differential-effects of captopril on regional hemodynamic-responses to angiotensin-I and bradykinin in conscious rats. *British J Pharm*. 108: 769–775.

Genz, J. and Grosell, M. 2011. *Fundulus heteroclitus* acutely transferred from seawater to high salinity require few adjustments to intestinal transport associated with osmoregulation. *Comp Biochem Physiol A*. 160(2): 156–165.

Genz, J., McDonald, D. M., and Grosell, M. 2011. Concentration of $MgSO_4$ in the intestinal lumen of *Opsanus beta* limits osmoregulation in response to acute hypersalinity stress. *Am J Physiol*. 300: R895–R909.

Genz, J., Taylor, J. R., and Grosell, M. 2008. Effects of salinity on intestinal bicarbonate secretion and compensatory regulation of acid-base balance in *Opsanus beta*. *J Exp Biol*. 211: 2327–2335.

Griffith, R. W. and Burdich, C. J. 1976. Sodium-potassium activated adenosine triphosphatase in coelacanth tissue: High activity in rectal glands. *Comp Biochem Physiol B*. 54: 557–559.

Griffith, R. W., Umminger, B. L., Grant, B. F., Pang, P. K. T., and Pickford, G. E. 1974. Serum composition of the coelacanth, *Latimeria chalumnae* Smith. *J Exp Zool*. 187: 87–102.

Grosell, M. 2007. Intestinal transport processes in marine fish osmoregulation. In: *Fish Osmoregulation*, B. Baldisserotto, J. M. Mancera, and B. G. Kapoor (Eds.), Science publishers, Enfield, pp. 332–357.

Grosell, M. 2006. Intestinal anion exchange in marine fish osmoregulation. *J Exp Biol*. 209: 2813–2827.

Grosell, M. 2011a. Intestinal anion exchange in marine teleosts is involved in osmoregulation and contributes to the oceanic inorganic carbon cycle. *Acta Physiol*. 202: 421–434.

Grosell, M. 2011b. The role of the gastrointestinal tract in salt and water balance. *Fish Physiol*. 30: 135–164.

Grosell, M., De Boeck, G., Johannsson, O., and Wood, C. M. 1999. The effects of silver on intestinal ion and acid-base regulation in the marine teleost fish, *Parophrys vetulus*. *Comp Biochem Physiol C*. 124: 259–270.

Grosell, M. and Genz, J. 2006. Ouabain sensitive bicarbonate secretion and acid absorption by the marine fish intestine play a role in osmoregulation. *Am J Physiol*. 291: R1145–R1156.

Grosell, M., Genz, J., Taylor, J. R., Perry, S. F., and Gilmour, K. M. 2009a. The involvement of H⁺-ATPase and carbonic anhydrase in intestinal HCO_3^- secretion on seawater-acclimated rainbow trout. *J Exp Biol*. 212: 1940–1948.

Grosell, M., Gilmour, K. M., and Perry, S. F. 2007. Intestinal carbonic anhydrase, bicarbonate, and proton carriers play a role in the acclimation of rainbow trout to seawater. *Am J Physiol*. 293: R2099–R2111.

Grosell, M. and Jensen, F. B. 1999. NO_2^- uptake and HCO_3^- excretion in the intestine of the European flounder (*Platichthys flesus*). *J Exp Biol*. 202: 2103–2110.

Grosell, M., Laliberte, C. N., Wood, S., Jensen, F. B., and Wood, C. M. 2001. Intestinal HCO_3^- secretion in marine teleost fish: Evidence for an apical rather than a basolateral Cl^-/HCO_3^- exchanger. *Fish Physiol Biochem*. 24: 81–95.

Grosell, M., Mager, E. M., Williams, C., and Taylor, J. R. 2009b. High rates of HCO_3^- secretion and Cl^- absorption against adverse gradients in the marine teleost intestine: The involvement of an electrogenic anion exchanger and H^+-pump metabolon? *J Exp Biol*. 212: 1684–1696.

Grosell, M., McDonald, M. D., Walsh, P. J., and Wood, C. M. 2004. Effects of prolonged copper exposure in the marine gulf toadfish (*Opsanus beta*). II. Drinking rate, copper accumulation and Na^+/K^+-ATPase activity in osmoregulatory tissues. *Aquat Tox*. 68: 263–275.

Grosell, M., Nielsen, C., and Bianchini, A. 2002. Sodium turnover rate determines sensitivity to acute copper and silver exposure in freshwater animals. *Comp Biochem Physiol C*. 133: 287–303.

Grosell, M. and Taylor, J. R. 2007. Intestinal anion exchange in teleost water balance. *Comp Biochem Physiol A*. 148: 14–22.

Grosell, M. and Wood, C. M. 2001. Branchial versus intestinal silver toxicity and uptake in the marine teleost *Parophrys vetulus*. *J Comp Physiol B*. 171: 585–594.

Grosell, M., Wood, C. M., Wilson, R. W., Bury, N. R., Hogstrand, C., Rankin, J. C., and Jensen, F. B. 2005. Bicarbonate secretion plays a role in chloride and water absorption of the European flounder intestine. *Am J Physiol*. 288: R936–R946.

Guffey, S., Esbaugh, A., and Grosell, M. 2011. Regulation of apical H^+-ATPase activity and intestinal HCO_3^- secretion in marine fish osmoregulation. *Am J Physiol*. 301: R1682–R1691.

Halm, D. R., Krasny, E. J., and Frizzell, R. A. 1985a. Electrophysiology of flounder intestinal mucosa.I. Conductance of cellular and paracellular pathways. *J Gen Physiol*. 85: 843–864.

Halm, D. R., Krasny, E. J., and Frizzell, R. A. 1985b. Electrophysiology of flounder intestinal mucosa.II. Relation of the electrical potential profile to coupled NaCl absorption. *J Gen Physiol*. 85: 865–883.

Hazon, N., Wells, A., Pillans, R. D., Good, J. P., Anderson, W. G., and Franklin, C. E. 2003. Urea based osmoregulation and endocrine control in elasmobranch fish with special reference to euryhalinity. *Comp Biochem Physiol B*. 136: 685–700.

Heuer, R. M., Esbaugh, A. J., and Grosell, M. 2012. Ocean acidification leads to counterproductive intestinal base loss in the gulf toadfish (*Opsanus Beta*). *Physiol Biochem Zool*. 85: 450–459.

Hickman, C. P. 1968. Ingestion, intestinal absorption, and elimination of seawater and salts in the southern flounder, *Paralichthys lethostigma*. *Can J Zool*. 46: 457–466.

Hirano, T. 1974. Some factors regulating water intake by the eel, *Anguilla japonica*. *J Exp Biol*. 61: 737–747.

Hirano, T. and Mayer-Gostan, N. 1976. Eel esophagus as an osmoregulatory organ. *Proc Nat Acad Sci*. 73: 1348–1350.

Hirano, T. and Utida, S. 1968. Effects of ACTH and cortisol on water movement insolated intestine of the eel, *Anguilla japonica*. *Gen Comp Endocrinol*. 11: 373–343.

Hogstrand, C., Ferguson, E. A., Galvez, F., Shaw, J. R., Webb, N. A., and Wood, C. M. 1999. Physiology of acute silver toxicity in the starry flounder (*Platichthys stellatus*) in seawater. *J Comp Physiol B*. 169: 461–473.

Holmes, W. N. and Donaldson, E. M. 1969. The body compartments and distribution of electrolytes. In: *Fish Physiology*, D. J. Randall (Ed.), Vol. 1, Academic Press, New York, pp. 1–90.

Huang, X., Jiao, B., Fung, C. K., Zhang, Y., Ho, W. K. K., Chan, C. B., Lin, H., Wang, D., and Cheng, C. H. K. 2007. The presence of two distinct prolactin receptors in seabream with different tissue distribution patterns, signal transduction pathways and regulation of gene expression by steroid hormones. *J Endocrinol*. 194: 373–392.

Hwang, P. P., Lee, T. H., and Lin, L. Y. 2011. Ion regulation in fish gills: Recent progress in the cellular and molecular mechanisms. *Am J Physiol Regul Integr Comp Physiol*. 301: R28–R47.

Jampol, L. M. and Epstein, F. H. 1970. Sodium-potassium-activated adenosine triphosphate and osmotic regulation by fishes. *Am J Physiol*. 218: 607–611.

Kajimura, M., Walsh, P. J., Mommsen, T. P., and Wood, C. M. 2006. The dogfish shark (*Squalus acanthias*) increases both hepatic and extra-hepatic ornithine-urea cycle enzyme activities for nitrogen conservation after feeding. *Physiol Biochem Zool*. 79: 602–613.

Kato, A., Chang, M. H., Kurita, Y., Nakada, T., Ogoshi, M., Nakazato, T., Doi, H., Hirose, S., and Romero, M. F. 2009. Identification of renal transporters involved in sulfate excretion in marine teleost fish. *Am J Physiol*. 297: R1647–R1659.

Kelly, S. P., Chow, I. N. K., and Woo, N. Y. S. 1999. Effects of prolactin and growth hormone on strategies of hypoosmotic adaptation in a marine teleost, *Sparus sarba*. *Gen Comp Endocrinol*. 113: 9–22.

Kirsch, R. and Meister, M. F. 1982. Progressive processing of ingested water in the gut of sea-water teleost. *J Exp Biol*. 98: 67–81.

Klinck, J. S., Green, W. W., Mirza, R. S., Nadella, S. R., Chowdhury, M. J., Wood, C. M., and Pyle, G. G. 2007. Branchial cadmium and copper binding and intestinal cadmium uptake in wild yellow perch (*Perca flavescens*) from clean and metal-contaminated lakes. *Aquat Tox*. 84: 198–207.

Kormanik, G. A. and Evans, D. H. 1982. The relation of Na and Cl extrusion in *Opsanus beta*, the Gulf toadfish, acclimated to seawater. *J Exp Zool*. 224: 187–194.

Kurita, Y., Nakada, T., Kato, A., Doi, H., Mistry, A. C., Chang, M. H., Romero, M. F., and Hirose, S. 2008. Identification of intestinal bicarbonate transporters involved in formation of carbonate precipitates to stimulate water absorption in marine teleost fish. *Am J Physiol*. 294: R1402–R1412.

Lemire, M. and Lagios, M. 1979. Ultrastructure of the secretory parenchyma of the postanal gland of the coelacanth, *Latimeria chalumnae* Smith. *Acta Anat*. 104: 1–15.

Lignot, J. H., Cutler, C. P., Hazon, N., and Cramb, G. 2002. Immunolocalisation of aquaporin 3 in the gill and the gastrointestinal tract of the European eel *Anguilla anguilla* (L.). *J Exp Biol*. 205: 2653–2663.

Lin, L.-Y., Weng, C.-F., and Hwang, P.-P. 2002. Regulation of drinking rate in euryhaline tilapia larvae (*Oreochromis mossambicus*) during salinity challenges. *Physiol Biochem Zool*. 74: 171–177.

Loretz, C. A. 1995. Electrophysiology of ion transport in the teleost intestinal cells. In: *Cellular and Molecular Approaches to Fish Ionic Regulation, Fish Physiology*, C. M. Wood and T. J. Shuttleworth (Eds.), Vol. 14, pp. 25–56.

Loretz, C. A. and Fourtner, C. R. 1988. Functional-characterization of a voltage-gated anion channel from teleost fish intestinal epithelium. *J Exp Biol*. 136: 383–403.

Loretz, C. A. and Takei, Y. 1997. Natriuretic peptide inhibition of intestinal salt absorption in the Japanese eel: Physiological significance. *Fish Physiol Biochem*. 17: 319–324.

Madsen, S. S., Mccormick, S. D., Young, G., Endersen, J. S., Nishioka, R. S., and Bern, H. S. 1994. Physiology of seawater acclimation in the striped bass, *Morone saxatilis* (Walbaum). *Fish Physiol Biochem*. 13: 1–11.

Mainoya, J. R. and Bern, H. A. 1984. Influence of vasoactive intestinal peptide and urotensin II on the absorption of water and NaCl by the anterior intestine of the tilapia *Sarotherodon mossambicus*. *Zool Sci*. 1: 100–105.

Malvin, R. L., Schiff, D., and Eiger, S. 1980. Angiotensin and drinking rates in the euryhaline killifish. *Am J Physiol*. 239: R31–R34.

Marshall, W. S. and Grosell, M. 2006. Ion transport, osmoregulation and acid-base balance. In: *Physiology of Fishes*, 3rd edn., D. Evans and J. B. Claiborne (Eds.), CRC Press, Boca Raton, FL.

Marshall, W. S., Howard, J. A., Cozzi, R. R. F., and Lynch, E. M. 2002a. NaCl and fluid secretion by the intestine of the teleost *Fundulus heteroclitus*: Involvement of CFTR. *J Exp Biol*. 205: 745–758.

Marshall, W. S., Lynch, E. M., and Cozzi, R. R. 2002b. Redistribution of immunofluorescence of CFTR anion channel and NKCC cotransporter in chloride cells during adaptation of the killifish *Fundulus heteroclitus* to sea water. *J Exp Biol*. 205: 1265–1273.

Martinez, A.-S., Cutler, C. P., Wilson, G. D., Phillips, C., Hazon, N., and Cramb, G. 2005a. Cloning and expression of three aquaporin homologues from the European eel (*Anguilla anguilla*): Effects of seawater acclimation and cortisol treatment on renal expression. *Biol Cell*. 97: 615–627.

Martinez, A. S., Cutler, C. P., Wilson, G. D., Phillips, C., Hazon, N., and Cramb, G. 2005b. Regulation of expression of two aquaporin homologs in the intestine of the European eel: Effects of seawater acclimation and cortisol treatment. *Am J Physiol*. 288: R1733–R1743.

Maxime, V., Pennec, J. P., and Peyraud, C. 1991. Effects of direct transfer from fresh-water to seawater on respiratory and circulatory variables and acid-base status in rainbow-trout. *J Comp Physiol B*. 161: 557–568.

McDonald, M. D. 2007. Renal contribution to teleost fish osmoregulation. *Fish Osmoregul*. Baldisserotto, B., Manara, J. M., Kapoor, B. G. (Eds.), Science Publishers, Enfield, NH. pp. 309–332.

McDonald, M. D. and Grosell, M. 2006. Maintaining osmotic balance with an aglomerular kidney. *Comp Biochem Physiol A*. 143: 447–458.

Mori, Y. and Ando, M. 1991. Regulation of ion and water transport across the eel intestine: Effects of acetylcholine and serotonin. *J Comp Physiol B*. 161: 387–392.

Morris, R. 1958. General problems of osmoregulation with special reference to cyclosomes. *Symp Zoo Soc Lond*. 1: 1–16.

Musch, M. W., Orellana, S. A., Kimberg, L. S., Field, M., Halm, D. R., Krasny, E. J., and Frizzell, R. A. 1982. Na$^+$-K$^+$-2Cl$^-$ co-transport in the intestine of a marine teleost. *Nature*. 300: 351–353.

Niv, Y. and Fraser, G. M. 2002. The alkaline tide phenomenon. *J Clin Gastroenterol*. 35: 5–8.

Nonnotte, G. and Truchot, J. P. 1990. Time course of extracellular acid-base adjustments under hypoosmotic or hyperosmotic conditions in the euryhaline fish *Platichthys-flesus*. *J Fish Biol*. 36: 181–190.

O'Grady, S. M. 1989. Cyclic nucleotide-mediated effects of ANF and VIP on flounder intestinal ion transport. *Am J Physiol*. 256: C142–C146.

O'Grady, S. M., Field, M, Nash, N. T., and Rao, M. C. 1985. Atrial natriuretic factor inhibits Na-K-Cl cotransport in the teleost intestine. *Am J Physiol*. 249: C531–C534.

Parmelee, J. T. and Renfro, J. L. 1983. Esophageal desalination of seawater in flounder: Role of active sodium transport. *Am J Physiol*. 245: R888–R893.

Part, P., Wright, P. A., and Wood, C. M. 1998. Urea and water permeability in dogfish (*Squalus acanthias*) gills. *Comp Biochem Physiol A*. 119: 117–123.

Pelis, R. M. and Renfro, J. L. 2003. Active sulfate secretion by the intestine of winter flounder is through exchange for luminal chloride. *Am J Physiol*. 284: R380–R388.

Perrot, M. N., Grierson, N., Hazon, N., and Balment, R. J. 1992. Drinking behavior in sea water and fresh water teleosts: The role of the renin-angiotensin system. *Fish Physiol Biochem*. 10: 161–168.

Perry, S. F., Braun, M. H., Genz, J., Vulesevic, B., Taylor, J. R., Grosell, M., and Gilmour, K. M. 2010. Acid-base regulation in the plainfin midshipman (*Porichthys notatus*), an aglomerular marine teleost. *J Comp Physiol*. 180: 1213–1225.

Perry, C. T., Salter, M. A., Harborne, A. R., Crowley, S. F., Jelks, H. L., and Wilson, R. W. 2011. Fish as major carbonate mud producers and missing components of the tropical carbonate factory. *Proc Natl Acad Sci*. 108: 3865–3869.

Pickering, A. D. and Morris, R. 1973. Localization of ion-transport in the intestine of the migrating river lamprey, *Lampetra fluviatilis* L. *J Exp Biol*. 58: 165–176.

Pierce, A. L., Fox, B. K., Davis, L. K., Visitacion, N., Kitahashi, T., Hirano, T., and Grau, E. G. 2007. Prolactin receptor, growth hormone receptor and putative somatolactin receptor in Mozambique tilapia: Tissue specific expression and differential regulation by salinity and fasting. *Gen Comp Endocrinol*. 154: 31–40.

Potts, W. and Evans, D. 1967. Sodium and chloride balance in the killifish, *Fundulus heteroclitus*. *Biol Bull*. 133: 411–425.

Raldua, D., Otero, D., Fabra, M., and Cerda, J. 2008. Differential localization and regulation of two aquaporin-1 homologs in the intestinal epithelia of the marine teleost *Sparus aurata*. *Am J Physiol*. 294: R993–R1003.

Salter, M. A., Perry, C. T., and Wilson, R. W. 2012. Production of mud-grade carbonates by marine fish: Crystalline products and their sedimentary significance. *Sedimentology*. 59: 2172–2198.

Sardella, B. A., Baker, D. W., and Brauner, C. J. 2009. The effects of variable water salinity and ionic composition on the plasma status of the Pacific hagfish (*Eptatretus stoutii*). *J Comp Physiol B*. 179: 721–728.

Sattin, G., Mager, E. M., and Grosell, M. 2010. Cytosolic carbonic anhydrase in the gulf toadfish is important for tolerance to hypersalinity. *Comp Biochem Physiol A*. 156: 169–175.

Seidelin, M., Madsen, S. S., Blenstrup, H., and Tipsmark, C. K. 2000. Time-course changes in the expression of the Na^+, K^+-ATPase in gills and pyloric caeca of brown trout (*Salmo trutta*) during acclimation to seawater. *Physiol Biochem Zool*. 73: 446–453.

Shehadeh, Z. H. and Gordon, M. S. 1969. The role of the intestine in salinity adaptation of the rainbow trout, *Salmo gairdneri*. *Comp Biochem Physiol*. 30: 397–418.

Skadhauge, E. 1974. Coupling of transmural flows of NaCl and water in the intestine of the eel (*Anguilla anguilla*). *J Exp Biol*. 60: 535–546.

Sleet, R. B. and Weber, L. J. 1982. The rate and manner of seawater ingestion by a marine teleost and corresponding seawater modification by the gut. *Comp Biochem Physiol*. 72A: 469–475.

Smith, H. W. 1930. The absorption and excretion of water and salts by marine teleosts. *Am J Physiol*. 93: 480–505.

Smith, C. P., Smith, P. L., Welsh, M. J., Frizzell, R. A., Orellana, S. A., and Field, M. 1980. Potassium transport by the intestine of the winter flounder *Pseudopleuronectes americanus*: Evidence for KCl co-transport. *Bull Mt Desert Island Biol Lab*. 20: 92–96.

Takei, Y., Hirano, T., and Kobayashi, H. 1979. Angiotensin and water intake in the Japanese eel, *Anguilla japonica*. *Gen Comp Endocrinol*. 38: 446–475.

Takei, Y. and Loretz, C. A. 2011. The gastrointestinal tract as an endocrine/neuroendocrine/paracrine organ: Organization, chemical messengers and physiological targets. *Fish Physiol*. 30: 261–317.

Takei, Y., Okubo, J., and Yamaguchi, K. 1988. Effect of cellular dehydration on drinking and plasma angiotensin II level in the eel, *Anguilla japonica*. *Zool Sci*. 5: 43–51.

Takei, Y and Tsuchida, T. 2000. Role of the renin-angiotensin system in drinking of the seawater-adapted eels *Anguilla japonica*: Reevaluation. *Am J Physiol*. 279: R1105–R1111.

Takei, Y., Tsuchida, T., Li, Z. H., and Conlon, J. M. 2001. Antidipsogenic effects of eel bradykinins in the eel *Anguilla japonica*. *Am J Physiol*. 281: R1090–R1096.

Takei, Y. and Yuge, S. 2007. The intestinal guanylin system and seawater adaptation in eels. *Gen Comp Endocrinol*. 152: 339–351.

Taylor, J. R. 2009. Intestinal HCO_3^- secretion in fish: A widespread mechanism with newly recognized physiological functions. PhD thesis, University of Miami.

Taylor, J. R. and Grosell, M. 2006a. Evolutionary aspects of intestinal bicarbonate secretion in fish. *Comp Biochem and Physiol A*. 143: 523–529.

Taylor, J. R. and Grosell, M. 2006b. Feeding and osmoregulation: Dual function of the marine teleost intestine. *J Exp Biol*. 209: 2939–2951.

Taylor, J. R. and Grosell, M. 2009. The intestinal response to feeding in seawater gulf toadfish, *Opsanus beta*, includes elevated base secretion and increased epithelial oxygen consumption. *J Exp Biol*. 212: 3873–3881.

Taylor, J. R., Mager E. M., and Grosell, M. 2010. Basolateral NBCe1 plays a rate-limiting role in transepithelial intestinal HCO_3^- secretion serving marine fish osmoregulation. *J Exp Biol*. 213: 459–468.

Taylor, J. R., Whittamore, J. M., Wilson, R. W., and Grosell, M. 2007. Postprandial acid-base balance in freshwater and seawater-acclimated European flounder, *Platichthys flesus*. *J Comp Physiol*. 177: 597–608.

Thomsen, J. et al. 2010. Calcifying invertebrates succeed in a naturally CO_2-rich coastal habitat but are threatened by high levels of future acidification. *Biogeosciences*. 7: 3879–3891.

Tipsmark, C. K., Sorensen, K. J., Hulgard, K., and Madsen, S. S. 2010. Claudin-15 and—25b expression in the intestinal tract of Atlantic salmon in response to seawater acclimation, smoltification and hormone treatment. *Comp Biochem Physiol A*. 155: 361–370.

Tomasso, J. R. and Grosell, M. 2004. Physiological basis for large differences in resistance to nitrite among freshwater and freshwater acclimated euryhaline fishes. *Environ Sci Technol*. 39: 98–102.

Tresguerres, M., Katoh, F., Fenton, H., Jasinska, E., and Goss, G. G. 2005. Regulation of branchial V-H$^+$-ATPase Na$^+$/K$^+$-ATPase and NHE2 in response to acid and base infusions in the Pacific spiny dogfish (*Squalus acanthias*). *J Exp Biol*. 208: 345–354.

Tresguerres, M., Levin, L. R., Buch, J., and Grosell, M. 2010. Modulation of NaCl absorption by HCO_3^- in the marine teleost intestine is mediated by soluble adenyl cyclase. *Am J Physiol*. 299: R62–R71.

Tresguerres, M., Parks, S. K., and Goss, G. G. 2007a. Recovery from blood alkalosis in the Pacific hagfish (*Eptatretus stoutii*): Involvement of gill V-H$^+$-ATPase and Na$^+$/K$^+$-ATPase. *Comp Biochem Physiol*. A148: 133–141.

Tresguerres, M., Parks, S. K., Katoh, F., and Goss, G. G. 2006. Microtubule-dependent relocation of branchial V-H$^+$-ATPase to the basolateral membrane in the Pacific spiny dogfish (*Squalus acanthias*): A role in base secretion. *J Exp Biol*. 209: 599–609.

Tresguerres, M., Parks, S. K., Wood, C. M., and Goss, G. G. 2007b. V-H$^+$-ATPase translocation during blood alkalosis in dogfish gills: Interaction with carbonic anhydrase and involvement in the postfeeding alkaline tide. *Am J Physiol*. 292: R2012–R2019.

Tsuchida, T. and Takei, Y. 1998. Effects of homologous atrial natriuretic peptide on drinking and plasma ANG II level in eels. *Am J Physiol*. 44: R1605–R1610.

Uesaka, T., Yano, K., Sugimoto, S., and Ando, M. 1996. Effects of eel neuropeptide Y on ion transport across the seawater eel intestine. *Zool Sci*. 13: 341–346.

Uesaka, T., Yano, K., Yamasaki, M., Nagashima, K., and Ando, M. 1994. Somatostatin-related peptides isolated from the eel gut: Effects on ion and water absorption across the intestine of the seawater eel. *J Exp Biol*. 188: 205–216.

Veillette, P. A., Sundell, K., and Specker, J. L. 1995. Cortisol mediates the increase in intestinal fluid absorption in Atlantic salmon during parr smolt transformation. *Gen Comp Endocrinol*. 97: 250–258.

Veillette, P. A. and Young, G. 2004. Temporal changes in intestinal Na$^+$, K$^+$-ATPase activity and in vitro responsiveness to cortisol in juvenile Chinook salmon. *Comp Biochem and Physiol A*. 138: 297–303.

Veillette, P. A. and Young, G. 2005. Tissue culture of sockeye salmon intestine: Functional response of Na$^+$-K$^+$-ATPase to cortisol. *Am J Physiol*. 288: R1598–R1605.

Walton Smith, F. G. 1931. *CRC Handbook of Marine Science*. Vol. 1, CRC Press, Cleveland, OH.

Watanabe, S., Hirano, T., Grau, E. G., and Kaneko, T. 2009. Osmosensitivity of prolactin cells is enhanced by the water channel aquaporin-3 in a euryhaline Mozambique tilapia (*Oreochromis mossambicus*). *Am J Physiol*. 296: R446–R453.

Whittamore, J. M., Cooper, C. A., and Wilson, R. W. 2010. HCO_3^- secretion and $CaCO_3$ precipitation play major roles in intestinal water absorption in marine teleost fish in vivo. *Am J Physiol*. 298: R877–R886.

Wilkes, P. R. H. and Mcmahon, B. R. 1986. Responses of a stenohaline fresh-water teleost (*Catostomus commersonii*) to hypersaline exposure.1. The dependence of plasma pH and bicarbonate concentration on electrolyte regulation. *J Exp Biol*. 121: 77–94.

Wilson, R. W., Gilmour, K., Henry, R., and Wood, C. 1996. Intestinal base excretion in the seawater-adapted rainbow trout: A role in acid-base balance? *J Exp Biol*. 199: 2331–2343.

Wilson, R. W. and Grosell, M. 2003. Intestinal bicarbonate secretion in marine teleost fish—Source of bicarbonate, pH sensitivity, and consequence for whole animal acid-base and divalent cation homeostasis. *Biochim Biophys Acta*. 1618: 163–193.

Wilson, R. W., Millero, F. J., Taylor, J. R., Walsh, P. J., Christensen, V., Jennings, S., and Grosell, M. 2009. Contribution of fish to the marine inorganic carbon cycle. *Science*. 323: 359–362.

Wilson, R. W., Wilson, J. M., and Grosell, M. 2002. Intestinal bicarbonate secretion by marine teleost fish–Why and how? *Biochim Biophys Acta*. 1566: 182–193.

Wood, C. M. and Bucking, C. 2010. The role of feeding in salt and water balance. In: *The Multifunctional Gut of Fish: Fish Physiology*, M. Grosell, A. P. Farrel, and C. J. Brauner (Eds.), Vol. 30, Academic Press, Elsevier, London, U.K., pp. 165–212.

Wood, C. M. and Bucking, C. 2011. The role of feeding in salt and water balance. *Fish Physiol*. 30: 165–212.

Wood, C. M., Bucking, C., and Grosell, M. 2010a. Acid-base responses to feeding and intestinal Cl⁻ uptake in freshwater- and seawater-acclimated killifish, *Fundulus heteroclitus*, an agastric euryhaline teleost. *J Exp Biol*. 213: 2681–2692.

Wood, C. M. and Grosell, M. 2012. Independence of net water flux from paracellular permeability in the intestine of *Fundulus heteroclitus*, a euryhaline teleost. *J Exp Biol*. 215: 508–517.

Wood, C. M., Kajimura, M., Bucking, C., and Walsh, P. J. 2007. Osmoregulation, ionoregulation and acid-base regulation by the gastrointestinal tract after feeding in the elasmobranch (*Squalus acanthias*). *J Exp Biol*. 210: 1335–1349.

Wood, C. M., Kajimura, M., Mommsen, T. P., and Walsh, P. J. 2005. Alkaline tide and nitrogen conservation after feeding in an elasmobranch (*Squalus acanthias*). *J Exp Biol*. 208: 2693–2705.

Wood, C. M., Matsuo, A. Y. O., Gonzalez, R. J., Wilson, R. W., Patrick, M. L., and Val, A. L. 2002. Mechanisms of ion transport in Potamotrygon, a stenohaline freshwater elasmobranch native to the ion-poor blackwaters of the Rio Negro. *J Exp Biol*. 205: 3039–3054.

Wood, C. M., Schultz, A. G., Munger, R. S., and Walsh, P. J. 2009. Using omeprazole to link the components of the post-prandial alkaline tide in the spiny dogfish, *Squalus acanthias*. *J Exp Biol*. 212: 684–692.

Wood, C. M., Walsh, P. J., Kajimura, M., McClelland, G. B., and Chew, S. F. 2010b. The influence of feeding and fasting on plasma metabolites in the dogfish shark (*Squalus acanthias*). *Comp Biochem Physiol A*. 155: 435–444.

Woosley, R. J., Millero, F. J., and Grosell, M. 2012. The solubility of fish-produced high magnesium calcite in seawater. *J Geophys Res-Oceans*. 117: C04018.

Yuge, S., Inoue, K., Hyodo, S., and Takei, Y. 2003. A novel guanylin family (guanylin, uroguanylin and renoguanylin) in eels: Possible osmoregulatory hormones in intestine and kidney. *J Biol Chem*. 278: 22726–22733.

Yuge, S. and Takei, Y. 2007. Regulation of ion transport in eel intestine by the homologous guanylin family of peptides. *Zool Sci*. 24: 1222–1230.

6 Gill Ionic Transport, Acid–Base Regulation, and Nitrogen Excretion

Pung-Pung Hwang and Li-Yih Lin

CONTENTS

6.1 INTRODUCTION

All vertebrates have to regulate their intracellular ionic compositions for the normal operation of cellular and biochemical reactions. Compared to terrestrial animals, aquatic vertebrates are faced with more challenging osmoregulatory environments with fluctuating ionic compositions and osmolarities, which directly affect the homeostasis of body fluids. Aquatic vertebrates have developed different strategies so that their body fluids can cope with the dramatic ionic and osmotic gradients found in aquatic environments. According to the fossil record, early vertebrates, hagfishes, are believed to have originated in a seawater (SW) environment, and some agnathan fish groups, lampreys, were the first invaders in freshwater (FW) (Halstead, 1985; Bartels and Potter, 2004; Evans and Claiborne, 2009). It was also proposed that lampreys and subsequent teleosts originally evolved in FW, and later some of them returned to SW (Halstead, 1985; Bartels and Potter, 2004; Evans and Claiborne, 2009). On the other hand, most extant chondrichthyan fishes are marine with very few FW species; however, no fossil record is available to clarify the early evolution of these cartilaginous species in FW (Halstead, 1985; Evans and Claiborne, 2009). Table 6.1 shows the concentrations of major solutes and osmolarity in plasma of different fishes in various aquatic environments. Following the early evolution of vertebrates, strategies dealing with body fluids also evolved from osmoconforming to osmo- and ionoregulating (Table 6.1). Hagfishes are osmoconformers with limited regulation of some divalent ions (Mg^{2+} and Ca^{2+}). Lampreys are pioneers in developing osmo- and ionoregulatory strategies, and teleosts are strict osmo- and ionoregulators. On the other hand, most marine chondrichthys are osmoconformers (slightly hyperosmotic to SW due to high plasma levels of urea and the counteracting solute, trimethylamine oxide) and ionoregulators, and the stenohaline FW stingrays (*Potamotrygon* sp.) are osmo- and ionoregulators with much reduced plasma levels of urea (Wood et al., 2002). Acid–base regulatory mechanisms achieved by apical Na^+/H^+ and

TABLE 6.1
Major Solutes (mM) and Osmolarity (mOsm) in Plasma of Different Fishes

	Na⁺	Cl⁻	K⁺	Ca²⁺	Mg²⁺	Urea	TMAO	Osmolarity
Seawater	439	513	9.3	9.6	50	—	—	1050
Hagfish	486	508	8.2	5.1	12	—	—	1035
Anadromous lamprey	156	159	5.8	3.5	7.0	—	—	333
Dogfish shark	255	241	6.0	5	3.0	441	72	1118
Euryhaline steelhead trout	153	135	4.0	1.4	0.7			325
Soft lake water[a]	0.17	0.03	—	0.22	0.15	—	—	<1[b]
River water[a]	0.39	0.23	0.04	0.52	0.21	—	—	<3[b]
Hard river water[a]	6.13	13.44	0.11	5.01	0.66	—	—	<30[b]
Landlocked lamprey	120	104	3.9	2.5	2.0	—	—	272
Euryhaline bull shark	221	220	4.2	3.0	1.3	151	19	595
Freshwater stingray	178	146	—	—	—	1.2	—	319
Euryhaline steelhead trout	153	133	3.8	1.4	0.5	—	—	260
Carp	130	125	2.9	2.1	1.2	—	—	274

Data from Evans and Claiborne (2009); anadromous lamprey, *Petromyzon marinus*; carp, *Cyprinus carpio*; euryhaline bull shark, *Carcharhinus leucas*; dogfish shark, *Scyliorhinus canicula*, *Squalus acanthias*; euryhaline steelhead trout, *Oncorhynchus mykiss*; freshwater stingray, *Potamotrygon* sp.; hagfish, *Myxine glutinosa*; landlocked lamprey, *Lampetra fluviatilis*.

[a] Data from Schmidt-Nielsen (1993).
[b] Sum of all solutes.

Cl^-/HCO_3^- exchangers in the transporting epithelia evolved early in primitive vertebrates, such as hagfishes. Lampreys were probably pioneers in evolving both ion absorption and secretory mechanisms other than the acid/base-linked Na^+/Cl^- uptake pathways that evolved in hagfishes.

Gills, a unique organ that evolved in early aquatic vertebrates, achieve acid–base regulation in all fish groups, from hagfishes to teleosts and chondrichthys. No information is available regarding the roles gills play in NaCl secretion or acid/base-unlinked NaCl uptake functions in lampreys. Gills might not be the organ that performs NaCl secretion function in marine chondrichthys (Takabe et al., 2012). Among fishes, ion uptake/secretion, acid–base regulation, and nitrogenous waste excretion mechanisms of gill functions of teleosts are the most studied. In addition, the majority of teleosts are ammonotelic, they excrete most of their waste nitrogen as ammonia, and the ammonia excretion is largely accomplished by their gills. Teleost gills have four paired gill arches, and for each arch, there are numerous gill filaments that in turn support thousands of lamellae. Teleost gill epithelium, which covers filaments and lamellae, is not only the barrier between body fluids and the external aquatic environment, but also the major site associated with ionic and acid–base regulation and nitrogenous waste excretion functions. The gill epithelium is composed of several distinct cell types: pavement cells (PVCs) (over 90% of the cell population), ionocytes (formerly, chloride cells or mitochondrion-rich cells, <10%), mucus cells, and stem cells. Ionocytes are the main cell types responsible for ion transport functions.

The topic of ionic and acid–base regulation in fish gills has been summarized in numerous detailed reviews (Hirose et al., 2003; Perry et al., 2003b; Evans et al., 2005; Marshall and Grosell, 2006; Perry and Gilmour, 2006; Hwang and Lee, 2007; Evans, 2008, 2011; Evans and Claiborne, 2009; Gilmour and Perry, 2009; Hwang, 2009; Hwang and Perry, 2010; Hwang et al., 2011). The majority of studies regarding fish iono- and osmoregulation were conducted in teleosts. Given the highly advanced approaches employing cell biology/molecular physiology and model animals, there has been much progress in knowledge related to teleost gill iono- and osmoregulatory mechanisms. Therefore, considering space limitations, the present review focuses on recent progress, particularly emphasizing details behind cellular and molecular pathways in teleost gill (and skin) iono-/osmoregulation, acid–base regulation, and ammonia excretion. For comparisons, some studies on fish species other than teleosts are also cited. Teleost embryonic and larval skin is similar to adult gills in terms of the morphology/function of ionocytes and related ion transporters. Some of the recent progress derived from studies on embryonic and larval skin is also discussed.

6.2 ION REGULATION IN SW GILLS

The current model for NaCl secretion by gill ionocytes of SW teleosts includes three critical transporters: the Na^+-K^+-ATPase (NKA) and Na^+-K^+-$2Cl^-$ cotransporter (NKCC) in the basolateral membrane and the cystic fibrosis transmembrane conductance regulator (CFTR) Cl^- channel in the apical membrane. The basolateral NKCC carries 1 Na^+, 1 K^+, and 2 Cl^- into the cell down the electrochemical gradient provided by the action of NKA. The accumulated intracellular Cl^- then exits the cell through the apical CFTR channel, and the interstitial Na^+ is pushed out of the gills through the paracellular pathway by the transepithelial potential (Hirose et al., 2003; Evans et al., 2005; Marshall and Grosell, 2006; Hwang and Lee, 2007; Hwang et al., 2011). This model was formed in the 1970s, and the involved proteins were further identified and clarified in the past decades.

6.2.1 NaCl Secretion

6.2.1.1 NKA

The NKA is the first and most comprehensively investigated transporter in the NaCl-secreting pathway. It provides an electrochemical gradient to drive the process of NaCl secretion (Takeyasu et al., 1990; Pressley, 1992). Changes in branchial NKA activities of euryhaline teleosts are thought to be necessary for salinity adaptation (Hwang and Lee, 2007). Increases in NKA activity upon salinity challenge are attributed to (a) the increase in NKA mRNA and/or protein abundance (Lee et al., 2000; Seidelin et al., 2000; Singer et al., 2002; Tipsmark et al., 2002; Lin et al., 2003; Scott et al., 2004) and (b) the modulation of the enzyme's hydrolytic rate (Crombie et al., 1996).

The NKA consists of two subunits, α and β, and several isoforms of each subunit were identified in fishes. In zebrafish (*Danio rerio*), 9 NKA α-subunit genes (7 α1, 1 α2, and 1 α3) were identified from genetic databases (Liao et al., 2009). Different isoforms of the α-subunit were also found in gills of euryhaline teleosts, including European eel (*Anguilla anguilla*) (Cutler et al., 1995), salmonids (Madsen et al., 1995; D'Cotta et al., 2000; Seidelin et al., 2000; Richards et al., 2003), Mozambique tilapia (*Oreochromis mossambicus*) (Hwang et al., 1998; Feng et al., 2002), and Antarctic nototheniid (*Trematomus bernachii*) (Guynn et al., 2002; Brauer et al., 2005). It is generally accepted that different NKA isoforms with spatial and/or temporal specificities are correlated with their kinetics or properties to match physiological demands. Regulation of branchial NKA α-isoforms (the catalytic subunit) was suggested to be crucial for salinity adaptation in euryhaline fishes. In tilapia, differential expressions of α1 and α3 mRNA/protein were found in gills of fish subjected to salinity changes (Lee et al., 1998; Feng et al., 2002). In milkfish (*Chanos chanos*), the protein abundances of α1 and α3 but not α2 isoforms in gills increased with environmental salinities (Tang et al., 2009). Similar phenomena were also found in salmonids, and it was suggested that NKAα1a and NKAα1b in gills have respective functions in fish residing in SW and FW (Bystriansky et al., 2006; Madsen et al., 2009; Nilsen et al., 2007; Richards et al., 2003). In addition to the expression level, McCormick et al. (2009) further used isoform-specific antibodies to reveal the salinity-dependent expression patterns of NKA α-isoforms (α1a and α1b) in ionocytes of Atlantic salmon (*Salmo salar*). With double immunohistochemistry, the two NKA isoforms were found in two distinct groups of ionocytes in gills of FW salmon (McCormick et al., 2009). In zebrafish, three isoforms of NKA α1 (ATP1a1.1, ATP1a1.2, and ATP1a1.5) were respectively localized to three subtypes of ionocytes (refer to Section 5.2) (Liao et al., 2009).

Recently, three NKA α1 isoforms (1a, 1b, and 1c) were identified in gills of the climbing perch (*Anabas testudineus*) (Ip et al., 2012a). The mRNA expression of NKA α1a was downregulated in gills of fish acclimated to SW; in contrast, SW acclimation led to an upregulation of α1b. Interestingly, high-ammonia exposure led to a significant upregulation of α1c, suggesting that α1c might be involved in active ammonia excretion.

6.2.1.2 NKCC

The NKCC includes two isoforms: NKCC1 and NKCC2. NKCC1 is considered to be the secretory isoform, as prominent expression of NKCC1 occurs in the basolateral membrane, and it is associated with epithelial Cl$^-$ secretion (Lytle et al., 1995). Two isoforms of NKCC1, NKCC1a and NKCC1b, were cloned from European eel, Mozambique tilapia, and brackish-adapted medaka (*Oryzias dancena*); however, NKCC1a is the major isoform in gills (Cutler and Cramb, 2002; Hiroi et al., 2008; Kang et al., 2010).

Using an antibody against human NKCC1 (T4), NKCC1 was localized to the basolateral membrane of SW ionocytes in the giant mudskipper (*Periophthalmodon schlosseri*) (Wilson et al., 2000), several salmonids (Hiroi and McCormick, 2007), killifish (Marshall et al., 2002), Hawaiian goby (*Stenogobius hawaiiensis*) (McCormick et al., 2003), Mozambique tilapia (Wu et al., 2003; Hiroi et al., 2005), and pufferfish (*Takifugu niphobles*) (Tang et al., 2011). Furthermore, using whole-mount in situ hybridization, NKCC1a mRNA was localized to ionocytes of medaka gills (Kang et al., 2010), supporting those immunocytochemical data in previous studies. Increases in mRNA and protein levels of branchial NKCC1 (NKCC1a) were found in euryhaline species subjected to SW transfer (Cutler and Cramb, 2002; Tipsmark et al., 2002; Scott et al., 2004; Lorin-Nebel et al., 2006; Tse et al., 2006; Hwang and Lee, 2007; Hiroi et al., 2008; Sardella and Kültz, 2009; Flemmer et al., 2010; Kang et al., 2010), supporting NKCC1a being involved in the NaCl secretion of ionocytes.

6.2.1.3   CFTR

Cl$^-$ secretion from apical membranes of SW ionocytes uses an anion channel with characteristics resembling those of the CFTR (Marshall et al., 1995). The first teleost CFTR was cloned from killifish (*Fundulus heteroclitus*) gills (Singer et al., 1998). Following SW exposure, upregulation of CFTR mRNA/protein was reported in gills of Atlantic salmon (Singer et al., 1998, 2002), killifish (Scott et al., 2004), Japanese eel (*Anguilla japonicus*) (Tse et al., 2006), and pufferfish (Tang and Lee, 2007). Immunohistochemistry revealed localization of the CFTR in apical membranes of SW ionocytes in several species, including the giant mudskipper (Wilson et al., 2000), Hawaiian goby (McCormick et al., 2003), killifish (Marshall et al., 2002; Katoh and Kaneko, 2003), tilapia (Hiroi et al., 2005), and grass pufferfish (*Tetraodon nigroviridis*) (Tang et al., 2011). Time-course studies in killifish subjected to salinity changes showed that the apical CFTR occurred 24 h after transfer from FW to SW, while it disappeared within 24 h of transfer from SW to FW (Marshall et al., 2002; Katoh and Kaneko, 2003; Scott et al., 2005). In tilapia, SW ionocytes with the apical CFTR appeared at 12 h and showed a remarkable increase in number between 24 and 48 h after transfer from FW to SW (Hiroi et al., 2005).

6.2.2   K$^+$ Secretion

The mechanism for NaCl regulation in SW fishes has been intensively investigated as the NaCl balance determines the greatest proportion of body fluid osmolarity. Other ions such as K$^+$ also play critical roles in animal physiology; however, their regulation in fishes has not been intensively investigated. Due to the actions of the NKA and NKCC in basolateral SW ionocytes, the concentration of intracellular K$^+$ is supposedly much higher than interstitial K$^+$, and the gradient probably drives K$^+$'s return to the interstitial fluid through a K$^+$ channel (eKir, an orthologue of the mammalian Kir5.1) located in basolateral membranes in eel gill ionocytes (Suzuki et al., 1999; Tse et al., 2006). In a recent study, Furukawa et al. showed that K$^+$ was excreted from gill ionocytes of SW-acclimated tilapia, using a newly developed technique that insolubilized and visualized excreted K$^+$ (Furukawa et al., 2012). They cloned a renal outer medullary K$^+$ channel (ROMK, Kir1.1) from tilapia gills and found that the ROMK mRNA level in gills increased in response to a high external K$^+$ concentration. In addition, immunohistochemistry revealed that ROMK was localized in apical membranes of ionocytes, and the immunosignals were most intense in fish acclimated to high-K$^+$ water. Their study suggested that SW fish secrete K$^+$ from gill ionocytes, and the apical ROMK is responsible for the egression of K$^+$. This new pathway needs to be studied in other species.

6.2.3   Acid–Base Regulation

Acid–base regulation in fishes is predominantly accomplished by the transfer of relevant acid/base ions (H$^+$ and HCO$_3^-$) in gills (Evans et al., 2005; Perry and Gilmour, 2006; Hwang and Perry, 2010). They cannot adjust the plasma pH by ventilation as in air breathers because of the low PCO$_2$ and bicarbonate concentration in the blood (Evans et al., 2005). It is generally accepted that SW fishes excrete metabolic acids through an Na$^+$/H$^+$ exchanger (NHE) in the apical membrane of ionocytes (Claiborne et al., 2002). The high external Na$^+$ concentration provides a strong gradient to drive H$^+$ secretion by the NHE in ionocytes. The molecular identity and cellular localization of NHE isoforms in marine and euryhaline fishes were reported in elasmobranches and teleosts. An earlier work using heterologous antibodies (against mammalian NHE2) demonstrated that NHE2 colocalized with Na-K-ATPase in gills of several elasmobranch species (Edwards et al., 2002) and increased protein detection following acid infusion (Tresguerres et al., 2005). In marine sculpin (*Myoxocephalus octodecemspinosus*), Catches et al. used a species-specific antibody to show that NHE2 is localized to apical membranes of ionocytes (Catches et al., 2006). Recently, Claiborne et al. cloned an NHE2-like gene from the dogfish (*Squalus acanthias*) and localized its mRNA and protein in gills (Claiborne et al., 2008a). Cytoplasmic and apical NHE2 immunoreactivity was observed in a subtype of ionocytes that was rich in NKA. Choe et al. cloned NHE3 from the gill of stingray (*Dasyatis sabina*) and used species-specific antibodies to label NHE3 in the gill of FW-adapted individuals (Choe et al., 2005). They demonstrated that apical NHE3 is colocalized with basolateral Na-K-ATPase and suggested that this subtype of ionocyte is involved in acid secretion.

In addition to the NHE, the role of H^+-ATPase in acid–base regulation of SW fishes was also reported in some elasmobranch species. In the marine dogfish, Tresguerres et al. found that H^+-ATPase was located in the basolateral portion of a subtype of ionocyte that was distinct from other ionocytes expressing NKA, and H^+-ATPase expression increased in fish exposed to metabolic alkalosis (Tresguerres et al., 2005, 2006b, 2007). In the euryhaline stingray (*D. sabina*), immuno-histochemical and in situ hybridization studies revealed two distinct ionocytes in gills (Piermarini and Evans, 2001; Piermarini et al., 2002; Choe et al., 2005): one that expresses apical NHE and basolateral NKA for H^+ secretion, and the other that uses apical pendrin-like Cl^-/HCO_3^- exchangers and basolateral H^+-ATPase to secrete HCO_3^-. These studies suggested that the H^+-ATPase is involved in base secretion instead of acid secretion in SW fish.

6.3   ION REGULATION IN FW GILLS

6.3.1   Na$^+$ Uptake/Acid Secretion

6.3.1.1   H$^+$-ATPase and ENaC

Pioneering work by Krogh (1937, 1938) initially suggested independent Na^+ and Cl^- uptake respectively linked to NH_4^+ and HCO_3^- in FW fish gills. Since then, at least two major pathways for apical Na^+ uptake/acid secretion—(a) H^+-ATPase driving the epithelial Na^+ channel (ENaC) and (b) the electroneutral NHE—were proposed and have been debated to the present. It is thermodynamically reasonable for gill cells to absorb Na^+ through the ENaC from a low-Na^+ FW environment down the intracellular negative potential created by apical H^+-ATPase. Pharmacological experiments (Reid et al., 2003; Preest et al., 2005; Esaki et al., 2007) and localization (Wilson et al., 2000) evidence support the existence of the ENaC in FW-acclimated teleost gills; however, no orthologues of the ENaC have been found in a teleost genome so far. ENaC orthologues were recently reported in the genome of the elephant shark (*Callorhinchus milii*) (Venkatesh et al., 2007), but their functions in gill ion regulatory mechanisms remain unknown. On the other hand, evidence for the involvement of apical H^+-ATPase is consistent and convincing. Pioneering work on rainbow trout (*Oncorhynchus mykiss*) provided protein and mRNA localization evidence of H^+-ATPase in gill lamellar PVCs and/or chloride cells (Lin et al., 1994; Sullivan et al., 1995, 1996). Subsequent evidence included stimulation of gill H^+-ATPase expression by hypercapnia or acidic treatment (Sullivan et al., 1995, 1996; Galvez et al., 2002; Al-Fifi, 2006; Yan et al., 2007) and inhibition of the Na^+ uptake function by H^+-ATPase inhibitors (Fenwick et al., 1999; Boisen et al., 2003). These were reinforced by recent molecular/cellular physiological studies. In rainbow trout gills, a group of isolated peanut lectin agglutinin-negative (PNA$^-$) cells revealed higher H^+-ATPase expression and bafilomycin (an H^+-ATPase inhibitor)-sensitive acid-activated (Chasiotis et al., 2012a) Na^+ uptake function, compared to a group of PNA$^+$ cells (Galvez et al., 2002; Reid et al., 2003). In zebrafish embryonic skin and gills, a group of H^+-ATPase-rich (HR) ionocytes, identified by in situ hybridization and immunocytochemistry of the enzyme, presented an in vivo acid-secreting function according to a scanning ion-selective electrode technique (SIET) (Lin et al., 2006). Loss-of-function of H^+-ATPase by injection with specific morpholinos suppressed the H^+ secretory function of HR cells and the whole-body Na^+ content in embryos acclimated to low-Na^+ FW (Horng et al., 2007). Simultaneously, HR cells in zebrafish were demonstrated to show an Na uptake function in vivo with an Na^+ green fluorescent probe, and Na^+ green accumulation in cells was suppressed by bafilomycin (Esaki et al., 2007). On the other hand, low-Na^+ or soft FW, which stimulates the Na^+ uptake function (Boisen et al., 2003; Chang et al., 2003), was reported to suppress the gill H^+-ATPase mRNA expression in rainbow trout (Al-Fifi, 2006) and zebrafish (Yan et al., 2007), implying that the environmental ionic strength may affect the involvement of H^+-ATPase in gill Na^+ uptake mechanisms.

6.3.1.2   NHE

Although the thermodynamics issue due to unfavorable chemical gradients was questioned (Avella and Bornancin, 1989; Parks et al., 2008), accumulating evidence and recent molecular physiological progress reinforce the notion of NHE-mediated Na^+ uptake/acid secretion mechanisms in FW teleost gills (Evans, 2011; Hwang et al., 2011). Detailed reviews of the localization, expression, and pharmacological responses of NHEs in FW gills were conducted (Evans, 2011; Kumai and Perry, 2012). Convincing molecular evidence of NHE3 initially emerged from the Osorezan dace (*Tribolodon hakonensis*) inhabiting a pH 3.5 lake (Hirata et al., 2003) and the euryhaline stingray (Choe et al., 2005). NHE3 was cloned and localized by immunocytochemistry and/or in situ hybridization in a group of gill ionocytes in both species, and gill mRNA expression was stimulated by acidic FW in the dace (Hirata et al., 2003), but was not affected by hypercapnia in the stingray (Choe et al., 2005). Among the eight members of the NHE family of zebrafish, NHE3b was identified by triple in situ hybridization/immunocytochemistry as the only isoform specifically expressed in apical membranes of HR-type ionocytes (Yan et al., 2007). This was supported by subsequent studies on tilapia (Hiroi et al., 2008; Inokuchi et al., 2008) and Japanese medaka (Wu et al., 2010; Lin et al., 2012). Impairments of H^+ activity (SIET) and Na^+ influx (Na^+ activity by SIET or Na^+ green accumulation) in HR cells by NHE3b knockdown or EIPA (an NHE3-specific inhibitor) treatment (Esaki et al., 2007; Shih et al., 2008, 2012) provided solid molecular physiological evidence for the functional role of NHE3 in Na^+ uptake/acid secretion in FW gill (skin) cells. Taken together, NHE3 appears to be the major member of the SLC9 family responsible for Na^+ uptake/acid secretion in FW teleosts gills as in mammalian kidneys (Choi et al., 2000). However, some species-specific differences were reported. NHE2 mRNA and NHE3 protein were colocalized in the same type of PNA^+ ionocytes in FW trout gills; and NHE2, but not NHE3, was stimulated by hypercapnia, suggesting a role of NHE2 in the acid-secreting function in trout PNA^+ ionocytes (Ivanis et al., 2008). Killifish studies also proposed a role of the NHE2-like gene in the acid-secreting function in FW gills (Edwards et al., 2005, 2010).

6.3.1.3   Driving Mechanism for Apical Na^+ Uptake

Recent emerging knowledge from Rhesus glycoproteins (Rh) provides some answers to long-term questions regarding the thermodynamics of apical NHEs in FW gills. Rhcg1 is co-expressed with the NHE and/or H^+-ATPase in apical membranes of specific groups of ionocytes in FW gills (and embryonic skin) (Nakada et al., 2007a; Wu et al., 2010) and was proposed to work with the NHE and H^+-ATPase (and others) as a functional metabolon (refer to the details in Section 6.4) (Wright and Wood, 2009). Subsequent molecular physiological evidence reinforced the notion that the H^+ gradient created by Rhcg1-mediated intracellular deprotonation and extracellular acid trapping may drive the operation of the electroneutral NHE (refer to details in Section 6.4) (Wu et al., 2010; Kumai and Perry, 2011; Shih et al., 2012). Rhcg1 appears to assist Na^+ uptake and/or the acid-secreting function, particularly in low-Na^+ and low-pH situations. Knockdown of Rhcg1 or acute high-NH_4^+ treatment simultaneously suppressed both NH_4^+ excretion and Na^+ uptake in zebrafish embryos acclimated to low-Na^+ (Shih et al., 2012) and low-pH FW (Kumai and Perry, 2011). Moreover, low-Na^+ or low-pH FW stimulated gill mRNA expression of Rhcg1 and NH_4^+ excretion in ionocytes in zebrafish (Shih et al., 2008, 2012) and medaka (Wu et al., 2010; Lin et al., 2012).

6.3.1.4   Carbonic Anhydrases and Basolateral Transporters

To achieve transepithelial Na^+ uptake and H^+ secretion (equivalently, HCO_3^- uptake) in FW gills, involvement of carbonate anhydrases and other basolateral transporters is essential as in mammalian kidney proximal tubules and collecting ducts (Purkerson and Schwartz, 2007). According to the proposed model for zebrafish HR cells (Hwang et al., 2011), apical NHE3b and H^+-ATPase transport H^+ out of cells, and H^+ combines with environmental HCO_3^- to generate H_2O and CO_2 by membrane carbonic anhydrase (CA15a; see later). CO_2 enters cells and is hydrated by cytosolic CA (CA2-like a; see later) to form H^+ and HCO_3^-. The basolateral anion exchanger (AE1b; see later) extrudes cytosolic HCO_3^- down the Cl^- gradient to fulfill the epithelial acid-secreting function, while basolateral NKA is responsible for the excretion of Na^+. The role of CA in Na^+ uptake and acid–base regulatory mechanisms had been proposed for a long time (Evans et al., 2005; Gilmour and Perry, 2009). The localization, and the physiological and pharmacological evidence for the role of cytosolic CA in FW gill acid secretion and/or Na^+ uptake are solid in many teleosts (for details, see review by Gilmour and Perry, 2009). The involvement of the membrane form of CA was unclear until a zebrafish study (Lin et al., 2008). In zebrafish, the membrane-form CA15a (which differs from the kidney CA4) and cytosolic CA2-like a were colocalized in HR cells, and loss-of-function experiments demonstrated their role in Na^+ uptake and H^+ secretion/HCO_3^- uptake (Lin et al., 2008). This was further supported by a very recent study, in which a proximity ligation assay was used to show NHE3b, Rhcg1, CA15a, and CA2-like a forming and functioning as a transport metabolon in zebrafish HR cells (Ito et al., 2013). With regard to basolateral transporters for transepithelial Na^+ uptake and acid secretion, the $Na^+ - HCO_3^-$ cotransporter (NBC) and NKA were repeatedly proposed as being major players (Hirose et al., 2003; Evans et al., 2005; Hwang and Lee, 2007; Parks et al., 2008; Hwang and Perry, 2010). Hypercapnia or acidic treatment was reported to stimulate gill NBCe1 mRNA expression in Osorezan dace (Hirata et al., 2003) and rainbow trout (Parks et al., 2007). This was further supported by a pharmacological experiment on isolated rainbow trout gill cells using DIDS (an NBC inhibitor) (Parks et al., 2007) and gill localization of NBCe1 in the dace (Hirata et al., 2003) and trout (Perry et al., 2003a). However, this notion of NBCe1 appears to require reevaluation because of recent convincing colocalization studies (Furukawa et al., 2011; Lee et al., 2011). In zebrafish and tilapia, double or triple in situ hybridization/immunocytochemistry was used to demonstrate the colocalization of NBCe1 and Na^+-Cl^- cotransporter (NCC) in a specific group of ionocytes (see Section 6.3.2), distinct from acid-secreting ionocytes, which express apical NHE3 and/or H^+-ATPase (Furukawa et al., 2011; Lee et al., 2011). Furthermore, AE1 was localized in the basolateral membrane of HR-type ionocytes and appeared to cooperate with apical NHE3b and H^+-ATPase to achieve transepithelial Na^+ uptake and H^+ secretion in zebrafish gill/skin secretions based on triple-labeling and loss-of-function experiments (Lee et al., 2011). More evidence from other species is necessary to reinforce this new insight. Direct evidence for the role of the basolateral NKA in FW gill Na^+ uptake and H^+ secretion functions is scanty. Acidic FW treatment did result in enhancement of NKA subunit mRNA in dace gills (Hirata et al., 2003). The diversity of NKA subunit subtypes raises the possibility of differential expression and functions of these subtypes in various types of ionocytes in FW gills; however, this issue has been overlooked so far (Liao et al., 2009). Among the four subtypes of the NKA α1 subunit, only ATP1a1a.5 was specifically localized in zebrafish HR ionocytes, and its mRNA expression was stimulated by low-Na^+ FW (Liao et al., 2009), suggesting NKA isoform-specific functions in FW gill ion and acid–base regulatory mechanisms.

6.3.2 Na⁺-Cl⁻ Cotransport

6.3.2.1 NCC

In mammalian kidneys, proximal tubules carry out the bulk reabsorption of solutes via NHEs and other associated transport pathways. Further fine-tuning of NaCl reabsorption is achieved by other redundant Na^+/Cl^- uptake pathways, and one of them is mediated by the electroneutral and thiazide-sensitive NCC in distal convoluted tubules (Reilly and Ellison, 2000). This pathway is conserved in gills of FW teleosts. An NCC-like protein (SLC12A10), distinct from the kidney orthologue (SLC12A3), was initially identified as a gill-specific member of the teleost SLC family (Hiroi et al., 2008). NCC-like protein and mRNA were respectively localized in a certain group of ionocytes in FW tilapia and zebrafish gills/embryonic skin (Hiroi et al., 2008; Inokuchi et al., 2008;

Wang et al., 2009), and its mRNA expression and cell number were stimulated by treatment with low-Cl^- artificial FW, suggesting a role of the NCC in FW gill Cl^- uptake function (Inokuchi et al., 2008, 2009; Wang et al., 2009). Recently, SIET was used to detect an inward metolazone-sensitive Cl^- current that occurs in convex ionocytes that express apical NCC immunoreactivity in tilapia embryonic skin, providing in vivo evidence for NCC's function in Cl^- uptake (Horng et al., 2009a). Subsequent studies on zebrafish provided further molecular physiological evidence to support this notion. Translational knockdown of the NCC impaired both Cl^- influx and content in zebrafish morphant embryos (Wang et al., 2009). All this evidence confirms a major function of the NCC in FW gill Cl^- uptake mechanisms. The NCC appears to serve as a redundant or supplementary player in FW gill Na^+ function. Acclimation to low-Na^+ FW increased the number of NCC-expressing ionocytes in tilapia gills (Inokuchi et al., 2009). Zebrafish morphants with NCC knockdown showed compensatory stimulation of Na^+ influx and content as well as NHE3b mRNA expression (Wang et al., 2009). Incubation with metolazone (a thiazide-like NCC-specific inhibitor) reduced both Cl^- and Na^+ influxes in zebrafish embryos (Wang et al., 2009). NBCe1b, a basolateral transporter that eliminates intracellular Na^+ from cells, was localized in NCC-expressing ionocytes of zebrafish (Lee et al., 2011) and tilapia (Furukawa et al., 2011). Furthermore, knockdown of GCM2, a transcriptional factor that specifically targets differentiation of HR cells, resulted in the disappearance of HR cells (thus impairing Na^+ uptake; refer to Section 6.3.1) (Chang et al., 2009), but also caused an increase in the number of NCC cells in zebrafish morphants (Shono et al., 2011). Taken together, the role of NCC and its relation with the NHE3 pathway in FW gill Na^+ uptake mechanisms require further exploration.

6.3.2.2 Basolateral Transporters

To achieve NCC-mediated NaCl absorption in FW gills, the basolateral Cl^- channel (ClC), $Na^+ - HCO_3^-$ cotransporter (NBC), and NAK may play some roles. ClC family members have different tissue distributions and subcellular localizations with broad physiological functions, including transepithelial Cl^- transport and cell-volume regulation in mammals (Jentsch et al., 2005). ClC3 was first cloned from tilapia and showed expression in various osmoregulatory organs including gills (Miyazaki et al., 1999). In recent studies on grass pufferfish, ClC3 was localized (with a heterologous antibody) in the basolateral membrane of all FW and SW gill NKA-rich ionocytes, and FW stimulated gill ClC3 protein levels (Tang et al., 2010, 2011). In tilapia, the ClC3 protein was only localized in a group of NCC-expressing ionocytes in FW gills, but showed weaker signals in all SW gill ionocytes; furthermore, acclimation to low-Cl^- FW or deionized water stimulated gill ClC3 protein expression (Tang and Lee, 2011). All these results support the possible role of ClC3, in collaboration with apical NCC, in the Cl^- uptake function in FW teleost gills. The role of NBCe1 in acid–base regulation appears to be convincing based on previous plentiful data (see Section 6.3.2). A recent striking finding of NBCe1 was its co-expression with apical NCC in the same type of ionocytes in tilapia and zebrafish (Furukawa et al., 2011; Lee et al., 2011). However, the role of NBCe1 in FW gill Na^+ uptake or other ion regulation needs reevaluation as its mRNA in zebrafish gills was stimulated by high-Na^+ FW compared to low-Na^+ FW (Lee et al., 2011). On the other hand, basolateral NKA in NCC-expressing ionocytes may establish a negative intracellular gradient to drive the electrogenic ClC and/or NBCe1 (see earlier) to eliminate Cl^- and/or Na^+ from cells. Immunocytochemical localization of NKA in tilapia NCC-expressing ionocytes (Hiroi et al., 2008; Inokuchi et al., 2008) and the presence of ATP1a1a.2 (a subtype of the NKA α1 subunit) mRNA by in situ hybridization in zebrafish NCC-type ionocytes (Liao et al., 2009) support this notion.

6.3.3 Cl^- Uptake/Base Secretion

Apical Cl^-/HCO_3^- exchange has long been proposed as a major pathway of Cl^- uptake/base secretion in FW gills (Krogh, 1937, 1938). Pharmacological evidence supported this notion (Perry et al., 2003a; Chang and Hwang, 2004; Parks et al., 2009; Hwang and Perry, 2010); however, early molecular evidence of the existence of AE1 (an SLC4 member) in gill cells by immunocytochemistry was scarce and debatable due to the antibody specificity (Wilson et al., 2000; Tresguerres et al., 2006a). A recent double and triple protein/mRNA-labeling study clearly demonstrated the basolateral localization of AE1 in Na⁺ absorption/acid secretion cells, HR-type ionocytes, in zebrafish skin/gills (refer to Section 6.3.1). On the other hand, pendrin (an SLC26 protein) initially emerged as another candidate for the apical Cl^-/HCO_3^- exchange function based on the colocalization of apical pendrin and basolateral H⁺-ATPase in a group of ionocytes, which differed from another group with apical NHE3 and basolateral NKA, in gills of the euryhaline stingray (Piermarini et al., 2002). Recently, pendrin (SLC26A4) and mRNAs of two paralogues (SLC26A3 and SLC26A6) were localized in certain groups of gill ionocytes in an FW teleost, zebrafish, and only a small portion (<10%) of SLC26A3-expressing cells co-expressed basolateral NKA (Bayaa et al., 2009; Perry et al., 2009). The role of these transporters in the Cl^- uptake mechanism was demonstrated by finding Cl^- uptake defects in zebrafish embryos injected with specific morpholinos of SLC26A3, SLC26A4, and SLC26A6 (Perry et al., 2009), and low-Cl^- FW stimulation of mRNA expression of SLC26 members in both embryos and adult gills (Bayaa et al., 2009; Perry et al., 2009). However, the driving mechanism for apical Cl^-/HCO_3^- exchange in FW fish gill cells is still questionable. Investigation of the stoichiometry of SLC26 members may give a possible answer to this puzzle. On the other hand, SLC26 member-expressing ionocytes appear to be analogous to mammalian kidney non-A-type intercalated cells in terms of the apical localization of pendrin (Wagner et al., 2009). The expression and functioning of cytosolic CA and other Cl^- transporters and H⁺-ATPase in basolateral membranes of cells are necessary to achieve epithelial Cl^- uptake/base secretion functions, and this remains to be studied in the future.

6.3.4   Ca²⁺ Uptake

In FW teleosts, besides dietary intake, the gills are responsible for the majority (>97% of the whole body) of Ca²⁺ uptake from the aquatic environment for body fluid Ca²⁺ homeostasis (Flik et al., 1995). The current mammalian model for transcellular Ca²⁺ transport was also proposed in FW teleost gills (Flik et al., 1995; Hoenderop et al., 2005): Ca²⁺ is transported through apical epithelial Ca²⁺ channels (ECaC, TRPV5, and/or TRPV6) into ionocytes, while intracellular Ca²⁺ is bound to calbindins that facilitate diffusion to the basolateral membrane, and is extruded via the basolateral plasma membrane Ca²⁺-ATPase (PMCA) and/or Na⁺/Ca²⁺ exchanger (NCX). However, only recently, molecular evidence became available for the existence and functional roles of relevant Ca²⁺ transporters in FW fish gills. ECaC, a highly Ca²⁺-selective channel, was first cloned from mammal kidneys (Hoenderop et al., 1999). Fish ECaC was initially cloned from fugu (Qiu and Hogstrand, 2004), and thereafter from zebrafish (Pan et al., 2005) and trout (Shahsavarani et al., 2006), and recently from a marine shark (*Triakis scyllium*) (Takabe et al., 2012). Two basolateral transporter isoforms (PMCA2 and NCX1b) and an NKA α1 subunit subtype (ATP1a1a.2) were identified as being co-expressed with ECaC in a subset of ionocytes in zebrafish (Liao et al., 2007, 2009; Pan et al., 2005). In rainbow trout, gill mitochondrion-rich cells and PVCs were proposed as sites of Ca²⁺ uptake on the basis of ECaC expression in both cell types (Shahsavarani et al., 2006); however, the Ca⁺ uptake capacity of isolated mitochondrion-rich cells (PNA⁺ cells, refer to Section 6.3.1) was about threefold higher than that of PNA⁻ cells (presumably PVCs) (Galvez et al., 2006). Upregulation of ECaC (but not PMCA2 or NCX1b) mRNA expression and the cell number of ECaC-expressing ionocytes were found to be induced by low-Ca⁺ FW (Liao et al., 2007; Pan et al., 2005), and similarly, acclimation to soft water or an infusion of CaCl₂ also upregulated ECaC mRNA and/or protein expressions in trout gills (Shahsavarani et al., 2006). Loss-of-function experiments on zebrafish ECaC further provided convincing molecular physiological evidence to support the role of ECaC in FW gill Ca²⁺ uptake (Tseng et al., 2009). Based on kinetics experiments, extrusion mechanisms by PMCA and NCX were proposed to operate far below their maximum capacity in fish gills (Flik et al., 1997). Therefore, the ECaC may serve as a rate-limiting step or as a major regulatory transporter of the

Ca^{2+} uptake mechanism in FW fish gills as in mammal kidneys (Hoenderop et al., 2005). This was supported by recent molecular physiological studies. Knockdown of stanniocalcin, a hypocalcemic hormone, was found to stimulate ECaC expression without effects on PMCA, an NCX, in zebrafish embryos (Tseng et al., 2009). Similarly, incubation with cortisol, a hypercalcemic hormone, and knockdown of the glucocorticoid receptor showed effects on ECaC expression but not on the other two Ca^{2+} transporters in zebrafish (Lin et al., 2011).

6.4 AMMONIA EXCRETION IN FW AND SW GILLS

Ammonia is a nitrogenous product of amino acid metabolism. Ammonia exists as two distinct chemical species, dissolved ammonia gas (NH$_3$) and ammonium ions $\left(NH_4^+\right)$ with ratios instantaneously variable depending on the pH. Here, we use the conventional term "ammonia" when the chemical is not specified, and chemical symbols $\left(NH_3 \text{ or } NH_4^+\right)$ when the chemical is specified. Ammonia causes a number of toxic effects in animal systems including fishes. Most of the severe toxic effects of ammonia are on the central nervous system, where it causes excessive activation of NMDA receptors on neurons, and also causes swelling and death of astrocytes (Ip and Chew, 2010). Due to its toxicity, ammonia must be rapidly excreted or metabolized to less toxic nitrogen metabolites (such as urea) to maintain low concentrations in animal bodies. The majority of teleosts are ammonotelic, that is, they excrete most of their waste nitrogen as ammonia.

Ammonia excretion by fishes is largely accomplished by gills and larval skin where gas exchange and ion transport take place. Although the mechanism underlying the excretion of ammonia has been investigated for several decades, with the recent discovery of Rh proteins that facilitate ammonia diffusion across cell membranes, this field of research has opened a new avenue to reveal the mechanism and regulation of ammonia excretion and its associated roles in branchial Na$^+$ uptake and acid secretion (refer to Section 6.3.1). In addition, with application of the SIET technique in fish embryos and larvae, ammonia excretion by specific types of ionocytes has been well investigated.

The human Rh blood type antigen has long been linked to destructive antibody production; however, the role of Rh proteins in ammonia transport of erythrocytes (Rhag) and non-erythrocytes (Rhbg and Rhcg) has only been clearly revealed in the past decade. The functional properties of these Rh proteins were examined and comprehensively reviewed in several reports (Weiner and Hamm, 2007; Wright and Wood, 2009). Some in vitro studies with *Xenopus* oocytes expressing Rh proteins provided evidence to support Rh proteins being NH$_3$ gas channels (Mak et al., 2006) or electrogenic NH$_4^+$ transporters (Nakhoul et al., 2006). X-ray crystallographic studies (Khademi et al., 2004; Li et al., 2007; Lupo et al., 2007) analyzed the molecular structures of AmtB and NeRh, bacterial homologues of Rh proteins, and suggested that these proteins are ammonia channel proteins. Based on the molecular structure, a model was proposed whereby a 20-Å-long hydrophobic pore serves as a channel that has NH$_4^+$-binding sites (Li et al., 2007; Lupo et al., 2007). The first site facing the extracellular medium serves as a vestibule that recruits ammonia binding predominantly as charged NH$_4^+$. Once NH$_4^+$ is recruited, H$^+$ is released, and NH$_3$ is conducted through the hydrophobic pore. On the intracellular side, NH$_3$ recruits an intracellular H$^+$ and is released as NH$_4^+$.

In teleosts, four homologues of Rh proteins (Rhag, Rhbg, Rhcg1, and Rhcg2) were first identified in gills of pufferfish (*Takifugu rubripes*) (Nakada et al., 2007b). Heterologous expression of pufferfish or rainbow trout Rh proteins in *Xenopus* oocytes showed that they mediate ammonia transport (Nakada et al., 2007b; Nawata et al., 2010b). Following this finding in the pufferfish, Rh proteins were identified in gills of several species, including mangrove killifish (*Kryptolebias marmoratus*) (Hung et al., 2007), rainbow trout (Nawata et al., 2007), marine sculpin (Claiborne et al., 2008b), toadfish (*Opsanus beta*) (Weihrauch et al., 2009), weatherloach (*Misgurnus anguillicaudatus*) (Moreira-Silva et al., 2009), zebrafish (Nakada et al., 2007a), and Japanese medaka (Wu et al., 2010). However, cellular distributions of Rh isoforms in these species were not as clear as those in the pufferfish. With isoform-specific antibodies, localization of Rh proteins in gills of pufferfish was clearly revealed (Nakada et al., 2007b). Rhbg and Rhcg2 were respectively localized to the basolateral and apical membranes of PVCs, which cover over 90% of the surface area of fish gills. Rhag was localized to pillar cells lining the blood channels of gill lamellae. Interestingly, Rhcg1 was localized to apical membranes of ionocytes, suggesting that ionocytes are also involved in ammonia excretion.

In some studies, the mRNA/protein level of Rh proteins was found to correlate with ammonia excretion in fishes, supporting their role in ammonia excretion. In rainbow trout, Rhcg2 and Rhbg mRNA levels in gill and skin increased after exposure to high-ammonia water (Nawata et al., 2007). A similar upregulation of some Rh isoforms was also found in gills of mangrove killifish (Hung et al., 2007) and pufferfish (Nawata et al., 2010a). However, downregulation of Rh isoforms (Rhag and Rhbg) was also found in pufferfish exposed to high-ammonia water (Nawata et al., 2010a). The downregulation of Rh proteins might prevent or decrease ammonia back flux in high-ammonia water.

In addition to correlation and in vitro evidence, genetic-based loss-of-function studies provide convincing support for the function of Rh proteins in vivo. In a zebrafish study, Rhcg1 was localized to the apical membrane of a subtype of ionocytes, HR cells (Nakada et al., 2007a; Shih et al., 2008). Shih et al. used a morpholino knockdown technique to block the translation of Rhcg1 in larvae and demonstrated that Rhcg1 is involved in ammonia excretion by HR cells (Shih et al., 2008). Braun et al. further demonstrated that knockdown of various Rh isoforms equally suppressed about 50% of ammonia excretion in zebrafish larvae (Braun et al., 2009).

It has been generally accepted that FW fishes predominately excrete NH_3 gas instead of ionic NH_4^+ as early studies found that acidification of the boundary layer adjacent to the gill surface could facilitate the excretion of ammonia (Wright et al., 1989; Wright and Wood, 2009). The acidic layer at the gill surface increases the conversion of NH_3 to NH_4^+ and thus maintains a favorable blood-to-water NH_3 gradient. This process is called "acid-trapping" ammonia excretion. A study by Wright et al. on rainbow trout suggested that CO_2 excretion by gills is the major source of the acidic layer (Wright et al., 1989). In a recent study on zebrafish, Shih et al. used morpholino knockdown and bafilomycin to demonstrate that H^+ secretion by H^+-ATPase in HR cells provides an alternate source of surface acidification to promote ammonia excretion (Shih et al., 2008). In air-breathing weather-loach, H^+-ATPase and Rhcg1 were also colocalized to a group of ionocytes. Treatment with bafilomycin decreased ammonia excretion, also supporting H^+-ATPase being involved in acid-trapping ammonia excretion (Moreira-Silva et al., 2009). Apparently, the colocalization of Rhcg1 and H^+-ATPase in apical membranes of ionocytes seems to play a critical role in ammonia excretion.

In addition to H^+-ATPase, the NHE was also found in apical membranes of ionocytes (see Section 6.3.1) and was suggested to mediate acid-trapping ammonia excretion. In zebrafish larvae, knockdown of the expression of NHE3 or treatment with EIPA partially suppressed ammonia excretion (Shih et al., 2008, 2012). In Japanese medaka, the NHE3 and Rhcg1 were also identified in a major group of ionocytes, which do not have apical H^+-ATPase (see Section 6.3.1). Treatment with EIPA was found to suppress ammonia excretion by ionocytes in larval skin (Wu et al., 2010).

Importantly, the coupling of NHE3 and Rhcg1 not only mediates acid-trapping ammonia excretion but also suggests to mediate Na^+ uptake by ionocytes. In both zebrafish and medaka, Rhcg1 and NHE3 mRNA expressions were induced by low-Na^+ FW acclimation, suggesting that these two proteins are involved in Na^+ uptake (Wu et al., 2010; Shih et al., 2012). Using SIET to probe Na^+, NH_4^+, and H^+ gradients of ionocytes in the larval skin of medaka, Wu et al. found that Na^+ uptake and ammonia excretion by ionocytes are tightly associated (Wu et al., 2010). In addition, both Na^+ uptake and ammonia excretion were blocked by EIPA. Those studies (Wu et al., 2010; Shih et al., 2012) suggested that Rhcg1 and NHE3 might form a protein complex to achieve "ammonium-dependent sodium uptake." Rhcg1 facilitates NH3 diffusion and thus generates an H^+ gradient across the apical membrane of ionocytes, and the H^+ gradient may drive Na^+ uptake via the NHE3. Meanwhile, excreted H^+ combines with external NH_3 to form NH_4^+ and thus maintains the NH_3 gradient across the apical membrane (the acid-trapping mechanism). Recently, this model was further tested with genetic knockdown experiments in zebrafish embryos. Morpholino knockdown of either Rhcg1 or NHE3b decreased both NH_4^+ excretion and Na^+ uptake in zebrafish larvae acclimated to low-Na^+ water or acidic water (Kumai and Perry, 2011; Shih et al., 2012). Those studies provided convincing loss-of-function evidence for the ammonium-dependent Na^+ uptake mechanism. As mentioned in Section 6.3.1, the driving force of the NHE in gills of FW fishes has been questioned for several decades. This model may be a solution for the long-debated question.

Although the mechanism of ammonia excretion has not been intensively investigated in SW fishes, it has long been considered that the mechanism in SW fishes differs from that in FW fishes. Early studies suggested that ionic NH_4^+ might be excreted via a paracellular pathway in the gill epithelium of SW fishes, as the junction of their gill epithelium is leaky for Na^+ secretions (Wilkie, 2002). Moreover, acid-trapping ammonia excretion was questioned in SW fishes (Weihrauch et al., 2009). As the buffering capability of SW is much higher than that of FW, acid secretion from gills of SW fishes might not form an effective acidic layer for ammonia trapping. In recent studies, however, expression of Rhcg mRNA in gills/skin was induced by high-ammonia exposure in sculpin (*M. octodecemspinosus*) (Claiborne et al., 2008b) and the mangrove killifish in brackish SW (Hung et al., 2007). In pufferfish exposed to high-ammonia SW, gill Rhcg, H$^+$-ATPase, and NHE3 mRNAs were also elevated (Nawata et al., 2010a). These studies implied that ammonia excretion in SW fishes is also mediated by SW-type ionocytes by a mechanism similar to that of FW fishes. In the euryhaline climbing perch (*A. testudineus*), gill mRNA levels of NKA, NKCC, and CFTR were induced by extremely high ammonia exposure $\left(100 \text{ mM } NH_4^+\right)$, suggesting that these transporters are also involved in active ammonia excretion by ionocytes (Ip et al., 2012a,b; Loong et al., 2012).

6.5 IDENTIFICATION AND COMPARISON OF IONOCYTE SUBTYPES

Teleost gills are in some respects analogous to mammalian kidneys in terms of the transporter expression patterns and functions. Similar to mammalian kidneys, teleost gills have long been proposed to develop pleomorphic subtypes of ionocytes to achieve different ion transport functions. In earlier studies, different populations or subtypes of ionocytes were identified based on their anatomic locations in gills and/or the ultrastructural characteristics of cytosol or apical openings by electron microscopy, and most of these identified ionocytes lacked convincing cellular (e.g., localization of the ion transporters) and physiological (ion transport) evidence to support the proposed ionoregulatory pathways (Hirose et al., 2003; Hwang and Lee, 2007). Given the advanced molecular/cellular/physiological approaches, identification and functional analysis of ionocytes subtypes in different species were recently accelerated, and it has become feasible to precisely study target ion transporters and related transport functions at the molecular/cellular physiological levels, thus largely enhancing our further understanding of and providing new insights into fish ionoregulatory mechanisms and their functional regulation (Hwang et al., 2011; Dymowska et al., 2012; Hiroi and McCormick, 2012).

6.5.1 SW FISHES

Figure 6.1 shows the model of ionocytes in SW teleosts. NaCl secretion mechanisms in SW teleost gills, as described earlier, are relatively better understood than ion uptake mechanisms in FW ones. It is generally accepted that SW fish gills have only 1 type of ionocytes expressing a similar set of ion transporters (refer to Section 6.2.1) among most teleosts (Figure 6.1). The only type of ionocytes in SW teleost gills, responsible for NaCl secretion, acid secretion, and ammonia excretion, express CFTR, NHE2/3, ROMK (Kir1.1), and Rhcg1 in apical membranes, and NKCC, NKA, Kir5.1 (eKir), and Rhbg in basolateral membranes of cells (Figure 6.1). Accessory cells (ACs) with a less developed tubular system and lower expression of NKA were identified to be adjacent to SW ionocytes by sharing an apical crypt. Although the identity of ACs is still being debated (Evans et al., 2005),

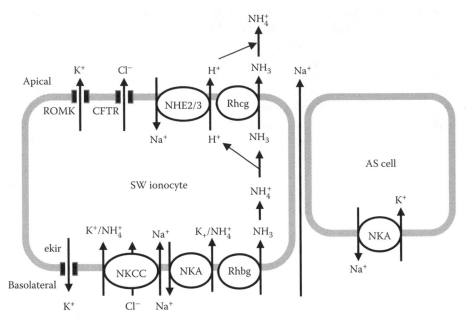

FIGURE 6.1 Model of ionocytes in seawater (SW) teleosts. For details, refer to the text (Sections 6.2.1 through 6.2.3). AC, accessory cell; CFTR, cystic fibrosis transmembrane conductance regulator; eKir, eel inwardly rectifying K+ channel (an orthologue of mammalian Kir5); NHE, Na+/H+ exchanger; NKA, Na+-K+-ATPase; NKCC 1, Na+-K+-2Cl− cotransporter; Rhcg/-bg, Rhesus glycoproteins; ROMK, an orthologue of the mammalian renal outer medullary K+ channel (Kir1.1); ion in dark gray, an unidentified transport pathway.

the multicellular complex of an AC and SW ionocyte is known to form apical membrane inter-digitations with leaky junctions, thus providing a paracellular route for Na+ extrusion in SW gills (Hootman and Philpott, 1980; Hwang and Hirano, 1985). Na+ extrusion through a paracellular route was recently demonstrated using SIET to detect Na+ outward currents only in a multicellular complex, but never in a single ionocyte without a neighboring accessory cell, in SW medaka (Shen et al., 2011). On the other hand, glycogen-rich cells, a recently identified gill cell type, were found to deposit a large amount of glycogen to provide the neighboring ionocyte with energy, and were proposed as a population of ACs based on the lack of NKA expression and sharing of an apical crypt with ionocytes (Chang et al., 2007; Tseng et al., 2007; Tseng and Hwang, 2008), although more detailed studies are required. Marine sculpin, another subtype of ionocyte, distinct from the traditional NKA-labeled SW ionocytes, expresses basolateral H+-ATPase, suggesting a possible role in base secretion (Catches et al., 2006); however, this needs to be clarified in other SW species.

6.5.2 FW FISHES

Ionoregulatory mechanisms in FW teleost gills appear to be more complicated than in SW ones as discussed earlier, and this is probably due to fish gills having to deal with FW environments, which are more diverse than SW ones in terms of ionic compositions, pH, hardness, and so on. Accordingly, it is not surprising to find more subtypes of ionocytes in FW gills than in SW ones. Given a variety of cell biological (localization of ion transporters and isolation of cells), molecular physiological (loss- or gain-of-function), electrophysiological, and pharmacological approaches, recent studies identified ionocyte subtypes according to criteria with greater physiological significance (Hwang et al., 2011; Dymowska et al., 2012; Hiroi and McCormick, 2012). In Figure 6.2, we summarized and compared the ionocyte subtypes in zebrafish, tilapia, and rainbow trout, which were selected because the related studies on these species are more comprehensive and convincing than those for other species.

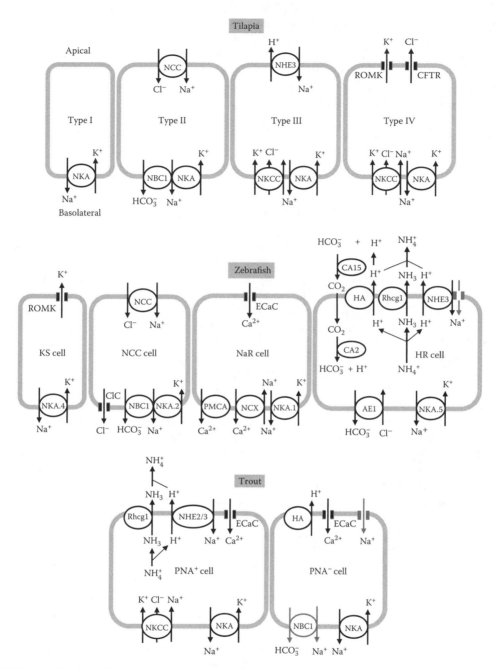

FIGURE 6.2 Models of ionocytes in different freshwater (FW) species. For details, refer to the text (Sections 6.3.1 through 6.3.4). Tilapia: type 1, II, III, and IV (i.e., SW ionocytes, refer to Figure 6.1) cells. Zebrafish: K⁺-secreting (KS), Na⁺-Cl⁻ cotransporter (NCC), H⁺-ATPase-rich (HR), and Na⁺-K⁺-ATPase-rich (NaR) cells. Rainbow trout: PNA⁺ and PNA⁻ cells. AE1, anion exchanger 1b; CA2, carbonic anhydrase 2-like a; CA15, carbonic anhydrase 2-like a (−15a); CFTR, cystic fibrosis transmembrane conductance regulator; ClC, Cl⁻ channel 3; ECaC, epithelial Ca^{2+} channel; HA, H⁺-ATPase; ROMK, an orthologue of the mammalian renal outer medullary K⁺ channel (Kir1.1); NBC1, $Na^+ - HO_3^-$ cotransporter (NBCe1b in zebrafish); NCX, Na^+/Ca^{2+} exchanger 1b; NHE, Na⁺/H⁺ exchanger (NHE3b in zebrafish and trout); NKA, Na⁺-K⁺-ATPase; NAK.15, NKA 1 subunit subtypes; NKCC, Na⁺-K⁺-2Cl⁻ cotransporter 1; PMCA, plasma membrane Ca^{2+}-ATPase 2; Rhcg1, Rhesus glycoprotein. Transporters indicated in dark gray had no convincing data of mRNA or protein localization.

In zebrafish, at least four subtypes of ionocytes were identified and functionally analyzed (Figure 6.2): HR cells, NKA-rich (NaR) cells, NCC-expressing (NCC) cells, and K$^+$-secreting (KS) cells. HR cells are responsible for Na$^+$ uptake/acid secretion/ammonia excretion/HCO$_3^-$ absorption by operations of apical H$^+$-ATPase, NHE3b, CA15a, and Rhcg1, cytosolic CA2-like a, and basolateral AE1b and NKA (ATP1a1a.5) (refer to Section 6.3.1). NaR cells mainly function in Ca^{2+} uptake via apical ECaC, and basolateral PMCA2/NCX1b/NKA (ATP1a1a.1) (refer to Section 6.3.4). NCC cells carry out Cl$^-$ and Na$^+$ uptake functions by NCC/NBCe1b/NKA (ATP1a1a.2) (refer to Section 6.3.2). On the other hand, KS cells show mRNA signals of the K$^+$ channel, Kir1.1 (ROMK). KS cells were proposed to function in K$^+$ secretion (Abbas et al., 2011); however, no direct evidence is available to support this function, and the physiological role of K$^+$ secretion in hyperosmoregulatory FW fishes is not known either. On the other hand, it is still unclear whether SLC26A3, SLC26A4, and SLC26A6 are expressed in the same type of cells or in one (or more) type of the identified ionocytes (HR, NaR, NCC, or KS cells) (refer to Section 6.3.3), and therefore the SL26-expressing cells are not presented in the proposed model (Figure 6.2).

Tilapia were also demonstrated to develop four subtypes (types I, II, III, and IV) of ionocytes (Figure 6.2). Type I ionocytes, expressing only basolateral NKA, were initially assumed to be immature ionocytes, but were recently confirmed to be an independent cell type with unknown function as they are distinguishable from newly emerging types II, III, and IV ionocytes (Hiroi and McCormick, 2012). Similar to zebrafish NCC cells, tilapia type II ionocytes mainly function in Cl$^-$/Na$^+$ uptake by operation of apical NCC and basolateral NBCe1, ClC3, and NKA (refer to Section 6.3.2). On the other hand, tilapia type III ionocytes are similar to zebrafish HR cells, as they carry out Na$^+$ uptake/acid secretion via apical NHE3 and basolateral NKCC1a and NKA (refer to Section 6.3.1). During transfer from FW to SW, type III ionocytes directly transform into type IV ionocytes, which are SW-type cells, as described earlier in terms of the ion transporter expression pattern and NaCl secretion function (refer to Section 6.2.1).

Rainbow trout are one of the most important model species and have provided basic knowledge of fish iono- and osmoregulatory mechanisms for many decades. Two subtypes of ionocytes were identified and functionally analyzed; however, there are some conflicts in the current model. In the generally accepted model, FW rainbow trout develop PNA$^+$ and PNA$^-$ subtypes of ionocytes, which can be distinguished from isolated gill cells using PNA (Galvez et al., 2002). PNA$^-$ cells are involved in bafilomycin-sensitive acid secretion/Na$^+$ uptake through apical H$^+$-ATPase and ENaC, and basolateral NBC and NKA, while PNA$^+$ cells carry out Cl$^-$ uptake/base secretion and Ca^{2+} uptake through apical AE and ECaC and basolateral H$^+$-ATPase and NKA (also refer to Sections 6.3.1, 6.3.3, and 6.3.4) (Reid et al., 2003; Galvez et al., 2006; Perry and Gilmour, 2006; Shahsavarani et al., 2006; Parks et al., 2007, 2008; Evans, 2008). In a recent review by Wright and Wood (2009), an apical "Na$^+$/NH$_4^+$ exchange complex" consisting of H$^+$-ATPase, NHE2/3, ENaC, and Rhcg working together as a metabolon achieved apical acid secretion/Na$^+$ uptake/ammonia excretion mechanisms in FW rainbow trout gill ionocytes. However, this proposed metabolon appears to conflict with a recent study demonstrating NHE2/3 localization only in PNA$^+$ cells that do not express H$^+$-ATPase (Ivanis et al., 2008). Furthermore, the ECaC protein is restricted to PNA$^+$ cells, as assumed by earlier models, but also extends to PNA$^-$ cells (Shahsavarani et al., 2006). In Figure 6.2, the proposed model of FW trout gill ionocytes mainly presents related transporters and pathways with convincing supporting localization evidence. Indeed, a very recent study by Hiroi and McCormick (Hiroi and McCormick, 2012) using triple labeling demonstrated two distinct subtypes of ionocytes in FW rainbow trout gills: one (presumably PNA$^+$ cells) expressed apical NHE3b and basolateral NKA and NKCC1, while the other showed only basolateral NKA signals. The roles of the two ionocyte subtypes in acid secretion/Na$^+$ uptake functions need to be reevaluated. Apparently, more detailed colocalization experiments on related transporters, including ENaC, AE, NBC, Rhcg, and others, are urgently required to clarify the proposed transport pathways in FW trout gill ionocytes.

The diversity of ionocyte subtypes among species has been reasonably ascribed to species differences (Dymowska et al., 2012; Hiroi and McCormick, 2012). As discussed earlier, some differences among species were noted to merely be due to different experimental approaches (e.g., limitations or inefficiencies of the experimental methods) used in those species. On the other hand, generalizing a comprehensive working model of FW teleost ionoregulatory mechanisms by comparing different species would not only be of physiological and evolutionary significance, but also helpful to accelerate our understanding of related issues by precisely studying the detailed transport mechanism of an ion or ionocyte subtype.

6.6 FUNCTIONAL MODIFICATION AND REGULATION OF IONOCYTES

6.6.1 REGULATION AT THE TRANSPORTER LEVEL

6.6.1.1 Regulation of NKA by FXYD

Regulation of NKA activity and kinetics to meet the demands of fish ionocytes is essential for an enzyme to execute its physiological functions. FXYD proteins, small single-transmembrane proteins, were proposed to be auxiliary regulatory subunits of NKA and were recently implicated in ionic regulation in teleost fishes (Hwang and Lee, 2007). Teleost FXYD (FXYD9) was first identified in gills of the euryhaline grass pufferfish, and its expression level was shown to be salinity-dependent (higher in FW than in SW) (Wang et al., 2008). Meanwhile, Tipsmark also identified eight FXYD isoforms (i.e., FXYD2, FXYD5, FXYD6, FXYD7, FXYD8, FXYD9, FXYD11, and FXYD12) in various tissues of Atlantic salmon (Tipsmark, 2008). Among these isoforms, FXYD11 was predominantly expressed in gills. Immunohistochemical and co-immunoprecipitation studies demonstrated protein interactions between FXYD and NKA in ionocytes (Wang et al., 2008; Tipsmark et al., 2010). In zebrafish, FXYD11 mRNA/protein was identified in NaR cells but not HR cells (Saito et al., 2010). FXYD11 mRNA expression was induced by diluted FW, and a close association between NKA and FXYD11 was demonstrated with an in situ proximity ligation assay (Saito et al., 2010). Knockdown of FXYD11 translation resulted in an increase in the number of NaR cells in larval skin, suggesting that impairment of NKA might lead to feedback upregulation of NKA expression and NaR cell differentiation (Saito et al., 2010).

6.6.1.2 Regulation of NKCC

Branchial NKCC1 plays a critical role in Cl^- secretion by SW fishes, and its expression level changes quickly upon salinity acclimation. It was suggested that short-term (non-genomic) regulation of NKCC was via phosphorylation and dephosphorylation of NKCC in cells responding to osmotic shocks (Hoffmann et al., 2007). Several reports found that branchial NKCC1 protein levels remained constant in teleosts after being transferred from SW to FW for 1 week (Tipsmark et al., 2002; Lorin-Nebel et al., 2006; Kang et al., 2010), implying that NKCC1 was inactivated but not degraded during functional modification. Kang et al. suggested that the inactive NKCC1 maintained in FW-acclimated brackish medaka was involved in the endurance of the hypo-osmoregulatory ability (Kang et al., 2010). Recently, Flemmer et al. found that within 1 h of transfer of killifish to higher- or lower-salinity water, phosphorylated NKCC1 respectively increased or decreased (Flemmer et al., 2010). They also suggested that the phosphorylation of NKCC is mediated by a cAMP–protein kinase A pathway.

6.6.1.3 Regulation of CFTR

In killifish subjected to SW challenge, CFTR protein levels and the short-circuit current measured in opercular membranes were upregulated within 1 h (Shaw et al., 2008). In contrast, when killifish were transferred from SW to FW, the CFTR immunostaining signals in gill ionocytes disappeared within 1 day (Katoh and Kaneko, 2003). In sea bass (*Dicentrarchus labrax*) transferred from SW to FW, branchial CFTR mRNA was downregulated within 1 day, followed by a decrease in the CFTR protein level over 7 days (Bodinier et al., 2009). Taken together, these studies suggest genomic regulation of CFTR in fish gills (Bodinier et al., 2009). On the other hand, Marshall et al. found that CFTR in opercular membranes of killifish was activated through focal adhesion kinase (FAK)-mediated protein phosphorylation, suggesting the short-term, non-genomic regulation of CFTR in response to osmotic stress (Marshall et al., 2009). A recent study reported that rapid acclimation of killifish to SW is mediated by trafficking of CFTR from intracellular vesicles to plasma membranes in opercular membranes within the first hour in SW (Shaw et al., 2008). Acute transfer of killifish to SW also caused increases in both mRNA and protein levels of serum glucocorticoid kinase 1 (SGK1) within 15 min of transfer. Using morpholinos to knock down SGK1 in adult gills, Notch et al. demonstrated that the increase in SGK1 protein was required for the trafficking of CFTR from intracellular vesicles to plasma membranes of ionocytes (Notch et al., 2011) and mitogen-activated protein kinase (MAPK) 14 was involved in the activation of SGK1 (Notch et al., 2012).

6.6.2 Regulation at the Ionocyte Level

6.6.2.1 Transformation of FW and SW Ionocytes

Although FW- and SW-type ionocytes share some morphological characteristics, their function and mechanism in ion transport greatly differ. Activating the salt secretion function of SW-type ionocytes and simultaneously suppressing the salt absorption function of FW-type cells, or vice versa, during acute salinity challenge in a timely and sufficient manner are critical for euryhaline teleosts to adapt to salinity changes. However, whether the rapid appearance of FW- and SW-type ionocytes originates from stem cells (or undifferentiated cells) or directly from the transformation of preexisting FW-type (or SW-type) cells has been a challenging and important issue. This issue remained a puzzle until some elegant and precise studies were carried out (Hiroi et al., 1999; Choi et al., 2010; Shen et al., 2011; Hiroi and McCormick, 2012; Inokuchi and Kaneko, 2012). Following sequential in vivo observations by confocal microscopy, 75% of skin ionocytes in tilapia larvae survived for 96 h after transfer from FW to SW, suggesting that the acute functional and morphological changes upon salinity challenge occurred in preexisting ionocytes, and they were mostly not associated with the recruitment of new ionocytes (Hiroi et al., 1999). This notion was supported by the appearance of changes in the cell size (Hiroi et al., 1999) and apical morphology (from convex/concave to pit types) (Choi et al., 2010) within 1 day of salinity challenge. The transformation from type III ionocytes (FW type, with NHE3 used as the marker transporter) to type IV (SW type, with NKCC used as the marker transporter) reflects a functional change from salt uptake to salt secretion during acute salinity challenge. A recent report on ionocytes in the skin of medaka larvae further provided convincing evidence for rapid functional changes by ionocytes (Shen et al., 2011). Shen et al. used SIET to monitor Na+ and Cl− transport by skin ionocytes. They distinguished multicellular-complex ionocytes from single ionocytes by microscopic observations of living animals. They showed that only multicellular-complex ionocytes secreted Na+ in SW-acclimated larvae, whereas both multicellular-complex and single ionocytes secreted Cl−. This direct evidence of differential transport activity adds physiological significance to the morphological distinctions among ionocytes. More importantly, they monitored ion transport by ionocytes in intact larvae following abrupt salinity changes and revealed that ionocytes displayed functional plasticity, altering Na+ and Cl− transport activities between FW and SW patterns within 6 h. Supporting the findings in medaka, acute transfer of tilapia from FW to 70% SW enhanced apoptosis of FW-type ionocytes and stimulated the recruitment of multicellular-complex SW ionocytes in gills 1–3 days after the transfer (Inokuchi and Kaneko, 2012).

6.6.2.2 Regulation of the Proliferation and Differentiation of Ionocytes

One strategy for teleosts to cope with a changing environment is to increase gill ionocyte numbers through cell renewal and proliferation, thus enhancing the overall ionoregulatory function. Early studies using cell biological approaches demonstrated stimulation of mitotic activity or turnover in gill ionocytes during salinity challenge (Conte and Lin, 1967; Chretien and Pisam, 1986; Wong and Chan, 1999). In tilapia embryonic skin examined by sequential in vivo observations, however, ionocytes showed no change in the turnover rate with 3 days of salinity treatment or with ~1–2 days of artificial FW with different Cl⁻ levels (Hiroi et al., 1999; Lin and Hwang, 2004). On the other hand, zebrafish embryos exhibited enhanced proliferation of ionocyte stem cells (using the transcription factor p63 as a marker) without an effect on cell apoptosis (according to a transferase dUTP nick end labeling [TUNEL] assay), resulting in increased HR cell numbers and their acid-secretion function, after 4 days of acclimation to acidic FW (Horng et al., 2009b). Regulation of ionocytes at the cellular level appears to be species- and/or environment-dependent.

Whether the increase in ionocytes during acclimation to a changing environment results from stimulation of stem cell proliferation, ionocyte differentiation (final maturation), or both was unclear until recent studies that dissected the molecular pathways for the differentiation and specification of ionocytes. In the proposed model, skin/gill ionocytes differentiate from the same stem cells by differential determinations of several transcription factors, FOXI3a/-b and GCM2 (Esaki et al., 2007, 2009; Hsiao et al., 2007; Janicke et al., 2007; Chang et al., 2009; Thermes et al., 2010; Chang and Hwang, 2011). This model opens a new avenue to further explore molecular and cellular mechanisms of the functional regulation of ionocytes. To cope with acidic environments, zebrafish embryos enhance their acid-secretion ability by stimulating acid-secreting functions (according to SIET) of single HR cells and also by increasing HR cell numbers that originate from stem cell proliferation and the terminal differentiation of HR cells by GCM2 (Chang et al., 2009; Horng et al., 2009b). On the other hand, zebrafish gill ionocytes enhance ionoregulatory functions to compensate for cold-induced ionic imbalances. This functional compensation is the result of an extended lifespan (by delaying apoptosis) and sustained ionocyte functions (by the FOXI3a-mediated stimulation of preexisting progenitor cells into ionocytes) (Chou et al., 2008). Apparently, fishes adopt different strategies, that is, adjusting the function of each ionocyte, stem cell proliferation, ionocyte differentiation/maturation, apoptosis, or all of these steps, to regulate overall gill ionoregulatory functions to cope with diverse environments.

6.6.3 REGULATION OF IONIC PERMEABILITY

Paracellular permeability characteristics of the fish gill epithelium are generally accepted to play a critical role in fish ion regulation. In SW fishes, the gill is "leaky" for paracellular Na⁺ secretion; in FW fishes, the gill is "tight" to limit passive ion loss (Hwang and Hirano, 1985; Hwang and Lee, 2007). Determination of the paracellular permeability of fish gill is generally linked to properties of the tight junction (TJ) protein complex located on the apical lateral sides of epithelial cells (Anderson and van Itallie, 2008). Increased permeability across the gill epithelium of fishes in SW is attributed to the presence of shallow "leaky" TJ connecting ACs with SW ionocytes. However, it should be noted that in the remainder of the SW fish gill epithelium, deep TJs link ionocytes and ACs with adjacent PVCs, and PVCs to adjacent PVCs, suggesting that the majority of the SW fish gill epithelium is not "leaky" (Sardet et al., 1979; Chasiotis et al., 2012a,b).

Occludin was the first transmembrane TJ protein identified and isolated from vertebrate epithelia (Furuse et al., 1993) and is the most comprehensively investigated TJ protein in teleosts. Expression patterns in fishes showed widespread occludin mRNA distributions among various tissues (Kumai and Perry, 2011; Chasiotis et al., 2010; Chasiotis and Kelly, 2011). However, gill tissues exhibit the highest levels of occludin mRNA, suggesting an important role for occludin in regulating branchial permeability (Chasiotis et al., 2010; Chasiotis and Kelly, 2011). Immunohistochemical experiments showed that occludin was distributed along the edges of lamellae (PVCs) and ionocytes of goldfish (*Carassius auratus*) gills (Chasiotis and Kelly, 2008). This observation was later confirmed by a goldfish gill cell separation study, in which comparable levels of occludin mRNA were demonstrated between isolated PVC and ionocyte fractions (Chasiotis et al., 2012b). Acclimation of goldfish to ion-poor water significantly increased occludin mRNA and protein abundance in gills (Chasiotis et al., 2009, 2012b), and the increased gill occludin abundance occurred in both PVCs and ionocytes (Chasiotis et al., 2012b). Moreover, Whitehead et al. used microarray analyses to show that the gill occludin mRNA abundance was upregulated in SW populations of killifish following hypoosmotic challenge (Whitehead et al., 2011). Those studies suggest that occludin probably contributes to a "tightening" of the gill epithelium.

The transmembrane protein, claudin, is believed to be the main structural and functional component of TJs (Kraemer et al., 2008). In mammals, about 24 claudin members were identified (van Itallie and Anderson, 2006). However, in fishes, about 57 members were identified because of tandem gene duplication and/or whole-genome duplication events, which were probably induced by the diversity of aquatic environments (Loh et al., 2004). Based on the "tight" versus "leaky" paradigm that has been proposed to exist between TJs of FW versus SW teleost gill epithelia, several studies provided new insights into potential roles of various claudins in euryhaline fish gills. For example, pufferfish claudin-3a, claudin-3c, and claudin-8d; killifish claudin-3 and claudin-4; and flounder and tilapia claudin-3-like and claudin-4-like mRNA and protein levels were found to be higher in FW than in SW (Bagherie-Lachidan et al., 2008, 2009; Tipsmark et al., 2008a,b; Duffy et al., 2011; Whitehead et al., 2011). SW-induced (10e) and FW-induced (27a and 30) claudins in salmon gills were all stimulated by cortisol (Tipsmark et al., 2009). These salinity-dependent expressions indicate that claudins might be vital to permeability changes associated with salinity adaptation and likely the formation of deeper TJs in FW gills of euryhaline teleosts.

So far, there has been no loss-of-function study conducted to demonstrate the roles of TJ proteins in fish gills. In the future, the functions of these proteins need to be further investigated with zebrafish genetic knockdown approaches.

6.7 CONCLUSIONS AND PERSPECTIVES

In SW teleosts, NaCl secretion and the related ion transporters (CFTR, NKCC, NKA, and eKir) have been clearly identified and functionally characterized. Compared to this, ionic regulatory mechanisms in FW teleost gills are still being explored, and many issues are being debated, because the mechanisms in FW gills are more complicated. With the availability of advanced cellular/ molecular physiological approaches and model animals, research on gill ionic uptake and acid–base regulation mechanisms in teleost gills is accelerating, and recent progress has boosted our knowledge in this field. As for ion transporters, related proteins, and transport pathways, there are K^+ excretion, a SW-specific NKA isoform, FXYD, FAK, SGK, and claudin, for example, in SW gills, and there are more in FW gills, including PMCA2, NCX1b, NCC, AE1, NBCe1, ClC3, SLC26a members, ROMK, Rh proteins, FOXI3, GCM2, and so on. More importantly, convincing evidence has emerged to answer several long-term debates or puzzling questions: for example, the metabolon of NHE/Rhcg1 providing chemical gradients to drive Na^+ uptake in FW gills, multiple labeling of transporters to precisely identify ionocyte subtypes, SIET real-time tracing of the functional plasticity of ionocytes during salinity changes, loss- (or gain-)of-function approaches providing direct evidence for the function of a transporter (or ionocyte), the molecular model of ionocyte differentiation opening a new window to study functional regulation, and so on. On the other hand, identification and comparison of ionocyte subtypes among species are not only of physiological and evolutionary significance, but also helpful in accelerating our understanding of related issues by precisely studying the detailed transport mechanisms of an ion or ionocyte subtype.

ACKNOWLEDGMENT

We thank Y. C. Tung for her technical and secretarial assistance.

REFERENCES

Abbas L, Hajihashemi S, Stead LF, Cooper GJ, Ware TL, Munsey TS, Whitfield TT, White SJ. 2011. Functional and developmental expression of a zebrafish Kir1.1 (ROMK) potassium channel homologue Kcnj1. *J Physiol* 589:1489–1503.

Al-Fifi ZIA. 2006. Studies of some molecular properties of the vacuolar H^+-ATPase in rainbow trout (*Oncorhynchus mykiss*). *Biotechnology* 5:455–460.

Anderson JM, van Itallie CM. 2008. Tight junctions. *Curr Biol* 18:R941–R943.

Avella M, Bornancin M. 1989. A new analysis of ammonia and sodium transport through the gills of the freshwater rainbow trout (*Salmo gairdneri*). *J Exp Biol* 142:155–175.

Bagherie-Lachidan M, Wright SI, Kelly SP. 2008. Claudin-3 tight junction proteins in *Tetraodon nigroviridis*: Cloning, tissue-specific expression, and a role in hydromineral balance. *Am J Physiol Regul* 294:R1638–R1647.

Bagherie-Lachidan M, Wright SI, Kelly SP. 2009. Claudin-8 and -27 expression in puffer fish (*Tetraodon nigroviridis*). *J Comp Physiol* 179B:419–431.

Bartels H, Potter IC. 2004. Cellular composition and ultrastructure of the gill epithelium of larval and adult lampreys implications for osmoregulation in fresh and seawater. *J Exp Biol* 207:3447–3462.

Bayaa MB, Vulesevic A, Esbaugh M, Braun ME, Ekker M, Perry SF. 2009. The involvement of SLC26 anion transporters in chloride uptake in zebrafish (*Danio rerio*) larvae. *J Exp Biol* 212:3283–3295.

Bodinier C, Boulo V, Lorin-Nebel C, Charmantier G. 2009. Influence of salinity on the localization and expression of the CFTR chloride channel in the ionocytes of *Dicentrarchus labrax* during ontogeny. *J Anat* 214:318–329.

Boisen AM, Amstrup J, Novak I, Grosell M. 2003. Sodium and chloride transport in soft water and hard water acclimated zebrafish (*Danio rerio*). *Biochim Biophys Acta* 1618:207–218.

Brauer PR, Sanmann JN, Petzel DH. 2005. Effects of warm acclimation on Na^+,K^+-ATPase alpha-subunit expression in chloride cells of antarctic fish. *Anat Rec* 285A:600–609.

Braun MH, Steele SL, Perry SF. 2009. Nitrogen excretion in developing zebrafish (*Danio rerio*): A role for Rh proteins and urea transporters. *Am J Physiol Renal Physiol* 296:F994–F1005.

Bystriansky JS, Richards JG, Schulte PM, Ballantyne JS. 2006. Reciprocal expression of gill Na^+/K^+-ATPase α1a and α1b during seawater acclimation of three salmonid fishes that vary in their salinity tolerance. *J Exp Biol* 209:1848–1858.

Catches JS, Burns JM, Edwards SL, Claiborne JB. 2006. Na^+/H^+ antiporter, V-H^+-ATPase and Na^+/K^+-ATPase immunolocalization in a marine teleost (*Myoxocephalus octodecemspinosus*). *J Exp Biol* 209:3440–3447.

Chang WJ, Horng JL, Yan JJ, Hsiao CD, Hwang PP. 2009. The transcription factor, glial cell missing 2, is involved in differentiation and functional regulation of H^+-ATPase-rich cells in zebrafish (*Danio rerio*). *Am J Physiol Regul* 296:R1192–R1201.

Chang IC, Hwang PP. 2004. Cl⁻ uptake mechanism in freshwater-adapted tilapia (*Oreochromis mossambicus*). *Physiol Biochem Zool* 77:406–414.

Chang WJ, Hwang PP. 2011. Development of zebrafish epidermis. *Birth Defects Res C* 93:205–214.

Chang IC, Wei YY, Chou FI, Hwang PP. 2003. Stimulation of Cl⁻ uptake and morphological changes in gill mitochondria-rich cells in freshwater tilapia (*Oreochromis mossambicus*). *Physiol Biochem Zool* 76:544–552.

Chang JCH, Wu SM, Tseng YC, Lee YC, Baba O, Hwang PP. 2007. Regulation of glycogen metabolism in gills and liver of the euryhaline tilapia (*Oreochromis mossambicus*) during acclimation to seawater. *J Exp Biol* 210:3494–3504.

Chasiotis H, Effendi JC, Kelly SP. 2009. Occludin expression in goldfish held in ion-poor water. *J Comp Physiol* 179B:145–154.

Chasiotis H, Kelly SP. 2008. Occludin immunolocalization and protein expression in goldfish. *J Exp Biol* 211:1524–1534.

Chasiotis H, Kelly SP. 2011. Permeability properties and occludin expression in a primary cultured model gill epithelium from the stenohaline freshwater goldfish. *J Comp Physiol* 181B:487–500.

Chasiotis H, Kolosov D, Bui P, Kelly SP. 2012a. Tight junctions, tight junction proteins and paracellular permeability across the gill epithelium of fishes: A review. *Respir Physiol Neurobiol* 184:269–281.

Chasiotis H, Kolosov D, Kelly SP. 2012b. Permeability properties of the teleost gill epithelium under ion-poor conditions. *Am J Physiol Regul* 302:R727–R739.

Chasiotis H, Wood CM, Kelly SP. 2010. Cortisol reduces paracellular permeability and increases occludin abundance in cultured trout gill epithelia. *Mol Cell Endocrinol* 323:232–238.

Choe KP, Kato A, Hirose S, Plata C, Sindic A, Romero MF, Claiborne JB, Evans DH. 2005. NHE3 in an ancestral vertebrate: Primary sequence, distribution, localization, and function in gills. *Am J Physiol Regul* 289:R1520–R1534.

Choi JH, Lee KM, Inokuchi M, Kaneko T. 2010. Morphofunctional modifications in gill mitochondria-rich cells of Mozambique tilapia transferred from freshwater to 70% seawater, detected by dual observations of whole-mount immunocytochemistry and scanning electron microscopy. *Comp Biochem Physiol* 158A:132–142.

Choi JY, Shah M, Lee MG, Schultheis PJ, Shull GE, Muallem S, Baum M. 2000. Novel amiloride-sensitive sodium-dependent proton secretion in the mouse proximal convoluted tubule. *J Clin Invest* 105:1141–1146.

Chou MY, Hsiao CD, Chen SC, Chen IW, Liu ST, Hwang PP. 2008. Hypothermic effects on gene expressions in zebrafish gills: Up-regulations in differentiation and function of ionocytes as compensatory responses. *J Exp Biol* 211:3077–3084.

Chretien M, Pisam M. 1986. Cell renewal and differentiation in the gill epithelium of fresh- or salt-water-adapted euryhaline fish (*Lebistes reticulatus*) as revealed by [H^3] thymidine radioautography. *Biol Cell* 56:137–150.

Claiborne JB, Choe KP, Morrison-Shetlar AI, Weakley JC, Havird J, Freiji A, Evans DH, Edwards SL. 2008a. Molecular detection and immunological localization of gill Na$^+$/H$^+$ exchanger in the dogfish (*Squalus acanthias*). *Am J Physiol Regul* 294:R1092–R1102.

Claiborne JB, Edwards SL, Morrison-Shetlar AI. 2002. Acid-base regulation in fishes: Cellular and molecular mechanisms. *J Exp Zool* 293:302–319.

Claiborne JB, Kratochvilova H, Diamanduros AW, Hall C, Phillips ME, Hirose S, Edwards S. 2008b. Expression of branchial Rh glycoprotein ammonia transporters in the marine longhorn sculpin (*Myoxocephalus octodecemspinosus*). *Bull Mt Desert Is Biol Lab Salisb Cove Maine* 47:67–68.

Conte FP, Lin DHY. 1967. Kinetics of cellular morphogenesis in gill epithelium during sea water adaptation of *Oncorhynchus* (walbaum). *Comp Biochem Pharmacol* 23:953–957.

Crombie HJ, Bell MV, Tytler P. 1996. Inhibition of sodium-plus-potassium stimulated adenosine triphosphatase (Na$^+$-K$^+$-ATPase) by protein kinase c activators in the gills of Atlantic cod (*Gadus morhua*). *Comp Biochem Physiol* 113B:765–772.

Cutler CP, Cramb G. 2002. Two isoforms of the Na$^+$/K$^+$/2Cl$^-$ cotransporter are expressed in the European eel (*Anguilla anguilla*). *Biochim Biophys Acta* 1566:92–103.

Cutler CP, Sanders IL, Hazon N, Cramb G. 1995. Primary sequence, tissue-specificity and expression of the Na$^+$,K$^+$-ATPase α-1 subunit in the European eel (*Anguilla anguilla*). *Comp Biochem Physiol* 111B:567–573.

D'Cotta H, Valotaire C, Le Gac F, Prunet P. 2000. Synthesis of gill Na$^+$-K$^+$-ATPase in Atlantic salmon smolts: Differences in α-mRNA and α-protein levels. *Am J Physiol Regul* 278:R101–R110.

Duffy NM, Bui P, Bagherie-Lachidan M, Kelly SP. 2011. Epithelial remodeling and claudin mrna abundance in the gill and kidney of puffer fish (*Tetraodon biocellatus*) acclimated to altered environmental ion levels. *J Comp Phsyiol B* 181:219–238.

Dymowska A, Hwang PP, Goss GG. 2012. Structure and function of ionocytes in the freshwater fish gill. *Respir Physiol Neurobiol* 184:249–257.

Edwards SL, Donald JA, Toop T, Donowitz M, Tse CM. 2002. Immunolocalisation of sodium/proton exchanger-like proteins in the gills of elasmobranchs. *Comp Biochem Physiol* 131A:257–265.

Edwards SL, Wall BP, Morrison-Shetlar A, Sligh S, Weakley JC, Claiborne JB. 2005. The effect of environmental hypercapnia and salinity on the expression of NHE-like isoforms in the gills of a euryhaline fish (*Fundulus heteroclitus*). *J Exp Zool* 303A:464–475.

Edwards SL, Weakley JC, Diamanduros AW, Claiborne JB. 2010. Molecular identification of Na$^+$-H$^+$ exchanger isoforms (NHE2) in the gills of the euryhaline teleost *Fundulus heteroclitus*. *J Fish Biol* 76:415–426.

Esaki M, Hoshijima K, Kobayashi S, Fukuda H, Kawakami K, Hirose S. 2007. Visualization in zebrafish larvae of Na$^+$ uptake in mitochondria-rich cells whose differentiation is dependent on FOXI3a. *Am J Physiol Regul* 292:R470–R480.

Esaki M, Hoshijima K, Nakamura N, Munakata K, Tanaka M, Ookata K, Asakawa K et al. 2009. Mechanism of development of ionocytes rich in vacuolar-type H$^+$-ATPase in the skin of zebrafish larvae. *Dev Biol* 329:116–129.

Evans DH. 2008. Teleost fish osmoregulation: What have we learned since August Krogh, Homer Smith, and Ancel Keys. *Am J Physiol Regul* 295:R704–R713.

Evans DH. 2011. Freshwater fish gill ion transport: August Krogh to morpholinos and microprobes. *Acta Physiologica* 202:349–359.

Evans DH, Claiborne JB. 2009. *Osmotic and Ionic Regulation in Fihses*. Evans DH, ed. Boca Raton, FL: CRC Press.

Evans DH, Piermarini PM, Choe KP. 2005. The multifunctional fish gill: Dominant site of gas exchange, osmoregulation, acid-base regulation, and excretion of nitrogenous waste. *Physiol Rev* 85:97–177.

Feng SH, Leu JH, Yang CH, Fang MJ, Huang CJ, Hwang PP. 2002. Gene expression of Na$^+$-K$^+$-ATPase α1 and α3 subunits in gills of the teleost *Oreochromis mossambicus*, adapted to different environmental salinities. *Mar Biotechnol* 4:379–391.

Fenwick JC, Wendelaar Bonga SE, Flik G. 1999. In vivo bafilomycin-sensitive Na$^+$ uptake in young freshwater fish. *J Exp Biol* 202:3659–3666.

Flemmer AW, Monette MY, Dyurisic M, Dowd B, Darman R, Gimenez I, Forbush B. 2010. Phosphorylation state of the Na$^+$-K$^+$-Cl$^-$ cotransporter (nkcc1) in the gills of Atlantic killifish (*Fundulus heteroclitus*) during acclimation to water of varying salinity. *J Exp Biol* 213:1558–1566.

Flik G, Kaneko T, Greco AM, Li J, Fenwick JC. 1997. Sodium dependent ion transporters in trout gills. *Fish Physiol Biochem* 17:385–396.

Flik G, Verbost PM, Wendelaar Bongar SE. 1995. *Calcium Transport Process in Fishes*. Wood CM, Shuttleworth TJ, eds. San Diego, CA: Academic Press.

Furukawa F, Watanabe S, Inokuchi M, Kaneko T. 2011. Responses of gill mitochondria-rich cells in Mozambique tilapia exposed to acidic environments (ph 4.0) in combination with different salinities. *Comp Biochem Physiol* 158A:468–476.

Furukawa F, Watanabe S, Kimura S, Kaneko T. 2012. Potassium excretion through ROMK potassium channel expressed in gill mitochondrion-rich cells of Mozambique tilapia. *Am J Physiol Regul* 302:R568–R576.

Furuse M, Hirase T, Itoh M, Nagafuchi A, Yonemura S, Tsukita S. 1993. Occludin: A novel integral membrane protein localizing at tight junctions. *J Cell Biol* 123:1777–1788.

Galvez F, Reid SD, Hawkings G, Goss GG. 2002. Isolation and characterization of mitochondria-rich cell types from the gill of freshwater rainbow trout. *Am J Physiol Regul* 282:R658–R668.

Galvez F, Wong D, Wood CM. 2006. Cadmium and calcium uptake in isolated mitochondria-rich cell populations from the gills of the freshwater rainbow trout. *Am J Physiol Regul* 291:R170–R176.

Gilmour KM, Perry SF. 2009. Carbonic anhydrase and acid-base regulation in fish. *J Exp Biol* 212:1647–1661.

Guynn SR, Scofield MA, Petzel DH. 2002. Identification of mrna and protein expression of the Na/K-ATPase $\alpha1$-, $\alpha2$- and $\alpha3$-subunit isoforms in Antarctic and New Zealand nototheniid fishes. *J Exp Mar Biol Ecol* 273:15–32.

Halstead LB. 1985. The vertebrates invasion of freshwater. *Philos Trans R Soc Lon Ser B Biol Sci* 309:243–258.

Hirata T, Kaneko T, Ono T, Nakazato T, Furukawa N, Hasegawa S, Wakabayashi S et al. 2003. Mechanism of acid adaptation of a fish living in a pH 3.5 lake. *Am J Physiol Regul* 284:R1199–R1212.

Hiroi J, Kaneko T, Tanaka M. 1999. In vivo sequential changes in chloride cell morphology in the yolk-sac membrane of Mozambique tilapia (*Oreochromis mossambicus*) embryos and larvae during seawater adaptation. *J Exp Biol* 202:3485–3495.

Hiroi J, McCormick SD. 2007. Variation in salinity tolerance, gill Na$^+$/K$^+$-ATPase, Na$^+$/K$^+$/2Cl$^-$ cotransporter and mitochondria-rich cell distribution in three salmonids *Salvelinus namaycush, Salvelinus fontinalis* and *Salmo salar*. *J Exp Biol* 210:1015–1024.

Hiroi J, McCormick SD. 2012. New insights into gill ionocyte and ion transporter function in euryhaline and diadromous fish. *Respir Physiol Neurobiol* 184:257–268.

Hiroi J, McCormick SD, Ohtani-Kaneko R, Kaneko T. 2005. Functional classification of mitochondrion-rich cells in euryhaline Mozambique tilapia (*Oreochromis mossambicus*) embryos, by means of triple immunofluorescence staining for Na$^+$/K$^+$-ATPase, Na$^+$/K$^+$/2Cl$^-$ cotransporter and cftr anion channel. *J Exp Biol* 208:2023–2036.

Hiroi J, Yasumasu S, McCormick SD, Hwang PP, Kaneko T. 2008. Evidences for an apical Na$^+$-Cl$^-$ cotransporter involved in ion uptake in a teleost fish. *J Exp Biol* 211:2584–2599.

Hirose S, Kaneko T, Naito N, Takei Y. 2003. Molecular biology of major components of chloride cells. *Comp Biochem Physiol* 136B:593–620.

Hoenderop JGJ, van der Kemp AW, Hartog A, van der Graaf SFJ, van Os CH, Willems PHGM, Bindels RJM. 1999. Molecular identification of the apical Ca^{2+} channel in 1,25-dihydroxyvitamin d3-responsive epithelia. *J Biol Chem* 274:8375–8378.

Hoenderop JG, Nilius B, Bindels RJ. 2005. Calcium absorption across epithelia. *Physiol Rev* 85:373–422.

Hoffmann EK, Schettino T, Marshall WS. 2007. The role of volume-sensitive ion transporter systems in regulation of epithelial transport. *Comp Biochem Physiol* 148A:29–43.

Hootman SR, Philpott CW. 1980. Accessory cells in teleost branchial epithelium. *Am J Physiol Regul* 238:R199–R206.

Horng JL, Hwang PP, Shih TH, Wen ZH, Lin CS, Lin LY. 2009a. Chloride transport in mitochondrion-rich cells of euryhaline tilapia (*Oreochromis mossambicus*) larvae. *Am J Physiol Regul* 297:C845–C854.

Horng JL, Lin LY, Huang CJ, Katoh F, Kaneko T, Hwang PP. 2007. Knockdown of v-ATPase subunit a (ATP6v1a) impairs acid secretion and ion balance in zebrafish (*Danio rerio*). *Am J Physiol Regul* 292:R2068–R2076.

Horng JL, Lin LY, Hwang PP. 2009b. Functional regulation of H$^+$-ATPase-rich cells in zebrafish embryos acclimated to an acidic environment. *Am J Physiol Regul* 296:C682–C692.

Hsiao CD, You MS, Guh YJ, Ma M, Jiang YJ, Hwang PP. 2007. A positive regulatory loop between FOXI3a and FOXI3b is essential for specification and differentiation of zebrafish epidermal ionocytes. *PLoS One* 2:e302.

Hung CYC, Tsui KNT, Wilson JM, Nawata CM, Wood CM, Wright PA. 2007. Rhesus glycoprotein gene expression in the mangrove killifish *Kryptolebias marmoratus* exposed to elevated environmental ammonia levels and air. *J Exp Biol* 210:2419–2429.

Hwang PP. 2009. Ion uptake and acid secretion in zebrafish (*Danio rerio*). *J Exp Biol* 212:1745–1752.

Hwang PP, Fang MJ, Tsai JC, Huang CJ, Chen ST. 1998. Expression of mRNA and protein of Na$^+$-K$^+$-ATPase alpha subunit in gills of tilapia (*Oreochromis mossambicus*). *Fish Physiol Biochem* 18:363–373.

Hwang PP, Hirano R. 1985. Effects of environmental salinity on intercellular organization and junctional structure of chloride cells in early stages of teleost development. *J Exp Zool* 236:115–126.

Hwang PP, Lee TH. 2007. New insights into fish ion regulation and mitochondrion-rich cells. *Comp Biochem Physiol A* 148:479–497.

Hwang PP, Lee TH, Lin LY. 2011. Ion regulation in fish gills: Recent progress in the cellular and molecular mechanisms. *Am J Physiol Regul* 301:R28–R47.

Hwang PP, Perry SF. 2010. *Ionic and Acid-Base Regulation.* Perry SF, Ekker M, Farrell AP, Brauner CJ, eds. San Diego, CA: Academic Press.

Inokuchi M, Hiroi J, Watanabe S, Hwang PP, Kaneko T. 2009. Morphological and functional classification of ion-absorbing mitochondria-rich cells in the gills of Mozambique tilapia. *J Exp Biol* 212:1003–1010.

Inokuchi M, Hiroi J, Watanabe S, Lee KM, Kaneko T. 2008. Gene expression and morphological localization of NHE3, NCC and NKCC1a in branchial mitochondria-rich cells of Mozambique tilapia (*Oreochromis mossambicus*) acclimated to a wide range of salinities. *Comp Biochem Physiol* 151A:151–158.

Inokuchi M, Kaneko T. 2012. Recruitment and degeneration of mitochondrion-rich cells in the gills of Mozambique tilapia *Oreochromis mossambicus* during adaptation to a hyperosmotic environment. *Comp Biochem Physiol A Mol Integr Physiol* 162(3):245–251.

Ip YK, Chew SF. 2010. Ammonia production, excretion, toxicity, and defense in fish: A review. *Front Physiol* 1:134.

Ip YK, Loong AM, Kuah JS, Sim EW, Chen XL, Wong WP, Lam SH, Delgado IL, Wilson JM, Chew SF. 2012a. Roles of three branchial Na+-K+-ATPase alpha-subunit isoforms in freshwater adaptation, seawater acclimation, and active ammonia excretion in *Anabas testudineus*. *Am J Physiol Regul* 303:R112–R125.

Ip YK, Wilson JM, Loong AM, Chen XL, Wong WP, Delgado IL, Lam SH, Chew SF. 2012b. Cystic fibrosis transmembrane conductance regulator in the gills of the climbing perch, *Anabas testudineus*, is involved in both hypoosmotic regulation during seawater acclimation and active ammonia excretion during ammonia exposure. *J Comp Physiol* 182B:793–812.

van Itallie CM, Anderson JM. 2006. Claudins and epithelia paracellular transport. *Annu Rev Physiol* 68:403–429.

Ito Y, Kobayashi S, Nakamura N, Esaki M, Hoshijima K, Hirose S. 2013. Close association of carbonic anhydrase (CA2a and CA15a), Na+/H+ exchanger (NHE3b), and ammonia transporter Rhcg1 in zebrafish ionocytes responsible for Na+ uptake. *Front Physiol* 4:59.

Ivanis G, Esbaugh AJ, Perry SF. 2008. Branchial expression and localization of SLC9a2 and SLC9a3 sodium/hydrogen exchangers and their possible role in acid-base regulation in freshwater rainbow trout (*Oncorhynchus mykiss*). *J Exp Biol* 211:2467–2477.

Janicke M, Carney TJ, Hammerschmidt M. 2007. FOXI3 transcription factors and notch signaling control the formation of skin ionocytes from epidermal precursors of the zebrafish embryo. *Dev Biol* 307:258–271.

Jentsch TJ, Poet M, Fuhrmann JC, Zdebik AA. 2005. Physiological functions of CLC Cl− channels gleaned from human genetic disease and mouse models. *Annu Rev Physiol* 67:779–807.

Kang CK, Tsai HJ, Liu CC, Lee TH, Hwang PP. 2010. Salinity-dependent expression of a Na+, K+, 2Cl− cotransporter in gills of the brackish medaka *Oryzias dancena*: A molecular correlate for hyposmoregulatory endurance. *Comp Biochem Physiol* 157A:7–18.

Katoh F, Kaneko T. 2003. Short-term transformation and long-term replacement of branchial chloride cells in killifish transferred from seawater to freshwater, revealed by morphofunctional observations and a newly established "time-differential double fluorescent staining" technique. *J Exp Biol* 206:4113–4123.

Khademi S, O'Connell Jr, Remis J, Robles-Colmenares Y, Miercke LJ, Stroud RM. 2004. Mechanism of ammonia transport by Amt/MEP/Rh: Structure of AmtB at 1.35 a. *Science* 305:1587–1594.

Kraemer AM, Saraiva LR, Korsching SI. 2008. Structural and functional diversification in the teleost s100 family of calcium-binding proteins. *BMC Evol Biol* 8:48.

Krogh A. 1937. Osmotic regulation in freshwater fishes by active absorption of chloride ions. *J Comp Physiol* 24A:656–666.

Krogh A. 1938. The active absorption of ions in some freshwater animals. *J Comp Physiol* 25A:335–350.

Kumai Y, Perry SF. 2011. Ammonia excretion via RHCG1 facilitates Na+ uptake in larval zebrafish *Danio rerio*, in acidic water. *Am J Physiol Regul* 301:R1517–R1528.

Kumai Y, Perry SF. 2012. Mechanisms and regulation of Na+ uptake by freshwater fish. *Respir Physiol Neurobiol* 184:249–256.

Lee TH, Hwang PP, Shieh YE, Lin CH. 2000. The relationship between "deep-hole" mitochondria-rich cells and salinity adaptation in the euryhaline teleost, *Oreochromis mossambicus*. *Fish Physiol Biochem* 23:133–140.

Lee TH, Tsai JC, Fang MJ, Yu MJ, Hwang PP. 1998. Isoform expression of Na+-K+-ATPase α-subunit in gills of the teleost *Oreochromis mossambicus*. *Am J Physiol Regul* 44:R926–R932.

Lee YC, Yan JJ, Cruz S, Horng JL, Hwang PP. 2011. Anion exchanger 1b, but not sodium-bicarbonate cotransporter 1b, plays a role in transport functions of zebrafish H+-ATPase-rich cells. *Am J Physiol Regul* 300:C295–C307.

Li X, Jayachandran S, Nguyen HH, Chan MK. 2007. Structure of the *Nitrosomonas europaea* Rh protein. *Proc Natl Acad Sci USA* 104:19279–19284.

Liao BK, Chen RD, Hwang PP. 2009. Expression regulation of Na+-K+-ATPase alpha1-subunit subtypes in zebrafish gill ionocytes. *Am J Physiol Regul* 296:R1897–R1906.

Liao BK, Deng AN, Chen SC, Chou MY, Hwang PP. 2007. Expression and water calcium dependence of calcium transporter isoforms in zebrafish gill mitochondrion-rich cells. *BMC Genomics* 8:354.

Lin YM, Chen CN, Lee TH. 2003. The expression of gill na, k-ATPase in milkfish, chanos chanos, acclimated to seawater, brackish water and fresh water. *Comp Biochem Physiol* 135A:489–497.

Lin LY, Horng JL, Kunkel JG, Hwang PP. 2006. Proton pump-rich cell secretes acid in skin of zebrafish larvae. *Am J Physiol Regul* 290:C371–C378.

Lin LY, Hwang PP. 2004. Mitochondria-rich cell activity in the yolk-sac membrane of tilapia (*Oreochromis mossambicus*) larvae acclimatized to different ambient chloride levels. *J Exp Biol* 207:1335–1344.

Lin TY, Liao BK, Horng JL, Yan JJ, Hsiao CD, Hwang PP. 2008. Carbonic anhydrase 2-like a and 15a are involved in acid-base regulation and Na+ uptake in zebrafish H+-ATPase-rich cells. *Am J Physiol Regul* 294:C1250–C1260.

Lin CC, Lin LY, Hsu HH, Thermes V, Prunet P, Horng JL, Hwang PP. 2012. Acid secretion by mitochondrion-rich cells of medaka (*Oryzias latipes*) acclimated to acidic fresh water. *Am J Physiol Regul* 302:R283–R291.

Lin H, Pfeiffer DC, Vogl AW, Pan J, Randall DJ. 1994. Immunolocalization of H+-ATPase in the gill epithelia of rainbow trout. *J Exp Biol* 195:169–183.

Lin CH, Tsai IL, Su CH, Tseng DY, Hwang PP. 2011. Reverse effect of mammalian hypocalcemic cortisol in fish: Cortisol stimulates Ca²⁺ uptake via glucocorticoid receptor-mediated vitamin d3 metabolism. *PLoS One* 6:e23689.

Loh YH, Christoffels A, Brenner S, Hunziker W, Venkatesh B. 2004. Extensive expansion of the claudin gene family in the teleost fish, *Fugu rubripes. Gen Res* 14:1248–1257.

Loong AM, Chew SF, Wong WP, Lam SH, Ip YK. 2012. Both seawater acclimation and environmental ammonia exposure lead to increases in mRNA expression and protein abundance of Na+: K+: 2Cl⁻ cotransporter in the gills of the climbing perch, *Anabas testudineus. J Comp Physiol* 182B:491–506.

Lorin-Nebel C, Boulot V, Bondinier C, Charmantier G. 2006. The Na+/K+/2Cl⁻ cotransporter in the sea bass *Dicentrarchus labrax* ontogeny: Involvement in osmoregulation. *J Exp Biol* 209:4908–4922.

Lupo D, Li XD, Durand A, Tomizaki T, Cherif-Zahar B, Matassi G, Merrick M, Winkler FK. 2007. The 1.3-a resolution structure of *Nitrosomonas europaea* Rh50 and mechanistic implications for NH₃ transport by rhesus family proteins. *Proc Natl Acad Sci USA* 104:19303–19308.

Lytle C, Xu JC, Biemesderfer D, Forbush BR. 1995. Distribution and diversity of Na-K-Cl cotransport proteins: A study with monoclonal antibodies. *Am J Physiol Regul* 269:C1496–C1505.

Madsen SS, Jensen MK, Nhr J, Kristiansen K. 1995. Expression of Na+-K+-ATPase in the brown trout, *Salmo trutta*: In vivo modulation by hormones and seawater. *Am J Physiol Regul* 269:R1339–R1345.

Madsen SS, Kiilerich P, Tipsmark CK. 2009. Multiplicity of expression of Na+, K+-ATPase α-subunit isoforms in the gill of Atlantic salmon (*Salmo salar*): Cellular localisation and absolute quantification in response to salinity change. *J Exp Biol* 212:78–88.

Mak DD, Dang B, Weiner ID, Foskett JK, Westhoff CM. 2006. Characterization of ammonia transport by the kidney Rh glycoproteins RhBG and RhCG. *Am J Physiol Regul* 290:F297–F305.

Marshall WS, Bryson SE, Midelfart A, Hamilton WF. 1995. Low conductance anion channel activated by camp in teleost Cl⁻ secreting cells. *Am J Physiol Regul* 37:R963–R969.

Marshall WS, Grosell M. 2006. *Ion Transport, Osmoregulation and Acid-Base Balance*. Evans DH, Claiborne JB, eds. Boca Raton, FL: CRC Press.

Marshall WS, Lynch EM, Cozzi RR. 2002. Redistribution of immunofluorescence of CFTR anion channel and NKCC cotransporter in chloride cells during adaptation of the killifish *Fundulus heteroclitus* to seawater. *J Exp Biol* 205:1265–1273.

Marshall WS, Watters KD, Hovdestad LR, Cozzi RRF, Katoh F. 2009. CFTR Cl⁻ channel functional regulation by phosphorylation of focal adhesion kinase at tyrosine 407 in osmosensitive ion transporting mitochondria rich cells of euryhaline killifish. *J Exp Biol* 212:2365–2377.

McCormick SD, Regish AM, Christensen AK. 2009. Distinct freshwater and seawater isoforms of Na+/K+-ATPase in gill chloride cells of Atlantic salmon. *J Exp Biol* 212:3994–4001.

McCormick SD, Sundell K, Bjornsson BT, Brown CL, Hiroi J. 2003. Influence of salinity on the localization of Na⁺/K⁺-ATPase, Na⁺/K⁺/2Cl⁻ cotransporter (NKCC) and CFTR anion channel in chloride cells of the Hawaiian goby (*Stenogobius hawaiiensis*). *J Exp Biol* 206:4575–4583.

Miyazaki H, Uchida S, Takei Y, Hirano T, Marumo F, Sasaki S. 1999. Molecular cloning of CLC chloride channels in *Oreochromis mossambicus* and their functional complementation of yeast CLC gene mutant. *Biochem Biophys Res Commun* 255:175–181.

Moreira-Silva J, Tsui TK, Coimbra J, Vijayan MM, Ip YK, Wilson JM. 2009. Branchial ammonia excretion in the Asian weatherloach *Misgurnusan guillicaudatus*. *Comp Biochem Physiol* 151C:40–50.

Nakada T, Hoshijima K, Esaki M, Nagayoshi S, Kawakami K, Hirose S. 2007a. Localization of ammonia transporter Rhcg1 in mitochondrion rich cells of yolk sac, gill, and kidney of zebrafish and its ionic strength-dependent expression. *Am J Physiol Regul* 293:R1743–R1753.

Nakada T, Westhoff CM, Kato A, Hirose S. 2007b. Ammonia secretion from fish gill depends on a set of Rh glycoproteins. *FASEB J* 21:1067–1074.

Nakhoul NL, Schmidt E, Abdulnour-Nakhoul SM, Hamm LL. 2006. Electrogenic ammonium transport by renal RhBG. *Transfus Clin Biol* 13:147–152.

Nawata CM, Hirose S, Nakada T, Wood CM, Kato A. 2010a. Rh glycoprotein expression is modulated in pufferfish (*Takifu gurubripes*) during high environmental ammonia exposure. *J Exp Biol* 213:3150–3160.

Nawata CM, Hung CC, Tsui TK, Wilson JM, Wright PA, Wood CM. 2007. Ammonia excretion in rainbow trout (*Oncorhynchus mykiss*): Evidence for Rh glycoprotein and H⁺-ATPase involvement. *Physiol Genomics* 31:463–474.

Nawata CM, Wood CM, O'Donnell MJ. 2010b. Functional characterization of rhesus glycoproteins from an ammoniotelic teleost, the rainbow trout, using oocyte expression and siet analysis. *J Exp Biol* 213:1049–1059.

Nilsen TO, Ebbesson LOE, Madsen SS, McCormick SD, Andersson E, Björnsson BT, Prunet P, Stefansson SO. 2007. Differential expression of gill Na⁺, K⁺-ATPase α- and β-subunits, Na⁺, K⁺, 2Cl⁻ cotransporter and CFTR anion channel in juvenile anadromous and landlocked Atlantic salmon *Salmo salar*. *J Exp Biol* 210:2885–2896.

Notch EG, Chapline C, Flynn E, Lameyer T, Lowell A, Sato D, Shaw JR, Stanton BA. 2012. Mitogen activated protein kinase 14–1 regulates serum glucocorticoid kinase 1 during seawater acclimation in Atlantic killifish, *Fundulus heteroclitus*. *Comp Biochem Physiol* 162A:443–448.

Notch EG, Shaw JR, Coutermarsh BA, Dzioba M, Stanton BA. 2011. Morpholino gene knockdown in adult *Fundulus heteroclitus*: Role of SGK1 in seawater acclimation. *PLoS One* 6:e29462.

Pan TC, Liao BK, Huang CJ, Lin LY, Hwang PP. 2005. Epithelial Ca²⁺ channel expression and Ca²⁺ uptake in developing zebrafish. *Am J Physiol Regul* 289:R1202–R1211.

Parks SK, Tresguerres M, Goss GG. 2007. Interactions between Na⁺ channels and Na⁺ – HCO₃⁻ cotransporters in the freshwater fish gill MR cell: A model for transepithelial Na⁺ uptake. *Am J Physiol Regul* 292:C935–C944.

Parks SK, Tresguerres M, Goss GG. 2008. Theoretical considerations underlying Na⁺ uptake mechanisms in freshwater fishes. *Comp Biochem Physiol* 148C:411–418.

Parks SK, Tresguerres M, Goss GG. 2009. Cellular mechanisms of Cl⁻ transport in trout gill mitochondrion-rich cells. *Am J Physiol Regul* 296:R1161–R1169.

Perry SF, Furimsky M, Bayaa M, Georgalis T, Shahsavarani A, Nickerson JG, Moon TW. 2003a. Integrated responses of Na⁺/HCO₃⁻ cotransporters and V-type H⁺-ATPases in the fish gill and kidney during respiratory acidosis. *Biochim Biophys Acta* 1618:175–184.

Perry SF, Gilmour KM. 2006. Acid-base balance and Co₂ excretion in fish: Unanswered questions and emerging models. *Respir Physiol Neurobiol* 154:199–215.

Perry SF, Shahsavarani A, Georgalis T, Bayaa M, Furimsky M, Thomas S. 2003b. Channels, pumps, and exchangers in the gill and kidney of freshwater fishes: Their role in ionic and acid-base regulation. *J Exp Zool* 300A:53–62.

Perry SF, Vulesevic B, Bayaa M. 2009. Evidence that SLC26 anion transporters mediate branchial chloride uptake in adult zebrafish (*Danio rerio*). *Am J Physiol Regul* 297:R988–R997.

Piermarini PM, Evans DH. 2001. Immunochemical analysis of the vacuolar proton-ATPase B-subunit in the gills of a euryhaline stingray (*Dasyatis sabina*): Effects of salinity and relation to Na⁺/K⁺-ATPase. *J Exp Biol* 204:3251–3259.

Piermarini PM, Verlander JW, Royaux IE, Evans DH. 2002. Pendrin immunoreactivity in the gill epithelium of a euryhaline elasmobranch. *Am J Physiol Regul* 283:R983–R992.

Preest MR, Gonzalez RJ, Wilson RW. 2005. A pharmacological examination of Na$^+$ and Cl$^-$ transport in two species of freshwater fish. *Physiol Biochem Zool* 78:259–272.

Pressley TA. 1992. Phylogenetic conservation of isoform-specific regions within α-subunit of Na, K-ATPase. *Am J Physiol Regul* 262:C743–C751.

Purkerson JM, Schwartz GJ. 2007. The role of carbonic anhydrases in renal physiology. *Kidney Int* 71:103–115.

Qiu A, Hogstrand C. 2004. Functional characterisation and genomic analysis of an epithelial calcium channel (ECAC) from pufferfish, *Fugu rubripes*. *Gene* 342:113–123.

Reid SD, Hawkings GS, Galvez F, Goss GG. 2003. Localization and characterization of phenamil-sensitive Na$^+$ influx in isolated rainbow trout gill epithelial cells. *J Exp Biol* 206:551–559.

Reilly RF, Ellison DH. 2000. Mammalian distal tubule: Physiology, pathophysiology, and molecular anatomy. *Physiol Rev* 80:277–313.

Richards JG, Semple JW, Bystriansky JS, Schulte PM. 2003. Na$^+$/K$^+$-ATPase A-isoform switching in gills of rainbow trout (*oncorhynchus mykiss*) during salinity transfer. *J Exp Biol* 206:4475–4486.

Saito K, Nakamura N, Ito Y, Hoshijima K, Esaki M, Zhao B, Hirose S. 2010. Identification of zebrafish Fxyd11a protein that is highly expressed in ion-transporting epithelium of the gill and skin and its possible role in ion homeostasis. *Front Physiol* 1:129.

Sardella BA, Kültz D. 2009. Osmo- and ionoregulatory responses of green sturgeon (*Acipenser medirostris*) to salinity acclimation. *J Comp Physiol* 179B:383–390.

Sardet C, Pisam M, Maetz J. 1979. The surface epithelium of teleostean fish gills. Cellular and junctional adaptations of the chloride cell in relation to salt adaptation. *J Cell Biol* 80:96–117.

Schmidt-Nielsen K. 1993. *Animal Physiology: Adaptation and Environment*. New York: Cambridge University Press.

Scott GR, Claiborne JB, Edwards SL, Schulte PM, Wood CM. 2005. Gene expression after freshwater transfer in gills and opercular epithelia of killifish: Insight into divergent mechanisms of ion transport. *J Exp Biol* 208:2719–2729.

Scott GR, Richards JG, Forbush B, Isenring P, Schulte PM. 2004. Changes in gene expression in gills of the euryhaline killifish *Fundulus heteroclitus* after abrupt salinity transfer. *Am J Physiol Regul* 287:C300–C309.

Seidelin M, Madsen SS, Blenstrup H, Tipsmark CK. 2000. Time-course changes in the expression of Na$^+$, K$^+$-ATPase in gills and pyloric caeca of brown trout (*Salmo trutta*) during acclimation to seawater. *Physiol Biochem Zool* 73:446–453.

Shahsavarani A, McNeill B, Galvez F, Wood CM, Goss GG, Hwang PP, Perry SF. 2006. Characterization of a branchial epithelial calcium channel (ECAC) in freshwater rainbow trout (*Oncorhynchus mykiss*). *J Exp Biol* 209:1928–1943.

Shaw JR, Sato D, VanderHeide J, LaCasse T, Stanton C, Lankowski A, Stanton S, Chapline C et al. 2008. The role of SGK and CFTR in acute adaptation to seawater in *Fundulus heteroclitus*. *Cell Physiol Biochem* 22:69–78.

Shen WP, Horng JL, Lin LY. 2011. Functional plasticity of mitochondrion-rich cells in the skin of euryhaline medaka larvae (*Oryzias latipes*) subjected to salinity changes. *Am J Physiol Regul* 300:R858–R868.

Shih TH, Horng JL, Hwang PP, Lin LY. 2008. Ammonia excretion by the skin of zebrafish (*Danio rerio*) larvae. *Am J Physiol Regul* 295:C1625–C1632.

Shih TH, Horng JL, Liu ST, Hwang PP, Lin LY. 2012. Rhcg1 and NHE3b are involved in ammonium-dependent sodium uptake by zebrafish larvae acclimated to low-sodium water. *Am J Physiol Regul* 302:R84–R93.

Shono T, Kurokawa D, Miyake T, Okabe M. 2011. Acquisition of glial cells missing 2 enhancers contributes to a diversity of ionocytes in zebrafish. *PLoS One* 6:e23746.

Singer TD, Clements KM, Semple JW, Schulte PM, Bystriansky JS, Finstad B, Fleming IA, McKinley RS. 2002. Seawater tolerance and gene expression in two strains of Atlantic salmon smolts. *Can J Fish Aquat Sci* 59:125–135.

Singer TD, Tucker SJ, Marshall WS, Higgins CF. 1998. A divergent cftr homologue: Highly regulated salt transport in the euryhaline teleost *F. heteroclitus*. *Am J Physiol* Regul 274:C715–C723.

Sullivan GV, Fryer JN, Perry SF. 1995. Immunolocalization of proton pumps (H$^+$-ATPase) in pavement cells of rainbow trout gill. *J Exp Biol* 198:2619–2629.

Sullivan GV, Fryer JN, Perry SF. 1996. Localization of mrna for proton pump (H$^+$-ATPase) and Cl$^-$/HCO$_3^-$ exchanger in rainbow trout gill. *Can J Zool* 74:2095–2103.

Suzuki Y, Itakura M, Kashiwagi M, Nakamura N, Matuki T, Sauta H, Naito N et al. 1999. Identification by differential display of a hypertonicity-inducible inward rectifier potassium channel highly expressed in chloride cells. *J Biol Chem* 274:11376–11382.

Takabe S, Teranishi K, Takaki S, Kusakabe M, Hirose S, Kaneko T, Hyodo S. 2012. Morphological and functional characterization of a novel Na$^+$/K$^+$-ATPase-immunoreactive, follicle-like structure on the gill septum of Japanese banded houndshark, *triakis scyllium*. *Cell Tissue Res* 348:141–153.

Takeyasu K, Lemas V, Fambrough DM. 1990. Stability of Na$^+$/K$^+$-ATPase α-subunit isoforms in evolution. *Am J Physiol Regul* 259:C619–C630.

Tang CH, Chiu YH, Tsai SC, Lee TH. 2009. Relative changes in the abundance of branchial Na$^+$/K$^+$-ATPase a-isoform-like proteins in marine euryhaline milkfish (*Chanos chanos*) acclimated to environments of different salinities. *J Exp Zool* 311A:522–530.

Tang CH, Hwang LY, Lee TH. 2010. Chloride channel CLC-3 in gills of the euryhaline teleost, *Tetraodon nigroviridis*: Expression, localization and the possible role of chloride absorption. *J Exp Biol* 213:683–693.

Tang CH, Hwang LY, Shen ID, Chiu YH, Lee TH. 2011. Immunolocalization of chloride transporters to gill epithelia of euryhaline teleosts with opposite salinity-induced Na$^+$/K$^+$-ATPase responses. *Fish Physiol Biochem* 37:709–724.

Tang CH, Lee TH. 2007. The effect of environmental salinity on the protein expression of Na$^+$/K$^+$-ATPase, Na$^+$/K$^+$/2Cl$^-$ cotransporter, cyctic fibrosis transmembrane conductance regulator, anion exchanger 1, and chloride channel 3 in gills of a euryhaline teleost, *Tetraodon nigroviridis*. *Comp Biochem Physiol* 147A:521–528.

Tang CH, Lee TH. 2011. Ion-deficient environment induces the expression of basolateral chloride channel, CLC-3-like protein, in gill mitochondrion-rich cells for chloride uptake of the tilapia *Oreochromis mossambicus*. *Physiol Biochem Zool* 84:54–67.

Thermes V, Lin CC, Hwang PP. 2010. Expression of Ol-foxi3 and Na$^+$/K$^+$-ATPase in ionocytes during the development of euryhaline medaka (*Oryzias latipes*) embryos. *Gene Expr Patterns* 10:185–192.

Tipsmark CK. 2008. Identification of Fxyd protein genes in a teleost: Tissue-specific expression and response to salinity change. *Am J Physiol Regul* 294:R1367–R1378.

Tipsmark CK, Baltzegar DA, Ozden O, Grubb BJ, Borski RJ. 2008a. Salinity regulates claudin mRNA and protein expression in the teleost gill. *Am J Physiol Regul* 294:R1003–R1014.

Tipsmark CK, Jørgensen C, Brande-Lavridsen N, Engelund M, Olesen JH, Madsen SS. 2009. Effects of cortisol, growth hormone and prolactin on gill claudin expression in Atlantic salmon. *Gen Comp Endocrinol* 163:270–277.

Tipsmark CK, Luckenbach JA, Madsen SS, Kiilerich P, Borski RJ. 2008b. Osmoregulation and expression of ion transport proteins and putative claudins in the gill of southern flounder (*Paralichthys lethostigma*). *Comp Biochem Physiol* 150A:265–273.

Tipsmark CK, Madsen SS, Ceidelin M, Christensen AS, Cutler CP, Cramb G. 2002. Dynamics of Na, K, 2Cl cotransporter and Na, K-ATPase expression in the branchial epithelium of brown trout (*Salmo trutta*) and Atlantic salmon (*Salmo salar*). *J Exp Zool* 293:106–118.

Tipsmark CK, Mahmmoud YA, Borski RJ, Madsen SS. 2010. Fxyd-11 associates with Na$^+$-K$^+$-ATPase in the gill of Atlantic salmon: Regulation and localization in relation to changed ion-regulatory status. *Am J Physiol Regul* 299:R1212–R1223.

Tresguerres M, Katoh F, Fenton H, Jasinska E, Goss GG. 2005. Regulation of branchial V-H$^+$-ATPase, Na$^+$/K$^+$-ATPase and NHE2 in response to acid and base infusions in the Pacific spiny dogfish (*Squalus acanthias*). *J Exp Biol* 208:345–354.

Tresguerres M, Katoh F, Orr E, Parks SK, Goss GG. 2006a. Chloride uptake and base secretion in freshwater fish: A transepithelial ion-transport metabolon? *Physiol Biochem Zool* 79:981–996.

Tresguerres M, Parks SK, Katoh F, Goss GG. 2006b. Microtubule-dependent relocation of branchial V-H$^+$-ATPase to the basolateral membrane in the Pacific spiny dogfish (*Squalus acanthias*): A role in base secretion. *J Exp Biol* 209:599–609.

Tresguerres M, Parks SK, Wood CM, Goss GG. 2007. V-H$^+$-ATPase translocation during blood alkalosis in dogfish gills: Interaction with carbonic anhydrase and involvement in the postfeeding alkaline tide. *Am J Physiol Regul Integr Comp Physiol* 292:R2012–R2019.

Tse WKF, Au DWT, Wong CKC. 2006. Characterization of ion channel and transporter mRNA expression in isolated gill chloride and pavement cells of seawater acclimating eels. *Biochem Biophys Res Commun* 346:1181–1190.

Tseng DY, Chou MY, Tseng YJ, Hsiao CD, Huang CJ, Kaneko T, Hwang PP. 2009. Effects of stanniocalcin 1 on calcium uptake in zebrafish (*Danio rerio*) embryo. *Am J Physiol Regul* 296:R549–R557.

Tseng YC, Huang CJ, Chang JCH, Teng WY, Baba O, Fann MJ, Hwang PP. 2007. Glycogen phosphorylase in glycogen-rich cells is involved in the energy supply for ion regulation in fish gill epithelia. *Am J Physiol Regul* 293:R482–R491.

Tseng YC, Hwang PP. 2008. Some insights into energy metabolism for osmoregulation in fish. *Comp Biochem Physiol* 148C:419–429.

Venkatesh B, Kirkness EF, Loh YH, Halpern AL, Lee AP, Johnson J, Dandona N et al. 2007. Survey sequencing and comparative analysis of the elephant shark (*Callorhinchus milii*) genome. *PLoS Biol* 5:e101.

Wagner CA, Devuyst O, Bourgeois S, Mohebbi N. 2009. Regulated acid-base transport in the collecting duct. *Pflügers Arch Eur J Physiol* 458:137–156.

Wang PJ, Lin CH, Hwang PP, Lee TH. 2008. Branchial Fxyd protein expression in response to salinity change and its interaction with Na+/K+-ATPase of the euryhaline teleost *Tetraodon nigroviridis*. *J Exp Biol* 211:3750–3758.

Wang YF, Tseng YC, Yan JJ, Hiroi J, Hwang PP. 2009. Role of SLC12a10.2, a Na-Cl cotransporter-like protein, in a Cl uptake mechanism in zebrafish (*Danio rerio*). *Am J Physiol Regul* 296:R1650–R1660.

Weihrauch D, Wilkie MP, Walsh PJ. 2009. Ammonia and urea transporters in gills of fish and aquatic crustaceans. *J Exp Biol* 212:1716–1730.

Weiner ID, Hamm LL. 2007. Molecular mechanisms of renal ammonia transport. *Annu Rev Physiol* 69:317–340.

Whitehead A, Roach JL, Zhang S, Galvez F. 2011. Genomic mechanisms of evolved physiological plasticity in killifish distributed along an environmental salinity gradient. *Proc Natl Acad Sci USA* 108:6193–6198.

Wilkie MP. 2002. Ammonia excretion and urea handling by fish gills: Present understanding and future research challenges. *J Exp Zool* 293:284–301.

Wilson JM, Laurent P, Tufts BL, Benos DJ, Donowitz M, Vogl AW, Randall DJ. 2000. NaCl uptake by the branchial epithelium in freshwater teleost fish: An immunological approach to ion-transport protein localization. *J Exp Biol* 203:2279–2296.

Wong CKC, Chan DKO. 1999. Chloride cell subtypes in the gill epithelium of Japanese eel *Anguilla japonica*. *Am J Physiol Regul Integr Comp Physiol* 277:R517–R522.

Wood CM, Matsuo AY, Gonzalez RJ, Wilson RW, Patrick ML, Val AL. 2002. Mechanisms of ion transport in *Potamotrygon*, a stenohaline freshwater elasmobranch native to the ion-poor blackwaters of the Rio Negro. *J Exp Biol* 205:3039–3054.

Wright PA, Randall DJ, Perry SF. 1989. Fish gill water boundary layer: A site of linkage between carbon dioxide and ammonia excretion. *J Comp Physiol* 158B:627–635.

Wright PA, Wood CM. 2009. A new paradigm for ammonia excretion in aquatic animals: Role of rhesus (Rh) glycoproteins. *J Exp Biol* 212:2303–2312.

Wu SC, Horng JL, Liu ST, Hwang PP, Wen ZH, Lin CS, Lin LY. 2010. Ammonium-dependent sodium uptake in mitochondrion-rich cells of medaka (*Oryzias latipes*) larvae. *Am J Physiol Regul* 298:C237–C250.

Wu YC, Lin LY, Lee TH. 2003. Na+, K+, 2Cl− cotransporter: A novelmarker for identifying freshwater- and seawater-type mitochondria-rich cells in gills of the euryhaline tilapia, *Oreochromis mossambicus*. *Zool Stud* 42:186–192.

Yan JJ, Chou MY, Kaneko T, Hwang PP. 2007. Gene expression of Na+/H+ exchanger in zebrafish H+-ATPase-rich cells during acclimation to low-Na+ and acidic environments. *Am J Physiol Regul* 293:C1814–C1823.

7 Endocrine Disruption

Heather J. Hamlin

CONTENTS

7.1 INTRODUCTION

The endocrine system is a chemical information network made up of certain glands, called endocrine glands, that produce and secrete chemical messages, or hormones. Hormones are important regulators of nearly every aspect of vertebrate life including metabolism, growth, reproduction, and development (Bern, 1990). Certain chemicals in the environment, both man-made and naturally occurring, have the ability to disrupt normal endocrine function, and are collectively referred to as endocrine-disrupting contaminants (EDCs) (Colborn, 1992). Kavlock et al. define an EDC as "an exogenous agent that interferes with the production, release, transport, metabolism, binding, action or elimination of natural hormones in the body responsible for the maintenance of homeostasis and the regulation of developmental processes" (Kavlock et al., 1996). Although other definitions of EDCs currently exist (Colborn, 1992, Chang et al., 2009), it is clear that any chemical with the ability to mimic or antagonize endogenous hormones has the potential to cause a variety of adverse health outcomes. The extent of harm an EDC can impart on an individual or population depends on a variety of factors: the concentration and duration of exposure to the EDC, the species and life stage of the individual, and environmental factors such as season, temperature, salinity, and the presence of other contaminants (Figure 7.1). Although potential EDCs are not usually released into the environment intentionally, most chemicals eventually end up in aquatic systems and waterways, which have been called the "ultimate sink" for environmental pollutants. Fishes are particularly susceptible to exposure to pollutants because they are continually exposed to the aquatic medium and can take up the chemicals through absorption or ingestion. In addition, many EDCs readily bioaccumulate in the fishes' fat stores, and can also bio-magnify up the food chain, and in this way cause appreciable body burdens of EDCs from relatively low environmental concentrations. Fishes are ideally suited to studies of EDC exposure because they have a demonstrated sensitivity to EDCs

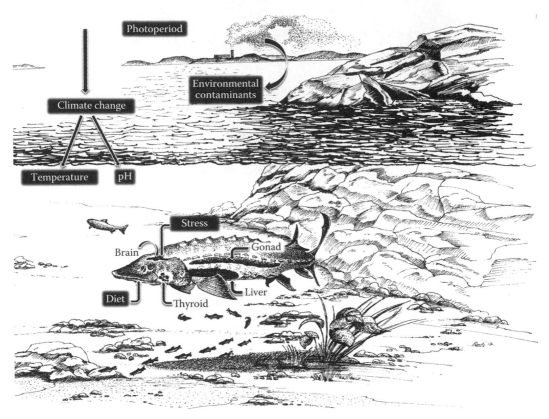

FIGURE 7.1 Environmental contaminants can disrupt normal endocrine function through a variety of pathways. Common targets of contaminant-induced dysfunction include the gonad, thyroid, brain, and clearance of steroid hormones by the liver. Other factors such as diet, stress, temperature, and photoperiod can also modulate endocrine function. (Drawing courtesy of David Basti.)

which has been confirmed in both field and laboratory studies, and many fishes have short life cycles conducive to multigenerational investigations (Porte et al., 2006).

Released in 1962, Rachel Carson's book *Silent Spring* marked a turning point in increasing public awareness that the largely unregulated use of pesticides could have severe and long-lasting consequences for wildlife (Carson, 1962). We now know that pesticides, such as dichlorodiphenyltrichloroethane (DDT), have caused eggshell thinning in birds, altered fecundity and fertility in fish, and developmental abnormalities in most taxa studied to date (Colborn et al., 1993). EDCs are comprised of a diverse group of industrial and agricultural chemicals, as well as several naturally occurring ones (Table 7.1). Many EDCs behave like estrogen and interact with the estrogen receptor. A number of EDCs found in sewage treatment effluent have been shown to feminize male fish living downstream (Tyler and Jobling, 2008), and indeed a number of effluent chemicals share a similar structural relationship with the phenolic ring of estrogen. Tens of thousands of man-made chemicals including plasticizers, industrial compounds, personal care products, pharmaceuticals, and a variety of others regularly make their way into aquatic systems. Following decades of field research and laboratory investigations, our knowledge and detection of EDC-induced dysfunction have increased dramatically (Pait and Nelson, 2002, Hoffmann et al., 2006, 2008, Hook et al., 2006, Samuelsson et al., 2006, Chandrasekar et al., 2011, Webb and Doroshov, 2011), and studies of EDC exposures are now experiencing a paradigm shift driven by several important discoveries: (1) EDC effects can manifest in decidedly different ways, often permanently, depending on whether the animal is exposed as an embryo, larva, or adult; (2) the effects of EDC exposure in embryos can be delayed,

TABLE 7.1

Effects of Endocrine-Disrupting Contaminants in Fish

Chemical[a]	Species	Exposure Concentration (Duration)	Effect	Reference
Bisphenol A	Atlantic cod (*Gadus morhua*)	50 µg/L (3 weeks)	Vitellogenin induction and production of zona radiata protein in plasma	Larsen et al. (2006)
Bisphenol A	Atlantic salmon (*Salmo salar*)	1000 µg/L (6 days)	Yolk sac edema and hemorrhage, histological abnormalities	Honkanen et al. (2004)
Bisphenol A	Brown trout (*Salmo trutta*)	1.75 µg/L (≈3.5 months)	Reduced semen volume and quality, delayed ovulation	Lahnsteiner et al. (2005)
DDE	Three-spine stickleback (*Gasterosteus aculeatus*)	50 µg/L (31 days)	Altered feeding behavior	Wibe et al. (2004)
EE$_2$	Three-spine stickleback (*G. aculeatus*)	10 ng/L (12 days)	Altered nest care behavior	Brian et al. (2006)
EE$_2$	Rare minnow (*Gobiocypris rarus*)	25 ng/L (7 days)	Inhibition of genes involved in the production of steroid hormones	Liu et al. (2012)
Glyphosate (Roundup®)	European eel (*Anguilla anguilla*)	17.9 µg/L (3 days)	Genotoxicity	Guilherme et al. (2012)
Glyphosate (Roundup®)	One-sided live bearer (*Jenynsia multidentata*)	0.5 mg/L (7 and 28 days)	Reduced male copulations	Hued et al. (2012)
4-Nonylphenol	Medaka (*Oryzias latipes*)	51.5 µg/L (≤104 days)	Skewed sex ratio toward female, intersex	Yokota et al. (2001)
4-Nonylphenol	Rainbow trout (*Oncorhynchus mykiss*)	30 µg/L (3 weeks)	Vitellogenin induction, increased gonadosomatic index in males	Jobling et al. (1996)
4-Nonylphenol	Spotted ray (*Torpedo marmorata*)	100 mg/kg injected (24 h)	Vitellogenin induction in males	Del Guidice et al. (2012)
PCBs[3] (Aroclor 1248)	Brown bullhead (*Ameiurus nebulosus*)	5 mg/kg injected (21 days)	Reduced plasma T3	Iwanowicz et al. (2009)
Vinclozolin (fungicide)	Fathead minnow (*Pimephales promelas*)	≤700 µg/L (21 days)	Complete reproductive failure, altered androgen production	Martinovic et al. (2008)

Source: Jobling, S. et al., *Environ. Toxicol. Chem.*, 15, 194, 1996; Yokota, H. et al., *Environ. Toxicol. Chem.*, 20, 2552, 2001; Honkanen, J.O. et al., *Chemosphere*, 55, 187, 2004; Wibe, A.E. et al., *Ecotoxicol. Environ. Saf.*, 57, 213, 2004; Lahnsteiner, F. et al., *Aquat. Toxicol.*, 75, 213, 2005; Larsen, B.K. et al., *Aquat. Toxicol.*, 78, S25, 2006; Brown, J.A. and Hazon, N., *The Renin-Angiotensin Systems of Fish and Their Roles in Osmoregulation*, Science Publishers, pp. 85–134, 2007; Martinovic, D. et al., *Environ. Toxicol. Chem.*, 27, 478, 2008; Iwanowicz, L.R. et al., *Aquat. Toxicol.*, 93, 70, 2009; Del Giudice, G. et al., *J. Fish Biol.*, 80, 2112, 2012; Guilherme, S. et al., *Ecotoxicology*, 21, 1381, 2012; Hued, A.C. et al., *Arch. Environ. Contam. Toxicol.*, 62, 107, 2012; Liu, S.Z. et al., *Aquat. Toxicol.*, 122, 19, 2012.

[a] Abbreviations used: DDE, dichlorodiphenyldichloroethylene (breakdown product of the pesticide DDT, dichlorodiphenyltrichloroethane); EE$_2$, 17α-ethinylestradiol; PCBs, polychlorinated biphenyls.

and not manifest until adulthood; and (3) high-dose exposures do not necessarily predict low-dose effects, and dose response curves that are not linear, or non-monotonic, are not uncommon in studies of EDC-induced dysfunction (Sumpter and Johnson, 2005).

7.2 ENDOCRINE SYSTEM: OVERVIEW

The endocrine system relies on the successful communication of cells sending chemical signals to other cells (or sometimes to themselves) and the ability of the target cells to receive those signals and respond appropriately. This is the basis for controlling virtually every aspect of physiology and behavior and is necessary to maintain homeostasis in a changing and harsh environment. While the term endocrine often refers to the glandular release of hormones into the blood, it is clear that chemical regulators also play a critical role in neural transmission. Because some neural regulators are released, like hormones, into the blood, it is often termed the neuroendocrine system, acknowledging the complex integration of these two systems that were once considered to be separate and distinct (Bern et al., 1985, Bern, 1990).

7.2.1 Pituitary and Endocrine Organs

In addition to the neural network, the endocrine system in fishes is made up of a number of glands, each regulating a host of biological processes. Although the gonad and thyroid axes will be described in greater detail, this section will provide a brief overview of some of the primary endocrine glands to give the reader an increased understanding of the extensive integration of this complex system (Hazon and Balment, 1998).

Considered the "master gland" of the endocrine system, the pituitary regulates many activities of other endocrine glands, and is comprised of two hormone-producing tissues, the *adenohypophysis* and the *neurohypophysis*. Hormone release from the adenohypophysis is controlled by the hypothalamus, and constitutes the hypothalamo–pituitary axis (HPA). The pituitary in fishes secretes a number of hormones that regulate a variety of physiologic functions including growth, reproduction, behavior, lipid metabolism, and others.

In mammals, the pancreas produces insulin and glucagon, which contribute to the regulation of blood glucose (often referred to as "blood sugar"). Insulin lowers blood glucose levels by causing it to enter cells, and glucagon raises blood glucose levels by stimulating glucose production from glycogen stores (Guyton, 1956). This process is relatively conserved among vertebrates, and it is thought that fishes follow a similar mechanism, although much less is known about glucose homeostasis in fishes as compared to mammals (Polakof et al., 2011). Calcium participates in many essential functions, and its tight regulation is important for many physiological processes including muscle contraction and maintenance of the potential difference across cell membranes. The ultimobranchial gland, often referred to as the ultimobranchial body, secretes calcitonin, which lowers blood calcium levels (Milhaud et al., 1977). The corpuscles of Stannius, small spherical bodies often lying on or within the kidneys, produce stanniocalcin, which also lowers blood calcium levels by inhibiting gill and intestinal calcium transport (Schein et al., 2012). Among the largest of the endocrine tissues, the fish gut contains endocrine cells that produce gastrin and cholecystokinin, which stimulate the release of gastric acids and pancreatic enzymes respectively (Kurokawa et al., 2003). Although on a far simpler scale, the urophysis in fishes has been likened to the neurohypophysis, due to the arrangement of axons in a urophyseal stalk at the caudal tip of the spinal cord. The urophysis produces urotensins, and although scientists still do not fully understand the extent of their actions, they appear to be involved in regulating blood pressure and osmoregulation, among other effects. First discovered in fish, urotensins were long thought to be "neurohormones from fish tails" (Bern et al., 1985) but we now know that urotensins are present in other vertebrates, including mammals, amphibians, and birds (Conlon, 2008).

Stress hormones in higher vertebrates are produced primarily by the adrenals, which have an inner medulla and an outer cortex, producing catecholamines (adrenaline and noradrenaline) and corticosteroids respectively. In fishes, the role of the medulla is played by chromaffin cells and the actions of the cortex are performed by the interrenals. These organs also play roles in osmoregulation, as do the juxtaglomerular (JG) cells of the kidney. The JG cells produce renin, a key component of the renin–angiotensin system that plays critical roles in osmoregulation and blood pressure regulation (Evans et al., 2005, Brown and Hazon, 2007).

7.2.2 HYPOTHALAMIC–PITUITARY–GONADAL AXIS

The hypothalamic–pituitary–gonadal (HPG) axis regulates many aspects of reproduction, and its disruption by EDCs has been well documented. In response to environmental cues, the hypothalamus releases the gonadotropin-releasing hormone (GnRH), which then triggers the release of gonadotropin hormones (GTHs) from the pituitary. Like other vertebrates, two main gonadotropins exist in fishes: follicle-stimulating hormone (FSH) and luteinizing hormone (LH). FSH is primarily responsible for early gonadal development and vitellogenesis, while LH is thought to be involved in final oocyte maturation and ovulation as well as spermiation (Peter and Yu, 1997, Weltzien et al., 2004). The activity of GTHs is controlled by the pituitary release of hypothalamic neuroendocrine factors, paracrine factors within the pituitary itself, and through feedback of gonadal hormones such as estrogen and testosterone. Although GnRH is thought to be the primary signal for FSH release, LH can be released in response to a number of factors including norepinephrine, neuropeptide Y, and inhibin/activin (Peter et al., 1991a,b, Ge et al., 1992). Dopamine released from the pituitary activates dopamine D_2 receptors on the gonadotroph, directly inhibiting LH secretion (Trudeau, 1997). It should be noted that the stimulation, inhibition, and overall regulation of the HPG axis can vary markedly between species, and with only a handful of the 30,000+ species of fishes having been investigated, there is still a great deal of interspecies variation left to be explored.

The release of pituitary GTHs into circulation stimulates the gonadal production of sex steroids. Steroid hormones, including estrogens and androgens, are synthesized from cholesterol that is produced primarily in the liver. The principal estrogen in fishes is 17β-estradiol (E_2) and is produced largely by the follicular cells of the ovary, although peripheral aromatization of testosterone (T), the precursor of E_2, can occur in other tissues (Diotel et al., 2010, Guiguen et al., 2010). Aromatization refers to the conversion of T to E_2 via the enzyme aromatase. The major androgens in fishes are T and 11-ketotestosterone (11-KT), although 11-KT is thought to be more biologically active, and is higher in males than in females (Borg, 1994). Although the testes are the major source of androgen production in males, castration does not eliminate androgens from the plasma entirely, indicating that other sources of androgen production are present. While we generally associate estrogens with females and androgens with males, both sexes produce and use these hormones (Mananos et al., 2009).

Because steroids are nonpolar compounds, with low solubility in water, they readily pass through cellular membranes and are rapidly destroyed by the liver and kidneys. The association of steroids with plasma proteins reduces their removal, and results in higher circulating concentrations, increasing the likelihood of reaching their intended target tissues. Once at the target cell, the steroid ligand can readily pass through the cell's phospholipid bilayer, and either bind to receptors on the nucleus (nuclear receptors) or interact with receptors or membrane-bound kinases on the plasma membrane (Hardy and Valverde, 1994). Often, the free sex steroid ligand binds to the nuclear receptor, and the ligand-bound receptor complex behaves like a transcription factor, binding to hormone-response elements or DNA sites, initiating mRNA synthesis. There are reports of non-genomic androgen receptors in fishes as well (Borg, 1994). For example, the intestinal uptake of L-leucine and glucose, which is stimulated by androgens, takes place so rapidly that it is very likely to be mediated by a non-genomic receptor (Habibi and Ince, 1984, Hazzard and Ahearn, 1992). Receptors for both androgens and estrogens are found in a variety of tissues including the pituitary, brain, gonads, accessory sex organs, and others. Although androgens and estrogens play many roles, they play critical roles in sexual differentiation and development. Steroids are removed from circulation primarily by the liver, which typically involves P450 enzymes and the conjugation or removal of side chains or attached groups. These conjugates are water-soluble, which renders them unsuitable for attachment to serum proteins and more easily excreted in urine. In fishes that lay eggs, the release of E_2 from the ovary triggers the production of vitellogenin (Vtg) by the liver. Vtg is a high-density lipoprotein and is the precursor of egg yolk, which provides critical nutrition to the developing embryo. In oviparous fishes, the tremendous growth of oocytes is due to the sequestering of Vtg, which passes through the surrounding follicle layers and is taken up by receptor-mediated endocytosis (Specker and Sullivan, 1994).

7.2.3 HYPOTHALAMIC–PITUITARY–THYROIDAL AXIS

In fishes, as in other vertebrates, the thyroid is composed of a group of follicles that are either scattered (most teleosts and agnathans) or present as a compact gland (parrotfish, swordfish, tuna, and elasmobranchs) in the ventral pharynx and possibly other organs (Kloas et al., 2009). Each follicle contains a single layer of follicular epithelial cells surrounding a colloid-filled space. The follicular cells synthesize thyroglobulin, a protein that acts as a substrate for the synthesis of thyroid hormones. These cells also sequester iodide from circulation that is incorporated into the tyrosyl groups of thyroglobulin to make thyroxine (T4) and to a lesser extent triiodothyronine (T3), the two major thyroid hormones (Orozco and Valverde-R, 2005). Fishes obtain iodide not only from the diet but also from the water via the gills (Brown et al., 2004). Like steroid hormones, thyroid hormones are usually carried in the bloodstream attached to binding proteins. Approximately 99% of plasma T4 is reversibly bound to plasma proteins, with less than 1% being free. In trout, about 10% of circulating T4 is transported in erythrocytes. While T4 is found at much higher circulating concentrations than T3, T3 is considered the biologically active form, with T4 being deiodinated at the target tissues to T3 (Orozco and Valverde, 2005). Although it appears that free T4 is the primary thyroid hormone responsible for feedback control of TSH secretion in teleosts, studies show that free T4 enters cells largely by transport systems, but can also enter by simple diffusion (Kawakami et al., 2006).

Similar to steroid hormones, THs produce their response by binding to specific nuclear thyroid receptors (TRs), which can act as transcription factors with a broad range of functional specificity, although some evidence suggests that TH could also use a direct non-genomic signaling pathway (Peter and Oommen, 1989). Nuclear TRs are part of a large superfamily of receptors that include the steroid receptors among others (Nelson and Habibi, 2009). Two thyroid receptor subtypes have been discovered in fish, TRα-1 and TRβ, and have been confirmed in several fish species including goldfish, zebrafish, Atlantic salmon, and Conger eels (Nelson and Habibi, 2009). When T3 binds to the nuclear receptors, they in turn bind in pairs to TH response elements. This complex then acts as a transcription factor to control RNA transcription. Although this process has not been well studied in fishes, it is assumed that they follow the vertebrate model as fish nuclei contain specific T3 sites that have high homology to sites seen in higher vertebrates. Although thyroid hormones do not appear to regulate metabolism in fishes as they do in endotherms, they play important roles in growth, development, osmoregulation, migratory behavior, and reproduction (Eales and Brown, 1993, Leatherland, 1993, 2000, Brown et al., 2004).

Regulation of thyroid hormone levels occurs via the hypothalamus–pituitary–thyroid (HPT) axis. In response to external stimuli, the hypothalamus acts on the pituitary to release the thyroid-stimulating hormone (TSH), which triggers the thyroid follicles to increase enzymatic uptake of iodide in preparation for thyroid hormone synthesis. Although the thyrotropin-releasing hormone (TRH) is the hypothalamic trigger for the pituitary release of TSH in mammals, the corticotrophin-releasing factor appears to be a more potent stimulator in fish (De Groef et al., 2006). For the synthesis of thyroid hormones, a glycoprotein located in the plasma membrane of the thyroid follicular cells called the sodium iodide symporter (NIS) mediates iodide transport into the follicles. Iodide is an essential component of thyroid hormones, so their production relies heavily on adequate supplies of iodide to the gland (Dohan et al., 2003) and a properly functioning NIS system to mediate the transfer. Also important are iodothyronine deiodinases, which control the conversion of T4 to the more biologically active T3. To date, three types of deiodinases have been discovered in fish (Orozco and Valverde-R, 2005).

7.3 EDC MODES OF ACTION

EDCs exert their effects by (1) mimicking or antagonizing endogenous hormones; (2) altering the numbers or availability of hormone-binding proteins; (3) modulating the synthesis or clearance of endogenous hormones; or (4) altering hormone receptor levels (Soto et al., 1995, Porte, et al., 2006, Ankley et al., 2009, Gore, 2010, James, 2011, Pelch et al., 2011). As discussed previously, the endocrine system relies on the ability of endocrine organs to produce hormones, but equally important is the ability of the body's tissues to receive the hormone signals and respond appropriately. Cells receive hormone signals through hormone-specific receptors. Hormones bind to their specific receptors, which are located on or within the cell, and the binding causes a conformational change in the receptor that then triggers an intracellular messaging system (Norris, 1997). For example, if a cell does not have receptors for estrogen, it won't be able to "see" and respond to estrogen, and will behave as if there is no estrogen present. Conversely, if a cell has receptors for estrogen and a chemical is present that is structurally similar enough to estrogen that it can bind to the estrogen receptor and cause a conformational change, then the cell will "see" estrogen and respond, even though estrogen is not actually present. As the estrogen mimics are not structurally identical to estrogen, the response could be weaker than its natural ligand, or it could bind to a part of the receptor in such a way as to not trigger a conformational change. When the latter occurs, although the binding will not trigger a cellular response, its occupation of the receptor prevents natural estrogen from binding, thereby reducing receptor availability. Receptors, such as the estrogen receptor, that have a tendency to bind to something other than their natural ligand are often referred to as promiscuous. The fish FSH receptor has also been described as promiscuous as it can also bind to LH, a phenomenon not yet observed in mammalian systems (Bogerd et al., 2005).

To complicate matters further, some estrogen-like compounds can produce effects similar to estrogen without interacting with the estrogen receptor at all. These mechanisms of action are diverse and can be mediated by a variety of signaling pathways including interactions with steroidogenic enzymes, binding proteins, and even growth factors (Gillesby and Zacharewski, 1998). Arguably the most well-studied EDCs are those that target estrogen receptor signaling by either interacting directly with estrogen receptors, indirectly through transcription factors, or by the modulation of enzymes critical for estrogen synthesis (Harris et al., 2011, Shanle and Xu, 2011). However, due to the conserved structural and functional nature of nuclear hormone receptors (NHRs) and the tremendous diversity of environmental contaminants, it is likely that all members of the NHR family are targets of EDCs (le Maire et al., 2010). Indeed, recent research shows a growing number of NHRs have been shown to be disrupted by EDC exposure (Ankley et al., 2009), including androgen and thyroid receptors.

EDCs also have the potential to have multigenerational effects through changes in the epigenome. The genome of an animal, comprised of DNA, is the building block of life and heritability. The epigenome directs how that DNA is read, and which genes will be expressed and which not. Therefore, epigenetics is the study of heritable changes in gene function (e.g., which genes are expressed) without altering the DNA sequence (Skinner et al., 2010, Uzumcu et al., 2012). This means that the regulation and expression of the fishes' DNA can be modified *in utero*, and those patterns of expression can be passed on to future generations even though the DNA itself isn't changed. EDCs have been shown to alter epigenetic regulation, and in this way can influence the genomic activity of future generations (Anway et al., 2005). Therefore, exposing fish to certain pollutants has the potential to result in heritable changes in their progeny. The epigenome tends to be more stable in somatic cells, but during early embryonic development, epigenetic programming can be particularly active and therefore especially vulnerable to reprogramming (Reik et al., 2001, Faulk and Dolinoy, 2011, Anderson et al., 2012). There are several mechanisms by which the epigenome is regulated, including DNA methylation, RNA interference, nuclear organization, and histone modification (Garcia-Gimenez et al., 2012). The role of epigenetic programming in causing reproductive and developmental impairment in fishes is not well understood, but as our knowledge increases in other vertebrate models, the importance of improving our understanding in fishes becomes increasingly clear.

7.4 EFFECTS OF EDC EXPOSURE IN FISHES

7.4.1 EDCs and the HPG Axis

Although EDC research is rapidly expanding, the effects of EDCs on reproduction, and end points directed by the HPG axis in particular, have been the focus of much of this research (Kavlock, et al., 1996, Lal, 2007, Rempel and Schlenk, 2008, Lister et al., 2011, Hachfi et al., 2012, Kraugerud et al., 2012). That many EDCs resemble or have the ability to disrupt the production of sex steroids makes the reproductive system especially vulnerable. Many EDCs found in aquatic environments are lipophilic and will readily bioaccumulate in exposed organisms, including polychlorinated biphenyls (PCBs), organochlorine pesticides, and alkylphenols (Hansen et al., 1985, Cheek et al., 2001, Sumpter and Johnson, 2005). Others such as bisphenol A (BPA) do not bioaccumulate, but are continually entering aquatic ecosystems through sources such as sewage treatment effluent and runoff from animal agriculture. BPA is used in the manufacture of plastics, the epoxy lining of metal food cans, building materials, adhesives, and other uses, and has been known to be an artificial estrogen since the 1930s (Flint et al., 2012). In fishes, BPA has been shown to reduce egg production, cause intersex gonads, reduce egg fertility, delay sexual maturation, atresia of oocytes in females, and decreased sexual behavior in males (Shioda and Wakabayashi, 2000, Metcalfe et al., 2001, Segner et al., 2003, Crain et al., 2007, Mandich et al., 2007). Although many studies use levels of BPA that are higher than would be found in the natural environment, reproductive impairments have been found in fishes at environmentally relevant concentrations as well (Flint et al., 2012).

Ethynylestradiol (EE_2) is a synthetic estrogen used in birth control pills, and is found in effluents and surface waters typically at concentrations of 0.5–7 ng/L, although concentrations as high as 50 ng/L have been reported. Life-long exposure of only 5 ng/L EE_2 to the F_1 generation of zebrafish has been shown to cause more than a 50% reduction in fecundity and complete population failure with no fertilization (Nash et al., 2004). In sand gobies, EE_2 exposure reduces breeding activity and fertility (Robinson et al., 2003).

The occurrence of altered sexual development has been extensively studied in roach (*Rutilus rutilus*), a freshwater carp species found commonly in European river systems. First noticed by sport fishermen, the fish living downstream from wastewater treatment works (WWTW) appeared to be neither male nor female, because they possessed both male and female gonads. While this is normal for some hermaphroditic fish species, the roach is normally gonochoristic with either a male or female phenotype. Scientists discovered that the chemicals in the effluent were estrogenic, causing dysfunctions in gonadal development, and altered sex steroid hormone dynamics and vitellogenin production in males (Figure 7.2). As discussed previously, vitellogenin is an egg protein precursor that is produced by the liver in response to circulating estrogen. The vitellogenin is then sequestered by the oocytes and will later support the developing embryo. Although vitellogenin is not normally produced by males, the presence of estrogenic pollutants causes some male fishes to produce it. The roach exposed to the wastewater treatment effluent (WWTE) had concentrations of vitellogenin that were higher than those found in fully mature females, and more than half of the blood protein content of these males consisted of vitellogenin (Tyler and Jobling, 2008).

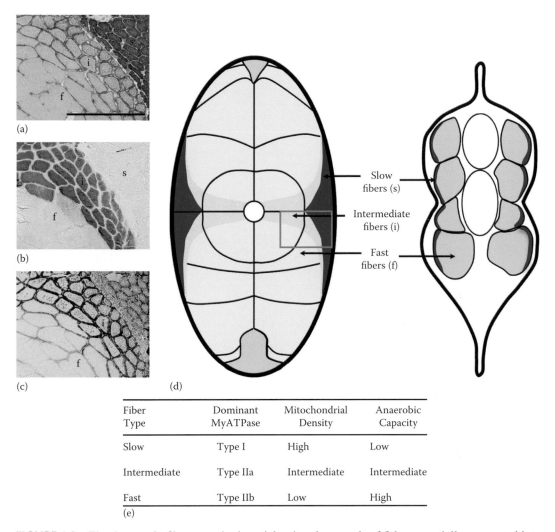

Fiber Type	Dominant MyATPase	Mitochondrial Density	Anaerobic Capacity
Slow	Type I	High	Low
Intermediate	Type IIa	Intermediate	Intermediate
Fast	Type IIb	Low	High

(e)

FIGURE 1.2 The three main fiber types in the axial swimming muscle of fish are spatially segregated into distinct regions. Slow oxidative fibers (s) are located at the lateral periphery of the muscle, and in the adult fish, they are most abundant near the horizontal septum. Fast oxidative (intermediate) fibers (i) are located next to the slow fibers, and fast glycolytic fibers (f) make up the remainder of the axial musculature. (a–c) Representative histological sections from an adult zebrafish (Modified from Scott, G.R. and Johnston, I.A., *Proc. Natl. Acad. Sci. USA*, 109, 14247, 2012); (a) Slow fibers are identified using immunohistochemistry with the S58 antibody (scale bar is 50 μm). (b) Intermediate fibers are identified based on the activity of alkaline-resistant myosin-ATPase activity. (c) Slow and intermediate fibers both contain abundance mitochondria, reflected by the activity of succinate dehydrogenase activity. (d) Diagram of a transverse section through an adult fish (left) and an embryo during late segmentation (right) showing the arrangement of fiber types (the blue box indicates the area of the muscle shown in a–c). Serial sections of the same muscle fiber are represented with letters (s, i, f) in a–c. (e) summarizes the major phenotypic differences.

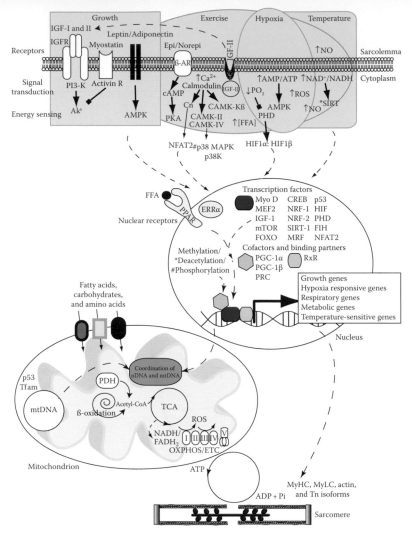

FIGURE 1.3 The mechanisms of muscle plasticity may overlap in response to exercise, hypoxia, and temperature, and through signals for growth. Signals interact with membrane receptors (R) for insulin-like growth factor (IGFR), myostatin (activin R), leptin and adiponectin, and epinephrine (Epi) and norepinephrine (Norepi) (beta-adrenergic receptor, β-AR). Signal transduction pathways may act through cAMP or PI3-K to activate or inhibit protein kinase A (PKA), serine/threonine-specific protein kinase (Akt), or AMPK. Changes in intracellular calcium act through calmodulin to activate calcineurin (Cn), Ca^{2+}/calmodulin-dependent protein kinases (CAMK), PI3-K, nuclear factor of activated T-cells (NFAT), and MAPK. Reduction in intracellular PO_2 inhibits the prolyl-hydroxylase domain (PHD), leading to the stabilization of the hypoxia inducible factor (HIF)-1α, which dimerizes with the constitutively expressed HIF-1β subunit. Energy perturbations are "sensed" by changes in intracellular AMP/ATP, NAD^+/NADH, reactive oxygen species (ROS), and nitric oxide (NO). These can activate AMPK and NAD^+-dependent protein deacetylase (SIRT), but can also increase cytosolic free fatty acid concentrations (FFA) that can interact with nuclear receptors. The signal transduction and energy sensing responses affect gene transcription through the actions of a number of transcription factors: cAMP response element binding (CREB) protein, estrogen-related receptors (ERR), myogenic regulatory factor (MyoD), myocyte enhancer factor (MEF), mammalian target of rapamycin (mTOR), forkhead box O (FOXO), nuclear respiratory factor (NRF), myogenic regulatory factor (MRF), tumor suppressor protein (p53), factor inhibiting HIF (FIH), peroxisome proliferator-activated receptor (PPAR), transcription factor A-mitochondrial (Tfam), in coordination with important cofactors: PPAR-gamma coactivator (PGC)1α, PGC-1-related coactivator (PRC) and PGC1β, and binding partner, retinoic acid receptor (R×R). These transcription factors change the expression of genes that alter fiber size, metabolic machinery (e.g., glycolysis), mitochondrial biogenesis (e.g., tricarboxylic acid cycle [TCA], electron transport chain [ETC]), and sarcomere structure (e.g., myosin heavy chain and light chain [MyHC, MyLC] and troponin [Tn] isoforms). Arrows indicate an activation and blunt-ended lines indicate an inhibitory effect. (Based on data from fish and mammals, modified from Hock, M.B. and Kralli, A., *Annu. Rev. Physiol.*, 71, 177, 2009; Gundersen, K., *Biol. Rev.*, 86, 564, 2011; Johnston, I.A. et al., Molecular biotechnology of development and growth in fish muscle, in: Tsukamoto, K., Takeuchi, T., Beard, Jr. T.D., and Kaiser, M.J., Eds., *Fisheries for Global Welfare and Environment*, 5th World Fisheries Congress, Terrapub., pp. 241–262, 2008; Johnston, I.A. et al., *J. Exp. Biol.*, 214, 1617, 2011; Saleem, A. et al., *Exerc. Sport Sci. Rev.*, 39(4), 199, 2011; Scarpulla, R.C. et al., *Trend. Endocrinol. Metabol.*, 23(9), 459, 2012.)

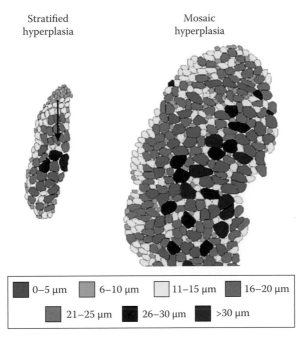

Stratified
hyperplasia

Mosaic
hyperplasia

| ■ 0–5 μm | ■ 6–10 μm | □ 11–15 μm | ■ 16–20 μm |
| ■ 21–25 μm | ■ 26–30 μm | ■ >30 μm | |

FIGURE 1.4 Myogenesis, the creation of new muscle fibers, occurs by multiple distinct processes. Stratified hyperplasia is the creation of new fibers from myogenic progenitor cells (MPC) located in distinct germinal zones. This leads to stratified variation in muscle fiber size between the younger and smaller fibers that are close to the germinal zones and the older and larger fibers located deep in the muscle. Myogenesis of this type begins in late embryogenesis and ends in the larval or early juvenile stage. Mosaic hyperplasia is the creation of new muscle fibers from MPCs that become interspersed throughout the fast muscle, which leads to a mosaic of muscle fiber sizes throughout the myotome (corresponding to differences in age). Camera lucida drawings of the fast muscle fibers from two zebrafish are shown: one at 7.5 mm total body length that is still undergoing myogenesis by stratified hyperplasia and one at 10 mm total body length that has begun mosaic hyperplasia. Fibers are color coded according to diameter size class. (Adapted from Johnston, I.A. et al., *J. Exp. Biol.*, 212, 1781, 2009. With permission.)

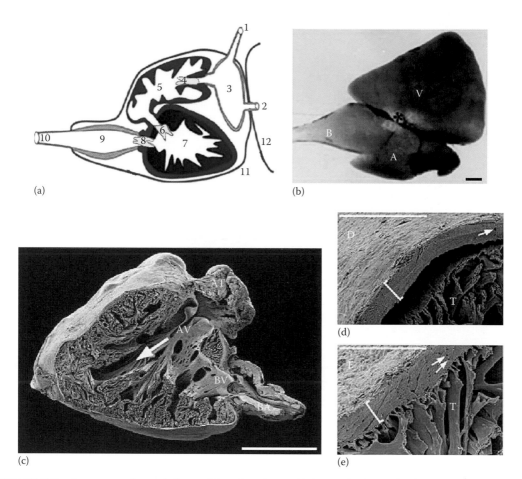

(a)

(b)

(c)

(d)

(e)

FIGURE 2.2 Anatomy and morphology of the teleost heart. (a) Anatomical organization of the heart including the cardiac chambers and pericardial cavity. (1) Ductus Cuvier, (2) hepatic vein, (3) sinus venosus, (4) sino atrial valve, (5) atrium, (6) atrio ventricular valve, (7) ventricle, (8) bulboventricular valve, (9) bulbus arteriosus, (10) ventral aorta, (11) pericardium, and (12) peritoneum. (Modified from Farrell and Pieperhoff, 2011. With permission.) (b) Picture of the heart of the Atlantic horse mackerel, *Trachurus trachurus.* A, atrium; V, ventricle; and B, bulbus arteriosus. (Reproduced from Icardo, J.M., *Anat. Rec. A,* 288, 900, 2006. With permission.) (c) Scanning electron microscopy (SEM) image showing the left side of the adult sockeye salmon (*Oncorhynchus nerka*) heart. AT, atrium; AV, atrioventricular valve; BA, bulbus arteriosus; and BV, bulboventricular valve. Scale bar = 2 mm. (d, e) High magnification SEM images of the apex of the sockeye salmon heart showing two distinct muscle layers (compact and spongy myocardium). *Note*: separation of the layers in (d) is a processing artifact. The brackets show the thickness of the compact myocardium, T is the trabeculae of the spongy myocardium, and the arrows indicate small coronary vessels within the compact myocardium. Scale bar = 200 μm. c, d and e. (Reproduced from Pieperhoff, S. et al., *J. Anat.* 215, 536, 2009. With permission.)

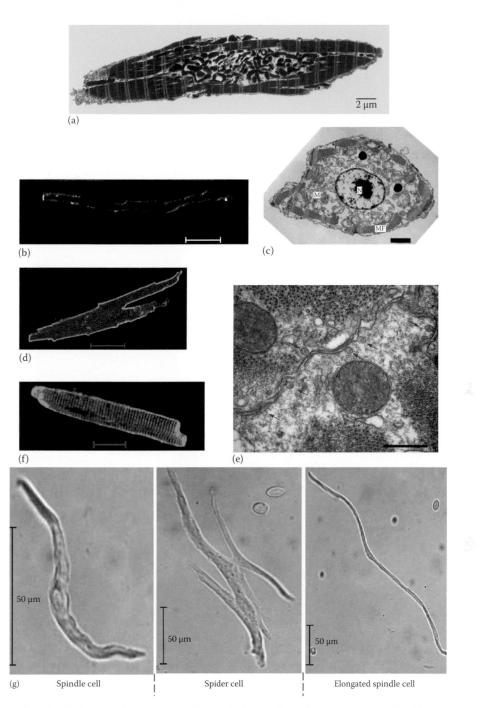

FIGURE 2.3 Morphology of fish myocytes. (a) Single isolated atrial myocyte from a Pacific bluefin tuna (*Thunnus orientalis*); note the sarcomeric pattern of the myofibrils and the internal organization of the organelles along the longitudinal axis of the cells. (Di Maio, A., unpublished observation, with permission). (b) Confocal image of an adult zebrafish (*Danio rerio*) ventricular myocyte with the sarcolemma visualized with di-8-anepps (bar 20 µm). (Luxan, G. and Shiels, H., unpublished observation, with permission). (c) Electron micrographs of a ventricular myocyte from rainbow trout (*Oncorhynchus mykiss*) heart in cross section (M, mitochondria; MF, myofibrils; N, nucleus; dark droplets are lipid). Scale bar 2 µm. (Adapted from Vornanen, M., *J. Exp. Biol.,* 201, 533, 1998. With permission.) (d) Sheet-like ventricular myocyte from the rainbow trout labeled with both fluo-4 AM (cytosol) and di-8-ANEPPS (SL membrane). Notice the absence of T-tubular invaginations. (Adapted from Shiels, H.A. and White, E., *Am. J. Physiol.* 288, R1756, 2005. With permission.) (e) Bluefin tuna ventricle with arrows showing caveolae—small flask-like inward projections of the SL (bar 0.5 µm). (Di Maio, A., unpublished data. With permission.) (f) Pacemaker cells of the trout heart showing three distinct morphologies. (Adapted from Haverinen, J. and Vornanen, M., *Am. J. Physiol.,* 292, R1023, 2007. With permission.)

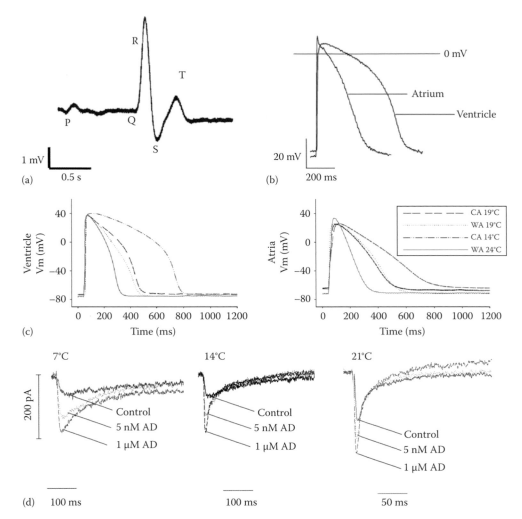

FIGURE 2.4 Excitation of the fish heart. (a) Electrocardiogram from an isolated rainbow trout (*Oncorhynchus mykiss*) heart at 11°C showing atrial depolarization (P), ventricular depolarization (QRS), and ventricular repolarization (T). (Patrick, S. and Shiels, H., unpublished data.) (b) Action potentials of the rainbow trout heart recorded with microelectrodes from intact cardiac tissue at 4°C. Note the resting membrane potential (I_{K1}), the upstroke (I_{Na}), the plateau phase (I_{Ca}), and repolarization ($I_{Ks,r}$). (Adapted from Haverinen, J. and Vornanen, M., *J. Exp. Biol.*, 209, 549, 2006. With permission.) (c) Effect of thermal acclimation and acute temperature exposure on action potential (AP) characteristics in bluefin tuna (*Thunnus orientalis*) cardiomyocytes. Left is ventricle and right is atria, as indicated. Note the prolongation with cooling and acceleration with warming in both myocardial tissues. Also, note the lack of thermal compensation (compare CA–cold acclimated 19°C with WA–warm acclimated 19°C). (Adapted from Galli, et al., *Am. J. Physiol.* 297, R502, 2009a. With permission.) (d) Effect of adrenergic stimulation (AD) and temperature (7°C, 14°C, and 21°C) on peak L-type Ca channel current (I_{Ca}) from trout atrial myocytes. Currents were elicited by 500 ms depolarizations from −70 to 0 mV using whole-cell patch clamp. Note the increase in current with increasing dose of AD. Also note the expanded timescale at 21°C. (Adapted from Shiels, H.A. et al., *Physiol. Biochem. Zool.*, 76, 816, 2003. With permission.)

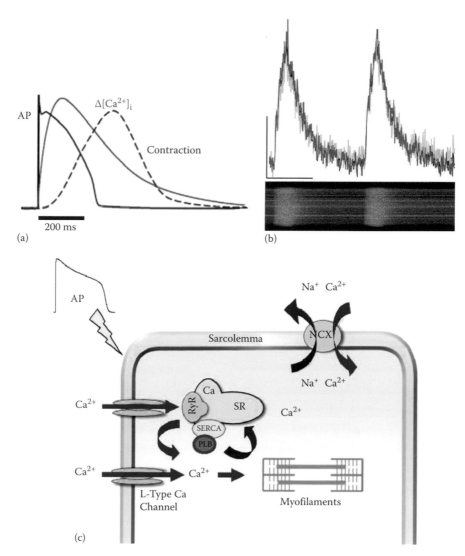

(a)

(b)

(c)

FIGURE 2.5   (a) Time course of excitation–contraction coupling. Data are from a rabbit ventricular myocyte at 37°C. (Adapted from Bers, D.M., *Nature*, 415, 2002. With permission.) (b) The Ca transient measured across the width of a 14°C ventricular myocyte from Pacific bluefin tuna (*Thunnus orientalis*). Representative time courses (top) and corresponding raw line scan images (below) show temporal and spatial characteristics of Ca flux. Scale is 100 nM [Ca] by 1 s. Line scans are 2500 lines, 512 pixels. (Adapted from Shiels, H.A. et al., *Proc. Biol. Sci.*, 278, 18, 2011. With permission.) (c) Schema for excitation–contraction coupling in the fish cardiac myocyte. The figure shows the SL being excited by AP that opens L-type Ca channels, allowing Ca influx (arrows) down its concentration gradient. Ca can also enter the cell via reverse-mode Na–Ca exchange (NCX). Ca influx can trigger Ca release from the sarcoplasmic reticulum (SR) through ryanodine receptors (RyR). Together, these Ca influxes cause a transient rise in Ca that initiates contraction of the myofilaments. Relaxation occurs when Ca is removed from the cytosol (arrows) either back across the SL via forward-mode NCX or back into the SR via the SR Ca pump (SERCA) whose activity is regulated by phospholamban (PLB). (Figure composed by Dr G.L.J. Galli.)

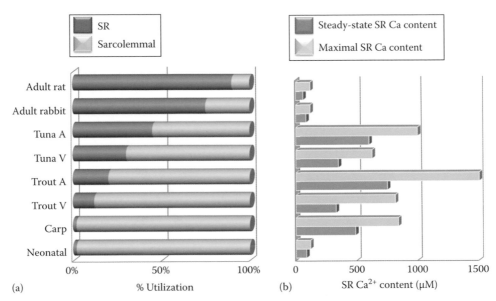

FIGURE 2.6 (a) SR and SL Ca utilization during contraction. Note how transsarcolemmal Ca flux contributes more Ca to contraction than SR Ca release in fishes as compared with mammals. (b) Maximal and steady-state SR Ca content expressed as µM Ca/nonmitochondrial cell volume. Notice the incongruence between SR Ca utilization during contraction (a) and the Ca content of the SR (b) with a species. Both figures are compiled from a number of studies (see following text), using various methods to assess the source of Ca for contraction and the SR Ca contents. Thus, the figure is illustrative rather than quantitative. Data adapted from: Adult rat ventricle (a) (Negretti et al., 1993) (b) (Delbridge et al., 1997); adult rabbit ventricle (a) (Bassani et al., 1994) (b) (Satoh et al., 1996); tuna atrium, tuna ventricle (a and b) (Galli et al., 2011); trout atrium (a) (Shiels et al., 2002) (b) (Haverinen and Vornanen, 2009a,b); trout and carp ventricle (a and b) (Haverinen and Vornanen, 2009a,b); neonatal rabbit ventricle (Haddock et al., 1999). (Figure compiled with help of Dr G.L.J. Galli.)

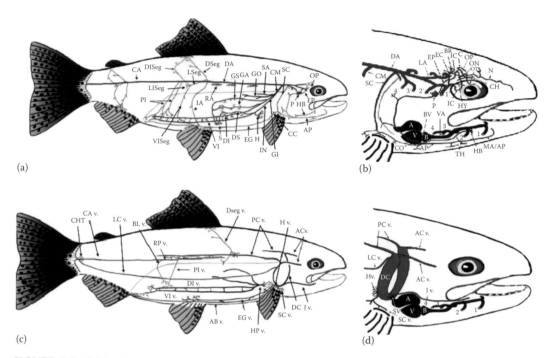

FIGURE 2.8 Major blood vessels in a teleost fish. Arterial (panels a and b) and venous (panels c and d) circulations in the rainbow trout (*Oncorhynchus mykiss*). A, atrium; AB v., abdominal vein; AC v., anterior cardinal vein; AP, afferent pseudobranch artery; B, bulbus arteriosus; BL v., bladder vein; BR, brain; BV, bulboventricular valve; C, commissure vessel; CA, caudal artery; CA v., caudal vein; CAP, capillary network; CH, choroid artery; CHT, caudal heart; CM, celiacomesenteric artery; CO, coronary artery; DA, dorsal aorta; DC, ductus Cuvier; DI, dorsal intestinal artery; DI v., dorsal intestinal vein; DISeg, dorsal intersegmental artery; DS, duodenosplenic artery; DSeg, dorsal segmental artery; DSeg v., dorsal segmental vein; EC, external carotid artery; EG, epigastric artery; EG v., epigastric vein; EP, efferent pseudobranch artery; GA, gastric artery; GI, gastrointestinal artery; GO, gonadal artery; GS, gastrosplenic artery; H, hepatic artery; HB, hypobranchial artery; H v., hepatic vein; HP v., hepatic portal vein; HY, hyoidean artery; IA, intercostal artery; IC, internal carotid artery; ICos, intercostal artery; ICos v., intercostal vein; IN, intestinal artery; J v., jugular vein; LA, lateral aorta; LC v., lateral cutaneous vein; LISeg, lateral intersegmental artery; LSeg, lateral segmental artery; MA, mandibular artery; N, nasal artery; ON, orbito-nasal artery; OP, ophthalmic artery; OT, optic artery; P, pseudobranch; PC v., posterior cardinal vein; PI, posterior intestinal artery; PI v., post-intestinal vein; RA, renal artery; RP, renal portal; RP v., renal portal vein; S, spleen; SA, swim bladder artery; SC, subclavian artery; SC v., subclavian vein; SV, sinus venosus; TH, thyroidean artery; V, ventricle; VA, ventral aorta; VI, ventral intestinal artery; VI v., ventral intestinal vein; VISeg, ventral intersegmental artery; 1–4 afferent branchial arteries to gill arches 1–4; 1′–4′ efferent branchial arteries from gill arches 1–4. (Modified From Olson, K.R. and Farrell, A.P., Secondary circulation and lymphatic anatomy, in Farrell, A.P., Ed., *Encyclopedia of Fish Physiology, from Genome to Environment*, 1st edn., Academic Press, London, U.K.; Waltham, MA, pp. 1161–1168, 2011. With permission.)

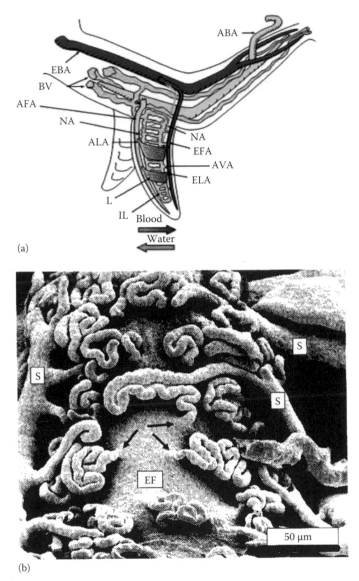

(a)

(b)

FIGURE 2.9 Blood vessels in the fish gill. (a) The afferent brachial artery (ABA) delivers blood to, and distributes it along, the gill arch. Afferent filamental arteries (AFAs) distribute blood to the respiratory lamellae (L) via afferent lamellar arterioles (ALAs). Oxygenated blood leaves the lamellae through short efferent lamellar arterioles (ELAs) and from there passes into the efferent filamental arteries (EFAs) and then into the efferent branchial artery (EBA). Small nutrient arteries (NAs) originate from the EFA near the base of the filament and reenter the filament. The intralamellar vasculature (IL) forms an extensive sinus with the filament that ultimately drains into the branchial veins (BV) in the gill arch. (Modified from Olson, K.R. and Farrell, A.P., Secondary circulation and lymphatic anatomy, in Farrell, A.P., Ed., *Encyclopedia of Fish Physiology, from Genome to Environment*, 1st edn, Academic Press, London, U.K./Waltham, MA, pp. 1161–1168, 2011. With permission.) (b) Scanning electron micrograph of a vascular corrosion replica showing the origin of the circulation in the gill filament of the climbing perch, *Anabas testudineus*. Tortuous narrow-bore arterioles (arrows) arise from an EFA and anastomose to form progressively larger secondary arterioles (S) that ultimately perfuse the core of the gill filament and the arch support tissue. (Adapted from Olson, 1986. *J Exp Zool* 275: 172–185. With permission.)

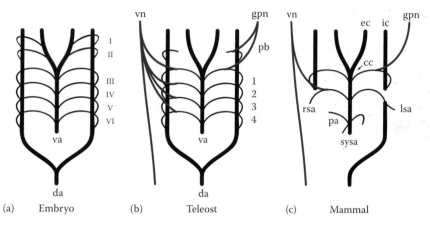

FIGURE 4.1 The aortic arches and the distribution of chemoreceptors in teleosts and mammals. Associated innervations from the glossopharyngeal (gpn, IX) and vagus (vn, X) nerves are shown in red. Innervation is bilateral but is displayed individually for clarity. (a) The embryonic condition, with six aortic arches (I–VI) connecting the ventral aorta (va) to paired dorsal aortae (da). Numerals correspond to homologous structures in (b) and (c). (b) In teleosts, the first aortic arch is absent, the second is modified into the nonrespiratory pseudobranch (pb), and aortic arches III–VI become gill arches 1–4, respectively. The pb and first gill arch receive gpn innervation, whereas all four arches receive vn innervation. (c) In mammals, the ventral aorta is modified to the common carotid (cc), which bifurcates to the external and internal carotids (ec, ic, respectively). The right da is lost. The va is modified to become the systemic aorta (sysa). The site of the carotid body, the cc, receives gpn innervation. (From Zachar, P.C. and Jonz, M.G., *Respir. Physiol. Neurobiol.*, 184, 301, 2012a; With permission.)

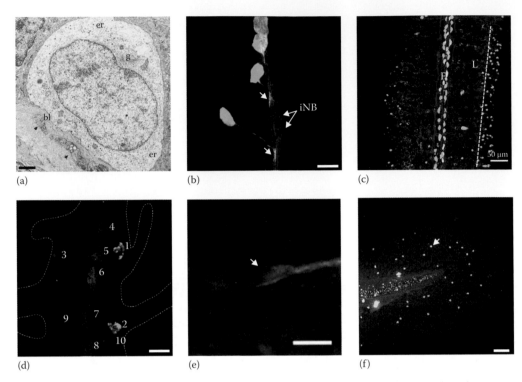

(a)   (b)   (c)

(d)   (e)   (f)

FIGURE 4.3 Morphology of neuroepithelial cells (NECs). (a) Electron micrograph showing ultrastructural characteristics of a single NEC of the filament in trout (*Salmo gairdneri*). The NEC is apposed to the basal lamina (bl). Small arrows indicate tonofilaments of surrounding cells. Several nerve profiles reside beneath the basal lamina (arrowheads). er, rough endoplasmic reticulum; g, Golgi apparatus. Scale bar indicates 1 μm. (From Dunel-Erb, S. et al., *J. Appl. Physiol. Resp. Environ. Exercise Physiol.*, 53, 1342, 1982. With permission.) (b) Confocal micrograph of NECs and associated innervation in the gill filament of zebrafish (*Danio rerio*). NECs and nerve fibers were labeled by serotonin (5-HT) and zn-12 immunohistochemistry, respectively. This tissue was taken from a specimen that had been acclimated to chronic hypoxia. Note the long neuron-like processes that formed and extended terminals (arrows) to form contact with the intrabranchial nerve bundle (iNB) following acclimation. Scale bar indicates 10 μm. (Modified from Jonz, M.G. et al., *J. Physiol.*, 560, 737, 2004. With permission.) (c) Confocal image showing NECs of gill filament (F) and lamellae (L) of goldfish (*Carassius auratus*) labeled by 5-HT and zn-12 immunohistochemistry. This tissue was taken from a specimen that had been acclimated to 7°C. Note that following acclimation, NECs of the lamellae were redistributed toward the distal regions of lamellae, as indicated by the dashed line, and maintained their innervation. (Modified from Tzaneva, V. and Perry, S.F., *J. Exp. Biol.*, 213, 3666, 2010. With permission.) (d) In zebrafish, cells containing the vesicular acetylcholine transporter (VAChT) do not contain serotonin (5-HT) or the synaptic vesicle protein SV2. Confocal micrographs from an adult gill filament demonstrate triple immunohistochemical labeling with antibodies against VAChT, 5-HT, and SV2. 10 cells are indicated: cells 1–3 are VAChT-positive, cells 4–8 are 5-HT and SV2-positive, and cells 9 and 10 are only SV2-positive. A dashed outline indicates the position of the lamellae. Scale bar is 10 μm. (From Shakarchi, K. et al., *J. Exp. Biol.*, 216(5), 869, 2013. With permission.) (e) By contrast, in trout intrabranchial neurons were immunolabeled with both 5-HT and VAChT (arrow), but NECs were immunonegative for these markers. Scale bar is 10 μm. (From Porteus, C.S. et al., *Acta Histochem.*, 115, 158, 2013. With permission.) (f) Confocal image of NECs of the skin in developing zebrafish. NECs and associated innervation were labeled by 5-HT and zn-12 immunohistochemistry. Scale bar indicates 100 μm. (Modified from Coccimiglio, M.L. and Jonz, M.G., *J. Exp. Biol.*, 215, 3881, 2012. With permission.)

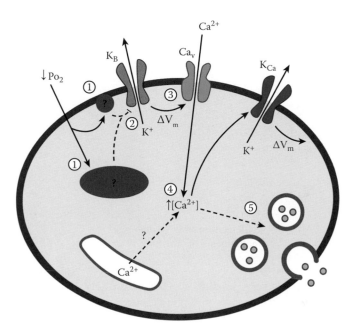

FIGURE 4.5 A current working model of O_2 sensing in gill neuroepithelial cells (NECs). (1) A decrease in P_{O2} is detected by an uncharacterized O_2 sensor that is membrane-delimited or cytosolic/mitochondrial. (2) Detection of hypoxia leads to inhibition of K^+ current through background K^+ (K_B) channels, leading to membrane depolarization (ΔV_m). (3) Voltage-dependent Ca^{2+} (Ca_V) channels are activated, permitting Ca^{2+} entry into the NEC. (4) Cytosolic Ca^{2+} increases, with intracellular stores possibly playing a role. (5) Ca^{2+} ions facilitate vesicle fusion and release of neurotransmitter. Ca^{2+} may also increase conductance through Ca^{2+}-activated K^+ (K_{Ca}) channels, as observed in goldfish NECs, modulating ΔV_m. Dotted lines represent presumptive mechanisms. (From Zachar, P.C. and Jonz, M.G., *Respir. Physiol. Neurobiol.*, 184, 301, 2012a; With permission.)

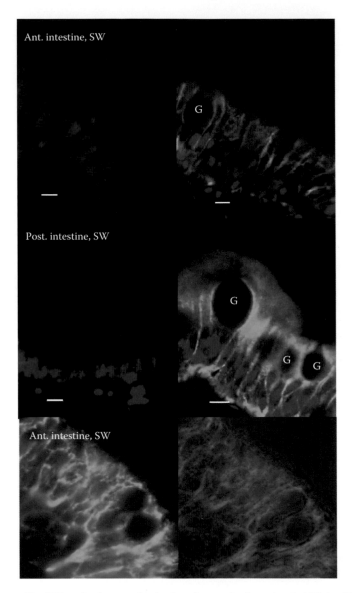

FIGURE 5.3 V-type H$^+$-ATPase in the anterior (top) and posterior intestine (middle) with negative controls (left panels) as well as Na$^+$/K$^+$-ATPase in the anterior intestine (bottom, left) and brightfield overlay (right) from toadfish held in seawater. Scale bar: 10 μ. (Images are from Guffey, S. et al., *Am. J. Physiol.*, 301, R1682, 2011.) (H$^+$ pump) and (From Tresguerres, M. et al., *Am. J. Physiol.*, 299, R62, 2010) (Na$^+$/K$^+$-ATPase).

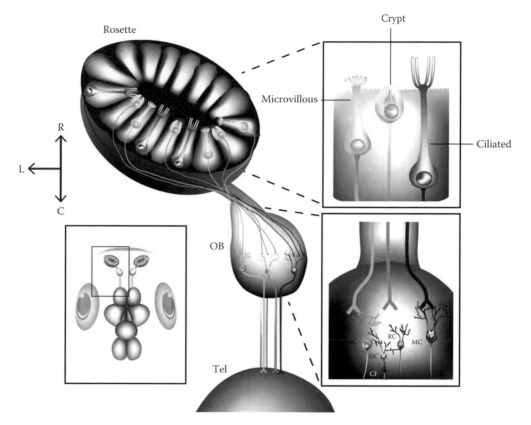

FIGURE 11.2 A representation of the organization of the ciliated, microvillous, and crypt cell olfactory sensory morphotypes in teleosts. The representation is shown on a goldfish-like arrangement of the olfactory system. In the peripheral olfactory organ, the three olfactory sensory neuron (OSN) morphotypes intermingle in the olfactory epithelium covering the rosette; however, the ciliated and microvillar morphotypes predominate. In the olfactory bulb (OB), the axons are organized according to morphotype. In general, the axons of the microvillous OSNs project laterally, the axons of the ciliated OSNs are medial, and the crypt cell axons are located in a small ventral region. Synaptic contacts are made in the olfactory bulb. The output neurons (mitral cells) project to the telencephalon (tel). The OB contains the mitral cells (mc), ruffed cells (rc), and granule cells (gc), as well as centrifugal fibers (cf) projecting from higher brain centers. R, rostral; C, caudal; L, lateral.

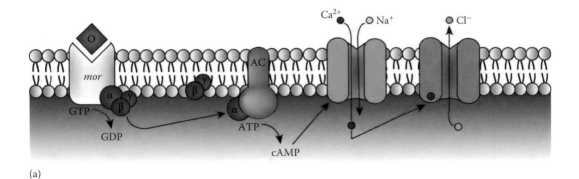

(a)

(b)

FIGURE 11.3 (a) The signal transduction cascade in ciliated olfactory sensory neurons (OSNs). The odorant (O) binds onto the *mor*-type G protein-coupled receptor (mor), the dissociated G protein subunits activate adenylate cyclase (AC), and ATP dephosphorylates, forming cAMP, and the cyclic nucleotide gated channels open, allowing for the inflow of cations (calcium: Ca^{2+}; sodium: Na^+). The intracellular calcium gates a chloride (Cl^-) channel. Chloride flows outward along its concentration gradient, further depolarizing the olfactory sensory neuron. (b) The signal transduction cascade in microvillar OSNs. The odorant (O) binds onto the *olfc*-type G protein-coupled receptor (*olfcs*), the dissociated G protein subunits activate phospholipase C (PLC), and phosphatidylinositol (3,4,5)-triphosphate (PI(3,4)P) dissociates to di-acyl glycerol (DAG) and inositol triphosphate (IP_3). The IP_3 gates the TRPM5 channels, allowing for the inward flow of cations. The intracellular calcium gates a chloride (Cl^-) channel. Chloride flows outward along its concentration gradient, further depolarizing the OSN.

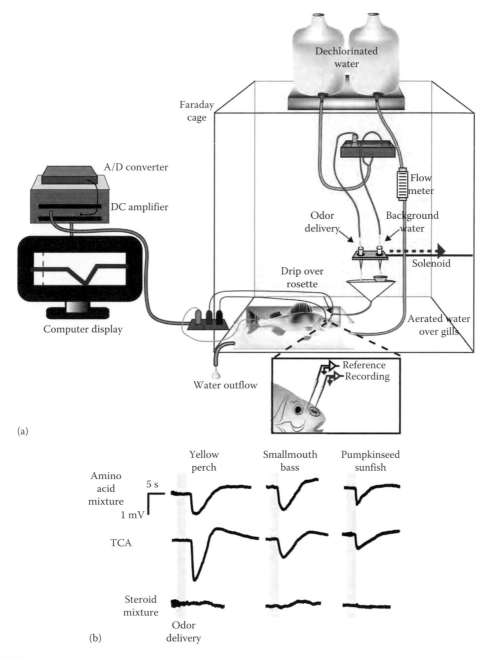

(a)

(b)

FIGURE 11.4 A representation of the delivery of test solutions to the olfactory epithelium of a teleost, and the recording of field potentials from the olfactory epithelium. (a) The recording of field potentials from the peripheral olfactory organ of a fish using the electro-olfactogram (EOG). Aerated water flows into the mouth and over the gills of an anesthetized fish. Background water flows into a small funnel into the nasal cavity and over the olfactory rosette. A second tube delivers water containing an odorant test solution. When activated, a computer-controlled solenoid allows for rapid switching between the background water and the odorant test solution being delivered over the rosette. A recording electrode is situated in the nasal cavity in close proximity to the olfactory epithelium, and a reference electrode is placed on the adjacent skin. The output of the electrodes is connected to a DC amplifier, the amplified signal is converted from analog to digital by an A/D converter, and the resulting signal is displayed on a computer. (b) EOG responses of three fish—a yellow perch, a smallmouth bass, and a pumpkinseed sunfish—to three classes of odorants: amino acids, a bile acid (taurocholic acid [TCA]), and a steroid mixture. EOG responses were observed in response to amino acids and the bile acid, but not to the steroids. The opaque bar indicated the duration of the odor delivery. 5 s; 1 mV.

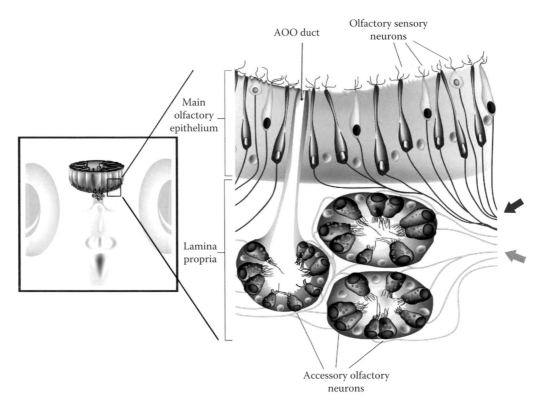

FIGURE 11.5 In the sea lamprey, the main olfactory epithelium contains ciliated olfactory sensory neurons (OSNs). There are tall (red), intermediate (yellow), as well as short (green) OSNs. Each form extends a single axon into the underlying lamina propria (red arrow). The main olfactory epithelium also contains ducts that lead to the accessory olfactory organ, small tubules containing rounded ciliated cells (blue). Axons extend from these accessory olfactory neurons (blue arrow). R, rostral; C, caudal. (Based on Laframboise, A.J. et al., *Neurosci. Lett.*, 414, 277, 2007; Ren, X. et al., *J. Comp. Neurol.*, 516, 105, 2009.)

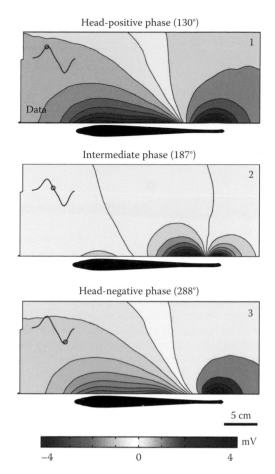

FIGURE 12.2 Electric organ discharge (EOD) voltage maps for *Apteronotus leptorhynchus* at three different phases of a single EOD cycle (130°, 187°, 288°) with EOD frequency (EODf) of 810 Hz. (Modified from Kelly, M. et al., *Biol Cybern*, 98, 479, 2008; Assad, C. et al., *J. Exp. Biol.*, 202, 1185, 1999.) Insert on each panel illustrates single EOD cycle with phase indicated by a circle. Color scale for voltage and spatial scale indicated at bottom.

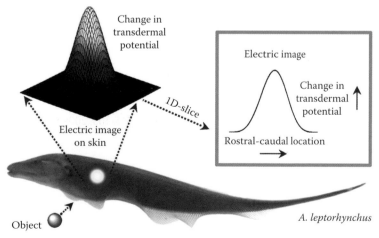

FIGURE 12.4 Schematic of *Apteronotus leptorhynchus* showing the electric image on the skin due to a nearby plastic sphere; voltage perturbation on the skin is color-coded, with light colors indicating larger values. The electric image is also illustrated in a 3D plot with the voltage perturbation represented by the same color code but also in the *z*-axis (height of the 2D Gaussian-shaped surface). As is the convention, a 1D slice of this surface is taken to illustrate the electric image in a single spatial dimension (i.e., rostral–caudal; see inset). (Photo credit: Robert Lacombe, University of Ottawa, Ottawa, Ontario, Canada.)

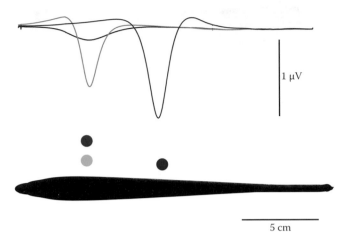

FIGURE 12.5 Overlays of electric images for prey-like objects at three different locations illustrating the decrease in amplitude and increase in width for increasing lateral distances (*y*-axis), and the increase in image amplitude in the caudal direction (*x*-axis). (Modified from Babineau, D. et al., *PLoS Comput. Biol.*, 3, e38, 2007.) Coordinates for the blue, green, and red objects respectively are $(x, y) = (5, 3), (5, 1.5), (10, 1.5)$ cm; origin is at the nose of a 21-cm fish. As is the convention, electric images are computed as the difference between the transdermal potentials, measured with and without the object present, along the rostral–caudal axis, that is, a 1D slice along the skin surface. Prey-like objects are modeled as 0.3-cm diameter discs with a conductivity of 0.0303 S/m (water conductivity: 0.023 S/m).

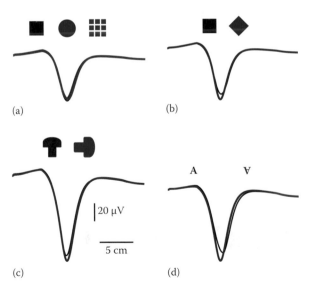

FIGURE 12.6 Electric images of complex objects, computed as in Figure 12.5 and by Babineau et al. (2006). Object shapes shown schematically in cross section (actual dimensions are as follows) and color-coded to the corresponding image in each panel. All objects are metal (conductivity = 3.7×10^7 S/m), are centered at $(x, y) = (6, 2)$ cm, and are uniform in the third dimension (dorsal–ventral axis of fish); that is, they are "rodlike," with the cross sections shown. Cross-sectional dimensions: (a) Cube (1×1 cm), disc (radius = 0.56 cm), fractionated cube (each fraction, 0.33×0.33 cm, with gap width 0.5 cm). All objects have equal volumes. (b) Cubes (1×1 cm), one rotated 45°. (c) Mushroom-shaped, one rotated 90° (length = 1.5 cm, large width = 1.2 cm, small width = 0.8 cm). (d) A-shaped, one rotated 180° (length = 1.6 cm).

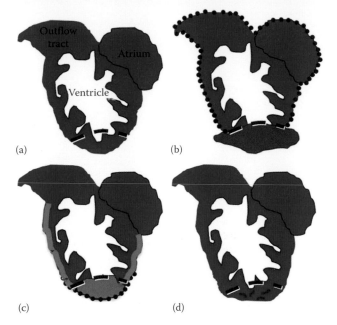

(a)   (b)

(c)   (d)

FIGURE 13.1   A model of heart regeneration. (a) Zebrafish hearts are comprised of a single atrium, ventricle, and an outflow tract. (b) In response to a resection injury (dashed black line), a blood clot (brown) quickly seals the wound. Within the first 1–3 days postamputation (dpa), there is organ-wide activation of the epicardium (black dots) and remodeling of the endocardium. (c) By 7 dpa, the blood clot is replaced with a collagen clot (light blue) and epicardium signals become localized to the injury site. At this time, *gata4+* cells (green) in the compact muscle undergo proliferation. (d) By 30–60 dpa, a contiguous myocardial wall has reformed, new cardiomyocytes are electrically coupled to the uninjured heart tissue, and new blood vessels penetrate the regenerate (dark red). Functionally, the regenerated heart is indistinguishable from the uninjured organ.

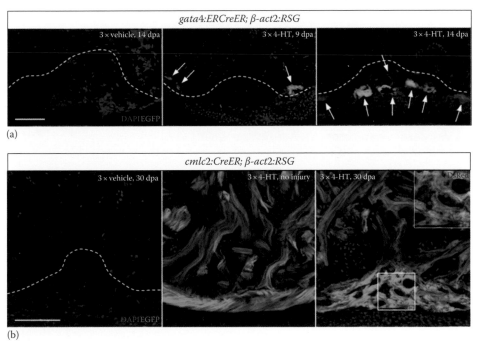

(a)

(b)

FIGURE 13.2   A *gata4+* subpopulation of cardiomyocytes contributes to heart regeneration. (a) *Tg(gata4:CreER; β-act2:RSG)* animals were injected with vehicle (left) or 4-HT to induce Cre-mediated recombination. EGFP+ cells (arrows) are present near the injury site at 9 days postamputation (dpa) and within the wound area by 14 dpa. (b) *Tg(cmlc2:CreER; β-act2:RSG)* animals were injected before injury with vehicle (left) or 4-HT. The majority of cardiomyocytes in the uninjured and at 30 dpa are labeled in the presence of 4-HT treatment. (Dashed line, approximate amputation plane; inset depicts DsRed expression that was used for determining labeling efficiency; scale bar, 50 µm.) (Reprinted from Macmillan Publishers Ltd. *Nature*, Kikuchi, K. et al., Primary contribution to zebrafish heart regeneration by gata4(+) cardiomyocytes, 464(7288), 601–605, Copyright 2010.)

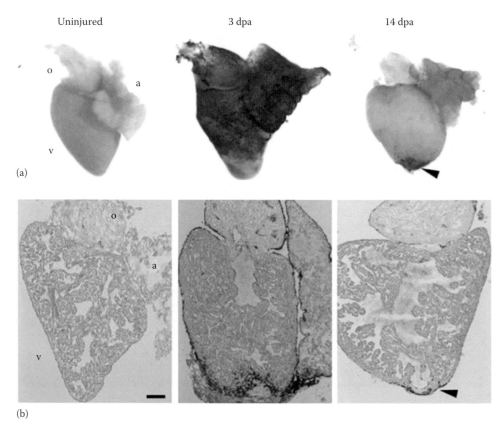

FIGURE 13.3 *Raldh2* expression is rapidly induced in the epicardium and endocardium. (a) Whole-mount in situ hybridizations show *raldh2* expression is absent in uninjured hearts but is detected organ-wide at 3 days postamputation (dpa) in the epicardium. Expression is confined to the injury site by 14 dpa in the epicardium and endocardium. (b) Sections of whole-mount stained hearts show the changing spatial expression of *raldh2* from uninjured to 14 dpa. (Arrowhead marks injury site; scale bar, 100 μm; o, outflow tract; a, atrium; v, ventricle.) (Reprinted from *Cell*, 127, Lepilina, A. et al., A dynamic epicardial injury response supports progenitor cell activity during zebrafish heart regeneration, 607–619, Copyright 2006, from Elsevier.)

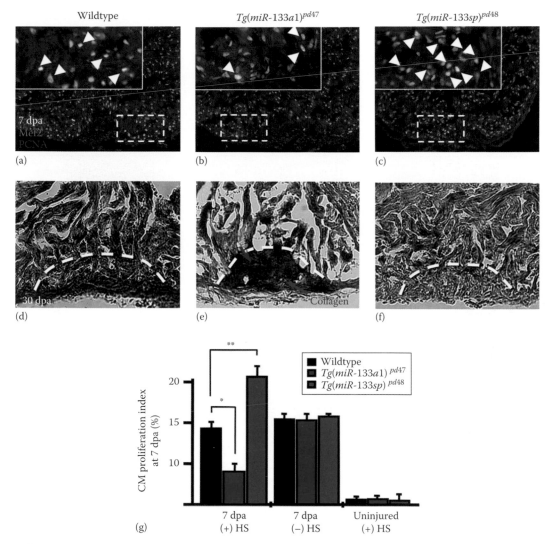

FIGURE 13.4 miR-133 controls cardiomyocyte proliferation and regeneration. (a–c) Wild-type, *Tg(hsp70:miR-133pre)*[pd47] and *Tg(hsp70:miR-133sp)*[pd48] animals were injured, heat-treated, and hearts were stained for Mef2 and PCNA at 7 days postamputation (dpa). (b) When miR-133 levels are overexpressed, cardiomyocyte proliferation is repressed by ~50%. (c) Conversely, depletion of miR-133 leads to elevated cardiomyocyte proliferation by ~45%. (d–f) Sustained miR-133 modulation and AFOG stains at 30 dpa reveal abnormal scar deposition and inhibition of regeneration in *Tg(hsp70:miR-133pre)*[pd47] or normal regeneration in control and *Tg(hsp70:miR-133sp)*[pd48] hearts. (g) Cardiomyocyte proliferation indices indicate miR-133 controls cardiomyocyte proliferation. (White arrowheads indicate proliferating cardiomyocytes; inset in (a–c) is an enlargement of dashed boxes; AFOG, acid fuchsin orange G; dashed white lines in (d–f) is approximate amputation plane; HS, heat shock.). (Modified from *Dev. Biol.*, 365, Yin, P. et al., Regulation of zebrafish heart regeneration by miR-133, 319–327, Copyright 2012, from Elsevier.)

Injury Type	Nonmyocardial Tissue Activation	Cardiomyocyte Activity	Duration of Regeneration	Necrotic Tissue	References
 Resection	Organ-wide activation of the epicardium and endocardium followed by restricted expression near injury site	*gata4* subpopulation of cardiomyocytes within the compact muscle layer proliferates	30–60 days for complete regeneration of missing myocardial tissue	Clean amputation of ventricular apex	Poss et al. (2002), Raya et al. (2003)
 Cryoinjury	Activation of the epicardium and endocardium mirrors that of a resection injury. Fibroblasts undergo rapid proliferation	Cardiomyocytes near injury dedifferentiate and re-enter the cell cycle	100–130 days to remove necrotic tissue and regeneration of new heart muscle	Necrotic patch of myocardium remains following injury	Schnabel et al. (2011), González-Rosa et al. (2011), Chablais et al. (2011)
 DTA ablation	Strong activation in both epicardium and endocardium layers despite localized damage to myocardium alone	The most robust cardiomyocyte proliferation among the three injury models	30 days to regenerate and integrate regions of newly created cardiomyocytes	Many patches of necrotic myocardial tissue	Wang et al. (2011)

FIGURE 13.5 Different heart injury models. Ventricular resection, cryoinjury, and diphtheria toxin A (DTA) ablation models of heart injury are compared with each other. Differences in regenerative responses are described within the following major categories: injury type, nonmyocardial tissue activation, cardiomyocyte activity, duration of regeneration, and presence or absence of necrotic tissue.

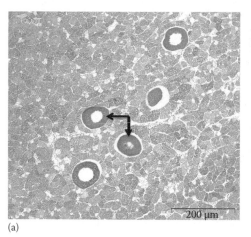
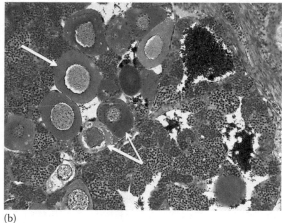

(a)   (b)

FIGURE 7.2   Histological sections showing intersex gonadal phenotypes. (a) Gonad in male roach exposed to wastewater treatment effluent in which the testis contains a large number of primary oocytes. (Photo courtesy of Anke Lange and Charles Tyler.) (b) Testes from male fathead minnows from Lake 260, an EE_2-exposed study lake. (Photo courtesy of Bradley Park and Karen Kidd.) Arrows in both images indicate primary-stage oocytes interspersed within spermatogenic seminiferous tubules in the testes.

Although banned in the United States for the past 40 years, the organochlorine insecticide DDT and its degradation product, dichlorodiphenyldichloroethylene (DDE), are still present in the environment and continue to bioaccumulate in animal tissues, including humans (Jaga and Dharmani, 2003). Like WWTE, DDT has been shown to have the ability to induce vitellogenesis in fish species, including Atlantic salmon (Tollefsen et al., 2003).

Virtually all living organisms need nitrogen to survive, but animals and plants need nitrogen in a fixed form. Most naturally occurring reactive nitrogen comes from nitrogen that has been fixed by bacteria; however, the past 20 years has seen a tremendous increase in the amount of reactive nitrogen produced by humans, through sources such as agricultural fertilizers, the burning of fossil fuels, and animal agriculture. Humans have now altered the nitrogen cycle more than any other natural cycle, and excess environmental nitrogen is now a leading global contaminant (Fields, 2004). Nitrate, a major form of reactive nitrogen, has increased dramatically in aquatic systems, including oceans and seas (Kim et al., 2011). Until recently, aquatic nitrate was thought to be relatively innocuous, but we now know nitrate has the ability to alter endocrine function in a variety of aquatic animals, including fishes and amphibians (Guillette and Edwards, 2005). In fishes, nitrate has been shown to alter circulating concentrations of sex steroids, alter uptake of iodine by the thyroid, and cause a number of reproductive impairments (Lahti et al., 1985, Hamlin et al., 2008). Male mosquito fish (*Gambusia holbrooki*) collected in nitrate contaminated springs show decreased sperm counts and increased length of the gonopodium (a modified fin used for sperm transfer), which correlate significantly with aquatic nitrate concentration (Edwards and Guillette, 2007). In addition, dry weight of developing embryos and the numbers of reproductive females were negatively correlated with nitrate concentration (Edwards et al., 2006). As discussed previously, not all EDCs elicit cellular responses through their interactions with hormone-specific receptors. The precise mechanism of action of nitrate in EDC-induced dysfunction is unclear, but it is not thought to interact with sex steroid receptors, but instead could interfere with key steroidogenic pathways. Nitrate can be converted to nitric oxide (NO) *in vivo*, and NO has been shown to interfere with a number of P450 enzymes essential for the production of steroid hormones, by binding to the heme group, which is the hallmark of all enzymes of the P450 superfamily (Zweier et al., 1995, Hannas et al., 2010, Ducsay and Myers, 2011). P450 enzymes are also used by the liver in the removal of hormones from circulation, as well as for toxin removal. Therefore, the conversion of nitrate to NO could serve as a mechanism to alter both steroidogenesis and the removal of sex steroids from circulation.

Although many teleosts are hermaphroditic, most species have separate sexes and it is generally accepted that sex determination is under genetic control, although few fish have sex chromosomes that can be differentiated from other chromosomes (Tyler et al., 1998). Although sex assignment is genetically mandated for most fish, sexual differentiation, or the phenotypic translation of genetic sex, can be altered by exposure to exogenous steroids and the enzymes controlling their synthesis (Piferrer et al., 1994, Donaldson, 1996, Devlin and Nagahama, 2002). Sex steroids play critical roles in regulating sexual differentiation, and treating fish with estradiol or testosterone can induce sex reversal, a technique sometimes used in aquaculture to skew a population toward males or females (Yamazaki, 1983, Piferrer, 2001). Japanese medaka show a skewed sex ratio in favor of females when exposed to certain EDCs, such as octylphenol, during development (Knorr and Braunbeck, 2002). Octylphenol belongs to a broader family called alkylphenols, which are used in the production of detergents, fuel additives, fragrances, and fire retardants, and are commonly released into the environment in sewage treatment and pulp mill effluent and textile industries (Giger et al., 1984, Ahel et al., 1994, 1996). DDT has also shown a similar effect in skewing sex ratios in favor of females (Cheek et al., 2001).

7.4.2 EDCs and the HPT Axis

Thyroid hormones play significant roles in reproduction, growth, and development, and disruptions in thyroid function can occur along any number of steps in the thyroid cascade (Brown et al., 2004, Liu et al., 2011, Chan and Chan, 2012). While T3 is the more biologically active of the thyroid hormones, interpretation of the action of T3 on target tissues is complicated because T3 and its receptor can occur as a heterodimer that includes a receptor for another ligand. Therefore, the activity of T3 could be altered by disruption of this second ligand as well as by interaction with its own receptor. A variety of chemicals can alter thyroid function, ranging from halogenated hydrocarbons to less complex anions or cations.

In addition to altering the HPG axis, organochlorine pesticides, including DDT, have the ability to alter thyroid function as well. While mullet (*Liza parsia*) exposed to DDT show a decrease in thyroid epithelial cell height, depletion of colloid, and degeneration of epithelial cells (Kumar Pandey et al., 1995), tilapia exposed to DDT show increased epithelial cell height as well as an increase in nuclear cell diameter. Other organochlorine pesticides such as endrin interfere with gamma-amino butyric acid (GABA) receptors and can cause reduced uptake of radio-labeled iodide into thyroid follicles as well as inhibit thyroid hormone synthesis (Grant and Mehrle, 1970, Bhattacharya et al., 1978). Endosulfan is an organochlorine pesticide used on food crops and as a wood preservative that antagonizes GABA-gated chloride channels. In addition to causing hypertrophy and hyperplasia of thyroid follicles, it can also lead to thyroidal degeneration and altered concentrations of thyroid hormones.

Organophosphorous pesticides are broadly used to combat insects by inhibiting acetylcholinesterase activity, which is essential for nerve function not only in insects, but in fish, humans, and other mammals as well. Commercial mixtures are sold under a variety of trade names such as malathion, parathion, chlorpyrifos, diazinon, and phosmet. Malathion is one of the most widely used pesticides in agriculture, landscaping, and public recreation areas in the United States. In the catfish *Heteropneustes fossilis*, malathion causes decreased TSH synthesis and secretion as well as lower plasma T4 levels and higher plasma T (Singh and Singh, 1980, Brown et al., 2004). Malathion has been shown to alter thyroid hormone levels in several other fish species including the walking catfish (*Clarias batrachus)* and the freshwater murrel (*Channa punctatus*).

PCBs are chemicals used mainly as insulating fluids and coolants in transformers, electric motors, and hydraulic fluids and can also be found in certain adhesives and paints. Approximately 209 PCB congeners exist. PCBs are highly stable and persistent in the environment, and readily bioaccumulate in aquatic food webs. Studies suggest that PCBs can alter indicators of thyroidal status in fishes, but their mode of action is poorly understood (Schnitzler et al., 2011). Rainbow trout exposed to PCBs show altered circulating concentrations of thyroid hormones, and increased liver weight and liver lipid content. Altered concentrations of thyroid hormones have been shown in European flounder (*Platichthys flesus*) and coho salmon (*Oncorhynchus kisutch*) as well (Leatherland and Sonstegard, 1978, 1980, Besselink et al., 1996).

Many other environmental contaminants have been shown to be or are at risk for altering thyroid function in fishes, including the rocket fuel additive perchlorate, flame retardants, pharmaceutical agents, environmental steroids, plasticizers, metals, and even ammonia (Chan and Chan, 2012). Although the thyroid cascade can act as a biomarker of EDC exposure, this system is extremely complex, and the thyroid has an unusual ability to compensate for physiological insult. Homeostatic adjustments can take place at the central level to regulate T4 production, or peripherally to regulate plasma T3. In addition, the thyroid itself can undergo morphological changes to compensate for inadequate production, and can store considerable amounts of thyroid hormone. Therefore, short-term exposure studies, even studies evaluating high doses of contaminant, might not deplete these stores. The thyroid hormone reserves are not well studied in fishes, and more work in this area is necessary to gain a more complete understanding of the effects of EDCs on thyroid function and how this system can be modulated by xenobiotic exposure. In addition, the effects of only a handful of EDCs have been investigated on very few species of fishes, so there is much to be learned regarding interspecies variation and the extent of responses to the diversity of chemicals present in the environment.

7.4.3   EDC Effects on Neuroendocrine Function and Behavior

The central neuroendocrine systems that regulate homeostatic processes such as growth, reproduction, and development are initiated by neuro-hormonal signals primarily in the hypothalamus, which are delivered first by neural and then by endocrine signals. The neuroendocrine system serves as a critical link between the fish and its environment, and is susceptible to perturbations by EDCs (Gore, 2010, Jin et al., 2011). Although not well studied in fishes, many neuroendocrine pathways are highly conserved among vertebrates, increasing the likelihood that responses to EDCs are equally conserved.

The hypothalamic GnRH neurons play significant roles in the control of reproduction for all vertebrates studied to date (Gore, 2002), and GnRH neurons have been shown to be direct targets of EDCs. In Atlantic croaker, PCBs decrease GnRH peptide content in the hypothalamus, numbers of GnRH receptors, and the amount of LH that is released in response to a GnRH challenge (Khan and Thomas, 2001). In terrestrial animals, developmental EDC exposure can disrupt GnRH systems later in life, and it is likely that fish follow a similar model (Gore, 2008).

It is well known that early-life exposure to hormones such as testosterone or estradiol can permanently modulate brain morphology in a sexually dimorphic manner. Meaning, exposure to EDCs can reorganize the brain and result in functional differences between males and females (Gore, 2008, Schwarz and McCarthy, 2008), although the mechanisms for these alterations are poorly understood. Reproductively active fish often show sex-specific behaviors that promote optimal mate choice and timing in gamete release, which can be critical for population survival and optimizing fitness (Munakata and Kobayashi, 2010). Alterations in brain function caused by EDCs can also lead to behavioral modifications that could compromise reproductive behaviors (Sebire et al., 2011, Soffker and Tyler, 2012). Sand gobies (*Pomatoschistus minutus*) have a polygynous mating system and exhibit male parental care (Forsgren, 1997). Normally, the male builds a nest and tends to the eggs until they hatch. Female gobies prefer males that show strong courting behavior and a high degree of parental care. Although male characteristics such as body size will contribute to her mate preference, the nest built by the male is a crucial resource important for egg development, and is thought to be an indication of male parental quality. When male gobies are exposed to EE_2, they are not able to acquire or defend a nest site and spend significantly less time courting females (Saaristo et al., 2009, 2010). Fathead minnows exposed to EE_2 show similar behavioral modifications (Salierno and Kane, 2009). These altered behaviors could greatly reduce the males' chances of mating successfully. EDCs can also affect other behaviors such as hierarchical or agonistic behaviors that can significantly affect breeding outcomes (Colman et al., 2009).

7.4.4 Improving Our Understanding of Low-Dose Effects

Thresholds of safety for a given contaminant are generally determined by exposing one or several species to a number of exposure concentrations, until a point is reached at which no negative effects can be seen, and this is determined to be the lowest observed adverse effect level (LOAEL). Regulators will often choose a concentration lower than the LOAEL (generally 5 to 10 times less) and deem it safe for the entire ecosystem. The dogma that "the dose makes the poison" has served US regulatory practice for decades. However, new studies of EDC exposure are challenging these traditional toxicology concepts because many EDCs have effects at low doses that are not predicted by higher dose exposures (Vandenberg et al., 2012). A variety of environmental contaminants including certain pesticides, plasticizers, and industrial chemicals have shown adverse health effects in fishes at low, ecologically relevant doses, or have effects following a non-monotonic dose response curve (NMDRC). Atrazine is a herbicide used primarily on corn and is the most commonly detected pesticide in ground- and surface water (Solomon et al., 1996). Exposure to low concentrations of atrazine induces aromatase expression in fishes, causing increased conversion of T to E_2. The Environmental Protection Agency (EPA) drinking water standard for atrazine in the United States is 3 ppb, and studies in zebrafish (*Danio rerio*) have shown that atrazine doses as low as 2.2 ppb significantly increased aromatase expression (Suzawa and Ingraham, 2008). Atrazine has also been shown to feminize males and cause such effects as degeneration of interstitial tissue in the testes, loss of male germ cells, and skewed sex ratios toward females (Hayes et al., 2011).

Additionally, negative effects can be observed at high and low doses, but may not be apparent at concentrations in between. These effects follow an NMDRC, in which the slope of the curve changes from negative to positive (or vice versa) during the range of doses examined. For example, lead nitrate stimulates the production of E_2 and T in catfish (*H. fossilis*) ovaries at low concentrations but inhibits their production at high doses (Chaube et al., 2010). Many NMDRCs have a U or inverted U shape and are often referred to as biphasic, from their descending or ascending dose response phases (Figure 7.3). Conventional regulatory toxicology assumes a monotonic dose response, justifying the use of high-dose testing and the LOAEL as the standard for chemical safety assessments.

How is it then that very low concentrations of EDCs can exert significant physiological change that higher doses might not? The endocrine system is especially tuned to respond to relatively low

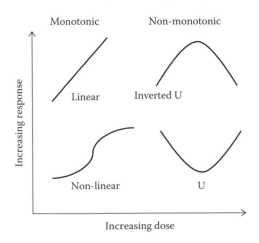

FIGURE 7.3   Monotonic versus non-monotonic dose response curves.

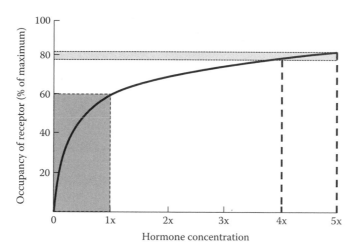

FIGURE 7.4 Theoretical example of the relationship between receptor occupancy and hormone concentration (or possibly hormone mimics). At low hormone concentrations an increase from 0 to 1× causes an increase in receptor occupancy of approximately 60% (dark gray box). However, at higher hormone concentrations, the same increase, in this case from 4× to 5×, causes an increase in receptor occupancy of only approximately 5% (pale gray box). (Figure redrawn from Vandenberg, L.N. et al., *Endocr. Rev.*, 33, 378, 2012.)

concentrations of hormones, generally in the picogram or nanogram per milliliter of plasma range for sex steroids or thyroid hormones. Receptor binding is often the initial step in endocrine disruption by compounds mimicking endogenous hormones (Welshons et al., 2003). The affinity of the ligand for the receptor must be great enough that a similar number of receptors are occupied as would be occupied if the natural ligand were present. The binding of hormones to their receptors shows a property of saturability, whereby as the receptors become saturated, there is no further change in response in spite of an increased concentration of hormone (or contaminant). Interestingly, the saturation of the response often occurs below a receptor occupancy of 100%. It is the change in receptor occupancy that is necessary to cause receptor-mediated response changes. Welshons and colleagues explain that the occupancy of receptors by the hormone ligand is not linear with hormone concentration, and saturation of the response (maximum biological response) can occur before saturation of the receptors. Therefore, lower doses of hormone (or hormone-mimicking compounds) can have a greater fold increase in receptor occupancy than high hormone doses, and cause marked changes in receptor occupancy, leading to considerable physiological change. Because it is the change in receptor occupancy that is necessary to cause receptor-mediated response changes, lower doses of hormone can cause greater degrees of changes in receptor occupancy, and hence a greater biological effect (Figure 7.4).

7.4.5 Population-Level Effects

There is clear evidence that fish exposed to EDCs in the laboratory exhibit a variety of endocrine dysfunctions that could lead to reduced reproductive fitness. These dysfunctions are also apparent in fish living in aquatic environments polluted with EDCs. Less clear is whether these effects can translate into meaningful disruptions at the population level, and whether low-dose exposures can contribute to species declines and impact the sustainability of wild populations (Mills and Chichester, 2005). Although studies at the population level are technically challenging and costly, several studies have been undertaken to address this question. An impressive 7-year study by Karen Kidd and colleagues examined the effects of low-dose EE_2 on a natural population of fathead minnows (*Pimephales promelas*) in north-western Ontario, Canada (Kidd et al., 2007). Fathead minnows are a common species in North America, and are an important food source for higher-level predators such as lake trout (*Salvelinus namaycush*) and others. These minnows generally live for about 2 years and often spawn for only one season. The primary estrogen used in birth control pills, EE_2, is a common contaminant in aquatic systems as it is not readily removed in municipal wastewater treatment processes. EE_2 was added to the lake at levels typically observed in municipal wastewaters. Within 7 weeks after the first addition of EE_2 to the lake, levels of vitellogenin began to rise in both male and female fathead minnows, with males having concentrations three orders of magnitude greater than reference samples. Levels of vitellogenin remained elevated throughout the study, with males having levels surpassing that of females for much of the study. Males also showed arrested testicular development and a condition called ova testes, in which oocytes are present in the testes (Figure 7.2). Females showed delayed ovarian development and a small number of atretic follicles, which were rarely observed in this population of fish prior to the EE_2 additions. The catch-per-unit effort decreased consistently following the EE_2 additions, and collapsed after the second season. The collapse was due to the loss of the young-of-the-year, and continued throughout the EE_2 additions for the next 2 years. Species collapse was not seen for the longer-lived pearl dace (*Margariscus margarita*), suggesting that life history traits are important factors in determining a species' risk, although the dace did show signs of decline during the third year (Palace et al., 2009). In addition, the pearl dace as well as other species investigated in the lake such as white sucker (*Catostomus commersonii*) and lake trout experienced vitellogenin inductions, whereas population declines were variable among species, illustrating that vitellogenin induction might not be sufficient to predict population-level effects.

EDCs can not only affect population dynamics through physiological disruptions, but could act through behavioral mechanisms, as described earlier for sand gobies, and could be extended to include changes in predator avoidance, locomotory performance, and habitat selection. Many marine species have a pelagic early life stage and are highly sensitive to alterations in dispersion and recruitment, although these alterations are often very difficult to detect (Sarria et al., 2011). Newborn pipefish exposed to concentrations of EE_2 within range of what has been reported in the environment shifted their vertical distribution toward the surface in a dose-dependent manner (Sarria et al., 2011). Changes such as this could have considerable impacts on feeding, predation, and overall population stability.

7.5   CONCLUSIONS

Although the past 15 years have been witness to an explosion in the number of studies covering many aspects of EDC exposure, contaminant-induced dysfunctions in wildlife have been documented for more than 70 years (Matthiessen, 2003). Many examples of EDC exposure that have been reported in wild fish populations pertain to estrogenic compounds, such as those released by sewage treatment and certain industrial effluents, which can cause a variety of endocrine disorders. These disorders can extend beyond the level of the receptor, and affect many aspects of the HPG or HPT axes. Although no animal appears to be immune to the effects of EDCs, aquatic wildlife, such as fish, appear to be particularly susceptible, which may be a consequence of the tendency of pollutants to accumulate in aquatic ecosystems, and their ability to biomagnify rapidly up aquatic food webs. While population studies are far less common than lab or correlational field studies, there is increasing evidence that low environmental doses of EDCs can have significant consequences for fish populations, and the extent of the impact depends on a number of variables, including fish life history traits. It is becoming increasingly clear that regulatory limits based on LOAELs are not sufficient to ascertain the potential harm imparted by EDCs, and new regulatory practices must take into account the unique nature of EDC-induced dysfunction that doesn't necessarily follow a monotonic dose response curve. As endocrine disruption research continues, and more tools are developed to expand our understanding of the physiological end points affected by EDCs, a clearer picture will emerge of the potential risks and the strategies necessary to effectively protect wild fish populations.

REFERENCES

Ahel M, Giger W, Schaffner C. 1994. Behavior of alkylphenol polyethoxylate surfactants in the aquatic environment. 2. Occurrence and transformation in rivers. *Water Research* 28: 1143–1152.

Ahel M, Schaffner C, Giger W. 1996. Behaviour of alkylphenol polyethoxylate surfactants in the aquatic environment. 3. Occurrence and elimination of their persistent metabolites during infiltration of river water to groundwater. *Water Research* 30: 37–46.

Anderson OS, Sant KE, Dolinoy DC. 2012. Nutrition and epigenetics: An interplay of dietary methyl donors, one-carbon metabolism and DNA methylation. *Journal of Nutritional Biochemistry* 23: 853–859.

Ankley GT et al. 2009. Endocrine disrupting chemicals in fish: Developing exposure indicators and predictive models of effects based on mechanism of action. *Aquatic Toxicology* 92: 168–178.

Anway MD, Cupp AS, Uzumcu M, Skinner MK. 2005. Epigenetic transgenerational actions of endocrine disruptors and mate fertility. *Science* 308: 1466–1469.

Bern HA. 1990. The new endocrinology—Its scope and its impact. *American Zoologist* 30: 877–885.

Bern HA, Pearson D, Larson BA, Nishioka RS. 1985. Neurohormones from fish tails: The caudal neurosecretory system. 1. 'Urophysiology' and the caudal neurosecretory system of fishes. *Recent Progress in Hormone Research* 41: 533–552.

Besselink HT, van Beusekom S, Roex E, Vethaak AD, Koeman JH, Brouwer A. 1996. Low hepatic 7-ethoxyresorufin-O-deethylase (EROD) activity and minor alterations in retinoid and thyroid hormone levels in flounder (*Platichthys flesus*) exposed to the polychlorinated biphenyl (PCB) mixture, Clophen A50. *Environmental Pollution* 92: 267–274.

Bhattacharya S, Kumar D, Das RH. 1978. Inhibition of thyroid-hormone formation by endrin in the head kidney preparation of a teleost *Anabas testudineus* Bloch. *Indian Journal of Experimental Biology* 16: 1310–1312.

Bogerd J, Granneman JCM, Schulz RW, Vischer HF. 2005. Fish FSH receptors bind LH: How to make the human FSH receptor to be more fishy? *General and Comparative Endocrinology* 142: 34–43.

Borg B. 1994. Androgen in teleost fishes. *Comparative Biochemistry and Physiology C-Pharmacology Toxicology & Endocrinology* 109: 219–245.

Brian JV, Augley JJ, Braithwaite VA. 2006. Endocrine disrupting effects on the nesting behavior of male three-spined stickleback *Gasterosteus aculeatus* L. *Journal of Fish Biology* 68:1883–1890.

Brown SB, Adams BA, Cyr DG, Eales JG. 2004. Contaminant effects on the teleost fish thyroid. *Environmental Toxicology and Chemistry* 23: 1680–1701.

Brown JA, Hazon N. 2007. *The Renin-Angiotensin Systems of Fish and Their Roles in Osmoregulation*. Science Publishers. Enfield, NH. pp. 85–134.

Carson R. 1962. *Silent Spring*. New York: Houghton Mifflin Company.

Chan WK, Chan KM. 2012. Disruption of the hypothalamic-pituitary-thyroid axis in zebrafish embryo-larvae following waterborne exposure to BDE-47, TBBPA and BPA. *Aquatic Toxicology* 108: 106–111.

Chandrasekar G, Arner A, Kitambi SS, Dahlman-Wright K, Lendahl MA. 2011. Developmental toxicity of the environmental pollutant 4-nonylphenol in zebrafish. *Neurotoxicology and Teratology* 33: 752–764.

Chang HS, Choo KH, Lee B, Choi SJ. 2009. The methods of identification, analysis, and removal of endocrine disrupting compounds (EDCs) in water. *Journal of Hazardous Materials* 172: 1–12.

Chaube R, Mishra S, Singh RK. 2010. In vitro effects of lead nitrate on steroid profiles in the post-vitellogenic ovary of the catfish *Heteropneustes fossilis*. *Toxicology In Vitro* 24: 1899–1904.

Cheek AO, Brouwer TH, Carroll S, Manning S, McLachlan JA, Brouwer M. 2001. Experimental evaluation of vitellogenin as a predictive biomarker for reproductive disruption. *Environmental Health Perspectives* 109: 681–690.

Colborn T, Saal FSV, Soto AM. 1993. Developmental effects of endocrine-disrupting chemicals in wildlife and humans *Environmental Health Perspectives* 101: 378–384.

Colborn T and Clement C. 1992. *Chemically-Induced Alterations in Sexual and Functional Development: The Wildlife/Human Connection*. Princeton, NJ: Princeton Scientific.

Colman JR, Baldwin D, Johnson LL, Scholz NL. 2009. Effects of the synthetic estrogen, 17alpha-ethinyl-estradiol, on aggression and courtship behavior in male zebrafish (*Danio rerio*). *Aquatic Toxicology* 91: 346–354.

Conlon JM. 2008. "Liberation" of urotensin II from the teleost urophysis: An historical overview. *Peptides* 29: 651–657.

Crain DA, Eriksen M, Iguchi T, Jobling S, Laufer H, LeBlanc GA, Guillette LJ. 2007. An ecological assessment of bisphenol-A: Evidence from comparative biology. *Reproductive Toxicology* 24: 225–239.

De Groef B, Van der Geyten S, Darras VM, Kuhn ER. 2006. Role of corticotropin-releasing hormone as a thyrotropin-releasing factor in non-mammalian vertebrates. *General and Comparative Endocrinology* 146: 62–68.

Del Giudice G, Prisco M, Agnese M, Verderame M, Rosati L, Limatola E, Andreuccetti P. 2012. Effects of nonylphenol on vitellogenin synthesis in adult males of the spotted ray *Torpedo marmorata*. *Journal of Fish Biology* 80: 2112–2121.

Devlin RH, Nagahama Y. 2002. Sex determination and sex differentiation in fish: An overview of genetic, physiological, and environmental influences. *Aquaculture* 208: 191–364.

Diotel N, Le Page Y, Mouriec K, Tong SK, Pellegrini E, Valliant C, Anglade I, Brion F, Pakdel F, Chung BC, Kah O. 2010. Aromatase in the brain of teleost fish: Expression, regulation and putative functions. *Frontiers in Neuroendocrinology* 31: 172–192.

Dohan O, De la Vieja A, Paroder V, Riedel C, Artani M, Reed M, Ginter CS, Carrasco N. 2003. The sodium/iodide symporter (NIS): Characterization, regulation, and medical significance. *Endocrine Reviews* 24: 48–77.

Donaldson EM. 1996. Manipulation of reproduction in farmed fish. *Animal Reproduction Science* 42: 381–392.

Ducsay CA, Myers DA. 2011. eNOS activation and NO function: Differential control of steroidogenesis by nitric oxide and its adaptation with hypoxia. *Journal of Endocrinology* 210: 259–269.

Eales JG, Brown SB. 1993. Measurement and regulation of thyroidal status in teleost fish. *Reviews in Fish Biology and Fisheries* 3: 299–347.

Edwards TM, Guillette LJ. 2007. Reproductive characteristics of male mosquitofish (*Gambusia holbrooki*) from nitrate-contaminated springs in Florida. *Aquatic Toxicology* 85: 40–47.

Edwards TM, Miller HD, Guillette LJ. 2006. Water quality influences reproduction in female mosquitofish (*Gambusia holbrooki*) from eight Florida springs. *Environmental Health Perspectives* 114: 69–75.

Evans DH, Piermarini PM, Choe KP. 2005. The multifunctional fish gill: Dominant site of gas exchange, osmoregulation, acid-base regulation, and excretion of nitrogenous waste. *Physiological Reviews* 85: 97–177.

Faulk C, Dolinoy DC. 2011. Timing is everything The when and how of environmentally induced changes in the epigenome of animals. *Epigenetics* 6: 791–797.

Fields S. 2004. Global nitrogen—Cycling out of control. *Environmental Health Perspectives* 112: A556–A563.

Flint S, Markle T, Thompson S, Wallace E. 2012. Bisphenol A exposure, effects, and policy: A wildlife perspective. *Journal of Environmental Management* 104: 19–34.

Forsgren E. 1997. Female sand gobies prefer good fathers over dominant males. *Proceedings of the Royal Society of London Series B-Biological Sciences* 264: 1283–1286.

Garcia-Gimenez JL, Sanchis-Gomar F, Lippi G, Mena S, Ivars D, Gomez-Cabrera MC, Vina J, Pallardo FV. 2012. Epigenetic biomarkers: A new perspective in laboratory diagnostics. *Clinica Chimica Acta; International Journal of Clinical Chemistry* 413: 1576–1582.

Ge W, Chang JP, Peter RE, Vaughan J, Rivier J, Vale W. 1992. Effects of porcine follicular-fluid, inhibin-A and activin-A on goldfish gonadotropin-release in-vitro *Endocrinology* 131: 1922–1929.

Giger W, Brunner PH, Schaffner C. 1984. 4-Nonylphenol in sewage-sludge—Accumulation of toxic metabolites from nonionic surfactants *Science* 225: 623–625.

Gillesby BE, Zacharewski TR. 1998. Exoestrogens: Mechanisms of action and strategies for identification and assessment. *Environmental Toxicology and Chemistry* 17: 3–14.

Gore AC. 2002. *GnRH: The Master Molecule of Reproduction*. Norwell, MA: Kluwer Academic Publishers.

Gore AC. 2008. Developmental programming and endocrine disruptor effects on reproductive neuroendocrine systems. *Frontiers in Neuroendocrinology* 29: 358–374.

Gore AC. 2010. Neuroendocrine targets of endocrine disruptors. *Hormones-International Journal of Endocrinology and Metabolism* 9: 16–27.

Grant BF, Mehrle PM. 1970. Chronic endrin poisoning in goldfish *Carassius auratus*. *Journal of the Fisheries Research Board of Canada* 27: 2225–2232.

Guiguen Y, Fostier A, Piferrer F, Chang CF. 2010. Ovarian aromatase and estrogens: A pivotal role for gonadal sex differentiation and sex change in fish. *General and Comparative Endocrinology* 165: 352–366.

Guilherme S, Santos MA, Barroso C, Gaivao I, Pacheco M. 2012. Differential genotoxicity of Roundup (R) formulation and its constituents in blood cells of fish (*Anguilla anguilla*): Considerations on chemical interactions and DNA damaging mechanisms. *Ecotoxicology* 21: 1381–1390.

Guillette LJ, Edwards TM. 2005. Is nitrate an ecologically relevant endocrine disruptor in vertebrates? *Integrative and Comparative Biology* 45: 19–27.

Guyton AC. 1956. *Textbook of Medical Physiology*. W. B. Saunders Co., Philadelphia, PA, 5.xiv+1030p. Illus.

Habibi HR, Ince BW. 1984. A study of androgen-stimulated L-leucine transport by the intestine of rainbow trout (*Salmo gairdneri* Richardson) invitro. *Comparative Biochemistry and Physiology a-Physiology* 79: 143–149.

Hachfi L, Couvray S, Simide R, Tarnowska K, Pierre S, Gaillard S, Richard S, Coupe S, Grillasca J-P, Prevot-D-Alvise N. 2012. Impact of endocrine disrupting chemicals (EDCs) on hypothalamic-pituitary-gonad-liver (HPGL) axis in fish. *World Journal of Fish and Marine Sciences* 4: 14–30.

Hamlin HJ, Moore BC, Edwards TM, Larkin ILV, Boggs A, High WJ, Main KL, Guillette LJ. 2008. Nitrate-induced elevations in circulating sex steroid concentrations in female Siberian sturgeon (*Acipenser baeri*) in commercial aquaculture. *Aquaculture* 281: 118–125.

Hannas BR, Das PC, Li H, LeBlanc GA. 2010. Intracellular conversion of environmental nitrate and nitrite to nitric oxide with resulting developmental toxicity to the crustacean *Daphnia magna*. *PloS One* 5: e12453.

Hansen PD, Vonwesternhagen H, Rosenthal H. 1985. Chlorinated hydrocarbons and hatching success in baltic herring spring spawners *Marine Environmental Research* 15: 59–76.

Hardy SP, Valverde MA. 1994. Novel plasma-membrane action of estrogen and antiestrogens revealed by their regulation of a large-conductance chloride channel. *FASEB Journal* 8: 760–765.

Harris CA, Hamilton PB, Runnalls TJ, Vinciotti V, Henshaw A, Hodgson D, Coe TS, Jobling S, Tyler CR, Sumpter JP. 2011. The Consequences of feminization in breeding groups of wild fish. *Environmental Health Perspectives* 119: 306–311.

Hayes TB et al. 2011. Demasculinization and feminization of male gonads by atrazine: Consistent effects across vertebrate classes. *Journal of Steroid Biochemistry and Molecular Biology* 127: 64–73.

Hazon N, Balment RJ. 1998. *Endocrinology*. Boca Raton, FL, CRC Press: pp. 441–463.

Hazzard CE, Ahearn GA. 1992. Rapid stimulation of intestinal D-glucose transport in teleosts by 19-alpha-methyltestosterone *American Journal of Physiology* 262: R412–R418.

Hoffmann JL, Thomason RG, Lee DM, Brill JL, Price BB, Carr GJ, Versteeg DJ. 2008. Hepatic gene expression profiling using GeneChips in zebrafish exposed to 17 alpha-methyldihydrotestosterone. *Aquatic Toxicology* 87: 69–80.

Hoffmann JL, Torontali SP, Thomason RG, Lee DM, Brill JL, Price BB, Carr GJ, Versteeg DJ. 2006. Hepatic gene expression profiling using Genechips in zebrafish exposed to 17 alpha-ethynylestradiol. *Aquatic Toxicology* 79: 233–246.

Honkanen JO, Holopainen IJ, Kukkonen JVK. 2004. Bisphenol A induces yolk-sac oedema and other adverse effects in landlocked salmon (*Salmo salar* m. sebago) yolk-sac fry. *Chemosphere* 55: 187–196.

Hook SE, Skillman AD, Small JA, Schultz IR. 2006. Gene expression patterns in rainbow trout, *Oncorhynchus mykiss*, exposed to a suite of model toxicants. *Aquatic Toxicology* 77: 372–385.

Hued AC, Oberhofer S, Bistoni MD. 2012. Exposure to a commercial glyphosate formulation (roundup (R)) alters normal gill and liver histology and affects male sexual activity of *Jenynsia multidentata* (Anablepidae, Cyprinodontiformes). *Archives of Environmental Contamination and Toxicology* 62: 107–117.

Iwanowicz LR, Blazer VS, McCormick SD, VanVeld PA, Ottinger CA. 2009. Aroclor 1248 exposure leads to immunomodulation, decreased disease resistance and endocrine disruption in the brown bullhead, *Ameiurus nebulosus*. *Aquatic Toxicology* 93: 70–82.

Jaga K, Dharmani C. 2003. Global surveillance of DDT and DDE levels in human tissues. *International Journal of Occupational Medicine and Environmental Health* 16: 7–20.

James MO. 2011. Steroid catabolism in marine and freshwater fish. *Journal of Steroid Biochemistry and Molecular Biology* 127: 167–175.

Jin YX, Shu LJ, Huang FY, Cao LM, Sun LW, Fu ZW. 2011. Environmental cues influence EDC-mediated endocrine disruption effects in different developmental stages of Japanese medaka (*Oryzias latipes*). *Aquatic Toxicology* 101: 254–260.

Jobling S, Sheahan D, Osborne JA, Matthiessen P, Sumpter JP. 1996. Inhibition of testicular growth in rainbow trout (*Oncorhynchus mykiss*) exposed to estrogenic alkylphenolic chemicals. *Environmental Toxicology and Chemistry* 15: 194–202.

Kavlock RJ et al. 1996. Research needs for the risk assessment of health and environmental effects of endocrine disruptors: A report of the US EPA-sponsored workshop. *Environmental Health Perspectives* 104: 715–740.

Kawakami Y, Shin DH, Kitano T, Adachi S, Yamauchi K, Ohta H. 2006. Transactivation activity of thyroid hormone receptors in fish (*Conger myriaster*) in response to thyroid hormones. *Comparative Biochemistry and Physiology B-Biochemistry & Molecular Biology* 144: 503–509.

Khan IA, Thomas P. 2001. Disruption of neuroendocrine control of luteinizing hormone secretion by Aroclor 1254 involves inhibition of hypothalamic tryptophan hydroxylase activity. *Biology of Reproduction* 64: 955–964.

Kidd KA, Blanchfield PJ, Mills KH, Palace VP, Evans RE, Lazorchak JM, Flick RW. 2007. Collapse of a fish population after exposure to a synthetic estrogen. *Proceedings of the National Academy of Sciences of the United States of America* 104: 8897–8901.

Kim TW, Lee K, Najjar RG, Jeong HD, Jeong HJ. 2011. Increasing N abundance in the northwestern Pacific Ocean due to atmospheric nitrogen deposition. *Science* 334: 505–509.

Kloas W et al. 2009. Endocrine disruption in aquatic vertebrates. In: Vaudry H, Roubos EW, Coast GM, Vallarino M Eds. *Trends in Comparative Endocrinology and Neurobiology*. Oxford, U.K.: Blackwell Publishing. pp. 187–200.

Knorr S, Braunbeck T. 2002. Decline in reproductive success, sex reversal, and developmental alterations in Japanese medaka (*Oryzias latipes*) after continuous exposure to octylphenol. *Ecotoxicology and Environmental Safety* 51: 187–196.

Kraugerud M, Doughty RW, Lyche JL, Berg V, Tremoen NH, Alestrom P, Aleksandersen M, Ropstad E. 2012. Natural mixtures of persistent organic pollutants (POPs) suppress ovarian follicle development, liver vitellogenin immunostaining and hepatocyte proliferation in female zebrafish (*Danio rerio*). *Aquatic Toxicology* 116: 16–23.

Kumar Pandey A, George KC, Peer Mohamed M. 1995. Effect of DDT on thyroid gland of the mullet *Liza parsia* (Hamilton-Buchanan). *Journal of the Marine Biological Association of India* 37: 287–290.

Kurokawa T, Suzuki T, Hashimoto H. 2003. Identification of gastrin and multiple cholecystokinin genes in teleost. *Peptides* 24: 227–235.

Lahnsteiner F, Berger B, Kletzl M, Weismann T. 2005. Effect of bisphenol A on maturation and quality of semen and eggs in the brown trout, *Salmo trutta* f. fario. *Aquatic Toxicology* 75: 213–224.

Lahti E, Harri M, Lindqvist OV. 1985. Uptake and distribution of radioiodine and the effect of ambient nitrate in some fish species *Comparative Biochemistry and Physiology A-Physiology* 80: 337–342.

Lal B. 2007. Pesticide-induced reproductive dysfunction in Indian fishes. *Fish Physiology and Biochemistry* 33: 455–462.

Larsen BK, Bjornstad A, Sundt RC, Taban IC, Pampanin DM, Andersen OK. 2006. Comparison of protein expression in plasma from nonylphenol and bisphenol A-exposed Atlantic cod (*Gadus morhua*) and turbot (*Scophthalmus maximus*) by use of SELDI-TOF. *Aquatic Toxicology* 78: S25–S33.

le Maire A, Bourguet W, Balaguer P. 2010. A structural view of nuclear hormone receptor: Endocrine disruptor interactions. *Cellular and Molecular Life Sciences* 67: 1219–1237.

Leatherland JF. 1993. Field observations on reproductive and developmental dysfunction in introduced and native salmonids from the great lakes. *Journal of Great Lakes Research* 19: 737–751.

Leatherland JF. 2000. *Contaminant-Altered Thyroid Function in Wildlife*. New York: Taylor & Francis. pp. 155–181.

Leatherland JF, Sonstegard RA. 1978. Lowering of serum thyroxine and triiodothyronine levels in yearling coho salmon, Oncorhynchus kisutch. *Journal of the Fisheries Research Board of Canada* 35: 1285–1289.

Leatherland JF, Sonstegard RA. 1980. Effect of dietary mirex and PCBs in combination with food-deprivation and testosterone administration on thyroid activity and bioaccumulation of organochlorines in rainbow trout *Salmo gairdneri* Richardson. *Journal of Fish Diseases* 3: 115–124.

Lister AL, Van der Kraak GJ, Rutherford R, MacLatchy D. 2011. *Fundulus heteroclitus*: Ovarian reproductive physiology and the impact of environmental contaminants. *Comparative Biochemistry and Physiology C-Toxicology & Pharmacology* 154: 278–287.

Liu SY, Chang JH, Zhao Y, Zhu GN. 2011. Changes of thyroid hormone levels and related gene expression in zebrafish on early life stage exposure to triadimefon. *Environmental Toxicology and Pharmacology* 32: 472–477.

Liu SZ, Qin F, Wang HP, Wu TT, Zhang YY, Zheng Y, Li M, Wang ZZ. 2012. Effects of 17 alpha-ethinylestradiol and bisphenol A on steroidogenic messenger ribonucleic acid levels in the rare minnow gonads. *Aquatic Toxicology* 122: 19–27.

Mananos E, Duncan N, Mylonas C. 2009. Reproduction and control of ovulation, spermiation and spawning in cultured fish. *Methods in Reproductive Aquaculture: Marine and Freshwater Species* CRC Press, Taylor and Francis Group, Boca Raton, FL.

Mandich A, Bottero S, Benfenati E, Cevasco A, Erratico C, Maggioni S, Massari A, Pedemonte F, Vigano L. 2007. In vivo exposure of carp to graded concentrations of bisphenol A. *General and Comparative Endocrinology* 153: 15–24.

Martinovic D, Blake LS, Durhan EJ, Greene KJ, Kahl MD, Jensen KM, Makynen EA, Villeneuve DL, Ankley GT. 2008. Reproductive toxicity of vinclozolin in the fathead minnow: Confirming an anti-androgenic mode of action. *Environmental Toxicology and Chemistry* 27: 478–488.

Matthiessen P. 2003. Historical perspective on endocrine disruption in wildlife. *Pure and Applied Chemistry* 75: 2197–2206.

Metcalfe CD, Metcalfe TL, Kiparissis Y, Koenig BG, Khan C, Hughes RJ, Croley TR, March RE, Potter T. 2001. Estrogenic potency of chemicals detected in sewage treatment plant effluents as determined by in vivo assays with Japanese medaka (*Oryzias latipes*). *Environmental Toxicology and Chemistry* 20: 297–308.

Milhaud G, Rankin JC, Bolis L, Benson AA. 1977. Calcitonin—Its hormonal action on gill. *Proceedings of the National Academy of Sciences of the United States of America* 74: 4693–4696.

Mills LJ, Chichester C. 2005. Review of evidence: Are endocrine-disrupting chemicals in the aquatic environment impacting fish populations? *Science of the Total Environment* 343: 1–34.

Munakata A, Kobayashi M. 2010. Endocrine control of sexual behavior in teleost fish. *General and Comparative Endocrinology* 165: 456–468.

Nash JP, Kime DE, Van der Ven LTM, Wester PW, Brion F, Maack G, Stahlschmidt-Allner P, Tyler CR. 2004. Long-term exposure to environmental concentrations of the pharmaceutical ethynylestradiol causes reproductive failure in fish. *Environmental Health Perspectives* 112: 1725–1733.

Nelson ER, Habibi HR. 2009. Thyroid receptor subtypes: Structure and function in fish. *General and Comparative Endocrinology* 161: 90–96.

Norris DO. 1997. *Vertebrate Endocrinology*, 3rd edn., Academic Press, Inc., San Diego, CA; Academic Press Ltd., London, U.K., xii + 634p.

Orozco A, Valverde-R C. 2005. Thyroid hormone deiodination in fish. *Thyroid* 15: 799–813.

Pait AS, Nelson JO. 2002. Endocrine disruption in fish an assessment of recent research and results. NOAA Technical Memorandum NOS NCOS CCMA 149. Silver Spring, MD: NOAA, NOS, Center for Coastal Monitoring and Assessment 55pp.

Palace VP, Evans RE, Wautier KG, Mills KH, Blanchfield PJ, Park BJ, Baron CL, Kidd KA. 2009. Interspecies differences in biochemical, histopathological, and population responses in four wild fish species exposed to ethynylestradiol added to a whole lake. *Canadian Journal of Fisheries and Aquatic Sciences* 66: 1920–1935.

Pelch KE, Beeman JM, Niebruegge BA, Winkeler SR, Nagel SC. 2011. *Endocrine-Disrupting Chemicals (EDCs) in Mammals.* Elsevier San Diego, CA. pp. 329–371.

Peter MCS, Oommen OV. 1989. Effects of thyroid and gonadal hormones invitro on hepatic succinate-dehydrogenase activity of teh teleost, *Anabas testudineus* (Bloch). *Zoological Science* 6: 185–189.

Peter RE, Trudeau VL, Sloley BD. 1991a. Brain regulation of reproduction in teleosts. *Bulletin of the Institute of Zoology Academia Sinica Monograph* 16: 89–118.

Peter RE, Trudeau VL, Sloley BD, Peng C, Nahorniak CS. 1991b. Actions of catecholamines, peptides and sex steroids in regulation of gonadotropin-II in the goldfish. *Proceedings of the Fourth International Symposium on the Reproductive Physiology of Fish*, Norwich, U.K., University of East Anglia: July 7–12, 1991. pp. 30–34.

Peter RE, Yu KL. 1997. Neuroendocrine regulation of ovulation in fishes: Basic and applied aspects. *Reviews in Fish Biology and Fisheries* 7: 173–197.

Piferrer F. 2001. Endocrine sex control strategies for the feminization of teleost fish. *Aquaculture* 197: 229–281.

Piferrer F, Zanuy S, Carrillo M, Solar, II, Devlin RH, Donaldson EM. 1994. Brief treatment with an aromatase inhibitor during sex-differentiation causes chromosomally female salmon to develop as normal, functional males. *Journal of Experimental Zoology* 270: 255–262.

Polakof S, Mommsen TP, Soengas JL. 2011. Glucosensing and glucose homeostasis: From fish to mammals. *Comparative Biochemistry and Physiology B-Biochemistry & Molecular Biology* 160: 123–149.

Porte C, Janer G, Lorusso LC, Ortiz-Zarragoitia M, Cajaraville MP, Fossi MC, Canesi L. 2006. Endocrine disruptors in marine organisms: Approaches and perspectives. *Comparative Biochemistry and Physiology C-Toxicology & Pharmacology* 143: 303–315.

Reik W, Dean W, Walter J. 2001. Epigenetic reprogramming in mammalian development. *Science* 293: 1089–1093.

Rempel MA, Schlenk D. 2008. Effects of environmental estrogens and antiandrogens on endocrine function, gene regulation, and health in fish. In: Jeon KW Ed. *International Review of Cell and Molecular Biology*, Vol. 267. San Diego, CA: Elsevier Academic Press Inc. pp. 207–252.

Robinson CD, Brown E, Craft JA, Davies IM, Moffat CF, Pirie D, Robertson F, Stagg RM, Struthers S. 2003. Effects of sewage effluent and ethynyl oestradiol upon molecular markers of oestrogenic exposure, maturation and reproductive success in the sand goby (*Pomatoschistus minutus*, Pallas). *Aquatic Toxicology* 62: 119–134.

Saaristo M, Craft JA, Lehtonen KK, Lindstrom K. 2009. Sand goby (*Pomatoschistus minutus*) males exposed to an endocrine disrupting chemical fail in nest and mate competition. *Hormones and Behavior* 56: 315–321.

Saaristo M, Craft JA, Lehtonen KK, Lindstrom K. 2010. Exposure to 17 alpha-ethinyl estradiol impairs courtship and aggressive behaviour of male sand gobies (*Pomatoschistus minutus*). *Chemosphere* 79: 541–546.

Salierno JD, Kane AS. 2009. 17 alpha-ethinylestradiol alters reproductive behaviors, circulating hormones, and sexual morphology in male fathead minnows (*Pimephales promelas*). *Environmental Toxicology and Chemistry* 28: 953–961.

Samuelsson LM, Forlin L, Karlsson G, Adolfsson-Eric M, Larsson DGJ. 2006. Using NMR metabolomics to identify responses of an environmental estrogen in blood plasma of fish. *Aquatic Toxicology* 78: 341–349.

Sarria MP, Santos MM, Reis-Henriques MA, Vieira NM, Monteiro NM. 2011. Drifting towards the surface: A shift in newborn pipefish's vertical distribution when exposed to the synthetic steroid ethinylestradiol. *Chemosphere* 84: 618–624.

Schein V, Cardoso JCR, Pinto PIS, Anjos L, Silva N, Power DM, Canario AVM. 2012. Four stanniocalcin genes in teleost fish: Structure, phylogenetic analysis, tissue distribution and expression during hypercalcemic challenge. *General and Comparative Endocrinology* 175: 344–356.

Schnitzler JG, Celis N, Klaren PHM, Blust R, Dirtu AC, Covaci A, Das K. 2011. Thyroid dysfunction in sea bass (*Dicentrarchus labrax*): Underlying mechanisms and effects of polychlorinated biphenyls on thyroid hormone physiology and metabolism. *Aquatic Toxicology* 105: 438–447.

Schwarz JM, McCarthy MM. 2008. Steroid-induced sexual differentiation of the developing brain: Multiple pathways, one goal. *Journal of Neurochemistry* 105: 1561–1572.

Sebire M, Katsiadaki I, Taylor NGH, Maack G, Tyler CR. 2011. Short-term exposure to a treated sewage effluent alters reproductive behaviour in the three-spined stickleback (*Gasterosteus aculeatus*). *Aquatic Toxicology* 105: 78–88.

Segner H, Caroll K, Fenske M, Janssen CR, Maack G, Pascoe D, Schafers C, Vandenbergh GF, Watts M, Wenzel A. 2003. Identification of endocrine-disrupting effects in aquatic vertebrates and invertebrates: Report from the European IDEA project. *Ecotoxicology and Environmental Safety* 54: 302–314.

Shanle EK, Xu W. 2011. Endocrine disrupting chemicals targeting estrogen receptor signaling: Identification and mechanisms of action. *Chemical Research in Toxicology* 24: 6–19.

Shioda T, Wakabayashi M. 2000. Effect of certain chemicals on the reproduction of medaka (*Oryzias latipes*). *Chemosphere* 40: 239–243.

Singh H, Singh TP. 1980. Thyroid-activity and TSH potency of the pituitary gland and blood serum in response to cythion and hexadrin treatment in the fresh-water catfish, *Heteropneustes fossilis* (Bloch). *Environmental Research* 22: 184–189.

Skinner MK, Manikkam M, Guerrero-Bosagna C. 2010. Epigenetic transgenerational actions of environmental factors in disease etiology. *Trends in Endocrinology and Metabolism* 21: 214–222.

Soffker M, Tyler CR. 2012. Endocrine disrupting chemicals and sexual behaviors in fish—A critical review on effects and possible consequences. *Critical Reviews in Toxicology* 42: 653–668.

Solomon KR et al. 1996. Ecological risk assessment of atrazine in North American surface waters. *Environmental Toxicology and Chemistry* 15: 31–74.

Soto AM, Sonnenschein C, Chung KL, Fernandez MF, Olea N, Serrano FO. 1995. The e-screen assay as a tool to identify estrogens—An update on estrogenic environmental pollutants. *Environmental Health Perspectives* 103: 113–122.

Specker JL, Sullivan CV. 1994. *Vitellogenesis in Fishes: Status and Perspectives*. National Research Council of Canada, Ottawa, pp. 304–315.

Sumpter JP, Johnson AC. 2005. Lessons from endocrine disruption and their application to other issues concerning trace organics in the aquatic environment. *Environmental Science & Technology* 39: 4321–4332.

Suzawa M, Ingraham HA. 2008. The herbicide atrazine activates endocrine gene networks via non-steroidal NR5A nuclear receptors in fish and mammalian cells. *Plos One* 3: e2117.

Tollefsen KE, Mathisen R, Stenersen J. 2003. Induction of vitellogenin synthesis in an Atlantic salmon (*Salmo salar*) hepatocyte culture: A sensitive in vitro bioassay for the oestrogenic and anti-oestrogenic activity of chemicals. *Biomarkers* 8: 394–407.

Trudeau VL. 1997. Neuroendocrine regulation of gonadotrophin II release and gonadal growth in the goldfish, *Carassius auratus*. *Reviews of Reproduction* 2: 55–68.

Tyler CR, Jobling S. 2008. Roach, sex, and gender-bending chemicals: The feminization of wild fish in English rivers. *Bioscience* 58: 1051–1059.

Tyler CR, Jobling S, Sumpter JP. 1998. Endocrine disruption in wildlife: A critical review of the evidence. *Critical Reviews in Toxicology* 28: 319–361.

Uzumcu M, Zama A, Oruc E. 2012. Epigenetic mechanisms in the actions of endocrine-disrupting chemicals: Gonadal effects and role in female reproduction. *Reproduction in Domestic Animals* 47(Suppl 4): 338–347.

Vandenberg LN et al. 2012. Hormones and endocrine-disrupting chemicals: Low-dose effects and nonmonotonic dose responses. *Endocrine Reviews* 33: 378–455.

Webb MAH, Doroshov SI. 2011. Importance of environmental endocrinology in fisheries management and aquaculture of sturgeons. *General and Comparative Endocrinology* 170: 313–321.

Welshons WV, Thayer KA, Judy BM, Taylor JA, Curran EM, vom Saal FS. 2003. Large effects from small exposures. I. Mechanisms for endocrine-disrupting chemicals with estrogenic activity. *Environmental Health Perspectives* 111: 994–1006.

Weltzien FA, Andersson E, Andersen O, Shalchian-Tabrizi K, Norberg B. 2004. The brain-pituitary-gonad axis in male teleosts, with special emphasis on flatfish (Pleuronectiformes). *Comparative Biochemistry and Physiology A-Molecular & Integrative Physiology* 137: 447–477.

Wibe AE, Fjeld E, Rosenqvist G, Jenssen BM. 2004. Postexposure effects of DDE and butylbenzylphthalate on feeding behavior in threespine stickleback. *Ecotoxicology and Environmental Safety* 57: 213–219.

Yamazaki F. 1983. Sex control and manipulation in fish. *Aquaculture* 33: 329–354.

Yokota H, Seki M, Maeda M, Oshima Y, Tadokoro H, Honjo T, Kobayashi K. 2001. Life-cycle toxicity of 4-nonylphenol to medaka (*Oryzias latipes*). *Environmental Toxicology and Chemistry* 20: 2552–2560.

Zweier JL, Wang PH, Samouilov A, Kuppusamy P. 1995. Enzyme-independent formation of nitric oxide in biological tissues. *Nature Medicine* 1: 804–809.

8 Thermal Stress

Suzanne Currie and Patricia M. Schulte

CONTENTS

8.1 INTRODUCTION

Most fishes, with the exception of a few large pelagic fish (e.g., tuna, mako sharks), cannot generate and retain sufficient endogenous heat to maintain a body temperature different from that of their external environment because of rapid heat exchange across the gills and the body surface (Hazel and Prosser, 1974; Stevens and Sutterlin, 1976). Thus, as has long been recognized (Glaser, 1929; Fry and Hart, 1948; Fry, 1958), temperature is one of the critical abiotic factors that affect fishes. Temperature directly impacts the rate of chemical reactions and the stability of weak chemical bonds, profoundly affecting the function of proteins and biological membranes (Schulte, 2011). Low temperatures tend to stabilize weak chemical bonds (including hydrogen bonds, ionic bonds, and hydrophobic interactions) while high temperatures tend to destabilize them. The properties of biological membranes also depend on these interactions. For example, biological membranes will tend to be more fluid at high temperatures and less fluid at low temperatures, which affects the permeability of these membranes and the functional properties of the proteins embedded within them. Similarly, high temperatures destabilize the bonds that support the secondary, tertiary, and quaternary structures of proteins, resulting in gradually increasing flexibility, which ultimately culminates in protein denaturation at high temperatures. In contrast, at low temperatures, proteins become less flexible, which can impede their biological function by reducing their ability to undergo the changes in conformation required to bind with substrates and catalyze reactions. Thus, biochemically catalyzed reactions tend to proceed more slowly at low temperatures and more quickly at higher temperatures, up to the point where the proteins begin to denature, at which point reaction rate decreases rapidly. In addition, temperature directly affects reaction rates by changing the thermal energy in the system. At higher temperatures, molecules move more swiftly and with higher energy, and are therefore more likely to undergo the high-energy molecular collisions that are required to overcome the activation energy barrier of the reaction. Together, the direct effects of temperature on reaction rates and on the stability of weak bonds result in an increase in reaction rates with temperature up to the point that proteins begin to denature or membranes become too fluid to function effectively. As most biological processes depend upon the proper functioning of proteins and membranes, these effects on protein flexibility, membrane fluidity, and reaction rate are integrated across all levels of biological organization to affect processes from the level of individual biochemical reactions up to population growth rates and ecosystem processes (Kingsolver, 2009). In this chapter, we review the short- and long-term effects on fishes due to both rapid and prolonged exposure to temperature change and how these effects may be interpreted as stressful.

8.1.1 CHARACTERIZING THE EFFECTS OF TEMPERATURE

The ability of a fish to tolerate a particular range of temperatures can be described using a thermal tolerance polygon (Figure 8.1), which displays the maximum and minimum acutely tolerated temperatures, and the effects of thermal acclimation on these temperatures. It is bounded at high and low temperatures by the maximum and minimum temperatures that can be tolerated for long periods. A thermal tolerance polygon is thus a complete characterization of the zone of tolerance of an organism, and the area enclosed by the polygon is a measure of the degree of eurythermality of a species. Many eurythermal fishes may experience seasonal temperature ranges greater than 30°C and in extreme examples such as in the viviparous *Poeciliopsis*, water temperatures may change by 22°C in 3 h (White et al., 1994). The annual killifish, *Austrofundulus limnaeus*, can experience daily changes in temperature of 20°C in its South American habitat (Podrabsky and Somero, 2004). The goby, *Gillichthys mirabilis*, experiences a wide seasonal temperature range of 5°C–37°C and may see changes of 4°C/h in the summer (Buckley and Hofmann, 2002). Similarly, the common killifish, *Fundulus heteroclitus*, experiences a greater than 20°C seasonal temperature range, and may experience temperature changes of as much as 10°C within a day in some seasons (Fangue et al., 2011). Tidepool sculpins (*Oligocottus maculosus*) also routinely experience dramatic daily tidal fluctuations in a water temperature of almost 15°C (Todgham et al., 2006). All of these species are extreme eurytherms, and would be expected to have exceptionally large thermal tolerance polygons. Although complete thermal tolerance polygons have been determined for relatively few of these species, the eurythermal fishes for which they have been determined (such as the sheepshead minnow, *Cyprinodon variegatus*, and the common killifish, *F. heteroclitus*) have thermal tolerance polygons with large areas (Beitinger et al., 2000; Fangue et al., 2006). In contrast, the highly stenothermal fish of the Southern Ocean (such as the Antarctic icefish of the suborder Notothenioidei) have relatively small thermal tolerance polygons that are centered at very low temperatures. It is noteworthy that these fish have evolved in the world's coldest and most thermally stable marine habitat (Eastman, 1991), while tropical warm water fishes tend to have small thermal tolerance polygons centered at high temperatures (Figure 8.1). Thermal tolerance polygons summarize the maximum and minimum temperatures that a species can tolerate for relatively short periods of time, but the thermal zone over which organisms can function, grow, and reproduce effectively are usually much smaller (Figure 8.1b). Temperatures outside of these regions but within the bounds of the thermal tolerance polygon are temperatures at which animals can survive, but which result in substantial thermal stress that reduces organismal performance. At these

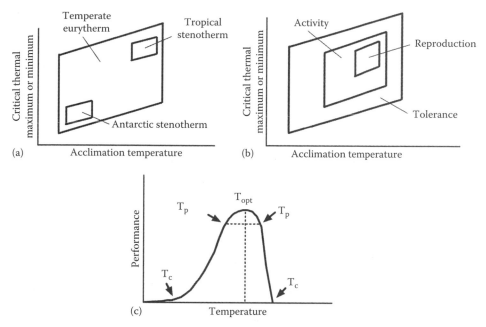

FIGURE 8.1   (a and b) Graphical representations of the relationship between temperature and thermal tolerance and (c) performance. (a and b) The "thermal tolerance polygon," which is obtained by plotting the temperatures at which a fish begins to demonstrate a specific behavior (which are termed the "critical thermal maximum or minimum") during an acute thermal ramp, for fish acclimated at a variety of temperatures. Typical end points include the onset of opercular spasms or loss of the ability to maintain equilibrium. (c) The shape of a typical thermal performance curve (TPC). Performance curves can be plotted for any biological rate process including cellular processes such as enzyme activity, organismal processes such as metabolic rate or swimming performance, or population processes such as the intrinsic rate of population growth. Temperature (on the x-axis) may be acute exposure temperature or acclimation temperature. The optimal temperature (T_{opt}) is the temperature at which performance is maximized. The *pejus* temperature (T_p) is the temperature at which performance declines below a specified percentage of the optimal performance. The critical temperature (T_c) is the point at which performance declines to zero.

temperatures, animals must divert energy away from growth, reproduction, and other activities in order to recruit processes that allow them to survive.

The effects of temperature on performance within the range of tolerated temperatures can be summarized in the form of a thermal performance curve (TPC) or reaction norm (Figure 8.1c) (Huey and Stevenson, 1979; Huey and Kingsolver, 1989; Hofmann and Todgham, 2010). TPCs can be used to identify a number of important characteristics of an organism, including the thermal optimum, the thermal breadth (or the range of temperatures over which performance is near optimal), and the thermal limits (or the threshold temperatures at which performance declines below some specified level). For example, Pörtner et al. have defined the *pejus* (Latin for "getting worse") temperature (T_p) as the temperature at which performance declines below the maximum level (usually arbitrarily set at ~80%–90% of maximum performance) and the critical temperature (T_c) at which performance is minimal. These parameters were originally defined in the context of the performance curves for aerobic scope (Pörtner et al., 2007 and see later), but similar parameters could be defined in the context of any organismal performance trait. It is possible to combine a thermal tolerance polygon and a TPC into a 3D graphic representation, by plotting performance in a third dimension on a tolerance polygon. For example, within the "activity" tolerance polygon in Figure 8.1b, one could plot swimming performance at a given acclimation temperature and at a series of acute exposure temperatures. This would yield a surface topology emerging out of the tolerance polygon. However, complex compound graphs of this nature have seldom been generated, as there are few species for which this type of comprehensive data on thermal tolerance and performance exists.

The ability to maintain performance in the face of thermal variation differs among fish species. Stenothermic species, such as tropical and Antarctic fishes, can perform well across relatively narrow thermal ranges, and thus have a narrow TPC, while eurythermic species (such as species found in salt marshes or in the intertidal zone) can maintain performance across a much wider range of temperatures and have much broader TPCs. But for both stenothermal and eurythermal fishes, temperatures that are higher than the upper *pejus* temperature or lower than the lower *pejus* temperature result in thermal stress.

At all levels of biological organization, empirically determined performance curves typically have a skewed normal distribution (see Figure 8.1c), with a much steeper slope at high temperatures than at low temperatures. As a result of this skewed shape, small changes in temperature above the optimum temperature can be very stressful. Indeed, many fishes die at temperatures only a few degrees above their optimal operating temperature (Somero, 2010). In contrast, most species can tolerate larger declines in temperature below the optimal temperature (T_{opt}) before reaching the T_c. However, as most TPCs have been determined for north temperate fishes, the extent to which this generalization holds for tropical fishes is unclear.

The slopes associated with various regions of a TPC are often summarized using the temperature coefficient (Q_{10}), which is a unitless quantity that describes the change in the rate of a process across a 10°C temperature interval. Q_{10} is calculated as

$$Q_{10} = \left(\frac{R_2}{R_1} \right)^{10/(T_2 - T_1)}$$

For many biochemical and physiological rate processes at biologically relevant temperatures, Q_{10} is approximately 2–3, which indicates that rate processes increase by about two- to threefold with every 10°C increase in temperature up to the thermal optimum. However, it is important to remember that for most processes, Q_{10} is not constant across the entire thermal range. Discontinuities in Q_{10} with temperature are indicators of physiological or biochemical perturbations, and are measures of inflection points in the TPC.

8.1.2 ANTHROPOGENIC THERMAL STRESS

With climate warming, eurythermal and stenothermal fishes will have to cope with both a gradual increase in habitat temperature and an increasing frequency of acute temperature fluctuations. Dramatic changes in temperature are more likely in freshwater systems, but certainly ocean temperatures have been warming for the last 50 years (Levitus et al., 2012) and are expected to become adverse for many marine species within the next 50–100 years (Donelson et al., 2012). To illustrate, in 2010, sea surface temperatures were the warmest on record for 30 years (National Oceanic and Atmospheric Administration). Freshwater river systems on both the Atlantic and Pacific coasts of North America have significantly warmed and fish now have a much higher probability of encountering river temperatures exceeding their traditional thermal limits (Morrison et al., 2002). The ultimate effects of global warming on fishes result from responses at multiple organizational levels, and it is these responses that will affect ecosystems (Pörtner and Farrell, 2008; Pörtner and Peck, 2010). Through the integration of responses at multiple levels of organization, climate change will thus negatively affect fish productivity by increasing the frequency of extreme temperature events that will have acute, physiological consequences (Pörtner and Farrell, 2008; Pörtner and Peck, 2010).

For stenothermal cold-water fishes such as salmonids, water temperatures now increase acutely on warm days in the summer months. For example, Atlantic salmon in the Miramichi River, New Brunswick, Canada, will experience temperature changes greater than 8°C/h (Lund et al., 2002).

Increases in water temperature may be particularly challenging for the extremely stenothermic Antarctic fishes. Notably, the West Antarctic Peninsula region in the Southern Ocean is experiencing some of the most rapid elevations in temperature of any marine environment. Add to this the fact that many animals in this habitat will perish only a few degrees above their normal ecological temperature (Somero, 2010). Thus, in these cold-water fishes, small changes in temperature may have profound effects on physiology and survival.

Fishes may also be exposed to acute *decreases* in water temperature as a result of natural or anthropogenic (e.g., effluents from power generation, which may also raise the temperature) thermoclines. Cold shock causes a rapid decrease in body temperature with behavioral and physiological consequences depending upon the animal's thermal range, acclimation history, and the magnitude of the "shock" (Donaldson et al., 2008). Compared to the response to heat stress, however, we have limited knowledge on the range of physiological effects of cold shock in fish.

8.1.3 BEHAVIORAL RESPONSES TO TEMPERATURE

One way of coping with thermal stress is simply to avoid stressful temperatures. For example, several species of elasmobranchs move to warmer waters when pregnant, presumably to reduce gestation times (Wallman and Bennett, 2006; Hight and Lowe, 2007). Many fishes engage in diel vertical migrations, whereby they feed during the night in relatively warm water and digest at lower temperatures during the day (Brett, 1971). Consistent with this hypothesis, in the Atlantic stingray (*Dasyatis sabina*), digestive efficiency increases in cooler waters after feeding (Di Santo and Bennett, 2011). That said, there are examples of fishes showing the reverse pattern—spending the day in warmer waters and the nights in cool, deep waters. This has recently been observed in brook charr (*Salvelinus fontinalis*; Bertolo et al., 2011) and has also been noted in female leopard sharks (*Triakissemi fasciata*) (Hight and Lowe, 2007), suggesting some degree of sexual dimorphism in behavioral thermoregulation. Thermoregulation has been postulated as an explanation for periodic deep-diving behavior in some shark species (e.g., blue sharks) and other large pelagic fishes (Campana et al., 2011). This behavior could be a means to warm up after a deep dive or to dive to cooler deep waters to cool down. However, to our knowledge, there are no clear data to indicate which, if any, strategy is used by these animals during diving.

Optimal temperatures, and thus behaviorally selected temperatures, may also vary among life stages in fishes. Ward et al. (2010) determined that in the presence of a thermal gradient, juvenile fishes positioned themselves higher in the water column than did adult conspecifics. This preference for warmer waters in juvenile fish likely facilitates growth; however, a position high in the water column may be costly in terms of predator risk (Ward et al., 2010). As adults, there is mounting evidence that female fish prefer warmer waters than males, possibly for optimal oogenesis (Podrabsky et al., 2008) and/or to maximize the growth rate of their larvae.

In salmonid fish (e.g., Atlantic salmon), when the temperature approaches or exceeds putative thermal limits, older, larger fish tend to move to cooler, deeper thermal refugia in the wild (Breau et al., 2007). Younger, smaller fish with higher metabolic rates can possibly satisfy their energetic requirements at these higher temperatures better than older fish and may also have greater thermal tolerance than their older counterparts (see later). Interestingly, juvenile salmon form aggregations of 8–85 fish in thermal refugia during the warmest periods of high temperature events/days (Breau et al., 2007). This aggregation may reduce oxygen consumption for some individuals in the group due to hydrodynamic, energetic savings in certain positions of the aggregation which could ultimately reduce overall physiological stress. Indeed, an individual fish's aerobic capacity will influence its position in a shoal (Killen et al., 2012). Together these data strongly suggest that fish exert some voluntary control over the thermal environment that they experience. However, it is not always possible to avoid exposure to temperatures outside the thermal optimum.

8.1.4 TIME SCALES OF RESPONSES TO THERMAL STRESS

When an ectothermic organism experiences a temperature outside the thermal optimum, it will experience these temperatures as stressful (i.e., these temperatures will perturb homeostasis unless compensatory mechanisms can be induced). These compensatory mechanisms result in changes in the shape of the TPCs or in the area of the thermal tolerance polygon, which eventually restores homeostasis. Ultimately, if a fish is unable to induce such mechanisms, performance will decline and the fish will be exposed to stress. Thermal exposures and the resulting compensatory processes can occur at a variety of time scales. Here, we will separate these adjustments across four time scales:

1. Acute thermal exposures resulting in a stress response
2. Longer-term exposures resulting in acclimation/acclimatization responses
3. Exposures to prior generations and/or at critical developmental windows, resulting in transgenerational or developmental plasticity
4. Long-term exposures that may drive adaptive change due to natural selection

Acute exposures may be due to hourly or daily fluctuations in temperature, and the compensatory adjustments associated with these exposures may be particularly important for surviving and recovering from brief, near-lethal temperature exposure. Longer-term exposures to a particular temperature regime may result in compensatory adjustments due to acclimation or acclimatization occurring over a period of weeks or months, for example, in response to seasonal changes in temperature. In the context of this review, we use the following definition of acclimation: a process of adjustment occurring in a laboratory setting, usually in response to a single environmental variable that is controlled by the experimenter. In contrast, acclimatization is a similar process of adjustment occurring in a natural setting, usually in response to multiple environmental variables. These adjustments due to acclimation or acclimatization are usually, but not always, reversible (Hazel and Prosser, 1974). In contrast, developmental plasticity is a form of adjustment that occurs due to thermal exposures early in development that often results in irreversible changes in phenotype that persist in the adult organism, and may even be passed on to subsequent generations in some cases— a process that is termed transgenerational plasticity. Transgenerational plasticity has been most clearly observed in plants (Herman and Sultan, 2011) and is an epigenetic, rather than a genetic, phenomenon. It is currently poorly understood in fishes. In contrast, genetic changes in populations due to thermal selection may constitute an adaptive response, which occurs over many generations. Responses at these various time scales can interact, and, for example, adaptive change or developmental plasticity could act to alter the acclimation response to temperature (Schulte et al., 2011).

In the following sections, we review what is known about the mechanisms that fishes use to respond to thermal variation at these time scales and across levels of biological organization to provide insight into the factors that allow fishes to cope with thermal stress.

8.2 RESPONSES TO ACUTE THERMAL STRESS

Fish, like all vertebrates, respond to acute thermal stress on multiple levels of organization. The initial response is detection by the nervous system, but details of the thermo-receptors in fishes remain enigmatic. These sensory receptors exist as free nerve endings and are present in the skin and hypothalamus. Several families of temperature-sensitive ion channels known as transient receptor potential (TRP) ion channels (Caterina et al., 1997) present on these free nerve endings facilitate the sensation of heat or cold in many vertebrates including fish (zebrafish, puffer fish; Saito and Shingai, 2006). To date, however, there is little known on the function of TRP channels in fish thermosensing, presenting a rich area for future research.

If possible, fish will avoid temperatures falling outside their ideal thermal range/niche. If this is not possible, there are a suite of behavioral, physiological, biochemical, and molecular reactions that are initiated in an attempt to allow the animal to cope and hopefully acclimate/acclimatize to the new thermal environment. These responses include increases in stress hormones such as cortisol and the catecholamines, adrenaline and noradrenaline, changes in cardiorespiratory function and metabolism, performance modifications, and upregulation of genes and proteins as part of an early cellular stress response (CSR). In this section, we present what is known about the effects of acute thermal change at the molecular, physiological, and behavioral level in fishes and how these responses are integrated in the whole animal.

8.2.1 CELLULAR STRESS RESPONSE TO ACUTE THERMAL STRESS

Molecular investigations of physiological stress have provided more detailed insights into the capacity of animals to cope with thermal stress (Kültz, 2005). The CSR is an early (i.e., minutes–hours) defense reaction resulting from strain or damage to macromolecules, and most often deoxyribonucleic acid (DNA) and protein (Kültz, 2005). Given that these macromolecules are particularly sensitive to temperature, it is not surprising that changes in gene and protein expression underpin physiological responses to both acute and chronic temperature change. There is a large suite of proteins involved in orchestrating the CSR, aimed at sensing, repairing, and minimizing protein damage arising as a result of an environmental insult. These cellular and molecular responses to temperature change have recently been described in an excellent review by Somero (2011). Here we will recap some of this information, with a focus on what we know about the CSR specifically in fish experiencing temperature stress.

In the last few years, physiologists interested in understanding thermal stress at the cellular and molecular levels have had access to a wealth of transcriptomic (Gracey et al., 2004; Podrabsky and Somero, 2004; Buckley et al., 2006; Kassahn et al., 2007; Hori et al., 2010; Lewis et al., 2010; Logan and Somero, 2011; Quinn et al., 2011) information. These valuable data have described hundreds to thousands of different temperature-dependent genes, targets that now form the basis of hypothesis-driven research questions. For example, recent microarray studies in several species of fish have revealed interesting patterns of gene expression in response to acute temperature change. In most microarray studies, molecular chaperones, particularly the heat shock proteins (HSPs), are clearly inducible in response to an acute heat stress (Buckley et al., 2006; Lewis et al., 2010). Indeed, proteins involved in protein folding, such as HSPs, have shown the highest level of inducible expression in microarray studies (Buckley et al., 2006; Kassahn et al., 2007).

Recently, Logan and Somero (2011) determined variation in the induction temperature among genes associated with different components of the CSR following an acute heat stress. For example, the onset temperature or T_{on} for protein rescue in the eurythermal goby (*G. mirabilis*), as evidenced by increases in ubiquitin-mediated proteasomal degradation genes, occurs before the cell reaches a temperature at which protein degradation is activated (Logan and Somero, 2011). Once protein-damaging temperatures are reached, HSP70, one of the most stress-inducible HSPs, is then induced. The mitochondrial paralog of the HSP70 family, HSPA9, occurs later and is thought to inhibit apoptosis (Logan and Somero, 2011). Apoptosis is programmed cell death essential for the removal of defective nucleated cells. There is some evidence that HSPs regulate components of the apoptotic pathway, such as the antiapoptotic NF-κB and its suppressor IκBα (Wong and Wispe, 1997). It is likely that apoptosis is not initiated until the repair of heat-damaged proteins is exceeded by molecular chaperones such as HSPs. There is genetic evidence of both inhibitory and stimulatory effects on apoptosis with acute heat stress in fish (Lewis et al., 2010; Logan and Somero, 2011) and it is possible that in some tissues, sufficient damage occurs to induce apoptotic genes (Buckley et al., 2006) and initiate apoptosis. More research is needed to understand the balance between cell repair/rescue and controlled death in the cells of fish and ectotherms, in general.

This gradation in the gene response to acute heat stress in fish indicates that as stress intensity increases, new classes of genes are activated that encode proteins playing a variety of functions in the repair of cellular damage, removal of irreversibly damaged cellular constituents, and regulation of cellular proliferation and life span (Somero, 2011). The plasma and intracellular membranes are substantially influenced by temperature; indeed, changes in membrane fluidity are one of the first cellular responses to heat stress (Königshofer et al., 2008). An increase in the expression of genes involved in membrane transport has also been reported in fish (Buckley et al., 2006) and likely reflects compromised cell and organelle membrane structure with acute heat stress. Furthermore, genes associated with cholesterol biosynthesis and storage and polyunsaturated fatty acid (PUFA) mobilization are correlated with acute and natural daily temperature cycling in the eurythermal *A. limnaeus*, implicating cholesterol in preserving membrane integrity (Podrabsky and Somero, 2004). Acute changes in temperature over the course of the day are too fast to allow for membrane acclimation (i.e., homeoviscous adaptation), and damage here could be fatal for the cell and possibly the organism.

Acute exposure to warm temperatures beyond the fish's thermal preference and niche can lead to reductions in cellular oxygen levels when oxygen demand exceeds oxygen supply. The fish is faced with rising oxygen demands for processes linked to the CSR (Somero, 2011) and aerobic metabolic rate increases, leading to oxidative stress and the creation of reactive oxygen species (ROS; Crockett, 2008). ROS are a natural by-product of metabolism, but if allowed to accumulate, they will trigger oxidative damage to lipids, proteins, nucleotides, and membranes. This ROS build-up creates a positive feedback loop in that more molecular chaperones are necessary to cope with the increased damage. In a further attempt to alleviate oxidative damage, the animal will synthesize and mobilize a suite of antioxidant enzymatic defenses such as superoxide dismutase (SOD), catalase, and glutathione peroxidase. Elevated temperature on its own may also increase ROS independently of effects arising from increased respiration (Somero, 2011).

Acute *decreases* in temperature may also lead to oxidative stress in fish as a result of an acceleration of mitochondrial respiration. Sudden and transient exposures to cold temperatures are not expected to result in the changes in mitochondrial densities that are observed with cold acclimation (see later). However, it is possible that these short-term decreases in temperature are partially dealt with by specific mitochondrial proteins. For example, the mRNA of uncoupling protein (UCP) isoforms increases upon acute cold exposure (28°C–18°C) in zebrafish brains (Tseng et al., 2011) and following several days of cold exposure in the brains of common carp (*Cyprinus carpio*; Jastroch et al., 2007). The UCPs are located in the mitochondrial inner membrane and catalyze proton conductance, reducing mitochondrial membrane potential and thus ROS production. The increase in UCP isoform mRNA expression in acutely cooled zebrafish brains paralleled increases in protein oxidation and also antioxidant (e.g., SOD, catalase) defenses (Tseng et al., 2011), indicating a cold-induced oxidative stress in fish that is at least partially countered by antioxidant enzymes and possible increases in UCPs. Increased expression of UCPs with acute cold will lead to decreases in ATP, leading to hypoxia. This is potentially very damaging to oxygen-sensitive neural tissue; thus, it may not be surprising that significant increases in hypoxia-inducible factor, HIF -1α, are also observed with cold shock (Tseng et al., 2011) as a hypoxic response and to ensure adequate energy supply to the brain through downstream glucose transport pathways. The limited data that are available would therefore suggest that fish are able to turn on specific molecular pathways to mitigate hypothermic oxidative damage.

Protein synthesis rates are also impacted by acute hypothermia. For example, protein synthesis rates of gill and heart tissue and cytosolic protein pools are depressed by over 80% in response to acute cold shock in *Tautogolabrus adspersus*, the saltwater cunner (Lewis and Driedzic, 2010). These animals metabolically depress during the cold winter months and this downregulation of protein synthesis is likely part of their physiological strategy to conserve energy during the cold. Despite these observed decreases in tissue and cytosolic protein synthesis, gill mitochondrial protein synthesis was defended following acute hypothermia (Lewis and Driedzic, 2010). These authors suggest that the fish may be maintaining oxidative capacity in gill lamellae upon emergence from a metabolically depressed state.

8.2.2 Effects of Behavior on the Cellular Stress Response

When resources are limited, animals in groups form dominance hierarchies with higher social status conferring fitness advantages while subordinate animals experience physiological costs such as chronic stress, elevated metabolism, and immunosuppression (Gilmour et al., 2005). On its own, the formation of social hierarchies in juvenile rainbow trout (*O. mykiss*) results in a CSR (i.e., induction of HSPs) and may influence how an animal copes with environmental stress such as acute heat shock (Currie et al., 2010). LeBlanc et al. (2011) determined that the induction of HSPs was inhibited in both dominant and subordinate fish experiencing an acute heat shock. Using critical thermal maximum (CTmax) as an index of thermal tolerance, they further observed that subordinate fish were more thermally sensitive than their dominant counterparts, but this response was decoupled from the induction of HSPs (LeBlanc et al., 2011). Thus, social stress affects how fish respond to acute increases in temperature with potentially deleterious consequences for overall fitness.

There is a rich literature on proactive and reactive stress-coping styles in mammals. Proactive individuals tend to form and follow routines, and reactive individuals are more flexible and respond to stress with decreased locomotion and freezing behavior. Fish also appear to possess these distinct stress-coping styles. Two lines of rainbow trout were bred from their high (HR) and low (LR) cortisol response to stress, with the LR fish behaving as proactive individuals and the HR fish acting in a more reactive manner (Pottinger and Carrick, 1999; Schjolden et al., 2005; Thomson et al., 2011). Initially, these divergent stress responses were characterized in confinement and handling stresses but these fish also have distinct responses to an acute thermal stress. The HR fish have a more pronounced response to an acute high temperature stress, responding with higher levels of cortisol, catecholamines, and HSPs and sustaining significantly less protein oxidative damage than their LR counterparts (LeBlanc et al., 2012). These physiological differences in these stress-coping behaviors may be, in part, dictated by environmental change/stress.

8.2.3 Metabolic Responses to Acute Thermal Stress

It has been suggested that, from the perspective of the whole animal, oxygen is the limiting factor for survival of short-term acute stressors such as temperature (Davenport and Davenport, 2007; Pörtner et al., 2007; Pörtner, 2010). This concept, which is known as the "oxygen limitation hypothesis" or "oxygen and capacity-limited thermal tolerance" (OCLT; Pörtner, 2001, 2010), implies that oxygen supply to tissues is optimal between the upper and lower *pejus* temperatures and that thermal effects on aerobic scope underpin consequences of climate change on distribution and abundance of aquatic animals (Pörtner and Farrell, 2008). Fry also classified environmental factors based on their influence on aerobic metabolism and aerobic scope (Fry, 1958). Beyond upper *pejus* limits, oxygen supply becomes limiting, while demand increases. This results in decreases in blood oxygen (hypoxemia), leading to declines in animal performance. Of course, *pejus* temperatures may be influenced by acclimation temperature, but only to a limited degree and cannot surpass the animal's thermal limits. Regardless, oxygen limitation at extreme temperatures may be the main determinant of temperature-dependent geographical distribution of marine fishes (Pörtner, 2010).

As one of us has pointed out previously (Schulte et al., 2011), sometimes assessing the physiological response of an animal to an acute temperature change is practically challenging if the time scale of the measurement of the response is long enough to allow the animal to physiologically adjust to the new temperature. A good example of this is the effect of acute temperature change on aerobic metabolic rate where the actual measurement of oxygen consumption may take up to an hour, at which point the animal may begin adjusting or acclimating. Despite these potential practical considerations, acute temperature change will affect metabolism in fishes, largely through changes in the regulation of metabolic enzymes and mitochondrial proteins. The metabolic responses of temperate fish species to cold acclimation are well described and we know that key enzymes of the oxidative electron transport chain (e.g., cytochrome c oxidase subunit (CCO) II, ccoII) and of the citric acid cycle (e.g., citrate synthase, CS) change with temperature. For example, CCO gene expression increases in response to acclimation to cold in fish, indicating enhanced aerobic capacity (Hardewig et al., 1999; McClelland et al., 2006). There are now a few microarray studies that have reported tissue-specific increases and decreases in CCO gene expression following acute heat stress (Buckley et al., 2006; Newton et al., 2012). However, on balance, it appears that aerobic metabolic gene expression generally decreases during acute warming events.

Compared to the wealth of data on the effects of acclimation to cold temperatures, we have limited information on the metabolic effects of *acute* cold exposure in fishes. Galloway and Kieffer (2003) did study metabolic recovery from exhaustive exercise in juvenile Atlantic salmon (*Salmo salar*) exercised at 12°C and recovered in either cold water at 6°C or warm water at 18°C. Recovery of plasma osmolality and lactate and metabolites in the anaerobic white muscle (e.g., PCr, glycogen, lactate) was delayed in fish that experienced cold shock relative to those that recovered at 18°C. Thus, acute exposure to cold temperatures following exhaustive exercise may retard metabolic recovery, whereas warm water may expedite this recovery (Galloway and Kieffer, 2003). Using functional magnetic resonance imaging (fMRI), Van den Burg et al. (2005) determined that a 10°C acute drop in temperature significantly decreased cerebral blood volume and increased Hb-O_2 affinity and plasma cortisol concentrations in the common carp (*C. carpio*). The drainage of blood from the brain may serve to temporarily isolate the brain from the circulation and protect it against cooling down too rapidly, allowing the fish to maintain important behaviors and sensorimotor function (Van den Burg et al., 2005). Collectively, these findings indicate significant fitness implications for fish if they are not able to seek out appropriate recovery temperatures following exercise.

It is well established that acute increases in water temperature increase aerobic metabolic rate in fish (Rodnick et al., 2004; Steinhausen et al., 2008; Kieffer and Wakefield, 2009; Clark et al., 2011; Gamperl et al., 2002). The Q_{10} for these rate increases is approximately 1.5–3 across most species (Beamish, 1964; Beamish and Mookherjii, 1964; Van Dijk et al., 1999), indicating an approximate doubling or tripling of aerobic metabolic rate for each 10°C increase in temperature. However, at the high end of the fish's thermal niche, oxygen consumption is less temperature-dependent and Q_{10} values decrease and may actually become negative as the fish approaches its thermal limits (Hardewig et al., 1999, 2004), reflecting a leveling and decrease in the TPC at and above the T_{opt}. The rate and magnitude of temperature increase affects metabolic rate; more acute, higher temperature exposures within the animal's thermal range cause a larger change in oxygen consumption compared to more gradual exposures, at least in fishes such as cold-water salmonids (Kieffer and Wakefield, 2009).

In contrast to aerobic activity, an acute temperature increase of 6°C did not affect anaerobic metabolism in juvenile *S. salar* (Galloway and Kieffer, 2003); however, this increase may not have been severe enough to increase activity levels. Furthermore, the energy status of cod (*Gadus morhua*) anaerobic white muscle appears to be relatively insensitive to thermal stress when compared to the thermal sensitivity of whole animal (Sartoris et al., 2003). Aerobic scope peaks at T_{opt} but decreases with increasing temperature and then anaerobic metabolism takes over at critical temperatures (T_c). Thus, anaerobic processes are likely recruited as the animal approaches T_c when temperatures become stressful and potentially damaging.

8.2.4 CARDIORESPIRATORY RESPONSES TO ACUTE THERMAL STRESS

Oxygen delivery limitations have been emphasized as a critical element in sublethal thermal stress, with cardiac failure as the ultimate cause of acute heat death. Circulatory rather than ventilatory performance may be the first process to cause oxygen limitation during acute heat stress. For example, arterial O_2 uptake/ventilation does not appear to be limiting during acute warming in Atlantic

North Sea cod (*G. morhua*), whereas the PO_2 of venous blood, which directly supplies the heart with oxygen, significantly decreases (Sartoris et al., 2003; Lannig et al., 2004). For North Sea cod at least, *pejus* temperatures are reached between 13°C and 16°C prior to a fatal drop in venous O_2 tension (Sartoris et al., 2003; Lannig et al., 2004). Similarly, when Chinook salmon (*Oncorhynchus tshawytscha*) experience acute warming, cardiac output increases through increases in heart rate, arterial PO_2 is maintained, but venous PO_2 and oxygen contents significantly decrease (Clark et al., 2008). At 25°C, cardiac arrhythmias are apparent and it is likely that venous oxygen depletion compromises the O_2 supply to systemic and cardiac tissues. Rainbow trout also increase cardiac output through increases in heart rate with decreases in stroke volume and venous capacitance, the latter preventing blood pooling in the venous periphery and providing a favorable pressure gradient for venous return and facilitating the increase in cardiac output (Sandblom and Axelsson, 2007) observed in most fish species with increases in temperature.

In Newfoundland populations of Atlantic cod, cardiac output, heart rate, and MO_2 all increase with acute temperature rise, up until a few degrees before the fish's critical thermal maximum (Gollock et al., 2006) and these three physiological variables are tightly correlated. As temperatures approach the thermal maximum for these fish, circulatory O_2 uptake is limited at least partly because of reductions in $Hb-O_2$ affinity and Hb-binding capacity. Similarly, in Pacific sockeye salmon (*Oncorhynchus nerka*), oxygen unloading to the tissues is favored at high temperatures when fish are swimming. Fatigue at high temperatures is due to cardiac limitation resulting from insufficient delivery of oxygen to working muscles and not oxygen uptake over the gills (Steinhausen et al., 2008). The Pacific pink salmon, *Oncorhynchus gorbuscha*, begins to suffer impaired cardiovascular function around its T_{opt} but is still able to perform, albeit likely briefly, at exceptionally high temperatures; however, temperatures slightly above their T_{opt} (~17°C) will likely lead to increased mortality (Clark et al., 2011). It is noteworthy that these authors observed sex-specific differences in the cardiorespiratory performance of these fish when swimming. Aerobic scope was significantly higher in males, suggesting that females of this species may be more susceptible to higher temperatures (Clark et al., 2011).

Collectively, these findings highlight the plasticity of the fish myocardium when faced with acute temperature change. In most species studied to date, it appears that an acute increase in temperature results in an increase in cardiac output, through increases in heart rate, with stroke volume either remaining constant or even decreasing (see Gamperl and Shiels, Chapter 2). It is not entirely clear why fish adopt this cardiovascular strategy but it is the subject of much discussion in the literature (Clark et al., 2008; Steinhausen et al., 2008; Mendonça and Gamperl, 2010).

As discussed earlier, an increase in water temperature can alter the oxygen affinity of hemoglobin, usually favoring off-loading to the tissues. Blood oxygen carrying capacity may also be affected by thermal stress, and with exposure to acute high temperatures there is a rapid release of stored red blood cells into the circulation (Houston and Murad, 1992) and an increase in younger cells (Lewis et al., 2012) in rainbow trout. Younger trout red blood cells are characterized with a more robust heat shock response (Lund et al., 2000) and as such are likely more thermal-tolerant. Younger, smaller fish, which presumably have a higher proportion of younger circulating red blood cells, have a larger heat shock response than do their larger, older counterparts (Fowler et al., 2009), which may contribute to the increased thermal tolerance of juvenile compared to adult fish. It is noteworthy that decreases in oxygen-limited tolerance to warm temperatures also occur with increasing body size (Pörtner and Peck, 2010). In addition to red blood cells appearing as a result of splenic contraction, there is evidence that acute thermal stress leads to the production of new red blood cells via erythropoiesis (Lewis et al., 2010, 2012). Ultimately, both strategies will lead to increased blood oxygen carrying capacity during acute warming events.

8.2.5 Effects of Acute Temperature Exposure on Swimming Performance

Exposure to acute temperature change will affect swimming performance and activity in fish. In salmonid fish, critical swimming speed or U_{crit} is maximized at temperatures below 18°C

(Taylor et al., 1996) and declines significantly above 23°C (Griffiths and Alderdice, 1972). An exception is in redband rainbow trout, where swimming performance was not negatively affected by acute temperature elevation from an acclimation temperature of 12°C–24°C (Gamperl et al., 2002). Temperature will also affect gait changes in fish. For example, as swimming speed increases, the bluegill sunfish (*Lepomis machrochirus*) changes gait from labriform swimming to more undulating movements of the body axis (Gibb et al., 1994), presumably to maximize locomotor activity. Both acute increases and decreases in temperature will affect this gait transition in bluegill (Jones et al., 2008).

8.3 ACCLIMATION AND ACCLIMATIZATION TO THERMAL STRESS

Many fish species have the capacity to acclimate to environmental temperatures that are stressful during acute exposures, but there is substantial variation among species in the extent of this capacity. In general, north temperate fishes have substantial capacity to acclimate (Beitinger et al., 2000), whereas tropical and Antarctic stenotherms generally show more limited capacity for acclimation than do fish from seasonally variable environments (Nilsson et al., 2010; Bilyk and DeVries, 2011; Muñoz et al., 2012). For example, Antarctic fish are typically found at environmental temperatures of approximately −1°C, and can be acclimated at temperatures up to about 4°C, and thus they have a thermal range of approximately 5°C. In contrast, *F. heteroclitus* can be acclimated to temperatures ranging from 0°C to 35°C. However, it is important to bear in mind that even though stenothermal fish have a relatively small range of temperatures over which they can acclimate, the actual process of acclimation itself may not always differ that much from that seen in more eurythermic fishes. For example, it is possible to compare the capacity for acclimation among species using a metric called the acclimation response ratio (or the change in CTmax divided by the difference in the acclimation temperatures). Using this metric, the Antarctic fish have acclimation response ratios ranging from 0.21 to 0.55 for CTmax (Bilyk and DeVries, 2011). These values can be compared to those of a highly eurythermal fish from the north temperate zone such as the common killifish, *F. heteroclitus*, which has an acclimation response ratio of 0.45 (Fangue et al., 2011). Thus, the acclimation response ratios of at least some Antarctic fish are similar to that of an extremely eurythermal north temperate zone fish, although the range of temperatures over which acclimation is possible is much narrower and temperate zone fishes can undergo a much larger absolute change in CTmax with acclimation.

The process of acclimation can be directly cued by temperature (Condon et al., 2010), but acclimation may also occur in anticipation of thermal stress cued by signals such as photoperiod (Guderley et al., 2001; Healy and Schulte, 2012a), although the majority of studies have focused on responses to thermal exposure *per se*. Changes as a result of thermal acclimation occur at a variety of levels of biological organization. In the following section, we summarize recent studies examining the capacity for acclimation of processes ranging from the molecular level to the whole organism level in fishes.

8.3.1 ACCLIMATION AT THE MOLECULAR AND CELLULAR LEVELS

As is the case in responses to acute temperature change, a picture of the general molecular and cellular response to thermal acclimation is emerging from transcriptomic studies that have examined gene expression responses to long-term acclimation in a variety of species (Castilho et al., 2009; Logan and Somero, 2010; Vergauwen et al., 2010; Windisch et al., 2011; Scott and Johnston, 2012). The nature of the gene expression response to thermal stress is strongly dependent on the acclimation time and the intensity of the thermal challenge. For example, gene expression can change on an hourly basis in response to temperature cycling conditions (Podrabsky and Somero, 2004), and acclimation to fluctuating and constant thermal environments can result in different physiological phenotypes (Todgham et al., 2006) and may elicit different transcriptional and likely physiological responses (Podrabsky and Somero, 2004).

As discussed with respect to acute thermal exposures, the early response to thermal transfer is consistent with a cellular and organismal stress response. For example, in zebrafish, the early response involves the upregulation of defense mechanisms, genes involved in tissue regeneration, and genes associated with the restructuring of oxygen supply through hematopoiesis (Vergauwen et al., 2010). Interestingly, this initial response was similar for both high and low temperature treatments, while the long-term response differed between increased and decreased temperatures (Vergauwen et al., 2010).

In contrast to the early response to thermal acclimation, after several weeks, the primary adjustments at the transcriptomic level appear to be shifts in the expression of metabolic genes consistent with alterations in fuel sources (Windisch et al., 2011) and other adjustments to the central metabolism (Seebacher et al., 2010). These changes have been suggested to be a compensation for the alteration in metabolic rate with temperature (Vergauwen et al., 2010). Alterations in the expression of genes associated with ion transport and protein biosynthesis have also been observed (Logan and Somero, 2010; Vergauwen et al., 2010). In addition, targeted studies of specific genes suggest that there are substantial changes in RNA processing and translation between warm- and cold-acclimated fish. For example, changes in the amount and activity of Cajal bodies, as indicated by changes in the expression of the p-80 coilin gene and by immunohistochemistry, have been detected between summer and winter acclimatized carp (Alvarez et al., 2007). Cajal bodies are nuclear organelles involved in the assembly of the RNA-processing machinery, among other functions. In general, however, the gene expression response after long-term acclimation is fairly modest. For example, in *G. mirabilis*, 9% of uniquely annotated genes differed in expression at 4 weeks of acclimation to 9°C, 19°C, or 28°C and fold changes were typically small. Similar patterns were observed in zebrafish acclimated to 18°C, 26°C, and 34°C for 4, 14, and 28 days, respectively (Vergauwen et al., 2010).

Despite the general similarity in responses to long-term acclimation across species, there appear to be some differences between species with respect to the role of the heat shock response during long-term acclimation to altered temperatures. In *G. mirabilis*, the signals of the acute CSR such as *hsp* upregulation were no longer evident after 4 weeks of acclimation (Logan and Somero, 2010). These observations suggest that, at least in this species, acclimation results in fundamental alterations in the cell that adjust thermal sensitivity such that the heat shock response is no longer required. However, this is not the case in all species. For example, in white muscle of bluefin tuna (*Thunnus orientalis*), *hsp*s remained upregulated in fish transferred from 20°C to 9°C after 90 days of acclimation (Castilho et al., 2009). Similarly, in zebrafish (*Danio rerio*), *hsp*s remained upregulated at high temperatures even after 28 days of acclimation (Vergauwen et al., 2010). These differences may reflect the particular temperatures chosen for each experiment. It is possible that *hsp* upregulation is maintained at temperatures that are outside of the acclimation capacity of a species, while within the normal range of acclimation temperatures, successful long-term acclimation would be expected to be associated with the eventual downregulation of HSPs. Consistent with this hypothesis, Vergauwen et al. (2010) observed that at the maximum acclimation temperature used in their experiments (34°C), the condition factor of zebrafish declined to the point that the fish would be unlikely to survive at this temperature over the long term. In contrast, *G. mirabilis* can easily be acclimated to 28°C and can survive at this temperature for long periods of time.

In comparison to transcriptomic studies, comprehensive studies of the acclimation response using proteomics are rare. The general message from these few studies is that proteomic changes appear to be less extensive than are transcriptomic changes. However, whether this lack of concordance reflects a difference in the sensitivity of the techniques or a true mismatch in the response between mRNA and protein levels remains unknown. The first study to utilize a proteomic approach to investigate the effects of thermal acclimation in fish was in *G. mirabilis* (Kültz and Somero, 1996). Of 602 protein spots separated by 2D gel electrophoresis, only two were induced after 2 months of acclimation to warm temperature (25°C) and three were induced by acclimation to low temperature (10°C), with one additional protein undergoing a small change in molecular weight and charge (possibly due to a posttranslational modification). At the time this study was conducted, techniques for identifying proteins from 2D gels were not very advanced for non-model species, so the actual proteins that changed in expression in response to thermal acclimation in *G. mirabilis* remain unknown (Kültz and Somero, 1996). Comparing this proteomic change to the transcriptomic changes observed under similar conditions in this species (Logan and Somero, 2010), less than 1% of the detected proteome changed in expression with acclimation versus a change in 9% of the transcriptome. Similarly, subtle changes in the proteome have been observed in white muscle in carp acclimated at 10°C and 30°C for 42 days (McLean et al., 2007). Only four protein bands demonstrated a significant effect of thermal acclimation as detected by 1D gel analysis, and only two of them could be identified (HSP90a and AMP deaminase), while 2D gel analysis detected very similar patterns between the acclimation temperatures. The only substantial change detected was a decrease in the amount of creatine kinase and an increase in the amount of creatine kinase degradation products with cold acclimation (McLean et al., 2007). To our knowledge, there is only one other published proteomic study of thermal acclimation in fish which examined changes in the cardiac proteome of rainbow trout (Klaiman et al., 2011). These authors did not report the total number of spots that varied in expression between acclimation treatments, but instead focused on examining variation in a small subset of contractile proteins that were identified using mass spectrometry, none of which were significantly affected by thermal acclimation. The only possible exceptions were a small change in the amount of one of the phosphorylated forms of the slow skeletal muscle isoform of troponin-T, and a modest change in the phosphorylation state of cardiac myosin-binding protein C.

Responses to thermal acclimation at the protein level have also been examined using biochemical approaches. Early studies of thermal compensation revealed great variety in the responses of specific enzymes to thermal acclimation, with some enzymes showing substantial changes in activity while others showed little or no change (Hazel and Prosser, 1974). More recently, a comprehensive examination of the acclimation response of the enzymes of glycolysis has been performed for several species of cyprinodontiform killifishes (*Fundulus* spp.; Pierce and Crawford, 1997a). Acclimation to lower temperature resulted in increases in the maximal activities of a variety of the glycolytic enzymes. Acclimation responses were greatest in the species that normally encounter low environmental temperatures, and were less pronounced in species from warmer habitats. Acclimation effects were also most pronounced for equilibrium enzymes (e.g., aldolase, lactate dehydrogenase), supporting a metabolic control theory perspective that suggests that control of metabolism is distributed across multiple steps in the pathway (Pierce and Crawford, 1997b). Taken together with the transcriptomic data, the proteomic and biochemical data point to the importance of adjustments in the central metabolism with long-term thermal acclimation.

8.3.2 Acclimation of the Cellular Stress Response

The CSR to acute high temperature stress, which is characterized by the induction of HSPs, is dependent on the thermal history of the cell and/or animal (Dietz and Somero, 1992; Dietz, 1994; Fangue et al., 2011). For example, DiIorio et al. (1996) showed that thermal history affected the expression of inducible HSP70 in *Poeciliopsis* spp. and hybrids. These responses of the CSR to acclimation may be tissue-specific as gene expression/protein levels and induction temperatures of components of the CSR may vary among tissues (Dietz and Somero, 1993; Currie et al., 2000; Buckley et al., 2006; Fowler et al., 2009), but together these data point to the important effects of acclimation on the CSR (Currie, 2011).

Acclimation particularly influences the onset temperature of the CSR. For example, when the eurythermal fish *G. mirabilis* is acclimated to seasonally warmer temperatures, the threshold induction temperature, or T_{on}, of HSPs (Dietz and Somero, 1992; Dietz, 1994) and other proteins involved in the CSR is higher than that of fish acclimated to cooler temperatures. Acclimation temperature also influences the *hsp* induction temperature for cold-water, stenothermal rainbow trout, *Oncorhynchus mykiss* (Currie et al., 2000), lamprey, *Petromyzon marinus* (Wood et al., 1999), and the fathead minnow (Dyer et al., 1991). The most comprehensive study of the effects of acclimation on the cellular stress response utilized transcriptomics to provide an unbiased scan of the acute response of gene expression to high temperature in the gills of *G. mirabilis* (Logan and Somero, 2011). Thermal acclimation was associated with an increase in T_{on} of the CSR, and at all acclimation temperatures, the CSR involved a three-step response. Mild heat stress caused the induction of *hsp70*, while moderate stress also induced genes in the ubiquitin proteasome degradation pathway. At very high temperatures, genes associated with cell cycle arrest and apoptosis were induced. Together, these data suggest that the CSR itself remains similar across acclimation temperatures, but that induction temperatures are shifted up or down with acclimation in this species. In addition, the T_{on} for HSPs may change over the course of the day. For example, the induction temperature for *hsp70* gene expression is lower at night than in the daytime in the diurnal fish *F. heteroclitus* (Healy and Schulte, 2012a) and the constitutive levels of *hsp70* mRNA are higher during the day than at night, suggesting that these fish may be anticipating warmer daytime temperatures by preproducing *hsp* transcripts.

The thermal plasticity of the CSR, and of the heat shock genes and proteins, at least, is probably related to plasticity in the activation of the heat shock transcription factor, HSF1 (Buckley and Hofmann, 2002). For example, the kinetics of HSF1 induction have been shown to change with acclimation in *G. mirabilis* (Lund et al., 2006), consistent with the observed changes in *hsp* gene expression.

8.3.3 Molecular and Cellular Responses to Low Temperature Acclimation

Many studies of thermal acclimation at the molecular and cellular levels have examined acclimation to low temperatures in north temperate zone fishes, and as these responses differ in some ways from the general response to thermal acclimation, we discuss them separately here. Transcriptomic studies of cold exposure have revealed a variety of common responses to low temperature (Gracey et al., 2004) including upregulation of stress proteins and chaperones, reorganization of metabolism, changes indicative of increased protein damage and catabolism, and changes in transport processes (Cossins et al., 2006). Interestingly, many similar changes are seen in yeast exposed to low temperatures (Cossins et al., 2006), suggesting a phylogenetically conserved common response to low temperature stress. However, in addition to this conserved response, tissue-specific and species-specific responses also occur. For example, changes in troponin expression occur in muscle fiber-type specific patterns in rainbow trout in response to cold acclimation (Alderman et al., 2012) and in tilapia (Schnell and Seebacher, 2008), which collectively have been interpreted as responses to the decreases in contraction efficiency associated with cold exposure. Furthermore, increases in ionocyte differentiation-related genes have been observed in cold-acclimated zebrafish that have been interpreted as compensation for the reductions in the efficiency of ion transport in the cold (Chou et al., 2008).

At the cellular level, one of the critical effects of acute exposure to low temperature is changes in the properties of cellular membranes, such as decreases in membrane fluidity. Compensatory responses to these effects of low temperature include changes in membrane phospholipid composition, causing changes in a variety of membrane properties such as membrane fluidity, membrane phase behavior, membrane thickness, and membrane permeability (Hazel, 1995; Crockett, 2008). The most widespread of the changes in membrane phospholipid composition with acclimation to low temperatures is an increase in the percentage of unsaturation of membrane phospholipids, particularly in the most common classes such as phosphatidylcholine and phosphatidylethanolamine, although the extent of these changes may vary among membrane compartments within a species. For example, mitochondrial membranes show substantial compensation to maintain fluidity, while little or no compensation is observed in sarcoplasmic reticulum membrane in cold-acclimated goldfish, and changes in the plasma membrane are intermediate in scope (Cossins and Prosser, 1982). Responses may also differ between raft and non-raft regions of the membrane (Zehmer and

Hazel, 2004). These changes in membrane saturation are driven in part by changes in the expression of desaturase enzymes, and in particular the Δ9 desaturase, which have been detected in both targeted and transcriptomic studies (Tiku et al., 1996; Gracey et al., 2004; Cossins et al., 2006).

Another common response to low temperature acclimation is an increase in mitochondrial enzyme amounts, which is often associated with an increase in the density of mitochondria in a tissue (O'Brien, 2011). In general, mitochondrial biogenesis is thought to convey three main benefits to fish in the cold. First, mitochondrial biogenesis helps to offset the decreases in mitochondrial activity in the cold. Second, increasing the number of mitochondria provides additional pathways for oxygen diffusion through lipids, which is much more efficient than diffusion of oxygen through the cytoplasm. Third, increasing the number of mitochondria may help to decrease the diffusion distance for oxygen within the tissue, depending on the subcellular location of these mitochondria (O'Brien, 2011). However, the extent of increased mitochondrial biogenesis in the cold varies among tissues (Grim et al., 2010), species (Guderley, 2004; Orczewska et al., 2010; Bremer and Moyes, 2011), and even among populations of a single species (Dhillon and Schulte, 2011).

The mechanisms associated with increased mitochondrial enzyme activity and mitochondrial density have been examined in a variety of species, but remain incompletely understood (O'Brien, 2011), possibly because of interspecific and intertissue variation in these mechanisms (Orczewska et al., 2010; Bremer et al., 2012). However, a general consensus is emerging that mitochondrial biogenesis in fish may be due to the actions of the transcription nuclear respiratory factor 1 (NRF-1) regulated by the transcriptional co-activator PGC1β, which is distinct from the mammalian pathway that involves PGC1α (Orczewska et al., 2010; Bremer et al., 2012), and that other regulatory factors may play a role in other parts of the response (McClelland et al., 2006). For example, the peroxisome-proliferator-activated receptors (PPARs) may be involved with increases in the expression of fatty acid oxidation genes with cold acclimation (Kondo et al., 2010). It is possible that the distributed control of different aspects of mitochondrial biogenesis, with different regulators responsible for changes in the electron transport chain, the enzymes of fatty acid oxidation, and the proliferation of mitochondrial membranes may allow fishes substantial flexibility to regulate these processes independently, depending on the specific metabolic needs (O'Brien, 2011).

Several lines of evidence suggest that both acute cold exposure (see Section 8.2.1) and cold acclimation may result in increased oxidative stress. For example, the change in mitochondrial membrane fluidity that occurs with acute cold exposure may result in a decrease in the efficiency of oxidative phosphorylation, causing an increase in the production of ROS (Kammer et al., 2011). In addition, changes in membrane composition that occur with cold acclimation may increase susceptibility to oxidative stress (Crockett, 2008). Oxidative stress may also increase at cold temperatures simply because oxygen is more soluble in aqueous fluids at low temperatures, increasing the oxygen concentration of the cytoplasm, although cold exposure has also been suggested to be associated with decreases in tissue oxygenation as a result of reduced cardiac performance (Pörtner and Farrell, 2008; Pörtner, 2010). Consistent with this second hypothesis, cold exposure and cold acclimation are associated with increases in the expression of the HIF-1α (Heise et al., 2007) and increases in the DNA-binding activity of this transcription factor (Rissanen et al., 2006), which ultimately result in increases in tissue capillarity that are thought to help restore oxygen delivery (Johnston, 1982; Egginton and Sidell, 1989) as well as increases in the amount of myoglobin in critical tissues such as the heart (Lurman et al., 2007).

Consistent with the possibility of increases in ROS at low temperatures, patterns consistent with increases in protein degradation have been observed in many species acclimated to low temperature (Todgham et al., 2007). For example, the activity of the 20S proteasome (measured at a common temperature) increases in wolffish acclimated to low temperature (Lamarre et al., 2009). Similarly, mRNA levels of 26S proteasome were upregulated at low acclimation temperatures in the annual killifish *A. limnaeus* (Podrabsky and Somero, 2004), and multiple genes in the ubiquitin–proteasome pathway are upregulated with cold exposure in carp (Gracey et al., 2004). Pathways that are protective against oxidative stress are also upregulated in the cold. For example, the activity of

SOD increases in stickleback with cold acclimation (Kammer et al., 2011), and in the North Sea eel pout *Zoarces viviparous* (Heise et al., 2007). These increases in SOD activity occurred independent of changes in SOD transcript levels in stickleback, suggesting that the regulation occurs at the protein level rather than at the level of transcription (Kammer et al., 2011). A less well-understood compensatory response to increases in ROS involves the upregulation of mitochondrial UCPs, which we discussed earlier in the context of the response to acute thermal stress. Upregulation of various genes in this family gene have been observed in a variety of species in response to cold exposure (Cossins et al., 2006; Mark et al., 2006; Tseng et al., 2011), and have been suggested to be important in reducing the production of ROS, although the exact function of these proteins in ectotherms remains relatively poorly understood.

One final response to exposure to extreme cold involves mechanisms to avoid or prevent the formation of ice in tissues, which can be extremely damaging to cellular structures. Ice formation is less of a problem for freshwater fishes, as ice is less dense than freshwater, and floats to the surface. Thus, freshwater fish can remain in the liquid water below the ice that is at temperatures above 0°C. However, in the marine environment, water freezes at about −1.8°C (depending on salinity), so marine fishes in cold water require mechanisms to avoid ice crystal formation. Ice formation is induced via a process of nucleation, in which existing ice crystals promote the formation of additional ice crystals at low temperature, so freezing can be prevented by removing ice-nucleating agents. Fishes have a variety of mechanisms to reduce ice crystal formation, including the production of small molecules such as glycerol that depress the freezing point of the tissues (Richards et al., 2010) and the production of antifreeze proteins (Ewart et al., 1999; Liebscher et al., 2006).

8.3.4 TISSUE AND ORGANISMAL ACCLIMATION RESPONSES

Acclimation to thermal stress also occurs at the level of tissue, organ, and whole organism. Changes in whole-organism maximum thermal tolerance with acclimation have been extensively studied in fishes (Beitinger and Bennett, 2000). Capacity for acclimation of maximum and minimum thermal tolerance varies among species, but can be extensive. Interestingly, the response to acclimation can differ from the response to repeated acute exposure to thermal stress. For example, in Atlantic killifish (*F. heteroclitus*), CTmax changes substantially with acclimation temperature (Fangue et al., 2011), but even repeated exposure to acute thermal stress does not result in any changes in CTmax. In general, loss of high temperature tolerance with acclimation to low temperature occurs more slowly than does acquisition of increased thermal tolerance with acclimation to high temperatures (Healy and Schulte, 2012a).

As discussed earlier, one system that may be a critical "weak link" during the acute response to thermal stress in fish is the cardiorespiratory system (Pörtner, 2010). As a result, substantial attention has been paid to the responses of the heart to thermal acclimation. For example, relative ventricular mass increases with cold acclimation in a variety of species (Klaiman et al., 2011; Young and Egginton, 2011), and in trout these changes are associated with changes in maximal rate of magnesium ATPase and modest changes in the phosphorylation state of cardiac contractile proteins (Klaiman et al., 2011). However, this increase in relative ventricular mass is not observed in all species or in all cases. For example, in cod, thermal acclimation had no effect on relative ventricular mass (Lurman et al., 2012). However, despite the lack of change in tissue mass, there was a large increase in maximum cardiac output and maximum power output in 4°C-acclimated hearts tested at 10°C, suggesting some kind of major change in cardiac function with an unknown mechanism. Changes in cardiac action potentials and other aspects of cardiac excitation contraction coupling have been observed in a variety of species in response to thermal acclimation (Galli et al., 2009a,b, 2011; Shiels et al., 2011; Korajoki and Vornanen, 2012) that may be involved in the compensation for the effects of temperature on cardiac function. These changes in cardiac performance with acclimation may help to restore aerobic scope at stressful temperatures (Kassahn et al., 2009; Pörtner, 2010).

At the organismal level, relatively few studies have comprehensively examined the effects of thermal acclimation on aerobic scope. In the eurythermal Atlantic killifish (*F. heteroclitus*), acclimation had little effect on the thermal window over which aerobic scope was greater than zero, but substantially increased aerobic scope across the preferred temperature range (Healy and Schulte, 2012b), resulting in a paradoxical decrease in the window over which aerobic scope was at least 80% of maximum with acclimation.

In contrast to the dearth of comprehensive studies on the effects of acclimation on aerobic scope, many studies have examined the effects of thermal acclimation on swimming performance. For example, thermal acclimation increases burst performance in a variety of species (Johnson and Bennett, 1995; Muñoz et al., 2012), although the extent of the change varies depending on the species. As a result, species interactions can be affected. For example, differences in the effects of acclimation on locomotor performance of a predator (Australian bass) and its prey (eastern mosquito fish) resulted in decreased predation pressure at high temperatures (Grigaltchik et al., 2012). Similarly, acclimation has been shown to affect both the thermal optimum for sustained swimming performance (Schnell and Seebacher, 2008) and the level of performance (Fangue et al., 2008) in some species. These changes even occur at the cellular level, as sperm from warm-acclimated mosquito fish swim faster than sperm from fish acclimated to cooler temperatures at all temperatures tested (Adriaenssens et al., 2012).

Thermal acclimation also affects the behavior of fishes. For example, acclimation to constant or variable environmental temperatures changed thermal preference in *A. limnaeus* (Podrabsky et al., 2008). Similarly, acclimation to low temperature in the northern subspecies of Atlantic killifish resulted in the fish being unable to detect or avoid lethally high temperatures, while fish acclimated to warmer temperatures could successfully perform this avoidance behavior (Fangue et al., 2009a). Although the physiological mechanisms underlying such behavioral changes remain largely unknown, a few studies have examined the effects of thermal acclimation on neurobiological processes involved in behavioral responses. For example, acclimation has been suggested to affect escape behavior and the properties of the Mauthner neuron circuit in goldfish (Szabo et al., 2008). However, this study did not include acute transfer trials, so it is difficult to distinguish between the effects of acclimation *per se* compared to the effects of acute temperature challenge, but by comparison to the literature, it is possible that at least some compensation occurred at lower acclimation temperatures. Hearing sensitivity has been shown to be affected by thermal acclimation in both *Ictalurus punctatus* and *Pimelodus pictus*, and this effect was smaller in the tropical *P. pictus* (Wysocki et al., 2009). However, much remains to be explored with respect to the effects of thermal acclimation on behavior and its underlying neural circuitry, and the role that these changes may play in ameliorating thermal stress.

8.4 EPIGENETIC EFFECTS IN RESPONSE TO TEMPERATURE

Fishes also demonstrate substantial phenotypic plasticity in response to temperature that is mediated by epigenetic effects. These changes are distinct from the effects of thermal acclimation in that these epigenetic mechanisms result in long-lasting, essentially irreversible, changes in the phenotype, whereas acclimation responses are typically fully reversible. Most work has focused on a specific type of epigenetic phenotypic plasticity called developmental plasticity, in which the temperatures experienced at a particular critical phase in the development result in changes in the adult phenotype. As with acclimation, developmental plasticity affects phenotypes at all levels of biological organization in fishes.

At the organismal level, temperature exposure during early development has been shown to affect growth rate in a variety of species of fish even when growth is compared at a common temperature (Finstad and Jonsson, 2012; Hurst et al., 2012). In cod, early exposure to differing temperatures induced developmental plasticity that influenced the shape of the TPC for growth rate in later life, suggesting an interaction between developmental plasticity and the capacity to acclimate (Hurst et al., 2012). In stickleback, low temperature exposure during early development resulted in rapid compensatory growth during the late juvenile period when fish were returned to a common temperature, but these changes resulted in reduced investment in reproductive capacity (Lee et al., 2012). In contrast, stickleback raised in higher temperatures during early development grew more slowly during the late juvenile phase when exposed to a common temperature and demonstrated increased reproductive investment. Interestingly, the differences between the treatments depended on the photoperiod, with the strongest effects being observed at photoperiods corresponding to the normal spring photoperiod immediately prior to the natural breeding season. Lee et al. (2012) interpret these results as suggesting that the fish alter their growth rates to attempt to achieve the largest possible size at the time of normal breeding, such that if the fish detects a photoperiod suggesting a limited amount of time prior to the breeding season, greater compensatory growth will occur, resulting in deleterious effects on fecundity.

Developmental temperature has been observed to affect adult metabolic rate and aerobic scope in a tropical reef fish *Acanthochromis polyacanthus* (Donelson et al., 2011), and these effects have persisted across generations—an example of an epigenetic phenomenon called transgenerational plasticity. However, despite the probable importance of metabolic rate and aerobic scope as an indicator of the thermal window of a species, and the variation in metabolic rates and aerobic scope that have been detected among fish population of a single species (Eliason et al., 2011; Healy and Schulte, 2012b), relatively few studies have examined the possible effects of developmental temperature on these phenotypes. Similarly, only a single study has comprehensively examined the effects of developmental temperature on thermal tolerance (Schaefer and Ryan, 2006), finding that zebrafish reared under a variable thermal regime had higher thermal tolerance than fish reared at a constant temperature.

Several studies have examined the effects of developmental temperature on adult swimming performance, and significant effects have been observed in species including the European sea bass *Dicentrarchus labrax* (Koumoundouros et al., 2009), zebrafish (Sfakianakis et al., 2011; Scott and Johnston, 2012), and stickleback (Lee et al., 2010). One important consideration in designing studies of developmental plasticity is to consider the environmental and evolutionary relevance of the temperatures tested. For example, exposure to low temperatures during development adversely affected the swimming performance of zebrafish as adults, but this may represent a pathological effect of stressful temperatures during development rather than a compensatory response that is beneficial to the organism (Sfakianakis et al., 2011). Interestingly, however, temperature-induced developmental plasticity not only affects swimming performance in zebrafish, but also affects the capacity for acclimation of swimming performance in this species (Scott and Johnston, 2012).

Development temperature has also been shown to affect adult body shape, body size, and meristic characteristics such as fin ray number in a variety of species (Georgakopoulou et al., 2007; Georga and Koumoundouros, 2010; Schmidt and Starck, 2010; Kawajiri et al., 2011). Some of these changes may be associated with changes in muscle development (Johnston, 2006). Development temperature has been shown to affect adult muscle composition in cod such that the number of fast mytomal fibers was greater in fish reared at high temperatures (Johnston and Andersen, 2008). In zebrafish, adult fiber number was highest at intermediate temperatures (26°C) and lower at 22°C and 31°C. (Johnston et al., 2009), whereas in muscle fiber number, maximum diameter, nuclear density, and size distribution were highest at 5°C compared to higher and lower treatment temperatures (Macqueen et al., 2008). Myogenic regulatory factors at the subcellular level that might be responsible for these differences have also been investigated (Fernandes et al., 2007; Macqueen et al., 2007; Johnston et al., 2009), and the expression of many of these factors appears to be affected by development temperature.

8.5 ADAPTATION TO THERMAL STRESS

Thermal stress at both high and low temperatures is likely to represent a strong selective pressure driving adaptation, but the nature, costs, and consequences of adaptive phenotypic variation associated with temperature remain poorly understood, not just in fishes but in all ectotherms

(Clarke, 2003; Angilletta, 2009). The extent to which thermally relevant phenotypes are heritable and can respond to natural selection is of more than academic interest, because of the potential effects of global climate change on fish habitats. Understanding the capacity for adaptation in thermally relevant traits may help in developing predictions about the likely responses of organisms to climate change (Somero, 2010). Interspecific comparative studies have documented extensive variation between species in thermally relevant traits that is correlated with habitat temperature (Hilton et al., 2010; Somero, 2010), which suggests that these traits may have adaptive significance, but few studies have taken a rigorous comparative approach (using, e.g., phylogenetically independent contrasts) to explicitly test for potential signatures of adaptive variation. One exception is a study of variation in glycolytic enzyme expression among species of killifish from different thermal habitats (Pierce and Crawford, 1997). These authors found that the maximal activities of three of the glycolytic enzymes (lactate dehydrogenase [LDH], pyruvate kinase, and glyceraldehyde 3-phosphate dehydrogenase [GAPDH]) correlated negatively with habitat temperature, even when phylogenetic relationships were factored out. This result strongly suggests that variation in the activity of these three glycolytic enzymes evolved adaptively in response to habitat temperature or some other correlated variable.

Examination of intraspecific variation in thermal phenotypes is another productive means of understanding thermal adaptation, because intraspecific trait variation is the material upon which natural selection acts. In order to provide strong support for a role for adaptation and natural selection in maintaining intraspecific variation, it is necessary to show that this variation is inherited and results in differences in fitness. Many studies have documented intraspecific variation in thermally relevant phenotypes in fish, including metabolic rate (Grabowski et al., 2009; Donelson and Munday, 2012; Healy and Schulte, 2012b), growth rate with temperature (Schultz et al., 1996; Conover et al., 2009; Kavanagh et al., 2010; Baumann and Conover, 2011), development rate and juvenile growth rate (Yamahira and Takeshi, 2008; Yamahira et al., 2007), the thermal dependence of swimming performance (Ohlberger and Staaks, 2008; Eliason et al., 2011), thermal preference (Fangue et al., 2009a), and thermal tolerance (Fangue et al., 2006). At the tissue and organ level, latitudinal variation in body shape and fin rays has been observed in medaka (*Oryzias latipes*), which is positively correlated with thermally induced developmental plasticity in these same traits (Kawajiri et al., 2011), as is latitudinal variation in vertebrae number (Yamahira and Nishida, 2009). At the subcellular level, latitudinal variation in enzyme expression and mRNA levels (Pierce and Crawford, 1996; Podrabsky et al., 2000; Oleksiak et al., 2002; Whitehead and Crawford, 2006) as well as mitochondrial properties (Lucassen et al., 2006; Fangue et al., 2009b; White et al., 2012) have all been observed.

Far fewer studies have examined whether this variation has a genetic component. Cold stress tolerance has been shown to be variable and heritable in the red drum *Sciaenops ocellatus* (Saillant et al., 2008), and CTmax has significant heritability in the least killifish *Heterandria formosa* (Doyle et al., 2011). Similarly, latitudinal variation in growth rate among populations of medaka persists in laboratory-raised animals, and demonstrates significant variation among families, suggesting a genetic component involved in this trait (Yamahira and Takeshi, 2008). Latitudinal variation in the number of abdominal vertebrae among medaka populations is also observed among laboratory-reared fish, but there was also substantial developmental plasticity in this trait, suggesting that the actual number of vertebrae observed in field populations represents a genotype by environment interaction (Yamahira and Nishida, 2009). Unfortunately, relatively few studies actually measure thermally relevant traits in laboratory-reared fishes, and many studies simply report the repeatability of a trait within an individual and suggest that this is a proxy for heritability, which is not actually the case.

Demonstrating heritable variation in a thermally relevant trait is only the first step in determining whether trait differences were established or are being maintained by natural selection. It is also necessary to show that these traits increase fitness, or to use pattern-based methods (such as phylogenetically independent contrasts) to detect the signatures of adaptation. Alternatively, it may also be possible to watch evolution "in action" over short time scales. Such studies may be particularly relevant in determining whether species may be able to adapt to the current rapid pace of climate change. An example of this approach is provided by studies using populations of grayling (*Thymallus thymallus*) in Norway (Kavanagh et al., 2010). In the 1880s, a man-made connection way made between the river Gudbrandsdalslågen and a nearby lake, Lesjaskogsvatnet, allowed grayling to colonize this new habitat. Subsequently, the construction of several dams limited additional colonization, allowing the lake populations to evolve more or less independently from the river populations. This time span represents about 22 generations for grayling. Once in the lake, multiple populations were established that spawned in various streams that enter the lake. These populations were shown to be partially genetically isolated, probably because of the natal homing behavior of grayling (Kavanagh et al., 2010). Populations from tributaries on the sunny side of the lake experience substantially higher temperatures in the spawning and rearing habitats than do populations on the shady side of the lake, establishing a system where natural selection might be expected to act on thermally relevant phenotypes. Kavanagh et al. (2010) conducted common garden experiments by rearing fish from two cold and two warm populations in the laboratory at high (12°C) and low (8°C) temperatures. Fish whose parents were from the shady side of the lake grew faster and had larger average muscle fiber cross-sectional area than did fish whose parents were from the sunny side of the lake, when compared at a common rearing temperature, despite the fact that starting egg size did not differ. Higher development rate in high-latitude populations is often observed in intraspecific comparisons, and this phenomenon has been termed "counter-gradient variation" (Conover et al., 2009). The data of Kavanagh et al. (2010) suggest that (at least in grayling) this pattern can evolve relatively quickly.

Another way to examine the potential for fishes to adapt to changes in temperature involves the use of the paradigm of "experimental evolution" (Garland and Rose, 2009). Studies utilizing an experimental evolution approach are relatively rare in fishes, but one study (Barrett et al., 2011) has examined the response of cold tolerance using this paradigm in threespine stickleback (*Gasterosteus aculeatus*). Threespine sticklebacks are found in freshwater and saltwater habitats in the northern hemisphere. Barrett et al. (2011) showed that freshwater populations of stickleback in British Columbia (Canada) have greater cold tolerance than do marine populations, which may be related to the lower winter temperatures in freshwater habitats compared to marine habitats in this region. Using fish from marine populations, Barrett et al. (2011) founded replicate populations in semi-natural experimental ponds and allowed them to breed naturally over multiple generations. By the third generation, the genetic composition of the populations in the ponds had shifted, as had their cold tolerance, such that the pond populations had significantly improved cold tolerance compared to the founder population, and their cold tolerance was similar to that of naturally evolved freshwater populations. Although it is not possible to eliminate the possibility that epigenetic effects caused the change in phenotype in the pond population using this experimental design, these data suggest that cold tolerance can shift rapidly in sticklebacks (either through genetic or epigenetic mechanisms) when they colonize a novel environment, and suggest the possibility of rapid adaptation in this thermally relevant trait.

8.5.1 Adaptation to Constant Cold

No review of the effects of thermal stress in fishes would be complete without a consideration of the Antarctic fishes. Water temperatures in the Antarctic Ocean can be as low as −1.9°C, and remain relatively stable throughout the year. Extremely low temperatures such as this would represent a substantial thermal stress and would even be lethal for most fishes. Thus, the type and extent of potential adaptations to cold in Antarctic fishes have been the subject of intense study for several decades (Petricorena and Somero, 2007). Compared to their subpolar or temperate relatives, which can be found in New Zealand and South America, Antarctic notothenoids differ in a number of respects that likely represent adaptations to their extremely cold environment, including the production of antifreeze proteins (DeVries, 1988), the loss of red blood cells, hemoglobin (Ruud, 1954), and myoglobin in many species of this suborder (Sidell et al., 1997); larger hearts and more extensive vasculature (Sidell and O'Brien, 2006); numerous cardiac mitochondria (O'Brien and Sidell, 2000); larger muscle fibers (O'Brien et al., 2003); low standard metabolic rates (Pörtner et al., 2007); and the existence of enzymes with properties that allow them to maintain function at low temperatures (Petricorena and Somero, 2007). Transcriptomic studies of the Antarctic notothenoid *Dissostichus mawsoni*, or the Antarctic toothfish, have revealed almost 200 gene families whose expression was highly upregulated compared to that in temperate fishes, and many of these gene families also exhibited Antarctic fish-specific gene duplications, suggesting that transcriptional upregulation may have been (at least in part) mediated by these duplications (Chen et al., 2008).

The lack of hemoglobin in Antarctic icefishes of the family Channichthyidae may cause them to be particularly sensitive to temperature elevation as a result of oxygen limitations at elevated temperatures (Thorne et al., 2010; Beers and Sidell, 2011). Further exacerbating the susceptibility of Antarctic fishes to high temperatures is the loss of the normally highly conserved heat shock response (Hofmann et al., 2000, 2005; Clark et al., 2008; Thorne et al., 2010). These fishes possess the genes for HSPs but have lost the capacity to increase *hsp* gene expression in response to acute heat stress (Hofmann et al., 2005). Transcriptome analysis of the notothenoid *Trematomus bernacchii* showed that heat-induced upregulation of HSP genes was absent compared with gene regulatory responses seen in temperate fish (Buckley and Somero, 2009). Instead, these fish have constitutively high levels of the (normally) inducible HSPs, which are likely part of the adaptive strategy this fish uses to cope with stable but extremely cold temperatures (Buckley et al., 2004). Interestingly, polar fish still increase whole-organism thermal tolerance following an acute thermal stress, a phenomenon known as heat-hardening (Bilyk et al., 2012), which suggests that the heat shock response and heat hardening at the whole organismal level may not be closely associated.

Polar fishes also have elevated levels of ubiquitin-conjugated proteins compared to their temperate zone relatives, indicative of higher levels of denatured proteins (Place et al., 2004), which may also explain elevated basal levels of HSPs in these fish. There is likely an energetic trade-off to maintaining constant levels of expensive, inducible HSPs to assist with protein folding and turnover in extreme cold (Hofmann et al., 2005). However, it appears that at least one species from this group, the Antarctic plunderfish (*Harpagifer antarcticus*), upregulate genes involved in transcription and translation, the acute inflammatory response oxidative stress, membrane fluidity, and β-oxidation (Thorne et al., 2010), representing a unique CSR to heat stress in these fish.

8.6   CONCLUSIONS AND PERSPECTIVES

Many years of research on the effects of thermal stress in fish have revealed certain common patterns in responses across taxa. At high temperatures, acute thermal stress may ultimately involve cardiac failure and the inability to provide oxygen to tissues, resulting in a cascade of effects at the cellular level, including the induction of the cellular stress response. Cardiac failure may be associated with changes in membrane properties with temperature that affect membrane potential (either at the plasma or mitochondrial membranes) or that might reduce the efficiency of neurotransmission, although the precipitating factors in aerobic collapse are still not fully understood. At low temperatures, failure may result from thermodynamic effects, causing these processes to simply slow down to the point where normal physiological function is not possible. But compared to our understanding of the effects of acute high temperature stress, the causes of failure at low temperatures in fishes remain unclear. Acclimation, developmental plasticity, and adaptation may alter the critical temperatures at which the effects of thermal stress occur, or provide organisms with protective responses that mitigate them, but ultimately all fishes have a specific range of temperatures over which they function optimally. Understanding the susceptibility of fishes to thermal stress may ultimately allow fish physiologists to provide some insight into whether fishes will be able to keep pace with global climate change.

ACKNOWLEDGMENTS

The authors acknowledge funding from the Natural Sciences and Engineering Research Council (NSERC) Canada and thank Sacha LeBlanc for editorial assistance.

REFERENCES

Adriaenssens B, van Damme R, Seebacher F, Wilson RS. 2012. Sex cells in changing environments: Can organisms adjust the physiological function of gametes to different temperatures? *Global Change Biol* 18: 1797–1803.

Alderman SL, Klaiman JM, Deck CA, Gillis TE. 2012. Effect of cold acclimation on troponin I isoform expression in striated muscle of rainbow trout. *Am J Physiol Regul Integr Comp Physiol* 303: R168–R176.

Alvarez M, Nardocci G, Thiry M, Alvarez R, Reyes M, Molina A, Vera MI. 2007. The nuclear phenotypic plasticity observed in fish during rRNA regulation entails Cajal bodies dynamics. *Biochem Biophys Res Comm* 360: 40–45.

Angilletta MJ, Jr. 2009. *Thermal Adaptation: A Theoretical and Empirical Synthesis.* Oxford University Press, Oxford, U.K.

Barrett RDH, Paccard A, Healy TM, Bergek S, Schulte PM, Schluter D, Rogers SM. 2011. Rapid evolution of cold tolerance in stickleback. *Proc R Soc B* 278: 233–238.

Baumann H, Conover DO. 2011. Adaptation to climate change: Contrasting patterns of thermal-reaction-norm evolution in Pacific versus Atlantic silversides. *Proc R Soc B* 278: 2265–2273.

Beamish FWH. 1964. Respiration of fishes with special emphasis on standard oxygen consumption: II. Influence of weight and temperature on respiration of several species. *Can J Zool* 42: 177–188.

Beamish FWH, Mookherjii PS. 1964. Respiration of fishes with special emphasis on standard oxygen consumption: I. Influence of weight and temperature on respiration of goldfish, *Carassius auratus* L. *Can J Zool* 42: 161–175.

Beers JM, Sidell BD. 2011. Thermal tolerance of Antarctic notothenioid fishes correlates with level of circulating hemoglobin. *Physiol Biochem Zool* 84: 353–362.

Beitinger TL, Bennett WA. 2000. Quantification of the role of acclimation temperature in temperature tolerance of fishes. *Environ Biol Fish* 58: 277–288.

Beitinger TL, Bennett WA, McCauley RW. 2000. Temperature tolerances of North American freshwater fishes exposed to dynamic changes in temperature. *Environ Biol Fish* 58: 237–275.

Bertolo A, Pépino M, Adams J, Magnan P. 2011. Behavioural thermoregulatory tactics in lacustrine brook charr, *Salvelinus fontinalis*. *PLoS One* 6: e18603.

Bilyk KT, DeVries AL. 2011. Heat tolerance and its plasticity in Antarctic fishes. *Comp Biochem Physiol A* 158: 382–390.

Bilyk KT, Evans CW, DeVries AL. 2012. Heat hardening in Antarctic notothenioid fishes. *Polar Biol* 35: 1447–1451.

Breau C, Cunjak RA, Bremset G. 2007. Age-specific aggregation of wild juvenile Atlantic salmon *Salmo salar* at cool water sources during high temperature events. *J Fish Biol* 71: 1179–1191.

Bremer K, Monk CT, Gurd BJ, Moyes CD. 2012. Transcriptional regulation of temperature-induced remodeling of muscle bioenergetics in goldfish. *Am J Physiol Regul Integr Comp Physiol* 303: R150–R158.

Bremer K, Moyes CD. 2011. Origins of variation in muscle cytochrome *c* oxidase activity within and between fish species. *J Exp Biol* 214: 1888–1895.

Brett JR. 1971. Energetic responses of salmon to temperature. A study of some thermal relations in the physiology and freshwater ecology of sockeye salmon (*Oncorhynchus nerka*). *Am Zool* 11: 99–113.

Buckley BA, Gracey AY, Somero GN. 2006. The cellular response to heat stress in the goby *Gillichthys mirabilis*: A cDNA microarray and protein-level analysis. *J Exp Biol* 209: 2660–2677.

Buckley BA, Hofmann GE. 2002. Thermal acclimation changes DNA-binding activity of heat shock factor 1 (HSF1) in the goby *Gillichthys mirabilis*: Implications for plasticity in the heat-shock response in natural populations. *J Exp Biol* 205: 3231–3240.

Buckley BA, Place SP, Hofmann GE. 2004. Regulation of heat shock genes in isolated hepatocytes from an Antarctic fish, *Trematomus bernacchii*. *J Exp Biol* 207: 3649–3656.

Buckley BA, Somero GN. 2009. cDNA microarray analysis reveals the capacity of the cold-adapted Antarctic fish *Trematomus bernacchii* to alter gene expression in response to heat stress. *Polar Biol* 32: 403–415.

Campana SE, Dorey A, Fowler M, Joyce W, Wang Z, Wright D, Yashayaev I. 2011. Migration pathways, behavioural thermoregulation and overwintering grounds of blue sharks in the northwest Atlantic. *PLoS One* 6: e16854.

Castilho PC, Buckley BA, Somero G, Block BA. 2009. Heterologous hybridization to a complementary DNA microarray reveals the effect of thermal acclimation in the endothermic bluefin tuna (*Thunnus orientalis*). *Mol Ecol* 18: 2092–2102.

Caterina MJ, Schumacher MA, Tominaga M, Rosen TA, Levine JD, Julius D. 1997. The capsaicin receptor: A heat-activated ion channel in the pain pathway. *Nature* 389: 816–824.

Chen Z, Cheng C-HC, Zhang J, Cao L, Chen L, Zhou L, Jin Y et al. 2008. Transcriptomic and genomic evolution under constant cold in Antarctic notothenioid fish. *Proc Natl Acad Sci USA* 105: 12944–12949.

Chou M-Y, Hsiao C-D, Chen S-C, Chen I-W, Liu S-T, Hwang P-P. 2008. Effects of hypothermia on gene expression in zebrafish gills: Upregulation in differentiation and function of ionocytes as compensatory responses. *J Exp Biol* 211: 3077–3084.

Clarke A. 2003. Costs and consequences of evolutionary temperature adaptation. *Trend Ecol Evol* 18: 573–581.

Clark MS, Fraser KPP, Burns G, Peck LS. 2008a. The HSP70 heat shock response in the Antarctic fish *Harpagifer antarcticus*. *Polar Biol* 31: 171–180.

Clark TD, Jeffries KM, Hinch SG, Farrell AP. 2011. Exceptional aerobic scope and cardiovascular performance of pink salmon (*Oncorhynchus gorbuscha*) may underlie resilience in a warming climate. *J Exp Biol* 214: 3074–3081.

Clark TD, Sandblom E, Cox GK, Hinch SG, Farrell AP. 2008b. Circulatory limits to oxygen supply during an acute temperature increase in the Chinook salmon (*Oncorhynchus tshawytscha*). *Am J Physiol Regul Integr Comp Physiol* 295: R1631–R1639.

Condon CH, Chenoweth SF, Wilson RS. 2010. Zebrafish take their cue from temperature but not photoperiod for the seasonal plasticity of thermal performance. *J Exp Biol* 213: 3705–3709.

Conover DO, Duffy TA, Hice LA. 2009. The covariance between genetic and environmental influences across ecological gradients: Reassessing the evolutionary significance of countergradient and cogradient variation. *Ann NY Acad Sci* 1168: 100–129.

Cossins A, Fraser J, Hughes M, Gracey A. 2006. Post-genomic approaches to understanding the mechanisms of environmentally induced phenotypic plasticity. *J Exp Biol* 209: 2328–2336.

Cossins AR, Prosser CL. 1982. Variable homeoviscous responses of different brain membranes of thermally-acclimated goldfish. *Biochim Biophys Acta* 687: 303–309.

Crockett EL. 2008. The cold but not hard fats in ectotherms: Consequences of lipid restructuring on susceptibility of biological membranes to peroxidation, a review. *J Comp Physiol B* 178: 795–809.

Currie S. 2011. Heat shock proteins and temperature. In: Farrell AP, Cech JJ, Stevens ED, Richards JG, Eds. *Encyclopedia of Fish Physiology: From Genome to Environment*. Academic Press, San Diego, CA. pp. 1688–1694.

Currie S, LeBlanc S, Watters MA, Gilmour KM. 2010. Agonistic encounters and cellular angst: Social interactions induce heat shock proteins in juvenile salmonid fish. *Proc R Soc B* 277: 905–913.

Currie S, Moyes CD, Tufts BL. 2000. The effects of heat shock and acclimation temperature on hsp70 and hsp30 mRNA expression in rainbow trout: In vivo and in vitro comparisons. *J Fish Biol* 56: 398–408.

Davenport J, Davenport JL. 2007. Interaction of thermal tolerance and oxygen availability in *Littorina littorea* and *Nucella lapillus*. *Mar Ecol Prog Ser* 332: 167–170.

DeVries AL. 1988. The role of antifreeze glycopeptides and peptides in the freezing avoidance of Antarctic fishes. *Comp. Biochem. Physiol. B* 90: 611–621.

Dhillon RS, Schulte PM. 2011. Intraspecific variation in the thermal plasticity of mitochondria in killifish. *J Exp Biol* 214: 3639–3648.

DiIorio PJ, Holsinger K, Schultz RJ, Hightower LE. 1996. Quantitative evidence that both Hsc70 and Hsp70 contribute to thermal adaptation in hybrids of the livebearing fishes *Poeciliopsis*. *Cell Stress Chap* 1: 139–147.

Di Santo V, Bennett WA. 2011. Effect of rapid temperature change on resting routine metabolic rates of two benthic elasmobranchs. *Fish Physiol Biochem* 37: 929–934.

Dietz TJ. 1994. Acclimation of the threshold induction temperatures for 70-kDa and 90-kDa heat shock proteins in the fish *Gillichthys mirabilis*. *J Exp Biol* 188: 333–338.

Dietz TJ, Somero GN. 1992. The threshold induction temperature of the 90-kDa heat shock protein is subject to acclimatization in eurythermal goby fishes (genus *Gillichthys*). *Proc Natl Acad Sci USA* 89: 3389–3393.

Dietz TJ, Somero GN. 1993. Species- and tissue-specific synthesis patterns for heat-shock proteins HSP70 and HSP90 in several marine teleost fishes. *Physiol Zool* 66: 863–880.

Donaldson MR, Cooke SJ, Patterson DA, Macdonald JS. 2008. Cold shock and fish. *J Fish Biol* 73: 1491–1530.

Donelson JM, Munday PL. 2012. Thermal sensitivity does not determine acclimation capacity for a tropical reef fish. *J Anim Ecol* 81: 1126–1131.

Donelson JM, Munday PL, McCormick MI, Nilsson GE. 2011. Acclimation to predicted ocean warming through developmental plasticity in a tropical reef fish. *Global Change Biol* 17: 1712–1719.

Donelson JM, Munday PL, McCormick MI, Pitcher CR. 2012. Rapid transgenerational acclimation of a tropical reef fish to climate change. *Nat Climate Change* 2: 30–32.

Doyle CM, Leberg PL, Klerks PL. 2011. Heritability of heat tolerance in a small livebearing fish, *Heterandria formosa*. *Ecotoxicology* 20: 535–542.

Dyer SD, Dickson KL, Zimmerman EG, Sanders BM. 1991. Tissue-specific patterns of synthesis of heat-shock proteins and thermal tolerance of the fathead minnow (*Pimephales promelas*). *Can J Zool* 69: 2021–2027.

Eastman JT. 1991. Evolution and diversification of Antarctic notothenioid fishes. *Am Zool* 31: 93–110.

Egginton S, Sidell BD. 1989. Thermal acclimation induces adaptive changes in subcellular structure of fish skeletal muscle. *Am J Physiol Regul Integr Comp Physiol* 256: R1–R9.

Eliason EJ, Clark TD, Hague MJ, Hanson LM, Gallagher ZS, Jeffries KM, Gale MK, Patterson DA, Hinch SG, Farrell AP. 2011. Differences in thermal tolerance among sockeye salmon populations. *Science* 332: 109–112.

Ewart KV, Lin Q, Hew CL. 1999. Structure, function and evolution of antifreeze proteins. *Cell Mol Life Sci* 55: 271–283.

Fangue NA, Hofmeister M, Schulte PM. 2006. Intraspecific variation in thermal tolerance and heat shock protein gene expression in common killifish, *Fundulus heteroclitus*. *J Exp Biol* 209: 2859–2872.

Fangue NA, Mandic M, Richards JG, Schulte PM. 2008. Swimming performance and energetics as a function of temperature in killifish *Fundulus heteroclitus*. *Physiol Biochem Zool* 81: 389–401.

Fangue NA, Osborne EJ, Todgham AE, Schulte PM. 2011. The onset temperature of the heat-shock response and whole-organism thermal tolerance are tightly correlated in both laboratory-acclimated and field-acclimatized tidepool sculpins (*Oligocottus maculosus*). *Physiol Biochem Zool* 84: 341–352.

Fangue NA, Podrabsky JE, Crawshaw LI, Schulte PM. 2009a. Countergradient variation in temperature preference in populations of killifish *Fundulus heteroclitus*. *Physiol Biochem Zool* 82: 776–786.

Fangue NA, Richards JG, Schulte PM. 2009b. Do mitochondrial properties explain intraspecific variation in thermal tolerance? *J Exp Biol* 212: 514–522.

Fernandes JMO, MacKenzie MG, Kinghorn JR, Johnston IA. 2007. FoxK1 splice variants show developmental stage-specific plasticity of expression with temperature in the tiger pufferfish. *J Exp Biol* 210: 3461–3472.

Finstad AG, Jonsson B. 2012. Effect of incubation temperature on growth performance in Atlantic salmon. *Mar Ecol Prog Ser* 454: 75–82.

Fowler SL, Hamilton D, Currie S. 2009. A comparison of the heat shock response in juvenile and adult rainbow trout (*Oncorhynchus mykiss*)-implications for increased thermal sensitivity with age. *Can J Fish Aquat Sci* 66: 91–100.

Fry FEJ. 1958. Temperature compensation. *Annu Rev Physiol* 20: 207–224.

Fry FEJ, Hart JS. 1948. The relation of temperature to oxygen consumption in the goldfish. *Biol Bull* 94: 66–77.

Galli GLJ, Lipnick MS, Block BA. 2009a. Effect of thermal acclimation on action potentials and sarcolemmal K^+ channels from Pacific bluefin tuna cardiomyocytes. *Am J Physiol Regul Integr Comp Physiol* 297: R502–R509.

Galli GLJ, Lipnick MS, Shiels HA, Block BA. 2011. Temperature effects on Ca^{2+} cycling in scombrid cardiomyocytes: A phylogenetic comparison. *J Exp Biol* 214: 1068–1076.

Galli GLJ, Shiels HA, Brill RW. 2009b. Temperature sensitivity of cardiac function in pelagic fishes with different vertical mobilities: Yellowfin tuna (*Thunnus albacares*), bigeye tuna (*Thunnus obesus*), mahimahi (*Coryphaena hippurus*), and swordfish (*Xiphias gladius*). *Physiol Biochem Zool* 82: 280–290.

Galloway BJ, Kieffer JD. 2003. The effects of an acute temperature change on the metabolic recovery from exhaustive exercise in juvenile Atlantic salmon (*Salmo salar*). *Physiol Biochem Zool* 76: 652–662.

Gamperl AK, Rodnick KJ, Faust HA, Venn EC, Bennett MT, Crawshaw LI, Keeley ER, Powell MS, Li HW. 2002. Metabolism, swimming performance, and tissue biochemistry of high desert redband trout (*Oncorhynchus mykiss* ssp.): Evidence for phenotypic differences in physiological function. *Physiol Biochem Zool* 75: 413–431.

Garland T Jr, Rose MR. 2009. *Experimental Evolution: Concepts, Methods, and Applications of Selection Experiments*. University of California Press, Berkeley, CA.

Georga I, Koumoundouros G. 2010. Thermally induced plasticity of body shape in adult zebrafish *Danio rerio* (Hamilton, 1822). *J Morphol* 271: 1319–1327.

Georgakopoulou E, Sfakianakis DG, Kouttouki S, Divanach P, Kentouri M, Koumoundouros G. 2007. The influence of temperature during early life on phenotypic expression at later ontogenetic stages in sea bass. *J Fish Biol* 70: 278–291.

Gibb AC, Jayne BC, Lauder GV. 1994. Kinematics of pectoral fin locomotion in the bluegill sunfish *Lepomis Macrochirus*. *J Exp Biol* 189: 133–161.

Gilmour KM, DiBattista JD, Thomas JB. 2005. Physiological causes and consequences of social status in salmonid fish. *Integr Comp Biol* 45: 263–273.

Glaser O. 1929. Temperature and heart-rate in *Fundulus* embryos. *J Exp Biol* 6: 325–339.

Gollock MJ, Currie S, Petersen LH, Gamperl AK. 2006. Cardiovascular and haematological responses of Atlantic cod (*Gadus morhua*) to acute temperature increase. *J Exp Biol* 209: 2961–2970.

Grabowski TB, Young SP, Libungan LA, Steinarsson A, Marteinsdóttir G. 2009. Evidence of phenotypic plasticity and local adaption in metabolic rates between components of the Icelandic cod (*Gadus morhua* L.) stock. *Environ Biol Fish* 86: 361–370.

Gracey AY, Fraser EJ, Li W, Fang Y, Taylor RR, Rogers J, Brass A, Cossins AR. 2004. Coping with cold: An integrative, multitissue analysis of the transcriptome of a poikilothermic vertebrate. *Proc Natl Acad Sci USA* 101: 16970–16975.

Griffiths JS, Alderdice DF. 1972. Effects of acclimation and acute temperature experience on the swimming speed of juvenile coho salmon. *J Fish Res Board Can* 29: 251–264.

Grigaltchik VS, Ward AJW, Seebacher F. 2012. Thermal acclimation of interactions: Differential responses to temperature change alter predator-prey relationship. *Proc R Soc B* 279: 4058–4064.

Grim JM, Miles DRB, Crockett EL. 2010. Temperature acclimation alters oxidative capacities and composition of membrane lipids without influencing activities of enzymatic antioxidants or susceptibility to lipid peroxidation in fish muscle. *J Exp Biol* 213: 445–452.

Guderley H. 2004. Metabolic responses to low temperature in fish muscle. *Biol Rev* 79: 409–427.

Guderley H, Leroy PH, Gagné A. 2001. Thermal acclimation, growth, and burst swimming of threespine stickleback: Enzymatic correlates and influence of photoperiod. *Physiol Biochem Zool* 74: 66–74.

Hardewig I, Pörtner HO, van Dijk P. 2004. How does the cold stenothermal gadoid Lota lota survive high water temperatures during summer? *J Comp Physiol B* 174: 149–156.

Hardewig I, van Dijk PLM, Moyes CD, Pörtner HO. 1999. Temperature-dependent expression of cytochrome-*c* oxidase in Antarctic and temperate fish. *Am J Physiol Regul Integr Comp Physiol* 277: R508–R516.

Hazel JR. 1995. Thermal adaptation in biological membranes: Is homeoviscous adaptation the explanation? *Annu Rev Physiol* 57: 19–42.

Hazel JR, Prosser CL. 1974. Molecular mechanisms of temperature compensation in poikilotherms. *Physiol Rev* 54: 620–677.

Healy TM, Schulte PM. 2012a. Factors affecting plasticity in whole-organism thermal tolerance in common killifish (*Fundulus heteroclitus*). *J Comp Physiol B* 182: 49–62.

Healy TM, Schulte PM. 2012b. Thermal acclimation is not necessary to maintain a wide thermal breadth of aerobic scope in the common killifish (*Fundulus heteroclitus*). *Physiol Biochem Zool* 85: 107–119.

Heise K, Estevez MS, Puntarulo S, Galleano M, Nikinmaa M, Pörtner HO, Abele D. 2007. Effects of seasonal and latitudinal cold on oxidative stress parameters and activation of hypoxia inducible factor (HIF-1) in zoarcid fish. *J Comp Physiol B* 177: 765–777.

Herman JJ, Sultan SE. 2011. Adaptive transgenerational plasticity in plants: Case studies, mechanisms, and implications for natural populations. *Front Plant Sci* 2: 102.

Hight BV, Lowe CG. 2007. Elevated body temperatures of adult female leopard sharks, *Triakis semifasciata*, while aggregating in shallow nearshore embayments: Evidence for behavioral thermoregulation? *J Exp Mar Biol Ecol* 352: 114–128.

Hilton Z, Clements KD, Hickey AJR. 2010. Temperature sensitivity of cardiac mitochondria in intertidal and subtidal triplefin fishes. *J Comp Physiol B* 180: 979–990.

Hofmann GE, Buckley BA, Airaksinen S, Keen JE, Somero GN. 2000. Heat-shock protein expression is absent in the antarctic fish *Trematomus bernacchii* (family Nototheniidae). *J Exp Biol* 203: 2331–2339.

Hofmann GE, Lund SG, Place SP, Whitmer AC. 2005. Some like it hot, some like it cold: The heat shock response is found in New Zealand but not Antarctic notothenioid fishes. *J Exp Mar Biol Ecol* 316: 79–89.

Hofmann GE, Todgham AE. 2010. Living in the now: Physiological mechanisms to tolerate a rapidly changing environment. *Annu Rev Physiol* 72: 127–145.

Hori TS, Gamperl AK, Afonso LOB, Johnson SC, Hubert S, Kimball J, Bowman S, Rise ML. 2010. Heat-shock responsive genes identified and validated in Atlantic cod (*Gadus morhua*) liver, head kidney and skeletal muscle using genomic techniques. *BMC Genomics* 11: 72.

Houston AH, Murad A. 1992. Erythrodynamics in goldfish, *Carassius auratus* L.: Temperature effects. *Physiol Zool* 65: 55–76.

Huey RB, Kingsolver JG. 1989. Evolution of thermal sensitivity of ectotherm performance. *Trend Ecol Evol* 4: 131–135.

Huey RB, Stevenson RD. 1979. Integrating thermal physiology and ecology of ectotherms: A discussion of approaches. *Am Zool* 19: 357–366.

Hurst TP, Munch SB, Lavelle KA. 2012. Thermal reaction norms for growth vary among cohorts of Pacific cod (*Gadus macrocephalus*). *Mar Biol* 159: 2173–2183.

Jastroch M, Buckingham JA, Helwig M, Klingenspor M, Brand MD. 2007. Functional characterisation of UCP1 in the common carp: Uncoupling activity in liver mitochondria and cold-induced expression in the brain. *J Comp Physiol B* 177: 743–752.

Johnson TP, Bennett AF. 1995. The thermal acclimation of burst escape performance in fish: An integrated study of molecular and cellular physiology and organismal performance. *J Exp Biol* 198: 2165–2175.

Johnston IA. 1982. Capillarisation, oxygen diffusion distances and mitochondrial content of carp muscles following acclimation to summer and winter temperatures. *Cell Tissue Res* 222: 325–337.

Johnston IA. 2006. Environment and plasticity of myogenesis in teleost fish. *J Exp Biol* 209: 2249–2264.

Johnston IA, Andersen Ø. 2008. Number of muscle fibres in adult Atlantic cod varies with temperature during embryonic development and pantophysin (*Pan*I) genotype. *Aquat Biol* 4: 167–173.

Johnston IA, Lee H-T, Macqueen DJ, Paranthaman K, Kawashima C, Anwar A, Kinghorn JR, Dalmay T. 2009. Embryonic temperature affects muscle fibre recruitment in adult zebrafish: Genome-wide changes in gene and microRNA expression associated with the transition from hyperplastic to hypertrophic growth phenotypes. *J Exp Biol* 212: 1781–1793.

Jones EA, Jong AS, Ellerby DJ. 2008. The effects of acute temperature change on swimming performance in bluegill sunfish *Lepomis macrochirus*. *J Exp Biol* 211: 1386–1393.

Kammer AR, Orczewska JI, O'Brien KM. 2011. Oxidative stress is transient and tissue specific during cold acclimation of threespine stickleback. *J Exp Biol* 214: 1248–1256.

Kassahn KS, Crozier RH, Pörtner HO, Caley MJ. 2009. Animal performance and stress: Responses and tolerance limits at different levels of biological organisation. *Biol Rev* 84: 277–292.

Kassahn KS, Crozier RH, Ward AC, Stone G, Caley MJ. 2007. From transcriptome to biological function: Environmental stress in an ectothermic vertebrate, the coral reef fish *Pomacentrus moluccensis*. *BMC Genomics* 8: 358.

Kavanagh KD, Haugen TO, Gregersen F, Jernvall J, Vøllestad LA. 2010. Contemporary temperature-driven divergence in a Nordic freshwater fish under conditions commonly thought to hinder adaptation. *BMC Evol Biol* 10: 350.

Kawajiri M, Fujimoto S, Yamahira K. 2011. Genetic and thermal effects on the latitudinal variation in the timing of fin development of a fish *Oryzias latipes*. *J Therm Biol* 36: 306–311.

Kieffer JD, Wakefield AM. 2009. Oxygen consumption, ammonia excretion and protein use in response to thermal changes in juvenile Atlantic salmon *Salmo salar*. *J Fish Biol* 74: 591–603.

Killen SS, Marras S, Ryan MR, Domenici P, McKenzie DJ. 2012. A relationship between metabolic rate and risk-taking behaviour is revealed during hypoxia in juvenile European sea bass. *Funct Ecol* 26: 134–143.

Kingsolver JG. 2009. The well-temperatured biologist. (American Society of Naturalists Presidential Address). *Am Nat* 174: 755–768.

Klaiman JM, Fenna AJ, Shiels HA, Macri J, Gillis TE. 2011. Cardiac remodeling in fish: Strategies to maintain heart function during temperature change. *PLoS One* 6: e24464.

Kondo H, Misaki R, Watabe S. 2010. Transcriptional activities of medaka *Oryzias latipes* peroxisome proliferator-activated receptors and their gene expression profiles at different temperatures. *Fisheries Sci* 76: 167–175.

Korajoki H, Vornanen M. 2012. Expression of SERCA and phospholamban in rainbow trout (*Oncorhynchus mykiss*) heart: Comparison of atrial and ventricular tissue and effects of thermal acclimation. *J Exp Biol* 215: 1162–1169.

Koumoundouros G, Ashton C, Sfakianakis DG, Divanach P, Kentouri M, Anthwal N, Stickland NC. 2009. Thermally induced phenotypic plasticity of swimming performance in European sea bass *Dicentrarchus labrax* juveniles. *J Fish Biol* 74: 1309–1322.

Königshofer H, Tromballa H-W, Löppert H-G. 2008. Early events in signalling high-temperature stress in tobacco BY2 cells involve alterations in membrane fluidity and enhanced hydrogen peroxide production. *Plant Cell Environ* 31: 1771–1780.

Kültz D. 2005. Molecular and evolutionary basis of the cellular stress response. *Annu Rev Physiol* 67: 225–257.

Kültz D, Somero GN. 1996. Differences in protein patterns of gill epithelial cells of the fish *Gillichthys mirabilis* after osmotic and thermal acclimation. *J Comp Physiol B* 166: 88–100.

Lamarre SG, Le François NR, Driedzic WR, Blier PU. 2009. Protein synthesis is lowered while 20S proteasome activity is maintained following acclimation to low temperature in juvenile spotted wolffish (*Anarhichas minor* Olafsen). *J Exp Biol* 212: 1294–1301.

Lannig G, Bock C, Sartoris FJ, Pörtner HO. 2004. Oxygen limitation of thermal tolerance in cod, *Gadus morhua* L., studied by magnetic resonance imaging and on-line venous oxygen monitoring. *Am J Physiol Regul Integr Comp Physiol* 287: R902–R910.

LeBlanc S, Höglund E, Gilmour KM, Currie S. 2012. Hormonal modulation of the heat shock response: Insights from fish with divergent cortisol stress responses. *Am J Physiol Regul Integr Comp Physiol* 302: R184–R192.

LeBlanc S, Middleton S, Gilmour KM, Currie S. 2011. Chronic social stress impairs thermal tolerance in the rainbow trout (*Oncorhynchus mykiss*). *J Exp Biol* 214: 1721–1731.

Lee W-S, Monaghan P, Metcalfe NB. 2010. The trade-off between growth rate and locomotor performance varies with perceived time until breeding. *J Exp Biol* 213: 3289–3298.

Lee W-S, Monaghan P, Metcalfe NB. 2012. The pattern of early growth trajectories affects adult breeding performance. *Ecology* 93: 902–912.

Levitus S, Antonov JI, Boyer TP, Baranova OK, Garcia HE, Locarnini RA, Mishonov AV et al. 2012. World ocean heat content and thermosteric sea level change (0–2000 m), 1955–2010. *Geophys Res Lett* 39: L10603.

Lewis JM, Driedzic WR. 2010. Protein synthesis is defended in the mitochondrial fraction of gill but not heart in cunner (*Tautogolabrus adspersus*) exposed to acute hypoxia and hypothermia. *J Comp Physiol B* 180: 179–188.

Lewis JM, Hori TS, Rise ML, Walsh PJ, Currie S. 2010. Transcriptome responses to heat stress in the nucleated red blood cells of the rainbow trout (*Oncorhynchus mykiss*). *Physiol Genomics* 42: 361–373.

Lewis JM, Klein G, Walsh PJ, Currie S. 2012. Rainbow trout (*Oncorhynchus mykiss*) shift the age composition of circulating red blood cells towards a younger cohort when exposed to thermal stress. *J Comp Physiol B* 182: 663–671.

Liebscher RS, Richards RC, Lewis JM, Short CE, Muise DM, Driedzic WR, Ewart KV. 2006. Seasonal freeze resistance of rainbow smelt (*Osmerus mordax*) is generated by differential expression of glycerol-3-phosphate dehydrogenase, phosphoenolpyruvate carboxykinase, and antifreeze protein genes. *Physiol Biochem Zool* 79: 411–423.

Logan CA, Somero GN. 2010. Transcriptional responses to thermal acclimation in the eurythermal fish *Gillichthys mirabilis* (Cooper 1864). *Am J Physiol Regul Integr Comp Physiol* 299: R843–R852.

Logan CA, Somero GN. 2011. Effects of thermal acclimation on transcriptional responses to acute heat stress in the eurythermal fish *Gillichthys mirabilis* (Cooper). *Am J Physiol Regul Integr Comp Physiol* 300: R1373–R1383.

Lucassen M, Koschnick N, Eckerle LG, Pörtner H-O. 2006. Mitochondrial mechanisms of cold adaptation in cod (*Gadus morhua* L.) populations from different climatic zones. *J Exp Biol* 209: 2462–2471.

Lund SG, Caissie D, Cunjak RA, Vijayan MM, Tufts BL. 2002. The effects of environmental heat stress on heat-shock mRNA and protein expression in Miramichi Atlantic salmon (*Salmo salar*) parr. *Can J Fish Aquat Sci* 59: 1553–1562.

Lund SG, Phillips MCL, Moyes CD, Tufts BL. 2000. The effects of cell ageing on protein synthesis in rainbow trout (*Oncorhynchus mykiss*) red blood cells. *J Exp Biol* 203: 2219–2228.

Lund SG, Ruberté MR, Hofmann GE. 2006. Turning up the heat: The effects of thermal acclimation on the kinetics of *hsp70* gene expression in the eurythermal goby, *Gillichthys mirabilis*. *Comp Biochem Physiol A* 143: 435–446.

Lurman GJ, Koschnick N, Pörtner H-O, Lucassen M. 2007. Molecular characterisation and expression of Atlantic cod (*Gadus morhua*) myoglobin from two populations held at two different acclimation temperatures. *Comp Biochem Physiol A* 148: 681–689.

Lurman GJ, Petersen LH, Gamperl AK. 2012. Atlantic cod (*Gadus morhua* L.) in situ cardiac performance at cold temperatures: Long-term acclimation, acute thermal challenge and the role of adrenaline. *J Exp Biol* 215: 4006–4014.

Macqueen DJ, Robb D, Johnston IA. 2007. Temperature influences the coordinated expression of myogenic regulatory factors during embryonic myogenesis in Atlantic salmon (*Salmo salar* L.). *J Exp Biol* 210: 2781–2794.

Macqueen DJ, Robb DHF, Olsen T, Melstveit L, Paxton CGM, Johnston IA. 2008. Temperature until the "eyed stage" of embryogenesis programmes the growth trajectory and muscle phenotype of adult Atlantic salmon. *Biol Lett* 4: 294–298.

Mark FC, Lucassen M, Pörtner HO. 2006. Thermal sensitivity of uncoupling protein expression in polar and temperate fish. *Comp Biochem Physiol D* 1: 365–374.

McClelland GB, Craig PM, Dhekney K, Dipardo S. 2006. Temperature- and exercise-induced gene expression and metabolic enzyme changes in skeletal muscle of adult zebrafish (*Danio rerio*). *J Physiol* 577: 739–751.

McLean L, Young IS, Doherty MK, Robertson DHL, Cossins AR, Gracey AY, Beynon RJ, Whitfield PD. 2007. Global cooling: Cold acclimation and the expression of soluble proteins in carp skeletal muscle. *Proteomics* 7: 2667–2681.

Mendonça PC, Gamperl AK. 2010. The effects of acute changes in temperature and oxygen availability on cardiac performance in winter flounder (*Pseudopleuronectes americanus*). *Comp Biochem Physiol A* 155: 245–252.

Morrison J, Quick MC, Foreman MGG. 2002. Climate change in the Fraser River watershed: Flow and temperature projections. *J Hydrol* 263: 230–244.

Muñoz NJ, Breckels RD, Neff BD. 2012. The metabolic, locomotor and sex-dependent effects of elevated temperature on *Trinidadian guppies*: Limited capacity for acclimation. *J Exp Biol* 215: 3436–3441.

Newton JR, De Santis C, Jerry DR. 2012. The gene expression response of the catadromous perciform barramundi *Lates calcarifer* to an acute heat stress. *J Fish Biol* 81: 81–93.

Nilsson GE, Ostlund-Nilsson S, Munday PL. 2010. Effects of elevated temperature on coral reef fishes: Loss of hypoxia tolerance and inability to acclimate. *Comp Biochem Physiol A* 156: 389–393.

Ohlberger J, Staaks G, Petzoldt T, Mehner T, Hölker F. 2008. Physiological specialization by thermal adaptation drives ecological divergence in a sympatric fish species pair. *Evol Ecol Res* 10: 1173–1185.

Oleksiak MF, Churchill GA, Crawford DL. 2002. Variation in gene expression within and among natural populations. *Nat Genet* 32: 261–266.

Orczewska JI, Hartleben G, O'Brien KM. 2010. The molecular basis of aerobic metabolic remodeling differs between oxidative muscle and liver of threespine sticklebacks in response to cold acclimation. *Am J Physiol Regul Integr Comp Physiol* 299: R352–R364.

O'Brien KM. 2011. Mitochondrial biogenesis in cold-bodied fishes. *J Exp Biol* 214: 275–285.

O'Brien KM, Sidell BD. 2000. The interplay among cardiac ultrastructure, metabolism and the expression of oxygen-binding proteins in Antarctic fishes. *J Exp Biol* 203: 1287–1297.

O'Brien KM, Skilbeck C, Sidell BD, Egginton S. 2003. Muscle fine structure may maintain the function of oxidative fibres in haemoglobinless Antarctic fishes. *J Exp Biol* 206: 411–421.

Petricorena ZLC, Somero GN. 2007. Biochemical adaptations of notothenioid fishes: Comparisons between cold temperate South American and New Zealand species and Antarctic species. *Comp Biochem Physiol A* 147: 799–807.

Pierce VA, Crawford DL. 1996. Variation in the glycolytic pathway: the role of evolutionary and physiological processes. *Physiol Zool* 69: 489–508.

Pierce VA, Crawford DL. 1997a. Phylogenetic analysis of glycolytic enzyme expression. *Science* 276: 256–259.

Pierce VA, Crawford DL. 1997b. Phylogenetic analysis of thermal acclimation of the glycolytic enzymes in the genus *Fundulus*. *Physiol Zool* 70: 597–609.

Place SP, Zippay ML, Hofmann GE. 2004. Constitutive roles for inducible genes: evidence for the alteration in expression of the inducible *hsp70* gene in Antarctic notothenioid fishes. *Am J Physiol Regul Integr Comp Physiol* 287: R429–R436.

Podrabsky JE, Clelen D, Crawshaw LI. 2008. Temperature preference and reproductive fitness of the annual killifish *Austrofundulus limnaeus* exposed to constant and fluctuating temperatures. *J Comp Physiol A* 194: 385–393.

Podrabsky JE, Javillonar C, Hand SC, Crawford DL. 2000. Intraspecific variation in aerobic metabolism and glycolytic enzyme expression in heart ventricles. *Am J Physiol Regul Integr Comp Physiol* 279: R2344–R2348.

Podrabsky JE, Somero GN. 2004. Changes in gene expression associated with acclimation to constant temperatures and fluctuating daily temperatures in an annual killifish *Austrofundulus limnaeus*. *J Exp Biol* 207: 2237–2254.

Pottinger TG, Carrick TR. 1999. Modification of the plasma cortisol response to stress in rainbow trout by selective breeding. *Gen Comp Endocr* 116: 122–132.

Pörtner HO. 2001. Climate change and temperature-dependent biogeography: Oxygen limitation of thermal tolerance in animals. *Naturwissenschaften* 88: 137–146.

Pörtner H-O. 2010. Oxygen- and capacity-limitation of thermal tolerance: A matrix for integrating climate-related stressor effects in marine ecosystems. *J Exp Biol* 213: 881–893.

Pörtner HO, Farrell AP. 2008. Physiology and climate change. *Science* 322: 690–692.

Pörtner HO, Peck MA. 2010. Climate change effects on fishes and fisheries: Towards a cause-and-effect understanding. *J Fish Biol* 77: 1745–1779.

Pörtner HO, Peck L, Somero G. 2007. Thermal limits and adaptation in marine Antarctic ectotherms: an integrative view. *Phil Trans R Soc B* 362: 2233–2258.

Quinn NL, McGowan CR, Cooper GA, Koop BF, Davidson WS. 2011. Identification of genes associated with heat tolerance in Arctic charr exposed to acute thermal stress. *Physiol Genomics* 43: 685–696.

Richards RC, Short CE, Driedzic WR, Ewart KV. 2010. Seasonal changes in hepatic gene expression reveal modulation of multiple processes in rainbow smelt (*Osmerus mordax*). *Mar Biotechnol* 12: 650–663.

Rissanen E, Tranberg HK, Sollid J, Nilsson GE, Nikinmaa M. 2006. Temperature regulates hypoxia-inducible factor-1 (HIF-1) in a poikilothermic vertebrate, crucian carp (*Carassius carassius*). *J Exp Biol* 209: 994–1003.

Rodnick KJ, Gamperl AK, Lizars KR, Bennett MT, Rausch RN, Keeley ER. 2004. Thermal tolerance and metabolic physiology among redband trout populations in south-eastern Oregon. *J Fish Biol* 64: 310–335.

Ruud JT. 1954. Vertebrates without erythrocytes and blood pigment. *Nature* 173: 848–850.

Saillant E, Wang X, Ma L, Gatlin III DM, Vega RR, Gold JR. 2008. Genetic effects on tolerance to acute cold stress in red drum, *Sciaenops ocellatus* L. *Aquac Res* 39: 1393–1398.

Saito S, Shingai R. 2006. Evolution of thermoTRP ion channel homologs in vertebrates. *Physiol Genomics* 27: 219–230.

Sandblom E, Axelsson M. 2007. Venous hemodynamic responses to acute temperature increase in the rainbow trout (*Oncorhynchus mykiss*). *Am J Physiol Regul Integr Comp Physiol* 292: R2292–R2298.

Sartoris FJ, Bock C, Serendero I, Lannig G, Pörtner HO. 2003. Temperature-dependent changes in energy metabolism, intracellular pH and blood oxygen tension in the Atlantic cod. *J Fish Biol* 62: 1239–1253.

Schaefer J, Ryan A. 2006. Developmental plasticity in the thermal tolerance of zebrafish *Danio rerio*. *J Fish Biol* 69: 722–734.

Schjolden J, Backström T, Pulman KGT, Pottinger TG, Winberg S. 2005. Divergence in behavioural responses to stress in two strains of rainbow trout (*Oncorhynchus mykiss*) with contrasting stress responsiveness. *Horm Behav* 48: 537–544.

Schmidt K, Starck JM. 2010. Developmental plasticity, modularity, and heterochrony during the phylotypic stage of the zebrafish, *Danio rerio*. *J Exp Zool (Mol Dev Ecol)* 314B: 166–178.

Schnell AK, Seebacher F. 2008. Can phenotypic plasticity facilitate the geographic expansion of the tilapia *Oreochromis mossambicus*? *Physiol Biochem Zool* 81: 733–742.

Schulte PM. 2011. Responses and adaptations to the effects of temperature: An introduction. In: Farrell AP, Cech JJ, Stevens ED, Richards JG, Eds. *Encyclopedia of Fish Physiology: From Genome to Environment*. Academic Press, San Deago, CA, pp. 1688–1694.

Schulte PM, Healy TM, Fangue NA. 2011. Thermal performance curves, phenotypic plasticity, and the time scales of temperature exposure. *Integr Comp Biol* 51: 691–702.

Schultz ET, Reynolds KE, Conover DO. 1996. Countergradient variation in growth among newly hatched *Fundulus heteroclitus*: Geographic differences revealed by common-environment experiments. *Funct Ecol* 10: 366–374.

Scott GR, Johnston IA. 2012. Temperature during embryonic development has persistent effects on thermal acclimation capacity in zebrafish. *Proc Natl Acad Sci USA* 109: 14247–14252.

Seebacher F, Brand MD, Else PL, Guderley H, Hulbert AJ, Moyes CD. 2010. Plasticity of oxidative metabolism in variable climates: Molecular mechanisms. *Physiol Biochem Zool* 83: 721–732.

Sfakianakis DG, Leris I, Kentouri M. 2011. Effect of developmental temperature on swimming performance of zebrafish (*Danio rerio*) juveniles. *Environ Biol Fish* 90: 421–427.

Shiels HA, Di Maio A, Thompson S, Block BA. 2011. Warm fish with cold hearts: Thermal plasticity of excitation-contraction coupling in bluefin tuna. *Proc R Soc B* 278: 18–27.

Sidell BD, O'Brien KM. 2006. When bad things happen to good fish: The loss of hemoglobin and myoglobin expression in Antarctic icefishes. *J Exp Biol* 209: 1791–1802.

Sidell BD, Vayda ME, Small DJ, Moylan TJ, Londraville RL, Yuan M-L, Rodnick KJ, Eppley ZA, Costello L. 1997. Variable expression of myoglobin among the hemoglobinless Antarctic icefishes. *Proc Natl Acad Sci USA* 94: 3420–3424.

Somero GN. 2010. The physiology of climate change: How potentials for acclimatization and genetic adaptation will determine "winners" and "losers." *J Exp Biol* 213: 912–920.

Somero GN. 2011. Comparative physiology: A "crystal ball" for predicting consequences of global change. *Am J Physiol Regul Integr Comp Physiol* 301: R1–R14.

Steinhausen MF, Sandblom E, Eliason EJ, Verhille C, Farrell AP. 2008. The effect of acute temperature increases on the cardiorespiratory performance of resting and swimming sockeye salmon (*Oncorhynchus nerka*). *J Exp Biol* 211: 3915–3926.

Stevens ED, Sutterlin AM. 1976. Heat transfer between fish and ambient water. *J Exp Biol* 65: 131–145.

Szabo TM, Brookings T, Preuss T, Faber DS. 2008. Effects of temperature acclimation on a central neural circuit and its behavioral output. *J Neurophysiol* 100: 2997–3008.

Taylor SE, Egginton S, Taylor EW. 1996. Seasonal temperature acclimatisation of rainbow trout: Cardiovascular and morphometric influences on maximal sustainable exercise level. *J Exp Biol* 199: 835–845.

Thomson JS, Watts PC, Pottinger TG, Sneddon LU. 2011. Physiological and genetic correlates of boldness: Characterising the mechanisms of behavioural variation in rainbow trout, *Oncorhynchus mykiss*. *Horm Behav* 59: 67–74.

Thorne MAS, Burns G, Fraser KPP, Hillyard G, Clark MS. 2010. Transcription profiling of acute temperature stress in the Antarctic plunderfish *Harpagifer antarcticus*. *Marine Genomics* 3: 35–44.

Tiku PE, Gracey AY, Macartney AI, Beynon RJ, Cossins AR. 1996. Cold-induced expression of delta 9-desaturase in carp by transcriptional and posttranslational mechanisms. *Science* 271: 815–818.

Todgham AE, Hoaglund EA, Hofmann GE. 2007. Is cold the new hot? Elevated ubiquitin-conjugated protein levels in tissues of Antarctic fish as evidence for cold-denaturation of proteins *in vivo*. *J Comp Physiol B* 177: 857–866.

Todgham AE, Iwama GK, Schulte PM. 2006. Effects of the natural tidal cycle and artificial temperature cycling on Hsp levels in the tidepool sculpin *Oligocottus maculosus*. *Physiol Biochem Zool* 79: 1033–1045.

Tseng Y-C, Chen R-D, Lucassen M, Schmidt MM, Dringen R, Abele D, Hwang P-P. 2011. Exploring uncoupling proteins and antioxidant mechanisms under acute cold exposure in brains of fish. *PLoS One* 6: e18180.

Van den Burg EH, Peeters RR, Verhoye M, Meek J, Flik G, Van der Linden A. 2005. Brain responses to ambient temperature fluctuations in fish: Reduction of blood volume and initiation of a whole-body stress response. *J Neurophysiol* 93: 2849–2855.

Van Dijk PLM, Tesch C, Hardewig I, Pörtner HO. 1999. Physiological disturbances at critically high temperatures: A comparison between stenothermal Antarctic and eurythermal temperate eelpouts (Zoarcidae). *J Exp Biol* 202: 3611–3621.

Vergauwen L, Benoot D, Blust R, Knapen D. 2010. Long-term warm or cold acclimation elicits a specific transcriptional response and affects energy metabolism in zebrafish. *Comp Biochem Physiol A* 157: 149–157.

Wallman HL, Bennett WA. 2006. Effects of parturition and feeding on thermal preference of Atlantic stingray, *Dasyatis sabina* (Lesueur). *Environ Biol Fish* 75: 259–267.

Ward AJW, Hensor EMA, Webster MM, Hart PJB. 2010. Behavioural thermoregulation in two freshwater fish species. *J Fish Biol* 76: 2287–2298.

White CR, Alton LA, Frappell PB. 2012. Metabolic cold adaptation in fishes occurs at the level of whole animal, mitochondria and enzyme. *Proc R Soc B* 279: 1740–1747.

White CN, Hightower LE, Schultz RJ. 1994. Variation in heat-shock proteins among species of desert fishes (Poeciliidae, *Poeciliopsis*). *Mol Biol Evol* 11: 106–119.

Whitehead A, Crawford DL. 2006. Neutral and adaptive variation in gene expression. *Proc Natl Acad Sci USA* 103: 5425–5430.

Windisch HS, Kathöver R, Pörtner H-O, Frickenhaus S, Lucassen M. 2011. Thermal acclimation in Antarctic fish: Transcriptomic profiling of metabolic pathways. *Am J Physiol Regul Integr Comp Physiol* 301: R1453–R1466.

Wong HR, Wispe JR. 1997. The stress response and the lung. *Am J Physiol Lung Cell Mol Physiol* 273: L1–L9.

Wood LA, Brown IR, Youson JH. 1999. Tissue and developmental variations in the heat shock response of sea lampreys (*Petromyzon marinus*): Effects of an increase in acclimation temperature. *Comp Biochem Physiol A* 123: 35–42.

Wysocki LE, Montey K, Popper AN. 2009. The influence of ambient temperature and thermal acclimation on hearing in a eurythermal and a stenothermal otophysan fish. *J Exp Biol* 212: 3091–3099.

Yamahira K, Kawajiri M, Takeshi K, Irie T. 2007. Inter- and intrapopulation variation in thermal reaction norms for growth rate: Evolution of latitudinal compensation in ectotherms with a genetic constraint. *Evolution* 61: 1577–1589.

Yamahira K, Nishida T. 2009. Latitudinal variation in axial patterning of the medaka (Actinopterygii: Adrianichthyidae): Jordan's rule is substantiated by genetic variation in abdominal vertebral number. *Biol J Linn Soc* 96: 856–866.

Yamahira K, Takeshi K. 2008. Variation in juvenile growth rates among and within latitudinal populations of the medaka. *Popul Ecol* 50: 3–8.

Young S, Egginton S. 2011. Temperature acclimation of gross cardiovascular morphology in common carp (*Cyprinus carpio*). *J Therm Biol* 36: 475–477.

Zehmer JK, Hazel JR. 2004. Membrane order conservation in raft and non-raft regions of hepatocyte plasma membranes from thermally acclimated rainbow trout. *Biochim Biophys Acta* 1664: 108–116.

9 Physiology of Social Stress in Fishes

Christina Sørensen, Ida Beitnes Johansen, and Øyvind Øverli

CONTENTS

9.1 STRESS BIOLOGY: AN INTEGRATIVE AND EVOLUTIONARY VIEW

Stress is a much used concept in physiology and medicine, yet, the definition of "stress" as a biological term has a long history of inconsistency and controversy. A consensus long remained where stress was typically seen as something negative to the organism, threatening homeostasis and survival (e.g., McEwen and Mendelson 1993; Moberg 2000). More recently, stress biology has encompassed the notion that maintaining complete stability is perhaps not the ultimate objective of organisms struggling to survive in a constantly changing world (McEwen and Wingfield 2003, 2010; Korte et al. 2005). Rather, efficient regulation requires anticipating varied challenges and executing physiological responses before the need actually arises. This process is referred to as allostasis or "stability through change" (Sterling and Eyer 1988; McEwen and Wingfield 2003, 2010). Allostasis thus refers to regulatory mechanisms mediating change through the prediction of an activity required to meet a new demand (Sterling 2012). In this concept, negative effects impairing health, fitness, and animal welfare emerge only when "allostatic overload" arises from chronic, unpredictable, or uncontrollable conditions which do not merit successful adjustment (McEwen and Stellar 1993; Korte et al. 2007). In their recent review, Koolhaas et al. (2011) suggest that the term "stress" should be restricted to conditions when an environmental demand exceeds the natural regulatory capacity of an organism, and in particular situations that include unpredictability and uncontrollability. Physiologically, such situations would be characterized by either absent or discordant anticipatory responses (unpredictability), or insufficient recovery (uncontrollability) of homeostasis. In this chapter, we will see how the unequal access to resources and aggressive interactions associated with the formation of dominance hierarchies in fishes can create a physiological situation which meets both of these criteria. We will present the physiological and behavioral responses typically observed in socially stressed animals. Although such situations undoubtedly infer some negative consequences for the subordinate animals, we will discuss the potential adaptive properties of socially induced stress responses.

Particular attention has recently been directed to possible cognitive and emotional processes in fishes (Rose 2002; Chandroo et al. 2004; Huntingford et al. 2006). The allostasis concept of coping with stress implies prediction of the activity required to meet new demand, a process which depends heavily on both experience and genetic factors affecting perception, learning, and memory. Thus, we find it timely to also review the effect of social status on contributors to the central processing of external events: the set of biochemical, molecular, and structural processes collectively referred to as neural plasticity.

Finally, we focus on the emerging realization that physiological and neurobiological mechanisms mediating responses on the organismal level are individually variable traits, manifested phenotypically as correlated trait clusters (alternatively termed: coping styles, behavioral syndromes, or personality traits), which are subject to selection (see reviews by Koolhaas et al. 1999; Gosling 2001; Sih et al. 2004; Dingemanse and Réale 2005; Korte et al. 2005; Koolhaas et al. 2007, 2010; Øverli et al. 2007). Notwithstanding some problems caused by the widely differing terminology, it is becoming increasingly clear that in order to elucidate the biology behind and the physiological consequences of stress, it is necessary to study the proximate mechanisms underlying linked behavioral and physiological trait characteristics. Studies on comparative models including teleost fishes have revealed a range of neuroendocrine–behavioral associations that appear to be conserved throughout the vertebrate subphylum (Øverli et al. 2005, 2007; Schjolden and Winberg 2007; Galhardo et al. 2011; Martins et al. 2011). For instance, has individual variation in stress-induced cortisol production been associated with a range of behaviors, with low-responsive (low cortisol response to stressors) individuals typically showing proactive behaviors such as enhanced aggression, social dominance, and rapid resumption of feed intake after stress (but see Ruiz-Gomez et al. (2008) and Schjolden et al. (2005a) for context-dependent exceptions).

Variation in such trait clusters can dramatically affect the outcome of physiological and behavioral studies. In particular, the influence of variation in stress responsiveness and related traits on the outcome of competitions for social dominance (in itself a powerful physiological modifier; see this chapter and Johnsson et al. (2005) for a comprehensive review) have been well documented in fishes (Pottinger and Carrick 2001; Øverli et al. 2004a; Dahlbom et al. 2011). In this chapter, we will therefore review how variation in stress-coping style relates to both the risk of being exposed to social stress and the effect of such stress. For definition, we rely on the terminology of Koolhaas et al. (1999), which states that coping style infers consistent correlations between physiological and behavioral traits (for instance, a high cortisol response to stress correlated with high aggression level and boldness; see also discussion in Section 9.4.2). Consistent behavioral patterns without knowledge of physiological parameters are often referred to as either personality traits (Gosling 2001) or behavioral syndromes. The latter term is typically used when several behavioral patterns are not only consistent, but also correlate to each other (Sih et al. 2004; see Section 9.4.2 for a more detailed discussion). Thus, in this context, the usage of terms such as "animal personality" in biology does not ascribe human characteristics to animals; it just means that an individual animal behaves in a consistent way, which distinguishes it from other members of the same species.

9.2   SOCIAL DOMINANCE AND SOCIAL STRESS

In social animals, that is, animals that live together and influence each other (Tinbergen 1953), variation in competitive ability implies that some animals will have preferential access over others to food or other commodities in demand. When such imbalance is a consistent feature of the interaction between two individuals, it is referred to as a dominance/subordination relationship (Drews 1993). In groups of more than two individuals, dominance/subordination relationships will often take the form of a dominance hierarchy or "pecking order," in which the outcome of overt or ritualized aggressive encounters is the main factor determining individual rank order (Huntingford and Turner 1987). Human investigators usually reveal the structure of dominance hierarchies by behavioral observations of who retreats and who wins during aggressive interactions. However, social hierarchy formation potentially affects almost every feature of animal life, so a range of other criteria including relative food consumption, mating frequency, or nonsexual affiliative behavior also indicate hierarchical relationships (Woolpy and Ginsburg 1967; Raab et al. 1986; Järvi 1990; Hill 1990; Winberg et al. 1993a,b).

In fishes (Abbott and Dill 1985; Franck and Ribowski 1993; Huntingford et al. 1993; Winberg and Nilsson 1993; Winberg et al. 1993a; Nakano 1994, 1995a,b; Gómez-Laplaza and Morgan 2003; Desjardins et al. 2012), as in other vertebrates (Raab et al. 1986; Blanchard et al. 1993; Albonetti and Farabollini 1994; Meerlo et al. 1997; Engh et al. 2005; Hsu et al. 2006; Korzan and Summers 2007), subordinate individuals show a general behavioral inhibition, characterized by suppressed aggressive and/or reproductive behavior, reduced feeding, and low spontaneous locomotor activity and exploration. Behavioral inhibition in subordinates can be viewed as a passive coping strategy to avoid costly interaction with dominants (Leshner 1980; Benus et al. 1991). In other words, it would be useless to engage in competition with an individual whose competitive skills are known to exceed your own; individuals with obvious superiority should not be challenged. Many of the behavioral changes seen in subordinates have also been reported to occur in response to other types of stress (McNaughton 1993; Martins et al. 2012). Thus, the behavioral characteristics of socially subordinate animals might in part reflect a general response to chronically stressful, unpredictable, and/or potentially dangerous situations where flight is not feasible.

Physiological stress responses can frequently be seen both in individuals that are subject to brief social defeat and in those who maintain a subordinate position in an established hierarchy. The degree of unpredictability and uncontrollability inflicted on the subordinate individual in such cases, along with the physiological and behavioral outcomes, clearly suggests that the restrictive criteria for a stressful condition are met (Koolhaas et al. 2011). Physiologically, subordinate fish show many of the signs of prolonged stress that are commonly observed in mammals, including elevated hypothalamus–pituitary–interrenal (HPI) axis activity, decreased hypothalamus–pituitary–gonadal (HPG) activity, and chronically increased brain serotonin (5-hydroxytryptamin, 5-HT) activity (Winberg et al. 1992; Francis et al. 1993; Winberg and Nilsson 1993; Fox et al. 1997; Winberg and Lepage 1998; Øverli et al. 1999a; Elofsson et al. 2000; Höglund et al. 2000; Bass and Grober 2001; Sloman and Armstrong 2002; Doyon et al. 2003; Sørensen et al. 2007; Filby et al. 2010a,b; Dahlbom et al. 2012; Jeffrey et al. 2012). Changes in other central signaling systems such as the nonapeptides arginine vasotocin (AVT) and isotocin (IT) have also been reported (Section 9.3.2.2, see also Godwin and Thompson 2012 for a review).

This social subordination paradigm, alternatively termed "social stress" or "defeat stress," has in mammals become a widely used model to understand the etiology and biological embedding of various stress-related physiological and psychological pathologies (for recent examples, see Scott et al. 2012; Shively and Willard 2012; Slattery et al. 2012; Turdi et al. 2012; Venzala et al. 2012). In this context, it should be noted that teleost fishes are rapidly emerging as an alternative to small mammals in biomedical and behavioral research for reasons such as easy maintenance, short generation times, and increasingly mapped genomes. Behavioral, genetic, and physiological screening of large numbers of individuals across treatments and/or over generations is one particularly attractive approach (Darland and Dowling 2001; Guo 2004; Lieschke and Currie 2007), typically utilizing zebrafish (*Danio rerio*), but important contributions are certainly not limited to this species (Thorgaard et al. 2002; Epstein and Epstein 2005; Terzibasi et al. 2007). Apart from biomedical application, models of social dominance and social stress provide excellent opportunity to gain fundamental knowledge about how the brain translates social information into physiological and behavioral responses on the neuroendocrine and transcriptomic levels (Øverli et al. 2004b; Burmeister et al. 2005; Summers and Winberg 2006; Fernald 2012; Sanogo et al. 2012;

Taborsky et al. 2013). The behavioral adjustments and fitness consequences seen in subordinate individuals also merit considerable attention in behavioral and evolutionary ecology (Weir et al. 2004; Fero et al. 2007; Edeline et al. 2010; Creel et al. 2013; Riebli et al. 2012; Taborsky and Oliveira 2012).

9.3 PHYSIOLOGICAL EFFECTS OF SOCIAL SUBORDINATION

With the exception of some phyla- and species-specific differences, the general neuroendocrine stress response appears to be well conserved between all vertebrates. The fish stress response has been thoroughly reviewed elsewhere (Barton and Iwama 1991; Wendelaar Bonga 1997; Barton 2002), so in the following sections we will just give a brief description of the physiological stress response in fishes. The physiological stress response is to a certain degree unspecific, in the sense that a variety of both perceived and real threats to the organism can trigger the same response (Wendelaar Bonga 1997). We will first present the general integrated stress response in fishes, and then move on to the specific case of social stress.

9.3.1 PHYSIOLOGICAL STRESS RESPONSE IN FISHES

The general stress response can be divided into a primary, a secondary, and a tertiary physiological response (Pickering 1981; Wedemeyer et al. 1990; Pickering and Pottinger 1995). The primary stress response represents the perception of threat and the subsequent and immediate neuroendocrine response to it. This response is dominated by the release of catecholamines and glucocorticoid stress hormones. The secondary response is characterized by various biochemical, physiological, and behavioral changes directly resulting from the actions of stress hormones, as well as pleiotropic effects of upstream control factors, such as corticotropin-releasing factor (CRF, also referred to as corticotropin-releasing hormone, CRH). The tertiary response pertains to whole animals and population dynamics and includes suppression of growth, immune function, or reproduction.

The hypothalamic–sympathetic–chromaffin axis is the fish equivalent of the mammalian sympathetic–adrenal–medullary axis. Chromaffin cells are functionally homologous to the mammalian adrenal medulla, and are scattered within or form small clusters in the fish head kidney (Chester Jones 1980). Within seconds of receiving stressful stimuli, the chromaffin cells start secreting the two catecholamines, adrenaline (epinephrine, E) and noradrenaline (norepinephrine, NE), of which E is the dominant species in teleost blood (McDonald and Milligan 1992; Randall and Perry 1992). Catecholamine release is predominantly triggered by preganglionic sympathetic cholinergic fibers of sympathetic nerves, although some non-cholinergic release can be triggered by other factors or conditions, such as circulating catecholamines themselves, cortisol, hyperkalemia, hypercapnia, and hypoxemia. As catecholamines are stored in vesicles inside the chromaffin cells, they can be rapidly released, and blood levels can increase from basal values of <5 nM to peak values of >1000 nM within 1–3 min in response to severe acute stressors. In most cases, the catecholamine response is of short duration as the biological half-life of catecholamines is <10 min. Nonetheless, during severe chronic stress, catecholamine levels may remain well above basal levels for several hours or days (for reviews, see McDonald and Milligan 1992; Randall and Perry 1992). Catecholamines are important for the cardiovascular, respiratory, and metabolic changes characterizing the acute stress response in fishes (Wendelaar Bonga 1997).

The fish equivalent of the mammalian hypothalamic–pituitary–adrenal (HPA) axis is referred to as the HPI axis. Interrenal cells are also found dispersed in the head kidney, and are functionally homologous to part of the mammalian adrenal cortex (*zona fasciculata*), and are the main source of corticosteroids in fishes (Wendelaar Bonga 1997). The main corticosteroid in ray-finned fishes is cortisol (Nandi and Bern 1965; Chester Jones 1980; Patiño et al. 1987), whereas 1α-hydroxycorticosterone is the major corticosteroid in elasmobranchs (Idler and Truscott 1967; Truscott and Idler 1972; for a review, see Anderson 2012). In this chapter, we will focus primarily on results from ray-finned fishes, and consequently cortisol. The endocrine control of cortisol secretion is complex, and appears to be mainly dominated by adrenocorticotropic hormone (ACTH) (Wendelaar Bonga 1997; Metz et al. 2005; Aluru and Vijayan 2008). CRH and urotensin I (UI) are produced and secreted in the preoptic area of the hypothalamus. They stimulate corticotropic cells in the anterior pituitary to release ACTH and other proopiomelanocortin (POMC)-derived pituitary hormones into the circulation. ACTH, in turn, stimulates the interrenal cells to produce and release cortisol. Unlike catecholamines, corticosteroids are hydrophobic and cannot be stored in vesicles, but are produced and released on demand. Therefore, a stress-induced rise in plasma levels occurs within minutes rather than seconds, which makes for easier measurement of cortisol without influence by handling stress. This is a major reason why elevation of plasma cortisol is the most commonly used stress indicator in fishes (for reviews, see Barton and Iwama 1991; Wendelaar Bonga 1997). There are considerable differences in the plasma cortisol response between different teleost species (Barton 2002), but a typical post-stress cortisol response falls between 30 and 300 ng/mL (Barton and Iwama 1991). In response to acute stress, plasma cortisol levels increase within a few minutes; however, it can take several hours or days for the plasma cortisol level to return to its basal condition. During chronic stress, plasma cortisol levels can remain well above control values for prolonged periods (for reviews, see Donaldson 1981; Barton and Iwama 1991; Wendelaar Bonga 1997). There is, however, considerable natural variation in circulating cortisol levels both at diurnal and seasonal scales, something which could in part be attributed to changes in light, temperature, and photoperiod (Ellis et al. 2012).

The effects of cortisol in teleost fishes are extensive. As the mineralcorticoid hormone aldosterone has only been demonstrated in minute amounts (if at all present), and without apparent physiological significance (Bern 1967; Chester Jones 1980; Sangalang and Uthe 1994), the general belief is that cortisol acts both as a mineralcorticoid and a glucocorticoid in teleost fishes (Stolte et al. 2008). In addition to the essential functions of cortisol (i.e., hydromineral balance and energy metabolism), several deleterious effects of prolonged and high cortisol exposure have been reported in socially stressed fish. These will be addressed in Section 9.3.2.

9.3.2 Physiological Effects of Social Subordination

In a range of vertebrate species, social defeat and social subordination are associated with physiological changes that may initially be adaptive but can ultimately threaten the animal's health or life (Louch and Higginbotham 1967; Noakes and Leatherland 1977; Golub et al. 1979; Ejike and Schreck 1980; Sapolsky 1990; Blanchard et al. 1993; Albeck et al. 1997; Shively et al. 1997; Winberg and Lepage 1998). These effects are mainly caused by the physiological stress response and, to a large extent, cortisol. As the effects are manifest in socially interacting animals, but not in the absence of external stressors during social isolation, they are referred to as social stress. This concerns free-living as well as captive animals, suggesting that social stress reactions, including hypersecretion of cortisol in subordinate individuals, are a general feature of animal life. Social stress is, however, likely to be enhanced under conditions of artificial rearing, where opportunities for social signalling and escape from fighting are limited (Sloman and Armstrong 2002). In mammals, a number of reports also infer stress in dominant individuals, especially in free-living animals where dominants are involved in high rates of aggressive encounters (Creel 2001; Honess and Marin 2006). This could probably be explained by taking the relative stability of social hierarchies into consideration. For example, in mammals living in a changing social environment with unstable hierarchies, top-ranking individuals frequently have to defend and reinforce their position (Sapolsky 1982, 1992). Thus, dominants would be frequently involved in aggressive encounters, which are of course stressful regardless of whether they are won or lost (but more so if they are lost, see next paragraph). In a stable social setting, dominant individuals are generally in a less stressful situation than their subordinate conspecifics. In this chapter, we primarily focus on the effects experienced by subordinate animals.

Elevated plasma cortisol levels are commonly found in subordinate fish confined in pairs (Laidley and Leatherland 1988; Pottinger and Pickering 1992; Øverli et al. 1999a; Elofsson et al. 2000; Sloman et al. 2000a, 2001a, 2002a; Pottinger and Carrick 2001; Höglund et al. 2002; Filby et al. 2010a; Alonso et al. 2012; Cammarata et al. 2012; Johansen et al. 2012; Sørensen et al. 2012). Cortisol levels increase in both dominant and subordinate individuals during initial stages of social interaction, but usually return to baseline in the dominant one within hours of establishment of the hierarchy (Øverli et al. 1999a). Still some studies have found persistently elevated cortisol in dominant individuals, for instance, in zebrafish (Pavlidis et al. 2011). It is unknown what may cause such differences, but it may relate to both species differences in social preference, and experimental timing and context. Species differ in group size preference (schooling vs. non-schooling), and in a schooling species such as the zebrafish, interacting in pairs may be relatively more stressful to both the dominant and subordinate individuals compared to control animals taken directly from a large group (Pavlidis et al. 2011). Stability of the hierarchy at the time of sampling may also affect relative cortisol levels, and species may differ markedly in time to establish a stable social hierarchy (Fox et al. 1997). In subordinate individuals, cortisol levels are typically elevated for days and possibly weeks after hierarchy establishment (Figure 9.1; Øverli et al. 1999a; Sloman et al. 2001a). In groups of fish, on the other hand, some studies report elevated plasma cortisol levels in subordinate individuals (Ejike and Schreck 1980; Winberg and Lepage 1998; Höglund et al. 2000), whereas other studies do not find this (Pottinger and Pickering 1992; Øverli et al. 1999b; Sloman et al. 2000b, 2001b, 2002b), and some report higher cortisol in dominant individuals (Buchner et al. 2004).

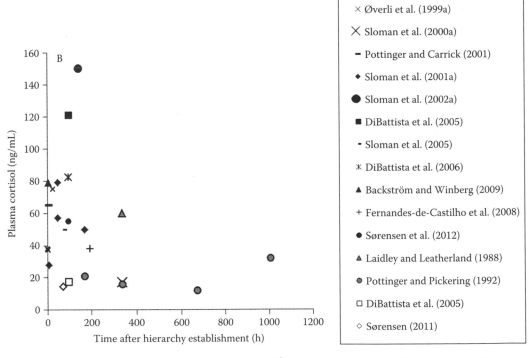

FIGURE 9.1 Reported plasma cortisol levels of subordinate rainbow trout in dyadic social hierarchies, plotted against time after hierarchy establishment. (Laidley and Leatherland (1988) reported cortisol values in nmol/L these values were converted to ng/mL.) All studies applied similar experimental paradigms; a dyadic fight between pairs of rainbow trout that had an equal previous residence time in the experimental setup (thus, all reports using resident–intruder paradigms were excluded). In 11 studies, the dyads consisted of individuals size-matched either by length or weight (black markers). Four studies reported cortisol values of non-matched pairs. Of these, two did not intentionally mismatch pairs (gray markers), while two studies had intentionally size-mismatched pairs (5%–20% difference in length or 50%–100% difference in weight (open markers).

This could be due to species differences, but could also be caused by differences in group composition and environmental context.

As will be discussed in Section 9.4.1, body size is important for competitive ability and capacity for social dominance (Miller 1964; Abbott et al. 1985; Knights 1987); however, relative size difference also influences the dynamics of hierarchies after establishment. Midas cichlids (*Cichlasoma citrinellum*) presented with lifelike dummies elicit most aggression against a perceived opponent of the same size, and aggression declines linearly with reduced opponent size (Barlow et al. 1984; Bond 1992). In pairs of European eel (*Anguilla anguilla*), aggression is highest in pairs of similar size, declining steeply at relative size differences between 1.5 and 2 (Knights 1987). The same general pattern is observed in groups of fish living in the wild (Wankowski and Thorpe 1979; Ang and Manica 2010). This size effect on hierarchy intensity could potentially be explained by fish of different sizes commonly occupying different microhabitats (Edmundson et al. 1968; Jenkins 1969), and not posing a direct threat to each other. Also, as both larger size and higher social status predict higher growth rate (discussed in Section 9.4.1, and see Yamagishi 1964; Li and Brocksen 1977), the relative size difference between dominant and subordinate individuals is likely to increase over time. Consequently, the perceived threat posed by a subordinate individual may be low at the outset and diminish over time. The dominant individual is likely to adjust its energy expenditure accordingly, leading to lower aggression and lower stress levels in the subordinate individual if the relative size difference is greater. In zebrafish, smaller subordinate individuals have higher reproductive success than subordinate individuals that are closer in size to the dominant individual, despite female zebrafish preferring larger males (Watt et al. 2011). One could speculate whether larger subordinate individuals receive more aggression and consequently have a higher stress level resulting in suppressed reproductive behavior, whereas smaller subordinate males receive less attention from the dominant individuals, allowing them to be more reproductively active and sire more offspring despite their apparent size disadvantage.

In salmonid fishes, cortisol levels appear to diminish with time spent in a social hierarchy (Figure 9.1). Although this could in theory reflect a gradual increase in growth rate divergence resulting in the subordinate individual posing less threat as time goes by, the reduction appears to take place on a time scale for which a significant change in relative size difference is unlikely to occur. It is therefore more likely to reflect other aspects of hierarchy dynamics. Indeed, in salmonid fishes a gradual reduction in aggression typically takes place on approximately the same time scale as the decline in cortisol levels (Figure 9.2; and see Winberg and Lepage 1998), and plasma cortisol levels may correlate with hierarchy intensity (Sloman et al. 2001a). On the other hand, data from studies on zebrafish suggest that aggression does not always decline during chronic social interaction (Paull et al. 2010; Dahlbom et al. 2012). Reports from zebrafish have also indicated persistently elevated cortisol levels for both dominant and subordinate individuals (Pavlidis et al. 2011). Whether this discrepancy reflects a difference in aggression and stress dynamics in this species relative to other species commonly studied is not known.

Social stress is also known to affect other key elements of the HPI axis (Pavlidis et al. 2011). In a study of rainbow trout (*Oncorhynchus mykiss*), social subordinates were found to have altered cortisol dynamics and reduced target tissue response to cortisol, while dominant fish had higher transcript levels for enzymes important in steroid biosynthesis, which could indicate an increased capacity for responding to subsequent stressors (Jeffrey et al. 2012). Social subordination also results in elevated CRH mRNA levels in the hypothalamic preoptic area of rainbow trout, suggesting increased synthesis and release of CRH, the main ACTH secretagogue (Winberg and Lepage 1998; Doyon et al. 2003; Bernier et al. 2008). Winberg and Lepage (1998) also demonstrated an increase in expression of pituitary POMC, the precursor peptide to ACTH and other biologically active peptides, in subordinate rainbow trout. These elevated mRNA levels are likely to reflect enhanced synthesis and release of ACTH and the other POMC-derived peptides.

Low social status is typically associated with reduced growth rate (e.g., Li and Brocksen 1977; Metcalfe 1986; Abbott and Dill 1989; Metcalfe et al. 1990; Pottinger and Pickering 1992; Ryer and

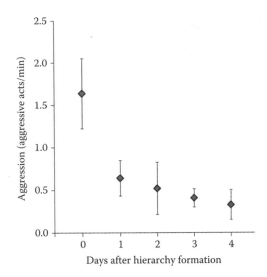

FIGURE 9.2 Aggression levels in dyadic hierarchies of rainbow trout. (From Sørensen, C., Regulation of brain cell proliferation in fish, in *Faculty of Mathematics and Natural Sciences*, University of Oslo, Oslo, Norway, 2011.) The pairs were filmed during and after hierarchy establishment, and then daily prior to feeding (between 12.00 and 16.00). Aggressive acts per minute were quantified for the 10 min interval immediately after hierarchy establishment (i.e., when aggressive activity turned unidirectional after an initial phase of bidirectional aggression), and for 10 min prior to feeding on the following days. An aggressive act was defined as a chase followed by a bite to any part of the subordinate individual's body.

Olla 1996; Sloman et al. 2000b, 2001b, 2002b; Filby et al. 2010a; Sørensen et al. 2012), as well as lower condition factor (the relationship between the weight and the cube of the length, commonly assumed to reflect nutritional status, i.e., "condition" of the animal; Ricker 1975; Sloman et al. 2000a,b). This is caused partially by monopolization of food sources by dominants (Metcalfe et al. 1989; McCarthy et al. 1992; Adams and Huntingford 1996; Adams et al. 1998; Maclean and Metcalfe 2001; Winberg et al. 1993b). Still, feeding dominant and subordinate individuals equal food rations does not rectify the difference in growth rate (Abbott and Dill 1989; Lee et al. 2011). This could potentially be explained by an increase in metabolic rate, and subordination, or the mere presence of a larger conspecific has been found to elevate standard metabolic rate in salmonids (Sloman et al. 2000c; Millidine et al. 2009). Social stress also causes down-regulation of digestive activity (Earley et al. 2004) and a general mobilization of energy reserves, which in the long term may lead to depletion of these reserves, and potentially weight loss or retarded growth (Ejike and Schreck 1980; Peters et al. 1988; Elofsson et al. 2000; Sloman et al. 2001b).

In the absence of aldosterone in teleosts, cortisol acts as the main mineralocorticoid hormone (Mommsen et al. 1999), and it is therefore not surprising that chronic stress affects ion- and osmoregulation. Sodium uptake is significantly elevated in subordinate fish (Sloman et al. 2002c, 2003a, 2004). Still plasma sodium concentrations are unaffected, which indicates a larger sodium excretion in these animals (Sloman et al. 2000a,b, 2004). Subordinate individuals are also more prone to take up copper and silver (Sloman et al. 2002c, 2003a), which can be explained by copper and silver being taken up in the fish body through sodium transport pathways. Cadmium, on the other hand, accumulates to a greater degree in gill tissue of dominant fish (Sloman et al. 2003b), and in contrast to silver and copper, cadmium enters the fish body through calcium transport pathways (Sloman et al. 2003c). It has also been shown that socially subordinate fish are less tolerant to hypoxia (Thomas and Gilmour 2012) and heat shock (LeBlanc et al. 2012); however, they appear to be able to regulate blood pH normally during respiratory acidosis (Mussa and Gilmour 2012).

Currie et al. (2010) demonstrated an induction of heat shock proteins that was correlated with interaction intensity in both dominant and subordinate rainbow trout. The induction of heat shock proteins in the brain was suggested to depend on cortisol. Cortisol is a potent modulator of immune function, and subordinate fish commonly show signs of impaired immune function (Peters and Schwarzer 1985; Peters et al. 1988, 1991; Kurogi and Lida 1999; Filby et al. 2010a; Cammarata et al. 2012). They are also more susceptible to bacterial infections (Peters et al. 1988)—something which could be explained by the aforementioned suppression of immune function, as well as injuries and skin damage caused by aggressive encounters. Indeed, subordinate fish sustain significantly more fin damage than other individuals (Abbott et al. 1985; Abbott and Dill 1985; Moutou et al. 1998). Social subordination has also been connected with a reduction in reproductive potential and function (Fox et al. 1997). In the African cichlid fish (*Astatotilapia burtoni*), for instance, subordinate status is associated with reduced levels of testosterone and 11-ketotestosterone, as well as reduced expression of gonadotropin-releasing hormone 1 (GnRH1) in the pituitary (Parikh et al. 2006). Ascending social position (i.e., going from subordinate status in one social context to dominant in a different social context) leads to rapid increases in production and release of the pituitary gonadotropins, luteinizing hormone (LH) and follicle-stimulating hormone (FSH) (Maruska et al. 2011). Subordinate male Arctic charr (*Salvelinus alpinus*) also have lower plasma levels of testosterone and 11-ketotestosterone (Elofsson et al. 2000), and male subordinate zebrafish have lower reproductive success than their dominant conspecifics (Paull et al. 2010). Interestingly, subordinate female zebrafish do not have lower overall reproductive success than dominant females; however, they sire fewer offspring with dominant males (Paull et al. 2010).

A considerable part of the detrimental effects of social subordination can be contributed to chronically elevated plasma cortisol levels. Cortisol treatment has, for instance, been shown to increase metabolic rate (Morgan and Iwama 1996; De Boeck et al. 2001), reduce food conversion efficiency (Gregory and Wood 1999), cause damage or morphological changes to the digestive tract (Barton et al. 1987), and increase energy mobilization (Barton et al. 1987; Morgan and Iwama 1996; De Boeck et al. 2001). These changes, together with suppression of appetite (Barton et al. 1987; Gregory and Wood 1999), lead to growth depression and reduced condition factor (McBride and Van Overbeeke 1971; Barton et al. 1987; Gregory and Wood 1999; De Boeck et al. 2001). Cortisol has also been shown to negatively affect immune function and increase susceptibility to infection, as well as increase mortality (Barton et al. 1987; Pickering and Pottinger 1989; Gregory and Wood 1999). Furthermore, blocking the cortisol-binding glucocorticoid receptor (GR) in subordinate fish partly inhibits the negative metabolic effects of social subordination (Gilmour et al. 2005; DiBattista et al. 2006).

Cortisol crosses the blood–brain barrier, and in mammals, corticosteroids have been found to inhibit brain cell proliferation (Gould et al. 1991; Cameron and Gould 1994; Ambrogini et al. 2002; Mayer et al. 2006; Montaron et al. 2006) and survival of newly formed neurons (Ambrogini et al. 2002; Wong and Herbert 2004), suppress long-term potentiation (Pavlides et al. 1995a), cause retraction and simplification of dendrites (Woolley et al. 1990), and even kill neurons (Starkman et al. 1992). In fishes, stress and cortisol treatment have been shown to reduce brain cell proliferation and affect the expression of genes and proteins involved in neural plasticity (Tognoli et al. 2010; Sørensen et al. 2011, 2012; Johansen et al. 2012).

9.3.2.1 Effects of Social Stress on Neural Plasticity

The behavioral and physiological response to social stress may be caused or supported by neural plasticity. Plasticity in neural circuits lets the animal respond to challenges according to its motivational state defined as "*endogenous, or internalized exogenous, factors that affect the probability of occurrence of a behavioural pattern*" (Zupanc and Lamprecht 2000). Motivation is thus an integration of factors such as experience and appraisal of the current environment, as well as internal state. Physiological changes in neural networks governing behavior are at the core of behavioural plasticity and there are at least two main forms of neural plasticity: structural reorganization and biochemical switching (for reviews, see Oliveira 2009; Zupanc and Lamprecht 2000). Both forms are influenced by stressful experiences in fishes (e.g., Winberg and Nilsson 1993; Øverli et al. 1999a;

Sørensen et al. 2012). Structural reorganization will be presented in Sections 9.3.2.3 and 9.3.2.4, whereas biochemical switching will be presented next (Section 9.3.2.2).

9.3.2.2 Neuromodulation

Through biochemical switching, the output of a neural network in response to a stimulus can be varied by the presence of a score of "neuromodulators." Neuromodulators affect their target neurons in a non-synaptic fashion, as they are not typically released in common "one-to-one" synapses, but rather diffuse over large distances and affect a great number of neurons. Neuromodulators exert their effects by binding to membrane receptors (typically G-protein coupled receptors), that may be located on all parts of the target neurons, thereby affecting the neurons' functional properties. Monoamine neurotransmitters, like dopamine, noradrenaline, and serotonin, as well as neuropeptides like somatostatin and CRH are well-known neuromodulators. The discovery of steroid hormone membrane receptors (Orchinik et al. 1991; Orchinik 1998) has led to the proposal that steroids should also be considered to be neuromodulators (Prager and Johnson 2009).

Neuromodulator function typically does not produce behaviors *per se*, but rather diffusely affects a number of neural circuits in an ongoing tuning of the animal's behavior. Neuromodulation and its role in stress coping has been extensively studied in fishes and other poikilothermic vertebrates (for reviews, see Winberg and Nilsson 1993; Summers and Winberg 2006; Korzan and Summers, 2007; Sørensen et al. 2013). Several neuromodulators are implied in social stress effects, among them the nonapeptides AVT and IT, which are homologues of mammalian arginine-vasopressin and oxytocin. For instance, dominant males of African cichlid fish have been found to have higher expression of AVT in the posterior preoptic area than subordinate males, whereas the subordinates had higher expression in the anterior preoptic area (Greenwood et al. 2008). Male zebrafish also had divergent AVT prevalence patterns, with dominant males displaying AVT immunoreactive neurons in the magnocellular preoptic area while subordinate individuals had AVT immunoreactive neurons in the parvocellular preoptic area (Larson et al. 2006). In the Mozambique tilapia (*Oreochromis mossambicus*), subordinate individuals had higher pituitary AVT concentrations, whereas dominant individuals had higher hindbrain IT levels (Almeida et al. 2012). Although it is clear that nonapeptides are involved in social behavior and social stress in fishes, the relationship appears to be complex, and there is significant interspecies variation (Godwin and Thompson 2012). Although increasing attention has been devoted to the nonapeptides recently, the most studied neuromodulators in relation to social stress in fishes are the monoamine neurotransmitters.

The monoamine neurotransmitters serotonin (5-HT), dopamine (DA), and noradrenaline (NE) are central in the coordination of physiological and behavioral stress responses, and a high degree of functional conservation has been indicated throughout the evolution of the vertebrate lineage (Blanchard et al. 1993; Winberg and Nilsson 1993; Stanford 1993; Höglund et al. 2005a; Maximino and Herculano 2010; Panula et al. 2010). Results from mammals indicate that these central signaling substances act to a large degree as neuromodulators, adjusting the responsiveness of target cells to other excitatory and inhibitory inputs (Huether 1996), which makes the monoamines very important in both biochemical switching and structural neural plasticity (Duman et al. 1999; Brezun and Daszuta 2000; Gaspar et al. 2003; Mattson et al. 2004; Williams and Undieh 2009; Christoffel et al. 2011). Prior to release, monoamines are stored in vesicles presynaptically, and upon release, they bind to receptors on all parts of the target neurons. Their effects are terminated by reuptake into presynaptic nerve terminals and possibly glial cells, followed by enzymatic catabolism. Monoamine metabolism is qualitatively identical in all vertebrates. DA is synthesized from the amino acid precursor tyrosine, and is further converted to NE in noradrenergic neurons (in certain neurons, NE is methylated to adrenaline (E), but a role for this catecholamine has not been identified in the teleost brain). After release and reuptake, DA is subjected to deamination by monoamine oxidase (MAO) to 3,4-dihydroxyphenylacetic acid (DOPAC) or methylation by catechol-O-methyl transferase (COMT) to 3-methoxytyramine (3-MT). Both 3-MT and DOPAC can be further converted to homovanillic acid (HVA), the importance of each pathway depending on species-specific distribution of the metabolizing enzymes.

The major metabolite of NE is considered to be 3-methoxy-4-hydroxyphenylglycol (MHPG), which is formed after deamination by MAO and methylation by COMT. 5-HT is synthesized from trypto-phan, and free 5-HT (i.e., not stored in vesicles) is converted to 5-hydroxyindoleacetic acid (5-HIAA) by MAO. For both 5-HT and the catecholamines, the ratio of the tissue concentration of their metabo-lites to that of the parent monoamine is frequently used as an index of neural activity, with increased concentration of the metabolite being taken to indicate increased release and turnover of the neu-rotransmitter (Shannon et al. 1986; Fillenz 1993).

Monoaminergic neurotransmission is strongly modified by social interaction and social stress in fishes and other vertebrates (reviewed by Winberg and Nilsson 1993; Johnsson et al. 2005). In a detailed time course study, Øverli et al. (1999a) found that brain monoaminergic systems were acti-vated very early during agonistic interaction in both eventual winners and losers, while in winners, monoaminergic activity rapidly returned to baseline levels. Subordinate rainbow trout displayed a substantial increase in 5-HT, DA, and NE activity after 3 and 24 h of interaction with a dominant (Figure 9.3). Telencephalic 5-HIAA/5-HT ratios were also elevated in dominant rainbow trout at 3 h after the termination of fights for social dominance. This effect was reversed in dominant fish within 24 h of social interaction, whereas in subordinates elevated 5-HIAA/5-HT ratios were observed in all brain regions by 24 h.

Numerous behaviors and physiological functions, including stress responses, mood, sleep, aggressiveness, fear, appetite, vascular function, pain, and reproduction are influenced by 5-HT in mammals (Jacobs and Azmitia 1992; Lucki 1998; Chaouloff 2000; Ursin 2002; Hensler 2010; Bardin 2011). In fishes, pharmacological and correlational data suggest an inhibitory role for 5-HT in aggression (Larson and Summers 2001; Winberg et al. 2001; Perreault et al. 2003; Höglund et al. 2005b; Lepage et al. 2005; Bell et al. 2007; Lynn et al. 2007), although this effect is clearly dependent on regional activation in the brain and time of exposure, with only long-term activation producing inhibitory effects on aggression (Summers and Winberg 2006). As in mammals, 5-HT has an appetite-suppressing effect in fishes, although both this effect and effects on locomotion depend on the interaction with other signaling systems, particularly those of the HPI axis, such as CRF (De Pedro et al. 1998; Bernier and Peter 2001; Clements et al. 2003).

As is the case for 5-HT, NE and DA have neuromodulatory actions. NE sets the level of activity in different CNS regions underlying different behavioral states. In mammals, NE plays an impor-tant role in the control of autonomic systems, but also influences diverse processes in the brain, including attention or arousal, memory, and the reward system (Ordway et al. 2007; Ramos and Arnsten 2007; Bush et al. 2010; Murchison et al. 2011), and NE is critically involved in the organ-ism's ability to rapidly respond to environmental changes through alterations in neuronal connectiv-ity and excitability (O'Donnell et al. 2012). Like 5-HT, NE appears to exert a widespread influence over arousal, sensory perception, emotion, and higher cognitive functions, as well as participating in the regulation of neuroendocrine-releasing factors at the level of the hypothalamus and pituitary. Abnormal function of these systems has been implicated in mental affective illnesses, anxiety, and depression (Coplan and Lydiard 1998; Delgado and Moreno 2000; Leonard 2001; O'Donnell et al. 2012). Our knowledge of behavioral and physiological roles for NE in the fish brain is limited (but see results referred in the next paragraph). Brain DA systems contribute to the control of motor activity, behavior, perception, and sleep and, at least in mammals, are involved in motivation, mood, reward, learning, and attention (Schultz 2007). In rainbow trout, Øverli et al. (1999a) demonstrated a substantial activation of brain catecholaminergic systems (DA and NE), as indicated by elevated MHPG/NE and DOPAC/DA ratios, after 24 h of social interaction in subordinate, but not in domi-nant, rainbow trout (Figure 9.3).

The activation of the 5-HT, NE, and DA systems in fishes shows that fights for social dominance only have limited and quickly reversible effects on brain monoaminergic activity in winners, while continued interaction in pairs induces substantial increases in the activity of catecholaminergic as well as serotonergic systems in subordinate individuals. Long-term effects of social interaction on brain monoaminergic activity have also been reported. In particular, subordinate fish continue

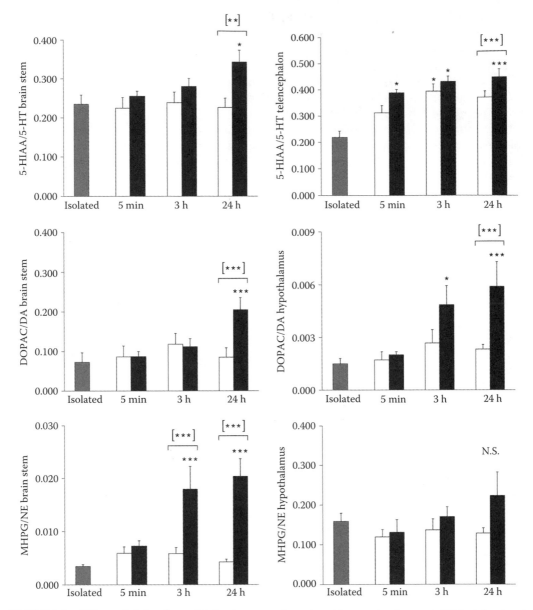

FIGURE 9.3 Effects of fights for social dominance and continued interaction in an established dominant–subordinate relationship on 5-HIAA/5-HT, DOPAC/DA, and MHPG/NE ratios in different brain regions of dominant and subordinate rainbow trout, as compared to isolated controls (mean + S.E.M.). F, df, and p-values are the result of one-way analysis of variance (ANOVA), followed by Tukey post hoc test. Post hoc significance levels are indicated by asterisks, where * is used to indicate a difference to controls and [*] indicates a difference between social ranks at a given time point (*$p < 0.05$, **$p < 0.01$, ***$p < 0.001$). (Data from Øverli, Ø. et al., *Brain Behav. Evol.*, 54, 263, 1999a.)

to show elevated brain 5-HT activity even after long-term social interaction in stable dominance hierarchies (Winberg et al. 1991, 1992; Winberg and Nilsson 1993; Cubitt et al. 2008). In rainbow trout interacting in groups of three, plasma cortisol in subordinates declined after 7 days of interaction, whereas brain 5-HIAA/5-HT remained elevated (Winberg and Lepage 1998). Regarding catecholaminergic neurotransmission, Höglund et al. (2000) reported that MHPG/NE was elevated in the optic tectum of subordinate Arctic charr as compared to isolated controls following 5 days of social interaction in groups of three. On the other hand, Winberg et al. (1991) reported that after an even longer period (2–11 weeks) of social interaction in groups, dominant Arctic charr had higher concentrations of the DA metabolite HVA in the telencephalon than subordinate fish. Recently, Dahlbom et al. (2012) reported that dyadic agonistic interaction resulted in elevated brain serotonergic activity in subordinate zebrafish, and also observed a sex difference in forebrain dopamine levels and forebrain 5-HIAA/5-HT ratios, with females displaying higher concentrations of dopamine but lower 5-HIAA/5-HT ratios than males (Dahlbom et al. 2012).

The organization of monoaminergic systems, especially that of 5-HT, appears to be remarkably constant throughout the vertebrate subphylum (Parent et al. 1984; Jacobs and Azmitia 1992; Maximino and Herculano 2010), and can be described as diffuse with very divergent projections patterns (5-HT: Kah and Chambolle 1983; Parent et al. 1984; Kaslin and Panula 2001; Jacobs and Azmitia 1992, see also review by Lillesaar 2011; catecholamines: Ekström et al. 1986, 1990; Meek et al. 1989, 1993; Ma 1994a,b, 1997, 2003; Holzschuh et al. 2001; Kaslin and Panula 2001; Wullimann and Rink 2002; Ma and Lopez 2003; McLean and Fetcho 2004). As mentioned previously, a substantial amount of data from mammals shows that 5-HT plays several fundamental roles both during development and in the adult, not only acting as a neurotransmitter and modulator, but also by regulating several aspects of structural plasticity such as cell proliferation, neuronal differentiation, neurite growth, and synaptogenesis. Direct evidence that 5-HT is involved in structural plasticity in fishes is, however, still lacking (Lillesaar 2011). Some recent reports show that anxiogenic or anxiolytic drugs acting on the serotonergic system in mammals act in a similar manner on behaviors taken to be indicative of fear and/or anxiety in fishes (Bencan et al. 2009; Maximino et al. 2011), so clearly fish models have the potential to increase our understanding of the role of monoamines in both regulation and functional effects of adult neuroplasticity.

9.3.2.3 Structural Plasticity

Structural reorganization encompasses different types of physical change in neural networks, including addition or removal of cells, modification of the connectivity within existing networks, or changes in the molecular constitution of the involved neurons (Zupanc and Lamprecht 2000; Oliveira 2009). Examples of this are long-term potentiation (Bliss and Lømo 1973), dendritic branching and debranching, axonal outgrowth and retraction, as well as alterations in the density of dendritic spines and synapses. In mammals, effects of stress and glucocorticoid hormones on dendritic arborization and synaptic density have been characterized in several brain regions (McEwen 1999; McKittrick et al. 2000; Sousa et al. 2000; Vyas et al. 2002; Radley et al. 2004, 2005, 2008; Stewart et al. 2005; Liston et al. 2006; Bennur et al. 2007). Such changes have been found to be reversible (Conrad et al. 1999; Radley et al. 2005), and accompanied by related behavioral changes (Vyas et al. 2002; Pawlak et al. 2003; Liston et al. 2006; Dias-Ferreira et al. 2009). Dendritic or synaptic change has not received much attention in fishes; much more studied is addition or removal of cells.

In mammals, new neurons are born throughout life in discrete parts of the brain: the subgranular zone and the subventricular zone (Figure 9.4a; Altman and Das 1965; Altman 1969; Kaplan and Hinds 1977; Cameron et al. 1993; Kuhn et al. 1996). Adult neurogenesis can be stimulated by a number of factors (Kempermann 2008) including physical exercise (van Praag et al. 1999; Ra et al. 2002) and environmental enrichment (Kempermann et al. 1997; Brown et al. 2003a). A range of acute (Gould et al. 1997, 1998; Tanapat et al. 2001; Heine et al. 2004) and chronic (Malberg and Duman 2003; Pham et al. 2003; Heine et al. 2004) stressors have been shown to reduce neurogenesis in mammals and the impact of stress on neurogenesis is believed to be mediated primarily by corticosteroid hormones (Gould et al. 1992; Cameron and Gould 1994; Ambrogini et al. 2002; Montaron et al. 2003; Wong and Herbert 2005). The role of neural plasticity in memory and learning is well established in mammals (Gould et al. 1999; Döbrössy et al. 2003; Ambrogini et al. 2004; Leuner et al. 2004; Morris 2006; Abrous and Wojtowicz 2008; Shors et al. 2012). Interestingly, considerable evidence suggests that stress and corticosteroids have a biphasic, inverted U-shaped effect on behavioral reactivity and cognition; brief, low intensity challenges are stimulatory, while severe, chronic

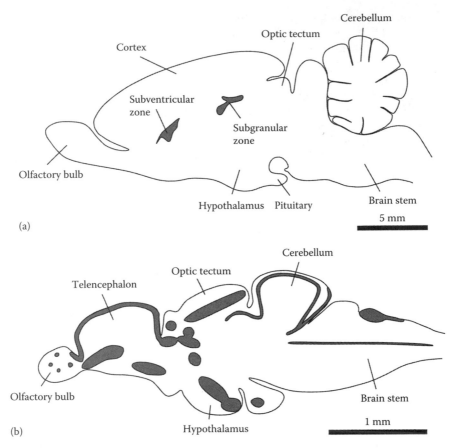

FIGURE 9.4 (a) Illustration of established proliferative zones (gray) in sagittal section of the rat brain; the subgranular zone of the dentate gyrus of the hippocampal formation, and the subventricular zone of the lateral ventricles (Gage 2000; Ming and Song 2011 and references therein). (b) Illustration of proliferative zones (gray) in sagittal section of the zebrafish brain. (After Grandel, H. et al., *Dev. Biol.*, 295, 263, 2006.) For complete description of the proliferative zones, see Grandel et al. (2006) and Zupanc et al. (2005). (From Sørensen et al. 2013. Reproduced with permission from *General and Comparative Endocrinology*.)

exposure is inhibitory (Figure 9.5; Sandi et al. 1996; Haller et al. 1997; Breuner et al. 1998; Øverli et al. 2002b; Jöels et al. 2006; Jöels 2007; Schwabe et al. 2012). The hypothesis has been raised that behavioral inhibition is the most adaptive strategy under threat when conditions are uncontrollable and unpredictable, or when repeated active coping attempts have failed (Wingfield 2003). It would then seem likely that the brain is influenced by the environment to create cognitive states affecting perceived predictability (e.g., reduced predictability with impaired neural plasticity). If this argument holds true, stress-induced behavioral inhibition and depressive states (often used as models for clinical depression in humans) may be viewed as adaptive strategies, rather than pathophysiological side effects of chronic stress. Thus, the physiological and behavioral effects of social defeat and stress, which may pose serious risks to the animal's health and life, may serve important adaptive purposes. Recent studies have addressed whether a similar relationship exists in fishes.

9.3.2.4 Neurogenesis in Fishes

Though not as extensively studied in fishes as in mammals, adult neurogenesis is still the most characterized form of structural neural plasticity in fishes. It has long been established that the brain of teleosts grows throughout life (Leyhausen et al. 1987; Zupanc and Horschke 1995). This growth is made possible by a pronounced proliferative activity taking place in discrete zones situated in all

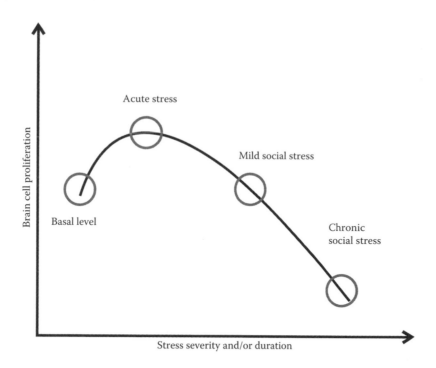

FIGURE 9.5 A suggested biphasic, inverted U-shaped effect of stress severity and/or duration on structural neural plasticity in rainbow trout. The suggested curve is based on results from the following studies: acute stress: Johansen et al. (2012); mild social stress: Sørensen (2011); chronic social stress: Johansen et al. (2012), Sørensen et al. (2012). (From Sørensen et al. 2013. Reproduced with permission from *General and Comparative Endocrinology*.)

regions of the fish brain (Figure 9.4b; Zupanc and Horschke 1995; Ekström et al. 2001; Zupanc et al. 2005; Grandel et al. 2006; Kuroyanagi et al. 2010; Maruska et al. 2012; Teles et al. 2012). Among these are areas homologous in structure and function to the neurogenic zones found in mammals (Northcutt and Davis 1983; Vargas et al. 2000, 2009; Rodríguez et al. 2002; Northcutt 2008; Broglio et al. 2010). Teleost fishes have strikingly high rates of adult brain cell proliferation that are orders of magnitude higher than those observed in rodents (Cameron and McKay 2001; Herculano-Houzel and Lent 2005; Hinsch and Zupanc 2007; Zupanc 2008). The initial number of new adult-born cells is reduced through apoptosis over the first few weeks after birth (Soutschek and Zupanc 1996; Ott et al. 1997), although the cells that do survive this period typically persist throughout the fish's life span (Ott et al. 1997; Zupanc et al. 2005; Hinsch and Zupanc 2007). A major proportion of these differentiate into neurons (Zupanc et al. 2005; Hinsch and Zupanc 2007). The functional significance of adult neurogenesis in teleosts is largely unknown, though adult-born neurons appear to integrate into already existing networks (Zupanc et al. 1996, 2005). As opposed to mammals, fish grow throughout life and increase the number of muscle fibers and sensory units (Johns and Easter 1977; Corwin 1981; Zakon 1984; Weatherley and Gill 1985; Rowlerson and Veggetti 2001). Therefore, it seems logical that adult-born neurons act as a central counterpart for the increasing number of peripheral structures (Raymond and Easter 1983). Adult-born neurons also contribute to the fish brain's astounding ability for regeneration after injury (Zupanc and Ott 1999; Clint and Zupanc 2001). A number of studies have indicated an effect of environmental type and complexity (e.g., hatchery rearing and environmental enrichment) on fish brain size and morphology (Kihslinger et al. 2006; Kihslinger and Nevitt 2006; Burns et al. 2009). Furthermore, it has been demonstrated that rearing in enriched or changing environments promotes learning in fishes (Kotrschal and Taborsky 2010; Strand et al. 2010, but see Brydges and Braithwaite 2009). On the cellular level, electric fish (*Brachyhypopomus gauderio*) living in the wild have higher brain cell proliferation than captive fish (Dunlap et al. 2011), and zebrafish that are kept in structurally enriched tanks have higher telencephalic cell proliferation (von Krogh et al. 2010). Thus, the general trend seems to be that an increase in environmental complexity leads to higher brain cell proliferation or neurogenesis, which in turn may be involved in enhanced cognitive performance observed in animals living in more complex environments.

The effect of social stress and genetic selection for stress responses on neural plasticity has recently been addressed in rainbow trout. In one experiment, socially subordinate individuals in pairs of socially interacting rainbow trout had a 40% reduction in the number of newborn cells in the telencephalon (Sørensen et al. 2012). Similarly, Maruska et al. (2012) found suppressed cell proliferation in the brains of subordinate African cichlid fish. In this study, the subordinate individuals were given an opportunity to rise in status, upon which the suppression was reversed within 1 day. This indicates that there is a rapid regulation of brain cell proliferation in response to social interaction in teleost fishes. Treatment of isolated rainbow trout with cortisol doses similar to those commonly seen in subordinate rainbow trout (Figure 9.1) caused a reduction in the number of actively proliferating telencephalic cells of approximately 50% (Sørensen et al. 2011). Thus, it is likely that cortisol mediates the reduction in cell proliferation observed in socially stressed fish. In a recent study, stress level was also linked with learning performance in African cichlid fish. High circulating cortisol levels were associated with inability to learn or lack of motivation to attempt learning a spatial task. The fish that did learn the task had the highest expression of immediate early genes in dorsolateral areas of the telencephalon, an area homologous to the mammalian hippocampus that is essential for spatial learning in fishes (Wood et al. 2011). This further strengthens the link between cognition, neuroplasticity, stress, and cortisol in fishes. The expression of neuroplasticity-related genes in response to acute and chronic stress has also been investigated in rainbow trout (Johansen et al. 2012). This study was performed on rainbow trout selected for differences in post-stress cortisol levels (Section 9.4.3). Regardless of selection line, the expression of proliferating cell nuclear antigen (PCNA), a marker of actively proliferating cells (Eisch and Mandyam 2007; Mandyam et al. 2007; Kuhn and Peterson 2008), was universally suppressed in the brain in response to chronic social stress (though this was just a trend in the telencephalon; $p = 0.08$; Johansen et al. 2012). On the other hand, PCNA expression was elevated in telencephalon, hypothalamus, and optic tectum during acute confinement stress. Expression of brain-derived neurotrophic factor (BDNF), a neurotrophic factor stimulating LTP, as well as migration and survival of young neurons, was also enhanced in hypothalamus and optic tectum of acutely stressed animals. In a different study, chronic mild social stress was induced by pairing rainbow trout with a larger conspecific. The dominant individuals displayed lower aggression levels than commonly seen in pairs of rainbow trout (Figure 9.2; Sørensen 2011), likely as a result from the size difference. Accordingly, the subordinate fish had only a modest increase in plasma cortisol levels (Figure 9.1; Sørensen 2011). In these fish, expression of PCNA and NeuroD, a transcription factor involved in neural differentiation and survival (Steiner et al. 2006), were unchanged. There was, however, increased expression of BDNF, suggesting a modest stimulatory effect of this mild social stress regime on neural plasticity (Sørensen 2011). Given that nonsocial acute stress has stimulatory effects and chronic social stress has inhibitory effects on neural plasticity, mild social stress could have an intermediate stress effect. Mild social stress could thus represent moving beyond the peak of the inverted U-shaped curve representing the relationship between stress severity and brain cell proliferation, though still not into the suppressive area (Figure 9.5). Nevertheless, these results suggest a biphasic effect of social or other types of stress on structural neural plasticity in rainbow trout.

This suggested biphasic relationship between stress intensity or duration and neuroplastic or cognitive effects is supported by emerging results from mammalian stress research. In a recent experiment, mice were subjected to social defeat, which caused approximately 50% of the mice to display social avoidance. These animals had higher neurogenesis than the mice that did not show this "appropriate" behavioral response (Lagace et al. 2010). Ablating neurogenesis by cranial irradiation inhibited social avoidance (Lagace et al. 2010), suggesting that neurogenesis is necessary for forming this appropriate behavioral response to social subordination. Similarly, Carpenter and Summers

(2009) reported that among subordinate rainbow trout given the option of escaping the dominant individual, only approximately half of the fish chose to do so, while the other half chose to stay with the dominant. The fish that learnt to escape the dominant individual had up-regulated expression of BDNF in the dorsolateral pallium, the telencephalic area thought to be homologous with the mammalian hippocampus (Carpenter et al. 2008). BDNF is well known to be intimately associated with learning, both at the cellular and the systemic levels (Patterson et al. 1992; Korte et al. 1995; Bramham and Messaoudi 2005), and is also an important permissive factor in the neurogenic process (Zigova et al. 1998; Benraiss et al. 2001; Rossi et al. 2006). There thus appears to be a link between stress coping and neural plasticity in fishes, as well as in mammals. Intermittent social stress (Lyons et al. 2010) and predictable, chronic, mild stress of a nonsocial nature (Parihar et al. 2011) have also been shown to stimulate neurogenesis and cognitive function in mammals. Ferragud et al. (2010), on the other hand, found a decrease in neurogenesis in subordinate mice, though the depression of neurogenesis was associated with a shift from use of spatial learning to habit-based learning. It thus appears that stress coping is associated with cognitive responses or strategies, and that beyond a certain threshold related to severity or duration of the stressor, a shift between cognitive strategies occurs. Before the threshold level is reached, a stimulation of cognitive activity (i.e., stimulation of neuroplasticity) takes place, and may be important for active coping attempts, whereas beyond the threshold, cognitive suppression as well as passive coping occurs.

Regardless of the direction of the change in each instance, neural plasticity appears to be involved in learning and behavioral responses related to stress coping, and in particular social subordination, in vertebrates including fishes. This view is further supported by results from mice suggesting a relationship between the level of neurogenic suppression and the intensity of a hierarchy as measured by aggression level (Yap et al. 2006), or as indicated by the behavioral response of the subordinate individuals (Mitra et al. 2006). We have also found such a relationship between aggression and cell proliferation in our studies in rainbow trout (Sørensen 2011; Sørensen et al. 2012). These results support the idea that regulation of neurogenesis is indeed part of an adaptive stress-coping response.

9.4 ANIMAL PERSONALITY AND STRESS-COPING STYLES: PHYSIOLOGICAL AND BEHAVIORAL INTERACTION

9.4.1 Predictors of Social Position

A caveat of all research on social interaction is the inherent difficulty in determining which traits are causes and which are consequences of social position. In competing for social dominance, it is not random which animal wins, and a number of factors have been shown to influence social hierarchy position. Higher innate aggressiveness increases the likelihood of attaining social dominance (Holtby et al. 1993; Adams and Huntingford 1996; Adams et al. 1998; Cutts et al. 1999b). Aggression level is positively correlated with metabolic rate in Atlantic salmon (Cutts et al. 1998), and having a higher metabolic rate prior to social interaction also predicts higher social status (Metcalfe et al. 1995; Yamamoto et al. 1998; Cutts et al. 1999a; McCarthy 2001). A higher metabolic rate is likely to be associated with higher feeding motivation, and having a higher feeding motivation has also been shown to increase the likelihood of attaining social dominance (Johnsson et al. 1996). Larger size is commonly considered a strong predictor of social dominance (Holtby et al. 1993; Rhodes and Quinn 1998; Cutts et al. 1999a), and relative size differences as small as approximately 5% have been found to ensure social dominance in rainbow trout (Abbott and Dill 1985). In other studies, however, no effect of size on tendency to become dominant has been demonstrated (Huntingford et al. 1990; Adams and Huntingford 1996; Yamamoto et al. 1998). In the case of *Cichlasoma dimerus*, size is a determining factor for social position in males, but not in females (Alonso et al. 2012).

Prior social experiences with winning or losing (winner/loser effects) are strong determinants of outcome of social interaction (Abbott and Dill 1985; Dugatkin 1997; Rhodes and Quinn 1998;

Hsu and Wolf 1999; Johnsson et al. 1999; Oliveira 2009; Oliveira et al. 2011). This is likely caused by neurochemical or endocrine changes brought on by prior social interaction (Winberg and Nilsson 1993; Øverli et al. 1999a; Höglund et al. 2001; Winberg et al. 2001). Furthermore, several studies report a change in the likelihood of attaining dominance after neurochemical or endocrinal modulations. Treatment with growth hormone has been shown to increase aggressive behavior and feeding motivation, which may in turn lead to social dominance (Johnsson and Björnsson 1994; Jönsson et al. 1998, 2003), and elevation of brain dopamine activity through treatment with the DA precursor L-3,4-dihydroxyphenylalanine (l-DOPA) ensured social dominance in Arctic charr (Winberg and Nilsson 1992). The role of brain monoamines in social interaction was discussed in Section 9.3.2.2.

Cortisol, on the other hand, has been shown to negatively influence performance in social competition, an effect that might also depend on regulation of brain monoamine activity (Øverli et al. 2002b). In one study, subordinate rainbow trout were found to have had higher basal cortisol levels 48 h prior to social interaction (Sloman et al. 2001a). This finding is supported by results from experiments where fish have been treated with cortisol. Cortisol-treated juvenile rainbow trout attained more fin damage, an indicator of hierarchy strength (Moutou et al. 1998), than untreated tank mates in mixed groups, something which indicates that the treated fish had lower social status (Gregory and Wood 1999). Within size-matched pairs of rainbow trout, cortisol treatment also increased the risk of becoming subordinate, an effect that even cancelled out the beneficial effect of being larger (DiBattista et al. 2005). The role of cortisol in determining social status is further strengthened by experiments on rainbow trout bred for high or low cortisol response to stress, where the animals responding to stress with higher plasma cortisol levels are more prone to becoming socially subordinate (Pottinger and Carrick 2001). These selection lines are presented in detail in Section 9.4.3. Many of the predicting factors for social position, including cortisol responses to stress, appear to be stable over time and characteristic for each individual. Such individual traits fall within the broader concept of animal personalities.

9.4.2 ANIMAL PERSONALITY AND STRESS-COPING STYLES

Severe and chronic stress is associated with prolonged elevation of cortisol levels. This can potentially compromise general health and the well-being of an organism. Still, post-stress cortisol levels are subject to great individual variation. As cortisol has such extensive somatic and central effects, it seems reasonable to expect that individual variation in cortisol responsiveness co-occurs with individual variation in other traits, such as behavior and cognition. Indeed, an increasing body of evidence suggests that individual variation in physiology is also associated with behavioral, cognitive, and emotional aspects of stress coping (for reviews, see Korte et al. 2005; Øverli et al. 2005, 2007; Koolhaas et al. 2010).

The concept that humans show consistent individual differences in behavior and personality is familiar to us all, and most people would agree with the definition suggested by Pervin and John (1997), who state that personality is *"those characteristics of individuals that describe and account for consistent patterns of feeling, thinking, and behaving."* In other words, if what a man does in a certain situation predicts his behavior the next time he is in a similar situation, or even his behavior in other contexts, he has a personality. Behavioral correlation across situations, the animal analogue of personality, has been extensively documented by ethologists and behavioral ecologists in a wide range of species, including mammals, birds, lizards, amphibians, fishes, molluscs, and arthropods (e.g., Budaev 1997; Gosling 2001; Bell and Stamps 2004; Dingemanse et al. 2004; Sinn and Moltschaniwskyj 2005; Réale et al. 2007; van Oers and Mueller 2010). Several influential publications have addressed the existence of consistent individual profiles and trait correlations within a population using different terminologies (see Koolhaas et al. 1999; Gosling 2001; Sih et al. 2004 for discussions of the terms "coping style," "animal personality," and "behavioral syndrome," respectively). Referring to the fact that suites of correlated traits are often described as syndromes, Sih et al. (2004) coined the term *"behavioral syndromes,"* and pointed out that this phenomenon can have important ecological and evolutionary implications. For instance, the notion of behavioral syndromes implies limited behavioral plasticity. Limited plasticity may in turn pose problems in rapidly changing environments, so why do animals not show more flexibility in behavior? Another question is how variation is maintained in a population. In other words, why does a population not simply drift toward a homogeneous group of the most successful phenotypes? Recently, such questions have sparked a surging interest in the evolution of consistent trait correlations (e.g., Sih et al. 2004; Wolf et al. 2007; van Oers and Mueller 2010).

In the following, we will adhere to the definition of coping style as "*a coherent set of behavioural and physiological stress responses which is consistent over time and which is characteristic to a certain group of individuals*" (Koolhaas et al. 1999). From a physiologist's point of view, it is important to convey the message that behavioral states have physiological correlates, and physiology may control behavior as much as behavior controls physiology. In some cases, neurobiological mechanisms that cause variation in behavior are plastic and modified by experience. In other cases, certain traits appear to be lifelong characteristics. Teleost fishes provide examples of both heritable and apparently consistent trait associations (see reviews by Øverli et al. 2005; Schjolden and Winberg 2007), as well as rapidly modifiable responses (Burmeister 2007). Different personalities, that is, consistent trait associations that diverge within the population, have been observed in Atlantic cod (*Gadus morhua*) (Meager et al. 2012), zebrafish (Moretz et al. 2007; Dahlbom et al. 2011), and Bluefin killifish (*Lucania goodei*) (McGhee and Travis 2010).

9.4.3 Stress-Coping Styles in Fishes

Animal personalities tend to become particularly apparent in extreme or stressful situations, and individual variation in how animals cope with stress has been the subject of many studies. Consistent and partially heritable variation in individual stress-coping styles raises a recurring problem in research on social hierarchies, namely the difficulty of discerning causes and consequences of social position (for reviews, see Gilmour et al. 2005; Øverli et al. 2007). It is evident that preexisting physiological and developmental differences are important in determining which animals become dominant or subordinate in social hierarchies. As discussed in Section 9.4.1, factors known to increase the chance of winning fights for social dominance include larger size, higher metabolic rate, and prior residency in the territory (Abbott and Dill 1985; Metcalfe et al. 1995; Yamamoto et al. 1998; Cutts et al. 1999a; Johnsson et al. 1999; McCarthy 2001). Experience with social subordination (Abbott and Dill 1985) and higher basal cortisol levels (Gregory and Wood 1999; Sloman et al. 2001a; DiBattista et al. 2005) increase the likelihood of becoming socially subordinate.

In rainbow trout, the magnitude of the cortisol response to stress shows both consistency over time and a moderate to high degree of heritability. High-responding (HR) and low-responding (LR) lines of rainbow trout have been generated by selection for consistently high or low post-stress plasma cortisol levels (Pottinger and Carrick 1999). Genetically determined differences in HPI axis activity affect social standing in such a way that LR fish (with low post-stress plasma cortisol) usually win fights for social dominance (Pottinger and Carrick 2001). Other studies suggest that LR fish are also characterized by a rapid recovery of food intake after transfer to a novel environment, and a reduced locomotor response in a territorial intrusion test (Øverli et al. 2001). Hence, some of the features of the LR trout line suggest that they represent selection for a proactive stress-coping style, whereas the HR trout line represents a reactive stress-coping style, as defined by Koolhaas et al. (1999). Proactive individuals are typically characterized by low corticosteroid production and high sympathetic output coupled with a fight–flight behavioral strategy, aggressive and bold emotional state, rigid and routine-like behavior, and high energy consumption. The reactive coping style incorporates opposite physiological and behavioral patterns, with high post-stress corticosteroid hormone levels, passive freeze-hide behavior, nonaggressive and cautious emotional states, behavioral flexibility, and energy conservation. It is debatable to what extent these categories are applicable to the diversity of species of fishes, but some examples of conservation have nevertheless been lately reported (e.g., Martins et al. 2011; Ruiz-Gomez et al. 2011). It should, however, also be pointed out that the behavior of the HR–LR lines of fish is highly context-dependent, and is influenced by factors such as novelty of the environment, group size, and prior history of stress (Schjolden et al. 2005a, 2006; Schjolden and Winberg 2007; Ruiz-Gomez et al. 2008).

Other experiments have addressed whether the LR–HR model reflects trait associations that are also present in nonselected hatchery populations of rainbow trout (Øverli et al. 2004a, 2006). It was found in these unselected populations that the duration of appetite inhibition after stress, a trait that is characteristic and different for the LR and HR fish, predicted social dominance with near 100% certainty, with rapid reversal of stress-induced anorexia being characteristic of individuals likely to become dominant in later contests (Øverli et al. 2004a). This means that unselected rainbow trout have similar trait associations to those seen in the LR and HR fish. The apparent parallel to genetically determined stress-coping styles in mammals, and the existence of similar trait associations in unselected fish populations suggest an evolutionarily conserved correlation among multiple traits. Thus, coping characteristics appear to be of importance in determining social rank and individual differences in aggressive behavior, which raises the question of whether observed physiological differences arise before, during, or after contests for social rank.

It has been proposed that LR fish demonstrate a different risk assessment than HR fish (Schjolden et al. 2005b), as they resume feeding earlier than HR fish after stress (Øverli et al. 2002a) and display greater initial effort to escape in a new environment (Schjolden et al. 2005b). Directly in line with what has been shown in mammals, Ruiz-Gomez et al. (2011) found that LR fish retained a learnt routine in a foraging experiment, whereas HR fish displayed a more flexible foraging behavior. In addition, Moreira et al. (2004) reported that LR fish retained a conditioned fear response for longer time than HR fish, indicating differences in fear extinction. Both results suggest more fixed behavioral patterns in LR fish and more flexibility in HR behavior, and neuro plasticity-related gene expression can explain this cognitive divergence (Johansen et al. 2012). Both socially stressed and control HR fish have significantly higher expression of NeuroD and doublecortin (DCX) in the brain than LR fish. DCX is a microtubule-associated protein involved in neural migration and neurite outgrowth, and both genes are intimately connected with neurogenesis (Brown et al. 2003b; Couillard-Despres et al. 2005). PCNA expression has also been found to be higher in the telencephalon of HR fish compared to LR fish (Johansen et al. 2012). Based on this, HR fish appear to have a larger pool of proliferating and/or differentiating neurons, and thus a higher degree of structural neural plasticity than LR fish. The HR and LR fish also differ in basal and stress-induced activity in the serotonin system: Despite having a lower basal serotonergic activity in both telencephalon and brain stem, HR fish have higher 5-HT concentrations in the brain stem in response to acute stress (Øverli et al. 2001). This suggests that there is a stronger stimulatory effect of stress on 5-HT neurons in HR than in LR fish. Serotonin is a potent modulator of adult hippocampal neurogenesis in mammals (Gould 1999; Brezun and Daszuta 2000; Dranovsky and Hen 2006), so different 5-HT dynamics could conceivably contribute to the difference in neural and behavioral plasticity between LR and HR fish.

In addition to the observed differences in neuromodulation and structural neural plasticity between the HR and LR fish, we have reported that the HR fish have lower expression of the mineralocorticoid receptor (MR) in the brain than LR fish (Johansen et al. 2011). There is, however, little difference between the lines in expression of either of the two GRs found in this species (GR1 and GR2). In mammals, there is an extensive literature on the role of corticosteroid receptors in the brain. Corticosterone (the main glucocorticoid in rodents) binds to MRs predominantly localized in limbic brain structures with a tenfold higher affinity than for GRs. GRs are more widely distributed in the brain (De Kloet et al. 2005). This divergence in both ligand affinity and receptor distribution allows for fine adjustment of the central responses to corticosteroids depending on the subregional MR/GR balance and circulating hormone levels (De Kloet et al. 2005). The high expression of MRs and GRs in limbic brain structures reflects their involvement in memory, learning, and general alertness, and MR and GR binding causes very different physiological effects

(De Kloet et al. 1998). MR activation leads to the maintenance of tonic HPA axis reactivity and neuronal excitability. GR activation, on the other hand, suppresses stress-induced HPA axis activation and restores disturbances in homeostasis through negative feedback (De Kloet et al. 1998). In addition, GR activation decreases neuronal excitation. In this way, corticosterone binding of two different receptor types is likely to contribute to the previously discussed biphasic effects on excitability of neurons, as well as cognition (Diamond et al. 1992; Pavlides et al. 1995b; Pugh et al. 1997; Joëls and Vreugdenhil 1998). The balance in hippocampal excitation and inhibition, accomplished through MRs and GRs, respectively, is therefore critical for attention and information processing during different stress intensities and durations (Chan-Palay and Kohler 1989). As MRs have a higher affinity for glucocorticoids than GRs, more than 80% of the MRs are occupied during basal conditions, whereas GR occupation varies greatly with changes in plasma corticosteroid levels (Reul and de Kloet 1985; Reul et al. 1987). The absolute ratio of MR to GR receptors might also provide a physiological basis for individual variation in the threshold value that shifts an animal between active and passive coping, as well as stress-induced cognitive and neural plasticity. Additionally, a role for non-genomic actions of corticosteroid receptors at the neuronal membrane in the coordination of rapid adaptive responses to stress has recently emerged in mammals (reviewed by Groeneweg et al. 2012).

The identity of GR and MR receptors varies between fish species, and in some species the different cortisol receptor types have been reported to differ in glucocorticoid affinity and sensitivity (Stolte et al. 2006; Bury and Sturm 2007). In rainbow trout, cortisol actions are mediated through GR1, GR2, and MR (Colombe et al. 2000; Bury et al. 2003), and rainbow trout MR has higher affinity for cortisol than GRs (Bury and Sturm 2007). Furthermore, the two different GR paralogs also differ in glucocorticoid sensitivity (Stolte et al. 2006; Li et al. 2012). As mentioned, LR fish have a higher expression of MR than HR fish (Johansen et al. 2011), which could contribute both to observed selection line differences in physiological and behavioral stress-coping style as well as cognition and neural plasticity.

9.5 CONCLUSIONS

In this chapter, we have given an overview of the physiological and behavioral responses to social stress in fishes, with emphasis on neural plasticity and individual variation in stress-coping styles. We have presented studies suggesting that physiology and neural plasticity is regulated in concert in response to social stress in fishes. A biphasic response pattern is apparent, with acute stress being stimulatory and chronic stress having suppressive effects on neural plasticity. These regulatory responses appear to be in line with results from mammals, suggesting that among the vertebrates, stress-coping mechanisms share common neural and behavioral components. Furthermore, we have presented results suggesting that there is great individual variation in physiological and behavioral stress responses, and that different stress-coping styles commonly co-occur within fish populations. Genetically determined individual variation in stress coping coincides with variation in capacity for neural plasticity. Yet much work remains before we can determine exactly how neural plasticity and stress coping are linked in fishes.

Knowledge about proximate and ultimate mechanisms involved in the control of neural plasticity and stress-coping styles in comparative models will be crucial in providing the necessary framework for research on behavioral stress responses as a substrate for natural selection. In this context, the hypothesis that behavioral inhibition is an adaptive strategy under uncontrollable and unpredictable conditions should be given particular attention. If this argument holds true, stress-induced behavioral inhibition and depressive states (often used as models for clinical depression in humans) may be viewed as adaptive stress-coping strategies, rather than pathophysiological side effects of stress. Such a revelation could have important consequences for research not only in fishes, but also in promoting a focus on prevention rather than cure in the increasingly extensive biomedical research on stress in humans.

REFERENCES

Abbott JC, Dill LM 1985. Patterns of aggressive attack in juvenile steelhead trout (*Salmo gairdneri*). *Can J Fish Aq Sci* 42:1702–1706.

Abbott JC, Dill LM 1989. The relative growth of dominant and subordinate juvenile steelhead trout (*Salmo gairdneri*) fed equal rations. *Behaviour* 108:104–113.

Abbott JC, Dunbrack RL, Orr CD 1985. The interaction of size and experience in dominance relationships of juvenile steelhead trout (*Salmo gairdneri*). *Behaviour* 92:241–253.

Abrous DN, Wojtowicz JM 2008. Neurogenesis and hippocampal memory system. In: *Adult Neurogenesis* (Gage FH, Kempermann G, Song H, eds.), pp. 445–461. New York: Cold Spring Harbor Laboratory Press.

Adams CE, Huntingford FA 1996. What is a successful fish? Determinants of competitive success in Arctic char (*Salvelinus alpinus*) in different social contexts. *Can J Fish Aq Sci* 53:2446–2450.

Adams CE, Huntingford FA, Turnbull JF, Beattie C 1998. Alternative competitive strategies and the cost of food acquisition in juvenile Atlantic salmon (*Salmo salar*). *Aquaculture* 167:17–26.

Albeck DS, McKittrick CR, Blanchard DC, Blanchard RJ, Nikulina J, McEwen BS, Sakai RR 1997. Chronic social stress alters levels of corticotrophin-releasing factor and arginine vasopressin mRNA in rat brain. *J Neurosci* 12:4895–4903.

Albonetti ME, Farabollini F 1994. Social stress by repeated defeat: Effects on social behaviour and emotionality. *Behav Brain Res* 62:187–193.

Almeida O, Gozdowska M, Kulczykowska E, Oliveira RF 2012. Brain levels of arginine-vasotocin and isotocin in dominant and subordinate males of a cichlid fish. *Horm Behav* 61:212–217.

Alonso F, Honji RM, Moreira RG, Pandolfi M 2012. Dominance hierarchies and social status ascent opportunity: Anticipatory behavioral and physiological adjustments in a Neotropical cichlid fish. *Physiol Behav* 106:612–618.

Altman J 1969. Autoradiographic and histological studies of postnatal neurogenesis. IV. Cell proliferation and migration in the anterior forebrain, with special reference to persisting neurogenesis in the olfactory bulb. *J Comp Neurol* 137:433–457.

Altman J, Das GD 1965. Autoradiographic and histological evidence of postnatal hippocampal neurogenesis in rats. *J Comp Neurol* 124:319–335.

Aluru N, Vijayan MM 2008. Molecular characterization, tissue-specific expression, and regulation of melanocortin 2 receptor in rainbow trout. *Endocrinology* 149:4577–4588.

Ambrogini P, Orsini L, Mancini C, Ferri P, Barbanti I, Cuppini R 2002. Persistently high corticosterone levels but not normal circadian fluctuations of the hormone affect cell proliferation in the adult rat dentate gyrus. *Neuroendocrinology* 76:366–372.

Ambrogini P, Orsini L, Mancini C, Ferri P, Ciaroni S, Cuppini R 2004. Learning may reduce neurogenesis in adult rat dentate gyrus. *Neurosci Lett* 359:13–16.

Anderson WG 2012. The endocrinology of 1α-hydroxycorticosterone in elasmobranch fish: A review. *Comp Biochem Physiol A* 162:73–80.

Ang TZ, Manica A 2010. Aggression, segregation and stability in a dominance hierarchy. *Proc R Soc B* 277:1337–1343.

Backström T, Winberg S 2009. Arginine–vasotocin influence on aggressive behavior and dominance in rainbow trout. *Physiol Behav* 96:470–475.

Bardin L 2011. The complex role of serotonin and 5-HT receptors in chronic pain. *Behav Pharmacol* 22:390–404.

Barlow GW, Rogers W, Bond AB 1984. Dummy-elicited aggressive behavior in the polychromatic Midas cichlid. *Biol Behav* 9:115–130.

Barton BA 2002. Stress in fishes: A diversity of responses with particular reference to changes in circulating corticosteroids. *Integr Comp Biol* 42:517–525.

Barton BA, Iwama GK 1991. Physiological changes in fish from stress in aquaculture with emphasis on the response and effects of corticosteroids. *Ann Rev Fish Dis* 1:3–26.

Barton BA, Schreck CB, Barton LD 1987. Effects of chronic cortisol administration and daily acute stress on growth, physiological conditions, and stress responses in juvenile rainbow trout. *Dis Aquat Org* 2:173–185.

Bass AH, Grober MS 2001. Social and neural modulation of sexual plasticity in teleost fish. *Brain Behav Evol* 57:293–300.

Bell AM, Backström T, Huntingford FA, Pottinger TG, Winberg S 2007. Variable neuroendocrine responses to ecologically-relevant challenges in sticklebacks. *Physiol Behav* 91:15–25.

Bell AM, Stamps JA 2004. Development of behavioural differences between individuals and populations of sticklebacks, *Gasterosteus aculeatus*. *Anim Behav* 68:1339–1348.

Bencan Z, Sledge D, Levin ED 2009. Buspirone, chlordiazepoxide and diazepam effects in a zebrafish model of anxiety. *Pharmacol Biochem Behav* 94:75–80.

Bennur S, Rao BSS, Pawlak R, Strickland S, McEwen BS, Chattarji S 2007. Stress-induced spine loss in the medial amygdala is mediated by tissue-plasminogen activator. *Neuroscience* 144:8–16.

Benraiss A, Chmielnicki E, Lerner K, Roh D, Goldman SA 2001. Adenoviral brain-derived neurotrophic factor induces both neostriatal and olfactory neuronal recruitment from endogenous progenitor cells in the adult forebrain. *J Neurosci* 21:6718–6731.

Benus RF, Bohus B, Koolhaas JM, van Oortmerssen GA 1991. Heritable variation for aggression as a reflection of individual coping strategies. *Experientia* 47:1008–1019.

Bern HA 1967. Hormones and endocrine glands of fishes. Studies of fish endocrinology reveal major physiologic and evolutionary problems. *Science* 158:455–462.

Bernier NJ, Alderman SL, Bristow EN 2008. Heads or tails? Stressor-specific expression of corticotropin-releasing factor and urotensin I in the preoptic area and caudal neurosecretory system of rainbow trout. *J Endocrinol* 196:637–648.

Bernier NJ, Peter RE 2001. The hypothalamic-pituitary-interrenal axis and the control of food intake in teleost fish. *Comp Biochem Physiol B* 129:639–644.

Blanchard DC, Sakai RR, McEwen B, Weiss SM, Blanchard RJ 1993. Subordination stress: Behavioral, brain, and neuroendocrine correlates. *Behav Brain Res* 58:113–121.

Bliss TV, Lømo T 1973. Long-lasting potentiation of synaptic transmission in the dentate area of the anaesthetized rabbit following stimulation of the perforant path. *J Physiol* 232:331–356.

Bond AB 1992. Aggressive motivation in the midas cichlid—evidence for behavioral efference. *Behaviour* 122:135–152.

Bramham CR, Messaoudi E 2005. BDNF function in adult synaptic plasticity: The synaptic consolidation hypothesis. *Prog Neurobiol* 76:99–125.

Breuner CW, Greenberg AL, Wingfield JC 1998. Noninvasive corticosterone treatment rapidly increases activity in Gambel's white-crowned sparrows (*Zonotrichia leucophrys gambelii*). *Gen Comp Endocrinol* 111:386–394.

Brezun JM, Daszuta A 2000. Serotonin may stimulate granule cell proliferation in the adult hippocampus, as observed in rats grafted with foetal raphe neurons. *Eur J Neurosci* 12:391–396.

Broglio C, Rodríguez F, Gómez A, Arias JL, Salas C 2010. Selective involvement of the goldfish lateral pallium in spatial memory. *Behav Brain Res* 210:191–201.

Brown J, Cooper-Kuhn CM, Kempermann G, Van Praag H, Winkler J, Gage FH, Kuhn HG 2003a. Enriched environment and physical activity stimulate hippocampal but not olfactory bulb neurogenesis. *Eur J Neurosci* 17:2042–2046.

Brown JP, Couillard-Després S, Cooper-Kuhn CM, Winkler J, Aigner L, Kuhn HG 2003b. Transient expression of doublecortin during adult neurogenesis. *J Comp Neurol* 467:1–10.

Brydges NM, Braithwaite VA 2009. Does environmental enrichment affect the behaviour of fish commonly used in laboratory work? *Appl Anim Behav Sci* 118:137–143.

Buchner AS, Sloman KA, Balshine S 2004. The physiological effects of social status in the cooperatively breeding cichlid *Neolamprologus pulcher*. *J Fish Biol* 65:1080–1095.

Budaev SV 1997. "Personality" in the guppy (*Poecilia reticulata*): A correlational study of exploratory behavior and social tendency. *J Comp Psychol* 111:399–411.

Burmeister SS 2007. Genomic responses to behavioral interactions in an African cichlid fish: Mechanisms and evolutionary implications. *Brain Behav Evol* 70:247–256.

Burmeister SS, Jarvis ED, Fernald RD 2005. Rapid behavioral and genomic responses to social opportunity. *PLoS Biol* 3(11):e363.

Burns JG, Saravanan A, Rodd FH 2009. Rearing environment affects the brain size of guppies: Lab-reared guppies have smaller brains than wild-caught guppies. *Ethology* 115:122–133.

Bury NR, Sturm A 2007. Evolution of the corticosteroid receptor signalling pathway in fish. *Gen Comp Endocrinol* 153:47–56.

Bury NR, Sturm A, Le Rouzic P, Lethimonier C, Ducouret B, Guiguen Y, Robinson-Rechavi M, Laudet V, Rafestin-Oblin ME, Prunet P 2003. Evidence for two distinct functional glucocorticoid receptors in teleost fish. *J Mol Endocrinol* 31:141–156.

Bush DEA, Caparosa EM, Gekker A, LeDoux J 2010. Beta-adrenergic receptors in the lateral nucleus of the amygdala contribute to the acquisition but not the consolidation of auditory fear conditioning. *Front Behav Neurosci* 4:154.

Cameron HA, Gould E 1994. Adult neurogenesis is regulated by adrenal steroids in the dentate gyrus. *Neuroscience* 61:203–209.

Cameron HA, McKay RD 2001. Adult neurogenesis produces a large pool of new granule cells in the dentate gyrus. *J Comp Neurol* 435:406–417.

Cameron HA, Woolley CS, McEwen BS, Gould E 1993. Differentiation of newly born neurons and glia in the dentate gyrus of the adult rat. *Neuroscience* 56:337–344.

Cammarata M, Vazzana M, Accardi D, Parrinello N 2012. Seabream (*Sparus aurata*) long-term dominant-subordinate interplay affects phagocytosis by peritoneal cavity cells. *Brain Behav Immunol* 26:580–587.

Carpenter RE, Sabirzhanov I, Arendt DH, Smith JP, Summers CH 2008. Brain derived neurotrophic factor mRNA in the hippocampus (dorsolateral pallium) of rainbow trout is differentially regulated by coping strategy, revealed in a new model of fear conditioning. *Soc Neurosci Abstr* 34, 292.211.

Carpenter RE, Summers CH 2009. Learning strategies during fear conditioning. *Neurobiol Learn Mem* 91:415–423.

Chan-Palay V, Köhler C (eds.) 1989. *The Hippocampus: New Vistas*. New York: AR. Liss.

Chandroo KP, Yue S, Moccia RD 2004. An evaluation of current perspectives on consciousness and pain in fishes. *Fish Fish* 5:281–295.

Chaouloff F 2000. Serotonin, stress and corticoids. *J Psychopharmacol* 14:139–151.

Chester Jones I 1980. The interrenal gland in Pisces. In: *General, Comparative and Clinical Endocrinology of the Adrenal Cortex* (Chester Jones I, Henderson IW, eds.), Vol. III, pp. 396–523. London, U.K.: Academic Press.

Christoffel DJ, Golden SA, Russo SJ 2011. Structural and synaptic plasticity in stress-related disorders. *Rev Neurosci* 22:535–549.

Clements S, Moore FL, Schreck CB 2003. Evidence that acute serotonergic activation potentiates the locomotor-stimulating effects of corticotropin-releasing hormone in juvenile chinook salmon (*Oncorhynchus tshawytscha*). *Horm Behav* 43:214–221.

Clint SC, Zupanc GK 2001. Neuronal regeneration in the cerebellum of adult teleost fish, *Apteronotus leptorhynchus*: Guidance of migrating young cells by radial glia. *Brain Res Dev Brain Res* 130:15–23.

Colombe L, Fostier A, Bury N, Pakdel F, Guiguen Y 2000. A mineralocorticoid-like receptor in the rainbow trout, *Oncorhynchus mykiss*: Cloning and characterization of its steroid binding domain. *Steroids* 65:319–328.

Conrad CD, Magariños AM, LeDoux JE, McEwen BS 1999. Repeated restraint stress facilitates fear conditioning independently of causing hippocampal CA3 dendritic atrophy. *Behav Neurosci* 113:902–913.

Coplan JD, Lydiard RB 1998. Brain circuits in panic disorder. *Biol Psychiatry* 44:1264–1276.

Corwin JT 1981. Postembryonic production and aging of inner ear hair cells in sharks. *J Comp Neurol* 201:541–553.

Couillard-Despres S, Winner B, Schaubeck S, Aigner R, Vroemen M, Weidner N, Bogdahn U, Winkler J, Kuhn HG, Aigner L 2005. Doublecortin expression levels in adult brain reflect neurogenesis. *Eur J Neurosci* 21:1–14.

Creel S 2001. Social dominance and stress hormones. *Trends Ecol Evol* 16:491–497.

Creel S, Dantzer B, Goymann W, Rubenstein DR 2013. The ecology of stress: Effects of the social environment. *Funct Ecol* 27:66–80.

Cubitt KF, Winberg S, Huntingford FA, Kadri S, Crampton VO, Øverli Ø 2008. Social hierarchies, growth and brain serotonin metabolism in Atlantic salmon (*Salmo salar*) kept under commercial rearing conditions. *Physiol Behav* 94:529–535.

Currie S, LeBlanc S, Watters MA, Gilmour KM 2010. Agonistic encounters and cellular angst: Social interactions induce heat shock proteins in juvenile salmonid fish. *Proc R Soc B* 277:905–913.

Cutts CJ, Brembs B, Metcalfe NB, Taylor AC 1999b. Prior residence, territory quality and life-history strategies in juvenile Atlantic salmon (*Salmo salar L.*). *J Fish Biol* 55:784–794.

Cutts CJ, Metcalfe NB, Taylor AC 1998. Aggression and growth depression in juvenile Atlantic salmon: The consequences of individual variation in standard metabolic rate. *J Fish Biol* 52:1026–1037.

Cutts CJ, Metcalfe NB, Taylor AC 1999a. Competitive asymmetries in territorial juvenile Atlantic salmon, *Salmo salar*. *Oikos* 86:479–486.

Dahlbom SJ, Backström T, Lundstedt-Enkel K, Winberg S 2012. Aggression and monoamines: Effects of sex and social rank in zebrafish (*Danio rerio*). *Behav Brain Res* 228:333–338.

Dahlbom SJ, Lagman D, Lundstedt-Enkel K, Sundström LF, Winberg S 2011. Boldness predicts social status in zebrafish (*Danio rerio*). *PLoS ONE* 6(8): e23565.

Darland T, Dowling JE 2001. Behavioral screening for cocaine sensitivity in mutagenized zebrafish. *Proc Natl Acad Sci USA* 98:11691–11696.

De Boeck G, Alsop D, Wood C 2001. Cortisol effects on aerobic and anaerobic metabolism, nitrogen excretion, and whole-body composition in juvenile rainbow trout. *Physiol Biochem Zool* 74:858–868.

De Kloet ER, Joëls M, Holsboer F 2005. Stress and the brain: From adaptation to disease. *Nat Rev Neurosci* 6:463–475.

De Kloet ER, Vreugdenhil E, Oitzl MS, Joëls M 1998. Brain corticosteroid receptor balance in health and disease. *Endocr Rev* 19:269–301.

De Pedro N, Pinillos ML, Valenciano AI, Alonso-Bedate M, Delgado MJ 1998. Inhibitory effect of serotonin on feeding behavior in goldfish: Involvement of CRF. *Peptides* 19:505–511.

Delgado PL, Moreno FA 2000. Role of norepinephrine in depression. *J Clin Psychiat* 61:5–12.

Desjardins JK, Hofmann HA, Fernald RD 2012. Social context influences aggressive and courtship behavior in a cichlid fish. *PLoS ONE* 7(7):e32781.

Diamond DM, Bennett MC, Fleshner M, Rose GM 1992. Inverted-U relationship between the level of peripheral corticosterone and the magnitude of hippocampal primed burst potentiation. *Hippocampus* 2:421–430.

Dias-Ferreira E, Sousa JC, Melo I, Morgado P, Mesquita AR, Cerqueira JJ, Costa RM, Sousa N 2009. Chronic stress causes frontostriatal reorganization and affects decision-making. *Science* 325:621–625.

DiBattista JD, Anisman H, Whitehead M, Gilmour KM 2005. The effects of cortisol administration on social status and brain monoaminergic activity in rainbow trout *Oncorhynchus mykiss*. *J Exp Biol* 208:2707–2718.

DiBattista JD, Levesque HM, Moon TW, Gilmour KM 2006. Growth depression in socially subordinate rainbow trout *Oncorhynchus mykiss*: More than a fasting effect. *Physiol Biochem Zool* 79:675–687.

Dingemanse NJ, Both C, Drent PJ, Tinbergen JM 2004. Fitness consequences of avian personalities in a fluctuating environment. *Proc R Soc B* 271:847–852.

Dingemanse NJ, Réale D 2005. Natural selection and animal personality. *Behaviour* 142:1159–1184.

Döbrössy MD, Drapeau E, Aurousseau C, Le Moal M, Piazza PV, Abrous DN 2003. Differential effects of learning on neurogenesis: Learning increases or decreases the number of newly born cells depending on their birth date. *Mol Psychiatr* 8:974–982.

Donaldson EM (ed.) 1981. *The Pituitary-Interrenal Axis as an Indicator of Stress in Fish*. London, U.K.: Academic Press.

Doyon C, Gilmour KM, Trudeau VL, Moon TW 2003. Corticotropin-releasing factor and neuropeptide Y mRNA levels are elevated in the preoptic area of socially subordinate rainbow trout. *Gen Comp Endocrinol* 133:260–271.

Dranovsky A, Hen R 2006. Hippocampal neurogenesis: Regulation by stress and antidepressants. *Biol Psychiatry* 59:1136–1143.

Drews C 1993. The concept and definition of dominance in animal behavior. *Behaviour* 125:283–313.

Dugatkin LA 1997. Winner and loser effects and the structure of dominance hierarchies. *Behav Ecol* 8:583–587.

Duman RS, Malberg J, Thome J 1999. Neural plasticity to stress and antidepressant treatment. *Biol Psychiatry* 46:1181–1191.

Dunlap KD, Silva AC, Chung M 2011. Environmental complexity, seasonality and brain cell proliferation in a weakly electric fish, *Brachyhypopomus gauderio*. *J Exp Biol* 214:794–805.

Earley RL, Blumer LS, Grober MS 2004. The gall of subordination: Changes in gall bladder function associated with social stress. *Proc R Soc B* 271:7–13.

Edeline E, Haugen TO, Weltzien FA, Claessen D, Winfield IJ, Stenseth NC, Vøllestad LA 2010. Body downsizing caused by non-consumptive social stress severely depresses population growth rate. *Proc R Soc B* 277:843–851.

Edmundson E, Everest FE, Chapman DW 1968. Permanence of station in juvenile chinook salmon and steelhead trout. *J Fish Res Bd Can* 25:1453–1464.

Eisch AJ, Mandyam CD 2007. Adult neurogenesis: Can analysis of cell cycle proteins move us "Beyond BrdU"? *Curr Pharm Biotechnol* 8:147–165.

Ejike C, Schreck CB 1980. Stress and social hierarchy rank in coho salmon. *Trans Am Fish Soc* 109:423–426.

Ekström P, Honkanen T, Steinbusch HWM 1990. Distribution of dopamine-immunoreactive neuronal perikarya and fibres in the brain of a teleost, *Gasterosteus aculeatus* L. Comparison with tyrosine hydroxylase- and dopamine-β-hydroxylase-immunoreactive neurons. *J Chem Neuroanat* 3:233–260.

Ekström P, Johnsson CM, Ohlin LM 2001. Ventricular proliferation zones in the brain of an adult teleost fish and their relation to neuromeres and migration (secondary matrix) zones. *J Comp Neurol* 436:92–110.

Ekström P, Reschke M, Steinbusch H, Van Veen T 1986. Distribution of noradrenaline in the brain of the teleost *Gasterosteus aculeatus* L.: An immunohistochemical analysis. *J Comp Neurol* 254:297–313.

Ellis T, Yildiz HY, Lopéz-Olmeda J, Spedicato MT, Tort L, Øverli Ø, Martins CIM 2012. Cortisol and finfish welfare. *Fish Physiol Biochem* 38:163–188.

Elofsson UOE, Mayer I, Damsgård B, Winberg S 2000. Intermale competition in sexually mature Arctic charr: Effects on brain monoamines, endocrine stress responses, sex hormone levels, and behavior. *Gen Comp Endocrinol* 118:450–460.

Engh AL, Siebert ER, Greenberg DA, Holekamp KE 2005. Patterns of alliance formation and postconflict aggression indicate spotted hyaenas recognize third-party relationships. *Anim Behav* 69:209–217.

Epstein FH, Epstein JA 2005. A perspective on the value of aquatic models in biomedical research. *Exp Biol Med* 230:1–7.

Fernald RD 2012. Social control of the brain. *Ann Rev Neurosci* 35:133–151.

Fernandes-de-Castilho M, Pottinger TG, Volpato GL 2008. Chronic social stress in rainbow trout: Does it promote physiological habituation? *Gen Comp Endocrinol* 155:141–147.

Fero K, Simon JL, Jourdie V, Moore PA 2007. Consequences of social dominance on crayfish resource use. *Behaviour* 144:61–82.

Ferragud A, Haro A, Sylvain A, Velázquez-Sánchez C, Hernández-Rabaza V, Canales JJ 2010. Enhanced habit-based learning and decreased neurogenesis in the adult hippocampus in a murine model of chronic social stress. *Behav Brain Res* 210:134–139.

Filby AL, Paull GC, Bartlett EJ, Van Look KJW, Tyler CR 2010a. Physiological and health consequences of social status in zebrafish (*Danio rerio*). *Physiol Behav* 101:576–587.

Filby AL, Paull GC, Hickmore TFA, Tyler CR 2010b. Unravelling the neurophysiological basis of aggression in a fish model. *BMC Genomics* 11:498.

Fillenz M 1993. Neurochemistry of stress: Introduction to techniques. In: *Stress from Synapse to Syndrome* (Stanford SC, Salmon P, eds.), pp. 247–279. London, U.K.: Academic Press.

Fox HE, White SA, Kao MHF, Fernald RD 1997. Stress and dominance in a social fish. *J Neurosci* 17:6463–6469.

Francis RC, Soma K, Fernald RD 1993. Social regulation of the brain pituitary gonadal axis. *Proc Natl Acad Sci USA* 90:7794–7798.

Franck D, Ribowski A 1993. Dominance hierarchies of male green swordtails (*Xiphophorus helleri*) in nature. *J Fish Biol* 43:497–499.

Gage FH 2000. Mammalian neural stem cells. *Science* 287:1433–1438.

Galhardo L, Vital J, Oliveira RF 2011. The role of predictability in the stress response of a cichlid fish. *Physiol Behav* 102:367–372.

Gaspar P, Cases O, Maroteaux L 2003. The developmental role of serotonin: News from mouse molecular genetics. *Nat Rev Neurosci* 4:1002–1012.

Gilmour KM, DiBattista JD, Thomas JB 2005. Physiological causes and consequences of social status in salmonid fish. *Integr Comp Biol* 45:263–273.

Godwin J, Thompson R 2012. Nonapeptides and social behavior in fishes. *Horm Behav* 61:230–238.

Golub MS, Sassenrath EN, Goo GP 1979. Plasma cortisol levels and dominance in peer groups of rhesus monkey weanlings. *Horm Behav* 12:50–59.

Gómez-Laplaza LM, Morgan E 2003. The influence of social rank in the angelfish, *Pterophyllum scalare*, on locomotor and feeding activities in a novel environment. *Lab Anim* 37:108–120.

Gosling SD 2001. From mice to men: What can we learn about personality from animal research? *Psychol Bull* 127:45–86.

Gould E 1999. Serotonin and hippocampal neurogenesis. *Neuropsychopharmacology* 21:S46–S51.

Gould E, Beylin A, Tanapat P, Reeves A, Shors TJ 1999. Learning enhances adult neurogenesis in the hippocampal formation. *Nat Neurosci* 2:260–265.

Gould E, Cameron HA, Daniels DC, Woolley CS, McEwen BS 1992. Adrenal hormones suppress cell division in the adult rat dentate gyrus. *J Neurosci* 12:3642–3650.

Gould E, McEwen BS, Tanapat P, Galea LA, Fuchs E 1997. Neurogenesis in the dentate gyrus of the adult tree shrew is regulated by psychosocial stress and NMDA receptor activation. *J Neurosci* 17:2492–2498.

Gould E, Tanapat P, McEwen BS, Flügge G, Fuchs E 1998. Proliferation of granule cell precursors in the dentate gyrus of adult monkeys is diminished by stress. *Proc Natl Acad Sci USA* 95:3168–3171.

Gould E, Woolley CS, McEwen BS 1991. Adrenal steroids regulate postnatal development of the rat dentate gyrus: I. Effects of glucocorticoids on cell death. *J Comp Neurol* 313:479–485.

Grandel H, Kaslin J, Ganz J, Wenzel I, Brand M 2006. Neural stem cells and neurogenesis in the adult zebrafish brain: Origin, proliferation dynamics, migration and cell fate. *Dev Biol* 295:263–277.

Greenwood AK, Wark AR, Fernald RD, Hofmann HA 2008. Expression of arginine vasotocin in distinct preoptic regions is associated with dominant and subordinate behaviour in an African cichlid fish. *Proc R Soc B* 275:2393–2402.

Gregory TR, Wood CM 1999. The effects of chronic plasma cortisol elevation on the feeding behaviour, growth, competitive ability, and swimming performance of juvenile rainbow trout. *Physiol Biochem Zool* 72:286–295.

Groeneweg FL, Karst H, de Kloet ER, Jöels M 2012. Mineralocorticoid and glucocorticoid receptors at the neuronal membrane, regulators of nongenomic corticosteroid signalling. *Mol Cell Endocrinol* 350:299–309.

Guo S 2004. Linking genes to brain, behavior and neurological diseases: What can we learn from zebrafish? *Genes Brain Behav* 3:63–74.

Haller J, Albert I, Makara GB 1997. Acute behavioural effects of corticosterone lack specificity but show marked context-dependency. *J Neuroendocrinol* 9:515–518.

Heine VM, Maslam S, Zareno J, Joëls M, Lucassen PJ 2004. Suppressed proliferation and apoptotic changes in the rat dentate gyrus after acute and chronic stress are reversible. *Eur J Neurosci* 19:131–144.

Hensler JG 2010. Serotonin in mood and emotion. In: *Handbook of Behavioral Neuroscience* (Müller CP, Jacobs BL, eds.), Vol. 21, pp. 367–378. Amsterdam, the Netherlands: Elsevier Science.

Herculano-Houzel S, Lent R 2005. Isotropic fractionator: A simple, rapid method for the quantification of total cell and neuron numbers in the brain. *J Neurosci* 25:2518–2521.

Hill DA 1990. Social relationships between adult male and female rhesus macaques: II. nonsexual affiliative behavior. *Primates* 31:33–50.

Hinsch K, Zupanc GK 2007. Generation and long-term persistence of new neurons in the adult zebrafish brain: A quantitative analysis. *Neuroscience* 146:679–696.

Höglund E, Balm PH, Winberg S 2000. Skin darkening, a potential social signal in subordinate Arctic charr (*Salvelinus alpinus*): The regulatory role of brain monoamines and pro-opiomelanocortin-derived peptides. *J Exp Biol* 203:1711–1721.

Höglund E, Bakke MJ, Øverli Ø, Winberg S, Nilsson GE 2005b. Suppression of aggressive behaviour in juvenile Atlantic cod (*Gadus morhua*) by L-tryptophan supplementation. *Aquaculture* 249:525–531.

Höglund E, Balm PHM, Winberg S 2002. Behavioural and neuroendocrine effects of environmental background colour and social interaction in Arctic charr (*Salvelinus alpinus*). *J Exp Biol* 205:2535–2543.

Höglund E, Kolm N, Winberg S 2001. Stress-induced changes in brain serotonergic activity, plasma cortisol and aggressive behavior in Arctic charr (*Salvelinus alpinus*) is counteracted by L-DOPA. *Physiol Behav* 74:381–389.

Höglund E, Weltzien FA, Schjolden J, Winberg S, Ursin H, Døving KB 2005a. Avoidance behavior and brain monoamines in fish. *Brain Res* 1032:104–110.

Holtby LB, Swain DP, Allan GM 1993. Mirror-elicited agonistic behavior and body morphology as predictors of dominance status in juvenile coho salmon (*Oncorhynchus kisutch*). *Can J Fish Aq Sci* 50:676–684.

Holzschuh J, Ryu S, Aberger F, Driever W 2001. Dopamine transporter expression distinguishes dopaminergic neurons from other catecholaminergic neurons in the developing zebrafish embryo. *Mech Dev* 101:237–243.

Honess PE, Marin CM 2006. Behavioural and physiological aspects of stress and aggression in nonhuman primates. *Neurosci Biobehav Rev* 30:390–412.

Hsu Y, Earley RL, Wolf LL 2006. Modulation of aggressive behaviour by fighting experience: Mechanisms and contest outcomes. *Biol Rev* 81:33–74.

Hsu Y, Wolf LL 1999. The winner and loser effect: Integrating multiple experiences. *Anim Behav* 57:903–910.

Huether G 1996. The central adaptation syndrome: Psychosocial stress as a trigger for adaptive modifications of brain structure and brain function. *Prog Neurobiol* 48:569–612.

Huntingford FA, Adams C, Braithwaite VA, Kadri S, Pottinger TG, Sandøe P, Turnbull JF 2006. Current issues in fish welfare. *J Fish Biol* 68:332–372.

Huntingford FA, Metcalfe NB, Thorpe JE 1993. Social status and feeding in Atlantic salmon *Salmo salar* parr: The effect of visual exposure to a dominant. *Ethology* 94:201–206.

Huntingford FA, Metcalfe NB, Thorpe JE, Graham WD, Adams CE 1990. Social dominance and body size in Atlantic salmon parr, *Salmo salar* L. *J Fish Biol* 36:877–881.

Huntingford FA, Turner A 1987. *Animal Conflict*. London, U.K.: Chapman & Hall.

Idler DR, Truscott B 1967. 1-alpha-hydroxycorticosterone: Synthesis in vitro and properties of an interrenal steroid in the blood of cartilaginous fish (genus Raja). *Steroids* 9:457–477.

Jacobs BL, Azmitia EC 1992. Structure and function of the brain serotonin system. *Physiol Rev* 72:165–229.

Järvi T 1990. The effects of male dominance, secondary sexual characteristics and female mate choice on the mating success of male Atlantic salmon *Salmo salar*. *Ethology* 84:123–132.

Jeffrey JD, Esbaugh AJ, Vijayan MM, Gilmour KM 2012. Modulation of hypothalamic-pituitary-interrenal axis function by social status in rainbow trout. *Gen Comp Endocrinol* 176:201–210.

Jenkins TMJ 1969. Social structure, position choice and microdistribution of two trout species (*Salmo trutta* and *Salmo gairdneri*) resident in mountain streams. *Anim Behav Monogr* 2:57–123.

Jöels M 2007. Role of corticosteroid hormones in the dentate gyrus. *Prog Brain Res* 163:355–370.

Jöels M, Pu Z, Wiegert O, Oitzl MS, Krugers HJ 2006. Learning under stress: How does it work? *Trends Cogn Sci* 10:152–158.

Joëls M, Vreugdenhil E 1998. Corticosteroids in the brain—Cellular and molecular actions. *Mol Neurobiol* 17:87–108.

Johansen IB, Sandvik GK, Nilsson GE, Bakken M, Øverli Ø 2011. Cortisol receptor expression differs in the brains of rainbow trout selected for divergent cortisol responses. *Comp Biochem Physiol D* 6:126–132.

Johansen IB, Sørensen C, Sandvik GK, Nilsson GE, Höglund E, Bakken M, Øverli Ø 2012. Neural plasticity is affected by stress and heritable variation in stress coping style. *Comp Biochem Physiol D* 7:161–171.

Johns PR, Easter SS 1977. Growth of adult goldfish eye.II. Increase in retinal cell number. *J Comp Neurol* 176:331–341.

Johnsson JI, Björnsson BT 1994. Growth hormone increases growth rate, appetite and dominance in juvenile rainbow trout, *Oncorhynchus mykiss*. *Anim Behav* 48:177–186.

Johnsson JI, Jonsson E, Björnsson BT 1996. Dominance, nutritional state, and growth hormone levels in rainbow trout (*Oncorhynchus mykiss*). *Horm Behav* 30:13–21.

Johnsson JI, Nöbbelin F, Bohlin T 1999. Territorial competition among wild brown trout fry: Effects of ownership and body size. *J Fish Biol* 54:469–472.

Johnsson JI, Winberg S, Sloman KA 2005. Social Interactions. In: *Fish Physiology* (Katherine A, Sloman RWW, Sigal B, eds.), Vol. 24, pp. 151–196. New York: Academic Press.

Jönsson E, Johnsson JI, Björnsson BT 1998. Growth hormone increases aggressive behavior in juvenile rainbow trout. *Horm Behav* 33:9–15.

Jönsson E, Johansson V, Björnsson BT, Winberg S 2003. Central nervous system actions of growth hormone on brain monoamine levels and behavior of juvenile rainbow trout. *Horm Behav* 43:367–374.

Kah O, Chambolle P 1983. Serotonin in the brain of the goldfish, *Carassius auratus*: An immunocytochemical study. *Cell Tissue Res* 234:319–333.

Kaplan MS, Hinds JW 1977. Neurogenesis in the adult rat: Electron microscopic analysis of light radioautographs. *Science* 197:1092–1094.

Kaslin J, Panula P 2001. Comparative anatomy of the histaminergic and other aminergic systems in zebrafish (*Danio rerio*). *J Comp Neurol* 440:342–377.

Kempermann G 2008. Activity dependency and aging in the regulation of adult neurogenesis. In: *Adult Neurogenesis* (Gage FH, Kempermann G, Song H, eds.), pp. 341–362. New York: Cold Spring Harbor Laboratory Press.

Kempermann G, Kuhn HG, Gage FH 1997. More hippocampal neurons in adult mice living in an enriched environment. *Nature* 386:493–495.

Kihslinger RL, Lema SC, Nevitt GA 2006. Environmental rearing conditions produce forebrain differences in wild chinook salmon *Oncorhynchus tshawytscha*. *Comp Biochem Phys A* 145:145–151.

Kihslinger RL, Nevitt GA 2006. Early rearing environment impacts cerebellar growth in juvenile salmon. *J Exp Biol* 209:504–509.

Knights B 1987. Agonistic behavior and growth in the European eel, *Anguilla anguilla* L, in relation to warm water aquaculture. *J Fish Biol* 31:265–276.

Koolhaas JM et al. 2011. Stress revisited: A critical evaluation of the stress concept. *Neurosci Biobehav Rev* 35:1291–1301.

Koolhaas JM, de Boer SF, Buwalda B, van Reenen K 2007. Individual variation in coping with stress: A multidimensional approach of ultimate and proximate mechanisms. *Brain Behav Evol* 70:218–226.

Koolhaas JM, de Boer SF, Coppens CM, Buwalda B 2010. Neuroendocrinology of coping styles: Towards understanding the biology of individual variation. *Front Neuroendocrinol* 31:307–321.

Koolhaas JM, Korte SM, de Boer SF, Van Der Vegt BJ, Van Reenen CG, Hopster H, De Jong IC, Ruis MAW, Blokhuis HJ 1999. Coping styles in animals: Current status in behavior and stress-physiology. *Neurosci Biobehav Rev* 23:925–935.

Korte M, Carroll P, Wolf E, Brem G, Thoenen H, Bonhoeffer T 1995. Hippocampal long-term potentiation is impaired in mice lacking brain-derived neurotrophic factor. *Proc Natl Acad Sci USA* 92:8856–8860.

Korte SM, Koolhaas JM, Wingfield JC, McEwen BS 2005. The Darwinian concept of stress: Benefits of allostasis and costs of allostatic load and the trade-offs in health and disease. *Neurosci Biobehav Rev* 29:3–38.

Korte SM, Olivier B, Koolhaas JM 2007. A new animal welfare concept based on allostasis. *Physiol Behav* 92:422–428.

Korzan WJ, Summers CH 2007. Behavioral diversity and neurochemical plasticity: Selection of stress coping strategies that define social status. *Brain Behav Evol* 70:257–266.

Kotrschal A, Taborsky B 2010. Environmental change enhances cognitive abilities in fish. *PLoS Biol* 8(4):e1000351.

Kuhn HG, Dickinson-Anson H, Gage FH 1996. Neurogenesis in the dentate gyrus of the adult rat: Age-related decrease of neuronal progenitor proliferation. *J Neurosci* 16:2027–2033.

Kuhn HG, Peterson DA 2008. Detection and phenotypic characterization of adult neurogenesis. In: *Adult Neurogenesis* (Gage FH, Kempermann G, Song H, eds.), pp. 25–47. New York: Cold Spring Harbor Laboratory Press.

Kurogi J, Lida T 1999. Social stress suppresses defense activities of neutrophils in tilapia. *Fish Pathol* 34:15–18.

Kuroyanagi Y, Okuyama T, Suehiro Y, Imada H, Shimada A, Naruse K, Takeda H, Kubo T, Takeuchi H 2010. Proliferation zones in adult medaka (*Oryzias latipes*) brain. *Brain Res* 1323:33–40.

Lagace DC, Donovan MH, DeCarolis NA, Farnbauch LA, Malhotra S, Berton O, Nestler EJ, Krishnan V, Eisch AJ 2010. Adult hippocampal neurogenesis is functionally important for stress-induced social avoidance. *Proc Natl Acad Sci USA* 107:4436–4441.

Laidley CW, Leatherland JF 1988. Cohort sampling, anesthesia and stocking density effects on plasma cortisol, thyroid hormone, metabolite and ion levels in rainbow trout, *Salmo gairdneri* Richardson. *J Fish Biol* 33:73–88.

Larson ET, O'Malley DM, Melloni RH, Jr. 2006. Aggression and vasotocin are associated with dominant-subordinate relationships in zebrafish. *Behav Brain Res* 167:94–102.

Larson ET, Summers CH 2001. Serotonin reverses dominant social status. *Behav Brain Res* 121:95–102.

LeBlanc S, Höglund E, Gilmour KM, Currie S 2012. Hormonal modulation of the heat shock response: Insights from fish with divergent cortisol stress responses. *Am J Physiol* 302R:184–192.

Lee G, Grant JWA, Comolli P 2011. Dominant convict cichlids (*Amatitlania nigrofasciata*) grow faster than subordinates when fed an equal ration. *Behaviour* 148:877–887.

Leonard BE 2001. Stress, norepinephrine and depression. *J Psych Neurosci* 26:S11–S16.

Lepage O, Larson ET, Mayer I, Winberg S 2005. Serotonin, but not melatonin, plays a role in shaping domi-nant-subordinate relationships and aggression in rainbow trout. *Horm Behav* 48:233–242.

Leshner AI 1980. The interaction of experience and neuroendocrine factors in determining behavioural adaptions to aggression. *Prog Brain Res* 53:427–438.

Leuner B, Mendolia-Loffredo S, Kozorovitskiy Y, Samburg D, Gould E, Shors TJ 2004. Learning enhances the survival of new neurons beyond the time when the hippocampus is required for memory. *J Neurosci* 24:7477–7481.

Leyhausen C, Kirschbaum F, Szabo T, Erdelen M 1987. Differential growth in the brain of the weakly electric fish, *Apteronotus leptorhynchus* (Gymnotiformes), during ontogenesis. *Brain Behav Evol* 30:230–248.

Li HW, Brocksen RW 1977. Approaches to analysis of energetic costs of intraspecific competition for space by rainbow trout (*Salmo gairdneri*). *J Fish Biol* 11:329–341.

Li Y, Sturm A, Cunningham P, Bury NR 2012. Evidence for a divergence in function between two glucocorti-coid receptors from a basal teleost. *BMC Evol Biol* 12:137.

Lieschke GJ, Currie PD 2007. Animal models of human disease: Zebrafish swim into view. *Nat Rev Genet* 8:353–367.

Lillesaar C 2011. The serotonergic system in fish. *J Chem Neuroanat* 41:294–308.

Liston C, Miller MM, Goldwater DS, Radley JJ, Rocher AB, Hof PR, Morrison JH, McEwen BS 2006. Stress-induced alterations in prefrontal cortical dendritic morphology predict selective impairments in perceptual attentional set-shifting. *J Neurosci* 26:7870–7874.

Louch CD, Higginbotham M 1967. The relation between social rank and plasma corticosterone levels in mice. *Gen Comp Endocrinol* 8:441–444.

Lucki I 1998. The spectrum of behaviors influenced by serotonin. *Biol Psychiatry* 44:151–162.

Lynn SE, Egar JM, Walker BG, Sperry TS, Ramenofsky M 2007. Fish on Prozac: A simple, noninvasive physiology laboratory investigating the mechanisms of aggressive behavior in *Betta splendens*. *Adv Physiol Educ* 31:358–363.

Lyons DM, Buckmaster PS, Lee AG, Wu C, Mitra R, Duffey LM, Buckmaster CL, Her S, Patel PD, Schatzberg AF 2010. Stress coping stimulates hippocampal neurogenesis in adult monkeys. *Proc Natl Acad Sci USA* 107:14823–14827.

Ma PM 1994a. Catecholaminergic systems in the zebrafish. I. Number, morphology, and histochemical characteristics of neurons in the locus coeruleus. *J Comp Neurol* 344:242–255.

Ma PM 1994b. Catecholaminergic systems in the zebrafish. II. Projection pathways and pattern of termination of the locus coeruleus. *J Comp Neurol* 344:256–269.

Ma PM 1997. Catecholaminergic systems in the zebrafish.III. Organization and projection pattern of medullary dopaminergic and noradrenergic neurons. *J Comp Neurol* 381:411–427.

Ma PM 2003. Catecholaminergic systems in the zebrafish. IV. Organization and projection pattern of dopaminergic neurons in the diencephalon. *J Comp Neurol* 460:13–37.

Ma PM, Lopez M 2003. Consistency in the number of dopaminergic paraventricular organ-accompanying neurons in the posterior tuberculum of the zebrafish brain. *Brain Res* 967:267–272.

Maclean A, Metcalfe NB 2001. Social status, access to food, and compensatory growth in juvenile Atlantic salmon. *J Fish Biol* 58:1331–1346.

Malberg JE, Duman RS 2003. Cell proliferation in adult hippocampus is decreased by inescapable stress: Reversal by fluoxetine treatment. *Neuropsychopharmacology* 28:1562–1571.

Mandyam CD, Harburg GC, Eisch AJ 2007. Determination of key aspects of precursor cell proliferation, cell cycle length and kinetics in the adult mouse subgranular zone. *Neuroscience* 146:108–122.

Martins CIM et al. 2012. Behavioural indicators of welfare in farmed fish. *Fish Physiol Biochem* 38:17–41.

Martins CIM, Silva PIM, Conceição LE, Costas B, Höglund E, Øverli Ø, Schrama JW 2011. Linking fearfulness and coping styles in fish. *PLoS ONE* 6(11):e28084.

Maruska KP, Carpenter RE, Fernald RD 2012. Characterization of cell proliferation throughout the brain of the African cichlid fish *Astatotilapia burtoni* and its regulation by social status. *J Comp Neurol* 520:3471–3491.

Maruska KP, Levavi-Sivan B, Biran J, Fernald RD 2011. Plasticity of the reproductive axis caused by social status change in an African cichlid fish: I. Pituitary gonadotropins. *Endocrinology* 152:281–290.

Mattson MP, Maudsley S, Martin B 2004. BDNF and 5-HT: A dynamic duo in age-related neuronal plasticity and neurodegenerative disorders. *Trends Neurosci* 27:589–594.

Maximino C, da Silva AWB, Gouveia A, Herculano AM 2011. Pharmacological analysis of zebrafish (*Danio rerio*) scototaxis. *Prog Neuro-Psychoph* 35:624–631.

Maximino C, Herculano AM 2010. A review of monoaminergic neuropsychopharmacology in zebrafish. *Zebrafish* 7:359–378.

Mayer JL, Klumpers L, Maslam S, de Kloet ER, Joëls M, Lucassen PJ 2006. Brief treatment with the glucocorticoid receptor antagonist mifepristone normalises the corticosterone-induced reduction of adult hippocampal neurogenesis. *J Neuroendocrinol* 18:629–631.

McBride JR, Van Overbeeke AP 1971. Effects of androgens, estrogens, and cortisol on the skin, stomach, liver, pancreas, and kidney in gonadectomized adult sockeye salmon (*Oncorhynchus nerka*). *J Fish Res Bd Can* 28:485–490.

McCarthy ID 2001. Competitive ability is related to metabolic asymmetry in juvenile rainbow trout. *J Fish Biol* 59:1002–1014.

McCarthy ID, Carter CG, Houlihan DF 1992. The effect of feeding hierarchy on individual variability in daily feeding of rainbow trout, *Oncorhynchus mykiss* (Walbaum). *J Fish Biol* 41:257–263.

McDonald DG, Milligan CL 1992. Chemical properties of the blood. In: *Fish Physiology* (Hoar WS, Randall DJ, Farrell AP, eds.), Vol. XII, pp. 55–133. San Diego, CA: Academic Press.

McEwen BS, Stellar E 1993. Stress and the individual. Mechanisms leading to disease. *Arch Intern Med* 153:2093–2101.

McEwen BS 1999. Stress and hippocampal plasticity. *Annu Rev Neurosci* 22:105–122.

McEwen BS, Mendelson S 1993. Effects of stress on the neurochemistry and morphology of the brain: Counterregulation versus damage. In: *Handbook of Stress: Theoretical and Clinical Aspects* (Breznitz S, Goldberger L, eds.), pp. 100–126. New York: The Free Press.

McEwen BS, Wingfield JC 2003. The concept of allostasis in biology and biomedicine. *Horm Behav* 43:2–15.

McEwen BS, Wingfield JC 2010. What is in a name? Integrating homeostasis, allostasis and stress. *Horm Behav* 57:105–111.

McGhee KE, Travis J 2010. Repeatable behavioural type and stable dominance rank in the bluefin killifish. *Anim Behav* 79:497–507.

McKittrick CR, Magariños AM, Blanchard DC, Blanchard RJ, McEwen BS, Sakai RR 2000. Chronic social stress reduces dendritic arbors in CA3 of hippocampus and decreases binding to serotonin transporter sites. *Synapse* 36:85–94.

McLean DL, Fetcho JR 2004. Ontogeny and innervation patterns of dopaminergic, noradrenergic, and serotonergic neurons in larval zebrafish. *J Comp Neurol* 480:38–56.

McNaughton N 1993. Stress and behavioural inhibition. In: *Stress from Synapse to Syndrome* (Stanford SC, Salmon P, eds.), pp. 191–206. London, U.K.: Academic Press.

Meager JJ, Fernö A, Skjæraasen JE, Järvi T, Rodewald P, Sverdrup G, Winberg S, Mayer I 2012. Multidimensionality of behavioural phenotypes in Atlantic cod, *Gadus morhua. Physiol Behav* 106:462–470.

Meek J, Joosten HWJ, Hafmans TGM 1993. Distribution of noradrenaline-immunoreactivity in the brain of the mormyrid teleost *Gnathonemus petersii*. *J Comp Neurol* 328:145–160.

Meek J, Joosten HWJ, Steinbusch HWM 1989. Distribution of dopamine immunoreactivity in the brain of the mormyrid teleost *Gnathonemus petersii*. *J Comp Neurol* 281:362–383.

Meerlo P, Overkamp GJF, Koolhaas JM 1997. Behavioural and physiological consequences of a single social defeat in Roman high- and low-avoidance rats. *Psychoneuroendocrinology* 22:155–168.

Metcalfe NB 1986. Intraspecific variation in competitive ability and food intake in salmonids: Consequences for energy budgets and growth rates. *J Fish Biol* 28:525–531.

Metcalfe NB, Huntingford FA, Graham WD, Thorpe JE 1989. Early social status and the development of life history strategies in Atlantic salmon. *Proc R Soc B* 236:7–19.

Metcalfe NB, Huntingford FA, Thorpe JE, Adams CE 1990. The effects of social status on life history variation in juvenile salmon. *Can J Zool* 68:2630–2636.

Metcalfe NB, Taylor AC, Thorpe JE 1995. Metabolic rate, social status and life history strategies in Atlantic salmon. *Anim Behav* 49:431–436.

Metz JR, Geven EJW, van den Burg EH, Flik G 2005. ACTH, alpha-MSH, and control of cortisol release: Cloning, sequencing, and functional expression of the melanocortin-2 and melanocortin-5 receptor in *Cyprinus carpio*. *Am J Physiol R* 289:814–826.

Miller RJ 1964. Studies on the social behavior of the blue gourami, *Trichogaster trichopterus* (Pisces, Belontiidae). *Copeia* 1964:469–496.

Millidine KJ, Metcalfe NB, Armstrong JD 2009. Presence of a conspecific causes divergent changes in resting metabolism, depending on its relative size. *Proc R Soc B* 276:3989–3993.

Ming GL, Song H 2011. Adult neurogenesis in the mammalian brain: Significant answers and significant questions. *Neuron* 70:687–702.

Mitra R, Sundlass K, Parker KJ, Schatzberg AF, Lyons DM 2006. Social stress-related behavior affects hippocampal cell proliferation in mice. *Physiol Behav* 89:123–127.

Moberg GP 2000. Biological response to stress: Implications for animal welfare. In: *The Biology of Animal Stress* (Mench JA, ed.), pp. 1–21. Wallingford, U.K.: CAB International.

Mommsen TP, Vijayan MM, Moon TW 1999. Cortisol in teleosts: Dynamics, mechanisms of action, and metabolic regulation. *Rev Fish Biol Fish* 9:211–268.

Montaron MF, Drapeau E, Dupret D, Kitchener P, Aurousseau C, Le Moal M, Piazza PV, Abrous DN 2006. Lifelong corticosterone level determines age-related decline in neurogenesis and memory. *Neurobiol Aging* 27:645–654.

Montaron MF, Piazza PV, Aurousseau C, Urani A, Le Moal M, Abrous DN 2003. Implication of corticosteroid receptors in the regulation of hippocampal structural plasticity. *Eur J Neurosci* 18:3105–3111.

Moreira PSA, Pulman KGT, Pottinger TG 2004. Extinction of a conditioned response in rainbow trout selected for high or low responsiveness to stress. *Horm Behav* 46:450–457.

Moretz JA, Martins EP, Robison BD 2007. The effects of early and adult social environment on zebrafish (*Danio rerio*) behavior. *Env Biol Fish* 80:91–101.

Morgan JD, Iwama GK 1996. Cortisol-induced changes in oxygen consumption and ionic regulation in coastal cutthroat trout (*Oncorhynchus clarki clarki*) parr. *Fish Physiol Biochem* 15:385–394.

Morris RG 2006. Elements of a neurobiological theory of hippocampal function: The role of synaptic plasticity, synaptic tagging and schemas. *Eur J Neurosci* 23:2829–2846.

Moutou KA, McCarthy ID, Houlihan DF 1998. The effect of ration level and social rank on the development of fin damage in juvenile rainbow trout. *J Fish Biol* 52:756–770.

Murchison CF, Schutsky K, Jin SH, Thomas SA 2011. Norepinephrine and ß1-adrenergic signaling facilitate activation of hippocampal CA1 pyramidal neurons during contextual memory retrieval. *Neuroscience* 181:109–116.

Mussa B, Gilmour KM 2012. Acid-base balance during social interactions in rainbow trout (*Oncorhynchus mykiss*). *Comp Biochem Physiol* A162:177–184.

Nakano S 1994. Variation in agonistic encounters in a dominance hierarchy of freely interacting red-spotted masu salmon (*Oncorhynchus masouishikawai*). *Ecol Freshw Fish* 3:153–158.

Nakano S 1995a. Individual differences in resource use, growth and emigration under the influence of a dominance hierarchy in fluvial red-spotted masu salmon in a natural habitat. *J Anim Ecol* 64:75–84.

Nakano S 1995b. Competitive interactions for foraging microhabitats in a size-structured interspecific dominance hierarchy of two sympatric stream salmonids in a natural habitat. *Can J Zool* 73:1845–1854.

Nandi J, Bern HA 1965. Chromatography of corticosteroids from teleost fishes. *Gen Comp Endocrinol* 5:1–15.

Noakes DLG, Leatherland JF 1977. Social dominance and interrenal cell activity in rainbow trout, *Salmo gairdneri* (Pisces, Salmonidae). *Env Biol Fish* 2:131–136.

Northcutt RG 2008. Forebrain evolution in bony fishes. *Brain Res Bull* 75:191–205.

Northcutt RG, Davis RE 1983. Telencephalic organization in ray-finned fishes. In: *Fish Neurobiology* (Davis RE, Northcutt RG, eds.), Vol. 2, Higher Brain Areas and Functions, pp. 203–236. Ann Arbor, MI: University of Michigan Press.

O'Donnell J, Zeppenfeld D, McConnell E, Pena S, Nedergaard M 2012. Norepinephrine: A neuromodulator that boosts the function of multiple cell types to optimize CNS performance. *Neurochem Res* 37:2496–2512.

Oliveira RF 2009. Social behavior in context: Hormonal modulation of behavioral plasticity and social competence. *Integr Comp Biol* 49:423–440.

Oliveira RF, Silva JF, Simões JM 2011. Fighting zebrafish: Characterization of aggressive behavior and winner-loser effects. *Zebrafish* 8:73–81.

Orchinik M 1998. Glucocorticoids, stress, and behavior: Shifting the timeframe. *Horm Behav* 34:320–327.

Orchinik M, Murray TF, Moore FL 1991. A corticosteroid receptor in neuronal membranes. *Science* 252:1848–1851.

Ordway GA, Schwartz MA, Frazer A 2007. *Brain Norepinephrine: Neurobiology and Therapeutics.* Cambridge, U.K.: Cambridge University Press. 1–642pp.

Ott R, Zupanc GK, Horschke I 1997. Long-term survival of postembryonically born cells in the cerebellum of gymnotiform fish, *Apteronotus leptorhynchus. Neurosci Lett* 221:185–188.

Øverli Ø et al. 2004a. Stress coping style predicts aggression and social dominance in rainbow trout. *Horm Behav* 45:235–241.

Øverli Ø, Harris CA, Winberg S 1999a. Short-term effects of fights for social dominance and the establishment of dominant-subordinate relationships on brain monoamines and cortisol in rainbow trout. *Brain Behav Evol* 54:263–275.

Øverli Ø, Korzan WJ, Larson ET, Winberg S, Lepage O, Pottinger TG, Renner KJ, Summers CH 2004b. Behavioral and neuroendocrine correlates of displaced aggression in trout. *Horm Behav* 45:324–329.

Øverli Ø, Kotzian S, Winberg S 2002b. Effects of cortisol on aggression and locomotor activity in rainbow trout. *Horm Behav* 42:53–61.

Øverli Ø, Olsen RE, Løvik F, Ringø E 1999b. Dominance hierarchies in Arctic charr, *Salvelinus alpinus* L.: Differential cortisol profiles of dominant and subordinate individuals after handling stress. *Aquac Res* 30:259–264.

Øverli Ø, Pottinger TG, Carrick TR, Øverli E, Winberg S 2001. Brain monoaminergic activity in rainbow trout selected for high and low stress responsiveness. *Brain Behav Evol* 57:214–224.

Øverli Ø, Pottinger TG, Carrick TR, Øverli E, Winberg S 2002a. Differences in behaviour between rainbow trout selected for high- and low-stress responsiveness. *J Exp Biol* 205:391–395.

Øverli Ø, Sørensen C, Nilsson GE 2006. Behavioral indicators of stress-coping style in rainbow trout: Do males and females react differently to novelty? *Physiol Behav* 87:506–512.

Øverli Ø, Sørensen C, Pulman KGT, Pottinger TG, Korzan WJ, Summers CH, Nilsson GE 2007. Evolutionary background for stress-coping styles: Relationships between physiological, behavioral, and cognitive traits in non-mammalian vertebrates. *Neurosci Biobehav Rev* 31:396–412.

Øverli Ø, Winberg S, Pottinger TG 2005. Behavioral and neuroendocrine correlates of selection for stress responsiveness in rainbow trout—A review. *Integr Comp Biol* 45:463–474.

Panula P, Chen YC, Priyadarshini M, Kudo H, Semenova S, Sundvik M, Sallinen V 2010. The comparative neuroanatomy and neurochemistry of zebrafish CNS systems of relevance to human neuropsychiatric diseases. *Neurobiol Dis* 40:46–57.

Parent A, Poitras D, Dube L 1984. Comparative anatomy of central monoaminergic systems. In: *Handbook of Chemical Neuroanatomy* (Björklund A, Hökfelt T, eds.), pp. 409–439. Amsterdam, the Netherlands: Elsevier Science Publishers.

Parihar VK, Hattiangady B, Kuruba R, Shuai B, Shetty AK 2011. Predictable chronic mild stress improves mood, hippocampal neurogenesis and memory. *Mol Psychiatr* 16:171–183.

Parikh VN, Clement T, Fernald RD 2006. Physiological consequences of social descent: Studies in *Astatotilapia burtoni. J Endocrinol* 190:183–190.

Patiño R, Redding JM, Schreck CB 1987. Interrenal secretion of corticosteroids and plasma cortisol and cortisone concentrations after acute stress and during seawater acclimation in juvenile coho salmon (*Oncorhynchus kisutch). Gen Comp Endocr* 68:431–439.

Patterson SL, Grover LM, Schwartzkroin PA, Bothwell M 1992. Neurotrophin expression in rat hippocampal slices: A stimulus paradigm inducing LTP in CA1 evokes increases in BDNF and NT-3 mRNAs. *Neuron* 9:1081–1088.

Paull GC, Filby AL, Giddins HG, Coe TS, Hamilton PB, Tyler CR 2010. Dominance hierarchies in zebrafish (*Danio rerio*) and their relationship with reproductive success. *Zebrafish* 7:109–117.

Pavlides C, Kimura A, Magariños AM, McEwen BS 1995a. Hippocampal homosynaptic long-term depression/depotentiation induced by adrenal steroids. *Neuroscience* 68:379–385.

Pavlides C, Watanabe Y, Magarinos AM, McEwen BS 1995b. Opposing roles of type I and type II adrenal steroid receptors in hippocampal long-term potentiation. *Neuroscience* 68:387–394.

Pavlidis M, Sundvik M, Chen YC, Panula P 2011. Adaptive changes in zebrafish brain in dominant-subordinate behavioral context. *Behav Brain Res* 225:529–537.

Pawlak R, Magarinos AM, Melchor J, McEwen B, Strickland S 2003. Tissue plasminogen activator in the amygdala is critical for stress-induced anxiety-like behavior. *Nat Neurosci* 6:168–174.

Perreault HAN, Semsar K, Godwin J 2003. Fluoxetine treatment decreases territorial aggression in a coral reef fish. *Physiol Behav* 79:719–724.

Pervin L, John OP 1997. *Personality: Theory and Research* (7th edn.). New York: Wiley & Sons.

Peters G, Faisal M, Lang T, Ahmed I 1988. Stress caused by social interaction and its effect on susceptibility to *Aeromonas hydrophila* infection in rainbow trout *Salmo gairdneri*. *Dis Aq Org* 4:83–89.

Peters G, Nüssgen A, Raabe A, Möck A 1991. Social stress induces structural and functional alterations of phagocytes in rainbow trout (*Oncorhynchus mykiss*). *Fish Shellfish Immunol* 1:17–31.

Peters G, Schwarzer R 1985. Changes in hemopoietic tissue of rainbow trout under influence of stress. *Dis Aq Org* 1:1–10.

Pham K, Nacher J, Hof PR, McEwen BS 2003. Repeated restraint stress suppresses neurogenesis and induces biphasic PSA-NCAM expression in the adult rat dentate gyrus. *Eur J Neurosci* 17:879–886.

Pickering AD 1981. *Stress and Fish*. London, U.K.: Academic Press.

Pickering AD, Pottinger TG 1989. Stress responses and disease resistance in salmonid fish—Effects of chronic elevation of plasma cortisol. *Fish Physiol Biochem* 7:253–258.

Pickering AD, Pottinger TG 1995. Biochemical effects of stress. In: *Environmental and Ecological Biochemistry* (Hochachka PW, Mommsen TP, eds.), pp. 349–379. Amsterdam, the Netherlands: Elsevier.

Pottinger TG, Carrick TR 1999. Modification of the plasma cortisol response to stress in rainbow trout by selective breeding. *Gen Comp Endocrinol* 116:122–132.

Pottinger TG, Carrick TR 2001. Stress responsiveness affects dominant-subordinate relationships in rainbow trout. *Horm Behav* 40:419–427.

Pottinger TG, Pickering AD 1992. The influence of social interaction on the acclimation of rainbow trout, *Oncorhynchus mykiss* (Walbaum) to chronic stress. *J Fish Biol* 41:435–447.

Prager EM, Johnson LR 2009. Stress at the synapse: Signal transduction mechanisms of adrenal steroids at neuronal membranes. *Sci Signal* 2: re5.

Pugh CR, Fleshner M, Rudy JW 1997. Type II glucocorticoid receptor antagonists impair contextual but not auditory-cue fear conditioning in juvenile rats. *Neurobiol Learn Mem* 67:75–79.

Ra SM, Kim H, Jang MH, Shin MC, Lee TH, Lim BV, Kim CJ, Kim EH, Kim KM, Kim SS 2002. Treadmill running and swimming increase cell proliferation in the hippocampal dentate gyrus of rats. *Neurosci Lett* 333:123–126.

Raab A, Dantzer R, Michaud B, Mormede P, Taghzouti K, Simon H, Le Moal M 1986. Behavioral, physiological and immunological consequences of social status and aggression in chronically coexisting resident intruder dyads of male rats. *Physiol Behav* 36:223–228.

Radley JJ, Rocher AB, Janssen WGM, Hof PR, McEwen BS, Morrison JH 2005. Reversibility of apical dendritic retraction in the rat medial prefrontal cortex following repeated stress. *Exp Neurol* 196:199–203.

Radley JJ, Rocher AB, Rodriguez A, Ehlenberger DB, Dammann M, McEwen BS, Morrison JH, Wearne SL, Hof PR 2008. Repeated stress alters dendritic spine morphology in the rat medial prefrontal cortex. *J Comp Neurol* 507:1141–1150.

Radley JJ, Sisti HM, Hao J, Rocher AB, McCall T, Hof PR, McEwen BS, Morrison JH 2004. Chronic behavioral stress induces apical dendritic reorganization in pyramidal neurons of the medial prefrontal cortex. *Neuroscience* 125:1–6.

Ramos BP, Arnsten AFT 2007. Adrenergic pharmacology and cognition: Focus on the prefrontal cortex. *Pharmacol Ther* 113:523–536.

Randall DJ, Perry SF 1992. Catecholamines. In: *Fish Physiology* (Hoar WS, Randall DJ, Farrell AP, eds.), Vol. XIIB, pp. 255–300. New York: Academic Press.

Raymond PA, Easter SS 1983. Postembryonic growth of the optic tectum in goldfish. I. Location of germinal cells and numbers of neurons produced. *J Neurosci* 3:1077–1091.

Réale D, Reader SM, Sol D, McDougall PT, Dingemanse NJ 2007. Integrating animal temperament within ecology and evolution. *Biol Rev* 82:291–318.

Reul JMHM, de Kloet ER 1985. Two receptor systems for corticosterone in rat brain: Microdistribution and differential occupation. *Endocrinology* 117:2505–2511.

Reul JMHM, van den Bosch FR, de Kloet ER 1987. Relative occupation of type-I and type-II corticosteroid receptors in rat brain following stress and dexamethasone treatment: Functional implications. *J Endocrinol* 115:459–467.

Rhodes JS, Quinn TP 1998. Factors affecting the outcome of territorial contests between hatchery and naturally reared coho salmon parr in the laboratory. *J Fish Biol* 53:1220–1230.

Ricker WE 1975. Computation and interpretation of biological statistics of fish populations. *Bull Fish Res Bd Can* 191:1–382.

Riebli T, Taborsky M, Chervet N, Apolloni N, Zürcher Y, Heg D 2012. Behavioural type, status and social context affect behaviour and resource allocation in cooperatively breeding cichlids. *Anim Behav* 84:925–936.

Rodríguez F, López JC, Vargas JP, Gómez Y, Broglio C, Salas C 2002. Conservation of spatial memory function in the pallial forebrain of reptiles and ray-finned fishes. *J Neurosci* 22:2894–2903.

Rose JD 2002. The neurobehavioral nature of fishes and the question of awareness and pain. *Rev Fish Sci* 10:1–38.

Rossi C et al. 2006. Brain-derived neurotrophic factor (BDNF) is required for the enhancement of hippocampal neurogenesis following environmental enrichment. *Eur J Neurosci* 24:1850–1856.

Rowlerson A, Veggetti A 2001. Cellular mechanisms of post-embryonic muscle growth in aquaculture species. In: *Fish Physiology* (Ian J, ed.), Vol. 18, pp. 103–140. Gulf Professional Publishing, Houston, Texas.

Ruiz-Gomez MD, Huntingford FA, Øverli Ø, Thörnqvist PO, Höglund E 2011. Response to environmental change in rainbow trout selected for divergent stress coping styles. *Physiol Behav* 102:317–322.

Ruiz-Gomez MD, Kittilsen S, Höglund E, Huntingford FA, Sørensen C, Pottinger TG, Bakken M, Winberg S, Korzan WJ, Øverli Ø 2008. Behavioral plasticity in rainbow trout (*Oncorhynchus mykiss*) with divergent coping styles: When doves become hawks. *Horm Behav* 54:534–538.

Ryer CH, Olla BL 1996. Growth depensation and aggression in laboratory reared coho salmon: The effect of food distribution and ration size. *J Fish Biol* 48:686–694.

Sandi C, Venero C, Guaza C 1996. Novelty-related rapid locomotor effects of corticosterone in rats. *Eur J Neurosci* 8:794–800.

Sangalang GB, Uthe JF 1994. Corticosteroid activity, *in vitro*, in interrenal tissue of Atlantic salmon (*Salmo salar*) part: 1. Synthetic profiles. *Gen Comp Endocrinol* 95:273–285.

Sanogo YO, Band M, Blatti C, Sinha S, Bell AM 2012. Transcriptional regulation of brain gene expression in response to a territorial intrusion. *Proc R Soc B* 279:4929–4938.

Sapolsky RM 1982. The endocrine stress response and social status in the wild baboon. *Horm Behav* 16:279–292.

Sapolsky RM 1990. Adrenocortical function, social rank, and personality among wild baboons. *Biol Psychiatry* 28:862–878.

Sapolsky RM 1992. Cortisol concentrations and the social significance of rank instability among wild baboons. *Psychoneuroendocrinology* 17:701–709.

Schjolden J, Backström T, Pulman KGT, Pottinger TG, Winberg S 2005a. Divergence in behavioural responses to stress in two strains of rainbow trout (*Oncorhynchus mykiss*) with contrasting stress responsiveness. *Horm Behav* 48:537–544.

Schjolden J, Pulman KGT, Metcalfe NB, Winberg S 2006. Divergence in locomotor activity between two strains of rainbow trout *Oncorhynchus mykiss* with contrasting stress responsiveness. *J Fish Biol* 68:920–924.

Schjolden J, Stoskhus A, Winberg S 2005b. Does individual variation in stress responses and agonistic behavior reflect divergent stress coping strategies in juvenile rainbow trout? *Physiol Biochem Zool* 78:715–723.

Schjolden J, Winberg S 2007. Genetically determined variation in stress responsiveness in rainbow trout: Behavior and neurobiology. *Brain Behav Evol* 70:227–238.

Schultz W 2007. Behavioral dopamine signals. *Trends Neurosci* 30:203–210.

Schwabe L, Joëls M, Roozendaal B, Wolf OT, Oitzl MS 2012. Stress effects on memory: An update and integration. *Neurosci Biobehav Rev* 36:1740–1749.

Scott KA, Melhorn SJ, Sakai RR 2012. Effects of chronic social stress on obesity. *Curr Obes Rep* 1:16–25.

Shannon NJ, Gunnet JW, Moore KE 1986. A comparison of biochemical indices of 5-hydroxytryptaminergic neuronal activity following electrical stimulation of the dorsal raphe nucleus. *J Neurochem* 47:958–965.

Shively CA, Laber-Laird K, Anton RF 1997. Behavior and physiology of social stress and depression in female cynomolgus monkeys. *Biol Psychiatry* 41:871–882.

Shively CA, Willard SL 2012. Behavioral and neurobiological characteristics of social stress versus depression in nonhuman primates. *Exp Neurol* 233:87–94.

Shors TJ, Anderson ML, Curlik DM, Nokia MS 2012. Use it or lose it: How neurogenesis keeps the brain fit for learning. *Behav Brain Res* 227:450–458.

Sih A, Bell A, Johnson JC 2004. Behavioral syndromes: An ecological and evolutionary overview. *Trends Ecol Evol* 19:372–378.

Sinn DL, Moltschaniwskyj NA 2005. Personality traits in dumpling squid (*Euprymna tasmanica*): Context-specific traits and their correlation with biological characteristics. *J Comp Psychol* 119:99–110.

Slattery DA, Uschold N, Magoni M, Bar J, Popoli M, Neumann ID, Reber SO 2012. Behavioural consequences of two chronic psychosocial stress paradigms: Anxiety without depression. *Psychoneuroendocrinology* 37:702–714.

Sloman KA, Armstrong JD 2002. Physiological effects of dominance hierarchies: Laboratory artefacts or natural phenomena? *J Fish Biol* 61:1–23.

Sloman KA, Baker DW, Ho CG, McDonald DG, Wood CM 2003c. The effects of trace metal exposure on agonistic encounters in juvenile rainbow trout, *Oncorhynchus mykiss*. *Aquat Toxicol* 63:187–196.

Sloman KA, Baker DW, Wood CM, McDonald G 2002c. Social interactions affect physiological consequences of sublethal copper exposure in rainbow trout, *Oncorhynchus mykiss*. *Env Toxicol Chem* 21:1255–1263.

Sloman KA, Gilmour KM, Metcalfe NB, Taylor AC 2000a. Does socially induced stress in rainbow trout cause chloride cell proliferation? *J Fish Biol* 56:725–738.

Sloman KA, Gilmour KM, Taylor AC, Metcalfe NB 2000b. Physiological effects of dominance hierarchies within groups of brown trout, *Salmo trutta*, held under simulated natural conditions. *Fish Physiol Biochem* 22:11–20.

Sloman KA, Lepage O, Rogers JT, Wood CM, Winberg S 2005. Socially-mediated differences in brain monoamines in rainbow trout: Effects of trace metal contaminants. *Aquat Toxicol* 71: 237–247.

Sloman KA, Metcalfe NB, Taylor AC, Gilmour KM 2001a. Plasma cortisol concentrations before and after social stress in rainbow trout and brown trout. *Physiol Biochem Zool* 74:383–389.

Sloman KA, Montpetit CJ, Gilmour KM 2002a. Modulation of catecholamine release and cortisol secretion by social interactions in the rainbow trout, *Oncorhynchus mykiss*. *Gen Comp Endocrinol* 127:136–146.

Sloman KA, Morgan TP, McDonald DG, Wood CM 2003a. Socially-induced changes in sodium regulation affect the uptake of water-bome copper and silver in the rainbow trout, *Oncorhynchus mykiss*. *Comp Biochem Physiol* C135:393–403.

Sloman KA, Motherwell G, O'Connor KI, Taylor AC 2000c. The effect of social stress on the standard metabolic rate (SMR) of brown trout, *Salmo trutta*. *Fish Physiol Biochem* 23:49–53.

Sloman KA, Scott GR, Diao ZY, Rouleau C, Wood CM, McDonald DG 2003b. Cadmium affects the social behaviour of rainbow trout, *Oncorhynchus mykiss*. *Aquat Toxicol* 65:171–185.

Sloman KA, Scott GR, McDonald DG, Wood CM 2004. Diminished social status affects ionoregulation at the gills and kidney in rainbow trout (*Oncorhynchus mykiss*). *Can J Fish Aq Sci* 61:618–626.

Sloman KA, Taylor AC, Metcalfe NB, Gilmour KM 2001b. Effects of an environmental perturbation on the social behaviour and physiological function of brown trout. *Anim Behav* 61:325–333.

Sloman KA, Wilson L, Freel JA, Taylor AC, Metcalfe NB, Gilmour KM 2002b. The effects of increased flow rates on linear dominance hierarchies and physiological function in brown trout, *Salmo trutta*. *Can J Zool* 80:1221–1227.

Sørensen C 2011. Regulation of brain cell proliferation in fish. In: *Faculty of Mathematics and Natural Sciences*. Oslo, Norway: University of Oslo.

Sørensen C, Bohlin LC, Øverli Ø, Nilsson GE 2011. Cortisol reduces cell proliferation in the telencephalon of rainbow trout (*Oncorhynchus mykiss*). *Physiol Behav* 102:518–523.

Sørensen C, Johansen IB, Øverli Ø 2013. Neural plasticity and stress coping in teleost fishes. *Gen Comp Endocrinol* 181:25–34.

Sørensen C, Nilsson GE, Summers CH, Øverli Ø 2012. Social stress reduces forebrain cell proliferation in rainbow trout (*Oncorhynchus mykiss*). *Behav Brain Res* 227:311–318.

Sørensen C, Øverli Ø, Summers CH, Nilsson GE 2007. Social regulation of neurogenesis in teleosts. *Brain Behav Evol* 70:239–246.

Sousa N, Lukoyanov NV, Madeira MD, Almeida OFX, Paula-Barbosa MM 2000. Reorganization of the morphology of hippocampal neurites and synapses after stress-induced damage correlates with behavioral improvement. *Neuroscience* 97:253–266.

Soutschek J, Zupanc GK 1996. Apoptosis in the cerebellum of adult teleost fish, *Apteronotus leptorhynchus*. *Brain Res Dev Brain Res* 97:279–286.

Stanford SC 1993. Monoamines in response and adaption to stress. In: *Stress from Synapse to Syndrome* (Stanford SC, Salmon P, eds.), pp. 282–331. London, U.K.: Academic Press.

Starkman MN, Gebarski SS, Berent S, Schteingart DE 1992. Hippocampal formation volume, memory dysfunction, and cortisol levels in patients with Cushing's syndrome. *Biol Psychiatry* 32:756–765.

Steiner B, Klempin F, Wang L, Kott M, Kettenmann H, Kempermann G 2006. Type-2 cells as link between glial and neuronal lineage in adult hippocampal neurogenesis. *Glia* 54:805–814.

Sterling P 2012. Allostasis: A model of predictive regulation. *Physiol Behav* 106:5–15.

Sterling P, Eyer J 1988. Allostasis: A new paradigm to explain arousal pathology. In: *Handbook of Life Stress, Cognition and Health* (Fisher S, Reason J, eds.), pp. 629–649. New York: Wiley.

Stewart MG, Davies HA, Sandi C, Kraev IV, Rogachevsky VV, Peddie CJ, Rodriguez JJ, Cordero MI, Donohue HS, Gabbott PLA, Popov VI 2005. Stress suppresses and learning induces plasticity in CA3 of rat hippocampus: A three-dimensional ultrastructural study of thorny excrescences and their postsynaptic densities. *Neuroscience* 131:43–54.

Stolte EH, de Mazon AF, Leon-Koosterziel KM, Jesiak M, Bury NR, Sturm A, Savelkoul HFJ, van Kemenade BMLV, Flik G 2008. Corticosteroid receptors involved in stress regulation in common carp, *Cyprinus carpio*. *J Endocrinol* 198:403–417.

Stolte EH, van Kemenade BMLV, Savelkoul HFJ, Flik G 2006. Evolution of glucocorticoid receptors with different glucocorticoid sensitivity. *J Endocrinol* 190:17–28.

Strand DA, Utne-Palm AC, Jakobsen PJ, Braithwaite VA, Jensen KH, Salvanes AGV 2010. Enrichment promotes learning in fish. *Mar Ecol Prog Ser* 412:273–282.

Summers CH, Winberg S 2006. Interactions between the neural regulation of stress and aggression. *J Exp Biol* 209:4581–4589.

Taborsky B, Oliveira RF 2012. Social competence: An evolutionary approach. *Trends Ecol Evol.* 27:679–688.

Taborsky B, Tschirren L, Meunier C, Aubin-Horth N 2013. Stable reprogramming of brain transcription profiles by the early social environment in a cooperatively breeding fish. *Proc R Soc B* 280:20122605.

Tanapat P, Hastings NB, Rydel TA, Galea LAM, Gould E 2001. Exposure to fox odor inhibits cell proliferation in the hippocampus of adult rats via an adrenal hormone-dependent mechanism. *J Comp Neurol* 437:496–504.

Teles MC, Sirbulescu RF, Wellbrock UM, Oliveira RF, Zupanc GK 2012. Adult neurogenesis in the brain of the Mozambique tilapia, *Oreochromis mossambicus*. *J Comp Physiol A* 198:427–449.

Terzibasi E, Valenzano DR, Cellerino A 2007. The short-lived fish *Nothobranchius furzeri* as a new model system for aging studies. *Exp Gerontol* 42:81–89.

Thomas JB, Gilmour KM 2012. Low social status impairs hypoxia tolerance in rainbow trout (*Oncorhynchus mykiss*). *J Comp Physiol B* 182:651–662.

Thorgaard GH et al. 2002. Status and opportunities for genomics research with rainbow trout. *Comp Biochem Physiol B* 133:609–646.

Tinbergen N 1953. *Social Behaviour in Animals*. London, U.K.: Chapman & Hall.

Tognoli C, Rossi F, Di Cola F, Baj G, Tongiorgi E, Terova G, Saroglia M, Bernardini G, Gornati R 2010. Acute stress alters transcript expression pattern and reduces processing of proBDNF to mature BDNF in *Dicentrarchus labrax*. *BMC Neurosci* 11:4.

Truscott B, Idler DR 1972. Corticosteroids in plasma of elasmobranchs. *Comp Biochem Physiol A* 42:41–50.

Turdi S, Yuan M, Leedy GM, Wu Z, Ren J 2012. Chronic social stress induces cardiomyocyte contractile dysfunction and intracellular Ca^{2+} derangement in rats. *Physiol Behav* 105:498–509.

Ursin R 2002. Serotonin and sleep. *Sleep Med Rev* 6:57–69.

van Oers K, Mueller JC 2010. Evolutionary genomics of animal personality. *Phil Trans R Soc B* 365:3991–4000.

van Praag H, Kempermann G, Gage FH 1999. Running increases cell proliferation and neurogenesis in the adult mouse dentate gyrus. *Nat Neurosci* 2:266–270.

Vargas JP, López JC, Portavella M 2009. What are the functions of fish brain pallium? *Brain Res Bull* 79:436–440.

Vargas JP, Rodríguez F, López, JC, Arias JL, Salas C 2000. Spatial learning-induced increase in the argyrophilic nucleolar organizer region of dorsolateral telencephalic neurons in goldfish. *Brain Res* 865:77–84.

Venzala E, García-García AL, Elizalde N, Delagrange P, Tordera RM 2012. Chronic social defeat stress model: Behavioral features, antidepressant action, and interaction with biological risk factors. *Psychopharmacology* 224:313–325.

von Krogh K, Sørensen C, Nilsson GE, Øverli Ø 2010. Forebrain cell proliferation, behavior, and physiology of zebrafish, *Danio rerio*, kept in enriched or barren environments. *Physiol Behav* 101:32–39.

Vyas A, Mitra R, Rao BSS, Chattarji S 2002. Chronic stress induces contrasting patterns of dendritic remodeling in hippocampal and amygdaloid neurons. *J Neurosci* 22:6810–6818.

Wankowski JWJ, Thorpe JE 1979. Spatial distribution and feeding in Atlantic salmon, *Salmo salar* L juveniles. *J Fish Biol* 14:239–247.

Watt PJ, Skinner A, Hale M, Nakagawa S, Burke T 2011. Small subordinate male advantage in the zebrafish. *Ethology* 117:1003–1008.

Weatherley AH, Gill HS 1985. Dynamics of increase in muscle fibers in fishes in relation to size and growth. *Experientia* 41:353–354.

Wedemeyer GA, Barton BA, McLeay DJ 1990. Stress and acclimation. In: *Methods for Fish Biology* (Schreck CB, Moyle PB, eds.), pp. 451–489. Bethesda, MD: American Fisheries Society.

Weir LK, Hutchings JA, Fleming IA, Einum S 2004. Dominance relationships and behavioural correlates of individual spawning success in farmed and wild male Atlantic salmon, *Salmo salar. J Anim Ecol* 73:1069–1079.

Wendelaar Bonga SE 1997. The stress response in fish. *Physiol Rev* 77:591–625.

Williams SN, Undieh AS 2009. Dopamine D1-like receptor activation induces brain-derived neurotrophic factor protein expression. *Neuroreport* 20:606–610.

Winberg S, Lepage O 1998. Elevation of brain 5-HT activity, POMC expression, and plasma cortisol in socially subordinate rainbow trout. *Am J Physiol R* 43:645–654.

Winberg S, Carter CG, Mccarthy JD, He ZY, Nilsson GE, Houlihan DF 1993b. Feeding rank and brain serotonergic activity in rainbow trout *Oncorhynchus mykiss. J Exp Biol* 179:197–211.

Winberg S, Nilsson GE 1992. Induction of social dominance by L-dopa treatment in Arctic charr. *Neuroreport* 3:243–246.

Winberg S, Nilsson GE 1993. Roles of brain monoamine neurotransmitters in agonistic behavior and stress reactions, with particular reference to fish. *Comp Biochem Physiol C* 106:597–614.

Winberg S, Nilsson GE, Olsén KH 1991. Social rank and brain levels of monoamines and monoamine metabolites in Arctic charr, *Salvelinus alpinus* (L). *J Comp Physiol A* 168:241–246.

Winberg S, Nilsson GE, Olsén KH 1992. Changes in brain serotonergic activity during hierarchical behavior in arctic charr (*Salvelinus alpinus* L) are socially induced. *J Comp Physiol A* 170:93–99.

Winberg S, Nilsson GE, Spruijt BM, Höglund U 1993a. Spontaneous locomotor activity in Arctic charr measured by a computerized imaging technique—Role of brain serotonergic activity. *J Exp Biol* 179:213–232.

Winberg S, Øverli Ø, Lepage O 2001. Suppression of aggression in rainbow trout (*Oncorhynchus mykiss*) by dietary L-tryptophan. *J Exp Biol* 204:3867–3876.

Wingfield JC 2003. Control of behavioural strategies for capricious environments. *Anim Behav* 66:807–815.

Wolf M, van Doorn GS, Leimar O, Weissing FJ 2007. Life-history trade-offs favour the evolution of animal personalities. *Nature* 447:581–584.

Wong EY, Herbert J 2004. The corticoid environment: A determining factor for neural progenitors' survival in the adult hippocampus. *Eur J Neurosci* 20:2491–2498.

Wong EY, Herbert J 2005. Roles of mineralocorticoid and glucocorticoid receptors in the regulation of progenitor proliferation in the adult hippocampus. *Eur J Neurosci* 22:785–792.

Wood LS, Desjardins JK, Fernald RD 2011. Effects of stress and motivation on performing a spatial task. *Neurobiol Learn Mem* 95:277–285.

Woolley CS, Gould E, McEwen BS 1990. Exposure to excess glucocorticoids alters dendritic morphology of adult hippocampal pyramidal neurons. *Brain Res* 531:225–231.

Woolpy JH, Ginsburg BE 1967. Social relationships in a group of captive wolves. *Am Zool* 7:305–311.

Wullimann MF, Rink E 2002. The teleostean forebrain: A comparative and developmental view based on early proliferation, Pax6 activity and catecholaminergic organization. *Brain Res Bull* 57:363–370.

Yamagishi H 1964. An experimental study on the effect of aggressiveness to the variability of growth in the juvenile rainbow trout, *Salmo gairdneri* Richardson. *Jap J Ecol* 14:228–232.

Yamamoto T, Ueda H, Higashi S 1998. Correlation among dominance status, metabolic rate and otolith size in masu salmon. *J Fish Biol* 52:281–290.

Yap JJ, Takase LF, Kochman LJ, Fornal CA, Miczek KA, Jacobs BL 2006. Repeated brief social defeat episodes in mice: Effects on cell proliferation in the dentate gyrus. *Behav Brain Res* 172:344–350.

Zakon HH 1984. Postembryonic changes in the peripheral electrosensory system of a weakly electric fish: Addition of receptor organs with age. *J Comp Neurol* 228:557–570.

Zigova T, Pencea V, Wiegand SJ, Luskin MB 1998. Intraventricular administration of BDNF increases the number of newly generated neurons in the adult olfactory bulb. *Mol Cell Neurosci* 11:234–245.

Zupanc GK 2008. Adult neurogenesis in teleost fish. In: *Adult Neurogenesis* (Gage FH, Kempermann G, Song H, eds.), pp. 571–592. New York: Cold Spring Harbor Laboratory Press.

Zupanc GK, Horschke I 1995. Proliferation zones in the brain of adult gymnotiform fish: A quantitative mapping study. *J Comp Neurol* 353:213–233.

Zupanc GK, Horschke I, Ott R, Rascher GB 1996. Postembryonic development of the cerebellum in gymnotiform fish. *J Comp Neurol* 370:443–464.

Zupanc GK, Hinsch K, Gage FH 2005. Proliferation, migration, neuronal differentiation, and long-term survival of new cells in the adult zebrafish brain. *J Comp Neurol* 488:290–319.

Zupanc GK, Lamprecht J 2000. Towards a cellular understanding of motivation: Structural reorganization and biochemical switching as key mechanisms of behavioral plasticity. *Ethology* 106:467–477.

Zupanc GK, Ott R 1999. Cell proliferation after lesions in the cerebellum of adult teleost fish: Time course, origin, and type of new cells produced. *Exp Neurol* 160:78–87.

10 Pain Perception

Victoria A. Braithwaite

CONTENTS

10.1 BACKGROUND

The capacity to respond and withdraw from a pain-inducing event, or injury, helps an animal protect itself and also decreases the risk of further damage occurring. Thus, a sensory system that supports the detection of tissue damage helps animals react appropriately to noxious events (Bateson 1991; Broom 2001). Animals that respond to tissue damage are more likely to survive and reproduce than those that are unable to detect when something harmful occurs. So being able to detect noxious events is an adaptive process and one that we should expect to be widespread across the animal kingdom (Kavaliers 1998).

Pain is a complex process that helps to offset adverse situations in life. It is often described as a "phenomenon" because there are aspects of pain that we still do not fully understand, particularly in our complex selves. A commonly quoted definition of human pain provided by Iggo (1984) describes it as "an unpleasant sensation or emotion associated with actual or potential tissue damage," and a version of this definition is the one adopted by the International Association for the Study of Pain (IASP 2012). In a review on the evolution of animal pain, however, Broom (2001) proposed that the wording should be modified to "an *aversive sensation and feeling* associated with actual or potential tissue damage" in order to make it more applicable to nonhuman animals. He suggested the use of "aversive," as opposed to "unpleasant" because aversion is a recognized and quantifiable process in animals. And he used "feeling" rather than "emotion," as he suggested the former is a physiological state that can be measured in terms of changes such as an increase in heart rate, or a change in stress hormone levels. This amended definition seems partially appropriate for the purpose of this review of pain in fishes as it gives us specific aspects of the pain process to consider. However, since Broom (2001) proposed the reworded definition, we have gained a deeper understanding of animal emotions and affective states (Paul et al. 2005; Rolls 2007; Braithwaite et al. 2013); thus, whether fishes have a capacity for an emotional response is not an impossible question to address, and it is a topic that will be considered in the second part of this review.

Over the last two decades, many advances have been made in terms of understanding animal pain, and this is particularly true for fishes where several systematic studies addressing the capacity for ray-finned, bony fishes to feel pain have now been published (Sneddon et al. 2003a,b; Dunlop and Laming 2005). The process of pain can be considered to be a series of stages. For example, in vertebrates (note, however, that some of these processes occur in invertebrates as well;

see Kavaliers 1998), the first stage involves the detection of a noxious event through specialized receptors known as nociceptors. Second, information about the noxious stimulation is conveyed through specialized fibers (A-delta and C fibers) within the nervous system. Third, the information reaches the spinal cord where reflexive responses may be triggered and some of these alter various physiological processes. Fourth, information associated with noxious stimulation may be transmitted to the brain where specific areas process it. And a final stage is associated with behavioral changes. Considering pain as a series of different stages in this way provides discrete events that can be studied in animals. This review will use this framework to determine which of these processes have been demonstrated in fishes.

Before considering the different stages of pain, it is important to recognize that the term "fishes" is very broad and it is too often collectively used to refer to many different distinct lineages. The specific nature of pain-processing systems and how sophisticated these are could reasonably be expected to vary across taxonomic groups that have evolved in different ways and under different selective pressures. Diverse groups such as agnathans, elasmobranchs, and ray-finned, bony fishes may therefore have different pain-processing capacities, or even mechanisms. A few studies have searched for specialized pain receptors and neurotransmitters associated with pain in jawless agnathans and in sharks and rays, and the findings from these studies will be summarized where appropriate (Mathews and Wickelgren 1978; Leonard 1985; Cameron et al. 1990; Snow et al. 1996). However, much of the recent research addressing pain in fishes has focused on ray-finned, bony fish species such as trout (*Oncorhynchus mykiss*), salmon (*Salmo salar*), and goldfish (*Carassius auratus*), and it is the advances that have been made in this particular group of fishes that will be the focus in this review (Chervova et al. 1994; Sneddon et al. 2003a; Dunlop and Laming 2005; Ashley et al. 2007; Nordgreen et al. 2007).

Determining whether any animal *feels* pain is difficult because a key component of pain is the feeling and sensation of hurt that it generates (Broom 2001). A subjective, personal feeling is something that only the individual can experience. For example, you will never be able to experience somebody else's pain, but through language you can be told in words what it feels like to experience the emotions associated with suffering. An accurate description may allow you to empathize with someone's situation and you can probably be persuaded that they are sore and hurting. However, with nonhuman animals, there is no common language to explain how an injured limb makes an animal feel. Although we cannot ask the animal directly, we might suspect something is wrong because of the way it limps, or protects itself, perhaps, it prefers to stay hidden and quiet, and if the injury is a severe one, it may have a suppressed appetite (Weary et al. 2006). All of these alterations to its normal behavior give some indication that the animal is affected by the injury, and whether it hurts can be explored in a number of ways. For example, we can look at how strongly the animal is motivated to protect the injury from further stimulation by determining if it will let us inspect the tissue damage. Does it vocalize or pull away when we touch the injured area? Approaches such as this can provide a way to assess the negative effects different animals associate with an injury.

10.2   NOCICEPTION

Nociceptors are free nerve endings found in tissues such as the skin, oral and nasal membranes, skeletal muscle, and visceral organs (Dubin and Patapoutian 2010). Their role is to detect when extreme events occur such as excessive temperatures, mechanical crushing, or noxious chemicals that will damage tissue. There are a number of receptor types and these respond to different forms of noxious stimulation (Dubin and Patapoutian 2010). Some receptors respond to more than one kind of noxious stimulus; for example, bimodal receptors such as mechanothermal nociceptors detect both mechanically damaging and excessive heat stimuli, whereas mechanochemical nociceptors respond to excessive mechanical and noxious chemical stimuli. Polymodal nociceptors are a further category of receptors and these respond to noxious mechanical, heat, and chemical stimuli (Dubin and Patapoutian 2010).

Nociceptors play a critical role in defense from stimuli that cause damage and they are found in both invertebrates and vertebrates (Kavaliers 1998). In fact, recent research examining the effects of noxious mechanical and thermal stimulation in *Drosophila* has now identified a particular form of invertebrate sensory neuron that has nociceptive properties—a multidendritic class IV neuron (Tracey et al. 2003; Hwang et al. 2007). With the many different molecular and genetic tools developed for *Drosophila*, identification of this nociceptive neuron is now enabling the mechanisms and even the genes that control nociception to be investigated (Zhong et al. 2010; Kim et al. 2012).

Within the vertebrates, specialized nociceptors have been extensively characterized in birds (Gentle et al. 2001) and mammals (Lynn 1994; Yeomans and Proudfit 1996), and, more recently, their presence in ray-finned, bony fishes was confirmed (Sneddon et al. 2003a; Dunlop and Laming 2005; Nordgreen et al. 2009). Before these papers were published, descriptions of the epidermis of several different species of fish such as minnow (*Phoxinus phoxinus*), stickleback (*Gasterosteus aculeatus*), sand goby (*Pomatoschistus minutus*), and gurnard (*Trigla lucerna*) had reported the presence of free nerve endings all over the fish body, and these were speculated to function as nociceptors (Whitear 1971, 1983). But it took more than a decade before this was confirmed for species such as trout, salmon, and goldfish, and different kinds of epidermal nociceptors were formally described (Chervova et al. 1994; Sneddon et al. 2003a; Dunlop and Laming 2005; Ashley et al. 2007; Nordgreen et al. 2007, 2009a).

After noxious stimuli are detected by the nociceptors, information about the damage is transmitted within the nervous system through specialized types of nerve fiber. In vertebrates, there are two kinds of fiber that relay nociceptive signals: lightly myelinated A-delta fibers and unmyelinated C fibers. Studies addressing nociceptive capacities often begin by looking for evidence of different classes of nociceptor and then look for the presence of A-delta and C fibers (Sneddon et al. 2003a). While nociceptors and their associated specialized fibers have been described in ray-finned bony fishes (Sneddon 2002; Sneddon et al. 2003a), their presence in agnathans and elasmobranch fishes has been harder to confirm. Physiological evidence of nociception by detecting intracellular activity in sensory neurons of the skin and mouth of sea lamprey (*Pteromyzon marinus*) was recorded after the skin had been punctured, crushed, or burned (Martin and Wickelgren 1971, 1978). These results suggest that nociception does occur in lampreys, but to date there have been no anatomical descriptions of cutaneous lamprey nociceptors. In elasmobranchs, some forms of nociception and associated neurotransmitters have been reported, but there are clearly differences between this group of fishes and ray-finned, bony fishes (Sneddon 2004). For instance, while A-delta fibers have been described in elasmobranch fishes, there appear to be many fewer C fibers and these are apparently completely absent in various species of ray (Coggeshall et al. 1978; Leonard 1985, Snow et al. 1993). However, neuromodulators associated with nociception, such as substance P and met-enkephalin, have been found in the spinal cord of brown stringray (*Dasyatis fluviorum*), the eagle ray (*Aetobatis narinari*), the shovelnose ray (*Rhinobatis battilum*), and black-tip shark (*Carcharhinus melanopterus*) (Cameron et al. 1990; Snow et al. 1996). Thus, the kinds and extent of nociception in elasmobranchs remain unclear, but with growing interest in this group of fishes, it seems likely that comparative studies will be undertaken in the near future.

In contrast to the variable nature of the reports concerning nociception in elasmobranchs, a more consistent pattern has emerged from studies with ray-finned bony fishes (Ashley and Sneddon 2008; Braithwaite and Boulcott 2008). Systematic studies of the main nerve innervating the head and face in rainbow trout (*O. mykiss*) have described different kinds of cutaneous nociceptor and also confirmed the presence of A-delta and C fibers in the maxillary, mandibular, and ophthalmic branches of the trigeminal nerve (Sneddon 2002, 2003a; Sneddon et al. 2003a; Ashley et al. 2006, 2007). Using deeply anesthetized trout restrained in a cradle, the receptive fields on the face and cornea were explored by linking specific kinds of stimulation to activity recorded in the trigeminal ganglion. Recordings from the nerves were achieved by removing the skin and bone of the head to provide access to the trigeminal ganglion (after removal of the olfactory lobes, optic lobes, and cerebellum). To prevent electrical activity associated with muscle twitching interfering with recordings made after receptor stimulation, a neuromuscular blocker, Pavulon, was used (Sneddon 2003a; Ashley et al. 2006, 2007).

The receptors were initially detected by applying a mechanical, glass probe onto different parts of the skin of the trout's head, face, and cornea (Sneddon 2003a; Ashley et al. 2006). When the probe activated a receptor, electrodes in the desheathed trigeminal ganglion recorded the neural activity. Once a receptor was located, the size of the receptor field was measured and its mechanical threshold could be determined with von Frey filaments (filaments of different length that apply specific amounts of force on a defined target—with shorter filaments requiring a greater force to bend compared to longer filaments). Sensitivity to thermal stimuli was quantified by placing a thermal simulator 1 mm above the area of the receptor field. A quartz-glass light bulb with a built-in reflector provided the heat source and this was increased in steps of 1°C to determine the threshold of the receptor (Sneddon 2003a). In later studies, Ashley et al. (2006, 2007) used a flat-based (1 mm diameter) thermode that made direct contact with the skin as this was found to give more sensitive thermal threshold data. To assess the chemical sensitivity, a single drop of 1% acetic acid (pH 2.8) was gently placed in the center of the receptive field (Sneddon 2003a). Any response that was detected within 5 ms of the acid solution being applied was considered a possible mechanical artifact and so was ignored. Stimulation with the acetic acid solution was subsequently repeated to check that the reaction to the chemical was a true response. As a control and to check that the mechanical action of placing the drop of acetic acid solution was not responsible for stimulating the receptor, a small drop of water was placed onto the receptive field. No responses to the water were ever recorded (Sneddon 2003a).

These different tests with rainbow trout revealed the presence of five different kinds of receptor. Two responded exclusively to mechanical stimulation: fast-adapting responses where the unit fired a few times but then ceased firing, and slow-adapting responses where the unit continuously fired during mechanical stimulation. Neither of these fast- or slow-adapting mechanoreceptors responded to the thermal or chemical assays. A third kind of receptor was polymodal and responded to noxious mechanical, thermal, and chemical stimuli. Two kinds of bimodal receptor were also found: one that responded to either mechanical or noxious thermal stimuli (mechanothermal nociceptors), and another that responded to mechanical and chemical stimuli (mechanochemical receptors) (Sneddon 2003a; Sneddon et al. 2003a; Ashley et al. 2007). Although the mechanochemical receptors were suspected to be nociceptors, this was only recently confirmed when their nociceptive characteristics were formally described (Mettam et al. 2012a).

The diameters of the receptive fields were generally similar across the different receptor types (~2.5 mm). The mechanical thresholds of polymodal receptors (1.02 ± 0.24 g) were higher than either the fast-adapting (0.37 ± 0.1 g) or slow-adapting (0.32 ± 0.14 g) mechanical receptors (Ashley et al. 2007). Apart from the mechanothermal nociceptors, all other receptors responded to an increase in strength of mechanical stimulation with a higher frequency of firing. There were no differences in threshold sensitivity to chemical stimulation for either polymodal or mechanochemical receptors (Ashley et al. 2007). However, the thermal thresholds of the mechanothermal receptors (33.16°C ± 0.97°C) were higher than those of polymodal receptors (28.87°C ± 1.32°C), and Ashley et al. (2007) proposed these bimodal nociceptors could be specialized for detecting noxious temperatures (Ashley et al. 2007). No cold-sensitive nociceptors were found on the head or cornea of rainbow trout (Ashley et al. 2006, 2007). The action potential amplitude, action potential duration, and conduction velocities were also found to be consistent across the different kinds of receptor (Ashley et al. 2007).

The trout fast-adapting units appear to be equivalent to touch receptors—in mammals, these typically have low mechanical thresholds (Leem et al. 1993). The slow-adapting receptors, however, appear to be units that detect pressure. The receptive field size of the trout receptors was similar to those described for mice and birds, but the mechanical thresholds were found to be lower for the fishes (i.e., more sensitive) than those recorded for birds and mammals (Gentle 1989; Cain et al. 2001). In fact, the mechanical thresholds for the trout polymodal and mechanothermal receptors were similar to those recorded for mammalian cornea nociceptors (López de Armentia et al. 2000). A more recent study of thermonociception in goldfishes used a heated belt wrapped around the body of the fishes to investigate thermal thresholds that elicited escape-type responses in the fishes (Nordgreen et al. 2009a). The mean nociceptive threshold for the goldfishes tested in this way was 38°C. In comparison, the mean threshold for trout polymodal receptors was 29°C and 33°C for mechanothermal receptors. While the goldfish threshold appears to be higher than those measured for the trout receptors, it should be noted that the techniques used to apply and quantify the noxious thermal threshold in the two studies are very different and so it is not clear if this is a true difference or a result of the techniques used (Ashley et al. 2007; Nordgreen et al. 2009a).

Once triggered, the information from nociceptors is conveyed through A-delta and C fibers. To confirm that these fiber types were present in trout, short lengths of all three branches of trout trigeminal nerve were fixed and embedded in resin. An ultratome was used to prepare sections that were then mounted on microscope slides and stained with toluidine blue. When viewed at 1000× magnification, it was possible to describe a range of fiber types, including both lightly myelinated A-delta fibers and unmyelinated C fibers. In terms of the number of different fiber types, these were similar across all three branches of the trigeminal nerve. Approximately a third of the fibers were identified as A-delta fibers with a diameter of 0.4–5.8 μm. A much smaller proportion, approximately 4%, was identified as small-diameter C fibers (0.1–1.2 μm). These were often found clustered in bundles and were sometimes associated with a Schwann cell. Trout A-delta and C fiber diameters were similar to those described for mammals (Lyn 1994).

A recent study with common carp (*Cyprinus carpio*) described the presence of A-delta and C fibers in the caudal fin (Roques et al. 2010). Bundles of fibers were found both within the lepido-trichia segments that create the fin rays and in the hypodermis, or soft tissue, between the rays. The nerves were systematically distributed across the tissue and the relative number and diameters of A-delta and C fibers were similar to those observed in the trout trigeminal nerve (Ashley et al. 2007; Roques et al. 2010). Dunlop and Laming (2005) also measured the conduction velocities of the nociceptive signals in trout and goldfish. They elicited a series of responses using a sharp pin attached to a solenoid that provided ten 30-ms pulses with 2-s interstimulus intervals. The pin was applied 5–10 mm behind the operculum cover and on the lateral line. Using this technique, they described three categories of responses. The initial fiber activity had a mean velocity of 4.2 ± 1.6 ms^{-1} in goldfishes and 5.3 ± 0.6 ms^{-1} in trout. These were concluded to be recordings from A-delta fibers. A second fiber group was also identified; activity in these fibers occurred after a short delay and were considered to be C fibers with conduction velocities in the range of 0.18–1.9 ms^{-1} (Dunlop and Laming 2005).

Together these various studies have described a nociceptive system for ray-finned bony fishes that is very similar to the nociceptive systems of mammals and birds, although one difference is that the fishes do not appear to have cold-sensitive nociceptors. The principal fiber types that are known to convey noxious stimuli information in birds and mammals (A-delta and C fibers) were also present in ray-finned, bony fishes, although the fishes have fewer C fibers. The significance of this difference in C fiber number is unclear; however, it could be speculated that it is associated with shorter-lived pain responses in fishes, because in birds and mammals, C fibers act over a longer time course compared to the responses detected through A-delta fiber stimulation.

10.3 PHYSIOLOGICAL RESPONSES TO NOXIOUS STIMULATION

Once the nociceptors have been stimulated and the A-delta and C fibers have conveyed the information about the tissue damage to the spinal cord, a series of responses are triggered. Some of these induce the animal to withdraw the affected area from the source of injury, whereas other responses trigger different stress-related physiological changes (Sumpter 1997; Barton 2002; Galhardo and Oliveira 2009). The elicited effects influence processes such as heart rate, ventilation rate, appetite, and attentional state (Gentle and Tilston 1999; Galhardo and Oliveira 2009). In goldfishes, while general arousal is associated with a decrease in heart rate and ventilation rate of the gills, fright responses associated with an object plunging into a tank housing the fishes caused tachycardia and an increased operculum beat rate (Laming and Savage 1980). Changes in operculum beat frequency provide an easy, noninvasive assay and have been used to monitor stress-related states in fishes (e.g. Lucas et al. 1993; Brown et al. 2005). Increased operculum beat rate associated with confinement stress is correlated with an increase in plasma cortisol (Barreto and Volpato 2004); however, change in operculum beat rate does not necessarily provide an accurate measure of the severity of a negative stimulus (Barreto and Volpato 2004).

Experiments with trout have monitored changes in physiological and motivational states that occur in response to noxious stimulation (Sneddon et al. 2003a; Dunlop et al. 2006; Reilly et al. 2008). In the study by Sneddon et al. (2003a), the responses of two noxious treatments were directly compared with those of two control groups. All groups were anesthetized and handled, and this was the extent of the experience for (1) the first control group, whereas (2) the second control was also injected with saline into the snout of the fishes. The two noxious treatment groups also received injections into the snout; these were either (3) bee venom or (4) acetic acid (0.1% in saline). All fishes were allowed to recover from the anesthesia before regular observations were made to assess motivation to feed, changes in operculum beat rate, and preference to be in an open, brightly lit area or a covered, darkened area of the tank. Although no group showed a consistent pattern of preferring to spend time in the open or sheltered part of the tank, there were clear differences in the motivation to feed and in the operculum beat rate. In contrast to the control treatments, fishes given either the bee venom or acetic acid had markedly increased operculum beat rates that remained elevated for over 2 h. Similarly, the two noxious stimuli groups showed little interest in food after they received their treatment. Motivation to feed was obvious in the two control groups after approximately 1 h, but the noxious treatment groups took closer to 3 h before they resumed feeding, indicating that noxious stimulation suppressed appetite (Sneddon et al. 2003a).

Fishes treated with the acetic acid solution have been shown to rub their snouts against the hard surfaces such as tank walls or the gravel substrate (Sneddon 2003b; Reilly et al. 2008). This rubbing response, however, was not observed in control fishes (Sneddon et al. 2003a). Rubbing an injured part of the body in humans and other mammals has been proposed to help relieve some of the negative effects of pain (Roveroni et al. 2001). The groups treated with bee venom and acetic acid also displayed a rocking response where the ventral surface of the fishes rested on the tank substrate with the fishes rocking from side to side using their pectoral fins to create the rocking motion (Sneddon et al. 2003a). In a more recent study, Newby and Stevens (2008) injected either 2% or 5% acetic acid into the snouts of rainbow trout that were not anesthetized prior to being given the noxious injection. These fishes were compared to control trout that were only handled and to saline-injected control fishes. The negative effects of the acetic acid were not as severe or prolonged as those reported by Sneddon et al. (2003a). Newby and Stevens (2008) found no decreased motivation to feed and no rocking or rubbing behavior, although more than half of the fishes treated with acetic acid lost equilibrium for several seconds. The authors did, however, observe a highly elevated operculum beat rate in the two acetic acid groups, and the response and duration of this effect were similar to the data reported by Sneddon et al. (2003a). Newby and Stevens (2008) concluded that trout do respond to noxious stimulation, but the ways in which the fishes react may be sensitive to the experimental conditions (i.e., use of anesthesia or not) and may also differ between trout strains.

Changes in ventilation rate associated with a subcutaneous injection of acetic acid solution have also been investigated in several other fish species including common carp and zebrafish (*Danio reiro*) (Reilly et al. 2008). The aversive effects of handling fishes out of water were also reported to induce stress-related physiological changes in species such as rainbow trout, threespined sticklebacks (*G. aculeatus*), and a tropical freshwater fish, the Panamanian bishop (*Brachyrhaphis episcopi*) (Brydges et al. 2009). While there are very few comparative studies explicitly addressing the effects of noxious or adverse treatments, those that have been done have all reported significant differences between species (Dunlop et al. 2006; Reilly et al. 2008; Brydges et al. 2009). Differences in how fishes respond to aversive stimuli should receive more attention in the future. At present, we tend to discuss "pain in fishes" and there is a tendency with this statement to expect all fish species to show similar responses; however, the little evidence that there is to date suggests that this is not the case (Dunlop et al. 2006; Reilly et al. 2008; Braithwaite et al. 2013).

Early experiments using conditioned electric shock avoidance in species such as goldfish demonstrated that some fishes can quickly learn to use the onset of a neutral cue such as a colored light to predict the delivery of an electric shock (Bitterman 1964; Agranoff et al. 1965). The notion that electric shocks are aversive stimuli for species such as trout and goldfish has been shown through electric shocks triggering reflexive Mauthner-initiated startle responses, and also because fishes will learn to avoid locations associated with the delivery of shocks (Gallon 1972; Eaton 1991; Dunlop et al. 2006; Millsopp and Laming 2008). Furthermore, exposure to electric shocks has been shown to increase plasma cortisol levels, but the cortisol effects were variable across fish species (e.g., goldfish and trout), and also variable across different experimental situations within a species (Dunlop et al. 2006). The ability of fishes to learn the association between the electric shock and a particular location has been found to vary with shock intensity in goldfishes (Gallon 1972; Dunlop et al. 2006). Thus, electric shocks have been concluded to be a noxious experience for fishes that can generate reflexive and physiological responses.

Overall, the experiments described in this section have shown that ray-finned, bony fishes exposed to different noxious stimuli express a number of reflexive and physiological changes. The fishes pay attention to the affected area and use solid structures in the local environment against which they rub the affected part of their body (Sneddon et al. 2003a; Reilly et al. 2008). Several of the responses that arise are associated with changes in stress physiology (Dunlop et al. 2006; Huntingford et al. 2006; Galhardo and Oliveira 2009). Treatment with noxious stimuli can also result in changes in motivational state such as a reduction in motivation to feed (Sneddon et al. 2003a).

Taking the different studies described in the first three sections together, there is considerable evidence that fishes have the necessary sensory receptors and nerve fibers to process nociceptive stimuli (Braithwaite and Boulcott 2007; Ashley and Sneddon 2008). All of these processes can occur unconsciously; that is, the brain and higher-order conscious processes do not need to be involved in order for nociceptive functions to work (Rose 2007; Braithwaite 2010). In animals that are sentient and do have higher-order mental processes capable of generating emotions and feelings, nociceptive information can affect behavior (Allen 2011; Braithwaite et al. 2011). Demonstrating conscious awareness of pain in any animal is difficult, but there is a general consensus that many animals do experience the unpleasant hurting sensation that accompanies pain. In the next section, I consider which areas of the brain are involved with higher-order processes that support emotions and sentience, and describe the evidence for these structures and their functions in ray-finned, bony fishes.

10.4 PAIN AND THE BRAIN

It has been argued that the fish brain is too simple to process emotions and feelings associated with pain (Rose 2002, 2007; Cabanac et al. 2009). Rose (2002), who accepts fishes are capable of nociception but does not consider them able to experience pain, argues that fishes lack key neural structures such as extensive frontal and parietal neocortical regions in their brains. Rose claims that the absence of these structures renders the fishes incapable of processing stimuli and interpreting them to be painful (Rose 2002, 2007). The brains of ray-finned, bony fishes are without a doubt less sophisticated than those of mammals, but growing evidence indicates that, both structurally and functionally, the brains of some fish species may support a capacity for sentience (Chandroo et al. 2004a,b; Braithwaite et al. 2013).

The brains of ray-finned, bony fishes have several of the same subdivisions seen in tetrapod vertebrates; a brain stem, the cerebellum, the mesencephalon (midbrain), a pair of optic lobes, the paired cerebral hemispheres of the telencephalon (forebrain), and the associated olfactory bulbs (Northcutt 2002). Specific brain regions, or a subset of these, within the brain of ray-finned, bony fishes have also been shown to process different kinds of information. Determining the functional role of different areas has benefitted considerably from the use of local lesions or ablations (Salas et al. 2006). By determining how the lesion impairs the behavior of the fishes, it has been possible to map specific sites within the brain that have a clearly defined role. For example, lesions to the cerebellum impair specific kinds of associative learning (Rodríguez et al. 2005). And while fishes do not possess a layered neocortex, there are specialized subdivisions within the telencephalon, such as the dorsal lateral region (Dl), that perform a role similar to the hippocampus found in mammalian and avian brains (Rodríguez et al. 2002; Broglio et al. 2010; Dúran et al. 2010). Dl is involved in learning and memory, and it is essential for the fishes to learn spatial and temporal associations (Portavella et al. 2002).

The position of Dl is different to the location of the hippocampus in mammals, but this appears to stem from contrasting processes during the early development of the mammalian and fish brain. In ray-finned, bony fishes, during the early development of the brain, the neural tube everts and pushes outward, whereas in mammals, the neural tube develops by inverting and pulling areas in toward the center of the developing brain (Salas et al. 2006). Such different developmental processes lead to key brain areas being in strikingly different locations. Such findings suggest that Rose's conclusion that fishes cannot experience the emotional aspects of pain because they are missing certain neural structures may not necessarily be justified. With such different developmental processes, it is difficult to know which parts of fish and mammalian brains are functionally homologous to one another. Brain lesion work using goldfishes has begun to provide a number of answers, but this is just a starting point, and we now need further studies using similar techniques and involving more fish species to get a better understanding of how the fish brain works and what kinds of cognitive and behavioral capacities it supports.

A further significant difference between mammalian and ray-finned, bony fish brains is found in the region next to Dl, the dorsomedial area (Dm). This has been found to be functionally similar to the mammalian amygdala (Portavella et al. 2002, 2004). The amygdala is involved in processing basic emotions such as fear, and ablation of Dm impairs fear conditioning in goldfishes (Portavella et al. 2004). The presence of Dm does not allow us to conclude that fishes have the emotional capacity for suffering, but its presence provides another piece of evidence that the fish brain can process more sophisticated kinds of information than Rose (2002, 2007) has indicated. Furthermore, the evidence to date reveals that brain structures with similar functions to the prefrontal structures in mammal brains are being found in fishes (Ito and Yamamoto 2009; Vargas et al. 2009).

In mammals, we find that information passes to the brain through two main tracts. The first conveys information from the head, the trigeminal tract. The second conveys signals from the rest of the body, the spinothalamic tract. Ascending projections then transmit information to the thalamus. Ray-finned, bony fishes have the same general ascending spinal projections and many of the main connections within the brain are similar to those found in terrestrial vertebrates (Goehler and Finger 1996; Finger 2000; Broglio et al. 2003). In mammals, ascending pathways are responsible for conveying pain-related information from the thalamus to the cortex, and then descending pathways relay information back to the thalamus. In fishes, connections between the thalamus and the pallial subdivisions of the telencephalon have been described in zebrafishes (Rink and Wullimann 2004), but it is not yet known whether these are involved with nociceptive activity.

Dunlop and Laming (2005) took electrophysiological recordings in the brains of goldfishes and trout and measured responses that were generated with either a sharp pin prod applied to the skin just behind the operculum (considered a noxious stimulus), or responses to stroking the skin with a brush along the lateral line (considered at non-noxious mechanical stimulation). Recording electrodes in the cerebellum, tectum, and telencephalon of both species detected activity in all brain regions when the pin or the brush stimuli were applied. In the goldfishes, the response frequencies to stimulation by the pin were significantly larger compared to those generated by the brush

(Dunlop and Laming 2005). The stimulus discrimination data were more variable for the trout, and the response frequencies measured in the telencephalon did not clearly distinguish the stimulus type (pin or brush). However, it was noted that when the pin or brush was applied to the contralateral side of the body, then, as with goldfishes, greater response frequencies were recorded in the brain when the trout were given the noxious pin prod (Dunlop and Laming 2005). Activity recorded in the telencephalon in response to the pin stimulus indicates that a nociceptive pathway does exist that takes information from the periphery to the telencephalon where higher-order brain functions (such as spatial cognition) are processed. However, studies specifically addressing which regions within the fish telencephalon process nociceptive information are now needed to further refine our understanding of what happens to these stimuli.

Summarizing the various examples highlighted in this section, ray-finned, bony fishes have a vertebrate brain with specialized regions, and several of these have been found to have similar functional roles to structures described in mammalian brains. The position of different structures in the fish brain appears to be very different in comparison to mammals, but contrasting processes in the early development of fish and mammalian brains explain these differences. Nociceptive information in fishes has been found to pass to the brain and reaches the telencephalon, where functionally similar regions to the mammalian hippocampus and amygdala have been described. Whether this information is processed in a way that leads to the negative feelings of hurt and suffering will now be discussed.

10.5   HIGHER-ORDER BEHAVIORAL RESPONSES TO NOXIOUS STIMULATION

Over the last decade, attempts have been made to demonstrate higher-order cognitive capacities that provide evidence for fishes having forms of mental awareness. To have some understanding of the hurt and aversive feelings that accompany pain, and so to be able to suffer from pain, requires that animals have a capacity for mental awareness. Several reviews have argued that certain species of fish have such mental capacities (Braithwaite and Huntingford 2004; Braithwaite 2005; Chandroo et al. 2004a,b; Braithwaite et al. 2013). In this section, I will highlight the kinds of change in cognitive processes that occur when species such as trout and goldfish are noxiously stimulated (Sneddon et al. 2003b; Millsopp and Laming 2008; Ashley et al. 2009), the kinds of information that fishes can learn, and how this can inform us about mental awareness (Nilsson et al. 2008; Nordgreen et al. 2010). However, to learn more about the breadth of cognitive capacities that fishes have, I recommend reading previously published papers (Braithwaite and Huntingford 2004; Braithwaite 2005; Chandroo et al. 2004a,b; Braithwaite et al. 2013).

Sneddon et al. (2003b) investigated the capacity of fishes to suffer by adopting a general paradigm that had previously tested changes in awareness in birds that were experiencing painful ankle joint inflammation (Gentle 2001). Earlier experiments with both mammals and birds have reported a change in sensitivity to pain when animals are put into a novel or fear-inducing environment (Kavaliers and Innes 1988; Gentle and Corr 1995; Gentle and Tilston 1999). It is proposed that the change to the environment can alter the animal's attention with regard to the pain so that it either shows hypoalgesia (a decreased sensitivity to pain) and thus focuses on its altered situation, or if the experience of the pain commands a significant part of the animal's attention, then the animal's response to the altered environment is impaired. Sneddon et al. (2003b) used this idea to investigate how trout given a noxious, acetic acid injection subcutaneously into the snout respond to a novel object (a brightly colored plastic column) placed into the fish's home tank. Trout normally express strong avoidance of novel objects and so under normal circumstances the fish would be expected to keep their distance from the novel column. Indeed, control fishes that were injected with a saline solution did stay away from it. Trout treated with an acetic acid solution, however, were found to move closer to the plastic tower, and thus appeared to have an impaired neophobic response. This was interpreted as the trout treated with acetic acid being distracted by their experience of pain which demanded the attention of the fish, and this decreased attention to the novel object thus impaired the usual avoidance response (Sneddon et al. 2003b).

To test whether the effects of the acetic acid were capturing the attention of the fishes, Sneddon et al. (2003b) repeated the experiment but this time they gave all fish an intramuscular injection of morphine sulfate. Fishes are known to have opiate receptors, and administration of an opioid in trout has been shown to cause decreased sensitivity to noxious mechanical stimulation (Vecino et al. 1992; Chervova et al. 1994). When Sneddon et al. (2003b) introduced the novel object this time, all fishes showed a similar response and stayed away from it. Thus, providing trout with an analgesic (morphine sulfate) apparently allowed the fishes in the acetic acid treatment group to now pay attention to the novel object and avoid it. This observation is similar to an earlier finding by Ehrensing et al. (1982), who showed that introducing morphine directly into the brain of goldfishes decreased their response to electric shock. However, when an opiate antagonist was used, the response to the electric shock was resumed. Although the study by Ehrensing et al. (1982) appeared to show an opioid generating an analgesic effect for a noxious electric shock, the results could also be explained by the morphine-reducing reactions to any kind of stimulus regardless of whether it is connected with pain. The study by Sneddon et al. (2003b), however, cannot be explained in this way because in their study the morphine revived the novel object avoidance behavior in the fishes, that is, treating the acetic acid treatment group with morphine allowed these fishes to actively move away from and stay away from the novel object.

Using a related approach, Ashley et al. (2009) monitored antipredator responses to determine how these may be modified in fishes given a noxious acetic acid injection. Their experiment assessed whether fishes would pay attention to their experience of noxious stimulation under the threatening situation of a predator's presence. Juvenile trout were monitored to obtain pretreatment measures of their operculum beat rate, their general activity, time spent swimming, and their use of a covered area of the tank. The fishes were then anesthetized and given an injection of either saline (control) or acetic acid (noxious treatment) into the snout. After a short recovery period, the fishes were then exposed to a predator stimulus (a pheromone extract made from the damaged skin of dead trout). Odor cues from damaged conspecifics have previously been shown to act as a trigger for antipredator behaviors (Mirza and Chivers 2003). The trout were then monitored at 30-min intervals over the next 3 h. The addition of the predator odor cues increased general activity in the saline-treated control trout, but it did not affect the acetic acid–treated fishes. Similarly, the time spent swimming was higher in control fishes compared to that in the acetic acid–treated fishes. The operculum beat rate was highly elevated in the noxiously treated acetic acid fishes compared to the controls. Finally, the use of the covered area changed in response to the addition of the predator cue; saline-injected fishes increased their use of cover, whereas the acetic acid–injected fishes either spent more time out of the cover or did not change their position at all. More detailed analysis revealed that it was bolder individuals treated with acetic acid that were more likely to move out into the open when the predator cue was introduced, but more timid individuals showed little change in their position with respect to cover. Interpreting these different results, Ashley et al. (2009) suggested that control fishes respond to the predator cue by trying to escape; they increased their swimming activity and moved into the sheltered area of the tank (Magurran et al. 1996). In contrast, the noxiously treated fishes often spent more time out in the open and were generally less likely to swim. Thus, similar to the effects reported by Sneddon et al. (2003b), it seems that the experience of noxious stimulation affects the ability of the fishes to pay appropriate attention to a threatening stimulus.

In a different kind of study, the decisions that trout and goldfishes make about how much to avoid exposure to a noxious stimulus (an electric shock) were found to vary under different kinds of social context (Dunlop et al. 2006). First, fishes were trained to avoid certain places in a 1.5-m-long tank; if they swam into these they experienced electric shocks. The tank was divided into quarters; two of these in one half of the tank never received an electric shock and could thus be regarded as "safe." In the next quarter, a low-intensity shock (3 V for goldfishes, 2.5 V for trout) was delivered at a specific rate, whereas in the final quarter a higher voltage was given (30 V for goldfishes and 25 V for trout).

The fishes had to learn that one half of the tank was unshocked and safe to spend time in, but if the fishes swam into the other half of the tank, they would receive low-voltage shocks, and if they continued further into the final quarter, they would receive higher-voltage shocks. Both species learned to avoid the electric shocks over the course of 2 days of trials, and were equally likely to be in either of the non-shock quarters.

In the next phase of the experiment, a single conspecific fish was placed in the quarter at the end of the tank that had previously been associated with the 30 V (or 25 V for trout) electric shocks. This quarter now no longer received aversive stimuli. Conspecific fish put into this section were restricted to this area by a Plexiglas wall, which allowed the fish to be seen but they could not physically interact. In this way, they acted as a visual social stimulus. The quarter immediately adjacent to the stimulus fish continued to have the low 3 V (or 2.5 V for trout) electric shocks delivered if the test fish moved into this section, and the remaining two quarters in the other part of the tank continued to be shock-free. The response of the test fishes to this new arrangement varied across species. Goldfishes spent most of their time in the unshocked quarter closest to the 3 V shock quarter. Thus, they were as close to the stimulus fish as they could be without receiving electric shocks. Trout, however, moved to be as close as they could to the stimulus fish, even though this exposed them to repeated low-level electric shocks. This series of experiments demonstrate that both goldfishes and trout prefer to avoid exposure to electric shocks, but the motivation to avoid the shocks is altered in the presence of a conspecific. The motivation to be close to another fish was higher in trout and they were prepared to experience repeated low-level electric shocks to be physically close to a companion fish. The study by Dunlop et al. (2006) demonstrates that the response of the fishes to the electric shocks is more than a reflex response; rather the behavior of the fishes appears to be modified by competing motivational demands such as being close to a conspecific. The cognitive mechanisms that affect changes in behavioral choices such as this are not known, but the results demonstrate that the fishes are processing the noxious stimuli in more sophisticated ways than a simple reflex response at the level of the spinal cord. Again, it should be noted that this study reports a species difference in how noxious stimulation alters behavior.

Demonstrating that an animal has some kind of awareness of its own state is difficult. One approach has been to determine when humans need awareness to perform a certain task and then see whether animals can achieve the same kind of task. One example of this is trace conditioning (the association of stimuli across a time gap). In humans, trace conditioning has been shown to be dependent on the hippocampus and the prefrontal cortex and post-task questioning has shown that only subjects that were aware of what the task was were able to form the association properly (Clark and Squire 1998). Rabbits also learn to associate stimuli despite there being an interval between the end of one stimulus and the onset of another, but only rabbits with an intact hippocampus can learn this kind of association (Clark and Squire 1998; Wallenstein et al. 1998). Trace conditioning has also been found to occur in fishes; for example, Nilsson et al. (2008) report that cod (*Gadus morhua*) can learn to associate a flashing light cue with food delivery even with an interstimulus gap of 2 min. Other experiments with goldfishes have reported that for trace conditioning to function, Dl and Dm must be intact (Broglio et al. 2005; Vargas et al. 2009). As we saw from some of the examples in the previous section, similar mental processes can occur in different species even though the underlying general structure of the brains is different. Thus, it has been argued that if trace conditioning only works when an animal is aware of the task, then the experiments described here indicate that fishes can form explicit memories and are aware of what they have remembered.

Taking this kind of learning one stage further with trout, Nordgreen et al. (2010) explored whether trout could be manipulated to alter the explicit memory of the trace conditioned task and so subsequently change the associative relationship. Specifically, they wanted to know if they could change a learned conditioned association between two stimuli by devaluing one of the stimuli in a separate situation and then observing whether this devaluation transferred back to the original task. If it did, this could be used to argue that the fishes have an explicit (declarative) mental representation of the stimuli in the association (Nordgreen et al. 2010). To investigate this, the authors trained trout to associate a green light switching on and then off with the delivery of a food reward. When the light was turned off, the food was delivered about 3.5 s later; thus, the trout learned to approach the feeder after the light switched off. The trace interval was relatively short, and it only took a few trials with the light followed by a short break and then food before almost all of the fishes had learned the trace conditioning task. After this initial training, the food (reward stimulus) was devalued in a series of separate trials. For these no light was used, but food pellets were now dropped into the tank from the feeder; however, if the fishes approached the feeder within 15 s of food being released, they received an electric shock. By moving away from the feeder in these trials, the fishes could escape exposure to the electric shocks. After 2 days of training with the food as a devalued stimulus (approximately 56 presentations of food paired with shock if the fishes approached the feeder), the fishes resumed the initial task. The green light was switched on to determine whether the fishes would approach the feeder, even though over the past 2 days the delivery of food had been aversive. Nordgreen et al. (2010) found that the aversive training with the feeder did indeed devalue the perceived value of food in the trace conditioning task. Prior to the devaluation, the trout approached the feeder and consumed the food within about 4 s. After the devaluation, this latency doubled so that once the green light switched off, it was closer to 8 s before the fishes fed. Control fishes that were trained in the same initial trace conditioning task, but never experienced the devaluation, continued to perform consistently throughout the experiment, feeding on the food about 4 s after it was released from the feeder (Nordgreen et al. 2010). The authors concluded that the trout had to have learned the association between the green light and the food, by explicitly understanding that the green light would lead to the delivery of food. This was not an unconsciously learned association, but rather one where the fishes were aware of the relationship between the light and the food reward. If the association had been acquired in an unconscious, reflex-like, way then the devaluation process should not have affected the response to the green light.

The different experiments described in this section suggest that fishes can form explicit memories of events, and that they understand the relationship between different stimuli presented in trace conditioning tasks. This cognitive capacity is important because without it, it is hard to argue that fishes have an understanding of the negative aspects of pain. Other examples showed that the experience of pain modulates the decisions fishes choose to make in different situations. For example, the experience of a noxious, acetic acid injection into the snout impaired the normal avoidance responses to something novel and potentially threatening. However, when fishes were given morphine sulfate as an analgesic, the normal, adaptive, avoidance response resumed. Together, these different kinds of empirical studies indicate that fishes have an awareness of noxious experiences.

10.6 ANALGESICS FOR FISHES

Given how recent research addressing fish pain has been, it is perhaps unsurprising that there is as yet little work addressing fish pain relief (Neiffer and Stamper 2009). Morphine is a standard form of pain relief against which other potential analgesics are assessed, and morphine has been demonstrated to have analgesic effects in a few different species of fishes (Sneddon et al. 2003b; Newby et al. 2009; Nordgreen et al. 2009b). Some research has addressed the pharmacokinetics of morphine. Newby et al. (2009) found that when morphine was delivered through the water surrounding goldfishes, uptake was very slow, and the authors suggested that this took too long to be an effective delivery technique. Other studies have looked at the effects of intramuscular injections of morphine in goldfishes and Atlantic salmon (Nordgreen et al. 2009b). When delivered in this way, the morphine increased quickly in the blood plasma, although the authors noted considerable individual variation in both species. While the degree of morphine metabolism could not be accurately calculated, it was considered to be relatively slow (Nordgreen et al. 2009b). The half-life of plasma morphine was similar in both goldfishes (12.5 h) and salmon (13.5 h).

Other pharmacokinetic work has compared morphine administered to winter flounder (*Pseudopleuronectes americanus*) and seawater-acclimated rainbow trout through either an intraperitoneal injection or an intravenous injection (Newby et al. 2006). The tests revealed differences depending on the delivery technique; there were also significant between-species differences, with winter flounder showing a slower disposition of morphine compared to the trout even though the fishes were held in similar temperatures. Newby et al. (2006) also reported that the disposition of morphine in fishes was in general considerably slower than that previously described in mammals.

A study with koi carp (*C. carpio*) ran a comparative assay to test the effects of saline (control) against ketoprofen (a nonsteroidal anti-inflammatory analgesic) and butorphanol (an opioid analgesic) when the fishes were given an incision into the body cavity (Harms et al. 2005). In comparison to saline or ketoprofen, only carp treated with butorphanol appeared to have any analgesic benefit in that their operculum beat rate and swimming behavior was the same both before and after surgery. A more recent comparative experiment in rainbow trout investigated the effects of the opioid buprenorphine, and a nonsteroidal anti-inflammatory drug (NSAID), carprofen (Mettam et al. 2012b). The buprenorphine had relatively little effect in terms of relieving pain associated with 0.1% acetic acid injection in the snout. The carprofen had more of an effect and fishes treated with this drug resumed feeding more rapidly. Mettam et al. (2012b) also investigated the effects of a local anesthetic, lidocaine, applied to the snout at the same time as they were injected with saline or acetic acid. The result of the local anesthetic was positive in that it helped to reduce several pain-related responses (e.g., swimming activity and operculum beat rate).

In comparison to other areas associated with fish pain research, comparative studies of the effects of anesthesia and analgesia are only just beginning (Mettam et al. 2012b; Zahl et al. 2012). The early studies within this area indicate there is still a lot to be learned and the study of different analgesic drugs and their effects on different species of fishes is likely to receive growing attention in the next few years.

10.7   GENERAL CONCLUSIONS

This review has summarized what we currently know and understand about pain in fishes. It has highlighted how similar the nociceptive system of ray-finned, bony fishes is to that found in birds and mammals, although there are some differences such as the lack of cold nociceptors and also the many fewer C fibers in fishes. What these differences signify is not yet clear. How other fish taxa such as the agnathans and elasmobranchs process noxious stimuli has also yet to be resolved.

The review also highlighted physiological changes that accompany pain responses in ray-finned, bony fishes and recognized that many of these effects are likely to occur without the fishes being directly aware of them. When and how fishes do have the cognitive capacity for recognizing and changing behavior in response to pain states and potential suffering was also reviewed. Current evidence indicates that some species of fish do create explicit mental representations of events, and as such they could be considered to have the capacity for suffering from the negative experiences associated with pain. There are still numerous challenges that need to be worked through. Our current understanding of species-specific responses to pain is poor. Similarly, we have limited knowledge of drugs that provide good analgesia for fishes, and studies quantifying the pharmacokinetics of pain-relieving drugs that induce appropriate levels of anesthesia in different fish species are now required. Despite these gaps in our understanding, the last decade has seen considerable progress in our understanding of fish nociception, and how pain induces changes to behaviors and affects decision making.

REFERENCES

Agranoff, B. W., Davis, R. E., and J. J. Brink. 1965. Memory fixation in the goldfish. *Proceedings of the National Academy of Sciences* 54: 788–793.
Allen, C. 2011. Fish cognition and consciousness. *Journal of Agricultural and Environmental Ethics* doi: 10.1007/s10806-011-9364-9.

Ashley, P. J., Ringrose, S., Edwards, K. L., Wallington, E., McCrohan, C. R., and L. U. Sneddon. 2009. Effects of noxious stimulation upon antipredator responses and dominance status in rainbow trout. *Animal Behaviour* 77: 403–410.

Ashley, P. J. and L. U. Sneddon. 2008. Pain and fear in fish. In *Fish Welfare*, ed. E. Branson, pp. 49–77. Wiley Blackwell Publishing, London, U.K.

Ashley, P. J., Sneddon, L. U., and C. R. McCrohan. 2006. Properties of corneal receptors in a teleost fish. *Neuroscience Letters* 410: 165–168.

Ashley, P. J., Sneddon, L. U., and C. R. McCrohan. 2007. Nociception in fish: Stimulus-response properties of receptors on head of trout (*Oncorhynchus mykiss*). *Brain Research* 1166: 47–54.

Barreto, R. E. and G. L. Volpato. 2004. Caution for using ventilatory frequency as an indicator of stress in fish. *Behavioural Processes* 66: 43–51.

Barton, B. A. 2002. Stress in fishes: A diversity of responses with particular reference to changes in circulating corticosteroids. *Integrative and Comparative Biology* 42: 517–525.

Bateson, P. 1991. Assessment of pain in animals. *Animal Behaviour* 42: 827–839.

Bitterman, M. E. 1964. Classical conditioning in the goldfish as a function of the CS-US interval. *Journal of Comparative & Physiological Psychology* 58: 356–366.

Braithwaite, V. A. 2005. Cognition in fish. *Behaviour and Physiology of Fish* 24: 1–37.

Braithwaite, V. A. 2010. *Do Fish Feel Pain?* Oxford University Press, Oxford, U.K.

Braithwaite, V. A. and P. Boulcott. 2007. Pain perception and fear in fish. *Diseases in Aquatic Organisms* 75: 131–138.

Braithwaite, V. A. and P. Boulcott. 2008. Can fish suffer? In *Fish Welfare*, ed. E. Branson, pp. 78–92. Wiley Blackwell Publishing, London, U.K.

Braithwaite, V. A. and Huntingford, F. A. 2004. Fish and welfare: Can fish perceive pain and suffering? *Animal Welfare* 13: S87–S92.

Braithwaite, V. A., Huntingford, F. A., and R. van den Bos. 2013. Variation in emotion and cognition among fishes. *Journal of Agricultural and Environmental Ethics* 26: 7–23.

Broglio, C., Gómez, A., Durán, E., Ocaña, F. M., Jiménez-Moya, F., Rodríguez, F., and C. Salas. 2005. Hallmarks of a common forebrain vertebrate plan: Specialized pallial areas for spatial, temporal and emotional memory in actinopterygian fish. *Brain Research Bulletin* 66: 277–281.

Broglio, C., Rodríguez, F., Gómez, A., Arias, J. L., and C. Salas. 2010. Selective involvement of the goldfish lateral pallium in spatial memory. *Behavioural Brain Research* 210: 191–201.

Broglio, C., Rodríguez, F., and C. Salas. 2003. Spatial cognition and its neural basis in teleost fishes. *Fish and Fisheries* 4: 247–255.

Broom, D. M. 2001. The evolution of pain. *Vlaams Diergeneeskundig Tijdschrift* 70: 17–21.

Brown, C., Gardner, C., and V. A. Braithwaite. 2005. Differential stress responses in fish from areas of high- and low predation pressure. *Journal of Comparative Physiology B* 175: 305–312.

Brydges, N. M., Boulcott, P., Ellis, T., and V. A. Braithwaite. 2009. Quantifying stress responses induced by different handling methods in three species of fish. *Applied Animal Behaviour Science* 116: 295–301.

Cabanac, M., Cabanac, A. J. and A. Parent 2009. The emergence of consciousness in phylogeny. *Behavioral Brain Research* 198: 267–272.

Cain, D. M., Khasabov, S. G., and D. A. Simone. 2001. Response properties of mechanoreceptors and nociceptors in mouse glabrous skin: An in vivo study. *Journal of Neurophysiology* 85: 1561–1574.

Cameron, A. A., Plenderleith, M. B., and Snow, P. J. 1990. Organization of the spinal cord in four species of elasmobranch fish: Cytoarchitecture and distribution of serotonin and selected neuropeptides. *Journal of Comparative Neurology* 297: 201–208.

Chandroo, K. P., Duncan, I. J. H., and R. D. Moccia. 2004a. Can fish suffer? Perspectives on sentience, pain, fear and stress. *Applied Animal Behaviour Science* 86: 225–250.

Chandroo, K. P., Yue, S., and R. D. Moccia. 2004b. An evaluation of current perspectives on consciousness and pain in fishes. *Fish and Fisheries* 5: 1–15.

Chervova, L. S., Lapshin, D. N., and A. A. Kamenskii. 1994. Pain sensitivity of trout and analgesia induced by intranasal administration of demorphine. *Doklady Biologikal Sciences* 338: 424–425.

Clark, R. E. and L. R. Squire. 1998. Classical conditioning and brain systems: The role of awareness. *Science* 280: 77–81.

Coggeshall, R. E., Leonard, R. B., Applebaum, M. L., and W. D. Willis. 1978. Organisation of peripheral nerves of the Atlantic stingray *Dasyatis sabina. Journal of Neurophysiology* 41: 97–107.

Dubin, A. E. and A. Patapoutian. 2010. Nociceptors: The sensors of the pain pathway. *Journal of Clinical Investigation* 120: 3760–3772.

Dunlop, R. and P. Laming. 2005. Mechanoreceptive and nociceptive responses in the central nervous system of goldfish (*Carassius auratus*) and trout (*Oncorhynchus mykiss*). *The Journal of Pain* 6: 561–568.

Dunlop, R., Millsopp, S., and P. Laming. 2006. Avoidance learning in goldfish (*Carassius auratus*) and trout (*Oncorhynchus mykiss*) and implications for pain perception. *Applied Animal Behaviour Science* 97: 255–271.

Dúran, E., Ocana, F. M., Broglio, C., Rodriguez, F. and C. Salas 2010. Lateral but not medial telencephalic pallium ablation impairs the use of goldfish spatial allocentric strategies in a "hole-board" task. *Behavioural Brain Research* 214: 480–487.

Eaton, R. C. 1991. Neuroethology of the Mauthner system. *Brain Behavior and Evolution* 37: 245–332.

Ehrensing, R. H., Michell, G. F., and A. J. Kastin. 1982. Similar antagonism of morphine analgesia by MIF-1 and naxolone in *Carassius auratus*. *Pharmacology Biochemistry and Behavior* 17: 757–761.

Finger, T. E. 2000. Ascending spinal systems in the fish *Prionotus carolinus*. *Journal of Comparative Neurology* 422: 106–122.

Galhardo, L. and R. F. Oliveira. 2009. Psychological stress and welfare in fish. *Annual Review of Biomedical Sciences* 11: 1–20.

Gallon, R. 1972. Effects of shock intensity on shuttlebox avoidance conditioning in goldfish. *Psychological Reports* 31: 855–858.

Gentle, M. J. 1989. Cutaneous sensory afferents recorded from the nervus intramandibularis of *Gallus gallus var domesticus*. *Journal of Comparative Physiology A* 164: 763–774.

Gentle, M. J. 2001. Attentional shifts alter pain perception in the chicken. *Animal Welfare* 10: S187–S194.

Gentle, M. J. and S. A. Corr. 1995. Endogenous analgesia in the chicken. *Neuroscience Letters* 201: 211–214.

Gentle, M. J. and V. L. Tilston. 1999. Reduction in peripheral inflammation by changes in attention. *Physiology and Behavior* 66: 289–292.

Gentle, M. J., Tilston, V., and D. E. F. McKeegan. 2001. Mechanothermal nociceptors in the scaly skin of the chicken leg. *Neuroscience* 106: 643–652.

Goehler, L. E. and T. E. Finger. 1996. Visceral afferent and efferent columns in the spinal cord of the teleost, *Ictalurus punctatus*. *Journal of Comparative Neurology* 371: 437–447.

Harms, C. A., Lewbart, G. A., Swanson, C. R., Kishimori, J. M., and S. M. Boylan. 2005. Behavioral and clinical pathology changes in koi carp (*Cyprinus carpio*) subjected to anesthesia and surgery with and without intra-operative analgesics. *Comparative Medicine* 55: 221–226.

Huntingford, F. A., Adams, C. E., Braithwaite, V. A., Kadri, S., Pottinger, T. G., Sandoe, P., and J. F. Turnbull. 2006. Current understanding on fish welfare: A broad overview. *Journal of Fish Biology* 68: 332–372.

Hwang, R. Y., Zhong, L., Xu, Y., Johnson, T., Zhang, F., Deisseroth, K., and W. D. Tracey. 2007. Nociceptive neurons protect Drosophila larvae from parasitoid wasps. *Current Biology* 17: 2105–1116.

IASP (last update May 2012). Taxonomy http://www.iasp-pain.org/Content/NavigationMenu/GeneralResourceLinks/PainDefinitions/default.htm

Iggo, A. 1984. *Pain in Animals*. Universities Federation for Animal Welfare, Hertfordshire, U.K.

Ito, H. and N. Yamamoto. 2009. Non-laminar cerebral cortex in teleost fishes? *Biology Letters* 5: 117–121.

Kavaliers, M. 1998. Evolutionary aspects of the neuromodulation of nociceptive behaviors. *American Zoologist* 29: 1345–1353.

Kavaliers, M. and D. G. L. Innes. 1988. Novelty-induced opioid analgesia in deer mice (*Peromyscus maniculatus*): Sex and population differences *Behavioral and Neural Biology* 49: 54–60.

Kim, S. E., Coste, B., Chadha, A., Cook, B., and A. Patapoutian. 2012. The role of *Drosophila* Piezo in mechanical nociception. *Nature* 483: 209–212.

Laming, P. and G. E. Savage. 1980. Physiological changes observed in the goldfish (*Carassius auratus*) during behavioural arousal and fright. *Behavioral and Neural Biology* 29: 255–275.

Leem, J. W., Willis, W. D., and J. M. Chung. 1993. Cutaneous sensory receptors in the rat foot. *Journal of Neurophysiology* 69: 1684–1699.

Leonard, R. B. 1985. Primary afferent receptive field properties and neurotransmitter candidates in a vertebrate lacking unmyelinated fibres. *Progress in Clinical Research* 176: 135–145.

López de Armentia, M., Cabanes, C., and C. Belmonte. 2000. Electrophysiological properties of identified trigeminal ganglion neurons innervating the cornea of the mouse. *Neuroscience* 101: 1109–1115.

Lucas, M., Johnstone, A. D. F., and I. G. Priede. 1993. Use of physiological telemetry as a method of estimating metabolism of fish in the natural environment. *Transactions of the American Fisheries Society* 122: 822–833.

Lynn, B. 1994. The fibre composition of cutaneous nerves and the classification and response properties of cutaneous afferents, with particular reference to nociception. *Pain Review* 1: 172–183.

Magurran, A. E., Irving, P. W., and P. A. Henderson. 1996. Is there a fish alarm pheromone? A wild study and critique. *Proceedings of the Royal Society of London, Series B* 263: 1551–1556.

Martin, A. R. and W. O. Wickelgren. 1971. Sensory cells in the spinal cord of the sea lamprey. *Journal of Physiology* 212: 65–83.

Mathews, G. and W. O. Wickelgren. 1978. Trigeminal sensory neurons of the sea lamprey. *Journal of Comparative Physiology* 123: 329–333.

Mettam, J. J., McCrohan, C. R., and L. U. Sneddon. 2012a. Characterisation of chemosensory trigeminal receptors in the rainbow trout, *Oncorhynchus mykiss*: Responses to chemical irritants and carbon dioxide. *Journal of Experimental Biology* 215: 685–693.

Mettam, J. J., Oulton, L. J., McCrohan, C. R., and Sneddon, L. U. 2012b. The efficacy of three types of analgesic drugs in reducing pain in the rainbow trout *Oncorhynchus mykiss*. *Applied Animal Behaviour Science* 133: 265–274.

Millsopp, S. and P. Laming. 2008. Trade-offs between feeding and shock-avoidance in goldfish (*Carassius auratus*). *Applied Animal Behaviour Science* 113: 247–254.

Mirza, R. S. and D. P. Chivers. 2003. Response of juvenile rainbow trout to varying concentrations of chemical alarm cue: Response thresholds and survival during encounters with predators. *Canadian Journal of Zoology* 81: 88–95.

Neiffer, D. L. and M. A. Stamper. 2009. Fish sedation, anaesthesia, analgesia and euthanasia: Considerations, methods, and types of drugs. *Institute for Laboratory Animal Research Journal* 50: 343–360.

Newby, N. C., Mendonça, P. C., Gamperl, K., and E. D. Stevens. 2006. Pharmacokinetics of morphine in fish: Winter flounder (*Pseudopleuronectes americanus*) and seawater-acclimated rainbow trout (*Oncorhynchus mykiss*). *Comparative Biochemistry and Physiology Part C* 143: 275–283.

Newby, N. C. and E. D. Stevens. 2008. The effects of the acetic acid "pain" test on feeding, swimming, and respiratory responses of rainbow trout (*Oncorhynchus mykiss*). *Applied Animal Behaviour Science* 114: 260–269.

Newby, N. C., Wilkie, M. P., and E. D. Stevens. 2009. Morphine uptake, disposition, and analgesic efficacy in the common goldfish (*Carassius auratus*). *Canadian Journal of Zoology* 87: 388–399.

Nilsson, J., Kristiansen, T. S., Fosseidengen, J. E., Ferno, A., and R. van den Bos. 2008. Learning in cod (*Gadus morhua*): Long trace interval retention. *Animal Cognition* 11: 215–222.

Nordgreen, J., Garner, J. P., Janczak, A. M., Ranheim, B., Muir, W. M., and T. E. Horsberg. 2009a. Thermonociception in fish: Effects of two different doses of morphine on thermal threshold and post test behaviour in goldfish (*Carassius auratus*). *Applied Animal Behaviour Science* 119: 101–107.

Nordgreen, J., Horsberg, T. E., Ranheim, B., and A. C. N. Chen. 2007. Somatosensory evoked potentials in the telencephalon of Atlantic salmon (*Salmo salar*) following galvanic stimulation of the tail. *Journal of Comparative Physiology A* 193: 1235–1242.

Nordgreen, J., Janczak, A. M., Hovland, A. L., Ranheim, B., and T. E. Horsberg. 2010. Trace classical conditioning in rainbow trout (*Oncorhynchus mykiss*): What do they learn? *Animal Cognition* 13: 303–309.

Nordgreen, J., Kolsrud, M. M., Ranheim, B., and T. E. Horsberg. 2009b. Pharmacokinetics of morphine after intramuscular injection in common goldfish *Carassius auratus* and Atlantic salmon *Salmo salar*. *Diseases in Aquatic Organisms* 88: 55–63.

Northcutt, R. G. 2002. Understanding vertebrate brain evolution. *Integrated and Comparative Biology* 42: 743–756.

Paul, E. S., Harding, E. J., and M. Mendl. 2005. Measuring emotional processes in animals: The utility approach. *Neuroscience and Biobehavioral Reviews* 29: 469–491.

Portavella, M., Torres, B., and C. Salas. 2004. Avoidance response in goldfish: Emotional and temporal involvement of medial and lateral telencephalic pallium. *Journal of Neuroscience* 24: 2335–2342.

Portavella, M., Vargas, J. P., Torres, B., and C. Salas. 2002. The effects of telencephalic pallial lesions on spatial, temporal, and emotional learning in goldfish. *Brain Research Bulletin* 57: 397–399.

Reilly, S. C., Quinn, J. P., Cossins, A. R., and L. U. Sneddon. 2008. Behavioral analysis of a nociceptive event in fish: Comparisons between three species demonstrate specific responses. *Applied Animal Behaviour Science* 114: 248–249.

Rink, E. and M. F. Wullimann 2004. Connections of the ventral telencephalon (subpallium) in the zebrafish (*Danio rerio*). *Brain Research* 1011: 204–220.

Rodríguez, F., Durán, E., Gómez, A., Ocaña, F. M., Álvarez, E., Jiménez-Moya, F., Broglio, C., and C. Salas. 2005. Cognitive and emotional functions of the teleost fish cerebellum. *Brain Research Bulletin* 66: 365–370.

Rodríguez, F., Lopez, J. C., Vargas, J. P., Gómez, Y., Broglio, C., and C. Salas. 2002. Conservation of spatial memory function in the pallial forebrain of reptiles and ray-finned fish. *Journal of Neuroscience* 22: 2894–2903.

Rolls, E. 2007. *Emotions Explained*. Oxford University Press, Oxford, U.K.

Roques, J. A. C., Abbink, W., Geurds, F., van de Vis, H., and G. Flik. 2010. Tailfin clipping, a painful procedure: Studies on Nile tilapia and common carp. *Physiology and Behavior* 101: 533–540.

Rose, J. D. 2002. The neurobehavioral nature of fishes and the question of awareness and pain. *Reviews in Fisheries Science* 10: 1–38.

Rose, J. D. 2007. Anthropomorphism and "mental welfare" of fishes. *Diseases of Aquatic Organisms* 75: 139–154.

Roveroni, R. C., Parada, C. A., Cecilia, M., Veiga, F. A., and C. H. Tambeli. 2001. Development of a behavioral model of TMJ pain in rats: The TMJ formalin test. *Pain* 94: 185–191.

Salas, C., Broglio, C., Durán, E., Gómez, A., Ocaña, F. M., Jiménez-Moya, F., and F. Rodríguez. 2006. Neuropsychology of learning and memory in teleost fish. *Zebrafish* 3: 157–171.

Sneddon, L. U. 2002. Anatomical and electrophysiological analysis of the trigeminal nerve in a teleost fish, *Oncorhynchus mykiss*. *Neuroscience Letters* 319: 167–171.

Sneddon, L. U. 2003a. Trigeminal somatosensory innervation of the head of the teleost fish with particular reference to nociception. *Brain Research* 972: 44–52.

Sneddon, L. U. 2003b. The evidence for pain in fish: The use of morphine as an analgesic. *Applied Animal Behaviour Science* 83: 153–162.

Sneddon, L. U. 2004. Evolution of nociception in vertebrates: Comparative analysis of lower vertebrates. *Brain Research Reviews* 46: 123–130.

Sneddon, L. U., Braithwaite, V. A., and M. J. Gentle. 2003a. Do fish have nociceptors: Evidence for the evolution of a vertebrate sensory system. *Proceedings of the Royal Society* 270: 1115–1121.

Sneddon, L. U., Braithwaite, V. A., and M. J. Gentle. 2003b. Novel object test: Examining nociception and fear in the rainbow trout. *Journal of Pain* 4: 431–440.

Snow, P. J., Plenderleith, M. B., and L. L. Wright. 1993. Quantitative study of primary sensory neurone populations of three species of elasmobranch fish. *Journal of Comparative Neurology* 334: 97–103.

Snow, P. J., Renshaw, G. M. C., and K. E. Hamlin. 1996. Localization of enkephalin immunoreactivity in the spinal cord of the long-tailed ray *Himantura fai*. *Journal of Comparative Neurology* 367: 264–273.

Sumpter, J. P. 1997. The endocrinology of stress. In: *Fish Stress and Health in Aquaculture*, eds. Iwama, G. K., Pickering, A. D., Sumpter, J. P., and C. B. Schreck. Cambridge University Press, Cambridge, U.K.

Tracey, W. D., Wilson, R. I., Laurent, G., and S. Benzer. 2003. Painless, a Drosophila gene essential for nociception. *Cell* 113: 261–273.

Vargas, J. P., López, J. C., and M. Portavella. 2009. What are the functions of fish brain pallium? *Brain Research Bulletin* 79: 436–440.

Vecino, E., Pinuela, C., Arévalo, R., Lara, J., Alonso, J. R., and J. Aijón. 1992. Distribution of enkephalin like immunoreactivity in the central nervous system of the rainbow trout: An immunocytochemical study. *Journal of Anatomy* 180: 435–453.

Wallenstein, G. V., Eichenbaum, H., and M. E. Hasselmo. 1998. The hippocampus as an associator of discontiguous events. *Trends in Neuroscience* 21: 317–323.

Weary, D. M., Niel, L., Flower, F. C., and D. Fraser. 2006. Identifying and presenting pain in animals. *Applied Animal Behaviour Science* 100: 64–76.

Whitear, M. 1971. The free nerve endings in fish epidermis. *Journal of Zoology* 163: 231–236.

Whitear, M. 1983. The question of free nerve endings in the epidermis of lower vertebrates, *Acta Biologica Hungarica* 34: 303–319.

Yeomans, D. C. and H. K. Proudfit. 1996. Nociceptive responses to high and low rates of noxious cutaneous heating are mediated by different nociceptors in the rat: Electrophysiological evidence. *Pain* 68: 141–150.

Zahl, I. H., Samuelsen, O., and A. Kiessling. 2012. Anaesthesia of farmed fish: Implications for welfare. *Fish Physiology and Biochemistry* 38: 201–218.

Zhong, L., Hwang, R. Y., and W. D. Tracey. 2010. Pickpocket is a DEG/ENaC protein required for mechanical nociception in Drosophila larvae. *Current Biology* 20: 429–434.

11 Chemoreception

Warren W. Green and Barbara S. Zielinski

CONTENTS

11.1 INTRODUCTION

For fishes, the surrounding chemicals of their environment provide information for feeding, defense, migration, and reproduction. Fishes have evolved strategies for sensing various compounds, and for conveying this information to specific brain regions for generating an appropriate response. The chemical structure of these molecules is diverse, and the responses range from attraction to avoidance, as well as various context-dependent responses.

In fishes, as in other vertebrates, the cells responding to chemical signals are exposed to the external environment. The cellular response is communicated to the brain, where neuronal integration ensues, and motor output is targeted to specific muscle groups. A central question in the study of the chemical senses relates to how chemical information is encoded peripherally and centrally, and subsequently how this leads to a behavioral response. There are two main chemosensory systems—taste and smell (gustation and olfaction)—as well as the common chemical sense, which conveys information regarding chemical irritants in the environment. For all, the coding of chemical information starts with the sensory cells, and continues in the brain, where neuronal groups are stimulated. For gustation and olfaction, there are examples of segregated processing, where chemicals activate responses along specific neural pathways, and responses are generated in the context of the original stimulus.

The purpose of this chapter is to review the current understanding of gustation and olfaction in fishes: the chemostimulatory compounds; how chemical information is encoded by sensory cells; and how this is transmitted and processed in the brain. Some aspects of chemoreception, including detailed discussion of odorants and tastants, were covered extensively by Michel (2006). We also introduce some unanswered questions relevant to ongoing studies of the chemical senses of fishes.

11.1.1 What Is the Difference between Gustation and Olfaction?

The fish's world is entirely in water. Odorous molecules diffuse through water, and tasting is not confined to the aqueous environment of the mouth; however, the senses of smell and taste utilize different neural pathways for responding to chemical compounds. For olfaction, chemosensory information is conveyed from the nose to the forebrain (olfactory bulb [OB]) by the olfactory nerve (cranial nerve I); and for taste, the cranial nerves VII (facial), IX (glossopharyngeal), and X (vagal) are used to transmit chemosensory information from epidermal taste buds to the hindbrain. These gustatory pathways are associated with the consumption of food.

When the olfactory system is stimulated, a fish's response can be movement, leading to attraction or aversion. For example, a food source may attract a fish. Olfactory cues in fishes also include pheromones, chemical signaling between individuals of the same species. Examples include attraction to reproductive pheromones, and escape or freezing response to released alarm pheromones. Olfactory cues have also been studied for a long time with regard to homing-in fishes, focusing largely on salmonids (e.g., Hasler et al., 1978). Pheromones utilized during migration and reproduction have been investigated in sea lampreys (e.g., Johnson et al., 2005; Sorensen et al., 2005), European eels (Huertas et al., 2006), goldfish (Sorensen et al., 1995), crucian carp (Olsen et al., 2006), and common carp (Lim and Sorensen, 2011).

11.1.2 What Compounds Stimulate the Chemical Senses?

Chemical compounds that are naturally present in the oceans, lakes, and rivers have benefited the survival and reproduction of fishes. Chemical cues associated with feeding derive from the food items. For example, amino acids indicate a source of protein, and bile acids released in the wastes of a prey may also indicate a food source.

As humans, we think of taste comprising of sweet, bitter, salty, sour, and umami (savory). In the fishes' world, palatability is conveyed by the taste of bile acids, amino acids, nucleotides, small peptides, quaternary ammonium compounds, organic acids, glucuronide conjugates, as well as quinine, strychnine, and tetrodotoxin (Caprio et al., 1993; Yamashita et al., 2006; Hara, 2007, 2011). For amino acids, the response threshold is normally 10^{-7}–10^{-8} M but decreases to 10^{-9} M for channel catfish (Kohbara et al., 1992; Ogawa and Caprio, 2010).

Just as bitter stimulates aversive responses to food in humans, there are aversive taste stimuli for fish. For example, algae may be protected against consumption by fish even when quinine and other alkaloids are tasted (Yamamori et al., 1988; Hara, 2011). In the predatory fish *Ariopsis felis*, the gustatory system mediates an aversive response to a potential prey, it, *Aplysia californica*, (Derby, 2007; Nusnbaum et al., 2012). This mollusk is chemically defended by releasing ink containing an unpalatable pigment originating from red seaweed consumed by the sea hare. This adaptation is also seen for the marine toxins tetrodotoxin, which defends the pufferfish (Tetraodontidae), and saxitoxin, a shellfish toxin originating from algal blooms. Both stimulate specialized gustatory receptors in fishes (Hara, 2011). Thus, gustation is an indicator of palatability, and has an adaptive component for the prey that emits unpalatable chemical signals. This ability to taste these compounds protects the fish from consuming prey and plants containing poisons.

Fish utilize olfaction to drive a variety of biologically important behaviors including feeding, mating, and predator avoidance. Both the olfactory and gustatory systems are involved in feeding, and therefore some amino acid and bile acid gustatory stimuli also stimulate the olfactory system. The olfactory repertoire of fishes is diverse and includes amino acids, bile salts, nucleotides, steroids, prostaglandins, and polyamines.

The fish olfactory system responds to many different L-amino acids which are associated with feeding responses. In cyprinids, gadids, and silurids, feeding behaviors such as biting and snapping are mediated by the lateral olfactory tract and are abolished when this tract is severed, thereby removing odor output from the lateral OB, which processes amino acid stimuli (Doving and Selset, 1980; Nikonov and Caprio, 2001, 2004). In sea lampreys, amino acid responsiveness is limited to basic amino acids (Li, 1994). In salmonids, olfactory imprinting to a home stream is needed for spawning migration and may involve an amino acid compound; however, the odors that are utilized for imprinting are still under investigation (Ueda, 2012). The olfactory system of cyprinids responds to nucleotides such as adenosine triphosphate (ATP) (Kang and Caprio, 1995; Hansen et al., 2003), which may signal fresh food to a predatory fish. In addition, polyamines, which are utilized in a variety of cellular functions and have been shown to be indicative of tissue decomposition, are olfactory stimuli to cyprinids (Rolen et al., 2003).

The olfactory system also responds to pheromones—species-specific signals for innate, intraspecific chemical communication (Wyatt, 2003). Danger is conveyed through alarm pheromones, compounds released by damaged skin cells from fish that have been preyed upon (e.g., Stebell and Vegusdal, 2010). These communicate predation to surrounding conspecifics (e.g., Wisenden, 2000).

Olfactory stimuli such as bile salts (or bile alcohols), prostaglandins, and steroids mediate migration and spawning behaviors in some fishes. Cyprinids communicate during mating by emitting and responding to prostaglandins and 17α, 20-β-dihydroxy-4-pregnen-3-one (Zheng and Stacey, 1997). The reproductive hormones and hormonal metabolites released into the water by conspecifics convey information during mating (Stacey et al., 2003). Some fishes, such as gobiids, respond to steroids such as etiocholanolone and its conjugates, but not to prostaglandins (Murphy et al., 2001; Laframboise and Zielinski, 2011). Lampreys synthesize and release bile alcohols such as petromyzonol sulfate (PZS), and its derivatives, which are utilized for migration and spawning (Li et al., 1995, 2002; Sorensen et al., 2005). A detailed examination of the role of steroids and bile alcohols in the reproductive behaviors of the round goby and the sea lamprey, respectively, is discussed in Sections 11.4.1 and 11.4.2.

Environmental calcium (Ca^{2+}) and sodium (Na^+) also elicit olfactory sensory responses in fish (Hubbard et al., 2000, 2002; Hubbard and Canário, 2007). In fishes that migrate from saltwater to freshwater and vice versa, or inhabit areas that experience shifts in salinity, it is necessary to detect changes in environmental concentrations of ions, such as Ca^{2+}, in order to regulate internal homeostasis of these ions and avoid detrimental osmotic conditions. The gilthead seabream (*Sparus aurata*), which inhabits estuarine waters, exhibits olfactory sensitivity to changes in environmental Ca^{2+} concentrations (Hubbard et al., 2002). Moreover, the goldfish (*Carassius auratus*), which lives in freshwater habitats, possesses a calcium-sensing receptor in the olfactory epithelium and exhibits olfactory sensitivity to changes in waterborne Ca^{2+} concentrations (Hubbard et al., 2002). In this freshwater fish, the observed olfactory sensitivity to Ca^{2+} may play a role in internal calcium homeostasis (Hubbard et al., 2002).

Clearly, fish are well aware of their chemical environment. The compounds that have been described represent diverse chemical structures, and originate from many components of the aquatic environment; yet fish have found a way to sense these, and to respond appropriately.

11.2 GUSTATION

A primary function of the gustatory system is to determine the palatability of food. While the sensory cells are located in epidermal taste buds, gustatory afferent neurons convey the sensory information to the brain, and neural projections within the brain integrate this information to several brain regions so that a coordinated response ensues. In the case of fish, motor responses to gustatory stimulation include mouth activity (e.g., Valentincic and Caprio, 1994). This section examines the function of the cells in the taste buds and in the brain regions receiving gustatory sensory input.

11.2.1 Taste Buds

The gustatory sensory cells (taste receptor cells) occur in clusters called taste buds in epidermal locations, including the oral cavity and lips. However, in silurids and ictalurids, taste buds are also located on the head, barbels, fins, and flanks. Gustatory responses and neuroanatomy have been examined extensively in these fishes, because of this highly developed taste system. Taste buds contain different cell types, based on ultrastructure and likely on function. The taste cells are categorized histologically as tubular or light cells, based on morphology and staining affinity. The ultrastructure and cellular composition of the taste cells have been described in detail by Michel (2006). The taste buds are innervated by nerve endings. Purigenic neurotransmission by ATP is likely important in teleost, elasmobranch, and lamprey taste cells (Kirino et al., 2013). In mammals, the ATP diffuses through channels and binds onto receptors on the nerve fibers, as well as neighboring taste cells, and the cells that released the ATP. This synaptic, paracrine, and autocrine regulation serves to amplify a chemosensory response. In mammalian taste buds, released serotonin inhibits receptor cells and apposes the positive feedback by the ATP receptors; and glutamate is a candidate neurotransmitter (reviewed by Chaudhari and Roper, 2010). Serotonin and glutamate are localized in some teleost taste cells, as is the inhibitory neurotransmitter gamma-aminobutyric acid (GABA) (Toyoshima et al., 1984; Eram and Michel, 2005). In *Ictalurus punctatus*, glutamate reactivity was found to be localized in afferent nerve fibers, and GABA-immunoreactivity was found in some taste cells with glia-like properties, basal cells, and cells with ensheathing morphology. GABA's role in these locations is unknown. It may function as an inhibitory neurotransmitter, or even modulate neural development in the taste bud (Semagor, 2010).

The cranial nerves innervating the taste buds are multimodal, with both sensory afferent and motor efferent fibers. Taste cells innervated by the facial nerve are crucial for food search (appetitive feeding behavior) (Atema, 1971), while those innervated by glossopharyngeal or vagal nerves determine if the food is ingested or rejected, and deal with the palatability of the food and ultimately if it will be swallowed.

11.2.1.1 Receptors and Sensory Transduction in Taste Cells

In mammals, three cell types are located within the taste bud. Type I taste cells have properties of glial cells, including neurotransmitter degradation and absorption; type II cells are known as receptor cells; and type III, as presynaptic cells (Chaudhari and Roper, 2010). Transduction for salty is initiated by the flow of Na^+ through ion channels into the cells; but the cell type responsible for salty transduction has not been identified. For sour, it is believed that organic acids permeating into in type III (presynaptic) cells dissociate, and the ensuing protons block a proton-sensitive potassium channel, thus depolarizing the membrane. For sweet, savory (umami), and bitter taste, ligand binding onto G protein-coupled receptors takes place on a single cell type known as type II (receptor) cells. In these cells, the G protein-coupled receptors are known as T1Rs and T2Rs (Mombaerts, 2004; Chandrashekar et al., 2006). The T1Rs mediate sweet and umami, while the T2Rs mediate bitter taste. The T1Rs function as heterodimers. For example, the T1R3 is an obligate partner for the umami receptor (T1R1+T1R3) and the sweet receptor (T1R2+T1R3). Fish also express T1Rs and T2Rs (Ishimaru et al., 2005). Oike et al. (2007) discovered T1R orthologues responding to amino acids, and the T2R orthologues responding to denatonium, a compound signaling bitter, in zebrafish (*Danio rerio*) and medaka (*Oryzias latipes*). By responding to amino acids, the T1R signals palatability, while the bitter responding T2Rs convey aversion in the fish taste system.

In mammals, the T1R and T2R taste receptors are coupled mainly to the G alpha i protein gustducin (McLaughlin et al., 1992; Wong et al., 1996; Chandrashekar et al., 2006). Gustducin lowers cellular cAMP levels inhibiting the activity of phosphodiesterase. An orthologue of mammalian gustducin has not been found in teleosts (Ohmoto et al., 2011; Oka and Korsching, 2011).

In mammalian taste receptor cells, the principal pathway for taste transduction includes G protein interaction with a phospholipase, consequent to synthesis of inositol-3-phosphate (IP_3), which opens ion channels on the endoplasmic reticulum, releasing Ca^{2+} into the cytosol and Ca^{2+}-dependent activation of the monovalent selective cation channel TrpM5 (Kinnamon, 2012). Zebrafish taste receptor cells also express the components of signal transduction that have been identified in the mammalian taste-signaling cascade, phospholipase-β-2 and the transient receptor potential channel TRPM5 (Yoshida et al., 2007). For this IP_3 second messenger transduction pathway, the G protein interaction is usually through a $G_{\alpha q}$ subunit. An example of this G protein subunit is present in both mammals and fishes (Kinnamon, 2010; Ohmoto et al., 2011). It is known as zfG14 in zebrafish (Ohmoto et al., 2011) and as $G_{\alpha 14}$ in mammals. However, the role of these G proteins in the fish gustatory transduction is not known yet.

11.2.2 SOMATOTOPY IN BRAIN STEM GUSTATORY CENTERS

The gustatory inputs carried by the facial, glossopharyngeal, and vagal cranial nerves stimulate regions in the medulla of the brain. In fishes, these cranial nerves project to ganglia in the medulla, the primary gustatory nucleus, located caudal to the cerebellum. This is equivalent to the nucleus of the solitary tract in mammals. There is a somatotopic organization of the gustatory projections onto the primary gustatory nucleus, so that gustatory input from a specific region on the fish's skin is processed in a similar location. The facial nerve innervates taste buds that are located on external appendages (such as barbels, and even the tail), and projects to the rostral part of the primary gustatory nucleus, in a region named the facial lobe. In fish such as channel catfish, with many taste buds and facial taste fibers, the facial lobe is prominent. Even within the facial lobe, gustatory input is mapped out in a somatotopic pattern (Finger and Morita, 1985).

The vagal nerve fibers project from oropharyngeal regions and the palatal organ to the vagal lobe, located caudal to the facial lobe. Thus, palatability decisions (food rejection or swallowing) conveyed by the glossopharyngeal and vagal nerves are conveyed to the same medullary region. In fish, such as goldfish and catfish, a three-neuron reflex connects vagal gustatory input to motor output, from the primary gustatory afferent (the vagal cranial nerve), to the vagal lobe interneuron, to the vagal motoneurons that innervate pharyngeal muscles. This reflex is specialized for food sorting, implying fine control of muscles in the palate (Finger, 2009). In the absence of vagal gustatory input, catfish will pick up, but not swallow, food (Atema, 1971).

11.3 OLFACTION

The olfactory system is a site of parallel neural processing, whereby different functional aspects of a sensory stimulus are conveyed along separate pathways. In this section, the details of fish olfactory information processing in the peripheral olfactory organ, the OB, and pathways to higher brain structures are examined within the context of parallel olfactory-processing pathways and olfactory-driven motor behaviors.

11.3.1 PERIPHERAL OLFACTORY ORGAN

The nares—the external openings to the environment—lead to the nasal cavity, which contains the peripheral olfactory organ, and within it the olfactory epithelium, a pseudo-stratified ciliated epithelium consisting of bipolar olfactory sensory neurons (OSNs), supporting cells, ciliated nonsensory cells and basal cells. In most teleost fishes, the peripheral olfactory organ is located beneath paired nostrils on either side of the head, and the olfactory epithelium is folded into a flower-like olfactory rosette. Lampreys and hagfish are monorhinic. A single nostril is located on the dorsal surface of the head and leads to a nasal cavity that contains a single multilamellar peripheral olfactory organ.

In some teleost species (especially in Acanthoptergii), the olfactory epithelium is flat or unilamellar (reviewed by Hansen and Zielinski, 2005; Zielinski and Hara, 2007). In many teleosts, accessory nasal sacs are utilized for ventilating water over the olfactory epithelium, so that fish are able to "sniff" (e.g., Doving et al., 1977; Nevitt, 1991; Belanger et al., 2003).

The axons of OSNs exit the olfactory epithelium by passing through the basal lamina into the underlying lamina propria, and fasciculate (i.e., bundle together) to form the olfactory nerve. An olfactory nerve extends from each of the pair of peripheral olfactory organs in teleost fishes, while an olfactory nerve extends from each side of the peripheral olfactory organ of the monorhinic lamprey. Olfactory information conducts along the olfactory nerves to a pair of OBs (located in the brain), where the OSN axon terminals form synaptic contacts. Second-order neurons project from the OB, largely into the telencephalon. The OBs can be located sessile (meaning that they are attached to the telencephalon) or pedunculated (meaning that they are located immediately caudal to the olfactory rosette), as in *Carassius* and *Ictalurus*. In the case of pedunculated OBs, a very short olfactory nerve extends between the peripheral olfactory organ and the OB, and an extended olfactory tract extends from the OB to the telencephalon (Figure 11.1).

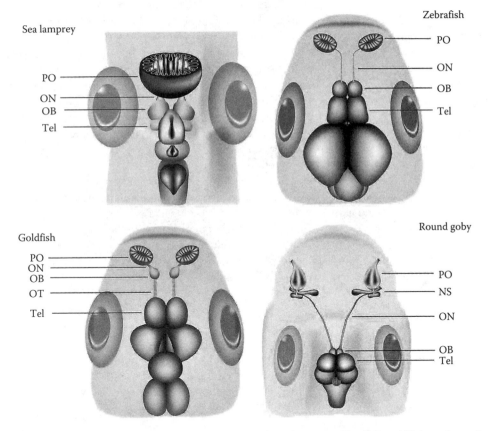

FIGURE 11.1 The anatomy of the olfactory system in the sea lamprey, zebrafish, goldfish, and round goby. In the sea lamprey, there is a single peripheral olfactory organ. There is bilateral symmetry, as two olfactory nerves extend into the olfactory bulb. The pallium is caudal to the olfactory bulb. In the zebrafish, the peripheral olfactory organ has a rosette shape. Elongate olfactory nerves extend to the sessile olfactory bulbs, located rostral to the telencephalon. In the goldfish, the peripheral olfactory organ also contains an olfactory rosette. The olfactory bulb is pedunculated, as the olfactory nerve is very short. An elongate olfactory tract extends from the olfactory bulb to the telencephalon. In the round goby, a single low fold is located on the floor of the peripheral olfactory organ. Nasal sacs help to ventilate the peripheral olfactory organ. Long olfactory nerves extend to the olfactory bulb, which is adjacent to the telencephalon. PO, peripheral olfactory organ; ON, olfactory nerve; OB, olfactory bulb; Tel, telencephalon; OT, olfactory tract; NS, nasal sacs.

11.3.1.1 Olfactory Sensory Neurons

The OSNs are able to respond to the odor molecules. Each OSN contains odor receptors that bind to specific odor ligands and are linked to G proteins, and then transduce this signal to effect membrane depolarization. The OSN axon conducts this information from the nasal cavity into the brain (the OB), and synaptically communicates this sensory input to secondary sensory neurons (the mitral cells). The OSNs are bipolar, with a single dendritic ending extending into the mucus covering the surface of the olfactory epithelium. The cell body resides in the olfactory epithelium, and a single axon extends into the OB. Odor receptors are located on the apical dendritic surface and each OSN expresses only a single type of receptor. Diverse odor information is communicated by the OSNs, largely because of the many different G protein-coupled odor receptor proteins expressed in the peripheral olfactory organ.

11.3.1.1.1 Morphotypes

In teleosts, the two principal OSN types are ciliated and microvillar, and there is also a third type known as crypt cells (Hansen and Finger, 2000; Belanger et al., 2003; Hansen et al., 2003) (Figure 11.2). These polymorphisms differ in relation to their shape, position within the olfactory epithelium, and functionality (summarized in Table 11.1). The ciliated OSN has a cell body

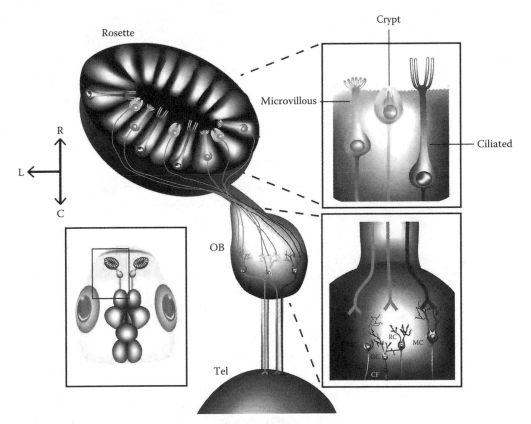

FIGURE 11.2 **(See color insert.)** A representation of the organization of the ciliated, microvillous, and crypt cell olfactory sensory morphotypes in teleosts. The representation is shown on a goldfish-like arrangement of the olfactory system. In the peripheral olfactory organ, the three olfactory sensory neuron (OSN) morphotypes intermingle in the olfactory epithelium covering the rosette; however, the ciliated and microvillar morphotypes predominate. In the olfactory bulb (OB), the axons are organized according to morphotype. In general, the axons of the microvillous OSNs project laterally, the axons of the ciliated OSNs are medial, and the crypt cell axons are located in a small ventral region. Synaptic contacts are made in the olfactory bulb. The output neurons (mitral cells) project to the telencephalon (tel). The OB contains the mitral cells (mc), ruffed cells (rc), and granule cells (gc), as well as centrifugal fibers (cf) projecting from higher brain centers. R, rostral; C, caudal; L, lateral.

TABLE 11.1

Summary of OSN Morphotypes and Their Various Properties

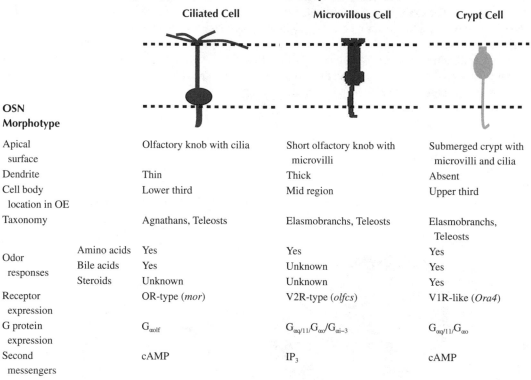

OSN Morphotype		Ciliated Cell	Microvillous Cell	Crypt Cell
Apical surface		Olfactory knob with cilia	Short olfactory knob with microvilli	Submerged crypt with microvilli and cilia
Dendrite		Thin	Thick	Absent
Cell body location in OE		Lower third	Mid region	Upper third
Taxonomy		Agnathans, Teleosts	Elasmobranchs, Teleosts	Elasmobranchs, Teleosts
Odor responses	Amino acids	Yes	Yes	Yes
	Bile acids	Yes	Unknown	Yes
	Steroids	Unknown	Unknown	Yes
Receptor expression		OR-type (*mor*)	V2R-type (*olfcs*)	V1R-like (*Ora4*)
G protein expression		$G_{\alpha olf}$	$G_{\alpha q/11}/G_{\alpha o}/G_{\alpha i-3}$	$G_{\alpha q/11}/G_{\alpha o}$
Second messengers		cAMP	IP_3	cAMP

Note: The dotted line in the first row designates the apical and basal surfaces of the olfactory epithelium.

located deep in the olfactory epithelium, near the basement membrane. Its dendrite is therefore long, and also quite thin, with a bulbous ending known as the olfactory knob, which bares cilia (reviewed by Hansen and Zielinski, 2005; Zielinski and Hara, 2007). These OSNs express cyclic nucleotide gated channels (Sato et al., 2005). In teleost fishes, ciliated OSNs are intermingled with the second morph, microvillous OSNs. The cell bodies of microvillous OSNs are located in the middle third of the olfactory epithelium, and they have mid-length, thick dendrites that end in microvilli. These cells express the transient receptor channel TRPC2 (Sato et al., 2005). The mouse orthologue of TRPC2 is found in sensory neurons of the vomeronasal organ of mammals (Liman et al., 1999). In addition to teleosts, both ciliated and microvillous OSNs are seen in a primitive extant ray-finned fish, the bichir (*Polypterus senegalus* and *Polypterus ornatipinnis*; Zeiske et al., 2009) and sturgeons of the genus Acipencer (Zeiske et al., 2003). While the ciliated OSNs are not found in elasmobranchs (Takami et al., 1994), in lampreys, all OSNs are ciliated, yet there are tall, intermediate, and short forms in the olfactory epithelium (Laframboise et al., 2007).

The final OSN morph is the crypt cell (Hansen and Zeiske, 1998). These cells lack a dendrite, and instead have a cell body located in the superficial olfactory epithelium and the longest axonal length of the OSN morphs. The crypt OSN gets its name from the crypt-like invagination at its apical surface which possesses both microvilli and cilia (Hansen and Zeiske, 1998). The crypt cell is found in teleost fishes (Hansen and Finger, 2000; Belanger et al., 2003; Castro et al., 2008; Bettini et al., 2012), sturgeons (Zeiske et al., 2003; Camacho et al., 2010), bichirs (Zeiske et al., 2009), and elasmobranchs

(Ferrando et al., 2007). The crypt cells show immunoreactivity to S100 calcium-binding protein and to nerve growth factor receptor TfkA (Catania et al., 2003; Germana et al., 2004).

The expression of OSN polymorphisms appears to follow an evolutionary pattern. In an ancient jawless fish (superclass Agnatha), the sea lamprey (*Petromyzon marinus*), only ciliated OSNs are seen (Vandenbossche et al., 1995); in elasmobranchs, these sensory cells only have microvilli (Takami et al., 1994; Schluessel et al., 2008; Ferrando et al., 2009), and crypt cells are present (Ferrando et al., 2007, 2009). By the divergence of the ray-finned fishes, the olfactory epithelium was populated by all three OSNs—ciliated, microvillous, and crypt (Zeiske et al., 2003, 2009; Camacho et al., 2010). Generally, crypt cells tend to be rare (e.g., Bettini et al., 2009). Considering their scarcity (e.g., Belanger et al., 2003), as well as the fact that their number may vary with season (Hamdani et al., 2008) and sex (Bettini et al., 2012), it is not surprising that the crypt cells remained undiscovered for so long.

11.3.1.1.2　Odor Receptors

The OSN receptor proteins are members of the seven-transmembrane G protein-coupled receptor superfamily (Buck and Axel, 1991). Niimura and Nei (2005) identified 102 intact zebrafish olfactory receptor genes; however, less than 50 olfactory receptor genes were identified in pufferfish (Alioto and Ngai, 2005; Niimur and Nei, 2005). In teleosts, the receptors do not exhibit topographic organization; rather they are randomly distributed throughout the olfactory epithelium (Ngai et al., 1993). Individual odorants generally bind to several receptors with different affinities, and individual receptors generally bind more than one odorant (Buck, 2000; Kajiya et al., 2001; Luu et al., 2004). Expression of a single olfactory receptor by each OSN is achieved by monoallelic exclusion; the olfactory receptor protein has an inhibitory function that prevents the activation of any other olfactory receptor gene (Serizawa et al., 2004). This one-neuron-one-receptor rule (monogenic expression) applies to the olfactory system in fishes, with the exception of very closely related receptor genes (Sato et al., 2007).

In teleost fishes, each of the previously mentioned OSN morphotypes (ciliated, microvillous, and crypt) expresses different types of receptors. Ciliated OSNs express the main olfactory *mor*-type receptors, thus named because they resemble the olfactory receptors identified in mammals (Figure 11.3a) (Cao et al., 1998). The receptors in the vomeronasal organ are termed V1R and V2R receptors (Dulac and Axel, 1995). As teleosts do not have a vomeronasal organ, different naming schemes for the orthologous V1Rs and V2Rs have been devised. The name *olfc* (Alioto and Ngai, 2006) was given to the genes coding for the teleost vomeronasal family 2 receptors (V2R). The microvillous OSNs express the *olfc* receptors (Figure 11.3b). The receptor OlfC a1 is the only de-orphanized fish odor receptor to date. In heterologous expression systems, the zebrafish OlfC a1 bind to acidic amino acids (Luu et al., 2004), and the goldfish OlfC a1 bind to basic amino acids (Aliota and Ngai, 2006). In both, OlfC a1 has relaxed specificity, and ligands with similar structures bind onto this particular receptor. The crypt cells express the vomeronasal family 1 receptor genes (V1R genes). These are called designated *ora*—olfactory receptor genes related to class A (Saraiva and Korsching, 2007; Oka et al., 2012). In zebrafish, the *ora* genes family is very small, with only six members (Saraiva and Korsching, 2007); and the gene for a single receptor, the *Ora-4*, is expressed by the majority of crypt cells (Oka et al., 2012).

Another olfactory receptor, the trace amine-associated receptor (TAAR), has been found in teleosts, including zebrafish and *Fugu* (Liberles and Buck, 2006). An ancestral form of the TAAR has been described in sharks (Hussain et al., 2009) and lampreys (Libants et al., 2009). TAAR expression is distributed sparsely in the olfactory epithelium, and not yet been associated with any particular OSN morph (Gloriam et al., 2005; Hussain et al., 2009). Over 100 TAAR genes have been found in zebrafish, and about half this number in stickleback (Hashiguchi and Nishida, 2007). In zebrafish, peak TAAR expression is seen at the border of sensory and nonsensory epithelium (Hussain et al., 2009). As the ligands for mammalian TAARs are amines (Lindemann and Hoene, 2005; Zhang et al., 2013) and amines are detected by the fish olfactory system (e.g., Rolen et al., 2003), the ligands for the fish TAARs may be amines.

Very little is known regarding olfactory receptors in non-teleost fishes. Few ancient fishes have the same published genetic resources as the sequenced genome for the sea lamprey (Smith et al., 2013). Chemosensory receptor genes detected in the sea lamprey included 27 OR-type genes, 28 TAARs, and 4 V1R-type (*ora*) genes; and all three gene families were expressed in the olfactory system of parasitic and adult life stages of the sea lamprey (Libants et al., 2009).

11.3.1.1.3 Signal Transduction

Once an odor binds to the G protein-coupled receptor, a signal transduction cascade involving second messengers opening ion channels takes place on the surface of the sensory neuron, leading to depolarization of the OSN and propagation of the signal along the OSN axon to its terminus in the OB. Ciliated OSNs express the G protein $G_{\alpha olf}$, which uses the cyclic adenosine monophosphate (cAMP) second messenger cascade and the cyclic nucleotide gated ion channel (Figure 11.3a). Localization of $G_{\alpha olf}$ has also been seen in sea lamprey OSNs, which are all ciliated (Frontini et al., 2003). Microvillous OSNs possess G proteins such as $G_{\alpha i}/G_{\alpha o}/G_{\alpha q}$, which use the phosphatidylinositol-3-phosphate (PIP$_3$) and diacylglycerol (DAG) second messenger cascade (Figure 11.3b) (Restrepo et al., 1996; Schild and Restrepo, 1998). The transduction of different odor classes has been associated by the two second messengers cAMP and IP$_3$ through experiments using various physiological and histological techniques. Findings regarding the association of odor class with second messenger and OSN type are summarized in Table 11.2.

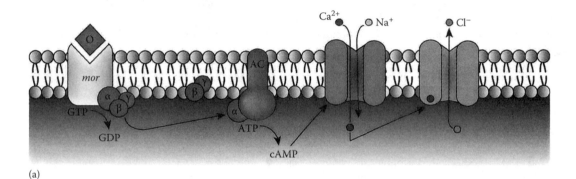

(a)

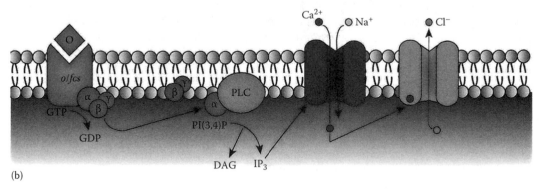

(b)

FIGURE 11.3 **(See color insert.)** (a) The signal transduction cascade in ciliated olfactory sensory neurons (OSNs). The odorant (O) binds onto the *mor*-type G protein-coupled receptor (mor), the dissociated G protein subunits activate adenylate cyclase (AC), and ATP dephosphorylates, forming cAMP, and the cyclic nucleotide gated channels open, allowing for the inflow of cations (calcium: Ca^{2+}; sodium: Na$^+$). The intracellular calcium gates a chloride (Cl$^-$) channel. Chloride flows outward along its concentration gradient, further depolarizing the olfactory sensory neuron. (b) The signal transduction cascade in microvillar OSNs. The odorant (O) binds onto the *olfc*-type G protein-coupled receptor (*olfcs*), the dissociated G protein subunits activate phospholipase C (PLC), and phosphatidylinositol (3,4,5)-triphosphate (PI(3,4)P) dissociates to di-acyl glycerol (DAG) and inositol triphosphate (IP$_3$). The IP$_3$ gates the TRPM5 channels, allowing for the inward flow of cations. The intracellular calcium gates a chloride (Cl$^-$) channel. Chloride flows outward along its concentration gradient, further depolarizing the OSN.

TABLE 11.2

Summary of Second Messengers and OSN Morphotypes Associated with Different Odor Classes Based on Electrophysiological, Calcium Imaging, and Biochemical Studies

Odor Class	Second Messenger		OSN Morph
	cAMP	IP$_3$	
Amino acids	Catfish (Hansen et al., 2003)	Catfish (Hansen et al., 2003)	Microvillous
	Goldfish (Rolen et al., 2003)	Goldfish (Rolen et al., 2003)	Elasmobranchs (Meredeth et al., 2012)
	Mackerel (Vielma et al., 2008)	Zebrafish (Michel et al., 2003; Michel, 1999)	Catfish (Hansen et al., 2003)
		Atlantic salmon (Lo et al., 1993)	Rainbow trout (Sato and Suzuki, 2001)
		Round goby (Laframboise and Zielinski, 2011)	Ciliated
			Sea lamprey
			Crypt
			Mackerel (Vielma et al., 2008)
			Rainbow trout (Bazaes and Schmachtenberg)
Bile salts	Catfish (Hansen et al., 2003)	Atlantic salmon (Lo et al., 1994)	Microvillous
	Goldfish (Rolen et al., 2003)		Elasmobranchs (Meredeth et al., 2012)
	Zebrafish (Ma and Michel, 1998; Michel, 1999)		Ciliated
	Round goby (Laframboise and Zielinski, 2011)		Sea lamprey
			Catfish (Nikonov and Caprio, 2001; Hansen et al., 2003)
			Goldfish (Rolen et al., 2003)
			Rainbow trout (Sato and Suzuki, 2001)
			Crypt
			Rainbow trout (Bazaes and Schmachtenberg)
Pheromones or steroids	Goldfish (Sorensen and Sato, 2005)	Round goby (Laframboise and Zielinski, 2011)	Crypt
	Round goby (Laframboise and Zielinski, 2011)		Rainbow trout (Bazaes and Schmachtenberg)

Microvillous OSN G protein expression is species-specific, and these neurons never express $G_{\alpha olf}$. In goldfish, microvillous OSNs express $G_{\alpha o}$, $G_{\alpha q}$, or $G_{\alpha i\text{-}3}$ (Hansen et al., 2004), while in catfish, this morph expresses $G_{\alpha q/11}$ (Hansen et al., 2003), and in the round goby $G_{\alpha o}$, it is associated with the microvillous OSNs (Belanger et al., 2003). In the shark (*Scyliorhinus canicula*) and rabbit fish (*Chimaera monstrosa*), which only have microvillous cells, OSNs express only $G_{\alpha o}$ (Ferrando et al., 2009, 2010). Crypt cells express $G_{\alpha o}$ in round gobies (Belanger et al., 2003) and catfish (Hansen et al., 2003), and in the goldfish, crypt cells can express both $G_{\alpha o}$ and $G_{\alpha q}$—the only published report of an OSN morphotype expressing more than one G protein (Hansen et al., 2004).

Generally speaking, amino acid odors are transduced via cAMP and/or IP$_3$. These responses appear to take place in microvillous OSNs and therefore involve G proteins other than $G_{\alpha olf}$ (e.g., $G_{\alpha o}$ or $G_{\alpha q}$).

A study of crypt cells found that responses to amino acids were mediated via cAMP (Vielma et al., 2008). These crypt cells also respond to bile salts, to gonadal extracts and hormones from the opposite sex, as well as to synthetic steroids 4-androstene-3, 17-dione, 17α-methyltesterone, 4-androsten-3-17-dione, and 17α-20-β–dihydroxy-4-pregnen-3-one (Bazaes and Schmachtenberg, 2012). With the exception of the Atlantic salmon, bile salt odors (proposed to function in social interactions) are processed by ciliated OSNs using cAMP (via $G_{\alpha olf}$). Pheromones (in this case, a blend of sex steroids) were transduced by cAMP in goldfish (Sorensen and Sato, 2005). In the round goby, a perciform, cAMP transduction is utilized for bile acids, IP_3 for amino acids, and both cAMP and IP_3 for steroid odors (Laframboise and Zielinski, 2011).

11.3.1.1.4 Olfactory Sensory Neurons: Summary

In the olfactory epithelium of fishes, there exists a relationship between form and function of the OSNs. The three OSN morphotypes (ciliated, microvillous, and crypt) vary in their shape and position within the epithelium. They also express different receptors (OR-type and V2R-type, as well as V1R-type and TAARs, though this last one has not been associated with a particular morph), and these are coupled to different G proteins. These OSN types also transduce different odor classes and utilize different second messengers to do so. Studies of OSNs have revealed these characteristics in several species of fishes, but additional information—particularly for more diverse taxonomic groups of fishes—is still necessary before a complete picture of OSN form and function can be drawn.

11.3.2 Olfactory Bulb

The OB is a forebrain structure with input from cranial nerve I (the olfactory nerve). It is the site of olfactory sensory input into the brain that allows for odor coding and also a location of neural integration, the sorting of sensory input toward appropriate locations in the higher brain centers via mitral cells (the bulbar output neurons), as well as regulatory input from centrifugal fibers (descending projections) from higher brain centers. An example of OB function is seen in the channel catfish. The lateral region of the catfish OB receives the axons of microvillous OSNs that respond to amino acid odors, and the mitral cells propagate neural activity along the lateral olfactory tract, to the lateral region of the telencephalon (Hansen et al., 2003; Nikonov et al., 2005). Although OB function is far from being completely understood, investigation of the OB in fishes has led to some understanding regarding olfactory sensory and integrative function of this brain region.

11.3.2.1 Parallel Processing Pathways in the Olfactory System

In teleost fishes, parallel olfactory processing pathways are delineated as medial and lateral subdivisions of the olfactory system; however, the sensory neuron subtypes, the ciliated and microvillar olfactory sensory neuron morphotypes, are distributed stochastically throughout the olfactory epithelium. The axons of the ciliated form extend to the medial region of the OB, and the microvillar forms project to the lateral region. From the OB, odor information is sent via the medial and lateral olfactory tracts to the telencephalon (reviewed by Laberge and Hara, 2001; Hansen and Reutter, 2004). The medial and lateral olfactory tracts contain the axons of mitral cells, the second-order neurons in the olfactory system. In cyprinid fishes, the ciliated, microvillous, and crypt-type OSNs are distributed throughout the epithelium. Ciliated OSNs project axons to the medial region of the OB, while the axons of microvillous OSNs project to the lateral region of the OB (Figure 11.2) (Morita and Finger, 1998; Hamdani et al., 2001a; Hansen et al., 2003; Sato et al., 2005; Hamdani and Doving, 2007). These are generalized pathways, and there are exceptions to these generalizations in which individual ciliated or microvillous neurons may not strictly adhere to the medial and lateral pathways, respectively (reviewed by Zielinski and Hara, 2007). The axons of crypt cells project into two small regions of the ventral portion of the OB in channel catfish (Hansen et al., 2003) and to the dorsomedial glomerular field in zebrafish (Gayoso et al., 2012).

In the cyprinids, it is clear that the primary output neurons (mitral cells) of the medial and lateral region of the OB project their axons along the medial and lateral olfactory tracts, respectively (Sheldon, 1912; Doving and Selset, 1980; Satou, 1990; Hamdani et al., 2000, 2001b; Hamdani and Doving, 2007), to the telencephalon, where these subdivisions are maintained (Nikonov et al., 2005). Generally, the medial olfactory pathway (ciliated OSNs) processes bile salt and pheromone odor information, while the lateral olfactory pathway (microvillar OSNs) processes amino acid and nucleotide odor information (Satou, 1990; Friedrich and Korsching, 1998; Hara and Zhang, 1998; Nikonov and Caprio, 2001; Hansen et al., 2003). Furthermore, within the medial and lateral subdivisions of the OB, specific subsets of output neurons process specific subclasses of odors (e.g., acidic, basic, or neutral amino acids in the lateral OB) (Nikonov and Caprio, 2004; Rolen and Caprio, 2007). It is clear that parallel processing pathways in the olfactory system are important for processing different classes of odors, and can be organized by their functional output.

11.3.2.2 Afferent Input and Olfactory Glomeruli

The OSN axons form small nerve bundles in the lamina propria beneath the olfactory epithelium and project along the olfactory nerve, entering the rostral portion of the OB. The OSN axons spread over the surface of the OB, forming the peripheral layer of this structure (olfactory nerve layer), then proceed deeper into the OB, and enter spherical regions of neuropil known as glomeruli. The OSN axons terminate in the olfactory glomeruli and form glutamatergic synaptic junctions onto postsynaptic partners. The dendritic endings of output neurons (mitral cells), interneurons (granule cells), and axonal endings of descending projections from higher brain regions are also located in the glomerular region, and participate in synaptic activity (Hildebrand and Shepherd, 1997; Gire and Schoppa, 2009).

11.3.2.3 Organization of Neurons within the Olfactory Bulb

The OB consists of several fundamental components that are conserved across phyla, even though there are some differences with respect to cell types, stratification, and organization. The OB of teleost fishes has four layers: the olfactory nerve layer, the glomerular layer, the mitral cell layer, and the granule cell layer.

The primary output neurons of the teleost OB are mitral cells. The cell bodies are in the mitral cell layer and dendrites extend into the glomerular layer where they synapse with the axon terminals of OSNs. A single mitral cell innervates more than one glomerulus in fish. Mitral cells also make inhibitory dendrodendritic synapses with granule cells (Satou, 1990; Laberge and Hara, 2001). The granule cell layer contains local interneurons known as granule cells, which form dendrodendritic synapses with mitral cells (Mori, 1987; Shepherd and Greer, 1998; Shepherd et al., 2007). The granule cells are GABAergic interneurons. The nuclei are located in the granule cell layer of the OB, and the processes extend throughout the OB.

Some species including the goldfish, catfish, and zebrafish have another form of output neuron known as ruffed cells, because they have a distinctive ruffle on the initial segment of the axon (Kosaka and Hama, 1979, 1980, 1981; Kosaka, 1980; Satou, 1990; Fuller and Byrd, 2005). The cell bodies of the ruffed cells are situated in the mitral cell layer and do not directly synapse with OSN axons or mitral cells, but instead form axodendritic connections with granule cells, which in turn synapse with mitral cells (Figure 11.2) (Kosaka and Hama, 1982). Ruffed cells project axons along the olfactory tract and may play a role in the oscillatory neural activity observed in the OB of fish. Zippel et al. (1999, 2000) observed that ruffed cells respond to a variety of odors in the opposite fashion to mitral cells (i.e., a given odor may elicit an excitatory response in a mitral cell and an inhibitory response in an associated ruffed cell). The role of ruffed cells in odor information processing is still relatively unknown.

11.3.2.4 Olfactory Glomeruli and Odor Coding

Chemotopy, the spatial patterning of activity in the OB related to odorant chemical features, is evident from studies of zebrafish (Friedrich and Korsching, 1997) and channel catfish (Nikonov and Caprio, 2001). OSN input to the glomeruli is needed for chemotopy to occur. This is very clear in the mammalian OB, where each glomerulus receives peripheral axonal input only from OSNs expressing a single odor receptor (Mombaerts et al., 1996). Consequently, a specific odor ligand activates synaptic activity within the glomerulus that receives the OSNs that express the receptor for an odor ligand (e.g., Tan et al., 2010). The OSNs that respond to structurally similar compounds project to glomeruli that are grouped together and those that are less similar are further apart, so that the glomeruli produce a chemotopic map arranged around the surface of the OB (Johnson and Leon, 2007). Imaging and electrophysiological experiments have shown that any given odor will activate a distinct group of glomeruli, with similar odors producing distinct but overlapping glomerular activity maps, and less similar odors producing activity maps that overlap less or not at all (mammals: Xu et al., 2000; Wachowiak and Cohen, 2001, 2003; Johnson and Leon, 2007). Activity distributions observed by Friedrich and Korsching (1998) in zebrafish are consistent with field potential recordings in salmonid species, showing that bile acids induce activity mainly in the dorsal and medial OB, whereas amino acids induce activity mainly in the lateral OB region (Døving et al., 1980; Hara and Zhang, 1998). Laberge and Hara (2004) showed that in brown trout and rainbow trout, bile acids elicit responses in the middorsal OB and amino acids elicit responses in the latero-posterior OB.

In this way, odors are spatially encoded on the surface of the OB and within the glomeruli. The activation of many glomeruli in the OB in response to a particular odor is known as combinatorial odor coding, due to the unique combination of glomeruli activated by a given odor. This spatial information regarding particular glomeruli that have been activated is conveyed to specific regions of higher brain centers, so that particular neurons with specific temporal qualities are activated. However, there are examples for the activation of only a single glomerulus by a given odor. This non-combinatorial spatial odor coding has been observed in the OB of zebrafish for sex pheromones (Friedrich and Korsching, 1998; Korsching, 2001). In zebrafish, a single glomerulus (or isolated glomerular complex) has been observed to be responsive to pheromones (Friedrich and Korsching, 1998). The pheromones 15-keto-prostaglandin $F_{2\alpha}$, 13, 14-dihydro-prostaglandin $F_{2\alpha}$, and prostaglandin $F_{2\alpha}$ stimulated the ventromedial posterior area linking the OB to the telencephalon in lake whitefish (Laberge and Hara, 2003). These spatial patterns of combinatorial, non-combinatorial, and chemotopic odor coding in the OB contribute to parallel processing pathways in the olfactory system.

Regional activation in the olfactory bulb is similar in catfish (Nikinov and Caprio, 2001; Hansen et al, 2003) and zebrafish (e.g. Friedrich and Korsching, 1998). Amino acids activate glomeruli in the rostral lateral region, nucleotides activate the caudal lateral region and bile acids stimulate the medial region. In zebrafish, the pheromones prostaglandin F2α and 17α, and 20-β-dihydroxy-4-pregnene-3-one-20-sulfate activate a glomerular complex in the ventral bulbar region (Friedrich and Korsching, 1998).

11.3.2.5 Temporal Coding in the Olfactory Bulb

Olfactory information is not only spatially encoded in the OB, but temporally encoded as well by the firing pattern of mitral cells (Wehr and Laurent, 1996; Friedrich and Laurent, 2001; Laurent, 2002; Friedrich and Laurent, 2004). Oscillations in the teleost OB, which are between 3 and 15 Hz, have been observed in many studies (reviewed by Satou, 1990). The mitral cells make dendrodendritic synapses with granule cells, inhibitory interneurons that shape the temporal structure of the primary output neuron response patterns through lateral inhibition, in both vertebrates and invertebrates (Laurent et al., 1996; MacLeod and Laurent, 1996; Shepherd and Greer, 1998; Urban and Sakmann, 2002), including zebrafish (Wiechert et al., 2010). The temporal coding of odor responses along the parallel processing pathways in the OB can be informative of odor identity in higher brain structures (Laurent, 2002; Wick et al., 2010). Odors activate several mitral cells at the same time, but each individual mitral cell has its own slow temporal firing pattern with increasing and decreasing firing rates over the course of the response. In addition, any given mitral cell may respond to several odors of similar structure (i.e., two different amino acids). Activity of the responding mitral cells gets redistributed over the group of responding mitral cells and becomes unique to that particular odor over the course of the response (decorrelation) (Friedrich and Laurent, 2001). This process of decorrelation makes the temporal response pattern of a group of mitral cells unique to a specific odor over time, and therefore different from similar odors (Friedrich and Laurent, 2001; Laurent, 2002). This allows for higher-order structures to interpret the overall slow temporal firing pattern of the collection of mitral cells to decode the stimulus identity (Friedrich and Laurent, 2001; Laurent, 2002; Wick et al., 2010).

Fast oscillatory responses are also present in the vertebrate OB and are generated in mitral cells by the dendrodendritic inhibitory interneurons and their circuit connections with primary output neurons. These fast oscillatory waveforms are believed to play a role in making odor representations different from one another and in aiding in odor identity in higher brain regions (Laurent and Davidowitz, 1994; Laurent, 2002; Kay and Stopfer, 2006; Kay et al., 2009).

11.3.2.6 Olfactory Bulb Projection Sites

The bulbar output neurons (mitral cells) transmit the spatially and temporally coded odor information from the OB to higher brain regions, where this information is integrated and ultimately effects behavioral responses. In teleosts, mitral cells project their axons along the lateral or medial olfactory tracts to the telencephalon (Doving and Selset, 1980; Kosaka and Hama, 1982; Satou, 1990; Hamdani and Doving, 2007), where a chemotopic map similar to that observed in the OBs persists (Nikonov et al., 2005). Moreover, in sturgeon, cod, and goldfish, mitral cells also project to midbrain structures such as the preoptic area, thalamus, habenula, and hypothalamus (Vonbartheld et al., 1984; Rooney et al., 1992; Huesa et al., 2000, 2003, 2006). In crucian carp, the lateral olfactory tract mediates feeding behaviors, while the medial portion of the medial olfactory tract and the lateral portion of the medial olfactory tract mediate alarm reaction and reproductive behaviors, respectively (Hamdani et al., 2000, 2001a,b; Hamdani and Doving, 2007).

11.4 OLFACTION IN WILD FISHES

Studies investigating olfactory sensory responses have only been reported for a small number of wild-caught fishes. These studies have recorded field potentials from the olfactory epithelium following the application of test odors. This technique is called the electro-olfactogram (EOG), shown in Figure 11.4. EOG studies have been utilized to examine responses to amino acids in wild black bullhead catfish (*Ameiurus melas*) (Dolensek and Valentincic, 2010) and olfactory response thresholds to amino acids in wild-caught hammerhead sharks (*Sphyrna lewini*) (Tricas et al., 2009). Moreover, the EOG responses and olfactory thresholds to amino acids and bile salts have been examined in several elasmobranch fishes including the Atlantic stingray (*Dasyatis sabina*) and the bonnethead shark (*Sphyrna tiburo*) (Meredith and Kajiura, 2010; Meredith et al., 2012). In addition, EOG recordings have been used to investigate putative pheromones in wild-caught field specimens of peacock blenny (*Salario pavo*) (Serrano et al., 2008), as well as to identify potential reproductive pheromones in the sea lamprey (Li et al., 1995; Siefkes et al., 2003) and steroidal pheromones in the round goby (*Neogobius melanostomus*) (Murphy et al., 2001; Laframboise and Zielinski, 2011).

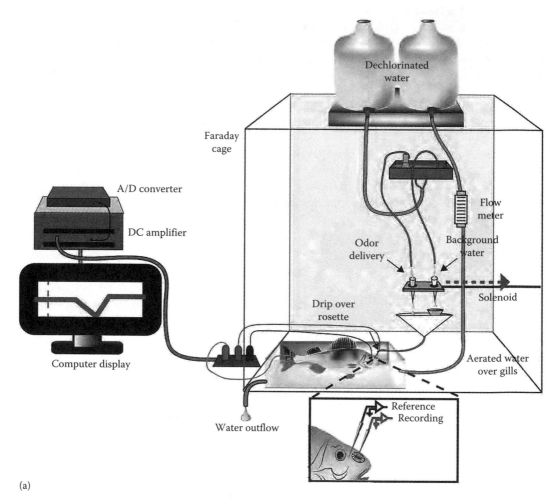

(a)

FIGURE 11.4   (See color insert.) A representation of the delivery of test solutions to the olfactory epithelium of a teleost, and the recording of field potentials from the olfactory epithelium. (a) The recording of field potentials from the peripheral olfactory organ of a fish using the electro-olfactogram (EOG). Aerated water flows into the mouth and over the gills of an anesthetized fish. Background water flows into a small funnel into the nasal cavity and over the olfactory rosette. A second tube delivers water containing an odorant test solution. When activated, a computer-controlled solenoid allows for rapid switching between the background water and the odorant test solution being delivered over the rosette. A recording electrode is situated in the nasal cavity in close proximity to the olfactory epithelium, and a reference electrode is placed on the adjacent skin. The output of the electrodes is connected to a DC amplifier, the amplified signal is converted from analog to digital by an A/D converter, and the resulting signal is displayed on a computer.

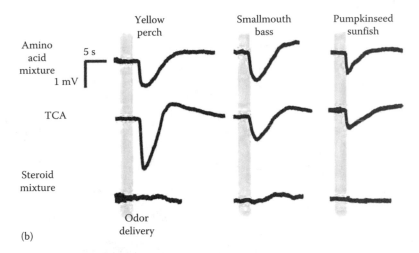

FIGURE 11.4 (continued) **(See color insert.)** A representation of the delivery of test solutions to the olfactory epithelium of a teleost, and the recording of field potentials from the olfactory epithelium. (b) EOG responses of three fish—a yellow perch, a smallmouth bass, and a pumpkinseed sunfish—to three classes of odorants: amino acids, a bile acid (taurocholic acid [TCA]), and a steroid mixture. EOG responses were observed in response to amino acids and the bile acid, but not to the steroids. The opaque bar indicated the duration of the odor delivery. 5 s; 1 mV.

11.4.1 OLFACTION, PHEROMONES, AND LOCOMOTOR RESPONSE IN THE SEA LAMPREY

The life cycle of the basal vertebrate, the sea lamprey, is composed of clearly defined life stages including the larva, seven stages of metamorphosis (Potter et al., 1982), parasitic feeding stage, migratory adult, and spawning adult. During the parasitic feeding stage, lampreys are thought to locate their prey using amino acids and amines released by the fish (Kleerekoper and Mogensen, 1963). In the spring, migrating adult lampreys move upstream and encounter a mixture of compounds that are released from larval sea lampreys in the streambed. This mixture includes PZS, allocholic acid (ACA), and two disulfated aminosterol derivatives known as petromyzonamine disulfate (PADS) and petromyzosterol disulfate (PSDS). This mixture acts as a pheromone that migrating adult lampreys detect and use as an indicator of suitable spawning habitat (Li et al., 1995; Bjerselius et al., 2000; Sorensen et al., 2005). Moreover, once in the spawning stream, spermiated male sea lampreys release the pheromones 3-keto petromyzonol sulfate (3kPZS) and 3-keto alcoholic acid (3kACA), which induce movement/searching behaviors and spawning behaviors in ovulating female sea lampreys (Li et al., 2002; Siefkes et al., 2003; Siefkes and Li, 2004; Johnson et al., 2005, 2009). After spawning, adult lampreys quickly deteriorate and die.

To detect and respond to these pheromones, as well as other odorants, sea lampreys utilize a unique olfactory system that consists of a peripheral olfactory organ containing both a main olfactory epithelium and tubular diverticula known as the accessory olfactory organ (Figure 11.5) (Hagelin and Johnels, 1955). The main olfactory epithelium of the sea lamprey consists of columnar cells that contain three morphotypes of OSNs. The three morphotypes have been termed tall, intermediate, and short OSN, with the tall morphotype being most prominent and the intermediate and short morphotype less prominent (Laframboise et al., 2007). The three morphotypes were also found to be present in both metamorphic and reproductive adult sea lampreys, indicating that these cells develop early and remain through adulthood (Laframboise et al., 2007). Interestingly, the density of OSNs present in the main olfactory epithelium is greatest during the larval life stage, when the cells are narrow and steadily decrease throughout maturation/development as the cells increase in width (VanDenBossche et al., 1995). Furthermore, a detailed examination of the peripheral olfactory organ revealed that drastic remodeling occurs during metamorphosis. Specifically,

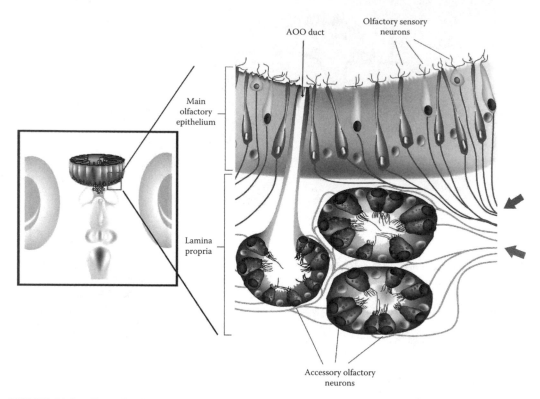

FIGURE 11.5 **(See color insert.)** In the sea lamprey, the main olfactory epithelium contains ciliated olfactory sensory neurons (OSNs). There are tall (red), intermediate (yellow), as well as short (green) OSNs. Each form extends a single axon into the underlying lamina propria (red arrow). The main olfactory epithelium also contains ducts that lead to the accessory olfactory organ, small tubules containing rounded ciliated cells (blue). Axons extend from these accessory olfactory neurons (blue arrow). R, rostral; C, caudal. (Based on Laframboise, A.J. et al., *Neurosci. Lett.*, 414, 277, 2007; Ren, X. et al., *J. Comp. Neurol.*, 516, 105, 2009.)

the entire olfactory organ enlarges in size and weight as it changes from an epithelial lined tube to a nasal sac containing lamellar folds with a partial midline septum that separates the nasal sac into two regions (VanDenBossche et al., 1997). Moreover, during metamorphosis, the accessory olfactory organ also becomes more prominent as two distinct cell types begin to form diverticula that are surrounded by blood vessels and nonmyelinated nerve bundles (Hagelin and Johnels, 1955; VanDenBossche et al., 1997).

Several electrophysiological studies have demonstrated that the main olfactory epithelium of the sea lamprey responds to amino acids as well as a variety of biologically important bile acids. Li et al. (1995) recorded extracellular field potentials (EOG) from the main olfactory epithelium of migratory adult sea lampreys in response to the bile acids PZS and ACA of larval sea lampreys. Interestingly, of 38 bile acids tested, migratory adult sea lampreys showed a very high level of olfactory sensitivity to PZS and ACA with threshold responses for both bile acids at approximately 10^{-12} M (Li et al., 1995). Furthermore, a subsequent study by Li and Sorensen (1997) demonstrated that there are at least four independent receptor sites within the main olfactory epithelium for bile acids including PZS and ACA. Due to the relatively inaccessible location and small size, there is currently no information regarding the physiological responses of the accessory olfactory organ to odorants. Interestingly, tract-tracing experiments in the sea lamprey have shown that olfactory sensory neurons from the main olfactory epithelium extend to the medial and lateral regions of the ipsilateral OB, and that the accessory olfactory organ contains sensory neurons that project axons solely to the medial region of the OB (Ren et al., 2009). Odorant stimulation and electrical stimulation of this medial region of the OB stimulates locomotion (Derjean et al., 2010). Moreover, primary output neurons in the medial region of the OB project their axons to the posterior tuberculum, which relays to the mesencephalic locomotor region. These neurons project to the reticulospinal neurons in the hindbrain, which are responsible for the activation of locomotion in the sea lamprey (Derjean et al., 2010). This pathway from the medial OB to the hindbrain constitutes the neural substrate for olfactory-driven locomotion in the sea lamprey (Derjean et al., 2010). Neural projections of primary output neurons from the lateral OB of the sea lamprey project to the lateral pallium as well as other forebrain regions (Derjean et al., 2010). Therefore, this other olfactory pathway is likely not directly involved in locomotion but may be involved in further integration of odor information (Derjean et al., 2010). This division of neural projections from the peripheral olfactory organ to the OB also indicates that there is a possibility of spatial organization within the OB of the sea lamprey. This further implies that the OB may be organized according to functional output rather than solely by input from the OSNs.

Currently, it is unknown how information regarding the identification of specific odorants is organized in the OB of the sea lamprey, and whether a chemotopic map, similar to that observed in teleost fish, allows for odorant discrimination. Frontini et al. (2003) demonstrated that the OB of the sea lamprey contain distinct glomerular territories and that all of these glomeruli, except the medial glomeruli, express G_{olf}—an olfactory regulatory G protein necessary for odorant reception. This indicates that the medial region of the OB is biochemically different from the other regions of the OB and lends evidence for the existence of spatial organization within the sea lamprey OB. In the lamprey, the glomerular layer and mitral cell layer comprise a single layer of glomeruli and mitral cells. This layer of the OB contains numerous aggregations of spherical or oval glomeruli that contain olfactory nerve fibers and mitral cell bodies and dendrites, as well as projections from both the granule and ependymal cells. Two morphologically distinct types of mitral cells exist in the lamprey OB, based on soma size and shape, and are both found in close proximity to OB glomeruli. Interestingly, Iwahori et al. (1987) also found that some mitral cell dendrites terminated in two or more glomeruli but that a majority of mitral cells terminate into a single glomerulus. The granule and ependymal cell layers comprise a large portion of the lamprey OB, and processes from these cells extend into the glomerular cell layer as well (Iwahori et al., 1987).

Currently, much is known about the olfactory repertoire of the sea lamprey, the anatomy and physiology of the olfactory organ, and the neural pathways involved in transforming olfactory inputs into a locomotor response. However, very little is known about how odor information is processed in the OB of the sea lamprey and what implications this odor processing has for olfactory-driven movement/behavioral responses in this species. Several studies are currently under way to elucidate the anatomical organization as well as the spatial and temporal coding mechanisms in the OB as they relate to the previously defined parallel olfactory pathways and olfactory-driven movement.

11.4.2 REPRODUCTIVE PHEROMONES IN A PERCIFORM: A STUDY OF THE ROUND GOBY

The round goby, *N. melanostomus*, is a small perciform native to the Black and Caspian seas of Eurasia. In the 1990s, round gobies were introduced into Lake St. Clair of the Laurentian Great Lakes through the transfer of ship ballast water. Breeding populations have spread and thrived in all five Great Lakes (Charlebois et al., 2001) and are starting to establish in the rivers that flow into the Great Lakes (Poos et al., 2010). Pheromones may be an important contributing factor to the reproductive success of this teleost species. Spawning occurs in nests occupied and guarded by males (MacInnis and Corkum, 2000). Strategies used by these reproductive males to attract females may include reproductive pheromones. Belanger et al. (2004), Gammon et al. (2006), and Corkum et al. (2006, 2009) observed that water conditioned by reproductively mature males attracts reproductively mature females. In order to determine the nature of the active compound (or compounds), two approaches were taken.

The first was to identify the steroids that are produced by reproductive male testes. Special attention was paid to the possible release of any androgens in which the A ring has a 5β- and 3α-reduced configuration by reproductive males. It was shown previously in the black goby, *Gobius jozo*, that one such steroid, 3α-hydroxy-5β-androstan-17-one 3-glucosiduronate (etiocholanolone glucuronide (ETIO-g), is produced by the testes of reproductive males and has the ability to attract reproductive females in this species (Colombo et al., 1980). Moreover, a range of steroids with a 5β- and 3α-reduced configuration, including free (i.e., unconjugated) ETIO and ETIO-g, stimulated olfactory epithelial field potentials (EOG) as well as gill ventilation responses in *N. melanostomus* (Murphy et al., 2001). Increased opercular activity has been utilized as an indicator of olfactory sensory responses to steroids in the round goby (Belanger et al., 2006). In these fish, gill ventilation increases when the fish are smelling steroids. This increased opercular movement increases the flow of water over the olfactory epithelium. Accessory nasal sacs caudal to the olfactory epithelium are utilized for gill ventilation during olfaction (reviewed by Hansen and Zielinski, 2005).

The second approach was to try and purify and then identify the active compound(s) in reproductive male–conditioned water extracts by high-performance liquid chromatography (HPLC)—initially using the technique of EOG to pinpoint those fractions that the fish are able to smell (Belanger et al., 2004). In relation to the first approach, it has been shown that testes and seminal vesicles of reproductive male *N. melanostomus* contain very distinct pockets of steroid-producing cells that synthesize 5β,3α-reduced steroids *in vitro*—in particular, 3α-hydroxy-5β-androstane-11,17-dione (11-oxo-etiocholanolone (11-O-ETIO)) (Arbuckle et al., 2005; Jasra et al., 2007). This steroid and conjugated derivatives, 3α,17β-dihydroxy-5β-androstan-11-one 17-sulfate; 3α-hydroxy-5β-androstan-11,17-dione 3-glucosiduronate (11-O-ETIO-3-g); 3α,17β-dihydroxy-5β-androstan-11-one 17-glucosiduronate and 3α-hydroxy-5β-androstan-11,17-dione 3-sulfate (11-O-ETIO-3-s), have been shown to be released into the water by reproductive males—especially following injection of the fish with gonadotropin-releasing hormone analogue (GnRHa) (Katare et al., 2011). This same study showed that conjugated steroids appear to be released mainly in the urine and free (unconjugated) 11-O-ETIO via the gills.

The second approach of coupling HPLC with EOG has worked well in isolating pheromones in studies on the sea lamprey *P. marinus* (Li et al., 2002) and Mozambique tilapia *Oreochromis mossambicus* (Barata et al., 2008), but has proven challenging in *N. melanostomus* (Belanger et al., 2004). In the *N. melanostomus* study, EOG responses were observed in many HPLC fractions and there was no clear "peak" that might have indicated which fractions contained an attractant pheromone. An alternative has been to use a behavioral bioassay to assess the fractions (Corkum et al., 2006, 2008, 2009; Gammon et al., 2006; Yavno and Corkum, 2010). A rapid assay of movement responses to small volumes of test solutions demonstrated that male urine attracted reproductively active females, and that when conditioned water from male *N. melanostomus* was fractionated by HPLC, movement responses occurred to the application of one block of fractions that contained conjugated 11-O-ETIO (Tierney et al., 2013). Fractions corresponding to unconjugated 11-O-ETIO evoked a dimorphic response—preference by nonreproductive females and avoidance by reproductive females. It must be stressed that the data so far fall short of proving that either the unconjugated or conjugated 11-O-ETIO are reproductive pheromones released by reproductive male round gobies in the wild.

When synthetic analogues of the released steroids were tested by EOG on *N. melanostomus*, it was clear that 11-O-ETIO and 11-O-ETIO-3-S are potent odors to round gobies, and that the 11-O-ETIO-17-sulfate did not elicit olfactory responses (Laframboise and Zielinski, 2011). As in goldfish and crucian carp some hormones released as sex pheromones can evoke EOG responses across species (Bjerselius and Olsen, 1993; Lim and Sorensen, 2011), an effort was made to determine if fish species sharing the same ecosystem as *N. melanostomus* in the Great Lakes region were able to smell the putative pheromones released by round goby males. Ochs et al. (2013) tested for EOG responses by rock bass (*Ambloplites rupestris*), bluegill sunfish (*Lepomis macrochirus*), pumpkinseed sunfish (*Lepomis gibbosus*), smallmouth bass (*Micropterus dolomieu*), and yellow perch (*Perca flavescens*) to positive controls for olfactory activity (amino acids and bile acids), as well as synthetic analogues of steroids released by male round gobies (unconjugated and conjugated 11-O-ETIO). When compared to round goby responses, amino acids and the bile acid consistently elicited field potential responses across all species, but only round gobies showed an EOG response to the putative pheromones. These studies point to the likelihood of nesting round goby males attracting females through the release of specific steroids.

Future elucidation of the exact chemical identity of these reproductive pheromones may become useful for biological control measure in areas of risk, such as sites with protected native species in danger of predation by the round gobies.

This approach, of pairing chemical isolation of pheromones with assay of biological function, by neural responses and behavioral responses has proven to be useful in identifying insect pheromones (more recently for arachnid chemical signals) (Olsen et al., 2011), sea lamprey pheromones (Li et al., 2002), and even nematode (*Caenorhabditis elegans*) pheromones (Kim et al., 2013). This strategy will likely become more useful for isolating pheromones from various other fish species as well.

11.5 SUMMARY

Although much is known regarding how chemosensory information is acquired by sensory cells and how neural information is organized in the brain, there are still many unanswered questions. Hopefully the future will bring answers regarding the matching of odor and taste receptors to specific ligands, an understanding of the chemical structure of pheromones, and the behavior responses elicited by these pheromones, particularly by fish in the wild. Regarding the coding of chemosensory information, we still have a long way to go before fully understanding temporal and spatial organization of information in the brain, the flow of information to output responses, the modulation of chemosensory activity, and the effect of centrifugal neurons projecting from other brain structures to regions that process chemosensory input. Although fishes have evolved exquisite chemosensory systems for thriving in the aquatic environment, the presence of chemicals brought in by industrial changes must also be considered for the future of fishes on earth.

REFERENCES

Alioto, T.S. and Ngai, J. 2005. The odorant receptor repertoire of teleost fish. *BMC Genomics* 6, 173.

Alioto, T.S. and Ngai, J. 2006. The repertoire of olfactory C family G protein-coupled receptors in zebrafish: Candidate chemosensory receptors for amino acids. *BMC Genomics* 7, 309.

Arbuckle, W.J., Belanger, A.J., Corkum, L.D., Zielinski, B.S., Li, W., Yun, S., Bachynski, S., and Scott, A.P. 2005. *In vitro* biosynthesis of novel 5β-reduced steroids by the testis of the round goby, *Neogobius melanostomus. Gen. Comp. Endocrinol.* 140(1), 1–13. doi: 10.1016/j.ygcen.2004.09.014

Atema, J. 1971. Structures and functions of the sense of taste in the catfish, *Ictalurus natilis. Brain Behav. Evol.* 4, 273–294.

Barata, E.N., Fine, J.M., Hubbard, P.C., Almeida, O.G., Frade, P., Sorensen, P.W., Canario, A.V.M. 2008. A sterol-like odorant in the urine of Mozambique tilapia males likely signals social dominance to females. *J Chem Ecol.* 34(4), 438–449.

Bazaes, A., Schmachtenberg, O. 2012. Odorant tuning of olfactory crypt cells from juvenile and adult rainbow trout. *J. Exp. Biol.* 215, 1740–1748.

Belanger, A.J., Arbuckle, W.J., Corkum, L.D., Gammon, D.B., Li, W., Scott, A.P., and Zielinski, B.S. 2004. Behavioural and electrophysiological responses by reproductive female *Neogobius melanostomus* to odors released by conspecific males. *J. Fish Biol.* 65(4), 933–946.

Belanger, R.M., Corkum, L.D., Li, W.M., Zielinski, B.S. 2006. Olfactory sensory input increases gill ventilation in male round gobies (*Neogobius melanostomus*) during exposure to steroids. *Comp Biochem Physiol A – Mol Integ Physiol.* 144(2), 196–202.

Belanger, R.M., Smith, C.M., Corkum, L.D., and Zielinski, B.S. 2003. Morphology and histochemistry of the peripheral olfactory organ in the round goby, *Neogobius melanostomus* (Teleostei: Gobiidae). *J. Morphol.* 257, 62–71.

Bettini, S., Lazzari, M., and Franceschini, V. 2012. Quantitative analysis of crypt cell population during postnatal development of the olfactory organ of the guppy, *Poecilia reticulate* (Teleostei Poecilidae), from birth to sexual maturity. *J. Exp. Biol.* 215, 2711–2715.

Bettini, S., Lazzari, M., Ciani, F., and Franceschini, V. 2009. Immunohistochemical and histochemical characteristics of the olfactory system of the guppy, *Poecilia reticulate* (Teleostei, Poecilidae). *Anat. Rec.* (Hoboken) 292(10), 1569–1576.

Bjerselius, R. and Olsen, K.H. 1993. A study of the olfactory sensitivity of crucian carp (*Carassius carassius*) and goldfish (*Carassius auratus*) to 17a,20b-dihydroxy-4-pregnen-3-one and prostaglandin F2a. *Chem. Senses.* 18, 427–436.

Bjerselius, R., Li, W.M., Teeter, J.H., Seelye, J.G., Johnsen, P.B., Maniak, P.J., Grant, G.C., Polkinghorne, C.N., and Sorensen, P.W. 2000. Direct behavioral evidence that unique bile acids released by larval sea lamprey (*Petromyzon marinus*) function as a migratory pheromone. *Can. J. Fish Aquat. Sci.* 57, 557–569.

Buck, L. and Axel, R. 1991. A novel multigene family may encode odorant receptors—A molecular basis for odor recognition. *Cell* 65(1), 175–187.

Buck, L.B. 2000. The molecular architecture of odor and pheromone sensing in mammals. *Cell* 100:611–618.

Camacho, S., Ostos-Garrido, M.V., Domezain, A., and Carmona, R. 2010. Study of the olfactory epithelium in the developing sturgeon. Characterization of the crypt cells. *Chem. Senses.* 35(2), 147–156.

Cao, Y., Oh, B.C., and Stryer, L. 1998. Cloning and localization of two multigene receptor families in goldfish olfactory epithelium. *Proc. Natl. Acad. Sci. USA* 95(20), 11987–11992.

Caprio, J., Brand, J.G., Teeter, J.H., Valentincic, T., Kalinoski, D.L., Kohbara, J., Kumazawa, T., Wegert, S. 1993. The taste system of the channel catfish—From biophysics to behavior. *Trends Neurosci.* 16, 192–197.

Castro, A., Becerra, M., Anadón, R., and Manso, M.J. 2008. Distribution of calretinin during development of the olfactory system in the brown trout, *Salmo trutta fario*: Comparison with other immunohistochemical markers. *J. Chem. Neuroanat.* 35(4), 306–316.

Catania, S., Germana, W., Laura, R., Gonzalez-Martinez, T., Ciriaco, E., and Vega, J.A. 2003. The crypt neurons in the olfactory epithelium of the adult zebrafish express TrkA-like immunoreactivity. *Neurosci. Lett.* 350:5–8.

Chandrashekar, J., Hoon, M.A., Ryba, N.J., and Zuker, C.S. 2006. The receptors and cells for mammalian taste. *Nature* 444, 288–294.

Charlebois, P.M., Corkum, L.D., Jude, D.J., and Knight, C. 2001. The round goby (*Neogobius melanostomus*) invasion: Current research and future needs. *J. Great Lakes Res.* 27, 263–266.

Chaudhari, N. and Roper, S.D. 2010. The cell biology of taste. *J. Cell Biol.* 190, 285–296.

Colombo, L., Marconato, A., Belvedere, P.C., and Friso, C. 1980. Endocrinology of teleost reproduction: A testicular steroid pheromone in the black goby, *Gobius jozo*. *Boll. Zool.* 47(3–4), 355–364. doi: 10.1080/11250008009438692.

Corkum, L.D., Arbuckle, W.J., Belanger, A.J., Gammon, D.B., Li, W., Scott, A.P. and Zielinski, B. 2006. Evidence of a male sex pheromone in the round goby (*Neogobius melanostomus*). *Biol. Invasions* 8(1), 105–112.

Corkum, L.D., Arbuckle, W.J., Belanger, A.J., Gammon, D.B., Li, W., Scott, A.P. and Zielinski, B. 2009. Can the invasive fish, the round goby, be controlled? In *Great Lakes Fisheries Commission Special Publication* 09–02. Edited by J.T. Tyson, R.A. Stein and J.M. Dettmers. pp. 71–78.

Corkum, L.D., Meunier, B., Moscicki, M., Zielinski, B.S., and Scott, A.P. 2008. Behavioral responses of female round gobies (*Neogobius melanostomus*) to putative steroidal pheromones. *Behaviour* 145, 1357–1365.

Derby, C.D. 2007. Escape by inking and secreting: Marine molluscs avoid predation through a rich array of chemical mechanisms. *Biol. Bull.* 213, 274–289.

Derjean, D., Moussaddy, A., Atallah, E., St-Pierre, M., Auclair, F., Chang, S., Ren, X., Zielinski, B., and Dubuc, R.J. 2010. A novel neural substrate for the transformation of olfactory inputs into motor output. *PLoS Biol.* 8, e1000567.

Dolensek, J.L. and Valentincic, T. 2010. Specificities of olfactory receptor neuron responses to amino acids in the black bullhead catfish (Amerius melas). *Pflugers Arch.* 459(3), 413–425.

Doving, K.B. and Selset, R. 1980. Behavior patterns in cod released by electrical-stimulation of olfactory tract bundlets. *Science* 207, 559–560.

Doving, K.B., Dubois-Dauphin, M., Holley, A., and Jourdain, F. 1977. Functional anatomy of the olfactory organ of fish and the ciliary mechanism of water transport. *Acta Zool.* 58, 245–255.

Doving, K.B., Selset, R., Thommesen, G. 1980. Olfactory sensitivity to bile acids in salmonid fishes. *Acta Physiol Scand.* 108, 123–131.

Dulac, C. and Axel, R. 1995. A novel family of genes encoding putative pheromone receptors in mammals. *Cell* 83, 195–206.

Eram, M. and Michel, W.C. 2005. Morphology and small metabolite profiles of facial and vagal innervated taste buds of the channel catfish, *Ictalurus punctatus. J. Comp. Neurol.* 486, 132–144.

Ferrando, S., Bottaro, M., Pedemonte, F., De Lorenze, S., Gallus, L., and Tagliafierro, G. 2007. Appearance of crypt neurons in the olfactory epithelium of the skate *Raja clavata* during development. *Anat. Rec.* 290, 1268–1272.

Ferrando, S., Gallus, L., Gambardella, C., Vacchi, M., Tagliafierro, G. 2010. G protein alpha subunits in the olfactory epithelium of the holocephalan fish Chimaera monstrosa. *Neurosci Letters.* 472(1), 65–67.

Ferrando, S., Gambardella, C., Ravera, S., Bottero, S., Ferrando, T., Gallus, L., Manno, V., Salati, A.P., Ramoino, P., and Tagliafierro, G. 2009. Immunolocalization of G-protein alpha subunits in the olfactory system of the cartilaginous fish *Scyliorhinus canicula. Anat. Rec.* (Hoboken) 292(11), 1771–1779.

Finger, T.E. 2009. Evolution of gustatory reflex systems in the brainstems of fishes. *Integrative Zoology.* 4(1), 53–63.

Finger, T.E. and Morita, Y. 1985. Two gustatory systems: Facial and vagal gustatory nuclei have different brainstem connections. *Science* 227, 776–778.

Friedrich, R.W. and Korsching, S.I. 1997. Combinatorial and chemotopic odorant coding in the zebrafish OB visualized by optical imaging. *Neuron* 18, 737–752.

Friedrich, R.W. and Korsching, S.I. 1998. Chemotopic, combinatorial, and noncombinatorial odorant representations in the OB revealed using a voltage-sensitive axon tracer. *J. Neurosci.* 18, 9977–9988.

Friedrich, R.W. and Laurent, G. 2001. Dynamic optimization of odor representations by slow temporal patterning of mitral cell activity. *Science* 291, 889–894.

Friedrich, R.W. and Laurent, G. 2004. Dynamics of OB input and output activity during odor stimulation in zebrafish. *J. Neurophysiol.* 91, 2658–2669.

Frontini, A., Zaidi, A.U., Hua, H., Wolak, T.P., Greer, C.A., Kafitz, K.W., Li, W.M., and Zielinski, B.S. 2003. Glomerular territories in the OB from the larval stage of the sea lamprey *Petromyzon marinus. J. Comp. Neurol.* 465, 27–37.

Fuller, C.L. and Byrd, C.A. 2005. Ruffed cells identified in the adult zebrafish OB. *Neurosci. Lett.* 379, 190–194.

Gammon, D.B., Li, W., Scott, A.P., Zielinski, B.S., and Corkum, L.D. 2006. Behavioural responses of female *Neogobius melanostomus* to odours of conspecifics. *J. Fish Biol.* 67, 615–626.

Gayoso, J., Castro, A., Anadon, R., and Manso, M.J. 2012. Crypt cells of the zebrafish *Danio rerio* mainly project to the dorsomedial glomerular field of the olfactory bulb. *Chem. Senses* 37, 357–369.

Germana, A., Montalbano, G., Laura, R., Ciriaco, E., del Valle M.E., and Vega, J.A. 2004. S100 protein-like immunoreactivity in the crypt olfactory neurons of the adult zebrafish. *Neurosci. Lett.* 37, 196–198.

Gire, D.H. and Schoppa, N.E. 2009. Control of on/off glomerular signaling by a local GABAergic microcircuit in the olfactory bulb. *J. Neurosci.* 29, 13454–13464.

Gloriam, D.E., Bjarnadottir, T.K., Yan, Y.L., Postlethwait, J.H., Schioth, H.B., and Fredriksson, R. 2005. The repertoire of trace amine G-protein—Coupled receptors: Large expansion in zebrafish. *Mol. Phylogenet. Evol.* 35, 470–482.

Hagelin, L. and Johnels, A.G. 1995. On the structure and function of the accessory olfactory organ in lampreys. *Acta Zool.* 36, 113–125.

Hamdani, E.H., Alexander, G., and Doving, K.B. 2001a. A projection of sensory neurons with microvilli to the lateral olfactory tract indicates their participation in feeding behaviour in crucian carp. *Chem. Senses* 26, 1139–1144.

Hamdani, E.H. and Doving, K.B. 2007. The functional organization of the fish olfactory system. *Prog. Neurobiol.* 82, 80–86.

Hamdani, E.H., Kasumyan, A., and Doving, K.B. 2001b. Is feeding behaviour in crucian carp mediated by the lateral olfactory tract? *Chem. Senses* 26, 1133–1138.

Hamdani, el H., Lastein, S., Gregersen, F., and Døving, K.B. 2008. Seasonal variations in olfactory sensory neurons—Fish sensitivity to sex pheromones explained? *Chem. Senses* 33(2), 119–123.

Hamdani, E.H., Stabell, O.B., Alexander, G., and Doving, K.B. 2000. Alarm reaction in the crucian carp is mediated by the medial bundle of the medial olfactory tract. *Chem. Senses* 25, 103–109.

Hansen, A. and Finger, T.E. 2000. Phyletic distribution of crypt-type olfactory receptor neurons in fishes. *Brain Behav. Evol.* 55, 100–110.

Hansen, A. and Reutter, K. 2004. Chemosensory systems in fish: structural, functional and ecological aspects. In: *The Senses of Fish Adaptations for the Reception of Natural Stimuli* (Von der Emide, G., Mogdans, J., Kapoor, B.G. Eds. pp. 55–89 Narosa Publishing House, New Delhi.

Hansen, A. and Zeiske, E. 1998. The peripheral olfactory organ of the zebrafish, *Danio rerio*: An ultrastructural study. *Chem. Senses* 23(1), 39–48.

Hansen, A. and Zielinski, B.S. 2005. Diversity in the olfactory epithelium of bony fishes: Development, lamellar development, lamellar arrangement, sensory neuron cell types and transduction components. *J. Neurocytol.* 34, 183–208.

Hansen, A., Anderson, K., and Finger, T.E. 2004. Differential distribution of olfactory receptor neurons in goldfish: Structural and molecular correlates. *J. Comp. Neurol.* 477, 347–359.

Hansen, A., Rolen, S.H., Anderson, K., Morita, Y., Caprio, J., and Finger, T.E. 2003. Correlation between olfactory receptor cell type and function in the channel catfish. *J. Neurosci.* 23, 9328–9339.

Hara, T.J. 2007. Gustation. In *Sensory Systems Neuroscience*, Vol. 25, Fish physiology, eds. A.P. Farrell and C.J. Brauner. Academic Press, New York, pp. 45–96.

Hara, T.J. 2011. Gustatory detection of tetrodotoxin and saxitoxin, and its competitive inhibition by quinine and strychnine in freshwater fishes. *Mar. Drugs* 9, 2283–2290.

Hara, T.J. and Zhang, C. 1998. Topographic bulbar projections and dual neural pathways of the primary olfactory neurons in salmonid fishes. *Neuroscience* 82, 301–313.

Hara, T.J. and Zhang, C.B. 1996. Spatial projections to the OB of functionally distinct and randomly distributed primary neurons in salmonid fishes. *Neurosci. Res.* 26, 65–74.

Hashiguchi, Y. and Nishida, M. 2007. Evolution and origin of trace amine associated receptor (TAAR) gene family in vertebrates: Lineage specific expansions and degradations of a second class of vertebrate chemosensory receptors expressed in the olfactory epithelium. *Mol. Biol. Evol.* 24, 2009–2107.

Hasler, A.D., Scholz, A.T., and Horrall, R.M. 1978. Olfactory imprinting and homing in salmon. *Am. Scientist* 66, 347–355.

Hildebrand, J.G. and Shepherd, G.M. 1997. Mechanisms of olfactory discrimination: Converging evidence for common principles across phyla. *Annu. Rev. Neurosci.* 20, 595–631.

Hubbard, P.C., Barata, E.N., Canario, A.V.M. 2000. Olfactory sensitivity to changes in environmental [ca^{2+}] in the marine teleost *Sparus aurata*. J. Exp. Biol. 203, 3821–3829.

Hubbard, P.C., Canario, A.V.M. 2007. Evidence that olfactory sensitivities to calcium and sodium are mediated by different mechanisms in the goldfish *Carassius auratus*. Neurosci Letters. 414, 90–93.

Hubbard, P.C., Ingleton, P.M., Bendell, L.A., Barate, E.N., and Canario, A.V.M. 2002. Olfactory sensitivity to changes in environmental [$Ca2^+$] in the freshwater teleost *Carassius auratus*: An olfactory role for the $Ca2^+$ sensing receptor. *J. Exp. Biol.* 205, 2755–2764.

Huertas, M., Scott, A.P., Hubbard, P.C., Canario, A.V., and Cerda, J. 2006. Sexually mature European eels (*Anguilla anguilla* L.) stimulate gonadal development of neighbouring males: Possible involvement of chemical communication. *Gen. Comp. Endocrinol.* 147, 304–313.

Huesa, G., Anadon, R., and Yanez, J. 2000. Olfactory projections in a chondrostean fish, *Acipenser baeri*: An experimental study. *J. Comp. Neurol.* 428, 145–158.

Huesa, G., Anadon, R., and Yanez, J. 2003. Afferent and efferent connections of the cerebellum of the chondrostean *Acipenser baeri*: A carbocyanine dye (DiI) tracing study. *J. Comp. Neurol.* 460, 327–344.

Huesa, G., Anadon, R., and Yanez, J. 2006. Topography and connections of the telencephalon in a Chondrostean, *Acipenser baeri*: An experimental study. *J. Comp. Neurol.* 497, 519–541.

Hussain, A., Saraiva, L.R., and Korsching, S.I. 2009. Positive Darwinian selection and the birth of an olfactory receptor clade in teleosts. *PNAS* 106, 4313–4318.

Ishimaru, Y., Okada, S., Naito, H., Nagai, T., Yasuoka, A., Matsumoto, I., and Abe, K. 2005. Two families of candidate taste receptors in fishes. *Mech. Dev.* 122, 1310–1321.

Iwahori, N., Kiyota, E., and Nakamura, K. 1987. A golgi-study on the olfactory-bulb in the lamprey, *Lampetra japonica*. *Neurosci. Res.* 5, 126–139.

Jasra, S.K., Arbuckle, W.J., Corkum, L.D., Li, W., Scott, A.P., and Zielinski, B. 2007. The seminal vesicle synthesizes steroids in the round goby *Neogobius melanostomus*. *Comp. Biochem. Physiol.* 148A(1), 117–123. doi: 10.1016/j.cbpa.2007.03.034.

Johnson, B.A. and Leon, M. 2007. Chemotopic odorant coding in a mammalian olfactory system. *J. Comp. Neurol.* 503, 1–34.

Johnson, N.S., Siefkes, M.J., and Li, W.M. 2005. Capture of ovulating female sea lampreys in traps baited with spermiating male sea lampreys. *North Am. J. Fish Manage* 25, 67–72.

Johnson, N.S., Yun, S.S., Thompson, H.T., Brant, C.O., and Li, W.M. 2009. A synthesized pheromone induces upstream movement in female sea lamprey and summons them into traps. *P Natl. Acad. Sci. USA* 106, 1021–1026.

Kajiya, K., Inaki, K., Tanaka, M., Haga, T., Kataoka, H., Touhara, K. 2001. Molecular bases of odor discrimination: Reconstitution of olfactory receptors that recognize overlapping sets of odorants. *J Neurosci.* 21(16), 6018–6025.

Kang, J. and Caprio, J. 1995. In vivo responses of single olfactory receptor neurons in the channel catfish, *Ictalurus punctatus. J. Neurophysiol.* 73, 172–177.

Katare, Y.K., Scott, A.P., Laframboise, A.J., Li, W., Alyasha'e, Z., Caputo, C.B., Loeb, S.J., and Zielinski, B. 2011. Release of free and conjugated forms of the putative pheromonal steroid 11-oxo-etiocholanolone by reproductively mature male round goby (*Neogobius melanostomus* Pallas, 1814). *Biol. Reprod.* 84(2), 288–298. doi: 10.1095/biolreprod.110.086546.

Kay, L.M. and Stopfer, M. 2006. Information processing in the olfactory systems of insects and vertebrates. *Semin. Cell Dev. Biol.* 17, 433–442.

Kay, L.M., Beshel, J., Brea, J., Martin, C., Rojas-Libano, D., and Kopell, N. 2009. Olfactory oscillations: The what, how and what for. *Trend. Neurosci.* 32, 207–214.

Kim, K.Y., Joo, H.J., Kwon, H.W., Kim, H., Hancock, W.S., and Pak, Y.D. 2013. Development of a method to quantitate nematode pheromone for study of small-molecule metabolism in *Caenorhabditis elegans. Anal. Chem.* 85, 2681–2688.

Kinnamon, S.C. 2010. T1R-mediated taste transduction mechanisms. *J. Dairy Sci.* 93, Supplement 1, 865–865.

Kinnamon, S.C. 2012. Taste receptor signalling—From tongues to lungs. *Acta Physiol.* 204, 158–168.

Kirino, M., Parnes, J., Hansen, A., Kiyohara, S., and Finger, T.E. 2013. Evolutionary origins of taste buds: Phylogeneic analysis of purinergic neurotransmission in epithelial chemosensors. *Open Biol.* 3, 1300015.

Kleerekoper, H. and Mogensen, J. 1963. Role of olfaction in the orientation of *Petromyzon marinus*, I, Response to a single amine in prey's body odor. *Physiol. Zool.* 36, 347–360.

Kohbara, J., Michel, W., and Caprio, J. 1992. Responses of single facial taste fibers in the channel catfish, *Ictalurus punctatus*, to amino acids. *J. Neurophysiol.* 68, 1012–1026.

Korsching, S.I. 2001. Odor maps in the brain: Spatial aspects of odor representation in sensory surface and OB. *Cell Mol. Life Sci.* 58, 520–530.

Kosaka, T. 1980. Ruffed cell—A new type of neuron with a distinctive initial unmyelinated portion of the axon in the olfactory-bulb of the goldfish (*Carassius auratus*).2. Fine-structure of the ruffed cell. *J. Comp. Neurol.* 193, 119–145.

Kosaka, T. and Hama, K. 1979. Ruffed cell—New type of neuron with a distinctive initial unmyelinated portion of the axon in the olfactory-bulb of the goldfish (*Carassius auratus*).1. Golgi impregnation and serial thin sectioning studies. *J. Comp. Neurol.* 186, 301–319.

Kosaka, T. and Hama, K. 1980. Presence of the ruffed cell in the olfactory-bulb of the catfish, parasilurus-asotus, and the sea eel, *Conger myriaster. J. Comp. Neurol.* 193, 103–117.

Kosaka, T. and Hama, K. 1981. Ruffed cell—A new type of neuron with a distinctive initial unmyelinated portion of the axon in the olfactory-bulb of the goldfish (*Carassius auratus*).3. 3-Dimensional structure of the ruffed cell dendrite. *J. Comp. Neurol.* 201, 571–587.

Kosaka, T. and Hama, K. 1982. Synaptic organization in the teleost OB. *Journal de Physiologie* 78, 707–719.

Laberge, F. and Hara, T.J. 2001. Neurobiology of fish olfaction: A review. *Brain Res. Rev.* 36, 46–59.

Laberge, F. and Hara, T.J. 2003. Non-oscillatory discharges of F-prostaglandin responsive neuron population in the olfactory bulb-telencephalon transition are in lake whitefish. *Neuroscience* 116, 1089–1095.

Laberge, F. and Hara, T.J. 2004. Electrophysiological demonstration of independent olfactory receptor types and associated neuronal responses in the trough olfactory bulb. *Comp. Biochem. Physiol. A. Mol. Integr. Physiol.* 137, 397–408.

Laframboise, A.J. and Zielinski, B.S. 2011. Responses of the round goby (*Neogobius melanostomus*) olfactory epithelium to steroids released by reproductive males. *J. Comp. Physiol. A Neuroethol. Sens. Neural. Behav. Physiol.* 10, 999–1008.

Laframboise, A.J., Ren, X., Chang, S., Dubuc, R., and Zielinski, B.S. 2007. Olfactory sensory neurons in the sea lamprey display polymorphisms. *Neurosci. Lett.* 414, 277–281.

Laurent, G. 2002. Olfactory network dynamics and the coding of multidimensional signals. *Nat. Rev. Neurosci.* 3, 884–895.

Laurent, G. and Davidowitz, H. 1994. Encoding of olfactory information with oscillating neural assemblies. *Science* 265, 1872–1875.

Laurent, G., Wehr, M., and Davidowitz, H. 1996. Temporal representations of odors in an olfactory network. *J. Neurosci.* 16, 3837–3847.

Li, W. 1994. The olfactory biology of adult sea lamprey (*Petromyzon marinus*). Doctoral dissertation, University of Minnesota, Ann Arbor, MI: ProQuest/UMI, (Publication No. 9514669).

Li, W. and Sorensen, P.W. 1997. Highly independent olfactory receptor sites for naturally occurring bile acids in the sea lamprey (*Petromyzon-marinus*). *J. Comp. Physiol. A-Sens. Neural. Behav. Physiol.* 180, 429–438.

Li, W.M., Scott, A.P., Siefkes, M.J., Yan, H.G., Liu, Q., Yun, S.S., and Gage, D.A. 2002. Bile acid secreted by male sea lamprey that acts as a sex pheromone. *Science* 296, 138–141.

Li, W.M., Sorensen, P.W., and Gallaher, D.D. 1995. The olfactory system of migratory adult sea lamprey (*Petromyzon marinus*) is specifically and acutely sensitive to unique bile-acids released by conspecific larvae. *J. Gen. Physiol.* 105, 569–587.

Libants, S., Carr, K., Wu, H., Teeter, J.H., Chung-Davidson, Y.W., Zhang, Z.P., Wilkerson, C., Li, W. July 2009. The sea lamprey *Petromyzon marinus* genome reveals the early origin of several chemosensory receptor families in the vertebrate lineage. *BMC Evol. Biol.* 9, 180. doi: 10.1186/1471-2148-9-180.

Liberles, S.D. and Buck, L.B. 2006. A second class of chemosensory receptors in the olfactory epithelium. *Nature* 442, 645–650.

Lim, H. and Sorensen, P.W. 2011. Polar metabolites synergize the activity of prostaglandin F2α in a species-specific hormonal sex pheromone released by ovulated common carp. *J. Chem. Ecol.* 37, 695–704.

Liman, E.R., Corey, D.P., and Dulac, C. 1999. TRP2: A candidate transduction channel for mammalian pheromone sensory signaling. *PNAS* 96, 5791–5796.

Lindemann, L. and Hoener, M.C. 2005. A renaissance in trace amines inspired by a novel GPCR family. *Trends Pharmacol. Sci.* 26, 274–281.

Lo, Y.H., Bellis, S.L., Cheng, L.J., Pang, J., Bradley, T.M., Rhoads, D.E. 1994. Signal transduction for taurocholic acid in the olfactory system of Atlantic salmon. *Chem. Senses.* 19(5), 371–380.

Lo, Y.H., Bradley, T.M., Rhoads, D.E. 1993. Stimulation of Ca(2+)-regulated olfactory phospholipase C by amino acids. *Biochemistry.* 32(46), 12358–12362.

Luu, P., Acher, F., Bertrand, H.O., Fan, J., and Ngai, J. 2004. Molecular determinants of ligand selectivity in a vertebrate odor receptor. *J. Neurosci.* 24, 10128–10137.

Ma, L., Michel, W.C. 1998. Drugs affecting phospholipase C-mediated signal transduction block the olfactory cyclic nucleotide-gated current of adult zebrafish. *J Neurophysiol.* 79(3), 1183–1192.

MacInnis, A.J. and Corkum, L.D. 2000. Fecundity and reproductive season of the round goby *Neogobius melanostomus* in the upper Detroit River. *T. Am. Fish. Soc.* 129, 136–144.

MacLeod, K. and Laurent, G. 1996. Distinct mechanisms for synchronization and temporal patterning of odor-encoding neural assemblies. *Science* 274, 976–979.

McLaughlin, S.K., McKinnon, P.J., and Margolskee, R.F. 1992. Gustducin is a taste-cell-specific G protein closely related to the transducins. *Nature* 357, 563–569.

Meredith, T.L., Caprio, J., Kajiura, S.M. 2012. Sensitivity and specificity of the olfactory epithelia of two elasmobranch species to bile salts. *J Exp Biol.* 215, 2660–2667.

Meredith, T.L., Kajiura, S.M. 2010. Olfactory morphology and physiology of elasmobranchs. *J Exp Biol.* 213, 3449–3456.

Michel, M. 2006. Chemoreception, Chapter 13. In. *The Physiology of Fishes*, 3rd edn. CRC Press, Boca Raton, pp. 471–498.

Michel, W.C. 1999. Cyclic nucleotide-gated channel activation is not required for activity-dependent labeling of zebrafish olfactory receptor neurons by amino acids. *Biol. Sig. Rec.* 8(6), 338–347.

Michel, W.C., Sanderson, M.J., Olson, J.K., Lipschitz, D.L. 2003. Evidence of a novel transduction pathway mediating detection of polyamines by the zebrafish olfactory system. *J Exp Biol.* 206(10), 1697–1706.

Mombaerts, P. 2004. Genes and ligands for odorant, vomeronasal and taste receptors. *Nat. Rev. Neurosci.* 5, 263–278.

Mombaerts, P., Wang, F., Dulac, C., Chao, S.K., Nemes, A., Mendelsohn, M., Edmondson, J., and Axel, R. 1996. Visualizing an olfactory sensory map. *Cell* 87, 675–686.

Mori, K. 1987. Membrane and synaptic properties of identified neurons in the olfactory-bulb. *Prog. Neurobiol.* 29, 275.

Morita, Y. and Finger, T.E. 1998. Differential projections of ciliated and microvillous olfactory receptor cells in the catfish, *Ictalurus punctatus*. *J. Comp. Neurol.* 398, 539–550.

Murphy, C.A., Stacey, N.E., and Corkum, L.D. 2001. Putative steroidal pheromones in the round goby, *Neogobius melanostomus*: Olfactory and behavioral responses. *J. Chem. Ecol.* 27(3), 443–470.

Nevitt, G.A. 1991. Do fish sniff? A new mechanism for olfactory sampling in pleuronectid flounders. *J. Exp. Biol.* 157, 1–18.

Ngai, J., Dowling, M., Buck, L., Axel, R., and Chess, A. 1993. The family of genes encoding odorant receptors in the channel catfish. *Cell* 72, 657–666.

Niimura, Y. and Nei, M. 2005. Evolutionary dynamic of olfactory receptor genes in fishes and tetrapods. *PNAS* 102, 6039.

Nikonov, A.A. and Caprio, J. 2001. Electrophysiological evidence for a chemotopy of biologically relevant odors in the OB of the channel catfish. *J. Neurophysiol.* 86, 1869–1876.

Nikonov, A.A. and Caprio, J. 2004. Odorant specificity of single OB neurons to amino acids in the channel catfish. *J. Neurophysiol.* 92, 123–134.

Nikonov, A.A., Finger, T.E., and Caprio, J. 2005. Beyond the olfactory bulb: An odotopic map in the forebrain. *P Natl. Acad. Sci. USA* 102, 18688–18693.

Nusnbaum, M., Aggio, J.F., and Derby, C.D. 2012. Taste-mediated behavioral and electrophysiological responses by the predatory fish *Ariopsis felis* to deterrent pigments from *Aplysia californica* ink. *J. Comp. Physiol. A. Neuroethol. Sens. Neural. Behav. Physiol.* 198(4), 283–294.

Ochs, C.L., Laframboise, A.J., Green, W.W., Basilious, A., Johnson, T.B., and Zielinski, B.S. 2013. Response of putative round goby (*Neogobius melanostomus*) pheromones by centrarchid and percid fish species in the Laurentian Great Lakes. *J. Great Lakes Res.* 39(1), 186–189.

Ogawa, K. and Caprio, J. 2010. Major differences in the proportion of amino acid fiber types transmitting taste information from oral and extraoral regions in the channel catfish. *J. Neurophysiol.* 103, 2062–2073.

Ohmoto, M., Okada, S., Nakamura, S., Abe, K., and Matsumoto, I. 2011. Mutually exclusive expression of Gaia and Ga14 reveals diversification of taste receptor cells in zebrafish. *J. Comp. Neurol.* 519, 1616–1629.

Oike, H., Nagai, T., Furuyama, A., Okada, S., Aihara, Y., Ishimaru, Y., Marui, T., Matsumoto, I., Misaka, T., Abe, K. 2007. Characterization of ligands for fish taste receptors. *J. Neurosci.* 27(21), 5584–5592.

Oka, Y. and Korsching, S.I. 2011. Shared and unique G alpha proteins in the zebrafish versus mammalian senses of taste and smell. *Chem. Sens.* 36, 357–365.

Oka, Y., Saraiva, L.R., and Korching, S.I. 2012. Crypt neurons express a single V1R-related ora gene. *Chem. Sens.* 37, 219–227.

Olsen, C.A., Kristennsen, A.S., and Stromgaard, K. 2011. Small molecules from spiders used as chemical probes. *Angew. Chem. Int. Ed. Engl.* 50, 11296–11311.

Olsen, K.H., Sawisky, G.R., and Stacey, N.E. 2006. Endocrine and milt responses of male crucian carp (*Carassius carassius L.*) to periovulatory females under field conditions. *Gen. Comp. Endocrinol.* 149, 294–302.

Poos, M., Dextrase, A.J., Schwalb, A.N., and Ackerman, J.D. 2010. Secondary invasion of the round goby into high diversity Great Lakes tributaries and species at risk hotspots: Potential new concerns for endangered freshwater species. *Biol. Inv.* 12, 1269–1284.

Potter, I.C., Hilliard, R.W., and Bird, D.J. 1982. Stages in metamorphosis. In: *The Biology of Lampreys*, Eds. Hardisty, M.W. and Potter, I.C. Academic Press, New York, pp. 137–164.

Ren, X., Chang, S., Laframboise, A., Green, W., Dubuc, R., and Zielinski, B. 2009. Projections from the accessory olfactory organ into the medial region of the OB in the sea lamprey (*Petromyzon marinus*): A novel vertebrate sensory structure. *J. Comp. Neurol.* 516, 105–116.

Restrepo, D., Teeter, J.H., Schild, D. 1996. Second messenger signaling in olfactory transduction. *J. Neurobiol.* 30(1), 37–48.

Rolen, S.H. and Caprio, J. 2007. Processing of bile salt odor information by single OB neurons in the channel catfish. *J. Neurophysiol.* 97, 4058–4068.

Rolen, S.H., Sorense, P.W., Mattison, D., and Caprio, J. 2003. Polyamines as olfactory stimuli in goldfish *Carassius auratus. J. Exp. Biol.* 206, 1683–1696.

Rooney, D., Doving, K.B., Ravailleveron, M., and Szabo, T. 1992. The central connections of the OBs in cod, *Gadus morhua* L. *J. Fur. Hirnforschung* 33, 63–75.

Saraiva, L.R. and Korsching, S.I. 2007. A novel olfactory receptor gene family in teleost fish. *Genome Re.* 17, 1448–1457.

Sato, K., Suzuki, N. 2001. Whole-cell response characteristics of ciliated and microvillous olfactory receptor neurons to amino acids, pheromone candidates and urine in rainbow trout. *Chem Senses.* 26(9), 1145–1156.

Sato, Y., Miyasaka, N., and Yoshihara, Y. 2005. Mutually exclusive glomerular innervation by two distinct types of olfactory sensory neurons revealed in transgenic zebrafish. *J. Neurosci.* 25, 4889–4897.

Sato, Y., Miyasaka, N., and Yoshihara, Y. 2007. Hierarchical regulation of odorant receptor gene choice and subsequent axonal projection of olfactory sensory neurons in zebrafish. *J. Neurosci.* 27, 1606–1615.

Satou, M. 1990. Synaptic organization, local neuronal circuitry, and functional segregation of the teleost olfactory-bulb. *Prog. Neurobiol.* 34, 115–142.

Schild, D., Restrepo, D. 1998. Transduction mechanisms in vertebrate olfactory receptor cells. *Physiol Rev.* 78(2), 429–466.

Schluessel, V., Bennett, M.B., Bleckmann, H., Blomberg, S., Collin, S.R. 2008. Morphometric and ultrastructural comparison of the olfactory system in elasmobranchs: The significance of structure-function relationships based on phylogeny and ecology. *J Morphology.* 269(11), 1365–1386.

Semagor, E., Chabrol, F., Bony, G., and Canedda, L. 2010. GABAergic control of neurite outgrowth and remodeling during development and adult neurogenesis: General rules and difference in diverse systems. *Frontiers Cell. Neurosci.* Doi: 10:3389.

Serizawa, S., Miyamichi, K., and Sakano, H. 2004. One neuron-one receptor rule in the mouse olfactory system. *Trends Gen.* 20, 648–653.

Serrano, R.M., Barata, E.N., Birkett, M.A., Hubbard, P.C., Guerreiro, P.S., and Canario, A.V. 2008. Behavioral and olfactory responses of female *Salaria pavo* (Pisces: Blenniidae) to a putative multi-component male pheromone. *J. Chem. Ecol.* 34, 647–658.

Sheldon, R.E. 1912. The olfactory tracts and centers in teleosts. *J. Comp. Neurol.* 22, 177–339.

Shepherd, G.M. and Greer, C.A. 1998. Olfactory bulb. In: *The Synaptic Organization of the Brain*, 4th edn., Ed. Shepherd, G.M. Oxford University Press, New York, pp. 159–204.

Shepherd, G.M., Chen, W.R., Willhite, D., Migliore, M., and Greer, C.A. 2007. The olfactory granule cell: From classical enigma to central role in olfactory processing. *Brain Res. Rev.* 55, 373–382.

Siefkes, M.J. and Li, W. 2004. Electrophysiological evidence for detection and discrimination of pheromonal bile acids by the olfactory epithelium of female sea lampreys (*Petromyzon marinus*). *J. Comp. Physiol. A Neuroethol. Sens. Neural. Behav. Physiol.* 190, 193–199.

Siefkes, M.J., Scott, A.P., Zielinski, B., Yun, S.S., and Li, W.M. 2003. Male sea lampreys, *Petromyzon marinus* L., excrete a sex pheromone from gill epithelia. *Biol. Reprod.* 69, 125–132.

Smith, J.J., Kuraku, S., Holt, C., Sauka-Spengler, T., Jiang, N., Campbell, M.S., Yandell, M.D., Manousaki, T., Meyer, A., Bloom, O.E., Morgan, J.R., Buxbaum, J.D., Sachidanandam, R., Sims, C., Garruss, A.S., Cook, M., Krumlauf, R., Wiedemann, L.M., Sower, S.A., Decatur, W.A., Hall, J.A., Amemiya, C.T., Saha, N.R., Buckley, K.M., Rast, J.P., Das, S., Hirano, M., McCurley, N., Guo, P., Rohner, N., Tabin, C.J., Piccinelli, P., Elgar, G., Ruffier, M., Aken, B.L., Searle, S.M., Muffato, M., Pignatelli, M., Herrero, J., Jones, M., Brown, C.T., Chung-Davidson, Y.W., Nanlohy, K.G., Libants, S.V., Yeh, C.Y., McCauley, D.W., Langeland, J.A., Pancer, Z., Fritzsch, B., de Jong, P.J., Zhu, B., Fulton, L.L., Theising, B., Flicek, P., Bronner, M.E., Warren, W.C., Clifton, S.W., Wilson, R.K., Li, W. 2013. Sequencing of the sea lamprey (*Petromyzon marinus*) genome provides insights into vertebrate evolution. *Nat Genet.* 45(4), 415–421.

Sorensen, P.W. and Sato, K. 2005. Second messenger systems mediating sex pheromone and amino acid sensitivity in goldfish olfactory receptor neurons. *Chem. Sens.* 30(suppl 1), i315–i316.

Sorensen, P.W., Fine, J.M., Dvornikovs, V., Jeffrey, C.S., Shao, F., Wang, J.Z., Vrieze, L.A., Anderson, K.R., and Hoye, T.R. 2005. Mixture of new sulfated steroids functions as a migratory pheromone in the sea lamprey. *Nat. Chem. Biol.* 1, 324–328.

Sorensen, P.W., Scott, A.P., Stacey, T.E., and Bowdin, L. 1995. Sulfated 17,20 beta-dihydroxy-4-pregnen-3-one functions as a potent and specific olfactory stimulant with pheromonal actions in the goldfish. *Gen. Comp. Endocrinol.* 100, 128–142.

Stacey, N., Chojnacki, A., Narayan, A., Cole, T.L., and Murphy, C. 2003. Hormonally derived sex pheromones in fish: Exogenous cues and signals from gonad to brain. *Can. J. Physiol. Pharmacol.* 81, 329–341.

Stebell, O.B. and Vegusdal, A. 2010. Socializing makes thick-skinned individuals: On the density of epidermal alarm substance cells in cyprinid fish, the crucian carp (*Carassius carassius*). *J. Comp. Physiol. A. Neurothol. Sens. Neural. Behav. Physiol.* 196, 639–647.

Takami, S., Luer, C.A., and Graziadei, P.P. 1994. Microscopic structure of the olfactory organ of the clearnose skate, *Raja eglanteria*. *Anat. Embryol.* (Berl). 190(3), 211–230.

Tan, J., Savigner, A., Ma, M.H., and Luo, M.M. 2010. Odor information processing by the olfactory bulb analyzed in gene-targeted mice. *Neuron* 65, 912–926.

Tierney, K., Kereliuk, M., Katare, Y.K., Scott, A.P., Loeb, S.J., and Zielinski, B. 2013. Invasive male round gobies (*Neogobius melanostomus*) release pheromones in their urine to attract females. *Can J Fish Aquat sci.* 70(3), 393–400.

Toyoshima, K., Nada, O., and Shimamura, A. 1984. Fine structure of monoamine-containing basal cells in the taste buds on the barbels of three species of teleosts. *Cell Tissue Res.* 235, 479–484.

Tricas, T.C., Kajiura, S.M., and Summers, A.P. 2009. Response of the hammerhead shark olfactory epithelium to amino acid stimuli. *J. Comp. Physiol. A. Neuroethol. Sens. Neural. Behav. Physiol.* 2009, 195, 947–954.

Ueda, H. 2012. Physiological mechanisms of imprinting and homing migration in Pacific salmon *Oncorhynchus* spp. *J. Fish Biol.* 81, 543–558.

Urban, N.N. and Sakmann, B. 2002. Reciprocal intraglomerular excitation and intra- and interglomerular lateral inhibition between mouse olfactory bulb mitral cells. *J. Physiol.—Lond.* 542, 355–367.

Valentincic, T. and Caprio, J. 1994. Consummatory feeding behavior to amino acids in intact and anosmic channel catfish *Ictalurus punctatus*. *Physiol. Behav.* 55, 857–863.

VanDenBossche, J., Seelye, J.G., and Zielinski, B.S. 1995. The morphology of the olfactory epithelium in larval, juvenile and upstream migrant stages of the sea lamprey, *Petromyzon-Marinus*. *Brain Behav. Evol.* 45, 19–24.

VanDenBossche, J., Youson, J.H., Pohlman, D., Wong, E., and Zielinski, B.S. 1997. Metamorphosis of the olfactory organ of the sea lamprey (*Petromyzon marinus L*): Morphological changes and morphometric analysis. *J. Morphol.* 231, 41–52.

Vielma, A., Ardiles, A., Delgado, L., and Schmachtenberg, O. 2008. The elusive crypt olfactory receptor neuron: Evidence for its stimulation by amino acids and cAMP pathway agonists. *J. Exp. Biol.* 211, 2417–2422.

Vonbartheld, C.S., Meyer, D.L., Fiebig, E., and Ebbesson, S.O.E. 1984. Central connections of the olfactory-bulb in the goldfish, *Carassius auratus. Cell Tissue Res.* 238, 475–487.

Wachowiak, M. and Cohen, L.B. 2001. Representation of odorants by receptor neuron input to the mouse olfactory bulb. *Neuron* 32, 723–735.

Wachowiak, M. and Cohen, L.B. 2003. Correspondence between odorant-evoked patterns of receptor neuron input and intrinsic optical signals in the mouse olfactory bulb. *J. Neurophysiol.* 89, 1623–1639.

Wehr, M. and Laurent, G. 1996. Odour encoding by temporal sequences of firing in oscillating neural assemblies. *Nature* 384, 162–166.

Wick, S.D., Wiechert, M.T., Friedrich, R.W., and Riecke, H. 2010. Pattern orthogonalization via channel decorrelation by adaptive networks. *J. Comput. Neurosci.* 28, 29–45.

Wiechert, M.T., Judkewitz, B., Riecke, H., and Friedrich, R.W. 2010. Mechanisms of pattern decorrelation by recurrent neuronal circuits. *Nat. Neurosci.* 13, U1003–U1132.

Wisenden, B.D. 2000. Olfactory assessment of predation risk in the aquatic environment. *Philos Tran. R Soc. Lond. B Biol. Sci.* 355, 1205–1208.

Wong, G.T., Gannon, K.S., and Margolskee, R.F. 1996. Transduction of bitter and sweet taste by gustducin. *Nature* 381, 796–800.

Wyatt, T.D. 2003. Pheromones and animal behavior: Communication by smell and taste. Cambridge University Press, Cambridge, U.K., p. 391.

Xu, F.Q., Greer, C.A., and Shepherd, G.M. 2000. Odor maps in the olfactory bulb. *J. Comp. Neurol.* 422, 489–495.

Xu, F.Q., Schaefer, M., Kida, I., Schafer, J., Liu, N., Rothman, D.L., Hyder, F., Restrepo, D., and Shepherd, G.M. 2005. Simultaneous activation of mouse main and accessory olfactory bulbs by odors or pheromones. *J. Comp. Neurol.* 489, 491–500.

Yamamori, K., Nakamura, M., Matsui, T., and Hara, T.J. 1988. Gustatory responses to tetrodotoxin and saxitoxin in fish, a possible mechanism for avoiding marine toxins. *Can. J. Fish. Aquat. Sci.* 45, 2182–2186.

Yamashita, S., Yamada, T., and Hara, T.J. 2006. Gustatory responses to feeding- and non-feeding-stimulant chemicals with special emphasis on amino acids in rainbow trout. *J. Fish Biol.* 68, 783–800.

Yavno, S. and Corkum, L.D. 2010. Reproductive female round gobies (*Neogobius melanostomus*) are attracted to visual male models at a nest rather than to olfactory stimuli in urine of reproductive males. *Behaviour* 147(1), 121–132.

Yoshida, Y., Saitoh, K., Aihara, Y., Okada, S., Misaka, T., and Abe, K. 2007. Transient receptor potential channel M5 and phospholipase C-beta2 colocalizing in zebrafish taste receptor cells. *Neuroreport* 18, 1517–1520.

Zeiske, E., Bartsch, P., and Hansen, A. 2009. Early ontogeny of the olfactory organ in a basal actinopterygian fish: Polypterus. *Brain Behav. Evol.* 73(4), 259–272.

Zeiske, E., Kasumyan, A., Bartsch, P., and Hansen, A. 2003. Early development of the olfactory organ in sturgeons of the genus Acipenser: A comparative and electron microscopic study. *Anat. Embryol.* (Berl). 206(5), 357–372.

Zhang, J., Pacifico, R., Cawley, D., Feinstein, P., and Bozza, T. 2013. Ultrasensitive detection of amines by a trace amine-associated receptor. *J. Neurosci.*, 33, 3228–3239.

Zheng, W. and Stacey, N.E. 1997. A steroidal pheromone and spawning stimuli act via different neuroendocrine mechanisms to increase gonadotropin and milt volume in male goldfish *Carassius auratus. Gen. Comp. Endocrinol.* 105, 228–238.

Zielinski, B.S. and Hara, T.J. 2007. Olfaction. In: *Sensory Systems Neuroscience*, Eds. Hara, T.J. and Zielinski, B.S., Vol. 25, Fish Physiology. Elsevier, San Diego, CA, pp. 1–43.

Zippel, H.P., Gloger, M., Luthje, L., Nassar, S., and Wilcke, S. 2000. Pheromone discrimination ability of olfactory bulb mitral and ruffed cells in the goldfish (*Carassius auratus*). *Chem. Sens.* 25, 339–349.

Zippel, H.P., Reschke, C., and Korff, V. 1999. Simultaneous recordings from two physiologically different types of relay neurons, mitral cells and ruffed cells, in the OB of goldfish. *Cell Mol. Biol.* 45, 327–337.

12 Active Electroreception
Signals, Sensing, and Behavior

John E. Lewis

CONTENTS

12.1 INTRODUCTION

What's in an image? Arguably, this question is most often considered in the context of vision. An observer can attempt an answer by considering the spatial pattern of color. Even in a static image or snapshot, the observer can extract a range of information through visual cues like perspective, figure–ground separation, and shading. Typically, though, vision is a dynamic process involving not only continuously changing images, but also active eye movements; these dynamics enable the extraction of additional visual information (e.g., Wexler and van Bostel 2005). Less intuitive, however, is the nature of information available to animals that sense with electricity (e.g., Heiligenberg 1993; Albert and Crampton 2005).

Most examples of electric sensing in nature are *passive*, in that the sensing animal detects the changing bioelectric fields produced by other animals. However, in so-called weakly electric fish, electric sensing is also *active*, in that information about the environment is provided through perturbations in a self-generated electric field. In both passive and active electric sensing, the spatiotemporal changes in electric field strength produced by a dynamic environment form the sensory basis for prey capture, object tracking, navigation, as well as communication (Lissmann 1951; Lissmann and Machin 1958; Heiligenberg 1991; Caputi and Budelli 2006); the difference between the two modes of electric sensing lies in the source of the electric field. This chapter will focus on active electroreception. First, I will provide a brief introduction to weakly electric fish, the electric organ discharge (EOD), and associated neural pathways. Then, I will consider what electric signals are available to weakly electric fish, and what information they provide about the environment. Next, I will summarize electric sensing, or rather some aspects of electrosensory information processing. Last, I will discuss some recent behavioral studies that will guide future work on both electrosensory signals and electrosensory processing.

12.1.1 WEAKLY ELECTRIC FISH

Weakly electric fish generate an electric field around their body using a specialized electric organ (EO). By sensing perturbations in this self-generated electric field with electroreceptors distributed over their skin, they can detect objects as well as the EODs of other fish (Lissmann 1951; Lissmann and Machin 1958; Caputi and Budelli 2006). Because this form of sensing involves probing the environment with a self-generated signal, it is a form of active sensing (e.g., Nelson and MacIver 1999, 2006). Active electric sensing has evolved independently in two groups of teleost fishes, the South and Central American gymnotiforms and the African mormyriforms (Moller 1995; Bullock et al., 2005). Ancestors of both groups were limited to passive electric sensing, an ability that also evolved independently in each case (e.g., Lavoué et al. 2012). In all active sensing species, the self-generated signal is weak (millivolt scale), with the exception of the gymnotiform *Electrophorus electricus*, which in addition to this weak active-sensing signal, also has the ability to produce a strong discharge (hundreds of volts) for prey capture and defense. The evolution and diversity of electric sensing has been the subject of several additional reviews (e.g., Alves-Gomes 1999; Crampton 2006; Crampton and Albert 2006; Lavoué et al. 2012), including previous editions of this book (Heiligenberg 1993; Albert and Crampton 2005).

12.1.1.1 Electric Organ Discharge (EOD)

EODs vary widely across species but in general come in two types: pulse type or wave type (Figure 12.1; Kramer 1990, 1996; Moller 1995; Caputi and Budelli 2006; Crampton and Albert 2006). A pulse type EOD comprises brief pulse discharges (pulse duration of about a millisecond), separated by much longer and often variable intervals (average pulse rate ranging from 1 to 150 Hz). On the other hand, a wave-type EOD is a quasi-sinusoidal periodic signal, oscillating at species-specific frequency ranges between 100 and 2000 Hz. Wave- and pulse-type discharges are found in both mormyriform and gymnotiform species. The spatial nature of the electric field resulting from the EOD is roughly dipolar for both pulse- and wave-type species. Over the EOD pulse time or cycle, the field changes polarity from a head-positive phase to a head-negative phase, resembling an oscillating dipole (Figure 12.2; see later section).

Electric fish species can also be differentiated by the nature of their EO (Bennett 1971; Moller 1995). In the Apteronotidae (gymnotiforms), the EO comprises the specialized terminals of spinal motor neurons, and thus the EOD is referred to as *neurogenic*. In all remaining gymnotiforms and in all mormyriforms, the EO comprises electrocytes derived from muscle tissue that are innervated by spinal motor neurons; thus, the EOD in these species is referred to as *myogenic*. In both cases, an action potential is fired by each cell of the EO for every EOD cycle. The timing of the EOD is controlled by a medullary pacemaker or command nucleus (Kramer 1990; Moortgat et al. 2000; Carlson 2002; Curti et al. 2006) whose action potential output drives the electrical pulses generated by the EO either directly in a one-to-one fashion through pacemaker relay neurons (gymnotiforms; Moortgat et al. 2000; Quintana et al. 2011) or indirectly via two relay nuclei (mormyriforms; Carlson 2002; Zhang and Kawasaki 2006).

The pacemaker nucleus in gymnotiforms comprises 100–200 neurons of two main classes: pacemaker neurons, which are intrinsic to the nucleus, and the relay neurons whose axons project down the spinal cord and innervate spinal motor neurons of the EO (Figure 12.3). Within the pacemaker, neurons are connected primarily by gap junctions (e.g., Moortgat et al. 2000). In *Apteronotus leptorhynchus*, an additional class of interneurons, the parvo-cells, has been identified (Smith et al.

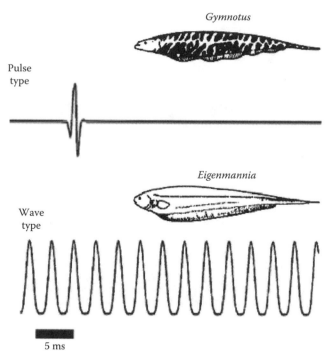

FIGURE 12.1 Electric organ discharges (EODs) for a pulse-type (a, *Gymnotus varzea*) and a wave-type (b, *Eigenmannia cf. virescens*) gymnotiform fish. (Modified from Heiligenberg, W.F., *Neural Nets in Electric Fish*, MIT Press, Cambridge, MA, 1991; Albert, J.S. and Crampton, W.G.R., Electroreception and electrogenesis. In: Evans D.H., Ed., *The Physiology of Fishes*, 3rd edn., CRC Press, New York, pp. 431–472, 2005.) Time bar applies to both recordings.

2000). The pacemaker nucleus receives inputs from two prepacemaker nuclei, the sublemniscal prepacemaker nucleus (sPPn) and the prepacemaker nucleus, which modulate EOD frequency (EODf) (Heiligenberg 1991; Metzner 1999).

The command nucleus of the mormyrid electromotor networks comprises about 20 cells that project to the 20–30 cells of the medullary relay nucleus, both directly and indirectly via the bulbar command-associated nucleus. In turn, neurons of the medullary relay nucleus project down the spinal cord to the EO. The command nucleus also receives inputs from the torus and mesencephalic precommand nucleus (Bell et al. 1983; Kramer 1990; Carlson 2002). Note that wave-type mormyrids have electromotor control of intermediate complexity, involving a direct multisynaptic pathway between the command (or pacemaker) nucleus, the lateral and medial relay nuclei, and the EO (Kawasaki 2009).

12.1.1.2 Electrosensory Pathways

The signals associated with electric sensing are encoded by two main classes of electroreceptors distributed over the fish's skin: ampullary and tuberous receptors (Heiligenberg 1993; Moller 1995). Ampullary-type electroreceptors are tuned to the low-frequency signals produced by other animals and underlie passive electric sensing. Tuberous electroreceptors are tuned to the high-frequency components of the fish's self-generated field and form the basis for active electric sensing. A large amount of work has described the electrosensory pathways in mormyriforms and gymnotiforms, both pulse and wave type. While the details differ between species, all have evolved dedicated pathways for timing (or phase) and amplitude information. This work has been reviewed extensively in past editions of this book (Heiligenberg 1993; Albert and Crampton 2005) as well as in many other

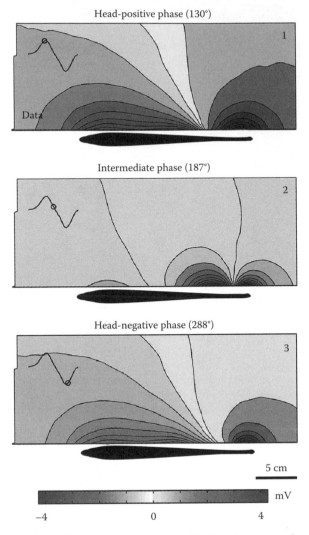

Head-positive phase (130°)

Intermediate phase (187°)

Head-negative phase (288°)

5 cm

−4   0   4   mV

FIGURE 12.2 **(See color insert.)** Electric organ discharge (EOD) voltage maps for *Apteronotus leptorhynchus* at three different phases of a single EOD cycle (130°, 187°, 288°) with EOD frequency (EODf) of 810 Hz. (Modified from Kelly, M. et al., *Biol Cybern*, 98, 479, 2008; Assad, C. et al., *J. Exp. Biol.*, 202, 1185, 1999.) Insert on each panel illustrates single EOD cycle with phase indicated by a circle. Color scale for voltage and spatial scale indicated at bottom.

reviews (e.g., Heiligenberg 1991; Moller 1995; Hopkins 1999; Rose 2004; Kawasaki 2009; Maler 2009). By way of example, here I will focus primarily on the electrosensory pathways (and processing in a subsequent section) in the wave-type gymnotiform *A. leptorhynchus*, one of the species for which most is known.

A. leptorhynchus (brown ghost knifefish) produce a wave-type EOD in the range of 600–1100 Hz. The spatiotemporal modulations of their high-frequency electric field are sensed by specialized tuberous electroreceptors (Carr et al. 1982). Each electroreceptor pore is innervated by a primary afferent neuron that projects to the electrosensory lateral line lobe (ELL) of the hindbrain, the first stage of central nervous system processing (Figure 12.3). These receptor afferents comprise two functional subgroups: T-units that fire action potentials phase-locked to the EOD cycle, and P-units that fire an action potential with a probability that depends on the amplitude of the EOD at the related skin location (Scheich 1973). T-units form the basis of the timing (or phase-coding) pathway, while P-units encode

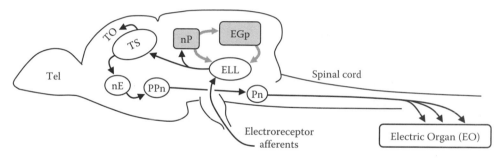

FIGURE 12.3 Schematic of electrosensory pathways of a wave-type gymnotiform, based on Heiligenberg (1991); details provided in the text. Abbreviations: ELL, electrosensory lateral line lobe; nP, nucleus praeminentialis; EGp, eminentia granularis posterior; TS, torus semicircularis; TO, optic tectum; nE, nucleus electrosensorius; PPn, prepacemaker nuclei (SPPn not shown); Pn, pacemaker nucleus. ELL feedback pathways are shown in gray. See Giassi et al. (2012) for connections with telencephalon (Tel), not shown in this figure.

EOD amplitude modulations (AMs; Heiligenberg 1991). The axons of T-units and P-units trifurcate in the ELL and terminate in three distinct somatotopic maps (Carr et al. 1982; Berman and Maler 1999; Metzner 1999; Maler 2009) called the centromedial segment, centrolateral segment, and lateral segment. A fourth segment of the ELL, the medial segment, receives inputs from ampullary receptors.

The ELL pyramidal neurons project to the midbrain torus semicircularis (TS), which then projects to the optic tectum (Figure 12.3; Berman and Maler 1999; Maler 2009). The ELL pyramidal neurons also project to the nucleus praeminentialis (nP), forming the basis for two important feedback pathways to the ELL (Figure 12.3, gray shading). The nP neurons project back to the ELL via the so-called direct feedback pathway; nP also projects to the eminentia granularis posterior (EGp) in the cerebellum. Granule cells in EGp then project back to ELL via the parallel fiber indirect feedback pathway (Berman and Maler 1999).

From the torus and tectum, there are multiple pathways to higher brain areas (e.g., Metzner 1999; Rose 2004; Giassi et al. 2012). The most direct pathway to the output network driving the EOD is from the torus to the diencephalic nucleus electrosensorius (nE), and then to two prepacemaker nuclei (the sPPn and the diencephalic prepacemaker nucleus (PPn)) that both innervate the pacemaker nucleus and directly control EODf.

12.2 SIGNALS: INFORMATION IN ELECTRIC FIELD PERTURBATIONS

Weakly electric fish actively probe their environment through their EOD. The fish sense how nearby objects or other fish distort the resulting electric field. While the spatial nature of these self-generated electric fields is roughly dipolar, the temporal details of the EOD vary significantly (see previous section; Assad et al. 1999). Despite the diversity of temporal properties, the physics underlying how the environment distorts the spatial aspects of the electric field is the same for all species. That said, differences in the spatial and temporal details of the dipole-like field due to, for example, different body shape or different EO properties lead to some quantitative differences that could be behaviorally relevant. In the interest of describing the basic concepts underlying electrosensory signals, I will ignore these subtle differences in the following discussion and instead focus on the fundamental similarities.

12.2.1 SPATIAL ASPECTS OF ELECTROSENSORY SIGNALS: THE ELECTRIC IMAGE

When an object with electrical conductivity different from that of the surrounding water is in the vicinity of the fish, the normal current flow will be diverted by the object, perturbing the fish's electric field. This perturbation results in a spatially extended change in voltage drop across the fish's skin,

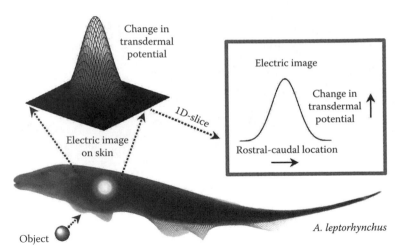

FIGURE 12.4 **(See color insert.)** Schematic of *Apteronotus leptorhynchus* showing the electric image on the skin due to a nearby plastic sphere; voltage perturbation on the skin is color-coded, with light colors indicating larger values. The electric image is also illustrated in a 3D plot with the voltage perturbation represented by the same color code but also in the z-axis (height of the 2D Gaussian-shaped surface). As is the convention, a 1D slice of this surface is taken to illustrate the electric image in a single spatial dimension (i.e., rostral–caudal; see inset). (Photo credit: Robert Lacombe, University of Ottawa, Ottawa, Ontario, Canada.)

called *the electric image* (Rasnow 1996; Babineau et al. 2006; Caputi and Budelli 2006). The electroreceptors (T- and P-units) transduce these voltage changes and transmit information through electrosensory afferents to the brain, forming the input stage of the electrosensory pathways (Figure 12.3).

To focus on the spatial details of the electric image, we consider only one point in time during the EOD. For simplicity, we choose a time when the amplitude of the dipole-like potential is maximally positive in the head region (with tail region negative); this corresponds to a time near the positive peak of the EOD (pulse or wave) if measured in the conventional manner, with one electrode near the head and one near the tail (Figure 12.2, top panel). For a simple object like a plastic sphere (insulator), the electric image on the skin surface resembles a Mexican hat or 2D Gaussian (e.g., Rasnow 1996; von der Emde et al. 1998; Chen et al. 2005; Babineau et al. 2006; Caputi and Budelli 2006), where the voltage perturbation is represented on the third dimension (Figure 12.4). A metal sphere (conductor) produces a similar-shaped electric image but with a voltage perturbation of opposite sign. To facilitate visualization and comparison, such skin surface electric images are typically viewed in cross section (e.g. along the rostral–caudal body axis; Figure 12.4, inset). Besides differences in image amplitude with object conductivity, note also that the electric image is a blurred representation of the sphere, being much broader than the sphere diameter.

12.2.1.1 Features of an Electric Image: Object Location

Given the simple situation just described, one can now attempt to answer the question "What's in an electric image?" In other words, what information about a sphere is available to the fish through the electric image it produces? We have already noted that the sign of the voltage perturbation provides information about electrical conductivity; we have also noted that this is relative conductivity (or *electrical contrast*, analogous to visual contrast), in that a "conductor" or an "insulator" is an object with conductivity greater or less (respectively) than that of the surrounding water. The consequence of this is that if an object has conductivity *equal* to that of the water, it will be electrically transparent, or "invisible" to the electric sense. This idea has been exploited experimentally by making agar walls with tank water to provide mechanical and physical isolation, while maintaining electrical transparency (e.g., Heiligenberg 1973). In this way, all sensory cues other than those related to the electric sense are eliminated and thus cannot influence behavior.

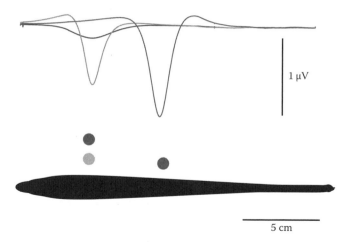

FIGURE 12.5 **(See color insert.)** Overlays of electric images for prey-like objects at three different locations illustrating the decrease in amplitude and increase in width for increasing lateral distances (y-axis), and the increase in image amplitude in the caudal direction (x-axis). (Modified from Babineau, D. et al., *PLoS Comput. Biol.*, 3, e38, 2007.) Coordinates for the blue, green, and red objects respectively are $(x, y) = (5, 3)$, $(5, 1.5)$, $(10, 1.5)$ cm; origin is at the nose of a 21-cm fish. As is the convention, electric images are computed as the difference between the transdermal potentials, measured with and without the object present, along the rostral–caudal axis, that is, a 1D slice along the skin surface. Prey-like objects are modeled as 0.3-cm diameter discs with a conductivity of 0.0303 S/m (water conductivity: 0.023 S/m).

A number of other object features are also represented in the electric image. For one, the location of the electric image peak on the fish skin provides information about the rostral–caudal object location (Figure 12.5, compare red and green traces). In addition, larger spheres will produce larger perturbations, resulting in electric images with larger amplitude. Thus, object size can be represented by the amplitude of the electric image. However, as pointed out by Rasnow (1996) and others, the electric image of a given object also has larger amplitude when it is closer to the fish; the image amplitude decreases with object distance (Figure 12.5, compare green and blue traces). This results in a size–distance ambiguity, such that an electric image with given amplitude could be due to a large object farther away or a relatively close smaller object. Rasnow (1996) proposed that a combination of image features could disambiguate object size and distance. The width of the electric image (at half-maximum for instance) increases linearly with distance; in other words, the images of faraway objects are increasingly blurry (Figure 12.5, compare green and blue traces). Rasnow (1996) showed that the ratio of image width and amplitude provided a measure of object distance that did not depend on object size. A related measure was suggested by von der Emde et al. (1998), namely the normalized slope of the image (which for a Gaussian-shaped image is proportional to the width); this quantity was shown to decrease with object distance.

12.2.1.2 Features of an Electric Image: Object Identification

Thus far, we have discussed image features that relate to the conductivity, rostral–caudal location, size, and distance of a simple sphere. Some small prey-like objects are well approximated electrically by small spheres, and thus allow extrapolation of these observations. However, to understand the limitations of object identification and discrimination, it is important to determine how the electric image changes with object geometry and context. Indeed, von der Emde et al. (1998) have shown that cubes and spheres can provide illusory cues for object distance (see later discussion). We can explore this issue using finite-element models of the electric field (Babineau et al. 2006, 2007; Kelly et al. 2008). We consider a number of objects with differently shaped cross sections. Note that these objects are uniform in the third dimension, that is, the dorsal–ventral axis of the fish.

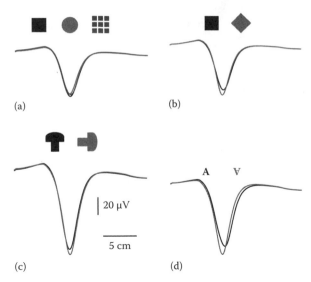

FIGURE 12.6 **(See color insert.)** Electric images of complex objects, computed as in Figure 12.5 and by Babineau et al. (2006). Object shapes shown schematically in cross section (actual dimensions are as follows) and color-coded to the corresponding image in each panel. All objects are metal (conductivity = 3.7×10^7 S/m), are centered at $(x, y) = (6, 2)$ cm, and are uniform in the third dimension (dorsal–ventral axis of fish); that is, they are "rodlike," with the cross sections shown. Cross-sectional dimensions: (a) Cube (1×1 cm), disc (radius = 0.56 cm), fractionated cube (each fraction, 0.33×0.33 cm, with gap width 0.5 cm). All objects have equal volumes. (b) Cubes (1×1 cm), one rotated 45°. (c) Mushroom-shaped, one rotated 90° (length = 1.5 cm, large width = 1.2 cm, small width = 0.8 cm). (d) A-shaped, one rotated 180° (length = 1.6 cm).

This allows us to simplify the analysis and limit the effects of object shape to only one plane. Figure 12.6 shows electric images for different objects and different orientations. In Figure 12.6a, images are shown for three different objects of the same size (i.e., volume) but varying cross section. All of these images are very similar in general shape, with a slightly asymmetric Mexican hat shape and only slight differences in amplitude. This holds true for other shapes as well, with the most obvious differences being in peak amplitude depending on object size (volume). Further, these images change relatively little under rotation, other than slight amplitude differences and asymmetries (Figure 12.6b through d). Overall, this suggests that such static electric images provide little detailed information about object shape.

12.2.1.3 Electric Images of Complex Scenes

While much intuition about electric imaging can be developed from considering the electric images of individual objects in isolation, it is also necessary to consider collections of objects and eventually understand the information available from complex electrosensory scenes. Again, electric field models have provided some clues. One very interesting observation that is likely of high adaptive value is that the image due to two objects is not the simple combination of the images from each object separately. Nearby objects influence the images each object produces, resulting in a nonlinear interaction, or "mutual polarization" effect (Rother et al. 2003; Kelly et al. 2008; Aguilera et al. 2012). The consequence of this effect is that electric imaging can allow detection of objects that are visually obscured, allowing fish to "see through" objects. One simple example of this was described earlier, where an object with conductivity equal to that of the surrounding water will be "transparent" electrically (though not necessarily visually transparent). Mutual polarization between objects can also allow such behavior, but through a slightly different mechanism. To illustrate this we consider a simple example: a spherical conductor positioned at a rostral–caudal location x_1 and a lateral distance y_1. As described earlier, this sphere will produce an electric image by perturbing the normal current flow around the fish. Previously, we were interested in the image on the fish skin as this is the relevant signal for electroreception. But we now consider the effect the object has on the entire field; in other words, the difference in the electric field with and without the object present. This difference, the so-called field perturbation (e.g., Caputi and Budelli 2006), would reveal another dipolar field, due to the fact that the fish's electric field induces a "new" dipole in the sphere. It is helpful to think about the electric image in this way when we consider the effects of an additional object, located farther away at x_1 and y_2 (where $y_2 > y_1$). This second object will be polarized by the fish electric field, but in addition it will be polarized by the first object. Thus, the second object is also polarized and will thus influence the polarization of the first object. All of these mutual effects will influence the electric image on the fish's skin. The interesting consequences of this phenomenon can be illustrated with another simple example. Imagine an object at a lateral distance y_2 that is out of detection range (i.e., y_2 is so large that the electric image on the skin is of negligible amplitude). Another conductor is now placed at a shorter distance y_1 (i.e., $y_2 > y_1$) that results in a larger amplitude and detectable image. From previous information, we know that these objects can influence each other, and thus influence the electric image on the fish's skin. Therefore, if the more distant object is now removed, the electric image can change by a potentially detectable amount. If the more distant object is returned, the electric image will change back. In this way, one object enhances the detectability of another. Another way of looking at this is that the fish can "see through" a visually opaque object. These ideas were first described by Rother et al. (2003), and their roles in electrosensory perception have been recently confirmed by Aguilera et al. (2012).

12.2.2 TEMPORAL ASPECTS OF ELECTROSENSORY SIGNALS: AMPLITUDE MODULATIONS

In the previous section, we dealt with spatial properties of the electric image at a fixed point in time. Much more attention, relatively, has been focused on the temporal nature of electrosensory signals. One of the classic case studies in neuroethology is the jamming avoidance response (JAR) exhibited by some wave-type fish (Heiligenberg 1991). Typically, fish within a species produce an EOD with individually specific frequency. However, when two fish of similar EODf encounter one another, the lower-frequency fish decreases its EODf and the higher-frequency fish increases its EODf. This behavior is exquisitely sensitive, resulting in reliable JARs to frequency differences of less than 1 Hz for EODf ranges of 300–600 Hz; this translates to sub-microsecond-level time discrimination (Rose and Heiligenberg 1985; Kawasaki et al. 1988). The signal features that enable the JAR are related to the interference of the two quasi-sinusoidal EODs, which results in an AM of both fish's EODs at a frequency equal to the frequency difference between the two fish (also referred to as the beat frequency). By comparing the amplitude and phase of this AM at different body locations, the fish can determine the sign of the frequency difference, allowing the appropriate change in EODf (Rose and Heiligenberg 1985; Heiligenberg 1991).

AMs not only arise from frequency differences in wave-type fish, but can also arise from objects in the environment. While we considered static electric images in an earlier section, more natural signals involve relative motion between fish and their environments, and will result in AMs for both pulse- and wave-type fish. For example, during the back-and-forth scan swimming typical of wave-type gymnotiforms, a nearby object would produce an electric image that would move back and forth along the fish's body. If measured at one point on the skin, the resulting signal would be a quasi-sinusoidal AM, similar in time scale to that resulting from another nearby fish with similar frequency. The difference lies in the spatial details. If the object was relatively small and nearby (prey-like), the AM would be spatially "local" (i.e., varies significantly between nearby skin locations); whereas an AM caused by another fish or a very large (fish-sized) object would be spatially "global" (i.e., very similar over a large fraction of the skin).

Another prominent behavior exhibited by a variety of fish species, both pulse- and wave-type, is tail-bending. Such curving of the body drastically changes the elongated dipole nature of the electric field, and will result in an AM if it is repeated in time, even in the absence of any surrounding object (e.g., Assad et al. 1999; Chen et al. 2005). Importantly, this AM is very large compared to that produced by a prey-sized object (mV compared to μV), suggesting that field distortions due to self-movements could swamp any prey-related signals. Recently though, tail-bending around a nearby object has been proposed as a means of generating cues for object distance (Sim and Kim, 2011). It remains to be seen whether fish actually use such information.

12.3   SENSING: NEURAL PROCESSING

Given the well-described neural pathways of *A. leptorhynchus*, much opportunity exists to test hypotheses on neural coding of electrosensory information. This task has been simplified by the relative ease in which certain behaviorally relevant stimuli can be generated under experimental conditions. The electric images due to objects and other fish are well approximated by an electric dipole, a stimulus that can be delivered using two electrodes. Depending on the voltage difference between electrodes and their geometry, different natural stimuli can be mimicked to varying degrees of accuracy. When the distance and voltage difference between electrodes is small ($\sim$μV, $\sim$cm), the stimulus is spatially "local" and mimics a prey-like electric image; when electrodes are separated by larger distances (tens of cm) and the voltage difference is larger (tens of mV), the electric image is spatially "global" (i.e., relatively uniform) and can mimic a conspecific fish (Chacron et al. 2003; Kelly et al. 2008; Chacron et al. 2011). In addition to the spatial aspects, stimulus frequencies can also be easily controlled experimentally to mimic natural signals: prey-like AM stimuli have most power in lower-frequency bands (<20 Hz), while conspecific stimuli can have power in a range of frequencies; depending on the difference between EODfs, the resulting beat frequencies can be small or large (up to 500 Hz). In addition, EODf modulations used by the fish for communication can alter the beat frequencies over brief time periods (10–100 ms) by hundreds of Hz, resulting in a spatially global, high-frequency signal specific to social contexts (Zakon et al. 2002; Zupanc et al. 2006).

The contrasting spatiotemporal signatures of prey-like AMs (low temporal frequency and spatially local) and communication-like (high temporal frequency and spatially global) have been exploited by many recent studies (see Chacron et al. 2011). The electroreceptors and associated sensory afferents are well suited to transduce and encode these signals. The P-unit afferents respond strongly (increase in firing rate) to a step change in transdermal potential, but then show marked adaptation (decreased firing rate) for a sustained, non-varying stimulus (Chacron et al. 2001). These afferents are thus particularly sensitive to changes in transdermal potential (Abbott and Regehr 2004). Quantitative measures of information coding have shown that P-units reliably transmit low-frequency information (<20 Hz) to the ELL through changes in their firing rate (Chacron et al. 2005; Gussin et al. 2007). High-frequency information, on the other hand, can be transmitted to ELL through changes in the degree of synchronization between multiple P-units (Benda et al. 2006). T-units are relatively scarce in *A. leptorhynchus* and are consequently poorly understood, though they are thought to play an important role in the coding of time-related information in other species (Heiligenberg 1991).

Tuberous P-units transmit information from their respective body locations to each of three segments in ELL, forming three distinct topographic maps of the body surface. The ELL maps are organized in functional columns, each comprising a variety of neuronal types (Maler 2009). One distinguishing feature of the primary output neurons (ELL pyramidal neurons) is that some respond to increases in transdermal potential and others respond to decreases (Saunders and Bastian 1984); these so-called E and I cells are thus functionally similar to the on- and off-type neurons found in many visual systems (Chacron et al. 2011). In addition, pyramidal neurons in the different ELL maps have receptive fields of different sizes, as well as different ion channel distributions, conferring a spatiotemporal selectivity to electrosensory inputs across maps (Krahe et al. 2008; Mehaffey et al. 2008; Maler 2009). Thus, the neural processing in ELL involves extracting specific features of incoming electrosensory stimuli. In other words, the transmission of information from the electrosensory afferents is separated into different streams by the ELL, depending on (among other things) whether the signal is spatially local or global, high- or low-frequency

(ELL maps), or whether it is due to a conductor or insulator (E and I cells). Such feature extraction or sparsification is a common element of sensory systems in general (Gabbiani et al. 1996; Metzner et al. 1998; Chacron et al. 2011).

The feedback pathways to ELL also play important roles in neural coding. One well-described function involves the indirect pathway from the cerebellum (EGp; Figure 12.3). Active sensing systems must be able to separate sensory inputs that are self-generated (due to active movements) from those that provide information about the environment. As mentioned earlier, the signals due to tail-bending are orders of magnitude larger than those due to prey, posing a significant constraint for prey detection. This problem is thought to be solved by cancelling out the effects of self-generated signals at the level of the ELL. In *A. leptorhynchus*, this is achieved through systematic changes in the strength of feedback from the indirect pathway through various forms of synaptic plasticity (Bastian et al. 2004; Lewis et al. 2007; Bol et al. 2011). These changes in synaptic strength result in the production of a "negative image" of the tail-bend (i.e., feedback inputs that are equal and opposite to those resulting from the bend). In this way, the effects of tail-bending are removed while prey signals are passed through to higher brain areas.

The pathways that allow the cancellation of self-generated signals are particularly sensitive to global, spatially diffuse inputs. Additional dynamics, arising from differential sensitivities to spatiotemporal inputs, are thought to be involved in the parsing of another set of signals. Spatially global inputs also arise from conspecific EODs. Thus, in the context of prey detection, these signals act as noise sources (similar to tail-bending). However, electrocommunication signals are also carried in these conspecific EODs, so a simple cancellation mechanism would not be ideal. It seems that the ELL solves this problem with an adaptive filtering mechanism, where the ELL neuron dynamics and frequency tuning (sensitivity) shifts with the spatial properties (local versus global) of the inputs (Chacron et al. 2003; Doiron et al. 2003).

Recent work has focused on the detection of stimulus features by neurons in the next processing stage, the midbrain torus (TS; Figure 12.3). The three maps of ELL converge to form one map in the TS, with TS neurons in general having a higher degree of spatial and temporal frequency selectivity (Chacron et al. 2011). In addition, some TS neurons show a preference for stimuli moving along the rostral–caudal body axis in one direction. This direction selectivity is coded in the burst response of these neurons and is thought to arise through a combination of synaptic depression in the ELL projection fibers and T-type calcium channels in the TS neurons (Chacron and Fortune 2010; Khosravi-Hashemi et al. 2011). This work represents a significant step forward, not only with respect to the focus on higher-order neurons in the electrosensory pathway, but also because of the use of stimuli that are dynamic not only in time but also in space. Relative motion is a critical aspect of electrosensory information processing and it will be important to further understand how related stimulus features are encoded by different neurons.

12.4 BEHAVIOR: ELECTROLOCATION

Thus far, our discussion has focused on possible sources of electrosensory information (features of the electric image and associated AMs) and how they may be encoded and processed, but ultimately we must rely on behavioral studies to determine how much of this information is used by the fish. Since the pioneering studies of Lissmann (1951), many studies have investigated various aspects of active electrosensing in weakly electric fish. The characterization of the JAR is a classic example of how to study the neural basis of animal behavior. These studies began with a thorough characterization of the behavior which, as mentioned earlier, involves wave-type fish changing their EODf to avoid jamming by another fish with similar frequency. The relative simplicity of this behavior (a decision to increase or decrease EODf) combined with its robustness under various experimental conditions allowed for a full description of the underlying neuronal pathways from electrosensory coding to electromotor output (Heiligenberg 1991). Further, this careful analysis revealed an extreme acuity in the time domain, in that these fish can discriminate time differences smaller than

1 μs (Rose and Heiligenberg 1985; Kawasaki et al. 1988). Current studies are focused on determining the limits of electrosensory processing in other contexts, relating to both spatial and temporal aspects of electrosensory inputs. Here, I review a selection of studies related to the particular features of the electric image discussed earlier.

12.4.1 Object Location and Identification

The task of locating and identifying prey, conspecifics, and other objects in the environment using the electric sense is generically referred to as electrolocation. Early studies focused on the sensitivity and range of electrolocation using sinusoidal electric dipole stimuli of varying frequencies. Knudsen (1974) found behavioral thresholds as low as 0.2 μV/cm in *Apteronotus* and *Eigenmannia* for stimulus frequencies similar to their own EODf. Based on these data, the detection range of a conspecific was predicted to be about five body lengths (30–120 cm depending on fish size and EOD field strength; Knudsen 1975). Detection thresholds increased for higher and lower frequencies. For low-frequency stimuli similar to prey items, thresholds were 0.5–1.0 μV/cm in *Apteronotus*, and around 50 μV/cm in *Eigenmannia* (Knudsen 1974). Similarly, the detection range for prey was estimated at about 3 cm (Knudsen 1975). Since then, more detailed analyses of prey capture behavior in *Apteronotus* have confirmed that the detection range is about 2–4 cm, depending on prey and water conductivity (MacIver et al. 2001). In pulse-type fish, similar detection thresholds have been found. A recent study by Pereira et al. (2012) took advantage of the "novelty response" in *Gymnotus* to assess object detection. These fish rapidly increase their EOD pulse rate when presented with a novel stimulus, and so provide a definitive behavioral response upon detection. Detection thresholds depended on object properties but were in the same range as found in wave-type fish (i.e., 2–5 cm). These ranges also depend on object location relative to the fish body and have been used to quantify a 3D sensing volume around the fish (Knudsen 1975; Snyder et al. 2007; Pereira et al. 2012). Note that this volume varies widely with the stimulus type, reaching many body lengths for conspecific signals and less than half a body length for prey-type objects.

In an elegant series of experiments, von der Emde et al. (1998) showed that pulse-type fish (*Gnathonemus petersii*) can use detailed features of the electric image to determine an object's distance. They first noted that the normalized slope (maximum slope over maximum amplitude) of the electric image of a sphere was smaller than that of a similar-sized cube at the same distance. Using a two-alternative forced-choice paradigm, fish were trained to choose the more distant of two identical objects. Next, when fish were provided with a sphere and a cube at the same distance, the authors predicted that fish would choose the object based on the image with smaller slope. Indeed, fish regularly chose the sphere as the "farther-away" object, suggesting that this electric image feature is used by fish to determine object distance.

In a previous section, we described electric images produced by objects of various shapes. At first glance, these images do not differ much relative to the differences in geometrical shape. Interestingly though, there is behavioral evidence that suggests that fish can in fact discriminate such objects (e.g., von der Emde and Fetz 2007; von der Emde et al. 2010). It is unclear, however, what particular image features enable this behavioral discrimination. Tail-bending or scan swimming may provide additional cues under these circumstances (e.g., Assad et al. 1999; Sim and Kim 2011). Fish may also take advantage of the different information available at different body locations. Recent studies have shown that body shape has an important influence on the electric image, and thus the image will vary with rostral–caudal location (Figure 12.5; Babineau et al. 2006, 2007), and is especially different in the nose region (Engelmann et al. 2008; Caputi et al. 2011).

12.4.2 Motion Cues

Thus far, I have focused primarily on static electric images, but as noted, relative motion between fish and object transforms such stimuli into time-varying stimuli. Active sensing behaviors, such as scan swimming (MacIver et al. 2001), chin-probing (Engelmann et al. 2008), and tail-bending

(Sim and Kim 2011), suggest that movement is critical for electrolocation. Using a modeling approach, Babineau et al. (2007) showed that the detection of small prey-like signals may be enhanced by motion. Recently, motion of a background has also been shown to increase object detection range (Fechler and von der Emde 2013). Interestingly, the AMs due to back-and-forth movements of a small sphere can produce a JAR (Carlson and Kawasaki 2007), suggesting that there are significant ambiguities involved in the discrimination of prey and conspecifics when the signals are small.

When larger objects move slowly, many electric fish will track the motion such that their relative position is constant (Heiligenberg 1973). Recent studies have quantified this tracking performance under various conditions and have identified back-and-forth "saccadic"-type movements under low signal-to-noise conditions (Stamper et al. 2012b). In other words, it appears that quick scanning movements are used to enhance tracking behavior; these movements cease in the presence of light when both visual and electrosensory inputs are available to provide localization cues.

In natural social contexts, involving multiple fish moving relative to one another, complex AMs can arise (Stamper et al. 2010; Yu et al. 2012). These so-called social envelopes can result in JAR-like changes in EODf even when individual EODfs are far outside those that would produce a JAR on their own (Stamper et al. 2012a). Further, these signals can be degraded by swimming movements, which act like a noise source, making EODf discrimination much more difficult (Yu et al. 2012). It will be interesting to see whether such movements can be used actively as a signal-cloaking or crypsis strategy.

12.5 SUMMARY

At this time, it is still not possible to determine the quantity and nature of the electrosensory information the fish extracts from the electric image. The robustness and relative simplicity of the JAR allowed for a thorough characterization of the sensory cues as well as the underlying neuronal networks. The key to this successful approach was a systematic study of the behavior and the development of techniques that allowed neural activity and behavioral response to be monitored simultaneously. Similar strategies must be developed to study the spatiotemporal aspects of electrolocation. We must better understand the dynamic electric images under more natural conditions. Promising work on this front has involved recordings from large electrode grids in streams inhabited by these fish (Henninger et al. 2012). In addition, we must diversify and improve electrosensory stimulation during electrophysiological experiments to more accurately reflect natural inputs (e.g., Kelly et al. 2008). Perhaps the biggest challenge will be in the study of motion-related active sensing. Movement is a necessary part of these behaviors and thus traditional experimental approaches in restrained animals will not suffice. The development of recording methods in freely swimming fish will be critical (Chen et al. 2005; Fotowat et al. 2012), as will the characterization of appropriate behavioral outputs (MacIver et al. 2001; Aguilera et al. 2012). Overall, much more work at the theoretical, behavioral, and neurophysiological levels is necessary before we understand all that is in the electric image.

ACKNOWLEDGMENTS

Supported by a Discovery Grant from NSERC of Canada and an Early Researcher Award from the Government of Ontario.

REFERENCES

Abbott LF and Regehr WG. 2004. Synaptic computation. *Nature* 431:796–803.
Aguilera PA, Pereira AC, and Caputi AA. 2012. Active electrolocation in pulse gymnotids: Sensory consequences of objects' mutual polarization. *J Exp Biol* 215:1533–1541.
Albert JS and Crampton WGR. 2005. Electroreception and electrogenesis. In: Evans DH, Ed. *The Physiology of Fishes*, 3rd edn. Boca Raton: CRC Press. pp. 431–472.
Alves-Gomes JA. 1999. Systematic biology of gymnotiform and mormyriform electric fishes: Phylogenetic relationships, molecular clocks and rates of evolution in the mitochondrial rRNA genes. *J Exp Biol* 202:1167–1183.

Assad C, Rasnow B, and Stoddard PK. 1999. Electric organ discharges and electric images during electrolocation. *J Exp Biol* 202:1185–1193.

Babineau D, Lewis JE, and Longtin A. 2007. Spatial acuity and prey detection in weakly electric fish. *PLoS Comput Biol* 3:e38.

Babineau D, Longtin A, and Lewis JE. 2006. Modeling the electric field of weakly electric fish. *J Exp Biol* 209:3636–3651.

Bastian J, Chacron MJ, and Maler L. 2004. Plastic and nonplastic pyramidal cells perform unique roles in a network capable of adaptive redundancy reduction. *Neuron* 41:767–779.

Bell CC, Libouban S, and Szabo T. 1983. Pathways of the electric organ discharge command and its corollary discharges in mormyrid fish. *J Comp Neurol* 216:327–338.

Benda J, Longtin A, and Maler L. 2006. A synchronization-desynchronization code for natural communication signals. *Neuron* 52:347–358.

Bennett MVL. 1971. Electrolocation in fish. *Ann NY Acad Sci* 188:242–269.

Berman N and Maler L. 1999. Neural architecture of the electrosensory lateral line lobe: Adaptations for coincidence detection, a sensory searchlight and frequency-dependent adaptive filtering. *J Exp Biol* 202:1243–1253.

Bol K, Marsat G, Harvey-Girard E, Longtin A, and Maler L. 2011. Frequency-tuned cerebellar channels and burst-induced LTD lead to the cancellation of redundant sensory inputs. *J Neurosci* 31:11028–11038.

Bullock TH, Hopkins CD, Popper AN, and Fay RR. 2005. *Electroreception*. New York: Springer.

Caputi AA, Aguilera PA, and Pereira AC. 2011. Active electric imaging: Body-object interplay and object's "electric texture." *PloS One* 6:e22793.

Caputi AA and Budelli R. 2006. Peripheral electrosensory imaging by weakly electric fish. *J Comp Physiol A* 192:587–600.

Carlson BA. 2002. Neuroanatomy of the mormyrid electromotor control system. *J Comp Neurol* 454:440–455.

Carlson BA and Kawasaki M. 2007. Behavioral responses to jamming and "phantom" jamming stimuli in the weakly electric fish Eigenmannia. *J Comp Physiol A* 193:927–941.

Carr CE, Maler L, and Sas E. 1982. Peripheral organization and central projections of the electrosensory nerves in gymnotiform fish. *J Comp Neurol* 211:139–153.

Chacron M, Doiron B, Maler L, Longtin A, and Bastian J. 2003. Non-classical receptive field mediates switch in a sensory neuron's frequency tuning. *Nature* 423:77–82.

Chacron MJ and Fortune ES. 2010. Subthreshold membrane conductances enhance directional selectivity in vertebrate sensory neurons. *J Neurophysiol* 104:449–462.

Chacron MJ, Longtin A, and Maler L. 2001. Negative interspike interval correlations increase the neuronal capacity for encoding time-dependent stimuli. *J Neurosci* 21:5328–5343.

Chacron MJ, Longtin A, and Maler L. 2011. Efficient computation via sparse coding in electrosensory neural networks. *Curr Opin Neurobiol* 21:752–760.

Chacron MJ, Maler L, and Bastian J. 2005. Feedback and feedforward control of frequency tuning to naturalistic stimuli. *J Neurosci* 25:5521–5532.

Chen L, House JL, Krahe R, and Nelson ME. 2005. Modeling signal and background components of electrosensory scenes. *J Comp Physiol A* 191:331–345.

Curti S, Comas V, Rivero C, and Borde M. 2006. Analysis of behavior-related excitatory inputs to a central pacemaker nucleus in a weakly electric fish. *Neuroscience* 140:491–504.

Crampton WGR. 2006. Evolution of electric signal diversity in gymnotiform fishes. II. Signal design. In: Ladich F, Collin SP, Moller P, and Kapoor BG, Eds. *Communication in Fishes*. Enfield, NH: Science Publishers. pp. 697–731.

Crampton WGR and Albert JS. 2006. Evolution of electric signal diversity in gymnotiform fishes. I. Phylogenetic systematics, ecology and biogeography. In: Ladich F, Collin SP, Moller P, and Kapoor BG, Eds. *Communication in Fishes*. Enfield, NH: Science Publishers. pp. 647–731.

Doiron B, Chacron M, Maler L, Longtin A, and Bastian J. 2003. Inhibitory feedback required for network oscillatory responses to communication but not prey stimuli. *Nature* 421:539–543.

Engelmann J, Bacelo J, Metzen M, Pusch R, Bouton B, Migliaro A, Caputi A, Budelli R, Grant K, and von der Emde G. 2008. Electric imaging through active electrolocation: Implication for the analysis of complex scenes. *Biol Cybern* 98:519–539.

Fechler K and von der Emde G. 2013. Figure-ground separation during active electrolocation in the weakly electric fish, *Gnathonemus petersii*. *J Physiol—Paris* 107:72–83.

Fotowat H, Harrison RR, and Krahe R. 2012. Wireless recording and computational modeling of natural elec-
trosensory input in freely swimming electric fish. *Front. Behav. Neurosci. Conference Abstract: Tenth International Congress of Neuroethology*. doi: 10.3389/conf.fnbeh.2012.27.00216.

Gabbiani F, Metzner W, Wessel R, and Koch C. 1996. From stimulus encoding to feature extraction in weakly electric fish. *Nature* 384:564–567.

Giassi ACC, Duarte TT, Ellis W, and Maler L. 2012. Organization of the gymnotiform fish pallium in relation to learning and memory: II. Extrinsic connections. *J Comp Neurol* 520:3338–3368.

Gussin D, Benda J, and Maler L. 2007. Limits of linear rate coding of dynamic stimuli by electroreceptor afferents. *J Neurophysiol* 97:2917–2929.

Heiligenberg W. 1973. Electrolocation of objects in the electric fish *Eigenmannia*. *J Comp Physiol* 87:137–164.

Heiligenberg WF. 1991. *Neural Nets in Electric Fish*. Cambridge, MA: MIT Press.

Heiligenberg WF. 1993. Electrosensation. In: Evans DH, Ed. *The Physiology of Fishes*. Boca Raton: CRC Press. pp. 137–160.

Henninger J, Benda J, and Krahe R. 2012. Undisturbed long-term monitoring of weakly electric fish in a small stream in Panama. *Front. Behav. Neurosci. Conference Abstract: Tenth International Congress of Neuroethology*. doi: 10.3389/conf.fnbeh.2012.27.00319.

Hopkins CD. 1999. Design features for electric communication. *J Exp Biol* 202:1217–1228.

Kawasaki M. 2009. Evolution of time-coding systems in weakly electric fishes. *Zoolog Sci* 26:587–599.

Kawasaki M, Rose G, and Heiligenberg W. 1988. Temporal hyperacuity in single neurons of electric fish. *Nature* 336:173–176.

Kelly M, Babineau D, Longtin A, and Lewis JE. 2008. Electric field interactions in pairs of electric fish: Modeling and mimicking naturalistic inputs. *Biol Cybern* 98:479–490.

Khosravi-Hashemi N, Fortune ES, and Chacron MJ. 2011. Coding movement direction by burst firing in electrosensory neurons. *J Neurophysiol* 106:1954–1968.

Knudsen E. 1974. Behavioral thresholds to electric signals in high frequency electric fish. *J Comp Physiol A* 91:333–353.

Knudsen E. 1975. Spatial aspects of the electric fields generated by weakly electric fish. *J Comp Physiol A* 99:103–118.

Krahe R, Bastian J, and Chacron MJ. 2008. Temporal processing across multiple topographic maps in the electrosensory system. *J Neurophysiol* 100:852–867.

Kramer B. 1990. *Electrocommunication in Teleost Fishes: Behavior and Experiments*. Berlin, Germany: Springer-Verlag.

Kramer B. 1996. *Electroreception and Communication in Fishes*. Progress in Zoology Series Vol. 42. Stuttgart, Germany: Gustav-Fischer.

Lavoué S, Miya M, Arnegard ME, Sullivan JP, Hopkins CD, and Nishida M. 2012. Comparable ages for the independent origins of electrogenesis in African and South American weakly electric fishes. *PloS One* 7:e36287.

Lewis JE, Lindner B, Laliberté B, and Groothuis S. 2007. Control of neuronal firing by dynamic parallel fiber feedback: Implications for electrosensory reafference suppression. *J Exp Biol* 210:4437–4447.

Lissmann H. 1951. Continuous electrical signals from the tail of a fish, *Gymnarchus niloticus Cuv*. *Nature* 167:201–202.

Lissmann H and Machin K. 1958. The mechanism of object location in *Gymnarchus niloticus* and similar fish. *J Exp Biol* 35:451–486.

MacIver MA, Sharabash NM, and Nelson ME. 2001. Prey-capture behavior in gymnotid electric fish: Motion analysis and effects of water conductivity. *J Exp Biol* 204:543–557.

Maler L. 2009. Receptive field organization across multiple electrosensory maps. I. Columnar organization and estimation of receptive field size. *J Comp Neurol* 516:376–393.

Mehaffey WH, Maler L, and Turner RW. 2008. Intrinsic frequency tuning in ELL pyramidal cells varies across electrosensory maps. *J Neurophysiol* 99:2641–2655.

Metzner W. 1999. Neural circuitry for communication and jamming avoidance in gymnotiform electric fish. *J Exp Biol* 202:1365–1375.

Metzner W, Koch C, Wessel R, and Gabbiani F. 1998. Feature extraction by burst-like spike patterns in multiple sensory maps. *J Neurosci* 18:2283–2300.

Moller P. 1995. *Electric Fishes: History and Behavior*. London, U.K.: Chapman & Hall. 583pp.

Moortgat KT, Bullock TH, and Sejnowski TJ. 2000. Precision of the pacemaker nucleus in a weakly electric fish: Network versus cellular influences. *J Neurophysiol* 83:971–983.

Nelson ME and MacIver MA. 1999. Prey capture in the weakly electric fish *Apteronotus albifrons*: Sensory acquisition strategies and electrosensory consequences. *J Exp Biol* 202:1195–1203.

Nelson ME and MacIver MA. 2006. Sensory acquisition in active sensing systems. *J Comp Physiol A* 192:573–586.

Pereira AC, Aguilera P, and Caputi AA. 2012. The active electrosensory range of *Gymnotus omarorum*. *J Exp Biol* 215:3266–3280.

Quintana L, Pouso P, Fabbiani G, and Macadar O. 2011. A central pacemaker that underlies the production of seasonal and sexually dimorphic social signals: Anatomical and electrophysiological aspects. *J Comp Physiol A* 197:75–88.

Rasnow B. 1996. The effects of simple objects on the electric field of *Apteronotus*. *J Comp Physiol A* 178:397–411.

Rose GJ. 2004. Insights into neural mechanisms and evolution of behaviour from electric fish. *Nature Rev Neurosci* 5:943–951.

Rose G and Heiligenberg W. 1985. Temporal hyperacuity in the electric sense of fish. *Nature* 318:178–180.

Rother D, Migliaro A, Canetti R, Gómez L, Caputi A, and Budelli R. 2003. Electric images of two low resistance objects in weakly electric fish. *Biosystems* 71:169–177.

Saunders J and Bastian J. 1984. The physiology and morphology of two types of electrosensory neurons in the weakly electric fish *Apteronotus leptorhynchus*. *J Comp Physiol A* 154:199–209.

Scheich H. 1973. Coding properties of two classes of afferent nerve fibers: High-frequency electroreceptors in the electric fish, *Eigenmannia*. *J Neurophysiol* 36:39–60.

Sim M and Kim D. 2011. Electrolocation based on tail-bending movements in weakly electric fish. *J Exp Biol* 214:2443–2450.

Smith GT, Lu Y, and Zakon HH. 2000. Parvocells: A novel interneuron type in the pacemaker nucleus of a weakly electric fish. *J Comp Neurol* 423:427–439.

Snyder JB, Nelson ME, Burdick JW, and Maciver MA. 2007. Omnidirectional sensory and motor volumes in electric fish. *PLoS Biol* 5:e301.

Stamper SA, Carrera, GE, Tan EW, Fugère V, Krahe R, and Fortune ES. 2010. Species differences in group size and electrosensory interference in weakly electric fishes: Implications for electrosensory processing. *Behav Brain Res* 207:368–376.

Stamper SA, Madhav MS, Cowan NJ, and Fortune ES. 2012a. Beyond the jamming avoidance response: Weakly electric fish respond to the envelope of social electrosensory signals. *J Exp Biol* 215:4196–4207.

Stamper SA, Roth E, Cowan NJ, and Fortune ES. 2012b. Active sensing via movement shapes spatiotemporal patterns of sensory feedback. *J Exp Biol* 215:1567–1574.

von der Emde G, Behr K, Bouton B, Engelmann J, Fetz S, and Folde C. 2010. Three-dimensional scene perception during active electrolocation in a weakly electric pulse fish. *Front Behav Neurosci* 4:26.

von der Emde G and Fetz S. 2007. Distance, shape and more: Recognition of object features during active electrolocation in a weakly electric fish. *J Exp Biol* 210:3082–3095.

von der Emde G, Schwarz S, Gomez L, Budelli R, and Grant K. 1998. Electric fish measure distance in the dark. *Nature* 395:890–894.

Wexler M and Van Boxtel JJA. 2005. Depth perception by the active observer. *Trends Cog Sci* 9:431–438.

Yu N, Hupé G, Garfinkle C, Lewis JE, and Longtin A. 2012. Coding conspecific identity and motion in the electric sense. *PLoS Comput Biol* 8:e1002564.

Zakon H, Oestreich J, Tallarovic S, and Triefenbach F. 2002. EOD modulations of brown ghost electric fish: JARs, chirps, rises, and dips. *J Physiol—Paris* 96:451–458.

Zhang Y and Kawasaki M. 2006. Interruption of pacemaker signals by a diencephalic nucleus in the African electric fish, *Gymnarchus niloticus*. *J Comp Physiol A* 192:509–521.

Zupanc GKH, Sîrbulescu RF, Nichols A, and Ilies I. 2006. Electric interactions through chirping behavior in the weakly electric fish, *Apteronotus leptorhynchus*. *J Comp Physiol A* 192:159–173.

13 Cardiac Regeneration

Viravuth P. Yin

CONTENTS

13.1 INTRODUCTION

The heart is a simple organ burdened with the pivotal task of maintaining circulation. Yet, despite its importance to life, the adult mammalian heart has little regenerative capacity. Not surprisingly, this deficiency plays a major role in making cardiovascular disease the leading cause of death in the United States as well as in the Western world. In fact, myocardial infarction ranks as the leading cause of death as well as the most costly medical condition to treat, killing over 750,000 Americans each year (Lloyd-Jones et al. 2010; Roger et al. 2012). Furthermore, common invasive procedures such as bypass surgery or angioplasty are temporary remedies, as the reoccurrence of a second myocardial infarction is approximately 14% within a year of the first incident (Holmes et al. 2005; Reeve et al. 2005; Bolli and Chaudhry 2010). In response to myocardial infarction, dead, necrotic heart tissue is replaced by dense, collagen-laden scar tissue (Holmes et al. 2005). While scarring is an essential cellular response to preserve remaining heart function, over time, this non-contractile, rigid fibrotic tissue causes arrhythmia. In contrast, teleosts, like the zebrafish, have a remarkable capacity to regenerate damaged cardiac tissues throughout their lifetime, thus making it an ideal model system to elucidate the cellular and genetic basis for cardiac tissue regeneration.

Regeneration is defined as a process that restores an injured tissue or organ to its uninjured state, such that both size and function are recapitulated. However, regenerative capacity is not distributed equally within the metazoan subkingdom (Sánchez-Alvarado and Tsonis 2006; Bely and Nyberg 2010). While mammals have poor regenerative capacity, certain vertebrates like newts and zebrafish harbor the remarkable capacity to regenerate multiple organs in response to acute injury (Dinsmore and American Society of Zoologists 1991; Poss et al. 2003). For instance, the adult zebrafish has an enhanced capacity to regenerate up to ~20% of damaged or lost heart tissue (Poss et al. 2002; Raya et al. 2003). In response to injury, the zebrafish heart quickly seals the wound with a temporary blood clot. Within the next week, a series of cellular changes is triggered, including cardiomyocyte proliferation and neovascularization of the new myocardium. Unlike the mammalian heart, little or no scar tissue remains after de novo creation of cardiomyocytes (Poss et al. 2002). While the majority of current cardiac regeneration studies center on stem cells and induced pluripotent stem cells, there is increasing evidence that newly regenerated cardiac muscle in the zebrafish emerges from an existing population of cardiomyocytes. Therefore, a more directed effort to elucidate the genetic regulatory networks that encourage natural regeneration from injured tissue itself may prove more promising for improving human cardiac health. In this chapter, I will discuss the recent advances in our understanding of natural heart regeneration in the adult zebrafish and how this model system is reshaping traditional views of the adult heart as a post-mitotic organ.

13.2   ZEBRAFISH AS A MODEL SYSTEM FOR TISSUE REGENERATION

The rise of the teleost zebrafish (*Danio rerio*) as a prominent vertebrate model system is attributed to its genetic tractability and transparency during embryonic development (Grunwald and Eisen 2002). Early genetic screens in the zebrafish not only identified key genetic factors underlining essential function and development but also solidified the zebrafish as an important vertebrate model system for cardiovascular research (Mullins et al. 1994; Driever et al. 1996; Malicki et al. 1996a; Pack et al. 1996; Schier et al. 1996; Solnica-Krezel et al. 1996; Stemple et al. 1996; Weinstein et al. 1996). With respect to cardiogenesis, these genetic screens led to the identification of key genetic factors that underscore heart field formation, morphogenesis, and the genetic foundation of numerous diseases (Stainier et al. 1993, 1996; Lee et al. 1994; Stainier and Fishman 1994).

The heart is the first organ that forms during vertebrate embryogenesis. However, in contrast to the four-chambered mammalian organ, the zebrafish heart is two-chambered, comprised of a single atrium and ventricle. Akin to mammals, the zebrafish heart is comprised of three major tissue layers (epicardium, myocardium, and endocardium). The myocardium is further subdivided into a compact and trabeculae layer, and is comprised of mainly cardiomyocytes, the contractile cells of the heart muscle. More recently, the popularity of the zebrafish model has encouraged and expanded studies into the adult animal.

The use of the zebrafish as a model system to address regeneration biology has garnered widespread recognition, support, and appreciation. Two features make the zebrafish an optimized model system to study organ regeneration. First, they are extraordinarily regenerative, equipped to regrow amputated fins, injured retinas, transected optic nerves and spinal cord, and resected heart muscle (Morgan 1901; Becker et al. 1997; Cameron 2000; Poss et al. 2002, 2003; Raya et al. 2003; Yin et al. 2008). Second, unlike other highly regenerative model systems like newts and axolotls, the zebrafish genome can be easily modified to overexpress or remove gene function, enabling functional dissection of regeneration. With regard to the adult heart, the zebrafish can rapidly regenerate lost cardiac muscle and restore cardiac function in as little as 30–60 days following injury (Poss et al., 2002).

13.3   HEART REGENERATION IN ADULT ZEBRAFISH

Following injury, the zebrafish heart responds with three defined cellular processes: wound healing and clot formation, cardiomyocyte proliferation, and reintegration of the regenerated tissue (Figure 13.1). Immediately following amputation, the wound is quickly sealed with a blood clot. A collection of genetic profiling studies has shown that cardiac wound healing has striking similarities with wound healing during appendage regeneration in the zebrafish and salamander (Lien et al. 2006; Sleep et al. 2010). Over the course of a week, the blood clot is replaced with a fibrin and collagen-rich clot, akin to scar tissue deposition within the mammalian heart.

Concomitant with this blood-to-fibrin clot transition, cardiomyocyte proliferation is initiated near or within the wound at 7 days postamputation (dpa), and continues at 30 dpa, mainly in the most apical cardiomyocytes of the regenerated wall. Peak cardiomyocyte proliferation activity is observed at 14 dpa and culminates in the reconstruction of a contiguous myocardial wall of electrically integrated cardiomyocytes by 30–60 dpa (Poss et al. 2002; Kikuchi et al. 2010). Remarkably, this process is executed with minimal residual scar tissue, suggesting that the zebrafish is able to

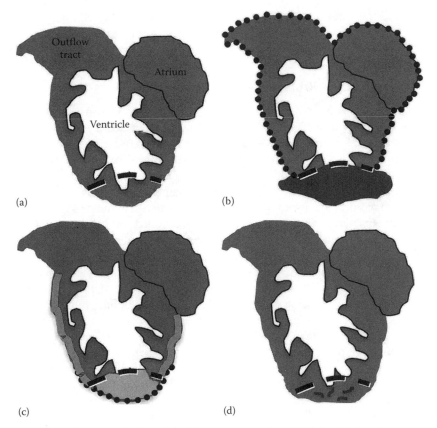

FIGURE 13.1 (See color insert.) A model of heart regeneration. (a) Zebrafish hearts are comprised of a single atrium, ventricle, and an outflow tract. (b) In response to a resection injury (dashed black line), a blood clot (brown) quickly seals the wound. Within the first 1–3 days postamputation (dpa), there is organ-wide activation of the epicardium (black dots) and remodeling of the endocardium. (c) By 7 dpa, the blood clot is replaced with a collagen clot (light blue) and epicardium signals become localized to the injury site. At this time, *gata4*+ cells (green) in the compact muscle undergo proliferation. (d) By 30–60 dpa, a contiguous myocardial wall has reformed, new cardiomyocytes are electrically coupled to the uninjured heart tissue, and new blood vessels penetrate the regenerate (dark red). Functionally, the regenerated heart is indistinguishable from the uninjured organ.

degrade the fibrin clot during cardiomyocyte replacement. When cardiomyocyte proliferation is disabled via a hypomorphic mutation in the cell cycle regulator, *mps1*, cardiac fibrosis heals the injury with little regeneration, analogous to a mammalian heart in response to myocardial infarction (Poss et al. 2002). Thus, injury-stimulated cardiomyocyte proliferation is crucial to prevent scarring and facilitate regeneration.

A key metric for cardiac regeneration is functional integration of electrical conduction between the regenerate and preexisting heart tissue. To address this question, optical voltage-mapping studies were conducted during the phases of regeneration following ventricular resection. At the onset of cardiomyocyte proliferation, newly created cardiomyocytes are uncoupled. However, by 14 dpa, recovery of conduction velocities is evident. By 30 dpa, a time when a continuous myocardial wall has formed, conduction velocities are nearly indistinguishable from optical maps of the uninjured heart (Kikuchi et al. 2010). By all measures, when challenged with an injury, adult zebrafish hearts activate a regeneration program that concludes with a repaired heart that is functionally identical with an uninjured counterpart.

Collectively, studies of cardiac regeneration in the zebrafish have been transformative for the cardiovascular field for many reasons: (1) the observations challenged the long-held notion that the adult vertebrate heart is a post-mitotic organ incapable of reentering the cell cycle; (2) they sparked in-depth analyses into the genetic and molecular mechanisms of natural heart regeneration in vertebrates; and (3) the results suggest a potential avenue to enhance regenerative potential in humans.

13.3.1 PRIMARY SOURCE OF REGENERATING CARDIOMYOCYTES

A long-standing question in heart regeneration concerns the source of regenerating cardiomyocytes. Studies in the newt and zebrafish have shown that cardiomyocytes are activated upon injury and reenter the cell cycle to replenish lost myocardial tissue (Becker et al. 1974; Bader and Oberpriller 1978, 1979; Poss et al. 2002; Raya et al. 2003). However, these new cardiomyocytes could emerge either from a progenitor or stem cell population, or from existing pools of contracting cardiomyocytes that dedifferentiate and reinitiate the cell cycle. Early studies suggested multiple populations of embryonic stem cells were contributing to regenerating myocardium. In particular, expression studies showed that embryonic cardiogenic transcription factors *nkx2.1* and *hand2* were reactivated within the first few weeks of ventricular resection (Lepilina et al. 2006). At the same time, fluorescent reporter strains demonstrated that the contractile gene, cardiac myosin light chain, *cmlc2*, was induced (Lepilina et al. 2006). However, in the absence of genetic labeling studies, the origin of regenerating cardiomyocytes remained unresolved.

More recently, two independent groups used genetic fate-mapping studies to conclusively ascertain if existing cardiomyocytes serve as a source for new tissue. In one study, embryonic cardiomyocytes were labeled using a *cmlc2*-driven floxed green fluorescent protein (GFP) reporter and an inducible Cre recombinase (*CreERT2*). When activated with 4-hydroxy-tamoxifen, all embryonic *cmlc2* positive cells were marked with GFP expression. In the adult, regeneration studies showed that newly created cardiomyocytes were marked with GFP, indicating they originated from existing cardiomyocytes (Jopling et al. 2010). However, as recombination was initiated during development, these results do not formally exclude the possibility that *cmlc2*-expressing progenitor cells contribute to newly regenerated cardiomyocytes. This is of particular consequence given that the new clonal analysis of cardiomyocytes during zebrafish development demonstrated that cardiomyocytes comprise a heterogeneous population (Gupta and Poss 2012).

A more conclusive set of studies was performed using tamoxifen-inducible fate mapping in a subset of cardiomyocytes in the adult animal. Using a similar approach as the aforementioned study, Kikuchi et al. (2010) first induced *cmlc2:CreER* recombination in adult hearts prior to ventricular amputation, confirming that indeed regenerated myocardial tissue emerged from an existing *cmlc2* population of cells (Figure 13.2). To further refine the source for regenerating cardiomyocytes, two transgenic strains, *Tg(gata4:GFP)* and *Tg(gata4:CreER)*, were used to label cells (Kikuchi et al. 2010). *gata4* encodes a transcription factor critical for embryonic development and vascularization (Heicklen-Klein and Evans 2004; Holtzinger and Evans 2005). With injury, *gata4* expression is induced by 7 dpa and limited to only the compact layer of the heart. By 14 dpa, proliferating *gata4* cells penetrate the wound apex, indicating that this subpopulation of cardiomyocytes plays a major role for myocardium regeneration (Figure 13.2; Kikuchi et al. 2010). Future functional studies to examine *gata4* roles are essential to further characterize this unique pool of cardiomyocytes.

Intact contractile machinery is critical for cardiomyocytes to generate the necessary contractile force. During myocardial regeneration, these sarcomeric structures are thought to disassemble as cardiomyocytes undergo division. A series of studies including transmission electron microscopy (EM) and histological staining indicate that indeed cardiomyocytes acquire a less organized sarcomeric structure during regeneration (Jopling et al. 2010; Kikuchi et al. 2010). Thus, zebrafish myocardial regeneration hinges on dedifferentiation and proliferation of existing cardiomyocytes.

13.3.2 ROLE OF THE EPICARDIUM DURING REGENERATION

Heart regeneration is an orchestrated interplay among all three major cardiac tissues. The epicardium is a tissue that outlines the chambers of the heart and functions as a mitogen source for

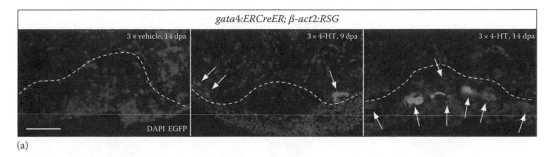

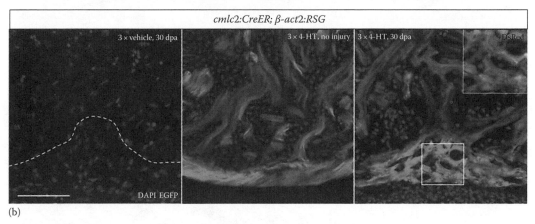

FIGURE 13.2 **(See color insert.)** A *gata4+* subpopulation of cardiomyocytes contributes to heart regeneration. (a) *Tg(gata4:CreER; β-act2:RSG)* animals were injected with vehicle (left) or 4-HT to induce Cre-mediated recombination. EGFP+ cells (arrows) are present near the injury site at 9 days postamputation (dpa) and within the wound area by 14 dpa. (b) *Tg(cmlc2:CreER; β-act2:RSG)* animals were injected before injury with vehicle (left) or 4-HT. The majority of cardiomyocytes in the uninjured and at 30 dpa are labeled in the presence of 4-HT treatment. (Dashed line, approximate amputation plane; inset depicts DsRed expression that was used for determining labeling efficiency; scale bar, 50 μm.) (Reprinted from Macmillan Publishers Ltd. *Nature*, Kikuchi, K. et al., Primary contribution to zebrafish heart regeneration by gata4(+) cardiomyocytes, 464(7288), 601–605, Copyright 2010.)

myocyte proliferation during heart development. In the adult, three events define the epicardium in response to cardiac injury: (1) reactivation of the embryonic genetic program, (2) differentiation into perivascular support cells, and (3) provision of mitogenic signals for cardiomyocyte proliferation. Reinitiation of the embryonic epicardium genetic circuit is evident by rapid induction of T-box transcription factor-18 (*tbx18*), Wilms tumor protein (*wt1*), and retinoic acid synthesizing enzyme retinaldehyde dehydrogenase 2 (*raldh2*) expression throughout the epicardium within 1–3 days following cardiac injury (Lepilina et al. 2006; Kikuchi et al. 2011). This expression becomes confined to the injured site by 7 dpa, concomitant with the elevated rate of epicardial cell proliferation (Figure 13.3; Poss et al. 2002; Lepilina et al. 2006).

A key role of the embryonic epicardium is to function as a reservoir for fibroblasts and vascular support cells during the process of epithelial-to-mesenchymal transition (EMT). In the adult zebrafish, EMT markers like *snail* and *twist*, and muscle markers *sm22α,β*, *α-smooth muscle actin*, and *pdgfr-B* are activated in response to injury in the adult heart (Kim et al. 2010). Activation of this genetic program precedes cell migration into the subepicardium and subsequent differentiation into perivascular support cells. Interestingly, studies using various mouse genetic lineage studies with *tbx18* and *wt1* promoters indicated a multipotency of the epicardium to transdifferentiate into coronary vessels and cardiomyocytes during development (Cai et al. 2008; Zhou et al. 2008; Smart et al. 2011). These observations were in contrast to earlier reports in vertebrate studies

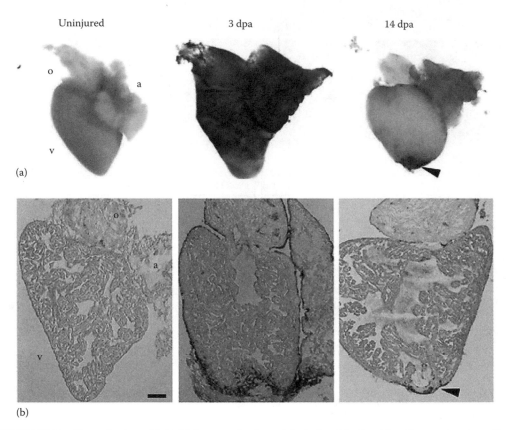

FIGURE 13.3 **(See color insert.)** *Raldh2* expression is rapidly induced in the epicardium and endocardium. (a) Whole-mount in situ hybridizations show *raldh2* expression is absent in uninjured hearts but is detected organ-wide at 3 days postamputation (dpa) in the epicardium. Expression is confined to the injury site by 14 dpa in the epicardium and endocardium. (b) Sections of whole-mount stained hearts show the changing spatial expression of *raldh2* from uninjured to 14 dpa. (Arrowhead marks injury site; scale bar, 100 μm; o, outflow tract; a, atrium; v, ventricle.) (Reprinted from *Cell*, 127, Lepilina, A. et al., A dynamic epicardial injury response supports progenitor cell activity during zebrafish heart regeneration, 607–619, Copyright 2006, from Elsevier.)

that indicated the epicardium was the primary source for perivascular and smooth muscle cells in the heart (Mikawa and Gourdie 1996; Dettman et al. 1998; Gittenberger-de Groot et al. 1998; Männer 1999; Merki et al. 2005; Lavine and Ornitz 2009).

In the zebrafish, however, *tbx18* and *wt1* expression included a subset of cardiomyocytes and intramyocardial cells, in addition to the strong epicardial activation (Kikuchi et al. 2011). Further in-depth analysis of mouse development subsequently showed that indeed both *tbx18* and *wt1* expression is more widespread than initially described, thus calling into question the capacity of epicardium-derived cells (EPDCs) to become cardiomyocytes (Christoffels et al. 2009). More recently, Kikuchi et al. (2011) utilized inducible Cre-mediated fate-mapping studies to examine the potential of EPDCs during heart regeneration. Using a *tcf21* reporter strain that expresses exclusively in the epicardium and subepicardium, the group followed the fate of EPDCs during embryonic development and during adult heart regeneration. In all studies, there was no evidence of *tcf21* cells adapting a myogenic fate. Instead, EPDCs gave rise to exclusively perivascular cells during neovascularization of the new regenerate.

Although EPDCs do not display the capacity to become cardiomyocytes, the confinement of epicardial *raldh2* expression to the injury site at 14 dpa suggested a potential to influence cardiomyocyte proliferation. Indeed, inactivation of retinoic acid signaling severely inhibited cardiomyocyte proliferation in response to ventricular resection (Kikuchi et al. 2011). It seems a major role for retinoic acid is to create a microenvironment conducive for cardiomyocyte proliferation. Notably, this retinoic acid-mediated enhancement of cardiomyocyte proliferation is also observed during mammalian heart development (Gittenberger-de Groot et al. 2010; Limana et al. 2010; Zhou et al. 2011). Collectively, these studies indicate that zebrafish epicardium plays an important role in creating a regenerative microenvironment for surrounding cardiomyocytes.

13.3.3 ENDOCARDIUM IS ALSO CRITICAL FOR CARDIOMYOCYTE PROLIFERATION

The endocardium is a thin tissue layer that lines the interior of the heart. In a similar manner to the epicardium, this tissue releases mitogenic factors that control myocardium proliferation during embryogenesis. Could this tissue display a similar pro-proliferative role in the context of adult heart regeneration? In response to cardiac injury, endocardial cells undergo morphological transformation becoming rounded and are detached from the myocardium within 6 hours postamputation (hpa) (Kikuchi et al. 2011). Accompanying these changes is the rapid induction of organ-wide *raldh2* expression in the endocardial tissue of both the atrium and the ventricle. This expression pattern becomes restricted to the ventricular apex at ~14 dpa (Figure 13.3; Kikuchi et al. 2011).

Although the importance and mechanism behind this refined expression need to be fully characterized, recent studies have provided some tantalizing possibilities. It has been demonstrated that exogenous introduction of lipopolysaccharide (LPS) triggers a similar organ-wide activation of *raldh2* expression, possibly resulting from inflammation and increased permeability of endocardial cells (Kikuchi et al. 2011). It has been hypothesized that cardiac injury triggers an inflammatory response and similar morphological changes to endocardial cells that precede *raldh2* expression activation in the heart. It is possible that restricted *raldh2* expression at the injury site recruits inflammatory cells to serve as a localized "homing" signal for cardiomyocyte proliferation. Consistent with this idea, transgenic knockdown of retinoic acid signaling with inducible dominant negative transgenes severely stunted cardiomyocyte proliferation during regeneration (Kikuchi et al. 2011). Interestingly, administration of retinoic acid agonists was not sufficient to promote cardiomyocyte proliferation in the adult zebrafish (Kikuchi et al. 2011). It is likely that additional regulatory mechanisms are in place to control the rate of retinoic acid-mediated cardiomyocyte proliferation in response to injury.

Retinoic acid is a potent mitogen during mammalian heart development and during zebrafish adult heart regeneration. Given the differences in regenerative potency, its induction and activity in the adult mammalian heart are predictably distinct from those in the zebrafish. For instance, *raldh2* expression is excluded from the endocardium and limited to epicardium cells. Whether combinatorial *raldh2* expression in the epicardium and endocardium in mammalian hearts could provide an avenue to enhance cardiomyocyte proliferative capacity remain to be investigated. Nonetheless, studies in the zebrafish indicate important roles for nonmyocardial tissues in crafting a pro-regenerative niche for the adult heart.

13.4 GENETIC CIRCUIT OF HEART REGENERATION

A major focus of cardiovascular research has been to elucidate the genetic hierarchy that promotes de novo creation of cardiomyocytes. The current consensus in the field is that regeneration is likely to reactivate a subset of the embryonic cardiac program in response to injury in the adult animal. Therefore, early approaches into elucidating the genetic components of heart regeneration relied on our understanding of embryonic developmental programs to identify candidate genes involved in promoting myocardial regeneration. For instance, fibroblast growth factor (Fgf) signaling, which enhances cardiomyocyte proliferation during development, has been shown to have a primary role in promoting epicardium EMT in response to cardiac injury in the adult zebrafish. A scan of the Fgf ligands and receptors identified *fgf17b*, *fgfr2*, and *fgfr4* expression in the epicardium as potential candidates to transduce Fgf signaling. Inactivation of the Fgf pathway with a dominant negative transgene culminated in defects in neovascularization and an increase in scar tissue (Lepilina et al. 2006).

As previously mentioned, *raldh2* expression localizes to epicardial and endocardial cells at the injury site to promote cardiomyocyte proliferation (Kikuchi et al. 2011). The crafting of a proliferative microenvironment is also exhibited by the TGFβ signaling pathway. In particular, TGFβ ligands are induced in both fibroblasts within the injury area and the surrounding cardiomyocytes. Interestingly, inhibition of TGFβ activity with small compounds inhibited regeneration due to repression of tissue remodeling in the injury site (Chablais and Jazwinska 2012). Thus, these studies demonstrate that the cardiac regenerative program is a genetic tug-of-war between scar formation and cardiomyocyte proliferation that is mediated in part by nonmyocardial tissues.

Along these parameters, regulators of cell cycle progression are likely to be important regulators of cardiomyocyte proliferation. In fact, the initial heart regeneration work demonstrated that a disruption in *mps1*, a cell cycle regulator with multiple roles, resulted in the enrichment of collagen-laden scar tissue in the wound site due to decreased cardiomyocyte proliferation (Poss et al. 2002). Remarkably, the *mps1* mutant was uncovered from a temperature-sensitive genetic screen for regulators of zebrafish fin regeneration, implicating conservation of the regenerative circuit across different organs (Poss et al. 2002). In a similar manner, pharmacological inhibition of the polo-like kinase 1 gene, which encodes a modulator of cell cycle progression, decreased the rate of cardiomyocyte proliferation (Jopling et al. 2010). Undoubtedly, focus on developmental genetic programs and cell cycle factors that promote cellular division has provided valuable insights into the genetic and cellular processes of heart regeneration.

With advances in microarray and sequencing technology, transcriptome analysis has emerged as a powerful tool to reveal potential regulators of heart regeneration. For example, using a microarray platform, Lien et al. (2006) showed that members of the platelet-derived growth factor (Pdgf) family were highly upregulated under various stages of regeneration. The contributions of Pdgf, however, are less clear. For instance, ex vivo studies on cultured cardiomyocytes showed that Pdgfs promote deoxyribonucleic acid (DNA) synthesis, suggesting a potential role in promoting cardiomyocyte proliferation (Lien et al. 2006). Subsequent in vivo studies, however, revealed that Pdgf expression is enriched in the epicardium with a likely role to promote vascular development in the regenerated tissue (Kim et al. 2010). Despite the uncertainty of Pdgf function, advances in genome sequencing promise to reveal intriguing factors that warrant functional studies in a model system.

13.4.1 MICRORNAS ARE IMPORTANT REGULATORS OF HEART REGENERATION

While transcriptional control is important to mount a regenerative response, recent evidence suggests that posttranscriptional regulation may also play prominent roles during heart regeneration. One key group of regulators is a class of small, noncoding ribonucleic acids (RNAs), termed microRNAs (miRNAs). miRNAs were originally identified from *Caenorhabditis elegans* genetic screens for defects in developmental transitions (Lee et al. 1993; Reinhart et al. 2000). Within the last 7–10 years, there has been a large collection of studies documenting the expanded roles of miRNAs, which now include organogenesis, stem cell maintenance, and brain and cardiac morphogenesis (Förstemann et al. 2005; Giraldez et al. 2005; Hatfield et al. 2005; Zhao et al. 2005; Shcherbata et al. 2006). miRNAs regulate gene expression at the posttranscriptional level by binding to the 3′ UTR of target mRNAs to inhibit protein translation (Krek et al. 2005; Hon and Zhang 2007; Friedman et al. 2009). miRNAs are predicted to control expression of >60% of human protein genes, and given that an individual miRNA may target multiple mRNAs, miRNAs offer unprecedented potential to rapidly modulate and integrate disparate cellular responses during development and homeostasis (Bartel 2004; Krek et al. 2005; Kloosterman and Plasterk 2006; Williams et al. 2009).

Recent miRNA studies employed microarray hybridizations to identify differentially expressed miRNAs between uninjured and regenerating hearts at 7 dpa (Yin et al. 2012). While a subset of miRNAs exhibited dynamical changes in expression patterns, miR-133, a highly conserved, cardiomyocyte-specific miRNA, was depleted in response to injury. Using heat-inducible transgenic overexpression *(Tg[hsp70:miR-133^{pre}])* and depletion *(Tg[hsp70:miR-133^{sp}])* strains, Yin et al. (2012) were able to reliably manipulate miR-133 expression in the heart at various stages following injury.

Remarkably, elevation of miR-133 at 7 dpa strongly shut down cardiomyocyte proliferation indices by ∼50%. Sustained overexpression of miR-133 for 30 days following injury produced prominent scar deposition in all treated animals when compared to control animals. Conversely, when levels of miR-133 were depleted with *Tg(hsp70:miR-133*sp*)* activation, cardiomyocyte proliferation indices were enhanced by 40%–45% (Figure 13.4; Yin et al. 2012). Whether this elevated level of cardiomyocyte proliferation will result in faster regeneration remains to be investigated.

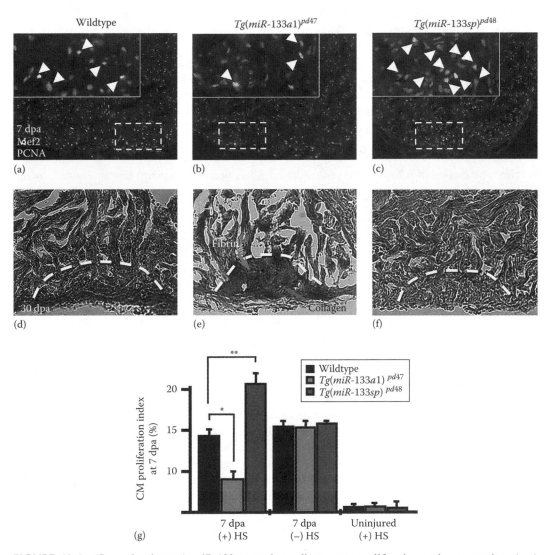

FIGURE 13.4 **(See color insert.)** miR-133 controls cardiomyocyte proliferation and regeneration. (a–c) Wild-type, *Tg(hsp70:miR-133pre)*^{pd47} and *Tg(hsp70:miR-133sp)*^{pd48} animals were injured, heat-treated, and hearts were stained for Mef2 and PCNA at 7 days postamputation (dpa). (b) When miR-133 levels are overexpressed, cardiomyocyte proliferation is repressed by ∼50%. (c) Conversely, depletion of miR-133 leads to elevated cardiomyocyte proliferation by ∼45%. (d–f) Sustained miR-133 modulation and AFOG stains at 30 dpa reveal abnormal scar deposition and inhibition of regeneration in *Tg(hsp70:miR-133pre)*^{pd47} or normal regeneration in control and *Tg(hsp70:miR-133sp)*^{pd48} hearts. (g) Cardiomyocyte proliferation indices indicate miR-133 controls cardiomyocyte proliferation. (White arrowheads indicate proliferating cardiomyocytes; inset in (a–c) is an enlargement of dashed boxes; AFOG, acid fuchsin orange G; dashed white lines in (d–f) is approximate amputation plane; HS, heat shock.). (Modified from *Dev. Biol.*, 365, Yin, P. et al., Regulation of zebrafish heart regeneration by miR-133, 319–327, Copyright 2012, from Elsevier.)

Nonetheless, the study was provocative in several ways. To date, depletion of miR-133 remains the only manipulation that enhanced muscle replacement. Second, it demonstrated roles for miRNAs during heart regeneration, and by virtue of miR-133 target gene identification, additional genetic regulators of heart regeneration were elucidated. From a list of ~150 potential miR-133 target genes that were identified with a combination of mRNA microarray and bioinformatics studies, the authors validated two target genes, the previously mentioned cell cycle regulator, *mps1*, and the gap junction protein, connexin-43 (*cx43*). As mentioned earlier, *mps1* function is to promote cardiomyocyte proliferation and limit scar formation (Poss et al. 2002). Cx43 is the major component of cardiac gap junctions and is essential for cardiac development (Reaume et al. 1995; Ya et al. 1998; Eckardt et al. 2006; Severs et al. 2008), and suppression by pharmacological inhibition stunted cardiomyocyte proliferation by 55% (Yin et al. 2012). Collectively, these studies revealed novel insights into the regenerative genetic network that now includes depletion of miR-133 and, as a result, elevation of *mps1* and *cx43* expression to promote heart regeneration. miRNAs are powerful genetic regulators that promise to be key factors in modulating cardiac regenerative capacity.

13.5   HEART REGENERATION INJURY MODELS

Surgical amputation of ~20% of the ventricular apex has been the standard injury zebrafish model for over a decade in studies of heart regeneration. In this assay, iridectomy scissors are used to open the pericardial cavity and an incision is made to remove a portion of the ventricular apex (Poss et al. 2002). This procedure is aggressive as it penetrates deep into the ventricular lumen and challenges the animal to undergo a rapid, dramatic regenerative response. Despite the significant blood loss, these animals display low levels of mortality, with greater than 90% survival. By and large, this approach has revealed fundamental genetic and cellular mechanisms that define natural heart regeneration.

Recently, three groups have used cryoinjury as an alternative injury model (Chablais et al. 2011; González-Rosa et al. 2011; Schnabel et al. 2011). Though variations exist among the different approaches, all models use either dry ice or liquid nitrogen cooled probes to inflict a local region of necrotic myocardium. An advantage with cryoinjury compared to ventricular resection is that the remaining dead cells activate pathology that mirrors the cellular programs activated by myocardial infarction events in mammals. Though the pace of regeneration is slower than a resection injury, cryoinjury nonetheless promotes removal of necrotic tissue, activates cardiomyocyte proliferation, and establishes a continuous myocardial wall.

More recently, chemical ablation with diphtheria toxin A (DTA) has been employed to destroy myocardial tissue. Using a double transgenic Cre/LoxP system with a tamoxifen-inducible CreER recombinase, DTA is activated in *cmlc2*-expressing cells, leading to death via programmed cell death. Remarkably, animals survive and display full recovery, even from the death of 60% of all cardiomyocytes in the adult heart (Wang et al. 2011). Cardiomyocyte proliferation and muscle replacement followed with full restoration of myocardial wall and cardiac function within 30 dpa. This approach is unique from the other injury models in that it enables tissue-specific programmed cell death of heart tissue and these studies show, for the first time, that cardiomyocyte death alone is sufficient to activate the regenerative program. A summary of the key findings from each injury model is depicted in Figure 13.5. Despite differences among the injury methods, common themes emerge during recovery. All three injury models trigger cardiomyocyte proliferation, muscle regeneration, and full electrical coupling between the regenerate and the preexisting heart tissue—processes that define heart regeneration.

13.6   PERSPECTIVES

The field of cardiac regeneration has been reinvigorated within the last 10 years due to recent advances in our understanding of regenerative potential afforded, in part, by model systems like the zebrafish. The zebrafish is uniquely suited for studies of heart regeneration given its combination of profound capacity to regenerate cardiac muscle and high conservation of genes with mammals.

Injury Type	Nonmyocardial Tissue Activation	Cardiomyocyte Activity	Duration of Regeneration	Necrotic Tissue	References
 Resection	Organ-wide activation of the epicardium and endocardium followed by restricted expression near injury site	*gata4* subpopulation of cardiomyocytes within the compact muscle layer proliferates	30–60 days for complete regeneration of missing myocardial tissue	Clean amputation of ventricular apex	Poss et al. (2002), Raya et al. (2003)
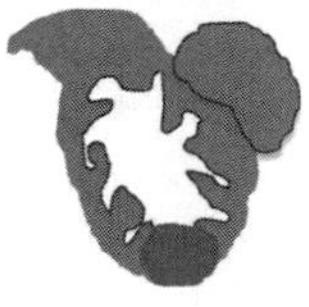 Cryoinjury	Activation of the epicardium and endocardium mirrors that of a resection injury. Fibroblasts undergo rapid proliferation	Cardiomyocytes near injury dedifferentiate and re-enter the cell cycle	100–130 days to remove necrotic tissue and regeneration of new heart muscle	Necrotic patch of myocardium remains following injury	Schnabel et al. (2011), González-Rosa et al. (2011), Chablais et al. (2011)
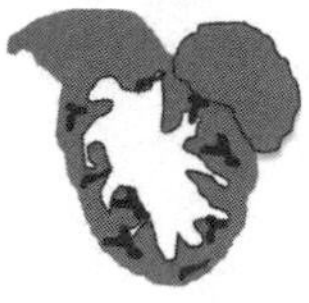 DTA ablation	Strong activation in both epicardium and endocardium layers despite localized damage to myocardium alone	The most robust cardiomyocyte proliferation among the three injury models	30 days to regenerate and integrate regions of newly created cardiomyocytes	Many patches of necrotic myocardial tissue	Wang et al. (2011)

FIGURE 13.5 **(See color insert.)** Different heart injury models. Ventricular resection, cryoinjury, and diphtheria toxin A (DTA) ablation models of heart injury are compared with each other. Differences in regenerative responses are described within the following major categories: injury type, nonmyocardial tissue activation, cardiomyocyte activity, duration of regeneration, and presence or absence of necrotic tissue.

While zebrafish regenerate heart muscle lost by injury, this capacity is muted in mammals, thus making human heart disease and heart failure the leading cause of death for the Western world. Given its profound impact on society, a multitude of disciplines have embarked upon the task of enhancing regenerative capacity, including developmental biologists, cardiologists, stem cell and induced-stem cell experts, and tissue engineers. While each strategy offers its own unique insight into this critical problem, stimulating heart regeneration is likely to require contributions from all areas of focus.

From studies in the zebrafish to newts to neonatal mice, it has become increasingly clear that regenerative capacity is not lost during evolution but rather lies latent in adult mammals. If we are to achieve the ultimate goal of enhancing cardiac regeneration in humans, future studies directed toward elucidating the genetic and molecular forces that define regenerative capacity will be instrumental. Given the recent advances that the zebrafish has made in the field of regenerative biology, its unprecedented regenerative capacity, and its advantages in genetic manipulation, it is without question that the zebrafish will continue to be a major contributor to our understanding of natural heart regeneration.

REFERENCES

Bader, D. and Oberpriller, J.O., 1978. Repair and reorganization of minced cardiac muscle in the adult newt (*Notophthalmus viridescens*). *Journal Morphology*, 155(3), 349–357.

Bader, D. and Oberpriller, J., 1979. Autoradiographic and electron microscopic studies of minced cardiac muscle regeneration in the adult newt, *Notophthalmus viridescens*. *Journal Experimental Zoology*, 208(2), 177–193.

Bartel, D.P., 2004. MicroRNAs: Genomics, biogenesis, mechanism, and function. *Cell*, 116(2), 281–297.

Becker, T. et al., 1997. Axonal regrowth after spinal cord transection in adult zebrafish. *Journal Comparative Neurology*, 377(4), 577–595.

Becker, R.O., Chapin, S., and Sherry, R., 1974. Regeneration of the ventricular myocardium in amphibians. *Nature*, 248(444), 145–147.

Bely, A.E. and Nyberg, K.G., 2010. Evolution of animal regeneration: Re-emergence of a field. *Trends in Ecology & Evolution*, 25(3), 161–170.

Bolli, P. and Chaudhry, H.W., 2010. Molecular physiology of cardiac regeneration. *Annals of the New York Academy of Sciences*, 1211, 113–126.

Cai, C.-L. et al., 2008. A myocardial lineage derives from Tbx18 epicardial cells. *Nature*, 454(7200), 104–108.

Cameron, D.A., 2000. Cellular proliferation and neurogenesis in the injured retina of adult zebrafish. *Visual Neuroscience*, 17(5), 789–797.

Chablais, F. et al., 2011. The zebrafish heart regenerates after cryoinjury-induced myocardial infarction. *BMC Developmental Biology*, 11, 21.

Chablais, F. and Jazwinska, A., 2012. The regenerative capacity of the zebrafish heart is dependent on TGFβ signaling. *Development (Cambridge, England)*, 139(11), 1921–1930.

Christoffels, V.M. et al., 2009. Tbx18 and the fate of epicardial progenitors. *Nature*, 458(7240), E8–9; discussion E9–10.

Dettman, R.W. et al., 1998. Common epicardial origin of coronary vascular smooth muscle, perivascular fibroblasts, and intermyocardial fibroblasts in the avian heart. *Developmental Biology*, 193(2), 169–181.

Dinsmore, C.E. and American Society of Zoologists, 1991. *A History of Regeneration Research: Milestones in the Evolution of a Science,* 1st edn., Cambridge, U.K.: Cambridge University Press.

Driever, W. et al., 1996. A genetic screen for mutations affecting embryogenesis in zebrafish. *Development*, 123, 37–46.

Eckardt, D. et al., 2006. Cardiomyocyte-restricted deletion of connexin43 during mouse development. *Journal of Molecular and Cellular Cardiology*, 41(6), 963–971.

Förstemann, K. et al., 2005. Normal microRNA maturation and germ-line stem cell maintenance requires Loquacious, a double-stranded RNA-binding domain protein. *PLoS Biology*, 3(7), e236.

Friedman, R.C. et al., 2009. Most mammalian mRNAs are conserved targets of microRNAs. *Genome Research*, 19(1), 92–105.

Giraldez, A.J. et al., 2005. MicroRNAs regulate brain morphogenesis in zebrafish. *Science*, 308(5723), 833–838.

Gittenberger-de Groot, A.C. et al., 1998. Epicardium-derived cells contribute a novel population to the myocardial wall and the atrioventricular cushions. *Circulation Research*, 82(10), 1043–1052.

Gittenberger-de Groot, A.C., Winter, E.M., and Poelmann, R.E., 2010. Epicardium-derived cells (EPDCs) in development, cardiac disease and repair of ischemia. *Journal of Cellular and Molecular Medicine*, 14(5), 1056–1060.

González-Rosa, J.M. et al., 2011. Extensive scar formation and regression during heart regeneration after cryoinjury in zebrafish. *Development (Cambridge, England)*, 138(9), 1663–1674.

Grunwald, D.J. and Eisen, J.S., 2002. Headwaters of the zebrafish—Emergence of a new model vertebrate. *Nature Reviews Genetics*, 3(9), 717–724.

Gupta, V. and Poss, K.D., 2012. Clonally dominant cardiomyocytes direct heart morphogenesis. *Nature*, 484(7395), 479–484.

Hatfield, S.D. et al., 2005. Stem cell division is regulated by the microRNA pathway. *Nature*, 435(7044), 974–978.

Heicklen-Klein, A. and Evans, T., 2004. T-box binding sites are required for activity of a cardiac GATA-4 enhancer. *Developmental Biology*, 267(2), 490–504.

Holmes, J.W., Borg, T.K., and Covell, J.W., 2005. Structure and mechanics of healing myocardial infarcts. *Annual Review of Biomedical Engineering*, 7, 223–253.

Holtzinger, A. and Evans, T., 2005. Gata4 regulates the formation of multiple organs. *Development (Cambridge, England)*, 132(17), 4005–4014.

Hon, L.S. and Zhang, Z., 2007. The roles of binding site arrangement and combinatorial targeting in microRNA repression of gene expression. *Genome Biology*, 8(8), R166.

Jopling, C. et al., 2010. Zebrafish heart regeneration occurs by cardiomyocyte dedifferentiation and proliferation. *Nature*, 464(7288), 606–609.

Kikuchi, K. et al., 2010. Primary contribution to zebrafish heart regeneration by gata4(+) cardiomyocytes. *Nature*, 464(7288), 601–605.

Kikuchi, K. et al., 2011a. tcf21+ epicardial cells adopt non-myocardial fates during zebrafish heart development and regeneration. *Development (Cambridge, England)*, 138(14), 2895–2902.

Kikuchi, K. et al., 2011b. Retinoic acid production by endocardium and epicardium is an injury response essential for zebrafish heart regeneration. *Developmental Cell*, 20(3), 397–404.

Kim, J. et al., 2010. PDGF signaling is required for epicardial function and blood vessel formation in regenerating zebrafish hearts. *Proceedings of the National Academy of Sciences of the United States of America*, 107(40), 17206–17210.

Kloosterman, W.P. and Plasterk, R.H.A., 2006. The diverse functions of microRNAs in animal development and disease. *Development Cell*, 11(4), 441–450.

Krek, A. et al., 2005. Combinatorial microRNA target predictions. *Nature Genetics*, 37(5), 495–500.

Lavine, K.J. and Ornitz, D.M., 2009. Shared circuitry: Developmental signaling cascades regulate both embryonic and adult coronary vasculature. *Circulation Research*, 104(2), 159–169.

Lee, R.K. et al., 1994. Cardiovascular development in the zebrafish. II. Endocardial progenitors are sequestered within the heart field. *Development (Cambridge, England)*, 120(12), 3361–3366.

Lee, R.C., Feinbaum, R.L., and Ambros, V., 1993. The *C. elegans* heterochronic gene lin-4 encodes small RNAs with antisense complementarity to lin-14. *Cell*, 75(5), 843–854.

Lepilina, A. et al., 2006. A dynamic epicardial injury response supports progenitor cell activity during zebrafish heart regeneration. *Cell*, 127, 607–619.

Lien, C.L. et al., 2006. Gene expression analysis of zebrafish heart regeneration. *PLoS Biology*, 4(8), e260.

Limana, F. et al., 2010. Myocardial infarction induces embryonic reprogramming of epicardial c-kit(+) cells: Role of the pericardial fluid. *Journal of Molecular and Cellular Cardiology*, 48(4), 609–618.

Lloyd-Jones, D. et al., 2010. Heart disease and stroke statistics—2010 update: A report from the American Heart Association. *Circulation*, 121(7), e46–e215.

Malicki, J. et al., 1996a. Mutations affecting development of the zebrafish retina. *Development (Cambridge, England)*, 123, 263–273.

Malicki, J. et al., 1996b. Mutations affecting development of the zebrafish ear. *Development (Cambridge, England)*, 123, 275–283.

Männer, J., 1999. Does the subepicardial mesenchyme contribute myocardioblasts to the myocardium of the chick embryo heart? A quail-chick chimera study tracing the fate of the epicardial primordium. *The Anatomical Record*, 255(2), 212–226.

Merki, E. et al., 2005. Epicardial retinoid X receptor alpha is required for myocardial growth and coronary artery formation. *Proceedings of the National Academy of Sciences of the United States of America*, 102(51), 18455–18460.

Mikawa, T. and Gourdie, R.G., 1996. Pericardial mesoderm generates a population of coronary smooth muscle cells migrating into the heart along with ingrowth of the epicardial organ. *Developmental Biology*, 174(2), 221–232.

Morgan, T.H., 1901. *Regeneration*, New York: Macmillan.

Mullins, M.C. et al., 1994. Large-scale mutagenesis in the zebrafish: In search of genes controlling development in a vertebrate. *Current Biology: CB*, 4(3), 189–202.

Pack, M. et al., 1996. Mutations affecting development of zebrafish digestive organs. *Development*, 123, 321–328.

Poss, K.D., Keating, M.T., and Nechiporuk, A., 2003. Tales of regeneration in zebrafish. *Developmental Dynamics: An Official Publication of the American Association of Anatomists*, 226(2), 202–210.

Poss, K.D. et al., 2002a. Mps1 defines a proximal blastemal proliferative compartment essential for zebrafish fin regeneration. *Development*, 129(22), 5141–5149.

Poss, K.D., Wilson, L.G., and Keating, M.T., 2002b. Heart regeneration in zebrafish. *Science (New York, N.Y.)*, 298(5601), 2188–2190.

Raya, A. et al., 2003. Activation of Notch signaling pathway precedes heart regeneration in zebrafish. *Proceedings of the National Academy of Sciences of the United States of America*, 100(Suppl 1), 11889–11895.

Reaume, A.G. et al., 1995. Cardiac malformation in neonatal mice lacking connexin43. *Science (New York, N.Y.)*, 267(5205), 1831–1834.

Reeve, J.L.V. et al., 2005, Don't lose heart—Therapeutic value of apoptosis prevention in the treatment of cardiovascular disease. *Journal of Cellular and Molecular Medicine*, 9(3), 609–622.

Reinhart, B.J. et al., 2000. The 21-nucleotide let-7 RNA regulates developmental timing in *Caenorhabditis elegans*. *Nature*, 403(6772), 901–906.

Roger, V.L. et al., 2012. Heart disease and stroke statistics—2012 update: A report from the American Heart Association. *Circulation*, 125(1), e2–e220.

Sánchez Alvarado, A. and Tsonis, P.A., 2006. Bridging the regeneration gap: Genetic insights from diverse animal models. *Nature Reviews Genetics*, 7(11), 873–884.

Schier, A.F. et al., 1996. Mutations affecting the development of the embryonic zebrafish brain. *Development (Cambridge, England)*, 123, 165–178.

Schnabel, K. et al., 2011. Regeneration of cryoinjury induced necrotic heart lesions in zebrafish is associated with epicardial activation and cardiomyocyte proliferation. *PLoS One*, 6(4), e18503.

Severs, N.J. et al., 2008. Remodelling of gap junctions and connexin expression in diseased myocardium. *Cardiovascular Research*, 80(1), 9–19.

Shcherbata, H.R. et al., 2006. The microRNA pathway plays a regulatory role in stem cell division. *Cell Cycle (Georgetown, Tex.)*, 5(2), 172–175.

Sleep, E. et al., 2010. Transcriptomics approach to investigate zebrafish heart regeneration. *Journal of Cardiovascular Medicine (Hagerstown, Md.)*, 11(5), 369–380.

Smart, N. et al., 2011. De novo cardiomyocytes from within the activated adult heart after injury. *Nature*, 474(7353), 640–644.

Solnica-Krezel, L. et al., 1996. Mutations affecting cell fates and cellular rearrangements during gastrulation in zebrafish. *Development (Cambridge, England)*, 123, 67–80.

Stainier, D.Y. et al., 1996. Mutations affecting the formation and function of the cardiovascular system in the zebrafish embryo. *Development (Cambridge, England)*, 123, 285–292.

Stainier, D.Y. and Fishman, M.C., 1994. The zebrafish as a model system to study cardiovascular development. *Trends in Cardiovascular Medicine*, 4(5), 207–212.

Stainier, D.Y., Lee, R.K., and Fishman, M.C., 1993. Cardiovascular development in the zebrafish. I. Myocardial fate map and heart tube formation. *Development (Cambridge, England)*, 119(1), 31–40.

Stemple, D.L. et al., 1996. Mutations affecting development of the notochord in zebrafish. *Development (Cambridge, England)*, 123, 117–128.

Wang, J. et al., 2011. The regenerative capacity of zebrafish reverses cardiac failure caused by genetic cardio-myocyte depletion. *Development (Cambridge, England)*, 138(16), 3421–3430.

Weinstein, B.M. et al., 1996. Hematopoietic mutations in the zebrafish. *Development (Cambridge, England)*, 123, 303–309.

Williams, A.H. et al., 2009. MicroRNA control of muscle development and disease. *Current Opinion in Cell Biology*, 21(3), 461–469.

Ya, J. et al., 1998. Heart defects in connexin43-deficient mice. *Circulation Research*, 82(3), 360–366.

Yin, V.P. et al., 2008. Fgf-dependent depletion of microRNA-133 promotes appendage regeneration in zebraf-ish. *Genes & Development*, 22(6), 728–733.

Yin, V.P. et al., 2012. Regulation of zebrafish heart regeneration by miR-133. *Developmental Biology*, 365(2), 319–327.

Zhao, Y., Samal, E., and Srivastava, D., 2005. Serum response factor regulates a muscle-specific microRNA that targets Hand2 during cardiogenesis. *Nature*, 436(7048), 214–220.

Zhou, B. et al., 2008. Epicardial progenitors contribute to the cardiomyocyte lineage in the developing heart. *Nature*, 454(7200), 109–113.

Zhou, B. et al., 2011. Adult mouse epicardium modulates myocardial injury by secreting paracrine factors. *The Journal of Clinical Investigation*, 121(5), 1894–1904.

14 Neuronal Regeneration

Ruxandra F. Sîrbulescu and Günther K.H. Zupanc

CONTENTS

14.1 INTRODUCTION

Teleost fishes have the highest regenerative potential of any vertebrate taxon studied. The capacity for structural and functional recovery during adulthood has been demonstrated in a number of species both in the central nervous system (CNS) and in the periphery. In addition to extensive studies examining the regeneration of fins (for review, see Akimenko et al. 2003) and cardiac muscle (for review, see Raya et al. 2004), most research has focused on the CNS, including brain, spinal cord, retina, and optic nerve (for reviews, see Anderson and Waxman 1985; Hitchcock and Raymond 1992; Zupanc 1999a, 2001, 2008a,b, 2009; Otteson and Hitchcock 2003; Zupanc and Zupanc 2006a; Chapouton et al. 2007; Kaslin et al. 2008; Zupanc 2011; Zupanc and Sîrbulescu 2011).

The first, although rather anecdotal, report of successful recovery after CNS injury was published by Koppányi and Weiss (1922). They described that 60 days after spinal cord transection, European carp (*Carassius vulgaris*) can regain normal swimming behavior. The first detailed analysis of such functional regeneration, combined with histological examination of the regrowing tissue, was performed by Tuge and Hanzawa (1935) in the spinal cord of Japanese rice minnows (*Oryzias latipes*).

Their work showed that the successful functional regeneration is closely linked to the structural repair of lesioned spinal cord tissue. A first indication that the structural and functional recovery of the CNS of teleost fishes may involve the generation of new neurons was obtained 25 years later. After the application of defined lesions to the optic tectum of crucian carps (*Carassius carassius*), Kirsche and Kirsche (1961) observed that successful regeneration critically depended on the preservation of "matrix zones," areas of high mitotic activity that give rise to new cells. If these are destroyed through injury, or depleted at more advanced ages of the fish, regeneration fails.

These early investigations laid the foundation for the study of adult neurogenesis (the generation of new neurons in the intact CNS) and neuronal regeneration (the regrowth of CNS tissue after injury through the generation of new neurons) in teleost fishes. Since then, research has focused on the exploration of a number of issues that are central to a better understanding of these phenomena: identification and mapping of the proliferation zones in the CNS, and analysis of their role in the supply of new cells that replace those lost to injury; characterization of the cellular processes involved in the development of the newly generated cells, in both the intact and injured CNSs; characterization of the progenitors that give rise to various types of new cells; identification of the molecular factors involved in the regulation of degenerative and regenerative processes after CNS injury; and employment of comparative approaches to elucidate the evolution of adult neurogenesis and neuronal regeneration. We have organized this chapter around these leitmotifs, starting with a review of adult neurogenesis in the intact CNS, followed by discussion of the involvement of adult-born cells in neuronal regeneration, and concluding with speculation about the evolution of these two intricately linked phenomena.

14.2   ADULT NEUROGENESIS IN THE INTACT CENTRAL NERVOUS SYSTEM

14.2.1   CELL PROLIFERATION

14.2.1.1   Rate of Cell Proliferation

Cell proliferation has been investigated in the CNS of several orders of teleost fishes; however, proliferation rates have only been determined in two species. In the brain of brown ghost knifefish (*Apteronotus leptorhynchus*), an average of 100,000 cells enter the S-phase of mitosis within any 2-h interval, as revealed by 5-bromo-2′-deoxyuridine (BrdU) labeling. This corresponds to approximately 0.2% of the total population of brain cells (Zupanc and Horschke 1995). Similar proliferation rates have been reported in the brain of zebrafish (*Danio rerio*) (Hinsch and Zupanc 2007). In the spinal cord of brown ghost knifefish, approximately 10,000 cells/mm^3 have been estimated to enter the S-phase of mitosis within any 2-h period (Sîrbulescu et al. 2009).

These estimates suggest that the rate of cell proliferation in the CNS of adult teleosts is at least one, if not two, orders of magnitude higher than in mammals, where the generation of new cells has been consistently reported only in two regions: the subgranular zone of the dentate gyrus from where new cells migrate a short distance into the granule cell layer of the hippocampus (Altman 1969; Luskin 1993; Lois and Alvarez-Buylla 1994; Lois et al. 1996; Pencea et al. 2001; Bédard and Parent 2004; Sanai et al. 2004; Curtis et al. 2007); and the anterior part of the subventricular zone of the lateral ventricle from where young cells migrate via the rostral migratory stream into the olfactory bulb (Altman and Das 1965; Kaplan and Bell 1984; Eriksson et al. 1998; Gould et al. 1999; Kornack and Rakic 1999; Seri et al. 2001).

14.2.1.2   Stem Cell Niches

The vast majority of new cells generated in the adult teleostean brain originate from specific proliferation zones. Many of these zones, particularly in the telencephalon and diencephalon, are located at or near ventricular surfaces (Figure 14.1). Although other proliferation zones, particularly in the cerebellum, are located in regions distant from any ventricle, some of them are thought to be derived from areas located at ventricular surfaces during embryonic stages of development. As a result of the eversion of the brain during development, the ventricular lumina associated with these areas can collapse or translocate (Pouwels 1978a,b; Kaslin et al. 2009).

In contrast to the mammalian brain, numerous proliferation zones have been found in the adult brain of all species of teleosts examined thus far, including guppy, *Poecilia reticulata* (Kranz and Richter 1970a,b; Richter and Kranz 1970a,b), brown ghost knifefish (Zupanc and Horschke 1995), gilt-head sea bream, *Sparus aurata* (Zikopoulos et al. 2000), three-spined stickleback, *Gasterosteus aculeatus* (Ekström et al. 2001), zebrafish (Zupanc et al. 2005; Grandel et al. 2006), annual killifish, *Austrolebias* sp. (Fernández et al. 2011), Mozambique tilapia, *Oreochromis mossambicus* (Teles et al. 2012), turquoise killifish, *Nothobranchius furzeri* (Tozzini et al. 2012), Burton's mouthbrooder, *Astatotilapia burtoni* (Maruska et al. 2012), and medaka, *O. latipes* (Kuroyanagi et al. 2010; Isoe et al. 2012).

An interesting characteristic feature of the adult teleostean brain is that the majority of new cells are generated in the cerebellum. For example, in brown ghost knifefish, approximately 75% of all adult-born cells in the brain originate in various subdivisions of the cerebellum (Zupanc and Horschke 1995).

Among the many proliferation zones in the brain, two are of particular interest from a comparative perspective: the olfactory bulb and the posterolateral region of the dorsal telencephalon. Based on neuroanatomical evidence (Northcutt and Braford 1980; Nieuwenhuys and Meek 1990; Braford 1995; Northcutt 1995; Butler 2000; Vargas et al. 2000) and results of functional studies

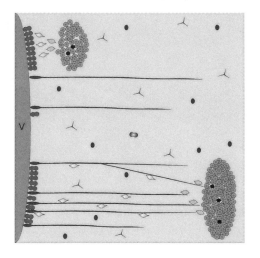

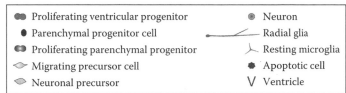

Proliferating ventricular progenitor	Neuron
Parenchymal progenitor cell	Radial glia
Proliferating parenchymal progenitor	Resting microglia
Migrating precursor cell	Apoptotic cell
Neuronal precursor	V Ventricle

FIGURE 14.1 Summary of the development of adult-born cells in the intact brain. Most cells generated in the intact brain are derived from progenitor cells harbored in specific proliferation zones, which are often associated with ventricular areas or derivatives of embryonic ventricular zones. In some areas of the brain, the progeny of the stem cells/progenitors reside near the proliferation zone where they were born. In other brain areas, the young cells migrate over relatively long distances from the proliferation zone to their target site. Radial glia have been implicated in the guidance of the migrating new cells. Differentiation into neurons or glia starts as early as during the migration of the young cells, and further maturation into specific subtypes of neuronal or glial cells becomes evident after arrival at the target site. At the target site, the number of new cells is regulated through apoptotic cell death. In addition to the mitotically active progenitor cells in the proliferation zones, quiescent progenitor cells exist throughout the parenchyma. In the intact brain, there is little mitotic activity among the latter population of progenitor cells. Similarly, most of the microglia remain in a resting state in the absence of injury. (Modified after Zupanc, G.K.H. and Sîrbulescu, R.F., *Curr. Top. Microbiol. Immunol.*, 367, 193, 2013.)

(Rodríguez et al. 2002; Portavella et al. 2004), part of the dorsolateral telencephalon of teleosts is thought to be homologous to the mammalian hippocampus. The existence of proliferation zones in these regions suggests that adult neurogenesis in the mammalian olfactory bulb and the hippocampus is a conserved trait of vertebrates (Zupanc 2006a).

In the spinal cord of teleost fishes, new cells are generated both in the ependymal layer and in the parenchyma (Anderson et al. 1983; Anderson and Waxman 1985; Reimer et al. 2008; Takeda et al. 2008; Sîrbulescu et al. 2009). Such a widespread distribution of proliferating cells is also observed in the spinal cord of mammals (Horner et al. 2000). The identity of the stem or progenitor cells in the mammalian spinal cord remains controversial: while some studies have indicated that neural stem cell activity is confined to ependymal cells (Meletis et al. 2008; Barnabé-Heider et al. 2010), others have failed to find evidence of proliferating ependymal cells (Spassky et al. 2005).

In the retina of teleost fishes, new neurons and photoreceptors are generated in two niches (Figure 14.2) (for review, see Otteson and Hitchcock 2003). One niche, commonly referred to as the ciliary marginal zone (CMZ), is located at the retinal periphery, where the retina meets the ciliary epithelium (Johns 1977). A second niche comprises rod progenitors in the central retina which give rise to rod photoreceptors in uninjured fish (Johns 1982). Recent studies have indicated that the rod progenitor cells are derived from Müller glia, which can act as stem cells (Bernardos et al. 2007).

In vitro examination of cells isolated from proliferation zones in the dorsal telencephalon and cerebellum has shown that they are capable of self-renewal and multipotency, thus exhibiting genuine

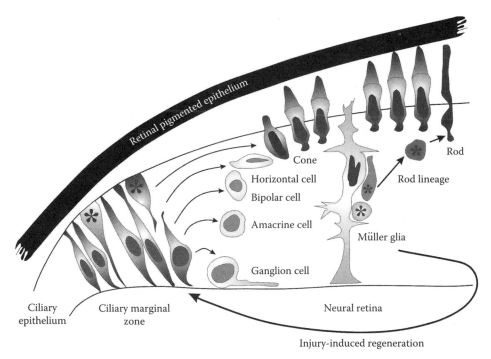

FIGURE 14.2 Development of adult-born cells in the intact and regenerating retina of teleost fish. The ciliary marginal zone located between the ciliary epithelium and the neural retina contains multipotent stem cells that span the width of the retinal epithelium. This zone is separated from the retinal pigmented epithelium by a narrow subretinal space. More restricted retinal progenitors (indicated by *asterisk*) within the ciliary marginal zone give rise to retinal ganglion cells, amacrine cells, bipolar cells, horizontal cells, and cone photoreceptors. Müller glia create a stem cell niche in the differentiated retina which can give rise to rod photoreceptors. Lesions induce the dedifferentiation of Müller glia, thus creating a population of progenitors that form a regenerative niche. The latter resembles the niche defined by the ciliary marginal zone and leads to the generation of all the cell types that have been lost to injury. (Modified after Raymond, P.A. et al., *BMC Dev. Biol.*, 6, 36, 2006.)

characteristics of stem cells (Hinsch and Zupanc 2006). In the adult teleostean brain, several lineages of stem cells have been identified. One subpopulation of progenitors displays the characteristics of radial glia (Pellegrini et al. 2007; Ganz et al. 2010; Chapouton et al. 2011; Rothenaigner et al. 2011). A second subpopulation of progenitors lacks the expression of canonical markers of glial cells (Kaslin et al. 2009; Alunni et al. 2010; Ganz et al. 2010; Rothenaigner et al. 2011). In the zebrafish cerebellum, these progenitors resemble neuroepithelial cells (Kaslin et al. 2009).

Few investigations have addressed the identity of the presumed stem cells in the intact spinal cord of adult teleosts. A study using transgenic zebrafish in which the oligodendrocyte transcription factor 2 (Olig2) was conjugated with the reporter enhanced green fluorescent protein (eGFP) revealed a discrete population of Olig2+ radial glia that persists into adulthood (Park et al. 2007). These radial glia can undergo asymmetrical divisions, characteristic of stem cells, to give rise to progeny that differentiate into oligodendrocytes, but not astrocytes or neurons. Interestingly, Olig2+ radial glia appear to have a widespread distribution within the spinal cord, further supporting the notion of a non-ependymal origin of the stem cells.

In the retina, cells of the CMZ and the Müller glia exhibit typical features of stem cells, including asymmetric mitotic division and self-renewal. It has been suggested that the neuroretinal stem cell population in the CMZ arises, together with a population of proliferative cells in the adjacent retinal pigmented epithelium, from a common stem cell in the larval eye (Wehman et al. 2005). The rod progenitors, which generate the rod receptor cells in the intact retina, appear to originate from the slowly dividing Müller glia scattered among differentiated retinal cells, as suggested by lineage tracing via transgenic expression of green fluorescent protein (GFP) under the control of a glia-specific promoter (Bernardos et al. 2007).

14.2.1.3 Control of Proliferative Activity

Stem cell activity in the zebrafish cerebellum can be regulated by growth factors. Inhibition of fibroblast growth factor (FGF) signaling in vivo results in a marked reduction of stem cell proliferation (Kaslin et al. 2009). Such an involvement of FGF in the control of mitotic activity has also been suggested by in vitro studies using stem cells isolated from the proliferation zones in the dorsal telencephalon and cerebellum of brown ghost knifefish (Hinsch and Zupanc 2006).

The transition of the adult stem cells from quiescence to an active state and vice versa is an important aspect in regulating proliferative activity. Upon proper stimulation, quiescent stem cells transition from the G_0 phase of the cell cycle to the G_1 phase and continue generating new cells. In ventricular zones of the zebrafish telencephalon, where radial glial cells act as stem cells (Adolf et al. 2006; Pellegrini et al. 2007; März et al. 2011), the transition between these two states is regulated by the transmembrane receptor protein Notch (Chapouton et al. 2010). Notch induction drives the stem cells into quiescence, whereas inhibition of Notch reinitiates cell division. Notch activation appears to arise from neighboring cells, presumably actively proliferating progenitors. It is thought that through this mechanism of lateral inhibition, equilibrium is maintained between quiescence and cellular proliferation, thus regulating the number of new cells produced.

The teleostean retina has been used as a comprehensive model system for investigating the mechanisms controlling adult stem cell activity. Several molecular characteristics commonly implicated in embryonic development have also been found to be involved in the regulation of stem cell activity in the adult retina (Raymond et al. 2006). Such characteristics include a diffuse distribution of N-cadherin on the basolateral plasma membranes of stem cells and progenitors in the CMZ; the activation of the Notch–Delta signaling pathway; and the expression of the transcription factors Pax6 and Rx1. Moreover, progenitor cells in the CMZ of adult fish respond to at least some of the extracellular signals that are known to influence embryonic retinal progenitors. For example, insulin-like growth factor-I (IGF-I) has been shown to stimulate proliferation of progenitors in the CMZ in organotypic cultures of intact eyecups of adult goldfish, *Carassius auratus* (Boucher and Hitchcock 1998). The mitogenic effect of IGF-I is consistent with the expression of both IGF-I and IGF-I receptors in the retina, and the in vivo modulation of cellular proliferation in the CMZ through the administration of recombinant growth hormone (Otteson et al. 2002). Interestingly, a recent study suggests that the tumor suppressor p53, which negatively regulates self-renewal of neural stem cells in mammals, has the opposite effect in the brain of adult medaka, positively regulating neurogenesis via cell proliferation (Isoe et al. 2012).

14.2.2 Development of New Cells

14.2.2.1 Migration

Two different migrational patterns of newly generated cells have been observed in the teleostean CNS (Figure 14.1). In certain instances, young cells migrate only a short distance and integrate proximally to their origin. In the retina, the new cells originate from the CMZ, an annulus of progenitor cells at the junction of the retina and the iris, and are continuously added appositionally to the margin of the retina (Johns and Easter 1977; Meyer 1978; Hagedorn and Fernald 1992; Marcus et al. 1999). Similarly, in the optic tectum—the projection target of the retinal ganglion cells in teleosts—the majority of new cells are generated at the caudal pole, where they remain during their subsequent development (Raymond and Easter 1983; Mansour-Robaey and Pinganaud 1990; Nguyen et al. 1999; Wullimann and Puelles 1999; Ekström et al. 2001; Candal et al. 2005; Zupanc et al. 2005; Grandel et al. 2006). As a result, the optic tectum grows asymmetrically by expanding primarily from its caudal end.

In other instances, newly generated cells migrate from their proliferation zones over relatively long distances to specific target areas. In the retina, rod progenitors arising from proliferating Müller glia in the inner nuclear layer migrate to the outer nuclear layer (Johns 1982; Julian et al. 1998). In the corpus cerebelli and the valvula cerebelli, new cells are generated in specific proliferation zones in the respective molecular layers. Subsequently, new cells migrate as far as several hundred micrometers into adjacent granular layers where they disperse uniformly (Zupanc et al. 1996, 2005, 2012; Grandel et al. 2006). Thus, both corpus cerebelli and valvula cerebelli grow in a rather symmetric fashion. A similar long-distance migration of young cells away from the corresponding proliferation zone has been observed in the preoptic area of zebrafish (Pellegrini et al. 2007).

Radial glial fibers appear to be involved in the guidance of newly formed cells during long-distance migration (Zupanc and Clint 2003; Pellegrini et al. 2007; Zupanc et al. 2012). In both the corpus cerebelli and the valvula cerebelli, such fibers, characterized by their immunoreactivity against vimentin and/or glial fibrillary acidic protein (GFAP), delineate the migratory path taken by the new cells. A few days after mitosis, many of these cells can be seen in close apposition to GFAP-labeled radial glial fibers (Zupanc et al. 2012). The nuclei of these cells are distinguished by their elongated morphology, a feature characteristic of migrating cells, as shown during both embryogenesis (Rakic 1971, 1972; Gregory et al. 1988) and adult stages of development (Alvarez-Buylla and Nottebohm 1988).

14.2.2.2 Regulation of Cell Numbers through Apoptosis

Little is known regarding mechanisms underlying the survival of new cells. In brown ghost knifefish, approximately half of the newborn cerebellar cells undergo apoptotic cell death within a few weeks after arriving at their target areas (Soutschek and Zupanc 1996; Zupanc et al. 1996; Ott et al. 1997). The number of apoptotic cells is significantly higher in the granule cell layers than in the corresponding molecular layers. Taken together, these findings suggest that apoptosis plays a critical role in regulating the number of young cells after they have reached their destination. Similarly, in the adult teleostean retina, apoptotic cell death occurs predominantly in areas where new cells differentiate and become integrated into visual circuits (Biehlmaier et al. 2001; Candal et al. 2005; Mizuno and Ohtsuka 2008, 2009).

Very little is known about the role of apoptosis during the development of adult-born cells in the intact spinal cord of teleosts. Immunolabeling against active caspase-3, an effector caspase critically involved in the late stages of apoptosis, has demonstrated the existence of low levels of constitutive apoptosis in the intact spinal cord of the brown ghost knifefish (Sîrbulescu et al. 2009).

14.2.2.3   Differentiation and Long-Term Persistence of New Cells

Experiments in zebrafish have shown that 9 months after administration of BrdU, approximately 50% of all adult-born cells express the neuron-specific protein Hu (Zupanc et al. 2005; Hinsch and Zupanc 2007). Such new neurons are particularly abundant in the dorsal telencephalon, including the region presumably homologous to the mammalian hippocampus, but are also found in other areas of the adult brain. Additional markers co-localizing with BrdU in the zebrafish brain include the neuronal markers parvalbumin, tyrosine hydroxylase, and serotonin (Grandel et al. 2006), acetylated tubulin (Pellegrini et al. 2007), and the glial marker S100β (Zupanc et al. 2005; Grandel et al. 2006). Co-localization of any of these markers with BrdU is, however, far less frequent than the co-localization of BrdU and Hu.

In the cerebellum, the vast majority of the new cells differentiate into granule cell neurons (Zupanc et al. 1996, 2005; Kaslin et al. 2009). In the corpus cerebelli, expression of the neuron-specific marker protein Hu commences when the migrating immature cells reach the granular layer (Zupanc et al. 2012). Retrograde tracing, combined with anti-BrdU immunohistochemistry, has demonstrated that the new granule cells develop proper axonal projections from the granular layer into the associated molecular layer (Zupanc et al. 1996, 2005), thus suggesting integration into the cerebellar neural network.

In the spinal cord of adult goldfish, new cells begin their neuronal differentiation as early as 24 h after their generation. A few weeks later, they express markers of mature neurons, indicating further specialization and possibly integration into functional circuits (Takeda et al. 2008). In the spinal cord of adult zebrafish, Olig2-expressing radial glia form a distinct stem cell population (Park et al. 2007). Using transgenic *olig2:egfp* zebrafish, these authors found that these cells differentiate into oligodendrocytes, but not into neurons or astrocytes (Park et al. 2007).

In the retina, the CMZ can generate all types of neurons and glia. The new neurons are incorporated into the retinal circuitry. Following migration from the inner nuclear layer to the outer nuclear layer, the rod progenitors differentiate into mature photoreceptors (Johns 1977, 1982; Raymond et al. 2006).

The lifespan of newly generated cells has been addressed in detail in two species: the brown ghost knifefish and the zebrafish. In each of these species, approximately half of the initially generated cells survive for at least several hundred days and likely for the rest of the individual's life (Ott et al. 1997; Zupanc et al. 2005; Hinsch and Zupanc 2007).

In the brown ghost knifefish, quantitative analysis has demonstrated that the long-term survival, together with the continuous production of new cells, leads to a permanent growth of the entire brain, except for very old fish in which the growth rate appears to reach a plateau. This parallels the continuous growth of the body in this species. While the body weight of the fish increases from 1 to 16 g, the total number of brain cells doubles from $5 \cdot 10^7$ to $1 \cdot 10^8$ (Zupanc and Horschke 1995).

In the intact spinal cord, newly generated neurons also exhibit considerable longevity. In adult goldfish, double-labeled BrdU$^+$/Hu C/D$^+$ neurons are still present in the gray matter and the leptomeninges 5 weeks after a single BrdU injection (Takeda et al. 2008).

14.2.3   Social and Endocrine Modulation of Cell Proliferation and Survival

Little is known about the effect of endocrine factors on adult neurogenesis. Modification of the estrogenic environment through inhibition of the brain estrogen-synthesizing enzyme aromatase B, blockade of nuclear estrogen receptors, and treatment with estrogens has demonstrated that estrogens modulate adult neurogenesis in zebrafish (Diotel et al. 2013). Specifically, estrogens decrease cell proliferation and migration of the young cells at the olfactory bulb/telencephalon junction and in the mediobasal hypothalamus. However, estrogens do not appear to have an effect on neurogenesis in the context of brain repair.

Similarly, the role of the social and behavioral context is largely unexplored in the regulation of adult neurogenesis. In rainbow trout (*Oncorhynchus mykiss*), experimental evidence has suggested a link between social hierarchy, plasma cortisol levels, and cellular proliferation in the brain (Sørensen et al. 2011, 2012). However, important aspects of this interplay, such as the effects of social factors and cortisol on differentiation and long-term survival of the new cells in specific brain areas, remain elusive.

In brown ghost knifefish, the generation and survival of cells in the dorsal thalamus can be modulated via social interaction with conspecifics (Dunlap et al. 2006; Dunlap and Chung 2013). Interestingly, a similar effect can be evoked by stimulation of the fish with electric signals mimicking the presence of a conspecific (Dunlap et al. 2008). This finding indicates that social signals conveyed in a single sensory modality may be sufficient to modulate cell proliferation in a specific brain area.

14.3 NEURONAL REGENERATION IN THE CENTRAL NERVOUS SYSTEM

Adult neurogenesis is closely related to neuronal regeneration, the ability to replace neurons lost to injury by newly generated ones. In teleosts, neuronal regeneration has been studied in detail in four CNS systems—the cerebellum, the dorsal telencephalon, the spinal cord, and the retina. Based on these model systems, we will discuss the major cellular events occurring during neuronal regeneration, the importance of this regenerative potential for functional recovery, and the molecular mechanisms involved in mediating these responses.

14.3.1 LESION PARADIGMS

A variety of lesion paradigms have been used to examine nervous tissue regeneration and functional recovery after CNS injuries in teleost fishes (Figure 14.3). To the best of our knowledge, the first mention of such a paradigm was published in 1922, in the form of a brief communication, by Theodor Koppányi and Paul Weiss, who transected the spinal cord of European carp, presumably in

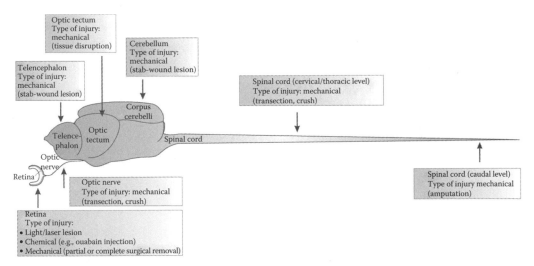

FIGURE 14.3 Lesion paradigms used in the study of CNS regeneration in teleost fishes. To indicate the sites of injury, a schematic drawing of the teleostean CNS is shown. Models of traumatic brain injury are mainly based on stab-wound lesions applied to the cerebellum and telencephalon, as well as more extensive tissue lesions in the optic tectum. Models of spinal cord injury involve transection or crush of the cord at cervical or thoracic levels, or amputation of the caudal part of the spinal cord. A variety of lesion paradigms have been applied to the retina, including light/laser-induced ablation, chemical lesions, and mechanical removal of patches of the retina. The effect of such lesions is examined in both the retina and the major projection area of retinal ganglion cells, the optic tectum. Injuries to the optic nerve are applied either through transection of the nerve or by crushing the nerve fibers. (Modified after Zupanc, G.K.H. and Sîrbulescu, R.F., *Curr. Top. Microbiol. Immunol.*, 367, 193, 2013.)

the cervical or thoracic region. Although the authors did not perform any histological experiments, and thus failed to provide evidence for structural regeneration, they claimed that after 2 months of recovery, the swimming movements of the regenerated fish differed "not in the slightest" from those of intact fish (Koppányi and Weiss 1922). Similar transection paradigms have been employed in a number of investigations, particularly to study axonal regeneration and recovery of locomotor function after spinal cord injury (for reviews, see Becker and Becker 2008; Sîrbulescu and Zupanc 2011).

A different lesion paradigm used to study regeneration after spinal cord injury is based on tail amputation, which completely removes the caudal part of the spinal cord (Figure 14.5; Waxman and Anderson 1980). This paradigm has been employed in two species of gymnotiform fishes, the black ghost knifefish (*Apteronotus albifrons*) and the brown ghost knifefish, primarily for the study of neuronal regeneration (for reviews, see Anderson and Waxman 1985; Waxman and Anderson 1986; Sîrbulescu and Zupanc 2011; Zupanc and Sîrbulescu 2011). As in apteronotids, including these two species, the modified axonal terminals of spinal motoneurons form a paired electric organ (de Oliveira-Castro 1955; Bennett 1971; Waxman et al. 1972), amputation of the tail, and thus part of the electric organ, leads to a reduction in the amplitude of the electric organ discharge. This characteristic enables the noninvasive monitoring of the partial loss of the electric behavior, and the subsequent regain of this behavioral function, which parallels structural regeneration (Sîrbulescu et al. 2009).

The majority of the studies addressing regeneration in the brain have focused on the optic tectum, the cerebellum, and the telencephalon. Based on earlier lesion studies that aimed to explore the function of the optic tectum, Walter Kirsche was the first to use this brain system to examine tissue regeneration and functional recovery in teleosts (Kirsche 1960; Kirsche and Kirsche 1961). Lesions are applied by mechanically disrupting varying amounts of tectal tissue.

In the cerebellum, lesions are generated with a scalpel in one hemisphere of the dorsalmost subdivision of the cerebellum, the corpus cerebelli (Zupanc et al. 1998; for reviews, see Zupanc and Zupanc 2006a; Zupanc 2008a,b, 2009, 2011; Zupanc and Sîrbulescu 2011).

In the telencephalon, lesions are applied by inserting a cannula (approximately 26–30 gauge) through one nostril and the olfactory bulb into the dorsal telencephalon (Ayari et al. 2010). Alternatively, lesions are generated vertically in the medial region (März et al. 2011) or laterally in the dorsolateral part (Ayari et al. 2010) of one hemisphere of the telencephalon by stabbing with a cannula through the skull.

In cerebellar, telencephalic, and tectal models of traumatic brain injury, control tissue can be obtained either from the region corresponding to the lesion site in intact or sham-operated animals, or from the contralateral hemisphere in lesioned animals.

In the retina, a multitude of approaches have been used to induce injuries, including surgical excision of a patch of retina (Hitchcock et al. 1992); intraocular injection of various toxins, such as ouabain (Maier and Wolburg 1979), kainic acid (Negishi et al. 1988), and tunicamycin (Negishi et al. 1991); thermal laser ablation (Braisted et al. 1994); and ultra-high-intensity light ablation (Bernardos et al. 2007) (for reviews, see Hitchcock and Raymond 1992; Hitchcock et al. 2004).

Severance of the optic nerve is used primarily to study axonal regeneration and recovery of visually guided behavior, but also to investigate the effect of such injury on the degenerative and regenerative processes in the region of the optic tectum innervated by the injured optic nerve (Sperry 1948; for reviews see Stuermer et al. 1992; Bernhardt 1999; Matsukawa et al. 2004; Beazley et al. 2006; Becker and Becker 2007a). A common technique to sever the optic nerve involves an incision in the dorsal conjunctiva and a downward rolling of the eyeball, thus making the optic nerve accessible to crushing or cutting.

14.3.2 Apoptotic Response

A characteristic response to CNS injury in teleost fishes is the rapid onset of cell death (Figure 14.4a). Using the terminal deoxynucleotidyl transferase dUTP nick end labeling (TUNEL) assay, the first labeled cells can be observed as early as 5 min after application of a stab-wound lesion to the

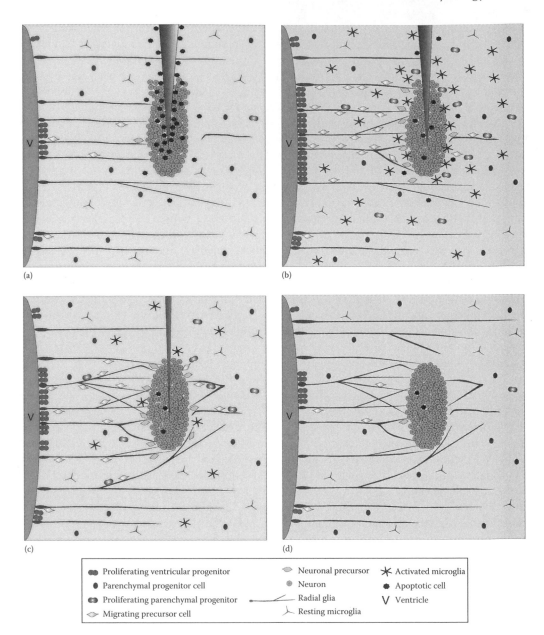

FIGURE 14.4   Sequence of major cellular events associated with brain injury and repair. (a) Within a few hours after injury, a large number of cells undergo apoptotic cell death at the lesion site. (b) Several days after the lesion, cell proliferation is markedly increased not only among progenitor cells within the ventricular proliferation zones which generate new cells constitutively, but also among the normally quiescent population of progenitor cells in the parenchyma. The new cells derived from these progenitors migrate toward the lesion site where they subsequently differentiate. At the same time, a pronounced increase in the number of activated microglia/macrophages is evident at and near the lesion site. They are thought to remove cellular debris through phagocytic activity. (c) Approximately 2 weeks after the lesion, the number of apoptotic cells has almost returned to baseline levels. The numbers of proliferating cells and activated microglia/macrophages have decreased, but they are still more numerous than in the intact brain. In addition, a meshwork of glial fibers has developed around the injury site. (d) Two to three months after the lesion, the cytoarchitecture of the brain area where the lesion occurred is restored. At this time point, only the persistence of the glial meshwork still marks the site of the injury. (Modified after Zupanc, G.K.H. and Sîrbulescu, R.F., *Curr. Top. Microbiol. Immunol.*, 367, 193, 2013.)

cerebellum of brown ghost knifefish (Zupanc et al. 1998). Thirty minutes after the lesion, the number of labeled cells reaches a plateau until 2 days post lesion, when the number of TUNEL-positive cells starts to decline, until background levels are reached approximately 20 days after injury. Further investigation of the morphological appearance of the TUNEL-positive cells using light and electron microscopy indicates that the vast majority of these cells undergo apoptosis, as opposed to necrosis. A similar transient increase in the number of cells undergoing apoptotic cell death has been observed in the dorsal telencephalon (Kroehne et al. 2011), the retina (Vihtelic and Hyde 2000; Yurco and Cameron 2005; Fimbel et al. 2007; Kassen et al. 2009; Bailey et al. 2010), and the spinal cord (Takeda et al. 2008; Sîrbulescu et al. 2009) of teleost fishes.

After amputation of the caudal spinal cord, at the interface between the intact tissue and the regenerating tissue, the number of apoptotic cells remains three to four times higher than baseline for up to 100 days post injury. The persistence of apoptosis in the area of the original lesion is remarkable, as it indicates that this type of cell death may also play a role in the restructuring of tissue and in the integration of new cells at this interface. This hypothesis is supported by the observation that the majority of apoptotic cells found at later stages of regeneration are new cells, and that many of the newly differentiated neurons and glia undergo apoptosis in this region of the regenerating spinal cord as late as 150–200 days post lesion (Sîrbulescu and Zupanc 2009).

The predominance of apoptosis in teleosts contrasts with necrosis as the dominant type of cell death in mammals after CNS injury (for reviews see Beattie et al. 2000; Vajda 2002; Liou et al. 2003). Unlike apoptosis, necrosis usually leads to inflammation at the site of the injury (for review, see Kerr et al. 1995). This inflammatory response initiates a cascade of events during a secondary (delayed) phase, which follows the primary (mechanical) phase of the injury. The events that take place during the secondary phase cause progressive cavitation and glial scarring, including upregulation of glial scar-associated molecules that lead to a retraction of axons from the site of the lesion (Balentine 1978; Zhang et al. 1997; Fitch et al. 1999; Horn et al. 2008; for reviews, see Reier et al. 1983; Fitch and Silver 2008; Rolls et al. 2009). By contrast, apoptosis is characterized by cell shrinkage, nuclear condensation, and production of membrane-enclosed particles that are digested by other cells. Most significantly, the side effects that accompany necrosis, such as inflammation of the surrounding tissue, are typically absent in apoptosis (for review, see Elmore 2007). Indeed, after traumatic brain injury in zebrafish, indication of acute, but not chronic, inflammation has been found (Kyritsis et al. 2012). Consequently, the elimination of damaged cells through apoptosis, instead of necrosis, is thought to be a key factor contributing to the enormous regenerative capability of the CNS of teleost fishes.

14.3.3 NEUROPROTECTION

It is likely that certain molecular factors exist in the microenvironment of the teleostean CNS which protect cells from dying. One such molecular candidate is calbindin-D_{28k}, a vitamin D-dependent calcium-binding protein, which is transiently increased in granular neurons of the cerebellum between 16 h and 7 days after injury (Zupanc and Zupanc 2006b). This upregulation of calbindin-D_{28k} might be involved in mitigating the effects of elevated levels of intracellular-free Ca^{2+} in the teleostean brain after injury. This hypothesis is supported by the findings that calbindin-D_{28k}-expressing neurons exhibit a relative resistance to neurotoxicity induced by glutamate, calcium ionophore, or acidosis (Mattson et al. 1991), and that the rate of survival of neurons can be increased after various types of insults by overexpression of the gene for calbindin-D_{28k} (D'Orlando et al. 2002; Ho et al. 1996; Monje et al. 2001; Phillips et al. 1999). Neurons can also survive axotomy by directly upregulating anti-apoptotic factors, such as B-cell lymphoma 2 (Bcl-2) and phospho-Akt, thus preventing the initiation of apoptosis (Ogai et al. 2012).

Another factor potentially involved in the promotion of cellular survival is glutamine synthetase. As revealed by 2D gel electrophoresis and mass spectrometry (see Section 14.3.9), the abundance of this astrocyte-specific enzyme is increased 3 days after cerebellar injury (Zupanc et al. 2006).

Under normal conditions, glutamine synthetase converts synaptically released glutamate into the nontoxic amino acid glutamine. However, under traumatic conditions, the extracellular level of glutamate is dramatically elevated (Faden et al. 1989; Katayama et al. 1990; Palmer et al. 1994), rendering the existing glutamine synthetase insufficient to catalyze the excessive amounts of glutamate released, and thus leading to a continuous overstimulation of glutamatergic synapses. This effect, commonly referred to as excitotoxicity (Olney 1969), is believed to be a major cause of cell death during the secondary phase of tissue damage in the CNS after traumatic injury (Hayes et al. 1992; Young 1992; Weber 2004; Lau and Tymianski 2010). Thus, the increase in the abundance of glutamine synthetase in the teleostean brain after traumatic injury is remarkable, as it may constitute another mechanism that furnishes regeneration-competent organisms with relative protection from cell death.

14.3.4 Activation of Microglia/Macrophages

Microglia/macrophages have been identified within a few days after lesions in several divisions of the CNS of teleost fishes (Figure 14.4b): the cerebellum (Zupanc et al. 2003), the dorsal telencephalon (Ayari et al. 2010; März et al. 2011; Kroehne et al. 2011), and the retina (Craig et al. 2008). In each of these regions, their numbers have been shown to return to background levels by 4 weeks post injury.

Macrophages/microglia are thought to mediate the removal of cellular debris through phagocytotic activity, although evidence for such a function in the teleostean CNS is limited. In the retina, phagocytosis has been shown to play a crucial role during regeneration. Normally, Müller glia engulf the cell bodies of apoptotic photoreceptors. However, if this process is disrupted by inhibiting phagocytosis, both proliferation of Müller cells in response to injury and regeneration of cone photoreceptors are significantly reduced (Bailey et al. 2010).

14.3.5 Reactive Cell Proliferation

The regenerative potential of teleost fishes is based on the ability to limit the degenerative effects of lesions and to generate new cells that replace those lost to injury. In the cerebellum, the dorsal telencephalon, and the spinal cord, the rate of mitosis starts to increase 1 day post lesion and peaks approximately 1 week after the injury, compared to controls (Figures 14.4b and c, and 14.5b; Zupanc and Ott 1999; Dervan and Roberts 2003; Reimer et al. 2008; Takeda et al. 2008; Sîrbulescu et al. 2009; Ayari et al. 2010; Kroehne et al. 2011; März et al. 2011; Kishimoto et al. 2012; Kyritsis et al. 2012). In the brain, the number of proliferating cells returns to baseline levels 3–4 weeks post lesion, whereas in the spinal cord, after tail amputation, control levels are still not reached 7 weeks after the injury.

In addition to the similarity in the time course of the transient upregulation of the proliferative response, a feature shared by different parts of the CNS is the diversity of sources that supply new cells for the repair of the injured tissue. After lesions applied to the cerebellum and the dorsal telencephalon, the cells generated in response to the injury originate from two major sources: the stem cell niches that generate new cells constitutively, and areas in the parenchyma near the injury site, which harbor quiescent progenitor cell populations in the intact brain (Zupanc and Ott 1999; Ayari et al. 2010; Kroehne et al. 2011; März et al. 2011; Kishimoto et al. 2012). Similarly, in the spinal cord, cells undergo mitosis after injury throughout the white and gray matters, as well as within the ependymal cell layer surrounding the central canal (Reimer et al. 2008; Takeda et al. 2008; Sîrbulescu et al. 2009). In the retina, mitotic cells appear to be primarily located proximal to the injured site (Stenkamp 2007). There is, however, some indication that enhanced proliferative activity can also occur distal to the lesion, suggesting the existence of diffusible factors that regulate this response (Yurco and Cameron 2005). Some of these factors appear to belong to the cysteinyl leukotriene signaling pathway as part of the acute inflammatory response in the brain (Kyritsis et al. 2012).

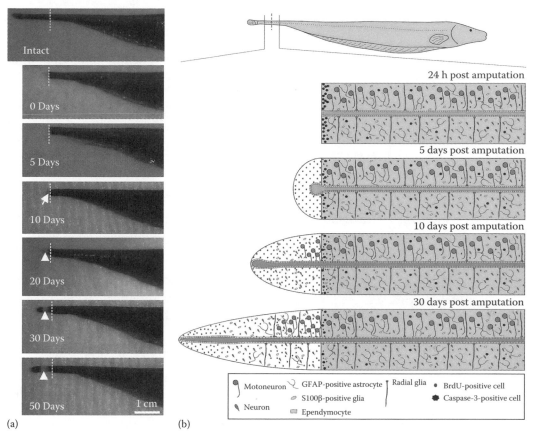

FIGURE 14.5 Regeneration of the tail and spinal cord in brown ghost knifefish. (a) Gradual regeneration of the tail over 50 days following a 1-cm amputation. Ten days after the transection, the caudal fin starts to regrow (arrow), followed a few days later by the tail stalk (arrowhead). The dashed line indicates the level of the cut. (b) Overview of some of the major processes involved in spinal cord regeneration in brown ghost knifefish. Amputation of the caudal part of the tail at the level indicated by the dashed line completely severs the spinal cord. Soon after the injury, numerous apoptotic active caspase-3-positive cells can be observed close to the lesion site. These high levels of apoptosis give way to cell proliferation, which increases rapidly and peaks 10 days after the injury. The massive levels of cell proliferation lead to the formation of an undifferentiated blastema at the tip of the regenerating tail, including the spinal cord. Subsequently, the caudal tip of the spinal cord, starting with the ependymal tube, extends into this blastema. The extension of the regenerating spinal cord appears to be supported by cell proliferation both at its caudal end and within the more rostral parenchyma. Differentiation of the newly generated cells into neurons and glia proceeds in a rostrocaudal direction. (Modified after Sîrbulescu, R.F. et al., *J. Comp. Physiol. A*, 195, 699, 2009; Sîrbulescu, R.F. and Zupanc, G.K.H., *Brain Res. Rev.*, 67, 73, 2011.)

Interestingly, in the cerebellum and dorsal telencephalon, it has been shown that a minor population of cells born as early as 2 days *before* the lesion also contributes to the restoration of the damaged tissue (Zupanc and Ott 1999; Kroehne et al. 2011). This observation suggests a direct relationship between the continued cell proliferation in the intact adult brain and the generation of new cells induced by injury. The continuous provision of a pool of undifferentiated cells in the intact brain appears to enable fish to recruit new cells more rapidly and in larger numbers in the event of injury than would be possible by only recruiting cells that are generated in response to a lesion.

In order to better understand the regenerative potential of teleost fishes, an important question concerns the source(s) of the regenerated neurons in the injured CNS. In several regeneration-competent organisms, dedifferentiation of cells in the immediate vicinity of the wound has been shown to represent a major mechanism contributing to the repair of tissues and organs, such as heart (Jopling et al. 2010; Kikuchi et al. 2010), bone (Knopf et al. 2011), and limbs (Kragl et al. 2009). By contrast, the role of dedifferentiation in the process of neuronal regeneration of brain and spinal cord tissue is less clear. Genetic lineage-tracing, combined with lesioning of the dorsal telencephalon, has suggested that most of the regenerating neurons are derived from radial glia-type progenitors, and that dedifferentiation of otherwise non-neurogenic cells plays only a minor role, if any (Kroehne et al. 2011). On the other hand, studies on the retina have shown that, in response to a lesion, the Müller glia can undergo dedifferentiation and give rise to a population of progenitors that serves as the basis for the regeneration of all cell types lost to injury (Yurco and Cameron 2005; Bernardos et al. 2007; Fimbel et al. 2007; Thummel et al. 2008).

14.3.6 Development of New Cells Produced in Response to Injury

Many of the new cells generated distal to the lesion site migrate, within the first few weeks after the injury, from the area where they were born to the area of the wound (Figure 14.4b and c). In the corpus cerebelli of brown ghost knifefish, a major source of such cells is the proliferation zone around the midline (Zupanc and Horschke 1995; Zupanc et al. 1996; see Section 14.2.2.1). After application of a stab wound to the dorsal part of one hemisphere of the corpus cerebelli, mitotic activity in this proliferation zone is markedly upregulated. As demonstrated by BrdU pulse-chase experiments, over the following 2 weeks, new cells migrate from this proliferation zone laterally within the dorsal molecular layer to the lesion site (Zupanc and Ott 1999). The migration of the young cells is paralleled by the appearance of GFAP- and vimentin-expressing radial glial fibers in the dorsal molecular layer, mainly between the injury site and the midline (Clint and Zupanc 2001, 2002). Among other possible functions, these radial glial fibers appear to provide a scaffolding for the migrating young cells, as inferred from the close apposition of BrdU-labeled cells to such fibers (Figure 14.4b through d; Clint and Zupanc 2001). On the other hand, there is no evidence that the GFAP- and vimentin-expressing cells form a permanent glial scar (Clint and Zupanc 2001, 2002; März et al. 2011).

A similar directed migration of new cells derived from adult stem cells residing in proliferation zones distal to the injury site has been found in the dorsal telencephalon of zebrafish. After lesioning the dorsolateral portion of one hemisphere, migrating new cells could be traced from the ventricular zone in the ventral telencephalon along a pathway in the subpallial and pallial regions to the injury site (Kishimoto et al. 2012), reaching their destination site within approximately 1 week. Consistent with the notion that the newly generated cells migrate to the injury site, it has been found that the BrdU-labeled cells express polysialylated-neural cell adhesion molecule (PSA-NCAM), a marker for migrating neurons, in the ventricular zone and its vicinity, but not in the area adjacent to the lesion site.

Differentiation of the new cells commences a few days after their generation. In the dorsal telencephalon, the young cells express the neuronal marker protein Hu as early as 3–4 days after the injury (Ayari et al. 2010; Kroehne et al. 2011; Kishimoto et al. 2012). These new neurons emerge both proximally and distally to the injury site (Figure 14.4). Over the following few days, the number of Hu-expressing new cells gradually increases, particularly at the injury site. This spatiotemporal pattern suggests that at least some of the new cells acquire immunological properties characteristic of neurons as early as during their migration toward the injury site.

Further examination of these cells at later stages of development has shown that they may differentiate into specific types of neurons, expressing marker proteins characteristic of mature neurons, such as parvalbumin or microtubule-associated protein 2a + 2b (MAP2a + 2b) (Kroehne et al. 2011), or Tbr1, a T-domain transcription factor characteristic of postmitotic glutamatergic neurons found in cortical areas of mammals (Englund et al. 2005) and in the pallium and medial–dorsal–lateral pallium of zebrafish (Kishimoto et al. 2012).

A similar differentiation of new cells into a specific subset of mature neurons has been demonstrated in the cerebellum after the application of stab-wound lesions (Zupanc et al. 1996). As the specific neuronal cell type—granular neurons—was identified by retrograde tracing from the molecular layer, this finding also suggests that the new granule cells have developed proper projections from the granular layer to the associated molecular layer.

In the retina, the progeny of dedifferentiated Müller glia redifferentiate into various cell types, including neurons, bipolar and amacrine cells, as well as Müller glia and cone photoreceptors (Wu et al. 2001; Yurco and Cameron 2005; Fausett and Goldman 2006; Raymond et al. 2006; Bernardos et al. 2007; Fimbel et al. 2007).

After spinal cord injury, the new cells develop into Hu C/D- and serotonin-expressing neurons, and S100β- and GFAP-expressing ependymocytes and glial cells (Figure 14.5b; Takeda et al. 2008; Sîrbulescu et al. 2009). After cervical spinal cord transection in zebrafish, new motor neurons are found in the proximity of the lesion site, with approximately 8% of the mitotic cells differentiating into HB9-positive or islet-1-positive neurons by 2 weeks after injury. By 6–8 weeks after the lesion, the newly formed cells express markers of mature neurons and are decorated with synaptic terminals, indicating successful integration into the spinal cord circuitry (Reimer et al. 2008). Interestingly, recent work in zebrafish has suggested that different classes of newly generated interneurons are derived from distinct progenitor cell domains localized around the central canal (Kuscha et al. 2012a).

14.3.7 IMPERFECT REGENERATION AND EXPERIMENTAL IMPROVEMENT OF REPAIR

Although apparent errors during regeneration in regeneration-competent organisms have rarely been examined systematically, review of the existing literature indicates that such deficiencies might occur more often than commonly assumed. After optic nerve transection in goldfish, new axonal sprouts emerge from the cut nerve stumps (Lanners and Grafstein 1980). Although the gross order of regenerating axons traveling to their synaptic target site, the optic tectum, is reminiscent of the order found in uninjured fish, the fascicle arrangement in the tectal fiber layer is quite erratic (Stuermer and Easter 1984). After destruction of the retina by ouabain injection, even more severe pathway aberrations occur in the course of retinal regeneration and of *de novo* formation of retinotectal axons (Stuermer et al. 1985). In the retina, these axons exhibit highly convoluted circuitry, including extensive fascicle crossings, hairpin loops, and circular routes. In the optic tectum, short and long fascicles are not neatly aligned, as observed during normal development, but intermingle and cross each other extensively.

Like the aberrations in the retinotectal projections, errors have also been observed in the regenerated retina itself. After neurotoxic lesions, or surgical excision of small patches of retina, laminar fusions that traverse the inner nuclear layer and the ganglion cell layer are the most noticeable type of error (Raymond et al. 1988; Hitchcock et al. 1992; Mensinger and Powers 2007). These fusions occur frequently at the interface between the regenerated and intact portions of the retina.

After amputation of the caudal part of the tail of brown ghost knifefish, in approximately 5%–10% of the individuals, the entire regenerate, including the spinal cord, is torsioned to various degrees (Figure 14.6; Sîrbulescu and Zupanc 2011). Perhaps related to this observation is the finding that apoptotic cells, identified by caspase-3 immunolabeling, persist at the transition zone between the rostrally located intact spinal cord and the more caudally situated regenerated tissue, at levels up to 10 times higher than baseline, for at least 110 days after spinal cord amputation (Sîrbulescu and Zupanc 2009). This phenomenon has been interpreted as an indication of a defective, or incomplete, organization of the transition zone between the intact tissue and the newly generated one. Similarly, in eels (*Anguilla anguilla*), it has been reported that the regenerated nervous tissue bridge that spans the gap between the caudal and rostral stumps of the transected spinal cord appears narrower and "more irregular in shape" (Doyle et al. 2001). An investigation of the regenerative response of the dopaminergic and serotonergic systems after spinal cord injury in zebrafish recorded long-term alterations of normal anatomical parameters, including hyperinnervation anterior to the lesion and reduced reinnervation caudal to the lesion site (Kuscha et al. 2012b).

Such observations indicate that, despite the impressive regenerative potential of teleost fishes, their ability to repair CNS tissue after injury has certain limits. This notion prompts the question

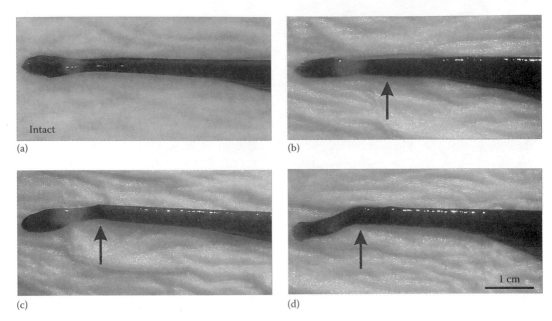

(a)   (b)

(c)   (d)

FIGURE 14.6 Successful and defective regeneration of the tail after amputation in brown ghost knifefish. (a) Intact tail. (b) After amputation of 1 cm of the tail (excluding the caudal fin), the majority of individuals successfully regrow the amputated part such that the regenerate is almost indistinguishable from intact tails. (c and d) In certain instances, however, the process of regeneration appears defective, resulting in various degrees of bending and torsion within the regenerate. *Arrows* indicate the site of the initial lesion. (Modified after Sîrbulescu, R.F. and Zupanc, G.K.H., *Brain Res. Rev.*, 67, 73, 2011.)

as to whether regeneration could be improved through experimental manipulation. In brown ghost knifefish, suppression of caspase-3 activation by administration of the apoptosis inhibitor 2,2′-methylenebis(1,3-cyclohexanedione) immediately after amputation of the caudal spinal cord results in a significant increase in the relative number of new cells that differentiate into neurons, in improved survival of the new neurons, and in significantly accelerated functional recovery, as assayed by monitoring the amplitude of the electric organ discharge (Sîrbulescu and Zupanc 2010a).

Improvement of regeneration has also been shown through a different type of experimental manipulation. Raising the ambient water temperature from 22°C to 30°C leads to a decrease in the number of apoptotic cells, an increase in the number of proliferating cells, and improved functional recovery (Sîrbulescu and Zupanc 2010b).

The deficits in regeneration, and the possibility to overcome them by experimental manipulation, enable investigators not only to better understand the mechanisms that mediate the regenerative potential of regeneration-competent organisms, but also to design therapeutic approaches that can be applied to regeneration-incompetent organisms, including humans. For example, the observation that administration of an apoptosis inhibitor immediately after the injury may have a long-term beneficial effect on the "quality" of the repair of neuronal tissue and on the regain of function prompts testing this effect in a mammalian model system. We propose such a comparative approach as a promising alternative to the traditional paradigms that focus exclusively on the study of regeneration-incompetent organisms to identify potential therapeutic targets.

14.3.8 FUNCTIONAL RECOVERY

Lesions of the corpus cerebelli in brown ghost knifefish do not cause any obvious behavioral defects. In the dorsal telencephalon, functional recovery after lesions has not been studied thus far.

In the retina, functional recovery as part of the regenerative process has been examined at both the physiological and behavioral levels, using mainly goldfish as a model system (for reviews, see Stenkamp 2007; Fleisch et al. 2011). Removal of parts of the retina by surgical and/or cytotoxic lesioning causes a reduction in the amplitude of different components of the electroretinogram, and a deviation from normal visually mediated reflexive behaviors as revealed through monitoring of the dorsal light reflex and the optokinetic nystagmus. The impairment of the dorsal light reflex is reflected by a tilting of the fish's vertical axis such that the intact eye is positioned away from downwelling light. The partial abolishment of the optokinetic nystagmus becomes evident through a marked reduction in the number of reset eye movements, compared to the untreated, control eye, after stimulation with high-contrast black-and-white square-wave gratings moving at a certain speed. As the retina regenerates, the components of the electroretinogram gradually reappear, and the two reflexive behaviors are progressively restored (Kästner and Wolburg 1982; Mensinger and Powers 1999, 2007; Lindsey and Powers 2007).

After transection of both optic nerves, fish fail to perform visually mediated behaviors, such as optokinetic reactions and startle responses, thus indicating total blindness. Functional vision is restored within a few weeks, as shown by the reappearance of these behaviors (Sperry 1948). Similarly, color discrimination, as demonstrated by the use of a conditioning paradigm, is fully reinstated within a few weeks after transection of the optic nerves, independent of whether the fish learned to discriminate certain colors pre- or postoperatively (Arora and Sperry 1963).

Experiments examining functional recovery after tectal lesions were initiated by Walter Kirsche in the early 1960s (Kirsche 1960; Kirsche and Kirsche 1961). Unilateral lesions of the optic tectum in crucian carps cause a number of motor aberrations, including sustained lateral body flexure and/or circling movements. The severity of these aberrations correlates positively with the extent of the lesion. In case of complete unilateral lesions of the tectum, tissue regeneration is absent, and functional recovery has not been observed. Partial lesions in young (3–6 cm) fish, leaving intact the proliferation zone located at the caudal–dorsal–medial borders of the tectum, result in a progressive recovery of the behavioral functions. The recovery becomes evident between 30 and 60 days post lesion, and the behavior typically returns to normal by 100 days after the lesion. The functional recovery parallels the reconstitution of the tectal cytoarchitecture, which is preceded by an increase in mitotic activity of the tectal proliferation zone (Richter 1968; Richter and Kranz 1970b, 1977).

In the olfactory system of zebrafish, chronic exposure to chemicals, such as detergents, damages olfactory sensory neurons and the resulting reduction in sensory input leads to degeneration of neurons in the glomeruli of the olfactory bulb, rendering the fish anosmic. Within 3 weeks after the chemical exposure is discontinued, the number of glomeruli returns to baseline, and olfactory-guided behaviors reappear (Paskin and Byrd-Jacobs 2012).

A large number of investigations have tested the capacity of teleost fishes to recover behavioral function after spinal cord lesions. The majority of the early studies have used the return to normal swimming behavior as a measure of spinal cord regeneration (Tuge and Hanzawa 1935, 1937; Bernstein and Bernstein 1967; Bernstein and Gelderd 1970; Coggeshall et al. 1982; Coggeshall and Youngblood 1983). In goldfish, Bernstein (1964) showed functional recovery 60 days after injury by evoking muscle contractions in myotomes located caudally to the lesion site through stimulation of spinal cord regions rostral to the injury. In the same species, the return of the highly stereotypical C-start escape response, as well as the regain of equilibrium and normal feeding behavior, have been used as measures of recovery after a crush lesion of the cervical spinal cord (Zottoli and Freemer 2003). However, the C-start response has not been observed to fully return to preoperative parameters, not even at more than 100 days post lesion. In minnows (*Phoxinus phoxinus*), neural control of the rapid skin color change in response to dark or light backgrounds returns to normal levels approximately 5 months after spinal cord transection (Healey 1962, 1967). In eels, after a similar injury, the tail beat frequency largely returns to baseline levels within 35 days. However, the amplitude of the tail beat is still lower than in intact fish by 45 days post lesion (Doyle et al. 2001). In this species, locomotor recovery can be significantly accelerated if the fish are forced to exercise by being placed in a tank with continuous water current (Doyle and Roberts 2006). After spinal cord transection, lesioned zebrafish can swim only approximately 5% of the distance covered by controls. However, by 6 weeks post lesion, they swim 57% of the distance measured in controls within a 5-min interval (Becker et al. 2004). When the zebrafish are challenged with forced swimming against a current, a markedly lower endurance becomes visible in fish that have regenerated, even 10–12 weeks after spinal cord lesions (van Raamsdonk et al. 1998). A possible explanation for this finding is that proximal and distal body regions in lesioned fish do not reach pre-lesion levels of coordination.

The majority of studies concern functional recovery after spinal cord transections at cervical or thoracic levels. Therefore, axonal regeneration, rather than *de novo* neurogenesis, has been assumed to be the basis of functional recovery. Nevertheless, neuronal regeneration is likely to play an important role in the recovery of functional spinal cord circuits, especially after ablations, as opposed to transection lesions. This has been demonstrated in brown ghost knifefish. Amputation of caudal spinal cord removes a portion of the neurons forming the electric organ, which leads to a proportional reduction in the amplitude of the electric organ discharge (Figure 14.7). As neurogenesis proceeds in the regenerating spinal cord, new motoneurons are generated, forming axons which reconstitute the electric organ. Thus, in this teleost, the process of neuronal differentiation after spinal cord lesion is paralleled by quantifiable behavioral recovery (Sîrbulescu et al. 2009; Sîrbulescu and Zupanc 2010a,b). The amplitude of the electric organ discharge recovers gradually after a 1-cm tail amputation, reaching baseline levels approximately 40 days after the lesion (Figure 14.7f; Sîrbulescu et al. 2009). The functional recovery becomes evident approximately 8–12 days after the lesion, which corresponds to the time when cell proliferation peaks and neuronal differentiation starts.

14.3.9 MOLECULAR FACTORS ASSOCIATED WITH REGENERATION

Our knowledge concerning the multitude of genes and proteins involved in the regenerative processes has been particularly advanced by two experimental approaches: microarrays and proteomic analysis.

Microarray studies have only been performed in zebrafish, and they have been instrumental in the identification of candidate genes that potentially underlie CNS regeneration. In the retina, such studies have examined changes that occur at various time points between 0 and 21 days after surgical lesions (Cameron et al. 2005), light injury (Kassen et al. 2007; Craig et al. 2008; Calinescu et al. 2009; Qin et al. 2009), optic nerve transection (Saul et al. 2010), or optic nerve crush (Veldman et al. 2007; McCurley and Callard 2010). After spinal cord transection, gene expression has been investigated in the spinal cord caudal to the lesion (Guo et al. 2011), as well as in the brain stem neurons of the nucleus of the medial longitudinal fascicle (Ma et al. 2012).

Although some variation exists between studies due to the range of methods and lesion paradigms, most of the altered transcripts can be grouped into several major functional categories that reflect remarkably well the cellular phenomena that describe regeneration. Within the first day after injury, a decrease is observed in factors associated with the normal function of affected cells, such as photoreceptor-specific genes encoding opsins, *phosphodiesterase 6c*, and *alpha* and *beta transducin* (Craig et al. 2008), and an increase in genes associated with an immune response and/or apoptosis, such as *tumor necrosis factor receptor 21*, *bcl2-associated X protein* (*bax*), *caspase 8*, *caspase 8/FADD-like apoptosis regulator*, and *bcl2-associated death promoter* (*bad*) (Kassen et al. 2007; McCurley and Callard 2010), *complement C7 precursor*, *chemokine C-X-C motif receptor 4b* (*cxcr4b*), *lymphocyte cytosolic plastin 1* (*L-plastin*), *perforin 1 precursor*, and *leukocyte surface antigen CD53* (Cameron et al. 2005).

At 1–2 days after injury, increases have been observed in the expression of genes encoding transcription factors, such as *sry-related high-mobility group box 11b* (*sox11b*), *fos*, *jun*, and *paired box 6a* (*pax6a*) (Craig et al. 2008), and proteins associated with cell cycle progression, such as *minichromosome maintenance 3, 4, 5*, and *7*, *cyclins B1, D1, F*, and *E*, and *proliferating cell nuclear antigen* (*pcna*) (Kassen et al. 2007).

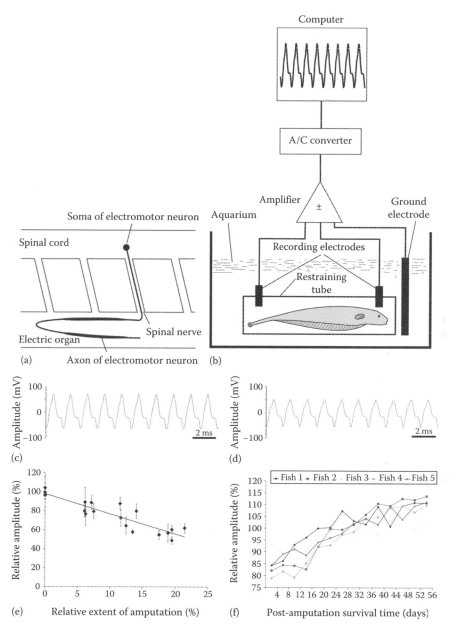

FIGURE 14.7 Behavioral recovery after spinal cord injury in brown ghost knifefish. (a) Anatomy of the electric organ. (Modified after Bennett, M.V.L., Electric organs, in *Fish Physiology, Vol. 5: Sensory Systems and Electric Organs*, eds., Hoar, W.S. and Randall, D.J., pp. 347–491, New York, Academic Press, 1971.) The somata of the electromotor neurons give rise to axons that leave the spinal cord via the spinal nerve to form the electric organ, located ventrally. Upon entering the electric organ, they run rostrally for several spinal segments before making a turn and extending in posterior direction to terminate near their entry level. Within the electric organ, the diameter of the axons greatly enlarges, up to 100 μm. (b) Experimental setup for recording the electric organ discharge in swimming fish. (c–f) Functional recovery after spinal cord injury. Waveforms of the electric organ discharges (c) before amputation and (d) 48 h after amputation of 1 cm of the tail. The extent to which the amplitude of the discharges drops is directly correlated with the relative extent of tail amputation. (e) Each data point represents one fish. After amputation, the amplitude of the electric organ discharges gradually recovers, until (f) it reaches baseline levels approximately 30 days later. (Modified after Sîrbulescu, R.F. et al., *J. Comp. Physiol. A*, 195, 699, 2009; Sîrbulescu, R.F. and Zupanc, G.K.H., *Brain Res. Rev.*, 67, 73, 2011.)

At later stages, changes in transcription regulation reflect the predominance of long-term regenerative processes. Some examples include genes associated with continued cell proliferation—*protein regulator of cytokinesis 1, proliferation associated protein 100 (p100), deoxycytidine kinase (dCK), class I c-tubulin, activating transcription factor 3 (atf3), cyclin B1,* and *tumor suppressor p53-binding protein* (Cameron et al. 2005; McCurley and Callard 2010; Saul et al. 2010); genes associated with cell growth and differentiation, including *engrailed 2b (eng2b)* and *zinc finger protein 2 (zic2)*; and genes involved in axonal genesis, including *growth-associated protein-43 (gap43), alpha* and *beta tubulin,* and the intermediate filament *plasticin,* which is expressed by regenerating ganglion cells (Cameron et al. 2005; Kassen et al. 2007). Increased tissue remodeling and cell migration is indicated by differential regulation of genes encoding enzymes involved in extracellular matrix processing, such as *matrix metalloproteinase 2 (mmp2), mmp9,* and *mmp13* (Cameron et al. 2005; Kassen et al. 2007). Interestingly, genes encoding some of the secreted growth factors upregulated during retina regeneration, including *midkine, progranulin,* and *galectin,* are also found after heart injuries in zebrafish, indicating shared mechanisms between various regenerative processes (Qin et al. 2009).

Differential proteome analysis has been performed in brown ghost knifefish on cerebellar tissue collected at two time points—30 min and 3 days after the lesion—reflecting different stages of the regenerative process (Zupanc et al. 2006; Ilieş et al. 2012). Thirty minutes after the lesion, the list of proteins whose abundance is significantly altered in lesioned tissue, compared to intact tissue, is comprised of two major categories (Ilieş et al. 2012). The first category includes proteins characteristic of degenerative processes, such as regulation of apoptotic cell death, destruction of the cytoskeleton, or remodeling of synaptic connections. Examples of upregulated proteins in this category include major histocompatibility complex (MHC) class I heavy chain, 26S proteasome non-ATPase regulatory subunit 8, and ubiquitin-specific protease 5, while levels of spectrin alpha 2 are decreased. The second category includes proteins associated with regenerative processes, such as regrowth of neurites and promotion of mitotic activity. Examples are alpha-internexin neuronal intermediate filament protein, erythrocyte membrane protein 4.1N, and tubulin alpha-1C chain. The finding of upregulation in the latter proteins is remarkable because it indicates that the course toward structural regeneration is set very early—within a few minutes after the injury—in the CNS of a regeneration-competent organism.

Three days after lesioning of the cerebellum, the list of proteins that exhibit significant changes in their abundance is dominated by cytoskeletal proteins (such as beta-actin and beta-tubulin), presumably reflecting the repair of injured axons, and proteins that mediate the correct assembly of structural proteins (such as chaperonin containing tailless-complex polypeptide 1, subunit epsilon, tropomodulins-3 and -4, and bullous pemphigoid antigen 1) (Zupanc et al. 2006). Furthermore, several proteins potentially involved in neuroprotection exhibit altered abundances at this time point. While the levels of 78,000-Da glucose-regulated protein and glutamine synthetase are increased, the level of cytosolic aspartate aminotransferase is decreased. Similar to the results of microarray studies, which have shown an upregulation in the expression of genes encoding transcription factors 1–2 days after injury (see earlier), proteomic analysis has also revealed an increase in the abundance of a potential transcriptional regulator, bone marrow zinc finger, 2–3 days after a cerebellar lesion. Taken together, these observations are congruent with the notion that, at this stage of regeneration, cells that have survived the primary phase of tissue damage are protected from cell death through a number of mechanisms, while at the same time processes are initiated that finally lead to the repair of injured cells and to the formation of new cells.

Large-scale approaches, such as microarray and proteomics studies, provide rich datasets comprising numerous genes and proteins potentially involved in regeneration. In order to verify and further examine the role that these candidates play, subsequent in-depth studies need to focus on the spatiotemporal dynamics of one or a few of such molecules. A summary of the molecular factors that have been the subject of such detailed investigations is given in Table 14.1. This growing list of factors, albeit far from comprehensive, begins to describe the special properties of a molecular microenvironment that supports, rather than inhibits, the complex process of regeneration in the CNS of vertebrates.

TABLE 14.1

Molecular Factors Involved in CNS Regeneration of Teleost Fish

| Class | Molecular Factor | Factor Regulation | | Model System | | |
		Modulation	Time Point/ Interval	Effect on Regeneration	Species	CNS Region	References
Apoptotic factors/ regulators	p-Akt	Increased expression	3–40 days	+	Goldfish (*Carassius auratus*)	Retina	Koriyama et al. (2006)
	p-Bad	Increased expression	3–40 days	±	Goldfish	Retina	Koriyama et al. (2006)
	Bax	No change	0–30 days	–	Goldfish	Retina	Koriyama et al. (2006)
	Bcl-2	Increased expression	10–20 days	+	Goldfish	Retina	Koriyama et al. (2006)
	Caspase–3	Increased expression	12 h to 2 days	±	Brown ghost knifefish (*Apteronotus leptorhynchus*)	Spinal cord	Sîrbulescu et al. (2009), Sîrbulescu and Zupanc (2009, 2010a)
Axonal growth inhibitors	Nogo-A	Decreased activity	10–30 days	±	Goldfish	Retina	Koriyama et al. (2006)
		N/A	N/A	Not inhibitory (lacks N-terminus)	Zebrafish (*Danio rerio*)	Spinal cord	Diekmann et al. (2005)
	Sema-3A	Decreased expression	2–7 days	–	Goldfish	Retina	Rosenzweig et al. (2010)
Cytoskeletal and associated structural proteins	MAP1B	Increased expression	N/A	+	Trout (*Oncorhynchus mykiss*)	Spinal cord	Alfei et al. (2004)
	GFAP	Increased expression	40 days; 4 months	±	Brown ghost knifefish; Black ghost knifefish (*Apteronotus albifrons*)	Spinal cord	Anderson et al. (1984), Sîrbulescu et al. (2009)
					Brown ghost knifefish	Brain	Clint and Zupanc (2001)
	Vimentin	Increased expression	15–100 days	+	Brown ghost knifefish	Brain	Clint and Zupanc (2002)
ECM components	CSPG	No change	N/A	–	Zebrafish	Spinal cord	Becker and Becker (2007b)
	Tenascin C	Increased expression	4 h, 2–11 days	+	Zebrafish	Spinal cord	Yu et al. (2011a)

(continued)

TABLE 14.1 (continued)
Molecular Factors Involved in CNS Regeneration of Teleost Fish

Class	Molecular Factor	Factor Regulation			Model System		References
		Modulation	Time Point/ Interval	Effect on Regeneration	Species	CNS Region	
Growth-associated proteins	GAP-43	Increased expression	7–14 days	+	Zebrafish	Spinal cord Retina	Becker et al. (1998) Kusik et al. (2010)
	Tuba1a	Increased expression	2–14 days	+	Zebrafish	Optic nerve	Goldman and Ding (2000), Senut et al. (2004), Veldman et al. (2010)
Growth factors/ receptors	BDNF	No change	1–20 days	+	Eel (Anguilla anguilla)	Spinal cord	Dalton et al. (2009)
	FGF3	Increased expression	14 days	+	Zebrafish	Spinal cord	Reimer et al. (2009)
	Retinoic acid receptor	Increased expression	14 days	+	Zebrafish	Spinal cord	Reimer et al. (2009)
	Trk B	No change	1–20 days	+	Eel	Spinal cord	Dalton et al. (2009)
Intra-/inter-cellular messengers	cAMP	N/A	N/A	+	Zebrafish	Spinal cord	Bhatt et al. (2004)
	NO	Increased expression	5–40 days	+	Goldfish	Retina	Koriyama et al. (2009)
Micro RNA	miR-133b	Increased expression	6 h to 7 days	+	Zebrafish	Spinal cord	Yu et al. (2011b)
mRNA editing enzymes	Apobec2a/2b	Increased expression	6 h to 8 days	+	Zebrafish	Retina	Powell et al. (2012)
Transcription regulators	ATF3	Increased expression	24 h to 7 days	+	Zebrafish	Retina	Veldman et al. (2007), Saul et al. (2010)
	KLF6a/KLF7a	Increased expression	1–12 days	+	Zebrafish	Retina	Veldman et al. (2007)
	Nkx6.1	Increased expression	14 days	+	Zebrafish	Spinal cord	Reimer et al. (2009)
	Pax6	Increased expression	14 days	+	Zebrafish	Spinal cord	Reimer et al. (2009)
	Sonic hedgehog a	Increased expression	14–42 days	+	Zebrafish	Spinal cord	Reimer et al. (2009)
	Sox11b	Increased expression	11 days	+	Zebrafish	Spinal cord	Guo et al. (2011)

Category	Molecule	Expression	Time		Species	Region	Reference
Transmembrane cell adhesion molecules	Contactin – 1a	Increased expression	6–14 days	+	Zebrafish	Spinal cord	Schweitzer et al. (2007)
	L 1.1	Increased expression	3–56 days	+	Zebrafish	Spinal cord	Becker et al. (1998), Becker et al. (2004)
	NCAM	No change	7–14 days	+	Zebrafish	Spinal cord	Becker et al. (1998)
	zfNLRR	Increased expression	24 h to 9 days	+	Zebrafish	Spinal cord, Retina	Bormann et al. (1999)
	P0	Increased expression	2–180 days	+	Zebrafish	Brain, Spinal cord	Schweitzer et al. (2003)
Other	Calbindin-D28k	Increased expression	16 h to 7 days	+	Brown ghost knifefish	Brain	Zupanc and Zupanc (2006b)
	Cysteine/Glycine-rich protein 1a	Increased expression	3–21 days	+	Zebrafish	Spinal cord	Ma et al. (2012)
	Somatostatin	Increased expression	24 h to 25 days	+	Brown ghost knifefish	Brain	Zupanc (1999b)
	Transglutaminase (neural)	Increased expression	10–40 days	+	Goldfish	Retina	Sugitani et al. (2006)

Immediately after injury, apoptotic cell death increases dramatically, but subsequent rises in the levels of apoptotic regulators, such as phospho-Akt (protein kinase B) and Bcl-2 (Koriyama et al. 2006; Ogai et al. 2012), facilitate a rapid decrease in the number of apoptotic cells (Zupanc et al. 1998; Sîrbulescu et al. 2009). In addition, heat shock proteins, such as heat shock factor-1 and HSP 70, show immediate induction, which can extend for several days after injury (Fujikawa et al. 2012). With a short lag, as compared to the initial wave of apoptosis, a marked increase in the expression of transcription factors, such as ATF3, Sox11b, Pax2, Pax6, Nkx6.1, Krüppel-like factor (KLF)6a, and KLF7a, and in growth-associated proteins, such as tubulin alpha 1A (Tuba1a) and GAP-43, indicates the transition toward regenerative processes, including cell proliferation, neuronal differentiation, and axonal regrowth (Veldman et al. 2007; Reimer et al. 2009; Saul et al. 2010; Parrilla et al. 2013). Elevated levels of growth factors, such as FGF-3, and several receptors of the retinoic acid pathway have been proposed to play a role in axonal regeneration (Reimer et al. 2009). During spinal cord regeneration in zebrafish, FGF signaling is necessary for the formation of glial bridges along which regenerating axons migrate (Goldshmit et al. 2012). Also in the retina, FGF signaling is required for the maintenance and regeneration of photoreceptors (Hochmann et al. 2012). MAP1B, a protein that has been associated in mammals with regeneration in the peripheral nervous system (Soares et al. 2002), is expressed in adult rainbow trout at high levels in both spinal and supraspinal neurons that show regenerative capacity, as well as in glial cells of the CNS (Alfei et al. 2004). In the regenerative phase following spinal cord transection, levels of major vault protein (MVP) increase; moreover, knockdown of MVP expression impairs structural and functional regeneration (Pan et al. 2013). Cellular coagulation factor XIII, a member of the transglutaminase family, is expressed in microglia, astrocytes, and retinal ganglion cells, and was found to promote neurite sprouting and elongation after optic nerve injuries in goldfish (Sugitani et al. 2012).

The excellent regeneration of teleost fishes after CNS injury is facilitated not only through the upregulation of growth-promoting molecules, but also through the absence or active inhibition of molecules that interfere with axonal regrowth. In zebrafish, Nogo-A, one of the main inhibitory proteins in mammalian myelin (Bradbury and McMahon 2006; Yiu and He 2006), lacks the inhibitory N-terminal region (Diekmann et al. 2005). Immunoreactivity to chondroitin sulfate proteoglycans, major components of the glial scar in mammals (Silver and Miller 2004), is not elevated after CNS injuries in zebrafish (Becker and Becker 2007b). A recent study has shown that the elevated expression of a regulatory microRNA, miR-133b, plays an important role in spinal cord regeneration in zebrafish by inhibiting the RhoA GTPase, which promotes growth cone collapse and would therefore prevent axonal regrowth (Yu et al. 2011b).

Taken together, the studies summarized here begin to sketch the biological mechanisms that enable teleost fishes to regenerate CNS tissue after injury. Nevertheless, considerable work will be required to advance our understanding of the molecular interactions which orchestrate this highly complex process.

14.4   CONCLUSIONS AND PERSPECTIVES

The study of regeneration-competent organisms provides us with the unique opportunity to gain a broad biological understanding of tissue repair in the adult CNS. As shown in this review, neuronal regeneration is intimately linked to adult neurogenesis. Every regeneration-competent organism examined thus far also generates new neurons constitutively, both in large numbers and in many regions of the adult CNS. Comparative analysis has suggested that adult neurogenesis is a primitive vertebrate trait (Zupanc 2006b). It is likely that the availability of all the cellular and molecular regulatory mechanisms necessary for the generation of new neurons in the intact CNS has greatly facilitated the repurposing of the cellular machinery for neuronal regeneration with only slight adaptive changes. We hypothesize that such a regenerative ability was also shared by many, if not all, early vertebrates.

One group of organisms that have shown extraordinary adaptation under the selective pressure caused by injury are the gymnotiforms; they include the black and brown ghost knifefish mentioned earlier in this review. Many species of this teleostean order are distinguished by their elongated, compressed caudal part of the body ("tail"). In their natural habitat, gymnotiforms often suffer from damage to, or loss of, parts of the tail because of predatory fish specialized in tail-eating (Mago-Leccia et al. 1985; Lundberg et al. 1996; de Santana and Vari 2010). Probably in response to this selective pressure, gymnotiforms have developed the extraordinary ability to regenerate tails, including large parts of the caudal spinal cord, rapidly and with high efficiency—even after repeated loss of the tail.

A broad understanding of the biology of adult neurogenesis and neuronal regeneration will also facilitate the analysis of the selective pressures that have caused the loss of the regenerative potential during the evolution of mammals. It has been hypothesized that in vertebrates with indeterminate growth, such as teleosts, the primary function of adult neurogenesis is to ensure a constant ratio of the number of central neuronal elements within motor/sensory pathways relative to the number of peripheral elements—muscle fibers and receptor cells—when the number of peripheral elements grows ("numerical matching hypothesis"; Figure 14.8; Zupanc 1999a, 2001, 2006a,b, 2008a,b, 2009). Such a function is consistent with the existence of an enormous neurogenic potential in the teleostean brain, which correlates with continuous peripheral growth and the constant increase in the number of individual muscle fibers or receptor cells (Johns and Easter 1977; Corwin 1981; Zakon 1984; Weatherley and Gill 1985; Koumans and Akster 1995; Zimmerman and Lowery 1999; Rowlerson and Veggetti 2001). By contrast, growth in mammals is primarily the result of an increase in the size, not in the number, of individual peripheral elements (Rowe and Goldspink 1969). As a corollary, the matching hypothesis predicts that, when the mode of muscle growth shifted from hyperplasia to hypertrophy in the course of the evolution of mammals, the neurogenic potential of CNS structures forming part of the motor pathway was reduced in parallel. Similarly, the number of neurogenic regions and the rate of neurogenesis decreased in brain areas associated with sensory processing when the continuous formation of new sensory cells in the sensory organs during adulthood was gradually abandoned.

Phylogenetic studies have shown that phenotypes are commonly lost by alteration of regulatory mechanisms without extensive modification of the individual developmental processes (such as cellular proliferation, migration, or differentiation) that had produced the lost form. On the other hand, novel environmental factors can initiate the expression of these ancestral elements in novel combinations through the process of developmental recombination to give rise to novel phenotypic traits, followed by the genetic accommodation of change (West-Eberhard 2003, 2005). If this hypothesis is correct, then the loss of adult neurogenesis and neuronal regeneration as phenotypic traits in many regions of the mammalian brain may have left intact some of the genetic or molecular pathways that induced the original phenotype. The existence of such ancestral elements has important implications from a biomedical point of view because then it should be possible to reactivate these elements under certain environmental conditions by turning off or on switches that control them. Indeed, quiescent stem cells have been discovered in the adult mammalian brain that can be activated by certain signals from the cellular environment, such as epidermal growth factor (Reynolds and Weiss 1992; Weiss et al. 1996; Palmer et al. 1999; Kondo and Raff 2000). Clearly, these adult stem cells could provide the substrate for the development of a cell replacement therapy based on the intrinsic ancestral potential of the adult human brain. The study of regeneration-competent organisms could play an important role in identifying the factors required to reactivate this potential.

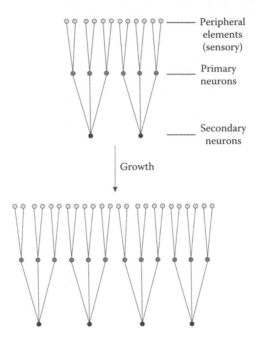

FIGURE 14.8 Schematic representation of the numerical matching of peripheral and central elements. The "matching hypothesis" makes predictions about the relationship between peripheral elements, such as muscle fibers or sensory receptor cells, and central elements in the brain involved in the motor control of these muscle fibers, or in the processing of sensory information received through these receptors, during growth. A number of observations suggest that the ratio of the corresponding peripheral elements and central elements is highly constant and maintained when the number of peripheral elements changes (here, a doubling in the number of sensory peripheral elements is shown). Hyperplasia of peripheral motor elements, or increase in the number of peripheral sensory elements, therefore, prompts a concomitant increase in the number of the corresponding central neurons. As such a growth pattern through addition of new peripheral elements to the population of existing ones is characteristic of teleost fishes, but largely absent in adult mammals, the matching hypothesis can explain the difference in the extent of adult neurogenesis between teleosts and mammals. (After Zupanc, G.K.H., *Semin. Cell Dev. Biol.*, 20, 683, 2009.)

LIST OF ABBREVIATIONS

Bcl-2	B-cell lymphoma 2
BrdU	5-bromo-2′-deoxyuridine
CMZ	ciliary marginal zone
CNS	central nervous system
eGFP	enhanced green fluorescent protein
FGF	fibroblast growth factor
GFAP	glial fibrillary acidic protein
GFP	green fluorescent protein
IGF-I	insulin-like growth factor I
KLF	Krüppel-like factor
MAP	microtubule-associated protein
MVP	major vault protein
Olig2	oligodendrocyte transcription factor-2
PSA-NCAM	polysialylated-neural cell adhesion molecule
TUNEL	terminal deoxynucleotidyl transferase dUTP nick end labeling

REFERENCES

Adolf, B., P. Chapouton, C.S. Lam, S. Topp, B. Tannhäuser, U. Strähle, M. Götz, and L. Bally-Cuif. 2006. Conserved and acquired features of adult neurogenesis in the zebrafish telencephalon. *Dev. Biol.* 295:278–293.

Akimenko, M.-A., M. Marí-Beffa, J. Becerra, and J. Géraudie. 2003. Old questions, new tools, and some answers to the mystery of fin regeneration. *Dev. Dyn.* 226:190–201.

Alfei, L., S. Soares, A. Alunni, M. Ravaille-Veron, Y. Von Boxberg, and F. Nothias. 2004. Expression of MAP1B protein and its phosphorylated form MAP1B-P in the CNS of a continuously growing fish, the rainbow trout. *Brain Res.* 1009:54–66.

Altman, J. 1969. Autoradiographic and histological studies of postnatal neurogenesis: IV. Cell proliferation and migration in the anterior forebrain, with special reference to persisting neurogenesis in the olfactory bulb. *J. Comp. Neurol.* 137:433–458.

Altman, J. and G.D. Das. 1965. Autoradiographic and histological evidence of postnatal hippocampal neurogenesis in rats. *J. Comp. Neurol.* 124:319–336.

Alunni, A., J.M. Hermel, A. Heuzé, F. Bourrat, F. Jamen, and J.S. Joly. 2010. Evidence for neural stem cells in the medaka optic tectum proliferation zones. *Dev. Neurobiol.* 70:693–713.

Alvarez-Buylla, A. and F. Nottebohm. 1988. Migration of young neurons in adult avian brain. *Nature* 335:353–354.

Anderson, M.J. and S.G. Waxman. 1985. Neurogenesis in adult vertebrate spinal cord in situ and *in vitro*: A new model system. *Ann. NY Acad. Sci.* 457:213–233.

Anderson, M.J., K.A. Swanson, S.G. Waxman, and L.F. Eng. 1984. Glial fibrillary acidic protein in regenerating teleost spinal cord. *J. Histochem. Cytochem.* 32:1099–1106.

Anderson, M.J., S.G. Waxman, and M. Laufer. 1983. Fine structure of regenerated ependyma and spinal cord in *Sternarchus albifrons*. *Anat. Rec.* 205:73–83.

Arora, H.L. and R.W. Sperry. 1963. Color discrimination after optic nerve regeneration in the fish *Astronotus ocellatus*. *Dev. Biol.* 7:234–243.

Ayari, B., K.H. El Hachimi, C. Yanicostas, A. Landoulsi, and N. Soussi-Yanicostas. 2010. Prokineticin 2 expression is associated with neural repair of injured adult zebrafish telencephalon. *J. Neurotrauma* 27:959–972.

Bailey, T.J., S.L. Fossum, S.M. Fimbel, J.E. Montgomery, and D.R. Hyde. 2010. The inhibitor of phagocytosis, O-phospho-L-serine, suppresses Müller glia proliferation and cone cell regeneration in the light-damaged zebrafish retina. *Exp. Eye Res.* 91:601–612.

Balentine, J.D. 1978. Pathology of experimental spinal cord trauma. I. The necrotic lesion as a function of vascular injury. *Lab. Invest.* 39:236–253.

Barnabé-Heider, F., C. Göritz, H. Sabelström, H. Takebayashi, F.W. Pfrieger, K. Meletis, and J. Frisén. 2010. Origin of new glial cells in intact and injured adult spinal cord. *Cell Stem Cell* 7:470–482.

Beattie, M.S., A.A. Farooqui, and J.C. Bresnahan. 2000. Review of current evidence for apoptosis after spinal cord injury. *J. Neurotrauma* 17:915–925.

Beazley, L.D., J. Rodger, C.A. Bartlett, A.L. Taylor, and S.A. Dunlop. 2006. Optic nerve regeneration: Molecular pre-requisites and the role of training—Restoring vision after optic nerve injury. *Adv. Exp. Med. Biol.* 572:389–395.

Becker, C.G. and T. Becker. 2007a. Growth and pathfinding of regenerating axons in the optic projection of adult fish. *J. Neurosci. Res.* 85:2793–2799.

Becker, C.G. and T. Becker. 2007b. Zebrafish as a model system for successful spinal cord regeneration. In *Model Organisms in Spinal Cord Regeneration*, eds. C. Becker and T. Becker, pp. 289–319. Weinheim, Germany: Wiley-VCH.

Becker, C.G. and T. Becker. 2008. Adult zebrafish as a model for successful central nervous system regeneration. *Restor. Neurol. Neurosci.* 26:71–80.

Becker, T., R.R. Bernhardt, E. Reinhard, M.F. Wullimann, E. Tongiorgi, and M. Schachner. 1998. Readiness of zebrafish brain neurons to regenerate a spinal axon correlates with differential expression of specific cell recognition molecules. *J. Neurosci.* 18:5789–5803.

Becker, C.G., B.C. Lieberoth, F. Morellini, J. Feldner, T. Becker, and M. Schachner. 2004. L1.1 is involved in spinal cord regeneration in adult zebrafish. *J. Neurosci.* 24:7837–7842.

Bédard, A. and A. Parent. 2004. Evidence of newly generated neurons in the human olfactory bulb. *Dev. Brain Res.* 151:159–168.

Bennett, M.V.L. 1971. Electric organs. In *Fish Physiology, Vol. 5: Sensory Systems and Electric Organs*, eds. W.S. Hoar and D. J. Randall, pp. 347–491. New York: Academic Press.

Bernardos, R.L., L.K. Barthel, J.R. Meyers, and P.A. Raymond. 2007. Late-stage neuronal progenitors in the retina are radial Müller glia that function as retinal stem cells. *J. Neurosci.* 27:7028–7040.

Bernhardt, R.R. 1999. Cellular and molecular bases of axonal regeneration in the fish central nervous system. *Exp. Neurol.* 157:223–240.

Bernstein, J.J. 1964. Relation of spinal cord regeneration to age in adult goldfish. *Exp. Neurol.* 9:161–174.

Bernstein, J.J. and M.E. Bernstein. 1967. Effect of glial-ependymal scar and teflon arrest on the regenerative capacity of goldfish spinal cord. *Exp. Neurol.* 19:25–32.

Bernstein, J.J. and J.B. Gelderd. 1970. Regeneration of the long spinal tracts in the goldfish. *Brain Res.* 20:33–38.

Bhatt, D.H., S.J. Otto, B. Depoister, and J.R. Fetcho. 2004. Cyclic AMP-induced repair of zebrafish spinal circuits. *Science* 305:254–258.

Biehlmaier, O., S.C. Neuhauss, and K. Kohler. 2001. Onset and time course of apoptosis in the developing zebrafish retina. *Cell Tissue Res.* 306:199–207.

Bormann, P., L.W. Roth, D. Andel, M. Ackermann, and E. Reinhard. 1999. zfNLRR, a novel leucine-rich repeat protein is preferentially expressed during regeneration in zebrafish. *Mol. Cell. Neurosci.* 13:167–179.

Boucher, S.E. and P.F. Hitchcock. 1998. Insulin-related growth factors stimulate proliferation of retinal progenitors in the goldfish. *J. Comp. Neurol.* 394:386–394.

Bradbury, E.J. and S.B. McMahon. 2006. Spinal cord repair strategies: Why do they work? *Nat. Rev. Neurosci.* 7:644–653.

Braford, M.R. 1995. Comparative aspects of forebrain organization in the ray-finned fishes: Touchstones or not? *Brain Behav. Evol.* 46:259–274.

Braisted, J.E., T.F. Essman, and P.A. Raymond. 1994. Selective regeneration of photoreceptors in goldfish retina. *Development* 120:2409–2419.

Butler, A.B. 2000. Topography and topology of the teleost telencephalon: A paradox resolved. *Neurosci. Lett.* 293:95–98.

Calinescu, A.A., T.S. Vihtelic, D.R. Hyde, and P.F. Hitchcock. 2009. Cellular expression of *midkine-a* and *midkine-b* during retinal development and photoreceptor regeneration in zebrafish. *J. Comp. Neurol.* 514:1–10.

Cameron, D.A., K.L. Gentile, F.A. Middleton, and P. Yurco. 2005. Gene expression profiles of intact and regenerating zebrafish retina. *Mol. Vis.* 11:775–791.

Candal, E., R. Anadón, W.J. DeGrip, and I. Rodríguez-Moldes. 2005. Patterns of cell proliferation and cell death in the developing retina and optic tectum of the brown trout. *Dev. Brain Res.* 154:101–119.

Chapouton, P., R. Jagasia, and L. Bally-Cuif. 2007. Adult neurogenesis in non-mammalian vertebrates. *Bioessays* 29:745–757.

Chapouton, P., K.J. Webb, C. Stigloher, A. Alunni, B. Adolf, B. Hesl, S. Topp, E. Kremmer, and L. Bally-Cuif. 2011. Expression of hairy/enhancer of split genes in neural progenitors and neurogenesis domains of the adult zebrafish brain. *J. Comp. Neurol.* 519:1748–1769.

Clint, S.C. and G.K.H. Zupanc. 2001. Neuronal regeneration in the cerebellum of adult teleost fish, *Apteronotus leptorhynchus*: Guidance of migrating young cells by radial glia. *Dev. Brain Res.* 130:15–23.

Clint, S.C. and G.K.H. Zupanc. 2002. Up-regulation of vimentin expression during regeneration in the adult fish brain. *Neuroreport* 13:317–320.

Coggeshall, R.E., S.G. Birse, and C.S. Youngblood. 1982. Recovery from spinal transection in fish. *Neurosci. Lett.* 32:259–264.

Coggeshall, R.E. and C.S. Youngblood. 1983. Recovery from spinal transection in fish: Regrowth of axons past the transection. *Neurosci. Lett.* 38:227–231.

Corwin, J.T. 1981. Postembryonic production and aging of inner ear hair cells in sharks. *J. Comp. Neurol.* 201:541–553.

Craig, S.E., A.A. Calinescu, and P.F. Hitchcock. 2008. Identification of the molecular signatures integral to regenerating photoreceptors in the retina of the zebrafish. *J. Ocul. Biol. Dis. Infor.* 1:73–84.

Curtis, M.A., M. Kam, U. Nannmark, M.F. Anderson, M.Z. Axell, C. Wikkelso, S. Holtas et al. 2007. Human neuroblasts migrate to the olfactory bulb via a lateral ventricular extension. *Science* 315:1243–1249.

Dalton, V.S., B.L. Roberts, and S.M. Borich. 2009. Brain derived neurotrophic factor and trk B mRNA expression in the brain of a brain stem-spinal cord regenerating model, the European eel, after spinal cord injury. *Neurosci. Lett.* 461:275–279.

Dervan, A.G. and B.L. Roberts. 2003. Reaction of spinal cord central canal cells to cord transection and their contribution to cord regeneration. *J. Comp. Neurol.* 458:293–306.

Diekmann, H., M. Klinger, T. Oertle, D. Heinz, H.M. Pogoda, M.E. Schwab, and C.A.O. Stuermer. 2005. Analysis of the reticulon gene family demonstrates the absence of the neurite growth inhibitor Nogo-A in fish. *Mol. Biol. Evol.* 22:1635–1648.

Diotel, N., C. Vaillant, C. Gabbero, S. Mironov, A. Fostier, M.M. Gueguen, I. Anglade, O. Kah, and E. Pellegrini. 2013. Effects of estradiol in adult neurogenesis and brain repair in zebrafish. *Horm. Behav.* 63:193–207.

D'Orlando, C., M.R. Celio, and B. Schwaller. 2002. Calretinin and calbindin D-28k, but not parvalbumin protect against glutamate-induced delayed excitotoxicity in transfected N18-RE 105 neuroblastoma-retina hybrid cells. *Brain Res.* 945:181–190.

Doyle, L.M. and B.L. Roberts. 2006. Exercise enhances axonal growth and functional recovery in the regenerating spinal cord. *Neuroscience* 141:321–327.

Doyle, L.M., P.P. Stafford, and B.L. Roberts. 2001. Recovery of locomotion correlated with axonal regeneration after a complete spinal transection in the eel. *Neuroscience* 107:169–179.

Dunlap, K.D., J.F. Castellano, and E. Prendaj. 2006. Social interaction and cortisol treatment increase cell addition and radial glia fiber density in the diencephalic periventricular zone of adult electric fish, *Apteronotus leptorhynchus*. *Horm. Behav.* 50:10–17.

Dunlap, K.D. and M. Chung. 2013. Social novelty enhances brain cell proliferation, cell survival, and chirp production in an electric fish, *Apteronotus leptorhynchus*. *Dev. Neurobiol.* 73:324–332.

Dunlap, K.D., E.A. McCarthy, and D. Jashari. 2008. Electrocommunication signals alone are sufficient to increase neurogenesis in the brain of adult electric fish, *Apteronotus leptorhynchus*. *Dev. Neurobiol.* 68:1420–1428.

Ekström, P., C.-M. Johnsson, and L.-M. Ohlin. 2001. Ventricular proliferation zones in the brain of an adult teleost fish and their relation to neuromeres and migration (secondary matrix) zones. *J. Comp. Neurol.* 436:92–110.

Elmore, S. 2007. Apoptosis: A review of programmed cell death. *Toxicol. Pathol.* 35:495–516.

Englund, C., A. Fink, C. Lau, D. Pham, R.A. Daza, A. Bulfone, T. Kowalczyk, and R.F. Hevner. 2005. Pax6, Tbr2, and Tbr1 are expressed sequentially by radial glia, intermediate progenitor cells, and postmitotic neurons in developing neocortex. *J. Neurosci.* 25:247–251.

Eriksson, P.S., E. Perfilieva, T. Bjork-Eriksson, A.M. Alborn, C. Nordborg, D.A. Peterson, and F.H. Gage. 1998. Neurogenesis in the adult human hippocampus. *Nat. Med.* 4:1313–1317.

Faden, A.I., P. Demediuk, S.S. Panter, and R. Vink. 1989. The role of excitatory amino acids and NMDA receptors in traumatic brain injury. *Science* 244:798–800.

Fausett, B.V. and D. Goldman. 2006. A role for alpha1 tubulin-expressing Müller glia in regeneration of the injured zebrafish retina. *J. Neurosci.* 26:6303–6313.

Fernández, A.S., J.C. Rosillo, G. Casanova, and S. Olivera-Bravo. 2011. Proliferation zones in the brain of adult fish *Austrolebias* (Cyprinodontiform[sic]: Rivulidae): A comparative study. *Neuroscience* 189:12–24.

Fimbel, S.M., J.E. Montgomery, C.T. Burket, and D.R. Hyde. 2007. Regeneration of inner retinal neurons after intravitreal injection of ouabain in zebrafish. *J. Neurosci.* 27:1712–1724.

Fitch, M.T., C. Doller, C.K. Combs, G.E. Landreth, and J. Silver. 1999. Cellular and molecular mechanisms of glial scarring and progressive cavitation: In vivo and in vitro analysis of inflammation-induced secondary injury after CNS trauma. *J. Neurosci.* 19:8182–8198.

Fitch, M.T. and J. Silver. 2008. CNS injury, glial scars, and inflammation: Inhibitory extracellular matrices and regeneration failure. *Exp. Neurol.* 209:294–301.

Fleisch, V.C., B. Fraser, and W.T. Allison. 2011. Investigating regeneration and functional integration of CNS neurons: Lessons from zebrafish genetics and other fish species. *Biochim. Biophys. Acta* 1812:364–380.

Fujikawa, C., M. Nagashima, K. Mawatari, and S. Kato. 2012. HSP70 gene expression in the zebrafish retina after optic nerve injury: A comparative study under heat shock stresses. *Adv. Exp. Med. Biol.* 723:663–668.

Ganz, J., J. Kaslin, S. Hochmann, D. Freudenreich, and M. Brand. 2010. Heterogeneity and Fgf dependence of adult neural progenitors in the zebrafish telencephalon. *Glia* 58:1345–1363.

Goldman, D. and J. Ding. 2000. Different regulatory elements are necessary for alpha1 tubulin induction during CNS development and regeneration. *Neuroreport* 11:3859–3863.

Goldshmit, Y., T.E. Sztal, P.R. Jusuf, T.E. Hall, M. Nguyen-Chi, and P.D. Currie. 2012. Fgf-dependent glial cell bridges facilitate spinal cord regeneration in zebrafish. *J. Neurosci.* 32:7477–7492.

Gould, E., A.J. Reeves, M. Fallah, P. Tanapat, C.G. Gross, and E. Fuchs. 1999. Hippocampal neurogenesis in adult old world primates. *Proc. Natl. Acad. Sci. USA* 96:5263–5267.

Grandel, H., J. Kaslin, J. Ganz, I. Wenzel, and M. Brand. 2006. Neural stem cells and neurogenesis in the adult zebrafish brain: Origin, proliferation dynamics, migration and cell fate. *Dev. Biol.* 295:263–277.

Gregory, W.A., J.C. Edmondson, M.E. Hatten, and C.A. Mason. 1988. Cytology and neuron-glial apposition of migrating cerebellar granule cells in vitro. *J. Neurosci.* 8:1728–1738.

Guo, Y., L. Ma, M. Cristofanilli, R.P. Hart, A. Hao, and M. Schachner. 2011. Transcription factor Sox11b is involved in spinal cord regeneration in adult zebrafish. *Neuroscience* 172:329–341.

Hagedorn, M. and R.D. Fernald. 1992. Retinal growth and cell addition during embryogenesis in the teleost, *Haplochromis burtoni*. *J. Comp. Neurol.* 321:193–208.

Hayes, R.L., L.W. Jenkins, and B.G. Lyeth. 1992. Neurotransmitter-mediated mechanisms of traumatic brain injury: Acetylcholine and excitatory amino acids. *J. Neurotrauma* 9:S173–S187.

Healey, E.G. 1962. Experimental evidence for regeneration following spinal section in the minnow (*Phoxinus phoxinus* L.). *Nature* 194:395–396.

Healey, E.G. 1967. Experimental evidence for the regeneration of nerve fibres controlling colour changes after anterior spinal section in the minnow (*Phoxinus phoxinus* L.). *Proc. R. Soc. Lond. B* 168:57–81.

Hinsch, K. and G.K.H. Zupanc. 2006. Isolation, cultivation, and differentiation of neural stem cells from adult fish brain. *J. Neurosci. Method* 158:75–88.

Hinsch, K. and G.K.H. Zupanc. 2007. Generation and long-term persistence of new neurons in the adult zebrafish brain: A quantitative analysis. *Neuroscience* 146:679–696.

Hitchcock, P., M. Ochocinska, A. Sieh, and D. Otteson. 2004. Persistent and injury-induced neurogenesis in the vertebrate retina. *Prog. Retin. Eye Res.* 23:183–194.

Hitchcock, P.F., K.J. Lindsey Myhr, S.S. Easter, Jr., R. Mangione-Smith, and D. Dwyer Jones. 1992. Local regeneration in the retina of the goldfish. *J. Neurobiol.* 23:187–203.

Hitchcock, P.F. and P.A. Raymond. 1992. Retinal regeneration. *Trend. Neurosci.* 15:103–108.

Ho, B.-K., M.E. Alexianu, L.V. Colom, A.H. Mohamed, F. Serrano, and S.H. Appel. 1996. Expression of calbindin-D_{28K} in motoneuron hybrid cells after retroviral infection with calbindin-D_{28K} cDNA prevents amyotrophic lateral sclerosis IgG-mediated cytotoxicity. *Proc. Natl. Acad. Sci. USA* 93:6796–6801.

Hochmann, S., J. Kaslin, S. Hans, A. Weber, A. Machate, M. Geffarth, R.H. Funk, and M. Brand. 2012. Fgf signaling is required for photoreceptor maintenance in the adult zebrafish retina. *PLoS One* 7:e30365.

Horn, K.P., S.A. Busch, A.L. Hawthorne, N. van Rooijen, and J. Silver. 2008. Another barrier to regeneration in the CNS: Activated macrophages induce extensive retraction of dystrophic axons through direct physical interactions. *J. Neurosci.* 28:9330–9341.

Horner, P.J., A.E. Power, G. Kempermann, H.G. Kuhn, T.D. Palmer, J. Winkler, L.J. Thal, and F.H. Gage. 2000. Proliferation and differentiation of progenitor cells throughout the intact adult rat spinal cord. *J. Neurosci.* 20:2218–2228.

Ilieş, I., M.M. Zupanc, and G.K.H. Zupanc. 2012. Proteome analysis reveals protein candidates involved in early stages of brain regeneration of teleost fish. *Neuroscience* 219:302–313.

Isoe, Y., T. Okuyama, Y. Taniguchi, T. Kubo, and H. Takeuchi. 2012. p53 Mutation suppresses adult neurogenesis in medaka fish (*Oryzias latipes*). *Biochem. Biophys. Res. Commun.* 423:627–631.

Johns, P.R. 1977. Growth of the adult goldfish eye. III. Source of the new retinal cells. *J. Comp. Neurol.* 176:343–357.

Johns, P.R. 1982. Formation of photoreceptors in larval and adult goldfish. *J. Neurosci.* 2:178–198.

Johns, P.R. and S.S. Jr. Easter. 1977. Growth of the adult goldfish eye: II. Increase in retinal cell number. *J. Comp. Neurol.* 176:331–342.

Jopling, C., E. Sleep, M. Raya, M. Martí, A. Raya, and J.C. Izpisúa Belmonte. 2010. Zebrafish heart regeneration occurs by cardiomyocyte dedifferentiation and proliferation. *Nature* 464:606–609.

Julian, D., K. Ennis, and J.I. Korenbrot. 1998. Birth and fate of proliferative cells in the inner nuclear layer of the mature fish retina. *J. Comp. Neurol.* 394:271–282.

Kaplan, M.S. and D.H. Bell. 1984. Mitotic neuroblasts in the 9-day-old and 11-month-old rodent hippocampus. *J. Neurosci.* 4:1429–1441.

Kaslin, J., J. Ganz, and M. Brand. 2008. Proliferation, neurogenesis and regeneration in the non-mammalian vertebrate brain. *Philos. Trans. R. Soc. Lond. B Biol. Sci.* 363:101–122.

Kaslin, J., J. Ganz, M. Geffarth, H. Grandel, S. Hans, and M. Brand. 2009. Stem cells in the adult zebrafish cerebellum: Initiation and maintenance of a novel stem cell niche. *J. Neurosci.* 29:6142–6153.

Kassen, S.C., V. Ramanan, J.E. Montgomery, T.C. Burket, C.G. Liu, T.S. Vihtelic, and D.R. Hyde. 2007. Time course analysis of gene expression during light-induced photoreceptor cell death and regeneration in albino zebrafish. *Dev. Neurobiol.* 67:1009–1031.

Kassen, S.C., R. Thummel, L.A. Campochiaro, M.J. Harding, N.A. Bennett, and D.R. Hyde. 2009. CNTF induces photoreceptor neuroprotection and Müller glial cell proliferation through two different signaling pathways in the adult zebrafish retina. *Exp. Eye Res.* 88:1051–1064.

Kästner, R. and H. Wolburg. 1982. Functional regeneration of the visual system in teleosts: Comparative investigations after optic nerve crush and damage of the retina. *Z. Naturforsch.* 37c:1274–1280.

Katayama, Y., D.P. Becker, T. Tamura, and D.A. Hovda. 1990. Massive increases in extracellular potassium and the indiscriminate release of glutamate following concussive brain injury. *J. Neurosurg.* 73:889–900.

Kerr, J.F.R., G.C. Gobé, C.M. Winterford, and B.V. Harmon. 1995. Anatomical methods in cell death. In *Cell Death*, eds. L. M. Schwartz and B. A. Osborne, pp. 1–27. San Diego, CA: Academic Press.

Kikuchi, K., J.E. Holdway, A.A. Werdich, R.M. Anderson, Y. Fang, G.F. Egnaczyk, T. Evans, C.A. Macrae, D.Y. Stainier, and K.D. Poss. 2010. Primary contribution to zebrafish heart regeneration by *gata4+* cardiomyocytes. *Nature* 464:601–605.

Kirsche, W. 1960. Zur Frage der Regeneration des Mittelhirnes der Teleostei. *Verh. Anat. Gesell.* 56:259–270.

Kirsche, W. and K. Kirsche. 1961. Experimentelle Untersuchungen zur Frage der Regeneration und Funktion des Tectum opticum von *Carassius carassius* L. *Z. Mikrosk. Anat. Forsch.* 67:140–182.

Kishimoto, N., K. Shimizu, and K. Sawamoto. 2012. Neuronal regeneration in a zebrafish model of adult brain injury. *Dis. Models Mech.* 5:200–209.

Knopf, F., C. Hammond, A. Chekuru, T. Kurth, S. Hans, C.W. Weber, G. Mahatma, S. Fisher, M. Brand, S. Schulte-Merker, and G. Weidinger. 2011. Bone regenerates via dedifferentiation of osteoblasts in the zebrafish fin. *Dev. Cell* 20:713–724.

Kondo, T. and M. Raff. 2000. Oligodendrocyte precursor cells reprogrammed to become multipotential CNS stem cells. *Science* 289:1754–1757.

Koppányi, T. and P. Weiss. 1922. Funktionelle Regeneration des Rückenmarkes bei Anamniern. *Anz. Akad. Wiss. Wien, Math.-Naturw. Kl.* 59:206.

Koriyama, Y., K. Homma, and S. Kato. 2006. Activation of cell survival signals in the goldfish retinal ganglion cells after optic nerve injury. *Adv. Exp. Med. Biol.* 572:333–337.

Koriyama, Y., R. Yasuda, K. Homma, K. Mawatari, M. Nagashima, K. Sugitani, T. Matsukawa, and S. Kato. 2009. Nitric oxide-cGMP signaling regulates axonal elongation during optic nerve regeneration in the goldfish *in vitro* and *in vivo*. *J. Neurochem.* 110:890–901.

Kornack, D.R. and P. Rakic. 1999. Continuation of neurogenesis in the hippocampus of the adult macaque monkey. *Proc. Natl. Acad. Sci. USA* 96:5768–5773.

Koumans, J.T.M. and H.A. Akster. 1995. Myogenic cells in development and growth of fish. *Comp. Biochem. Physiol.* 110A:3–20.

Kragl, M., D. Knapp, E. Nacu, S. Khattak, M. Maden, H.H. Epperlein, and E.M. Tanaka. 2009. Cells keep a memory of their tissue origin during axolotl limb regeneration. *Nature* 460:60–65.

Kranz, D. and W. Richter. 1970a. Autoradiographische Untersuchungen über die Lokalisation der Matrixzonen des Diencephalons von juvenilen und adulten *Lebistes reticulatus* (Teleostei). *Z. mikrosk.-anat. Forsch.* 82:42–66.

Kranz, D. and W. Richter. 1970b. Autoradiographische Untersuchungen zur DNS-Synthese im Cerebellum und in der Medulla oblongata von Teleostiern verschiedenen Lebensalters. *Z. mikrosk.-anat. Forsch.* 82:264–292.

Kroehne, V., D. Freudenreich, S. Hans, J. Kaslin, and M. Brand. 2011. Regeneration of the adult zebrafish brain from neurogenic radial glia-type progenitors. *Development* 138:4831–4841.

Kuroyanagi, Y., T. Okuyama, Y. Suehiro, H. Imada, A. Shimada, K. Naruse, H. Takeda, T. Kubo, and H. Takeuchi. 2010. Proliferation zones in adult medaka (*Oryzias latipes*) brain. *Brain Res.* 1323:33–40.

Kuscha, V., S.L. Frazer, T.B. Dias, M. Hibi, T. Becker, and C.G. Becker. 2012a. Lesion-induced generation of interneuron cell types in specific dorsoventral domains in the spinal cord of adult zebrafish. *J. Comp. Neurol.* 520:3604–3616.

Kuscha, V., A. Barreiro-Iglesias, C.G. Becker, and T. Becker. 2012b. Plasticity of tyrosine hydroxylase and serotonergic systems in the regenerating spinal cord of adult zebrafish. *J. Comp. Neurol.* 520:933–951.

Kusik, B.W., D.R. Hammond, and A.J. Udvadia. 2010. Transcriptional regulatory regions of *gap43* needed in developing and regenerating retinal ganglion cells. *Dev. Dyn.* 239:482–495.

Kyritsis, N., C. Kizil, S. Zocher, V. Kroehne, J. Kaslin, D. Freudenreich, A. Iltzsche, and M. Brand. 2012. Acute inflammation initiates the regenerative response in the adult zebrafish brain. *Science* 338:1353–1356.

Lanners, H.N. and B. Grafstein. 1980. Early stages of axonal regeneration in the goldfish optic tract: An electron microscopic study. *J. Neurocytol.* 9:733–751.

Lau, A. and M. Tymianski. 2010. Glutamate receptors, neurotoxicity and neurodegeneration. *Pflügers Arch.* 460:525–542.

Lindsey, A.E. and M.K. Powers. 2007. Visual behavior of adult goldfish with regenerating retina. *Vis. Neurosci.* 24:247–255.

Liou, A.K.F., R.S. Clark, D.C. Henshall, X.-M. Yin, and J. Chen. 2003. To die or not to die for neurons in ischemia, traumatic brain injury and epilepsy: A review on the stress-activated signaling pathways and apoptotic pathways. *Prog. Neurobiol.* 69:103–142.

Lois, C. and A. Alvarez-Buylla. 1994. Long-distance neuronal migration in the adult mammalian brain. *Science* 264:1145–1148.

Lois, C., J.M. Garcia-Verdugo, and A. Alvarez-Buylla. 1996. Chain migration of neuronal precursors. *Science* 271:978–981.

Lundberg, J.G., C. Cox-Fernandes, J.S. Albert, and M. Garcia. 1996. *Magosternarchus*, a new genus with two new species of electric fishes (Gymnotiformes: Apteronotidae) from the Amazon River Basin, South America. *Copeia* 1996:657–670.

Luskin, M.B. 1993. Restricted proliferation and migration of postnatally generated neurons from the forebrain subventricular zone. *Neuron* 11:173–189.

Ma, L., Y.M. Yu, Y. Guo, R.P. Hart, and M. Schachner. 2012. Cysteine- and glycine-rich protein 1a is involved in spinal cord regeneration in adult zebrafish. *Eur. J. Neurosci.* 35:353–365.

Mago-Leccia, F., J.G. Lundberg, and J.N. Baskin. 1985. Systematics of the South American freshwater fish genus *Adontosternarchus* (Gymnotiformes, Apteronotidae). *Contrib. Sci. (Los Angel, CA.)* 358:1–19.

Maier, W. and H. Wolburg. 1979. Regeneration of the goldfish retina after exposure to different doses of ouabain. *Cell Tissue Res.* 202:99–118.

Mansour-Robaey, S. and G. Pinganaud. 1990. Quantitative and morphological study of cell proliferation during morphogenesis in the trout visual system. *J. Hirnforsch.* 31:495–504.

Marcus, R.C., C.L. Delaney, and S.S. Easter. 1999. Neurogenesis in the visual system of embryonic and adult zebrafish (*Danio rerio*). *Vis. Neurosci.* 16:417–424.

Maruska, K.P., R.E. Carpenter, and R.D. Fernald. 2012. Characterization of cell proliferation throughout the brain of the African cichlid fish *Astatotilapia burtoni* and its regulation by social status. *J. Comp. Neurol.* 520:3471–3491.

März, M., R. Schmidt, S. Rastegar, and U. Strähle. 2011. Regenerative response following stab injury in the adult zebrafish telencephalon. *Dev. Dyn.* 240:2221–2231.

Matsukawa, T., K. Arai, Y. Koriyama, Z. Liu, and S. Kato. 2004. Axonal regeneration of fish optic nerve after injury. *Biol. Pharm. Bull.* 27:445–451.

Mattson, M.P., B. Rychlik, C. Chu, and S. Christakos. 1991. Evidence for calcium-reducing and excito-protective roles for the calcium-binding protein calbindin-D_{28k} in cultured hippocampal neurons. *Neuron* 6:41–51.

McCurley, A.T. and G.V. Callard. 2010. Time course analysis of gene expression patterns in zebrafish eye during optic nerve regeneration. *J. Exp. Neurosci.* 2010:17–33.

Meletis, K., F. Barnabé-Heider, M. Carlén, E. Evergren, N. Tomilin, O. Shupliakov, and J. Frisén. 2008. Spinal cord injury reveals multilineage differentiation of ependymal cells. *PLoS Biol.* 6:e182.

Mensinger, A.F. and M.K. Powers. 1999. Visual function in regenerating teleost retina following cytotoxic lesioning. *Vis. Neurosci.* 16:241–251.

Mensinger, A.F. and M.K. Powers. 2007. Visual function in regenerating teleost retina following surgical lesioning. *Vis. Neurosci.* 24:299–307.

Meyer, R.L. 1978. Evidence from thymidine labeling for continuing growth of retina and tectum in juvenile goldfish. *Exp. Neurol.* 59:99–111.

Mizuno, T.A. and T. Ohtsuka. 2008. Quantitative analysis of activated caspase-3-positive cells and a circadian cycle of programmed cell death in the adult teleost retina. *Neurosci. Lett.* 446:16–19.

Mizuno, T.A. and T. Ohtsuka. 2009. Quantitative study of apoptotic cells in the goldfish retina. *Zoolog. Sci.* 26:157–162.

Monje, M.L., R. Phillips, and R. Sapolsky. 2001. Calbindin overexpression buffers hippocampal cultures from the energetic impairments caused by glutamate. *Brain Res.* 911:37–42.

Negishi, K., K. Sugawara, S. Shinagawa, T. Teranishi, C.H. Kuo, and Y. Takasaki. 1991. Induction of immunoreactive proliferating cell nuclear antigen (PCNA) in goldfish retina following intravitreal injection with tunicamycin. *Dev. Brain Res.* 63:71–83.

Negishi, K., T. Teranishi, S. Kato, and Y. Nakamura. 1988. Immunohistochemical and autoradiographic studies on retinal regeneration in teleost fish. *Neurosci. Res. Suppl.* 8:S43–S57.

Nguyen, V., K. Deschet, T. Henrich, E. Godet, J.S. Joly, J. Wittbrodt, D. Chourrout, and F. Bourrat. 1999. Morphogenesis of the optic tectum in the medaka (*Oryzias latipes*): A morphological and molecular study, with special emphasis on cell proliferation. *J. Comp. Neurol.* 413:385–404.

Nieuwenhuys, R. and J. Meek. 1990. The telencephalon of actinopterygian fishes. In *Comparative Structure and Evolution of the Cerebral Cortex*, eds. E. G. Jones and A. Peters, pp. 31–73. New York: Plenum.

Northcutt, R.G. 1995. The forebrain of gnathostomes: In search of a morphotype. *Brain Behav. Evol.* 46:275–318.

Northcutt, R.G. and M.R. Braford. 1980. New observations on the organization and evolution of the telencephalon of actinopterygian fishes. In *Comparative Neurology of the Telencephalon*, ed. S. O. E. Ebbesson, pp. 41–98. New York: Plenum.

Ogai, K., S. Hisano, K. Mawatari, K. Sugitani, Y. Koriyama, H. Nakashima, and S. Kato. 2012. Upregulation of anti-apoptotic factors in upper motor neurons after spinal cord injury in adult zebrafish. *Neurochem. Int.* 61:1202–1211.

de Oliveira-Castro, G. 1955. Differentiated nervous fibers that constitute the electric organ of *Sternarchus albifrons*, Linn. *An. Acad. Bras. Cienc*. 27:557–564.

Olney, J.W. 1969. Brain lesions, obesity, and other disturbances in mice treated with monosodium glutamate. *Science* 164:719–721.

Ott, R., G.K.H. Zupanc, and I. Horschke. 1997. Long-term survival of postembryonically born cells in the cerebellum of gymnotiform fish, *Apteronotus leptorhynchus*. *Neurosci. Lett*. 221:185–188.

Otteson, D.C. and P.F. Hitchcock. 2003. Stem cells in the teleost retina: Persistent neurogenesis and injury-induced regeneration. *Vision Res*. 43:927–936.

Otteson, D.C., P.F. Cirenza, P.F. Hitchcock. 2002. Persistent neurogenesis in the teleost retina: Evidence for regulation by the growth-hormone/insulin-like growth factor-I axis. *Mech Dev*. 117:137–149.

Palmer, A.M., D.W. Marion, M.L. Botscheller, D.M. Bowen, and S.T. DeKosky. 1994. Increased transmitter amino acid concentration in human ventricular CSF after brain trauma. *Neuroreport* 6:153–156.

Palmer, T.D., E.A. Markakis, A.R. Willhoite, F. Safar, and F.H. Gage. 1999. Fibroblast growth factor-2 activates a latent neurogenic program in neural stem cells from diverse regions of the adult CNS. *J. Neurosci*. 19:8487–8497.

Pan, H.C., J.F. Lin, L.P. Ma, Y.Q. Shen, and M. Schachner. 2013. Major vault protein promotes locomotor recovery and regeneration after spinal cord injury in adult zebrafish. *Eur. J. Neurosci*. 37:203–211.

Park, H.C., J. Shin, R.K. Roberts, and B. Appel. 2007. An *olig2* reporter gene marks oligodendrocyte precursors in the postembryonic spinal cord of zebrafish. *Dev. Dyn*. 236:3402–3407.

Parrilla, M., C. Lillo, M.J. Herrero-Turrión, R. Arévalo, J. Aijón, J.M. Lara, and A. Velasco. 2013. Pax2+ astrocytes in the fish optic nerve head after optic nerve crush. *Brain Res*. 1492:18–32.

Paskin, T.R. and C.A. Byrd-Jacobs. 2012. Reversible deafferentation of the adult zebrafish olfactory bulb affects glomerular distribution and olfactory-mediated behavior. *Behav. Brain Res*. 235:293–301.

Pellegrini, E., K. Mouriec, I. Anglade, A. Menuet, Y. Le Page, M.M. Gueguen, M.H. Marmignon, F. Brion, F. Pakdel, and O. Kah. 2007. Identification of aromatase-positive radial glial cells as progenitor cells in the ventricular layer of the forebrain in zebrafish. *J. Comp. Neurol*. 501:150–167.

Pencea, V., K.D. Bingaman, L.J. Freedman, and M.B Luskin. 2001. Neurogenesis in the subventricular zone and rostral migratory stream of the neonatal and adult primate forebrain. *Exp. Neurol*. 172:1–16.

Phillips, R.G., T.J. Meier, L.C. Giuli, J.R. McLaughlin, D.Y. Ho, and R.M. Sapolsky. 1999. Calbindin D_{28K} gene transfer via herpes simplex virus amplicon vector decreases hippocampal damage in vivo following neurotoxic insults. *J. Neurochem*. 73:1200–1205.

Portavella, M., B. Torres, and C. Salas. 2004. Avoidance response in goldfish: Emotional and temporal involvement of medial and lateral telencephalic pallium. *J. Neurosci*. 24:2335–2342.

Pouwels, E. 1978a. On the development of the cerebellum of the trout, *Salmo gairdneri*: I. Patterns of cell migration. *Anat. Embryol*. 152:291–308.

Pouwels, E. 1978b. On the development of the cerebellum of the trout, *Salmo gairdneri*: III. Development of neuronal elements. *Anat. Embryol*. 153:37–54.

Powell, C., F. Elsaeidi, and D. Goldman. 2012. Injury-dependent Müller glia and ganglion cell reprogramming during tissue regeneration requires Apobec2a and Apobec2b. *J. Neurosci*. 32:1096–1109.

Qin, Z., L.K. Barthel, and P.A. Raymond. 2009. Genetic evidence for shared mechanisms of epimorphic regeneration in zebrafish. *Proc. Natl. Acad. Sci. USA* 106:9310–9315.

van Raamsdonk, W., S. Maslam, D.H. de Jong, M.J. Smit-Onel, and E. Velzing. 1998. Long term effects of spinal cord transection in zebrafish: Swimming performances, and metabolic properties of the neuromuscular system. *Acta Histochem*. 100:117–131.

Rakic, P. 1971. Neuron-glia relationship during granule cell migration in developing cerebellar cortex: A Golgi and electronmicroscopic study in *Macacus rhesus*. *J. Comp. Neurol*. 141:283–312.

Rakic, P. 1972. Mode of cell migration to the superficial layers of fetal monkey neocortex. *J. Comp. Neurol*. 145:61–83.

Raya, Á., A. Consiglio, Y. Kawakami, C. Rodriguez-Esteban, and J. C. Izpisúa-Belmonte. 2004. The zebrafish as a model of heart regeneration. *Cloning Stem Cells* 6:345–351.

Raymond, P.A., L.K. Barthel, R.L. Bernardos, and J.J. Perkowski. 2006. Molecular characterization of retinal stem cells and their niches in adult zebrafish. *BMC Dev. Biol*. 6:36.

Raymond, P.A. and S.S. Easter, Jr. 1983. Postembryonic growth of the optic tectum in goldfish. I. Location of germinal cells and numbers of neurons produced. *J. Neurosci*. 3:1077–1091.

Raymond, P.A., M.J. Reifler, and P.K. Rivlin. 1988. Regeneration of goldfish retina: Rod precursors are a likely source of regenerated cells. *J. Neurobiol*. 19:431–463.

Reier, P.J., L.J. Stensaas, and L. Guth. 1983. The astrocytic scar as an impediment to regeneration in the central nervous system. In *Spinal Cord Reconstruction*, eds. C. C. Kao, R. P. Bunge, and P. J. Reier, pp. 163–195. New York: Raven Press.

Reimer, M.M., V. Kuscha, C. Wyatt, I. Sörensen, R.E. Frank, M. Knüwer, T. Becker, and C.G. Becker. 2009. Sonic hedgehog is a polarized signal for motor neuron regeneration in adult zebrafish. *J. Neurosci.* 29:15073–15082.

Reimer, M.M., I. Sörensen, V. Kuscha, R.E. Frank, C. Liu, C.G. Becker, and T. Becker. 2008. Motor neuron regeneration in adult zebrafish. *J. Neurosci.* 28:8510–8516.

Reynolds, B.A. and S. Weiss. 1992. Generation of neurons and astrocytes from isolated cells of the adult mammalian central nervous system. *Science* 255:1707–1710.

Richter, W. 1968. Regeneration im Tectum opticum bei adulten *Lebistes reticulatus* (Peters 1859) (Poeciliidae, Cyprinodontiformes, Teleostei). *J. Hirnforsch.* 10:173–186.

Richter, W. and D. Kranz. 1970a. Autoradiographische Untersuchungen über die Abhängigkeit des ^{3}H-Thymidin-Index vom Lebensalter in den Matrixzonen des Telencephalons von *Lebistes reticulatus* (Teleostei). *Z. mikrosk.-anat. Forsch.* 81:530–554.

Richter, W. and D. Kranz. 1970b. Die Abhängigkeit der DNS-Synthese in den Matrixzonen des Mesencephalons vom Lebensalter der Versuchstiere (*Lebistes reticulatus*—Teleostei): Autoradiographische Untersuchungen. *Z. mikrosk.-anat. Forsch.* 82:76–92.

Richter, W. and D. Kranz. 1977. Über die Bedeutung der Zellproliferation für die Hirnregeneration bei niederen Vertebraten: Autoradiographische Untersuchungen. *Verh. Anat. Gesell.* 71:439–445.

Rodríguez, F., J.C. López, J.P. Vargas, Y. Gómez, C. Broglio, and C. Salas. 2002. Conservation of spatial memory function in the pallial forebrain of reptiles and ray-finned fishes. *J. Neurosci.* 22:2894–2903.

Rolls, A., R. Shechter, and M. Schwartz. 2009. The bright side of the glial scar in CNS repair. *Nat. Rev. Neurosci.* 10:235–241.

Rosenzweig, S., D. Raz-Prag, A. Nitzan, R. Galron, M. Paz, G. Jeserich, G. Neufeld, A. Barzilai, and A.S. Solomon. 2010. Sema-3A indirectly disrupts the regeneration process of goldfish optic nerve after controlled injury. *Graefes Arch. Clin. Exp. Ophthalmol.* 248:1423–1435.

Rothenaigner, I., M. Krecsmarik, J.A. Hayes, B. Bahn, A. Lepier, G. Fortin, M. Gotz, R. Jagasia, and L. Bally-Cuif. 2011. Clonal analysis by distinct viral vectors identifies bona fide neural stem cells in the adult zebrafish telencephalon and characterizes their division properties and fate. *Development* 138:1459–1469.

Rowe, R.W.D. and G. Goldspink. 1969. Muscle fibre growth in five different muscles in both sexes of mice. *J. Anat.* 104:519–530.

Rowlerson, A. and A. Veggetti. 2001. Cellular mechanisms of post-embryonic muscle growth in aquaculture species. In *Muscle Development and Growth*, ed. I. A. Johnston, pp. 103–140. San Diego, CA: Academic Press.

Sanai, N., A.D. Tramontin, A. Quinones-Hinojosa, N.M. Barbaro, N. Gupta, S. Kunwar, M.T. Lawton et al. 2004. Unique astrocyte ribbon in adult human brain contains neural stem cells but lacks chain migration. *Nature* 427:740–744.

de Santana, C.D. and R.P. Vari. 2010. Electric fishes of the genus *Sternarchorhynchus* (Teleostei, Ostariophysi, Gymnotiformes); phylogenetic and revisionary studies. *Zool. J. Linn. Soc.-Lond.* 159:223–371.

Saul, K.E., J.R. Koke, and D.M. García. 2010. Activating transcription factor 3 (ATF3) expression in the neural retina and optic nerve of zebrafish during optic nerve regeneration. *Comp. Biochem. Physiol. A* 155:172–182.

Schweitzer, J., T. Becker, C.G. Becker, and M. Schachner. 2003. Expression of protein zero is increased in lesioned axon pathways in the central nervous system of adult zebrafish. *Glia* 41:301–317.

Schweitzer, J., D. Gimnopoulos, B.C. Lieberoth, H.M. Pogoda, J. Feldner, A. Ebert, M. Schachner, T. Becker, and C.G. Becker. 2007. Contactin1a expression is associated with oligodendrocyte differentiation and axonal regeneration in the central nervous system of zebrafish. *Mol. Cell. Neurosci.* 35:194–207.

Senut, M.C., A. Gulati-Leekha, and D. Goldman. 2004. An element in the α1-tubulin promoter is necessary for retinal expression during optic nerve regeneration but not after eye injury in the adult zebrafish. *J. Neurosci.* 24:7663–7673.

Seri, B., J.M. García-Verdugo, B.S. McEwen, and A. Alvarez-Buylla. 2001. Astrocytes give rise to new neurons in the adult mammalian hippocampus. *J. Neurosci.* 21:7153–7160.

Silver, J. and J.H. Miller. 2004. Regeneration beyond the glial scar. *Nat. Rev. Neurosci.* 5:146–156.

Sîrbulescu, R.F., I. Ilieş, and G.K.H. Zupanc. 2009. Structural and functional regeneration after spinal cord injury in the weakly electric teleost fish, *Apteronotus leptorhynchus*. *J. Comp. Physiol. A* 195:699–714.

Sîrbulescu, R.F. and G.K.H. Zupanc. 2009. Dynamics of caspase-3-mediated apoptosis during spinal cord regeneration in the teleost fish, *Apteronotus leptorhynchus*. *Brain Res.* 1304:14–25.

Sîrbulescu, R.F. and G.K.H. Zupanc. 2010a. Inhibition of caspase-3-mediated apoptosis improves spinal cord repair in a regeneration-competent vertebrate system. *Neuroscience* 171:599–612.

Sîrbulescu, R.F. and G.K.H. Zupanc. 2010b. Effect of temperature on spinal cord regeneration in the weakly electric fish, *Apteronotus leptorhynchus*. *J. Comp. Physiol. A* 196:359–368.

Sîrbulescu, R.F. and G.K.H. Zupanc. 2011. Spinal cord repair in regeneration-competent vertebrates: Adult teleost fish as a model system. *Brain Res. Rev.* 67:73–93.

Soares, S., Y. von Boxberg, M.C. Lombard, M. Ravaille-Veron, I. Fischer, J. Eyer, and F. Nothias. 2002. Phosphorylated MAP1B is induced in central sprouting of primary afferents in response to peripheral injury but not in response to rhizotomy. *Eur. J. Neurosci.* 16:593–606.

Sørensen, C., L.C. Bohlin, Ø. Øverli, and G.E. Nilsson. 2011. Cortisol reduces cell proliferation in the telencephalon of rainbow trout (*Oncorhynchus mykiss*). *Physiol. Behav.* 102:518–523.

Sørensen, C., G.E. Nilsson, C.H. Summers, and Ø. Øverli. 2012. Social stress reduces forebrain cell proliferation in rainbow trout (*Oncorhynchus mykiss*). *Behav. Brain Res.* 227:311–318.

Soutschek, J. and G.K.H. Zupanc. 1996. Apoptosis in the cerebellum of adult teleost fish, *Apteronotus leptorhynchus*. *Dev. Brain Res.* 97:279–286.

Spassky, N., F.T. Merkle, N. Flames, A.D. Tramontin, J.M. García-Verdugo, and A. Alvarez-Buylla. 2005. Adult ependymal cells are postmitotic and are derived from radial glial cells during embryogenesis. *J. Neurosci.* 25:10–18.

Sperry, R.W. 1948. Patterning of central synapses in regeneration of the optic nerve in teleosts. *Physiol. Zool.* 21:351–361.

Stenkamp, D.L. 2007. Neurogenesis in the fish retina. *Int. Rev. Cytol.* 259:173–224.

Stuermer, C.A.O., M. Bastmeyer, M. Bahr, G. Strobel, and K. Paschke. 1992. Trying to understand axonal regeneration in the CNS of fish. *J. Neurobiol.* 23:537–550.

Stuermer, C.A.O. and S.S. Easter, Jr. 1984. A comparison of the normal and regenerated retinotectal pathways of goldfish. *J. Comp. Neurol.* 223:57–76.

Stuermer, C.A.O., A. Niepenberg, and H. Wolburg. 1985. Aberrant axonal paths in regenerated goldfish retina and tectum opticum following intraocular injection of ouabain. *Neurosci. Lett.* 58:333–338.

Sugitani, K., T. Matsukawa, Y. Koriyama, T. Shintani, T. Nakamura, M. Noda, and S. Kato. 2006. Upregulation of retinal transglutaminase during the axonal elongation stage of goldfish optic nerve regeneration. *Neuroscience* 142:1081–1092.

Sugitani, K., K. Ogai, K. Hitomi, K. Nakamura-Yonehara, T. Shintani, M. Noda, Y. Koriyama, H. Tanii, T. Matsukawa, and S. Kato. 2012. A distinct effect of transient and sustained upregulation of cellular factor XIII in the goldfish retina and optic nerve on optic nerve regeneration. *Neurochem. Int.* 61:423–432.

Takeda, A., M. Nakano, R.C. Goris, and K. Funakoshi. 2008. Adult neurogenesis with 5-HT expression in lesioned goldfish spinal cord. *Neuroscience* 151:1132–1141.

Teles, M.C., R.F. Sîrbulescu, U.M. Wellbrock, R.F. Oliveira, and G.K.H. Zupanc. 2012. Adult neurogenesis in the brain of the Mozambique tilapia, *Oreochromis mosambicus*. *J. Comp. Physiol. A* 198:427–449.

Thummel, R., S.C. Kassen, J.M. Enright, C.M. Nelson, J.E. Montgomery, and D.R. Hyde. 2008. Characterization of Müller glia and neuronal progenitors during adult zebrafish retinal regeneration. *Exp. Eye Res.* 87:433–444.

Tozzini, E.T., M. Baumgart, G. Battistoni, and A. Cellerino. 2012. Adult neurogenesis in the short-lived teleost *Nothobranchius furzeri*: Localization of neurogenic niches, molecular characterization and effects of aging. *Aging Cell* 11:241–251.

Tuge, H. and S. Hanzawa. 1935. Physiology of the spinal fish, with special reference to the postural mechanism. *Sci. Rep. Tohoku Imp. Univ., Biol.* 10:589–606.

Tuge, H. and S. Hanzawa. 1937. Physiological and morphological regeneration of the sectioned spinal cord in adult teleosts. *J. Comp. Neurol.* 67:343–365.

Vajda, F.J. 2002. Neuroprotection and neurodegenerative disease. *J. Clin. Neurosci.* 9:4–8.

Vargas, J.P., F. Rodríguez, J.C. López, J.L. Arias, and C. Salas. 2000. Spatial learning-induced increase in the argyophilic nucleolar organizer region of dorsolateral telencephalic neurons in goldfish. *Brain Res.* 865:77–84.

Veldman, M.B., M.A. Bemben, and D. Goldman. 2010. *Tuba1a* gene expression is regulated by KLF6/7 and is necessary for CNS development and regeneration in zebrafish. *Mol. Cell. Neurosci.* 43:370–383.

Veldman, M.B., M.A. Bemben, R.C. Thompson, and D. Goldman. 2007. Gene expression analysis of zebrafish retinal ganglion cells during optic nerve regeneration identifies KLF6a and KLF7a as important regulators of axon regeneration. *Dev. Biol.* 312:596–612.

Vihtelic, T.S. and D.R. Hyde. 2000. Light-induced rod and cone cell death and regeneration in the adult albino zebrafish (*Danio rerio*) retina. *J. Neurobiol.* 44:289–307.

Waxman, S.G. and M.J. Anderson. 1980. Regeneration of spinal electrocyte fibers in *Sternarchus albifrons*: Development of axon-Schwann cell relationships and nodes of Ranvier. *Cell Tissue Res.* 208:343–352.

Waxman, S.G. and M.J. Anderson. 1986. Regeneration of central nervous system structures: *Apteronotus* spinal cord as a model system. In *Electroreception*, eds. T. H. Bullock and W. Heiligenberg, pp. 183–208. New York: Wiley & Sons.

Waxman, S.G., G.D. Pappas, and M.V.L. Bennett. 1972. Morphological correlates of functional differentiation of nodes of Ranvier along single fibers in the neurogenic electric organ of the knife fish *Sternarchus*. *J. Cell Biol.* 53:210–224.

Weatherley, A.H. and H.S. Gill. 1985. Dynamics of increase in muscle fibres in fishes in relation to size and growth. *Experientia* 41:353–354.

Weber, J.T. 2004. Calcium homeostasis following traumatic neuronal injury. *Curr. Neurovasc. Res.* 1:151–171.

Wehman, A.M., W. Staub, J.R. Meyers, P.A. Raymond, and H. Baier. 2005. Genetic dissection of the zebrafish retinal stem-cell compartment. *Dev. Biol.* 281:53–65.

Weiss, S., C. Dunne, J. Hewson, C. Wohl, M. Wheatly, A.C. Peterson, and B.A. Reynolds. 1996. Multipotent CNS stem cells are present in the adult mammalian spinal cord and ventricular neuroaxis. *J. Neurosci.* 16:7599–7609.

West-Eberhard, M.J. 2003. *Developmental Plasticity and Evolution*. Oxford/New York: Oxford University Press.

West-Eberhard, M.J. 2005. Developmental plasticity and the origin of species differences. *Proc. Natl. Acad. Sci. USA* 102:6543–6549.

Wu, D.M., T. Schneiderman, J. Burgett, P. Gokhale, L. Barthel, and P.A. Raymond. 2001. Cones regenerate from retinal stem cells sequestered in the inner nuclear layer of adult goldfish retina. *Invest. Ophthalmol. Vis. Sci.* 42:2115–2124.

Wullimann, M.F. and L. Puelles. 1999. Postembryonic neural proliferation in the zebrafish forebrain and its relationship to prosomeric domains. *Anat. Embryol. (Berl.)* 199:329–348.

Yiu, G. and Z. He. 2006. Glial inhibition of CNS axon regeneration. *Nat. Rev. Neurosci.* 7:617–627.

Young, W. 1992. Role of calcium in central nervous system injuries. *J. Neurotrauma* 9:S9–S25.

Yu, Y.M., M. Cristofanilli, A. Valiveti, L. Ma, M. Yoo, F. Morellini, and M. Schachner. 2011a. The extracellular matrix glycoprotein tenascin-C promotes locomotor recovery after spinal cord injury in adult zebrafish. *Neuroscience* 183:238–250.

Yu, Y.M., K.M. Gibbs, J. Davila, N. Campbell, S. Sung, T.I. Todorova, S. Otsuka, H.E. Sabaawy, R.P. Hart, and M. Schachner. 2011b. MicroRNA miR-133b is essential for functional recovery after spinal cord injury in adult zebrafish. *Eur. J. Neurosci.* 33:1587–1597.

Yurco, P. and D.A. Cameron. 2005. Responses of Müller glia to retinal injury in adult zebrafish. *Vision Res.* 45:991–1002.

Zakon, H.H. 1984. Postembryonic changes in the peripheral electrosensory system of a weakly electric fish: Addition of receptor organs with age. *J. Comp. Neurol.* 228:557–570.

Zhang, Z., C.J. Krebs, and L. Guth. 1997. Experimental analysis of progressive necrosis after spinal cord trauma in the rat: Etiological role of the inflammatory response. *Exp. Neurol.* 143:141–152.

Zikopoulos, B., M. Kentouri, and C.R. Dermon. 2000. Proliferation zones in the adult brain of a sequential hermaphrodite teleost species *(Sparus aurata)*. *Brain Behav. Evol.* 56:310–322.

Zimmerman, A.M. and M.S. Lowery. 1999. Hyperplastic development and hypertrophic growth of muscle fibers in the white seabass (*Atractoscion nobilis*). *J. Exp. Zool.* 284:299–308.

Zottoli, S.J. and M.M. Freemer. 2003. Recovery of C-starts, equilibrium and targeted feeding after whole spinal cord crush in the adult goldfish *Carassius auratus*. *J. Exp. Biol.* 206:3015–3029.

Zupanc, G.K.H. 1999a. Neurogenesis, cell death and regeneration in the adult gymnotiform brain. *J. Exp. Biol.* 202:1435–1446.

Zupanc, G.K.H. 1999b. Up-regulation of somatostatin after lesions in the cerebellum of the teleost fish *Apteronotus leptorhynchus*. *Neurosci. Lett.* 268:135–138.

Zupanc, G.K.H. 2001. Adult neurogenesis and neuronal regeneration in the central nervous system of teleost fish. *Brain Behav. Evol.* 58:250–275.

Zupanc, G.K.H. 2006a. Neurogenesis and neuronal regeneration in the adult fish brain. *J. Comp. Physiol. A* 192:649–670.

Zupanc, G.K.H. 2006b. Adult neurogenesis and neuronal regeneration in the teleost fish brain: Implications for the evolution of a primitive vertebrate trait. In *The Evolution of Nervous Systems in Non-Mammalian Vertebrates*, eds. T. H. Bullock and L. R. Rubenstein, pp. 485–520. Oxford, U.K.: Academic Press.

Zupanc, G.K.H. 2008a. Adult neurogenesis and neuronal regeneration in the brain of teleost fish. *J. Physiol. Paris* 102:357–373.

Zupanc, G.K.H. 2008b. Adult neurogenesis in teleost fish. In *Adult Neurogenesis*, eds. F. H. Gage, G. Kempermann, and H. Song, pp. 571–592. New York: Cold Spring Harbor Laboratory Press.

Zupanc, G.K.H. 2009. Towards brain repair: Insights from teleost fish. *Semin. Cell Dev. Biol.* 20:683–690.

Zupanc, G.K.H. 2011. Adult neurogenesis in teleost fish. In *Neurogenesis in the Adult Brain*, Vol. 1, eds. T. Seki, K. Sawamoto, J. M. Parent, and A. Alvarez-Buylla, pp. 137–168. Tokyo, Japan: Springer-Verlag.

Zupanc, G.K.H. and S.C. Clint. 2003. Potential role of radial glia in adult neurogenesis of teleost fish. *Glia* 43:77–86.

Zupanc, G.K.H., S.C. Clint, N. Takimoto, A.T.L. Hughes, U.M. Wellbrock, and D. Meissner. 2003. Spatiotemporal distribution of microglia/macrophages during regeneration in the cerebellum of adult teleost fish, *Apteronotus leptorhynchus*: A quantitative analysis. *Brain Behav. Evol.* 62:31–42.

Zupanc, G.K.H., K. Hinsch, and F.H. Gage. 2005. Proliferation, migration, neuronal differentiation, and long-term survival of new cells in the adult zebrafish brain. *J. Comp. Neurol.* 488:290–319.

Zupanc, G.K.H. and I. Horschke. 1995. Proliferation zones in the brain of adult gymnotiform fish: A quantitative mapping study. *J. Comp. Neurol.* 353:213–233.

Zupanc, G.K.H., I. Horschke, R. Ott, and G.B. Rascher. 1996. Postembryonic development of the cerebellum in gymnotiform fish. *J. Comp. Neurol.* 370:443–464.

Zupanc, G.K.H., K.S. Kompass, I. Horschke, R. Ott, and H. Schwarz. 1998. Apoptosis after injuries in the cerebellum of adult teleost fish. *Exp. Neurol.* 152:221–230.

Zupanc, G.K.H. and R. Ott. 1999. Cell proliferation after lesions in the cerebellum of adult teleost fish: Time course, origin, and type of new cells produced. *Exp. Neurol.* 160:78–87.

Zupanc, G.K.H. and R.F. Sîrbulescu. 2011. Adult neurogenesis and neuronal regeneration in the central nervous system of teleost fish. *Eur. J. Neurosci.* 34:917–929.

Zupanc, G.K.H. and R.F. Sîrbulescu. 2013. Teleost fish as a model system to study successful regeneration of the central nervous system. *Curr. Top. Microbiol. Immunol.* 367:193–233.

Zupanc, G.K.H., R.F. Sîrbulescu, and I. Ilieş. 2012. Radial glia in the cerebellum of adult teleost fish: Implications for the guidance of migrating new neurons. *Neuroscience* 210:416–430.

Zupanc, M.M., U.M. Wellbrock, and G.K.H. Zupanc. 2006. Proteome analysis identifies novel protein candidates involved in regeneration of the cerebellum of teleost fish. *Proteomics* 6:677–696.

Zupanc, G.K.H., and M.M. Zupanc. 2006a. New neurons for the injured brain: Mechanisms of neuronal regeneration in adult teleost fish. *Regen. Med.* 1:207–216.

Zupanc, M.M. and G.K.H. Zupanc. 2006b. Upregulation of calbindin-D_{28k} expression during regeneration in the adult fish cerebellum. *Brain Res.* 1095:26–34.

Index